D1599479

DISCARDED

PLANT DISTURBANCE ECOLOGY

THE PROCESS AND THE RESPONSE

PLANT DISTURBANCE ECOLOGY
THE PROCESS AND THE RESPONSE

Edited by

Edward A. Johnson
University of Calgary
Calgary, Alberta, Canada

Kiyoko Miyanishi
University of Guelph
Guelph, Ontario, Canada

ELSEVIER

AMSTERDAM • BOSTON • HEIDELBERG • LONDON
NEW YORK • OXFORD • PARIS • SAN DIEGO
SAN FRANCISCO • SINGAPORE • SYDNEY • TOKYO
Academic Press is an imprint of Elsevier

Front cover photos (clockwise from top left): Fire in Manuka heath in the foothills of the New Zealand Alps west of the Canterbury Plains (courtesy Bushfire Cooperative Research Centre); parabolic dune field advancing over vegetation in the Newcastle Bight, New South Wales, Australia (R. G. Davidson-Arnott); iced tree near Fort Drum, New York, March 1991 (David Fisk); and wind-snapped tree at Tarn Hows, Lake District, England (C. P. Quine).

Back cover photo: Mountain pine beetle outbreak, view from Avalanche Peak near the eastern boundary of Yellowstone National Park (courtesy Jeff Hicke).

Academic Press is an imprint of Elsevier
30 Corporate Drive, Suite 400, Burlington, MA 01803, USA
525 B Street, Suite 1900, San Diego, California 92101-4495, USA
84 Theobald's Road, London WC1X 8RR, UK

This book is printed on acid-free paper. ♾

Library of Congress Cataloging-in-Publication Data
Application submitted

British Library Cataloguing-in-Publication Data
A catalogue record for this book is available from the British Library.

ISBN 13: 978-0-12-088778-1
ISBN 10: 0-12-088778-9

For information on all Academic Press publications
visit our Web site at www.books.elsevier.com

Printed in the United States of America
07 08 09 10 11 9 8 7 6 5 4 3 2 1

« A qui s'adresse ce livre?
Il est pour ceux qui aiment comprendre.»

Jean-Philippe Derenne,
L'Amateur de Cuisine (Éditions Stock)

Contents

Contributors

Numbers in parentheses indicate the pages on which the authors' contributions begin.

BARRY J. COOKE (487), Natural Resources Canada, Canadian Forest Service, Northern Forestry Centre, 5320 122nd St., Edmonton, AB T6H 3S5, Canada (bcooke@nrcan.gc.ca)

NOBLE T. DONKOR (579), Canadian University College, 5415 College Ave., Lacombe, AB T4L 2E5, Canada (ndonkor@cauc.ca)

JOHN J. FINNIGAN (15), CSIRO Centre for Complex Systems Science, P.O. Box 821, Canberra, ACT 2601, Australia (John.Finnigan@csiro.au)

RÉJEAN GAGNON (555), Départemente des Sciences Fondamentales, Université du Québec à Chicoutimi, Chicoutimi, QC G7H 2B1, Canada (rgagnon@uqac.ca)

BARRY A. GARDINER (103), Forestry Commission Research Agency, Roslin, Midlothian EH25 9SY, UK (Barry.Gardiner@forestry.gsi.gov.uk)

DAVID F. GREENE (181), Department of Geography, Concordia University, Montreal, PQ H3G 1M8, Canada (greene@alcor.concordia.ca)

SHERI L. GUTSELL (441), Department of Biological Sciences, University of Calgary, Calgary, AB T2N 1N4, Canada (gutsell@ucalgary.ca)

MASAKI HAYASHI (311), Department of Geology and Geophysics, University of Calgary, Calgary, AB T2N 1N4, Canada (hayashi@ucalgary.ca)

JUSTIN HEAVILIN (527), Department of Mathematics and Statistics, Utah State University, Logan, UT 84322-3900 (jheavilin@gmail.com)

PATRICK A. HESP (215), Department of Geography and Anthropology, Louisiana State University, Baton Rouge, LA 70803-4105 (pahesp@lsu.edu)

KAZ HIGUCHI (157), Air Quality Research Division, Atmospheric Science and Technology Directorate, Science and Technology Branch, Environment Canada, 4905 Dufferin St., Toronto, ON M3H 5T4, Canada (kaz.higuchi@ec.gc.ca)

MARK R. HJELMFELT (59), Institute of Atmospheric Sciences, South Dakota School of Mines and Technology, 501 E. St. Joseph St., Rapid City, SD 57701 (Mark.Hjelmfelt@sdsmt.edu)

SATOMI INAHARA (283), 3-5-14 Hatanaka, Niiza, Saitama 352-0012, Japan (inaharas@mbi.nifty.com)

YVES JARDON (555), Université du Québec à Chicoutimi, 33 Chemin Boudreau, Mulgrave et Derry, Québec, QC J8L 2W9, Canada (yjardon@nationalschool.ca)

EDWARD A. JOHNSON (1, 249, 441), Department of Biological Sciences, University of Calgary, Calgary, AB T2N 1N4, Canada (johnsone@ucalgary.ca)

KATHLEEN F. JONES (181), U.S. Army Corps of Engineers, Snow and Ice Branch, Cold Regions Research & Engineering Laboratory, 72 Lyme Rd, Hanover, NH 03755 (Kathleen.F.Jones@erdc.usace.army.mil)

JESSE A. LOGAN (527), USDA Forest Service, Rocky Mountain Research Station, Logan Forestry Science Lab, 860 North 1200 East, Logan, UT 84321 (logan.jesse@gmail.com)

M. LUISA MARTÍNEZ (215), Depto. de Ecologia Funcional, Instituto de Ecologia, A.C., km 2.5 Antigua Carretera a Coatepec, Congregacion El Haya, Xalapa, Ver. 91070, Mexico (marisa.martinez@inecol.edu.mx)

GEOFFRY N. MERCER (371), School of Physical, Environmental and Mathematical Sciences, University of New South Wales, Australian Defence Force Academy, Canberra ACT 2600, Australia (g.mercer@adfa.edu.au)

KIYOKO MIYANISHI (1, 249), Department of Geography, University of Guelph, Guelph, ON N1G 2W1, Canada (kmiyanis@uoguelph.ca)

HUBERT MORIN (555), Départemente des Sciences Fondamentales, Université du Québec à Chicoutimi, Chicoutimi, QC G7H 2B1, Canada (hmorin@uqac.ca)

FUTOSHI NAKAMURA (283), Department of Forest Science, Graduate School of Agriculture, Hokkaido University, Sapporo 060-8589, Japan (nakaf@for.agr.hokudai.ac.jp)

VINCENT G. NEALIS (487), Natural Resources Canada, Canadian Forest Service, 506 West Burnside Road, Victoria, BC V8Z 1M5, Canada (vnealis@nrcan.gc.ca)

JAMES POWELL (527), Department of Mathematics and Statistics, Utah State University, Logan, UT 84322-3900 (powell@math.usu.edu)

OLGA J. PROULX (181), Department of Geography, Concordia University, Montreal, PQ H3G 1M8, Canada (oj_prou@alcor.concordia.ca)

CHRISTOPHER P. QUINE (103), Forestry Commission Research Agency, Roslin, Midlothian EH25 9SY, UK (Chris.Quine@forestry.gsi.gov.uk)

JACQUES RÉGNIÈRE (487), Natural Resources Canada, Canadian Forest Service, 1055 P.E.P.S. Street, PO Box 10380 Stn. Ste-Foy, QC G1V 4C7, Canada (JRegniere@cfl.forestry.ca)

AMIR SHABBAR (157), Climate Research Division, Atmospheric Science and Technology Directorate, Science and Technology Branch, Environment Canada, 4905 Dufferin St., Toronto, ON M3H 5T4, Canada (amir.shabbar@ec.gc.ca)

J. DUNGAN SMITH (603), U.S. Geological Survey, 3215 Marine St., Suite E-127, Boulder, CO 80303, USA (jdsmith@usgs.gov)

GARTH VAN DER KAMP (311), Environment Canada, National Hydrology Research Centre, 11 Innovation Blvd., Saskatoon, SK S7N 3H5, Canada (garth.vanderkamp@ec.gc.ca)

ARNOLD G. VAN DER VALK (341), Department of Ecology, Evolution and Organismal Biology, Iowa State University, Ames, IA 50011 (valk@iastate.edu)

RODNEY O. WEBER (371), School of Physical, Environmental and Mathematical Sciences, University of New South Wales at the Australian Defence Force Academy, Canberra ACT 2600, Australia (r.weber@adfa.edu.au)

PAUL H. ZEDLER (397), Nelson Institute for Environmental Studies and UW Arboretum, University of Wisconsin–Madison, 550 N. Park St., Madison, WI 53706 (phzedler@wisc.edu)

Preface

The role of natural disturbances in community dynamics has not always been clearly recognized, even though ecologists have thought of communities as dynamic and undergoing succession since the early years of the science. However, by the 1970s, there were a number of important studies showing that communities were subject to a variety of natural disturbances that occurred often enough to have significant ecological effects. This understanding of disturbances as recurrent natural events required a rethinking of succession, because it was becoming obvious that the idea of directional succession toward some stable endpoint was no longer as realistic as it once seemed. Out of this concern arose the gap phase dynamics argument which, in its simplest version, had succession occurring in gaps of varying sizes or, in its more complex form, that the community consisted of a spectrum of life histories adapted to different-sized gaps. In 1985, the Pickett and White book *Ecology of Natural Disturbance and Patch Dynamics* summarized these emerging viewpoints. This book was immensely influential because it came at the right time, calling attention to the disparate evidence for the role of disturbance in communities and providing an alternative way of thinking about community dynamics from traditional succession.

Considerable literature has been generated since the Pickett and White book by ecologists interested in elucidating disturbance dynamics as both a community process and a metapopulation process. However, most ecologists still have only a vague understanding of *how* natural disturbances operate and, consequently, have only an incomplete understanding of the ecological effects of these disturbances. The approach taken by ecologists has generally been to describe some aspects of a disturbance (e.g., size, frequency, severity, or season) and then to correlate this with the vegetation response (e.g., composition of the regenerating plant community). There have generally been few attempts to understand the disturbance processes and how they affect ecological processes. Because of this, the study of disturbance ecology has lost some of its forward momentum and has been reduced to simply giving the disturbance regime in terms of descriptive rather than causal variables. Furthermore, with only a vague understanding of the disturbance process, the variables used to describe disturbance regimes are sometimes arbitrary and often are unclear if the variable is of the disturbance or of the ecological effects (e.g., severity).

A way to overcome this descriptive tradition is to ask how the disturbance actually *causes* an ecological effect. The key is to understand those parts of the disturbance mechanism that are causally connected to the specific ecological process of interest. For example, if we are interested in tree mortality caused by wind, what we need first is an understanding of the specific wind phenomenon of interest (e.g., gusts, downbursts, hurricanes, etc.). Then we need to know how the wind will apply pressure to the tree canopy. This pressure will create a moment arm that will cause the trunk to bend quite rapidly (dynamic loading). The tree may then fail by breakage of the stem or uprooting or not, depending on the loading and on the physical properties of the tree. Of course, a local population consists of individuals with differences in their physical properties affecting their responses of bending, breaking, and uprooting given specific loadings.

This book will be useful to both physical scientists who want to know how their expertise could be useful in ecological and environmental areas and plant ecologists, environmental scientists, environmental managers, and foresters who want to have an up-to-date understanding of the physical processes involved in natural disturbances. The book is organized into sections, each of which deals with a particular type of disturbance process: wind, ice storm, geomorphic, hydrologic, combustion, and biotic. Each section includes one or two chapters providing background on the physical or biotic processes involved in the disturbance coupled with one or two chapters on how the disturbance causes necrosis or death to individuals and their effects on population or community processes. We have not tried to cover every type of disturbance affecting plant communities. One will immediately notice that there is a much more sophisticated understanding of the disturbances than of how the disturbance is causing specific ecological effects.

Edward A. Johnson
Kiyoko Miyanishi

Acknowledgments

We gratefully acknowledge the following people who contributed to this book by reviewing and providing very helpful comments on individual chapters of this book: G. T. Auble, J. K. Bailey, J. W. Bartolome, J. M. Briggs, Y. Brunet, P. E. Busher, R. G. D. Davidson-Arnott, P. J. Gerla, R. Hardy, M. Hayashi, A. Hopkin, C. Johnson, M. Kent, G. M. Lackmann, F. Lorenzetti, I. D. Lunt, D. A. MacLean, Y. E. Martin, M. A. Maun, S. Mitchell, J. R. Moore, K. F. Nordstrom, J. S. Olson, A. Potapov, A. G. Rhoads, D. Rosenberry, J.-C. Ruel, M.-H. Ruz, Michael L. Scott, D. J. Stensrud, B. R. Sturtevant, R. J. Trapp, D. X. Viegas, J. Volney, B. W. Webb, J. D. Wilson, S. Wilson, and S. S. Ziegler.

Marie Puddister in the Department of Geography, University of Guelph, prepared all of the final figures, converting all of the contributed graphics from various formats into uniformly formatted figures. We greatly appreciate her skill and patience in working with all of the contributing authors scattered across Canada, the United States, Scotland, Australia, and Japan.

Finally, we thank Charles Crumly and Kelly Sonnack, formerly of Academic Press/Elsevier, for their encouragement and incredible patience through the proposal and preparation stages of this book and Meg Day and Jonathan Cornwell for their help with the final stages.

1

Disturbance and Succession

Edward A. Johnson
University of Calgary

Kiyoko Miyanishi
University of Guelph

INTRODUCTION

Natural or anthropogenic disturbance was traditionally viewed as an event that initiated primary or secondary succession, and succession explained the development of vegetation in the absence of disturbance. Thus, the concepts of disturbance and succession are inextricably linked in plant ecology.

Succession has been used in so many different ways and situations that it is almost useless as a precise idea. However, no matter whether succession has been considered a population (Peet and Christensen, 1980), community (Cooper, 1923a; 1923b; Clements, 1916), or ecosystem (Odum, 1969) phenomenon or process, it has contained certain common ideas. Succession is an orderly unidirectional process of community change in which communities replace each other sequentially until a stable (self-reproducing) community is reached (see definitions in Abercrombie *et al.* 1973; Small and Witherick, 1986; Allaby, 1994). The explanation of why and how succession is directed has changed over its more than hundred-year history, but most arguments share the notion that species are adapted to different stages in successions and in some way make the environment unsuited for themselves and more suited for the species in the next stage. This group selection argument was first instilled into succession in the Lamarckian ideas of Warming, Cowles, and Clements.

Succession arose at the end of the 1800s and early 1900s out of a naturalist observation tradition when quantitative methods were almost nonexistent, Aristotelean essentialism (Hull, 1965a, 1965b; Nordenskiöld, 1928) still had a firm grip on how nature should be understood, and meteorology, soil science, biology, and geology were very poorly developed. Further, and equally important, spatial and temporal scales of observation were limited to the scale of a naturalist's sight.

DISTURBANCE AS THE NEMESIS OF SUCCESSION

By the beginning of 2000, most of the original classical examples of succession (e.g., Cowles, 1899; Shelford, 1911; Cooper, 1923b) given in textbooks had been restudied and found not to support the original arguments.

The first example in North America of primary succession was that on sand dunes (Cowles, 1899). The spatial sequence of plant communities as one moves away from the lake was interpreted by Cowles (1899; 1901) and Clements (1916) as representing a temporal succession of communities from dune grasses to cottonwoods, then pines and oaks to the climax beech-sugar maple forest (see Fig. 1 in Chapter 8). Olson's (1958) study of the same dunes using techniques that allowed actual dating of the dunes produced a much more complex picture of community changes than the previously proposed simple sequence from grasses to mesophytic forest. Olson found that dunes of similar age supported a wide range of plant communities, depending on the location as well as the disturbance history of the site.

A second classic example of primary succession was that on glacial till left by the retreating glacier at Glacier Bay, Alaska (Cooper, 1923a; 1923b; 1926; 1931; 1939). Again, the spatial pattern of vegetation on areas deglaciated at varying times was interpreted as representing the temporal stages of communities through which each site would pass from herbaceous *Dryas* and *Epilobium* to shrubby willow and alder thickets, then Sitka spruce forest, and finally the spruce-hemlock climax forest. Subsequently, Crocker and Major's (1955) study of soil properties at the different aged sites concluded that occupation of each site by the shrubs, particularly the nitrogen-fixing alders, allowed subsequent establishment of the later successional tree species through soil alteration (changes in pH and addition of carbon and nitrogen). However, Cooper's original study sites were reexamined by Fastie (1995), who found that the tree ring record from spruces in the oldest three sites did *not* indicate early suppression of growth with subsequent release once the spruces had exceeded the height of the alder or willow canopy. In other words, these oldest sites apparently had not experienced a succession from a community dominated by alders and willows to one dominated by spruce. Furthermore, the oldest sites showed a much more rapid colonization by a dense stand of trees soon after the sites were deglaciated compared to the younger sites. The differences between the different aged sites in their vegetation history (i.e., the order and rate of species establishment) as shown by Fastie's reconstructions were explained primarily by the availability of propagules (distance to seed source) at the time the retreating ice exposed the bare substrate. Interestingly, Cooper (1923b) had also noted that "establishment of the climax does not depend upon previous dominance of alder, for in the areas of pure willow thicket the spruces were found to be invading with equal vigor," "almost any plant of the region may be found among the vanguard," and "even the climax trees make their first appearance with the pioneers." Despite such observations, the lasting legacy of the early studies of primary succession at Glacier Bay has been the classic successional idea of sequential invasion and replacement of dominants driven by facilitation. As Colinvaux stated in his 1993 textbook *Ecology 2:* "The record from Glacier Bay shows that a spruce-hemlock forest cannot grow on the raw habitat left by the glacier, but that spruce trees and hemlock can claim habitats that have first been lived on by pioneer plants and alder bushes. . . . [I]t is undeniable that primary succession on glacial till at Glacier Bay is driven by habitat modification."

A third example of primary succession was the hydrarch succession of bogs and dune ponds. As with both of the previous examples, the spatial pattern of vegetation outward from the edge of bogs was interpreted as representing successional stages, leading to the conclusion by Clements (1916) that the open water would eventually become converted to a mesophytic forested site. However, the paleoecological reconstruction by Heinselman (1963) of the Myrtle Lake bog in Minnesota indicated that, despite deposition of organic matter and mineral sediments into the bog since deglaciation, the open water has persisted and has not been filled in and invaded by the surrounding forest because of the rising water table with the accumulation of peat.

Shelford (1911; 1913) used the spatial sequence of ponds in the Indiana Dunes of Lake Michigan to develop a model of temporal change in vegetation resulting from hydrarch succession. Jackson *et al.* (1988) tested this classic hydrosere model by using paleoecological data spanning 3000 years and found no evidence of significant change in vegetation until the early 1800s, when rapid change occurred following European settlement. They concluded that the spatial differences in vegetation along the chronosequence reflected differential effects of disturbance rather than any temporal successional pattern.

A fourth example, this time of secondary succession in forests, by Stephens (1955) and Oliver and Stephens (1977) concerned whether forest canopy composition resulted from the continuous recruitment of new stems of more shade-tolerant species. What they found in the old, mixed-species, northern hardwoods Harvard Forest in Massachusetts was the overriding influence of small- and large-scale disturbances, both natural and anthropogenic. While small disturbances allowed release of suppressed understory trees that might otherwise never make it to the canopy, large disturbances resulted in seedling establishment of new trees. Thus, the canopy composition was determined by disturbance processes. Foster (1988) came to the same conclusions about the old-growth Pisgah Forest in New Hampshire.

Poulson and Platt's (1996) long-term study of Warren Woods, the classic example of a climax beech-maple forest (Cain, 1935), led them to conclude that natural disturbances were chronic, occurring dependably on an ecological time scale and producing continual changes in light regimes. Because tree species respond differentially to the changing light conditions, different species are favored under different light regimes. Thus, the relative abundance of tree species in the understory at any given time *cannot* be used to predict the composition of the canopy at some later time. Poulson and Platt (1996) presented data to show that the relative abundance of beech and maple (as well as other species) in the canopy fluctuates in response to spatial and temporal fluctuations in frequency and sizes of treefall gaps. This is despite Cain's (1935) tentative conclusion, based on the abundant maple reproduction he observed, that "maple seems destined to increase in importance." In other words, the system is neither in, nor tending toward, an equilibrium climax community dominated by the most shade-tolerant species growing from the understory into the canopy.

Finally, the study by Forcier (1975) proposed a climax microsuccession with yellow birch replacing beech, sugar maple replacing yellow birch, and beech replacing sugar maple. This microsuccession model was based on a static study of trees less than 2.0 cm diameter at breast height (dbh) at Hubbard Brook Experimental Forest in New Hampshire to determine the pattern, structure, and population dynamics of the seedling layer. Major problems with this study include the assumption that this forest is in a climax state of equilibrium, despite its history of logging (1906–1920) and hurricane disturbance in 1938 (Merrens and Peart, 1992), and the lack of aging of the canopy trees to show that they had not established concurrently following disturbance. In fact, Foster (1988) cites numerous studies that indicate repeated disturbances of the forests of central New England by windstorms, ice storms, pathogens, fire, and short-term climate changes. For example, the patterns of growth response and establishment of the canopy trees in the Pisgah Forest in New Hampshire showed the impact of 12 historically recorded storms between 1635 and 1938.

THE CHRONOSEQUENCE BASIS OF SUCCESSION

Despite the evidence presented in the preceding and other empirical studies that do not support the traditional ideas of succession, the tendency for ecologists to see vegetation changes as stages of succession has persisted (see Egler, 1981). As noted by Burrows (1990), "the basic concept of sequential development of vegetation on bare surfaces (first a colonizing phase, followed by immature 'seral' phases and culminating in a mature and stable 'climax' phase) is firmly embedded in the literature of vegetation ecology and in the minds of many plant ecologists."

One reason for this persistence may lie in the chronosequence method typically used to study succession. This method involves a space-for-time substitution; that is, a chronosequence assumes that different sites, which are similar except in age since some initiating disturbance, can be considered a time sequence (Salisbury, 1952; Pickett, 1988). The key assumptions of chronosequences are that each of the sites representing different developmental stages had the same initial conditions and has traced the same sequence of changes. This assumption is rarely, if ever, carefully tested. In fact, the validity of this assumption is highly unlikely, given our increasing understanding of the temporal changes in environment and species availability over the time span represented by the chronosequences. As indicated in the previous section, studies that can "see" back in time through the pollen record or forest reconstructions (e.g., Stephens, 1955; Heinselman, 1963; Walker, 1970; Oliver and Stephens 1977; Jackson *et al.*, 1988; Johnson *et al.*, 1994; Fastie, 1995) have not shown the classic successional changes hypothesized from chronosequence studies. Instead, communities are found to be constantly changing, with species reassembling in different, often unfamiliar, combinations (Davis, 1981). These changes are often caused by changes in the physical environment.

COUPLING DISTURBANCE AND VEGETATION PROCESSES

Natural disturbances and often previously unappreciated human disturbances became a serious challenge to traditional succession beginning as early as the 1940s (e.g., Stearns, 1949; Raup, 1957) but really solidifying in the 1970s. Traditional succession viewed disturbances as infrequent and anomalous occurrences that initiated succession, which then proceeded in the absence of further disturbance. However, Raup (1957) commented that the ideas of succession and climax were "based largely upon the assumption of long-term stability in the physical habitat. Remove this assumption and the entire theoretical structure becomes a shambles."

The increasing recognition of the pervasiveness of disturbances (White, 1979) led to the idea that ecological systems consisted of patches of different times since the last disturbance. This approach was reviewed in the patch dynamics book of Pickett and White (1985). The early development of patch dynamics focussed on wind-created gaps in deciduous forests (e.g., Barden, 1981; Runkle, 1981; 1982) but has since spread to almost all types of vegetation, whether appropriate and supported by evidence or not. Here again we often see the chronosequence approach being used without testing the assumptions. Remarkably, the notion of patch dynamics did not overthrow the traditional concept of succession because many ecologists simply saw patch dynamics as representing microscale successions (Forcier, 1975). Thus, the contemporary concept of succession has become a strange combination of traditional ideas of succession and patch dynamics.

Part of the reason for this strange, often inconsistent, idea of succession has been a rather poor understanding of the disturbance itself. One way to make progress in the study of dynamics of ecological systems and disturbance is to connect the disturbance processes to specific ecological processes. By process, we mean a natural phenomenon composed of a series of operations, actions, or mechanisms that explain (cause) a particular effect. This research approach (Fig. 1) has at least three parts:

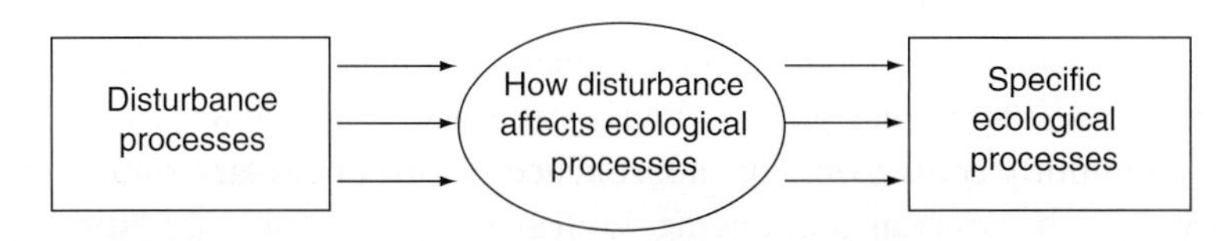

FIG. 1 Diagram illustrating the process-response model or approach to studying ecological effects of disturbance.

1. The ecological processes that will be affected by the disturbance must be precisely defined.
2. The parts of the disturbance processes that cause the ecological effect must be defined.
3. The ecological and disturbance processes must be brought together either as a coupling or a forcing.

Disturbance ecology has usually been approached in a much more informal manner than this program suggests. This has been due in part to the tools used by community ecologists (the main group of ecologists who have studied disturbances). Community ecologists in the last 50 years have taken a statistical–case study approach and have been less interested in physical environmental processes. The statistical–case study approach uses correlations or step-wise elimination curve-fitting (regressions) between variables that community ecologists hope are relevant. It is, in many ways, simply exploratory data analysis, although often used to develop predictive models. While such exploratory analysis plays a role in any research, the selection of variables is often haphazard, arbitrary, and guided by past usage or convenience (see example given by Miyanishi, 2001). No dimensional analysis is used to test either the selection of variables or the manner in which the variables are combined in the statistical model. Sometimes the variables are politically motivated (e.g., ecological integrity) and, after being chosen, the scientific models and units are then sought. Mechanisms by which cause and effect are defined are not explored formally. This is not to say that a statistical approach is not valid (e.g., population genetics) but that community ecologists have tended to use statistics to describe patterns. Rarely are statistical models seen as processes or mechanisms.

The study of earth surface interactions has been a flourishing research area that involves a wide range of disciplines, including atmospheric physics, hydrology, geomorphology, and biogeochemistry. The emerging field of biogeosciences has developed as the various disciplines have tried to integrate geophysical, geochemical, and biological (ecological) processes that are coupled to make up earth surface systems at all spatial and temporal scales (Hedin *et al.*, 2002). These developments appear to have had little effect on community ecologists and their approach to studying disturbances and vegetation dynamics (but see Waring and Running, 1998).

Community ecologists have approached disturbances largely as a multivariate set of variables that describe a disturbance regime. The axes of the multivariate space consist of general descriptive variables, such as frequency of the disturbance and its severity, intensity, and size (Fig. 2).

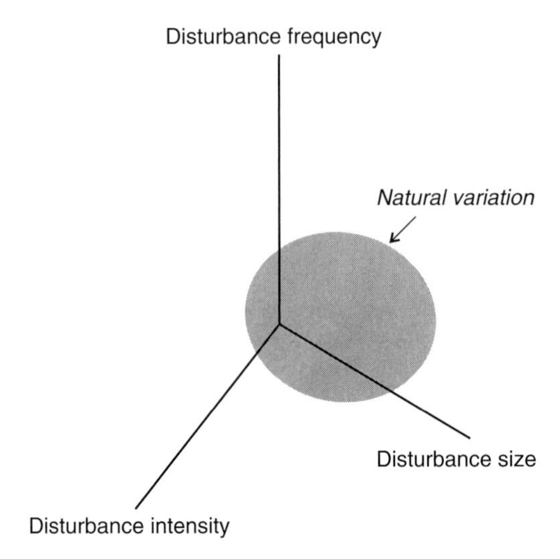

FIG. 2 Disturbance regime diagram illustrating the commonly used examples of disturbance descriptor variables such as frequency, intensity, and size. The shaded area attempts to indicate the multivariate space within which natural variation in a particular disturbance type occurs.

This approach does not clearly define the disturbance or ecological processes of interest. The variables are often themselves vague in what they measure (e.g., frequency of what process of the disturbance) or about whether it is the disturbance or ecological effect that is being considered (e.g., severity or intensity). The coupling or forcing is almost never clearly defined. Finally, this multivariate regime model has been used more as an informal idea and rarely tested with empirical data.

We now give an example of connecting snow avalanches to tree populations (cf. Johnson, 1987) to illustrate how the approach of coupling disturbance processes and ecological processes might be used to study vegetation dynamics. Trees that grow on avalanche paths are subject to breakage and uprooting from the recurrent avalanches. Avalanche frequency changes down the slope following an extreme-value distribution whose slope depends on the tangent of the slope angle (Fig. 3). The impact pressure (k Nm^{-2}) of avalanches also increases down slope (Fig. 4). The breakage of trees by avalanches can be determined by calculating the bending stress as the tree is deflected (Fig. 5). Bending stress (F) is determined by the applied load (P) and its lever arm (a), tree radius (r), and moment of inertia (I). Bending is determined for the deflection of the tree from its center of gravity by using a nonlinear differential equation for a tapered cantilever beam. Determining when

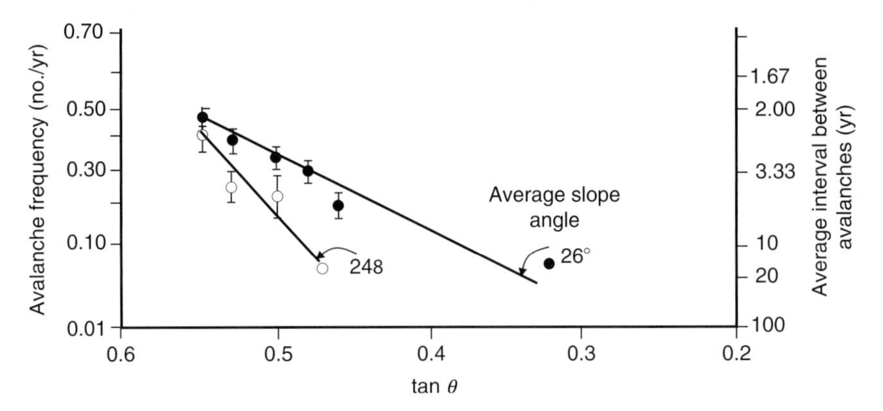

FIG. 3 Extreme-value distribution of avalanche frequency changes down slope.

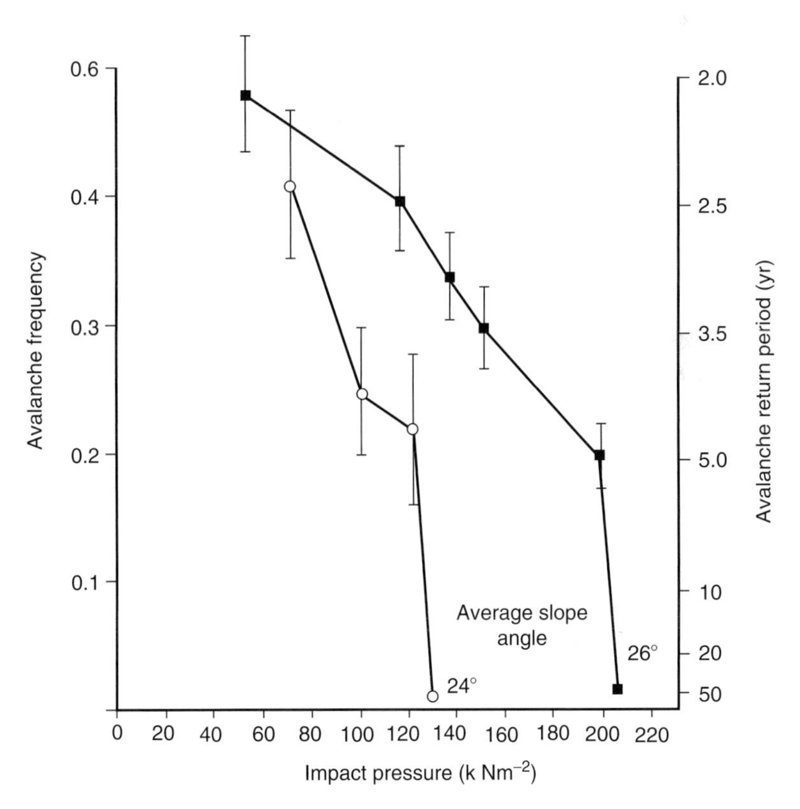

FIG. 4 Changes in impact pressure (k Nm^{-2}) of avalanches down slope.

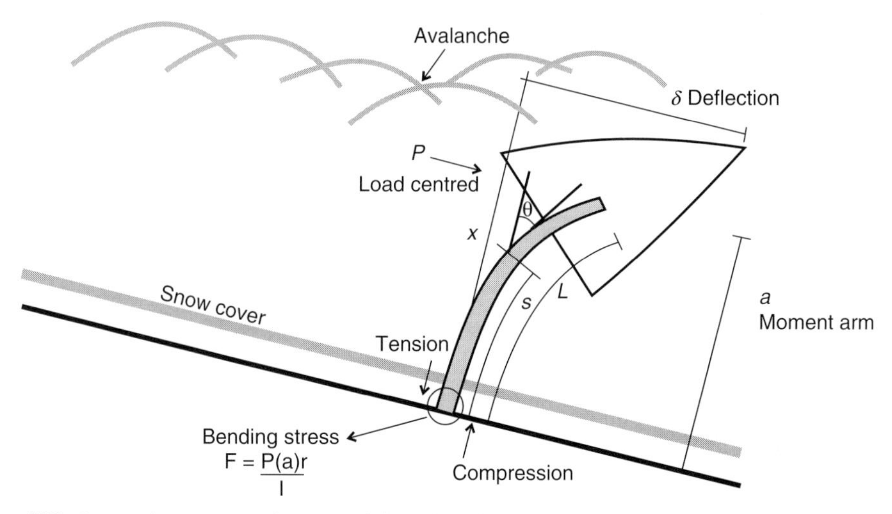

FIG. 5 Bending stress as the tree is deflected by the avalanche.

the bending stress of the deflected beam at different avalanche loadings surpasses the modulus of rupture tells when the tree will break. Thus, by using the ages of trees at different diameters, one can determine the age-specific mortality of trees from avalanches at different avalanche frequencies. This mortality can then be compared to the age-specific mortality from other causes, such as thinning. Remember, trees must be above a certain size to break, which usually means they are big enough to start competing.

CONCLUSION

In organizing this book, it became apparent to us that there was abundant literature in the physical sciences (e.g., atmospheric physics, meteorology, hydrology, geomorphology) on the physical processes involved in natural disturbances, indicating a reasonably good understanding of these disturbance processes. However, the literature addressing the ecological effects of these disturbances generally showed a lack of awareness of this literature and hence made little or no attempt to couple any of the ecological processes to the disturbance processes. As a result, most ecologists have clung to the concept of succession, despite not only the lack of empirical validation (beyond the flawed chronosequence studies) but also the abundant empirical evidence disproving it as "a pervasive and fundamental phenomenon in nature" (Pickett *et al.*, 1992).

The problem is that succession is often not well defined by users of the concept, and its meaning has undergone a shift from the clear definition given by its original proponents (and as stated in numerous technical dictionaries and textbooks of biology, ecology, and geography through the 1990s) to a vague and all-inclusive concept of simply vegetation change over time. A good example of this shift can be seen in the glossary definition of succession between the 1998 and 2004 editions of Christopherson's textbook *Elemental Geosystems*; the 1998 edition states that "changes apparently move toward a more stable and mature condition," while the 2004 edition replaces this with "communities are in a constant state of change as each species adapts to conditions; ecosystems do not exhibit a stable point or successional climax condition as previously thought."

We believe it is time we stop standing on our heads trying to make the concept fit our empirical observations and simply accept that the concept does not reflect a real phenomenon in nature and should therefore be abandoned. On one field trip in Colorado in the 1980s, after being informed by the trip leader that 224 different plant associations had been identified in the area, Grant Cottam wondered aloud, "In how many different ways can the number of species present be combined? I suspect it's very close to 224."

What we propose here is a change in viewpoint of vegetation dynamics that not only accepts the pervasive nature of disturbance in ecosystems but also incorporates the understanding that has developed of the physical disturbance processes. All disturbances have differential impacts on different populations within communities and also on different ecological processes. Therefore, to advance our understanding of the ecological effects of disturbances, we must couple the disturbance processes with ecological processes. In fact, as explained by Johnson *et al.* (2003), concepts such as metapopulations provide us with a way to incorporate disturbances into population processes.

The following chapters attempt to introduce this change in viewpoint by providing an introduction to the physical processes involved in a sampling of natural disturbances (wind, ice storms, hydrologic and fluvial disturbances, and so forth) and an attempt to couple these disturbance processes to their effects on individual plants, populations, and communities.

This book does not address all natural disturbances or all important ecological effects; for example, we do not discuss the effects of disturbances on nutrient cycling processes. However, we hope the book does provide a guide or blueprint for a different, more interdisciplinary, approach to plant disturbance ecology.

REFERENCES

Abercrombie, M., Hickman, C. J., and Johnson, M. L. (1973). *A Dictionary of Biology.* Penguin Books Ltd., Harmondsworth, UK.

Allaby, M. (Ed.) (1994). *The Concise Oxford Dictionary of Ecology.* Oxford University Press, Oxford.

Barden, L. A. (1981). Forest development in canopy gaps of a diverse hardwood forest of the southern Appalachian Mountains. *Oikos* 37, 205–209.

Burrows, C. J. (1990). *Processes of Vegetation Change.* Unwin Hyman, London.

Cain, S. A. (1935). Studies on virgin hardwood forest: III. Warren's Woods, a beech-maple climax forest in Berrien County, Michigan. *Ecology* 16, 500–513.

Christopherson, R. W. (1998). *Elemental Geosystems.* Pearson Education, Upper Saddle River, NJ.

Christopherson, R. W. (2004). *Elemental Geosystems.* 4th edition. Pearson Education, Upper Saddle River, NJ.

Clements, F. E. (1916). *Plant Succession: An Analysis of the Development of Vegetation.* Publication 242, Carnegie Institute of Washington, Washington, DC.

Colinvaux, P. (1993). *Ecology* 2. John Wiley & Sons, Inc., New York.

Cooper, W. S. (1923a). The recent ecological history of Glacier Bay, Alaska: Permanent quadrats at Glacier Bay: an initial report upon a long-period study. *Ecology* 4, 355–365.

Cooper, W. S. (1923b). The recent ecological history of Glacier Bay, Alaska: The present vegetation cycle. *Ecology* 4, 223–246.

Cooper, W. S. (1926). The fundamentals of vegetational change. *Ecology* 7, 391–413.

Cooper W. S. (1931). A third expedition to Glacier Bay, Alaska. *Ecology* 12, 61–95.

Cooper, W. S. (1939). A fourth expedition to Glacier Bay, Alaska. *Ecology* 20, 130–155.

Cowles, H. C. (1899). The ecological relations of the vegetation on the sand dunes of Lake Michigan. *Bot Gaz* 27, 95–117, 167–202, 281–308, 361–391.

Cowles, H. C. (1901). The physiographic ecology of Chicago and vicinity. *Bot Gaz* 31, 73–108, 145–182.

Crocker, R. L., and Major, J. (1955). Soil development in relation to vegetation and surface age at Glacier Bay, Alaska. *J Ecol* 43, 427–228.

Davis, M. B. (1981). Quaternary history and the stability of forest communities. *In Forest Succession: Concepts and Applications* (D. C. West, H. H. Shugart, and D. B. Botkin, Eds.). Springer-Verlag, New York, pp. 132–153.

Egler, F. E. (1981). Letter to the editor. *B Ecol Soc Am* 62, 230–232.

Fastie, C. L. (1995). Causes and ecosystem consequences of multiple pathways of primary succession at Glacier Bay, Alaska. *Ecology* 76, 1899–1916.

Forcier, L. K. (1975). Reproductive strategies and the co-occurrence of climax tree species. *Science* 189, 808–810.

Foster, D. R. (1988). Disturbance history, community organization and vegetation dynamics of the old-growth Pisgah Forest, south-western New Hampshire, U.S.A. *J Ecol* 76, 105–134.

Hedin, L., Chadwick, O., Schimel, J., and Torn, M. (2002). *Linking Ecological Biology & Geoscience: Challenges for Terrestrial Environmental Science.* Report to the National Science Foundation August 2002. Workshop held at the Annual Meeting of the Ecological Society of America, Madison, WI, August 4–5, 2001.

Heinselman, M. L. (1963). Forest sites, bog processes, and peatland types in the glacial Lake Agassiz region, Minnesota. *Ecol Monogr* 33, 327–374.

Hull, D. L. (1965a). The effect of essentialism on taxonomy—two thousand years of stasis, Part I. *Br J Philos Sci* 15, 314–326.

Hull, D. L. (1965b). The effect of essentialism on taxonomy-two thousand years of stasis, Part II. *Br J Philos Sci* 16, 1–18.

Jackson, S. T., Futyma, R. P., and Wilcox, D. A. (1988). A paleoecological test of a classical hydrosere in the Lake Michigan dunes. *Ecology* 69, 928–936.
Johnson, E. A. (1987). The relative importance of snow avalanche disturbance and thinning on canopy plant populations. *Ecology* 68, 43–53.
Johnson, E. A., Miyanishi, K., and Kleb, H. (1994). The hazards of interpretation of static age structures as shown by stand reconstructions in a *Pinus contorta–Picea engelmannii* forest. *J Ecol* 82, 923–931.
Johnson, E.A., Morin, H., Miyanishi, K., Gagnon, R., and Greene, D.F. (2003). A process approach to understanding disturbance and forest dynamics for sustainable forestry. In: V. Adamowicz, P. Burton, C. Messier, and D. Smith (eds) *Towards Sustainable Management of the Boreal Forest.* Ottawa: NRC Press. pp. 261–306.
Merrens, E. J., and Peart, D. R. (1992). Effects of hurricane damage on individual growth and stand structure in a hardwood forest in New Hampshire, USA. *J Ecol* 80, 787–795.
Miyanishi, K. (2001). Duff consumption. *In Forest Fires: Behavior and Ecological Effects* (E. A. Johnson and K. Miyanishi, Eds.). Academic Press, San Diego, pp. 437–475.
Nordenskiöld, E. (1928). *The History of Biology: A Survey* (Translated from the Swedish by L. B. Eyre). Tudor Publishing Co., New York.
Odum, E. P. (1969). The strategy of ecosystem development. *Science* 164, 262–270.
Oliver, C. D., and Stephens, E. P. (1977). Reconstruction of a mixed-species forest in central New England. *Ecology* 58, 562–572.
Olson, J. S. (1958). Rates of succession and soil changes on southern Lake Michigan sand dunes. *Bot Gaz* 119, 125-170.
Peet, R. K., and Christensen, N. L. (1980). Succession: a population process. *Vegetatio* 43, 131–140.
Pickett, S. T. A. (1988). Space-for-time substitution as an alternative to long term studies. *In Long-term Studies in Ecology: Approaches and Alternatives* (G. E. Likens, Ed.) Springer, New York, pp. 110–135.
Pickett, S. T. A. and White, P. S. (1985). *The Ecology of Natural Disturbance and Patch Dynamics.* Academic Press, Orlando, FL.
Pickett, S. T. A., Parker, V. T., and Fiedler, P. L. (1992). The new paradigm in ecology: implications for conservation biology above the species level. *In Conservation Biology: the Theory and Practice of Nature Conservation Preservation and Management* (P. L. Fiedler and S. K. Jain, Eds.). Chapman and Hall, New York, pp. 66–88.
Poulson, T. L., and Platt, W. J. (1996). Replacement patterns of beech and sugar maples in Warren Woods, Michigan. *Ecology* 77, 1234–1253.
Raup, H. M. (1957). Vegetational adjustment to the instability of the site. *In* Proceedings and Papers, 6th Technical Meeting, International Union for Conservation of Nature and Natural Resources, Edinburgh, June 1956, pp. 36–48.
Runkle, J. R. (1981). Gap regeneration in some old-growth forests of eastern United States. *Ecology* 62, 1041–1051.
Runkle, J. R. (1982). Patterns of disturbance in some old-growth mesic forests of eastern North America. *Ecology* 63, 1533–1546.
Salisbury, E. J. (1952). *Downs and Dunes: Their Plant Life and Environment.* G. Bell and Sons Ltd., London.
Shelford, V. E. (1911). Ecological succession. II. Pond fishes. *Biol Bull* 21, 127–151.
Shelford, V. E. (1913). *Animal Communities in Temperate America as Illustrated in the Chicago Region.* University of Chicago Press, Chicago.
Small, J., and Witherick, M. (1986). *A Modern Dictionary of Geography.* Edward Arnold, London.
Stearns, F. W. (1949). Ninety years change in a northern hardwood forest in Wisconsin. *Ecology* 30, 350–358.

Stephens, E. P. (1955). The Historical-Developmental Method of Determining Forest Trends. Ph.D. dissertation, Harvard University, Cambridge, MA.

Walker, D. (1970). Direction and rate in some British post-glacial hydroseres. *In Studies in the Vegetational History of the British Isles* (D. Walker and R. G. West, Eds.). Cambridge University Press, Cambridge, pp. 117–139.

Waring, R. H., and Running, S. W. (1998). *Forest Ecosystems: Analysis at Multiple Scales.* Second edition. Academic Press, San Diego.

White, P. S. (1979). Pattern, process and natural disturbance in vegetation. *Bot Rev* 45, 229–299.

2

The Turbulent Wind in Plant and Forest Canopies

John J. Finnigan

CSIRO Centre for Complex Systems Science

INTRODUCTION

Anyone who has walked in a tall forest on a windy day will have been struck by how effectively the trees shelter the forest floor from the wind. Tossing crowns and creaking trunks, and the occasional crash of a falling tree, attest to the strength of the wind aloft, but only an occasional weak gust is felt near the ground. An attentive observer will also notice that these gentle gusts seem to precede the most violent blasts aloft. Clearly, the presence of the foliage absorbs the strength of the wind very effectively before it can reach the ground. It is surprising, therefore, that only in the last two decades have we arrived at a satisfactory understanding of the way that interaction with the foliage changes the structure of the wind in tall canopies and leads to significant differences from normal boundary-layer turbulence. In this chapter I explore these differences, focusing especially on what the special structure of canopy turbulence implies for disturbance ecology. Gentle gusts at the ground foreshadowing the strong blasts aloft are but one example of the many curious and counterintuitive phenomena we will encounter.

While we now have a consistent theory of the turbulent wind in uniform canopies, the translation of this understanding to disturbed canopy flows is still in its infancy. Nevertheless, important recent advances have yielded general principles for the effect of hills and the proximity of forest edges and windbreaks on the mean flow, although this new understanding does not yet have much to say about effects on turbulence structure. My focus will generally be on strong winds, for which the effects of buoyancy on the turbulence are negligible, at least at canopy scale (although buoyancy can have significant effects at large scale, generating destructive downslope winds and microbursts as detailed in Chapter 3). However, buoyancy can be relevant at canopy scale in two contexts.

The first is fire. The heat released by burning fuel, whether the fire is confined to the ground cover or becomes a crown fire, can generate strong anabatic flow, with fires burning vigorously to hilltops and ridge crests. The second is cold air drainage, where the presence of the canopy exacerbates and concentrates gravity currents, which strongly influence carbon dioxide concentrations and ambient temperature patterns at night. Both these areas are poorly understood, although I make a few remarks at the end of this chapter based on some very recent analysis.

Returning to the central topic of strong winds and the damage they cause, phenomena that generate exceptional winds, such as cyclones, tornados, downslope windstorms, and microbursts, occur on scales much larger than the canopy height. Their ability to break or blow down trees depends on the way their effects are manifested at the canopy scale. We find that even very strong winds rarely exert static loads large enough to break or uproot mature trees. Instead, we must address phenomena such as resonance and the interaction between plant motion and turbulent eddies that can generate peak loads sufficient to cause the damage we observe. These questions are intimately involved with understanding the turbulent structure of canopies.

Let us begin by setting the scene for this discussion of canopy flow by a brief survey of atmospheric boundary layer flow in general, in which tall canopies form only one kind of a range of rough surfaces commonly encountered on land.

Notation

Most symbols are introduced when they first appear. Standard meteorological notation is used throughout based on a right-handed rectangular Cartesian coordinate system, x_i $\{x ,y, z\}$, with x_1, (x) aligned with the mean velocity, x_2, (y) parallel to the ground and normal to the wind, and x_3 (z) normal to the ground surface. Velocity components are denoted by u_i {u, v, w}, with u_i,(u) the streamwise, u_2,(v) the cross-stream horizontal, and u_3,(w) the vertical component. The time-average operator is denoted by an overbar and departures from the time-average by a prime, thus $c(t) = \overline{c} + c'(t)$. Within the canopy it is also assumed that a spatial average has been performed over thin slabs parallel to the ground so that the large point-to-point variation in velocity and other properties caused by the foliage is smoothed out. This spatial averaging process is an important formal step in deriving conservation equations for atmospheric properties in a canopy, but I do not discuss it further here. The interested reader should refer to Finnigan (2000) and references therein.

THE STRUCTURE OF THE ATMOSPHERIC BOUNDARY LAYER OVER LAND

Throughout most of its depth and for most of the time, the atmosphere is stably stratified with the virtual potential temperature increasing with height. The virtual potential temperature θ_v is the temperature corrected for the natural expansion of the atmosphere as pressure falls with increasing height and for the effect on buoyancy of the water vapor content of the air (Garratt, 1992). If the rate of change of θ_v with height is zero, then the atmosphere is neutrally stratified and a parcel of air displaced vertically will experience no buoyancy forces. For $\partial\theta_v/\partial z > 0$, the air is stably stratified and displaced parcels will tend to return to their origins. For $\partial\theta_v/\partial z < 0$, the air is unstably stratified and parcels will spontaneously accelerate vertically, generating convective turbulence.

Only in the atmospheric boundary layer during daytime is this state of stable stratification regularly overturned through the heating of the layers of air that are in contact with the surface. Conventionally, we divide the atmospheric boundary layer (ABL) into a relatively shallow surface layer (ASL) around 50 to 100 m deep and a convective boundary layer (CBL) extending some kilometers in height, depending on time of day. In the ASL, the wind and turbulence structure directly reflect the character of the surface, while in the CBL the main role of surface processes is to provide a buoyancy flux that determines the depth of the CBL and is the source of the kinetic energy of the turbulent eddies within it.

The ABL is turbulent for most of the time, and the source of this turbulence in the CBL is the flux of buoyancy produced at the warm surface. Heated parcels of air rise until they reach a level in the atmosphere at which they experience no further vertical accelerations. This level, z_i, increases rapidly from a few hundred meters or less at sunrise to 1 to 2 km or more around noon on sunny days when the growth in z_i slows and ceases. The top of the CBL at z_i is usually marked by a sharp discontinuity in temperature or "capping inversion" (Fig. 1A and B). At sundown there is a rapid collapse in z_i (Fig. 1B). The key scaling parameters in the CBL are z_i and the velocity scale, w^*,

$$w^* = \left[\frac{g}{T_0} \overline{w'\theta_v'}\, z_i \right]^{1/3} \tag{1}$$

where g is the acceleration of gravity, T_0 is a reference temperature in degrees K, and $\overline{w'\theta_v'}$ is the kinematic turbulent flux of buoyancy. The overbar denotes a time average, and the prime denotes a turbulent fluctuation around this average so that the kinematic buoyancy flux is the average covariance between turbulent fluctuations in θ_v and the vertical wind component, w. These scales remind us that the typical size of convectively-

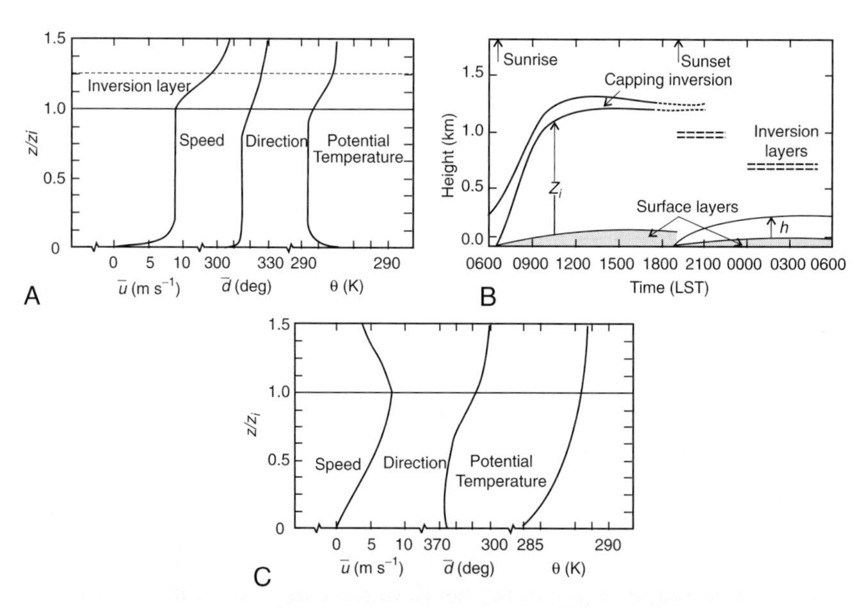

FIG. 1 Atmospheric boundary layer structure and profiles: (**A**) velocity and temperature profiles in the daytime convective boundary layer; (**B**) daily evolution of boundary layer structure; and (**C**) velocity and temperature profiles in the night-time stable boundary layer. (From Kaimal and Finnigan [1994], reprinted with permission from Oxford University Press © 1994.)

driven eddies in the CBL is its depth, z_i, while their velocity is set by the flux of buoyancy from the surface.

The vigorous mixing generated by convective turbulence in the CBL ensures that the average velocity and scalar gradients in the CBL are small as we see in Fig. 1A. In the daytime, strong wind shear and large gradients in potential temperature are confined to the ASL, which typically occupies 10% of the fully developed ABL. There, the turbulent eddies derive their energy from both the unstable stratification and the wind shear. In fact, the wind shear can maintain the ASL in a turbulent state even when the stratification is stable at night (Fig. 1C). These two contributions to turbulence are captured in the gradient Richardson number, Ri, the parameter we use to characterize the local stability of the atmosphere,

$$Ri = \frac{\dfrac{g}{T_0}\dfrac{\partial \overline{\theta_v}}{\partial z}}{[\partial \overline{u}/\partial z]^2} \tag{2}$$

Positive values of Ri denote stable stratification and negative, unstable. When Ri reaches positive values between 0.2 and 0.25, turbulence will decay and the boundary layer will become laminar except in some special circumstances. When Ri is negative, turbulence generated by the unstable stratification augments the "mechanical" turbulence produced by shear (Finnigan *et al.*, 1984).

Within the ASL, the atmosphere responds much more directly to surface conditions than in the CBL, and this is reflected in the length and velocity scales used to describe the turbulence there. The appropriate length scale is the height z so that the distance to the ground controls the eddy size. The velocity scale is u^*, a parameter called the friction velocity, which is formed from the frictional force τ exerted by the wind on the ground.

$$u^* = \sqrt{\tau/\rho} = \sqrt{-\overline{u'w'}} \tag{3}$$

where ρ is air density, and over flat ground in steady winds, the kinematic surface stress, τ/ρ, is equal to (minus) the average covariance between streamwise, u', and vertical, w', wind fluctuations in the ASL, $\tau/\rho = -\overline{u'w'}$. Combining the two scales above and introducing a further lengthscale, the Obukhov length, L_{MO}, to represent the effects of buoyancy in the surface layer, a comprehensive similarity theory that describes profiles of wind, temperature, and other scalars, as well as their turbulent moments, has been developed for the ASL. This is the eponymous Monin-Obukhov scaling.

With

$$L_{MO} = \frac{u^{*3}}{\left[\kappa \frac{g}{T_0} \overline{w'\theta'_v}\right]} \tag{4}$$

Monin-Obukhov theory predicts that the velocity and scalar profiles in the surface layer will be logarithmic, taking the form:

$$\overline{u}(z) = \frac{u^*}{\kappa} \ln\left(\frac{z-d}{z_0}\right) - \psi_m\left(\frac{z-d}{L_{MO}}\right) \tag{5}$$

$$\overline{\theta}(z) - \overline{\theta}_0 = \frac{\theta^*}{\kappa} \ln\left(\frac{z-d}{z_\theta}\right) - \psi_\theta\left(\frac{z-d}{L_{MO}}\right) \tag{6}$$

where $\kappa \approx 0.4$ is an empirical constant called von Karman's constant, d is an effective origin for the logarithmic profile that must be included when Monin-Obukhov theory is applied over tall vegetation, and z_0 and z_θ are the "roughness lengths" for momentum and scalar θ, respectively. They characterize the exchange properties of the surface for the particular species. The reference concentration of the scalar at the canopy surface is $\overline{\theta}_0$, while the "no-slip" condition ensures that the surface reference value for velocity is zero. The empirical functions ψ_m and ψ_θ of the dimensionless stability parameter $(z - d)/L_{MO}$ embody the influence of stability on the velocity and scalar profiles, respectively. The effect on the logarithmic velocity profile of stable and unstable stratification can be seen in Fig. 2.

The canonical state of the daytime ABL in light to moderate winds is taken to be a convective well-mixed CBL with negligible gradients in wind and scalars above an ASL where profiles obey the logarithmic Monin-Obukhov forms of Equations (5) and (6). In reality, we rarely observe a truly well-mixed CBL, and measured wind and scalar profiles in the CBL usually have some vertical structure but, on average, the gradients there are much smaller than within the ASL. When synoptic winds are very strong, however, significant shear can exist through most of the boundary layer while the vertical variation in θ_v is small. The influence of the surface buoyancy flux then becomes secondary, and a balance between surface friction and Coriolis effects sets the depth of the ABL (Garratt, 1992). Similarly, at night, as the ground cools by radiation and the boundary layer is stably stratified, turbulent mixing is weak and the nocturnal boundary depth is seldom more than 200 m.

Over rough surfaces, which for our purposes are all land surfaces, we distinguish a further subdivision of the lowest part of the ASL into the canopy layer and the roughness sublayer (RSL); in the "canopy" category

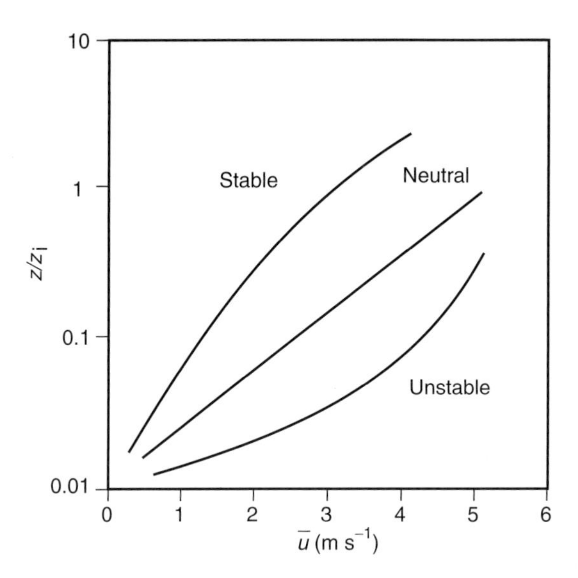

FIG. 2 Changes to the logarithmic velocity profile in the surface layer caused by stable and unstable stratification. (From Kaimal and Finnigan [1994], reprinted with permission from Oxford University Press © 1994.)

we include all tall roughness. The RSL extends from the ground to around two to three canopy heights, while the canopy layer itself occupies the bottom half or third of the RSL. What distinguishes the RSL from the ASL as a whole is that, within the RSL, the vertical and horizontal distributions of sources and sinks of scalars and momentum affect the average profiles, causing them to depart from the Monin-Obukhov forms (Fig. 3). For example, as the canopy is approached from above even in neutral conditions, the velocity and scalar profiles adopt forms similar to those found above the RSL in unstable conditions, where the shear-generated turbulence is augmented by buoyancy.

Above the roughness sublayer, boundary layer turbulence over any kind of flat rough surface is essentially indistinguishable from that over a smooth surface (Raupach *et al.*, 1991). Within the RSL and canopy layers, however, the differences are profound. Those differences are the main concern of the rest of this chapter. Detailed accounts of the behavior of the ABL, its turbulent structure, and the characterization of its several states may be found in many textbooks. To sample some different perspectives, see Kaimal and Finnigan (1994), Garratt (1992), Panofsky and Dutton (1984), or Fleagle and Businger (1963).

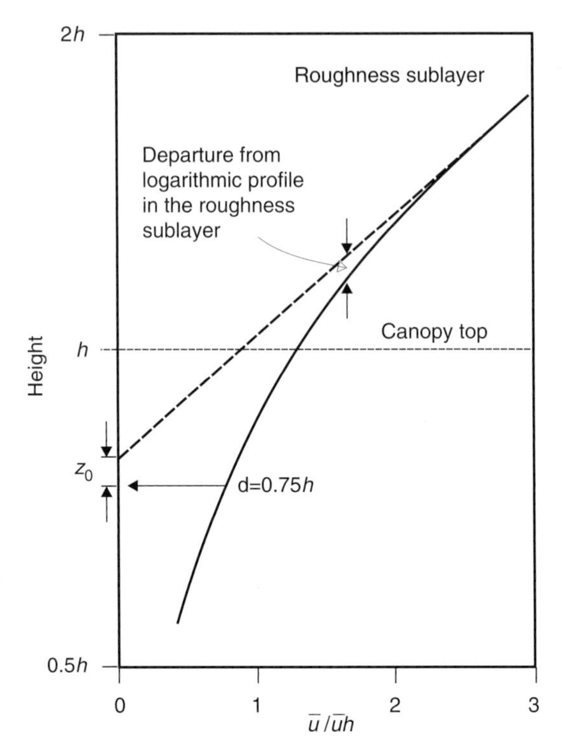

FIG. 3 Modifications to the logarithmic velocity profile in the roughness sublayer and the canopy. (From Kaimal and Finnigan [1994], reprinted with permission from Oxford University Press © 1994.)

CHARACTERISTICS OF TURBULENT FLOW IN AND ABOVE PLANT CANOPIES

The first theories of canopy turbulence supposed that the high turbulence intensities measured amid and just above the foliage were simply the result of the superposition of eddies shed from leaves and stalks onto standard boundary layer turbulence. However, this did not explain the observations that were available from the earliest two-point turbulence measurements (e.g., Allen, 1968) that canopy turbulence was correlated over distances similar to the canopy height—a much larger lengthscale than any associated with eddies shed from the foliage. Through the 1970s and 1980s, solid evidence accumulated to show that canopy turbulence was dominated by large, energetic eddies of whole-canopy scale and that the turbulence structure and dynamics were distinctly different from those of boundary layer turbulence.

This evidence was a mixture of the circumstantial, such as the failure of diffusive models of turbulent transport in the canopy (Denmead and Bradley, 1987), and the direct, such as that obtained from the application of conditional sampling techniques borrowed from wind tunnels to measurements in the field. These revealed the strongly intermittent nature of eddy activity in the canopy (Finnigan and Mulhearn, 1978; Finnigan, 1979a; 1979b; Shaw *et al.*, 1983) and showed that momentum transfer was accomplished by the infrequent penetration of strong gusts into the foliage from the ASL above. Time-height reconstructions of the turbulent velocity field in the x-z plane, obtained from multiple sensors in forests (e.g., Gao *et al.*, 1989; Gardiner, 1994), were the first to link directly the spatial structure of the eddies with their temporal intermittency and confirmed earlier inferences from space–time correlations (Shaw *et al.*, 1995).

Finally, a phenomenological theory—the "mixing layer" hypothesis—that explained the origin and scale of these large eddies was advanced by Raupach *et al.* (1996). Since then, a range of measurements in uniform canopies has strengthened its predictions, while limits to its range of validity have been explored by measurements in sparse canopies (Bohm *et al.*, 2000; Novak *et al.*, 2000). A recent survey of canopy turbulence (Finnigan, 2000) deals with the nature and dynamics of canopy turbulence in great detail. Here I concentrate on characteristics especially relevant to the forces exerted on plants in the canopy and hence to canopy damage and disturbance.

Velocity Moments

Fig. 4 (A–I) consists of a set of "family portraits" of single-point turbulence statistics measured in near-neutral flow in 12 canopies on flat ground. These plots are taken from Raupach *et al.* (1996) and range from wind tunnel models through cereal crops to forests. Details of the various canopies included in these plots are given in Table 1. The canopies span a wide range of roughness density λ (defined as the total frontal area of canopy elements per unit ground area) and a 500-fold height range. The vertical axis and length scales are normalized with canopy height h and velocity moments with either $U_h = \overline{u}(h)$, the value of the time averaged "mean" velocity at the canopy height, or the friction velocity u^*, where u^* is measured in the constant shear stress layer above the canopy. Details for each experiment are given in Table 1. The observations in Fig. 4 have many common features, their differences being mainly attributable to their differing foliage area distribution. Fig. 4J shows that the main differences in this parameter lie in the extent to which the foliage is clustered in a crown at the canopy top.

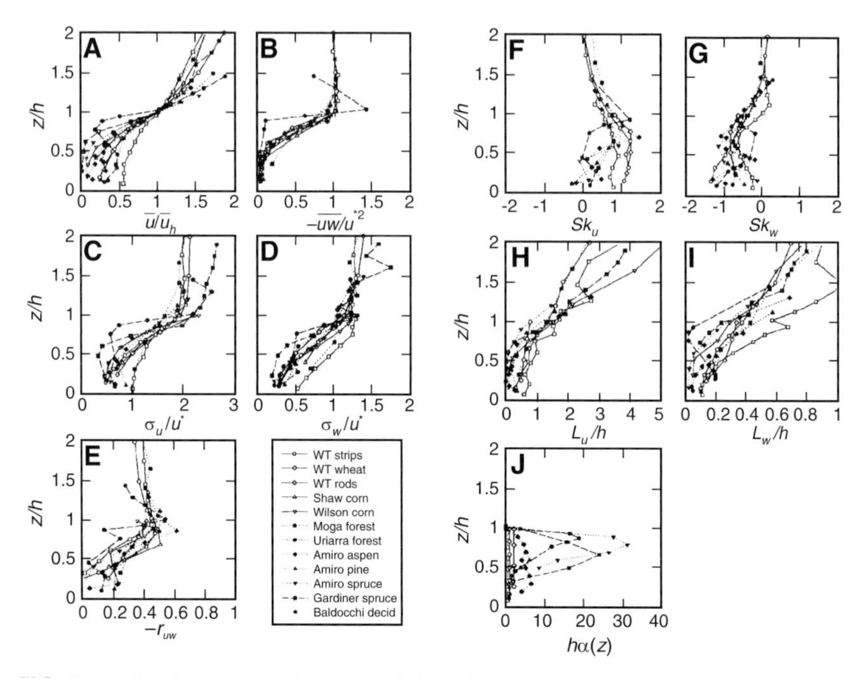

FIG. 4 A "family portrait" of canopy turbulence for canopies A to L in Table 1, showing profiles with normalized height z/h of (A) $U/U_h = \bar{u}(z)/\bar{u}(h)$; (B) $-\overline{u'w'}/u^{*2}$; (C) σ_u/u^*; (D) σ_w/u^*; (E) $-r_{uw} = -\overline{u'w'}/\sigma_u\sigma_w$; (F) *Sku*; (G) *Skw*; (H) *Lu*; (I) *Lw* ; (J) $h\alpha(z)$, where $\alpha(z)$ is leaf area per unit volume. *Lu* and *Lw* are integral length scales derived by applying Taylor's hypothesis to the point-valued integral time scales of the fluctuating velocity (Finnigan, 2000). (From Raupach *et al.* [1996], reprinted with permission of Springer Science and Business Media.)

All normalized mean velocity profiles $\bar{u}(z)/U_h$ shown in Fig. 4A have a characteristic inflection point at $z = h$. The shear is maximal at $z = h$, and its strength can be described by the length scale, $L_h = \bar{u}(h)/(\partial\bar{u}(h)/\partial z)$. Above the canopy, we observe a standard boundary layer profile; within the canopy space, the profile can be described as roughly exponential. As we shall see later, the inflexion point plays a critical role in canopy dynamics.

In Fig. 4B, the downward momentum flux $-\overline{u'w'}$ normalized by u^*, we see a standard constant stress layer above the canopy[1] and then the rapid decay of $-\overline{u'w'}$ as streamwise momentum is absorbed as aerodynamic drag on the foliage. Note that in all the cases of Fig. 4, $-\overline{u'w'}$ is practically zero

[1] A constant stress layer is a signal that the mean velocity is steady in time and varies only in strength in the vertical direction. Micrometeorologists take this as the simplest useful reference state of the planetary boundary layer against which more complex situations, such as flow over hills or across forest edges, can be compared.

TABLE 1 Physical and Aerodynamic Parameters of the Canopies in Fig. 4*

Canopy	Identifier	h(m)	λ	Uh/u*	Ls/h	Reference
WT strips	A	0.06	0.23	3.3	0.85	Raupach *et al.* (1986)
WT wheat	B	0.047	0.47	3.6	0.57	Brunet *et al.* (1994)
WT rods	C	0.19	1	5.0	0.49	Seginer *et al.* (1976)
Shaw corn	D	2.6	1.5	3.6	0.39	Shaw *et al.* (1974)
Wilson corn	E	2.25	1.45	3.2	0.46	Wilson *et al.* (1982)
Moga eucalypt	F	12	0.5	2.9	0.58	Unpublished
Uriarra pine	G	20	2.05	2.5	0.29	Denmead and Bradley (1987)
Amiro aspen	H	10	1.95	2.6	0.58	Amiro (1990)
Amiro pine	I	15	1	2.2	0.50	Amiro (1990)
Amiro spruce	J	12	5	2.4	0.44	Amiro (1990)
Gardiner spruce	K	12	5.1	4.0	0.30	Gardiner (1994)
Baldocchi deciduous	L	24	2.5	2.8	0.12	Baldocchi and Meyers (1988a; b)

*The roughness density or frontal area index, λ, is assumed to be half the single-sided leaf area index for field canopies.

by ground level, indicating that all the horizontal momentum has been absorbed by the canopy elements and does not reach the ground, even though some of the canopies in the ensemble are not particularly dense. The normalized standard deviations of streamwise and vertical fluctuations, pictured in Fig. 4C and D, are more scattered than $-\overline{u'w'}$ within the canopy but are also strongly inhomogeneous in the vertical. Values of σ_u/u^* seem to group around 0.75 in the lower canopy, indicating that there is a good deal of horizontal "sloshing" motion there but that this motion is *inactive* in the sense that it transfers little momentum.

In Fig. 4E, the correlation coefficient $r_{uw} = \overline{u'w'}/\sigma_u\sigma_w$ can be interpreted as the efficiency of momentum transport. Well above the canopy, r_{uw} assumes its standard ASL value of about –0.32, corresponding to $\sigma_u/u^* = 2.5$, $\sigma_w/u^* = 1.25$, values typical of a constant stress surface layer (Garratt, 1992). At the canopy top, in contrast, the stress ratios fall to $\sigma_u/u^* = 2.0$, $\sigma_w/u^* = 1.0$, producing a value of $r_{uw} = -0.5$. Although there is a good deal of scatter in the data, it is clear that the turbulence around the canopy top transports substantially more momentum per unit variance than in the surface layer above, suggesting that, in some way, the character of the turbulence has changed.

Moving from second to third moments, we observe in Figs. 4F and G that the skewnesses of streamwise and vertical velocity fluctuations are of order +1 and –1, respectively, in the canopy space, in sharp contrast to the near-zero values these statistics assume in the ASL. Positive skewness tells us that large positive excursions in velocity (gusts) are much more likely than negative excursions (lulls). Gaussian velocity distributions have skewnesses of zero. Large positive Sk_u and negative Sk_w values are associated with the intermittent penetration of the canopy by strong streamwise gusts from the ASL.

Length and Time Scales

In Figs. 4H and I, the streamwise L_u and vertical L_w Eulerian integral length scales obtained from integral time scales using Taylor's "frozen turbulence" hypothesis are plotted,

$$L_u = \frac{\overline{u}}{\sigma_u^2} \int_0^\infty \overline{u_1'(t) u_1'(t+\tau)}\, d\tau; \qquad L_w = \frac{\overline{u}}{\sigma_w^2} \int_0^\infty \overline{u_3'(t) u_3'(t+\tau)}\, d\tau \qquad (7)$$

Both these scales are much larger than the size of individual canopy elements, reaching $L_u \cong h$; $L_w \cong h/3$ in the upper canopy. Other measures of the scale of turbulent eddies, both indirectly from spectra (Kaimal and Finnigan, 1994) and directly from two-point velocity measurements (Shaw *et al.*, 1995), confirm that typical sizes of turbulent eddies in tall canopy flows are of order the canopy height.

Large Eddy Structure in Canopy Turbulence

Until the 1960s, the limitations of turbulence sensors in wind tunnels and in the field meant that we had only a sketchy idea of the detailed structure of turbulent flows. Turbulence was assumed to be a superposition of "eddies"—coherent patterns of velocity and pressure—spanning a wide range of space and time scales. More sophisticated measurement and visualization techniques applied from the mid 1960s onward, however, began to reveal that different turbulent flows were dominated by eddies of quite different character and that the average properties of the different flows, such as their ability to transport momentum and scalars, were linked to these differences in eddy structure. In shear flows, such as boundary layers, it quickly became clear that large coherent eddies that spanned the whole width of the shear layers contained most of the kinetic energy of the turbulence and were responsible for most of the transport.

Most of the early information on large eddy structure in canopy turbulence was circumstantial inasmuch as what was really being observed was intermittency in time series measured at single points. Conditional sampling techniques, such as "quadrant-hole" analysis, revealed that transport of momentum and heat to the canopy is primarily accomplished by sweeps (downward incursions of fast-moving air) in contrast to the surface layer above, where transfer is dominated by ejections (upward movement of slow-moving air). Moreover, transfer by these sweeps is very intermittent. For example, Finnigan (1979b) found in the upper levels of a wheat crop that more than 90% of the momentum transfer occurred only 5% of the time. More sophisticated techniques, such as wavelet analysis (e.g., Collineau and Brunet, 1993; Lu and Fitzjarrald, 1994), clarified the sequence of ejections and sweeps in a typical transfer event, but these point measurements alone could not shed light on the spatial structure of the eddies responsible.

Time-height plots of data from vertical arrays of anemometers were the first to show unequivocally that the transfer events were coherent through the roughness sublayer and that the sweep events corresponded to displacement of canopy air by the coherent airmass of a canopy scale eddy. Fig. 5, from Gao *et al.* (1989), shows a time-height plot from an 18-m tall mixed forest in southern Canada. Both velocity vectors and temperature contours are shown. The velocity vectors demonstrate that an ejection precedes the sweep event and that it is weaker than the sweep within the canopy but stronger above. The compression of temperature contours into a microfront at the leading edge of the sweep is also characteristic and in a time series is signaled by a "ramp" structure, where a slow increase in temperature over periods of minutes is followed by a sudden temperature drop that takes only seconds.

Time-height plots such as that shown in Fig. 5 are usually composites of many events, with the sharply changing front of the ramp structure being used as a sampling trigger so that individual events can be aligned. Since they also rely on the wind to advect the turbulent field past the sensor [so that *minus* time can be taken as a surrogate for the streamwise (x) space axis], they reveal only a slice through the eddy on the x-z plane and not its three-dimensional structure. An extension and refinement of the time-height compositing technique by Marshall *et al.* (2002) has used wavelet transforms to produce pseudo-instantaneous flow maps coincident with the maximum bending force on a model tree in a wind tunnel. The use of the maximum force on the tree as a sampling trigger is more directly relevant to the subject of tree damage, of course, and it also confirms that the sweep events are indeed those responsible for exerting the maximum bending moment, thereby establishing a direct connection that allows earlier information on large eddy structure to be interpreted in terms of tree damage.

A more objective approach to deducing the spatial structure of the dominant eddies has been taken by Finnigan and Shaw (2000) and Shaw *et al.* (2004), who have applied the techniques of empirical orthogonal function

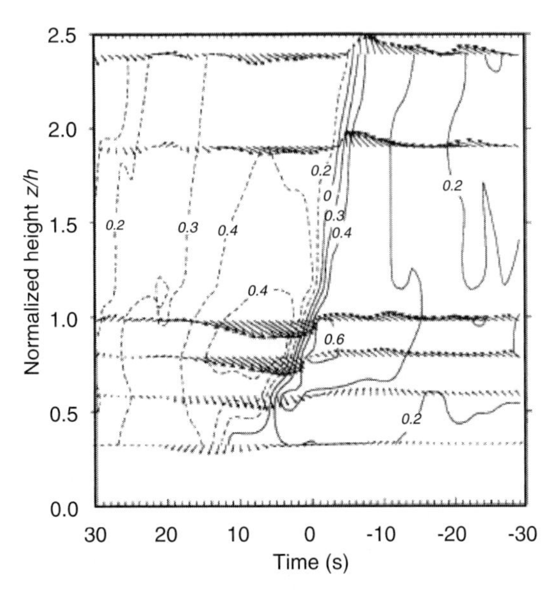

FIG. 5 Time-height plot of ensemble averaged temperature and fluctuating velocity fields measured in moderately unstable conditions ($L_{MO} = -138$ m) in an 18-m tall mixed forest. Dashed lines are isotherms below the mean and solid lines are isotherms above the mean. Contour interval is 0.2°C and the maximum arrow length represents a wind magnitude of 1.9 m s^{-1}. (From Gao *et al.* [1989], reprinted with permission of Springer Science and Business Media.)

(EOF) analysis to a wind tunnel model canopy and to the output of a large eddy simulation model of canopy flow. EOF analysis in the context of turbulent flows was introduced by Lumley (1967; 1981). It consists of finding the sequence of orthogonal eigenfunctions and associated eigenvalues that converges optimally fast when the variance or kinetic energy of the turbulent flow is represented as the sum of this sequence. The spatial structure of the turbulent field is contained in the eigenfunctions, and the rate of convergence of the sequence of eigenvalues is a sensitive indicator of the presence and relative importance of coherent structures. In shear flows dominated by coherent motions, a large fraction of the variance or turbulent kinetic energy is captured in just the first few eigenmodes, whereas in flows with no dominant structure the rate of convergence is slower. A great strength of the method is that only trivial assumptions about the form of the dominant structure need be made a priori so that we obtain essentially objective knowledge about the spatial structure of the coherent motion. Lumley (1981), in introducing the application of the technique to turbulent flows, thoughtfully discusses the relative merits and drawbacks of conditional sampling and the EOF method.

Despite the advantages of the EOF method, it has not been widely used in turbulent flows because the required empirical eigenfunctions are those of the two-point velocity covariance tensor $R_{ij}(\mathbf{x}, \mathbf{x}') = \overline{u_i'(\mathbf{x})u_j'(\mathbf{x}')}$ with spatial separation in three dimensions, a data set whose collection requires a particularly intensive experiment. The technique has been applied to numerical simulations as much as to real data. Moin and Moser (1989), for example, performed an EOF analysis of a numerical simulation of channel flow; their paper is an excellent introduction to the technique. Finnigan and Shaw (2000) used as their data set a two-point covariance field obtained in a wind tunnel model canopy as described by Shaw *et al.* (1995). Shaw *et al.* (2004) applied the method to a large eddy simulation[2] of the velocity and scalar fields of a canopy with the same physical characteristics as the wind tunnel model.

With some simple and plausible assumptions about the distribution of the coherent eddies in space, EOF analysis enables us to reconstruct the instantaneous velocity and scalar fields of a "typical" large eddy without any a priori assumptions, such as the coincidence of the eddy with the scalar microfront or the maximum tree bending moment. Fig. 6 shows the velocity

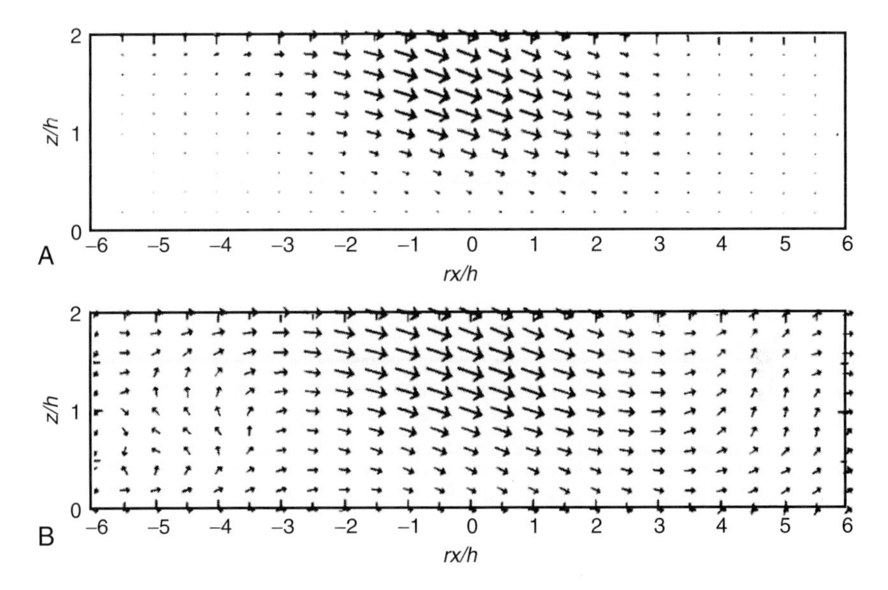

FIG. 6 The u', w' vector field of the characteristic eddy projected onto the $(rx,0,z)$ plane, where the horizontal coordinate rx is relative to the notional center of the eddy. (**A**) Arrow lengths are proportional to the magnitude of the velocity vector. (**B**) Arrow lengths in regions of low velocity are magnified to show the flow direction more clearly. (From Finnigan and Shaw [2000], reprinted with permission of Springer Science and Business Media.)

[2] Large eddy simulation is a computer-intensive modeling technique that solves the Navier Stokes equations directly for the large, energy-containing turbulent eddies at the expense of parameterizing the fine scales. It works especially well in canopy turbulence. See, for example, Shaw and Patton (2003).

field of the characteristic eddy deduced from wind tunnel data by Finnigan and Shaw (2000) projected onto the $(x,0,z)$ plane so it can be compared with the time-height composite reconstructions of Gao *et al.* (1989) shown in Fig. 5. If anything, the dominance of the sweep motion is clearer in the EOF reconstruction than in the time-height plot. Fig. 7, also from Finnigan and Shaw (2000), shows a cross-section of the eddy at its midpoint ($x/h = 0$ in Fig. 6) projected onto the y-z plane. This vector plot reveals information that cannot be found in time-height plots. It shows that: the eddy consists of a double roller vortex pair, with the strong downward sweep motion being concentrated on the plane of symmetry; the blocking action of the ground produces strong lateral outflows in the upper canopy; and the recirculation around the vortex centers, which are situated at about $z/h \approx \pm 1.5$; $y/h \approx \pm 1$, occurs in the upper part of the roughness sublayer.

Of particular interest in the context of forces on the plants is the spatial distribution of momentum transfer by the characteristic eddy, which is shown in Fig. 8 as a contour plot on the y-z plane at $x/h = 0$. Here we see that the sweep motion, which is the part of the eddy responsible for strong aerodynamic forces on the plants, is confined to a width of $y/h \approx \pm 0.5$

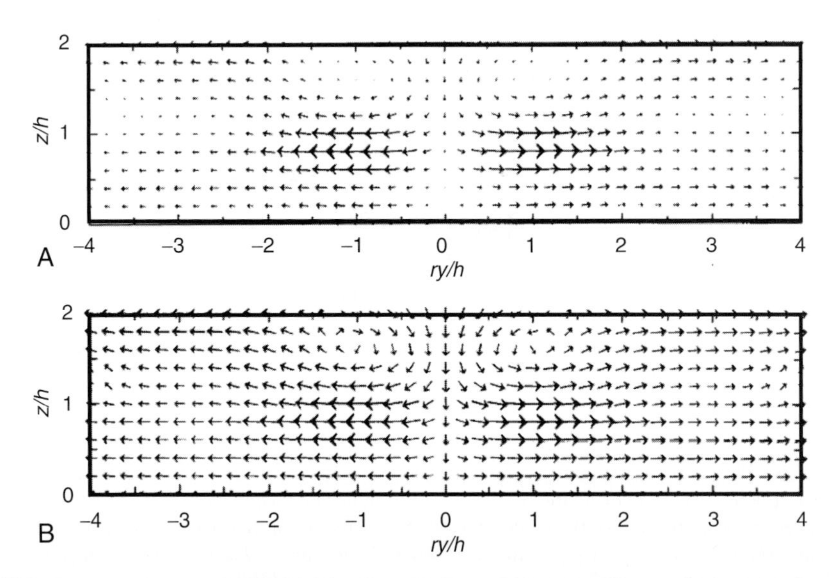

FIG. 7 The v', w' vector field of the characteristic eddy projected onto the $(ry,0,z)$ plane, where the horizontal coordinate ry is relative to the notional center of the eddy. (**A**) Arrow lengths are proportional to the magnitude of the velocity vector. (**B**) Arrow lengths in regions of low velocity are magnified to show the flow direction more clearly. (From Finnigan and Shaw [2000], reprinted with permission of Springer Science and Business Media.)

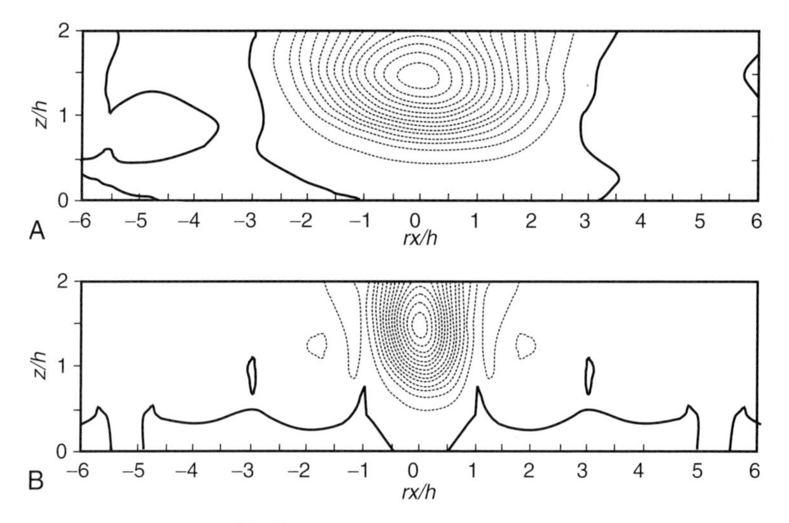

FIG. 8 Spatial contours of $\overline{\{u',w'\}}$ in the characteristic eddy. (**A**) Projection onto the (*rx,0,z*) plane (contour interval, 0.5; minimum value, –7.34). (**B**) Projection onto the (*0,ry,z*) plane (contour interval, 0.5; minimum value, –7.55). Solid contour lines denote positive covariance, dotted lines negative covariance.

while the circulation of the pair of vortices that make up the eddy ensure that the sweep is flanked laterally by two ejection regions as the upward motion of the vortices carries low momentum (and warmer/colder air in the daytime/night-time) out of the canopy. Combining Figs. 6, 7, and 8 indicates that although the characteristic canopy eddies are coherent over about 8 hours in the streamwise and 4 hours in the lateral directions, the intense sweep region is much smaller, being about 3 hours in the streamwise and 1 hour in the lateral directions. This three-dimensional information has obvious relevance when we come to consider windthrow and crop lodging later in this chapter.

A Dynamic Model for the Large Eddies—the Mixing Layer Hypothesis

It is evident not only that the statistics of turbulence in the RSL are quite different from those in the surface layer above but the large-eddy structure deduced by conditional sampling and EOF analysis also has distinct differences. A decade ago, Raupach *et al.* (1986; 1996) proposed that these differences might be explained by taking the plane-mixing layer rather than the boundary layer as a pattern for RSL turbulence.

The Plane-Mixing Layer

The plane-mixing layer is the free shear layer that forms when two airstreams of different velocity, initially separated by a splitter plate, merge downstream of the trailing edge of the plate. Flows of this kind are common in engineering applications and have been intensively studied. Conventionally, the splitter plate occupies the horizontal half plane ($x < 0$; $z = 0$) and the difference between the free stream velocities above and below the plate is ΔU. The width of the mixing layer can be characterized by the vorticity thickness $\delta_\omega = \Delta U / (\partial \bar{u}/\partial z)_{\max}$ as shown in Fig. 9. Even if the boundary layers on each

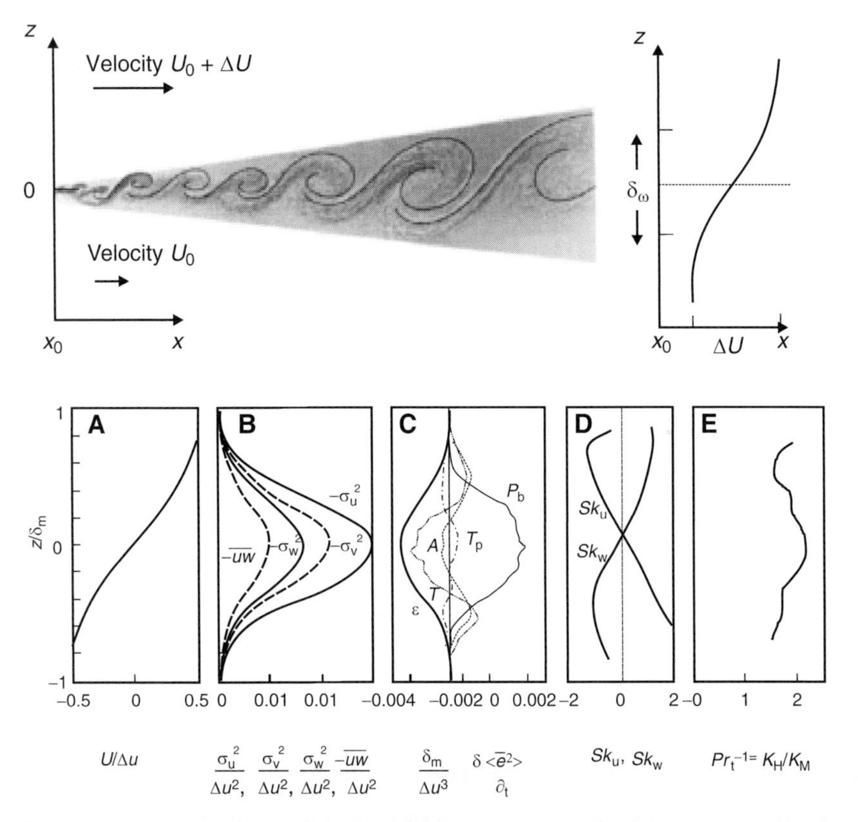

FIG. 9 Composite plot defining the parameters and flow statistics in the self-preserving region of a plane mixing layer. Variables are normalized with vorticity thickness δ_ω and velocity difference ΔU. (**A**) U = $\bar{u}$; (**B**) $\sigma_u, \sigma_v, \sigma_w$, and $-\overline{u'w'}$; (**C**) the budget of turbulent kinetic energy: P_b is shear production, ε is viscous dissipation, A is advection, T is turbulent transport, and T_p is pressure transport [for further explanation, see Finnigan (2000)]; (**D**) Sku, Skw; (**E**) inverse turbulent Prandtl number, Pr_t^{-1}. (From Raupach *et al.* [1996], reprinted with permission of Springer Science and Business Media.)

side of the splitter plate are laminar at $x = 0$, the mixing layer rapidly becomes turbulent and "self-preserving" whereupon $\delta_\omega \propto (x - x_0)$, where x_0 is a virtual origin (Townsend, 1976).

In Fig. 9A–E, various properties of fully developed, self-preserving mixing layers are plotted with the vertical coordinate z scaled by δ_ω and the velocity moments by ΔU. In Fig. 9A we see the inflection point in the mean velocity profile that is characteristic of mixing layers. The turbulent velocity variances σ_u^2, σ_v^2 and σ_w^2 and the shear stress $-\overline{u'w'}$ (Fig. 9B) all peak at the position of maximum shear, $z = 0$, and then decay toward the edge of the layer. Finally, in Fig. 9D we observe that streamwise and vertical skewnesses are of order one in the outer parts of the mixing layer, reversing sign as the centerline is crossed. If we compare these single-point moments with their counterparts in the RSL that were plotted in Fig. 4A–D, we see close correspondence between the canopy layer and the low speed side of the mixing layer.

Mixing layers are also populated with coherent motions that span the depth of the layer. The inflected velocity profile of the mixing layer is inviscidly unstable to small perturbations and growing, unstable modes of Kelvin-Helmholtz type emerge in the early stages of mixing layer development. The initial Kelvin-Helmholtz waves rapidly evolve into distinct transverse vortices or rollers connected by "braid" regions. Streamwise vorticity initially present in the flow is strongly amplified by strain in the braid regions while, at the same time, the transverse rollers become kinked and are less organized. As the mixing layer develops further, the transverse vorticity weakens and most of the total vorticity becomes concentrated in streamwise vortices in the braid regions and the streamwise component of kinked rollers (Rogers and Moser, 1992). The streamwise wavelength of the initial Kelvin-Hemholtz instability, Λ_x, is preserved through the stages of transition to fully developed turbulence and becomes the effective average spacing of the ultimate coherent structures. In a fully developed turbulent mixing layer, observations of Λ_x/δ_ω range from 3.5 to 5. A more detailed account of this process can be found in Raupach *et al.* (1996).

The Canopy-Mixing Layer Analogy

The obvious feature of plane-mixing layers that first prompted the comparison between them and canopy flows was the inflected mean velocity profile. Since it was known that this was inviscidly unstable to small perturbations, unlike the uninflected boundary layer profile (which becomes unstable only if viscosity is present), it was seen as a possible source of the high turbulence intensities in the RSL. Given the fact that the unstable eigenmodes were the same scale as

δ_ω, a bulk measure of the profile also seemed to offer a clue to the origin of the large eddies in the RSL. I have already noted the similarities among first, second, and third velocity moments in mixing layers and the RSL. Comparing integral length scales, terms in the turbulent kinetic energy budgets and other properties, such as the turbulent Prandtl number, confirm that RSL turbulence shares the character of mixing layer rather than boundary layer turbulence to a compelling degree (Raupach *et al.*, 1996; Finnigan, 2000).

We can go further and make a direct comparison between features of the coherent structures in the RSL and the mixing layer if we first state the relationship between the shear scale L_S and the vorticity thickness, δ_ω,

$$L_S = \frac{\overline{u}(h)}{(\partial \overline{u}/\partial z)_{z=h}} \cong \frac{1}{2}\delta_\omega = \frac{1}{2}\frac{\Delta U}{(\partial \overline{u}/\partial z)_{\max}} \tag{8}$$

Equation (8) is exact if the maximum shear is at $z = h$ (which can serve as a dynamic definition of h) and if we take the velocity deep within the canopy, where the shear is small (see Fig. 4A), as equivalent to the free stream velocity on the low-speed side of the mixing layer. Observations of the spacing of coherent eddies in fully developed mixing layers then fall in the range $10 > \Lambda_x/L_S > 7$. Raupach *et al.* (1996) compared estimates of Λ_x obtained from nine of the canopies in Fig. 4 against the shear length scale L_S. They found that the data over a very wide range of canopy types and scales closely fit the relationship $\Lambda_x = 8.1L_S$, which falls in the middle of the range for fully developed mixing layers. This is compelling evidence that the process controlling the generation of large coherent structures in the RSL is very similar to that in the plane-mixing layer; further supporting argument can be found in Raupach *et al.* (1996).

We can summarize the canopy-mixing layer analogy by reference to the schematic diagram, Fig. 10:

1. The first stage is the emergence of the primary Kelvin-Helmholtz instability. We suppose that this occurs when a large-scale gust or sweep from the boundary layer well above the canopy raises the shear at $z = h$ above some threshold level at which the instability can grow fast enough to emerge from the background before it is smeared out by the ambient turbulence. The growth rate of the mixing layer instability is proportional to the *magnitude* of the shear at the inflection point whereas, in the high Reynolds number canopy flow, the *scale* of the shear L_S and therefore the scale of the instability is independent of windspeed, depending only on canopy height and aerodynamic drag.
2. The second stage is the clumping of the vorticity of the Kelvin-Helmholtz waves into transverse vortices or rollers connected by braid

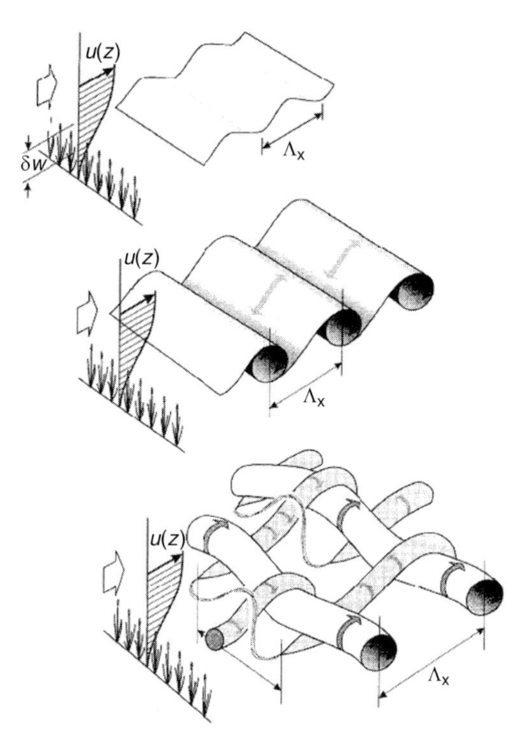

FIG. 10 Diagram of stages in the development of the mixing-layer type instability in the roughness sublayer. (From Finnigan [2000], reprinted with permission from *Annual Review of Fluid Mechanics*, Volume 32, 2000 by Annual Reviews.)

regions of highly strained fluid. The spacing of the rollers is similar to the wavelength of the initial instability.

3. Finally, secondary instabilities in the rollers lead to their kinking and pairing, while any ambient streamwise vorticity in the braid regions is strongly amplified by strain, resulting in coherent structures whose transverse and streamwise dimensions are of the same order and most of whose vorticity is aligned in the streamwise direction just as found for the "characteristic eddy" revealed by EOF analysis in Fig. 7. Note that the schematic "coherent structure" shown as the third panel of Fig. 10 is based on diagrams in Rogers and Moser (1992).

This conceptual picture has some further consequences. Because we expect that the primary instability will emerge from the background only when the shear at the canopy top exceeds some threshold, the convection velocity of the ultimate coherent structures will be larger than the mean velocity. We imagine that the footprint of the sweep that raises the shear above the threshold will be of larger scale than Λ_x, so we should expect to see some streamwise periodicity in canopy eddies. Short sections of periodicity are revealed by short-time two-point space-time correlations, and by using this technique in a wheat canopy, Finnigan (1979b) found that canopy eddies arrived in groups of three or four. Each group has a common convection velocity $U_c \approx 1.8\bar{u}(h)$ that presumably corresponds to the velocity of the large sweep that initiated the instability. The mixing layer analogy also has obvious consequences for scaling and modeling canopy flow. A single length scale δ_ω and single velocity scale u^* or $\bar{u}(h)$ determine the primary instability and ultimate coherent structures in the RSL. Within the RSL, δ_ω replaces $(z-d)$, the height-dependent length scale of the surface layer.

EFFECTS OF TOPOGRAPHY AND HETEROGENEITY

If we take the horizontally homogeneous quasi-steady flow described previously as the canonical state, then in the context of disturbance ecology the most important kinds of perturbations that we need to consider are those caused by hilly terrain and by forest edges or clearings. In trying to draw general conclusions about the way that canopy flows depart from horizontally homogeneous behavior, it is useful to break the problem into two parts: the nature of the forcing and the response of the canopy. We can identify two kinds of forcing. The first occurs when we accelerate the canopy flow by applying a pressure field $\bar{p}(x,y,z)$ that varies over a horizontal scale L much larger than the canopy height (i.e., $L >> h$). This is the kind of pressure field that results when the canopy is on a hill. Alternatively, such a pressure field could be generated diabatically by a strong contrast in surface energy balance, such as occurs in a true sea breeze or an "inland sea breeze." The second case occurs when the scale of the pressure gradient is comparable to the canopy height and this kind of forcing occurs around local obstructions, such as clearings or forest edges. Diabatic forcing can also occur on this scale, as in the case of anabatic and katabatic flows, where density gradients and accompanying pressure gradients develop initially at the scale of the canopy height. The most obvious case of this that interests us is fire.

A key measure of the response of the canopy is the distance needed to reestablish horizontally homogeneous flow within the canopy after a sudden change in the external forcing. By considering the idealized case of a one-dimensional canopy flow, it is easy to show that the flow responds exponentially to a change in the driving pressure gradient with a distance constant L_C (Finnigan and Brunet, 1995; Finnigan and Belcher, 2004) and that,

$$L_C = (C_d\, a)^{-1} \tag{9}$$

where C_d is the dimensionless drag coefficient of the foliage and a is the foliage area per unit volume of space, so we can write the aerodynamic drag force as $F_D = u|u|/L_C$. Most of the canopy drag is transmitted as a pressure force on the foliage, so the drag is proportional to the square of the windspeed while the modulus sign recognizes that the drag force is always directed in the instantaneous wind direction. L_C is the fundamental lengthscale of adjustment of the within-canopy flow to external forcing and appears in many contexts, such as in the changes that must be made to Kolmogorov inertial sublayer spectral forms within canopies (Finnigan, 2000). In the discussion below, we will see in each case that the flow response is a balance between the strength of the forcing and the tendency of the within-canopy flow to return to equilibrium with a response distance L_C.

Flow over Hills

Before considering canopy flows in complex topography, it is useful to set the scene by describing flow over rough hills without canopies. I will confine this discussion to hills sufficiently small that the flow perturbations they cause are confined within the boundary layer. In practice, this means that the hill height H and the hill horizontal lengthscale L satisfy $H << z_i$ and $L << h^*$, where h^* is the "relaxation length" of the boundary layer defined as $h^* = z_i U_0 / u_*$ or $z_i U_0 / w_*$ according to whether the flow is neutrally stratified or convectively unstable. U_0 is the velocity characteristic of the ABL above the surface layer; thus, in the case of the convective boundary layer portrayed in Fig. 1A, it would be the constant windspeed above $z/z_i \approx 0.2$. The horizontal length scale L is defined as the distance from the hillcrest to the half-height point. In continuously hilly terrain it can be more appropriate to use a characteristic wavelength λ as the horizontal length scale and in sinusoidal terrain, $L = \lambda/4$.

Fig. 11 shows the main features of the velocity field about an isolated hill. The figure could represent flow approaching an axisymmetric hill or a two-dimensional ridge at right angles. Close to the surface, the flow decelerates slightly at the foot of the two-dimensional ridge before accelerating to the summit. In the case of an axisymmetric hill, the deceleration is replaced by a region of lateral flow divergence at the foot of the hill. The wind reaches its maximum speed above the hilltop and then decelerates on the lee side. If the hill is steep enough, a separation bubble forms in which the mean flow reverses direction. Whether the flow separates or not, a wake region forms behind the hill with a marked velocity deficit extending for at least $10H$ downwind. The same information is made more concrete

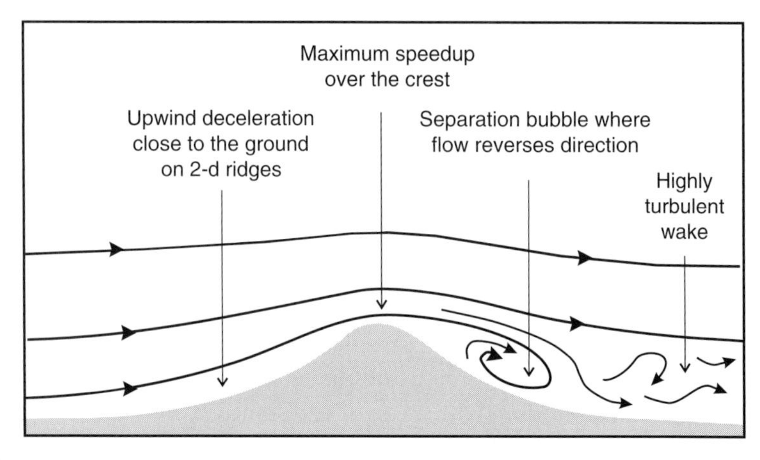

FIG. 11 Drawing of the flow over a two-dimensional ridge showing the formation of a downstream separation region when the ridge is steep enough. On an axisymmetric hill, the upwind deceleration region is replaced by a region of lateral flow divergence. (From Kaimal and Finnigan [1994], reprinted with permission from Oxford University Press © 1994.)

in Fig. 12, which plots velocity profiles well upwind, over the hilltop and in the wake. The vertical coordinate z measures height above the local surface. In Fig. 12, it is made dimensionless with the inner layer height, l, defined below. Upwind we have a standard logarithmic profile [Equation (5)], but on the hill top the profile is accelerated with the maximum relative speed-up occurring quite close to the surface at $z/l \sim 0.3$. In the wake we see a substantial velocity deficit extending to at least $z = H$.

Much of the understanding we now have about the dynamics of flow over hills derives from linear theory, which assumes that the mean flow perturbations caused by the hill are small in comparison to the upwind flow. Although linear theory is limited to hills of low slope, $H/L \ll 1$, its insights are applicable to much steeper hills. Linear theory supposes a division of the flow field into two main regions, an *inner* region of depth l and an *outer* region above, which are distinguished by essentially different dynamics (Fig. 13). The flow dynamics are governed by the balance between advection, streamwise pressure gradient, and the vertical divergence of the shear stress. Over low hills this balance can be expressed in an approximate linearized momentum equation,

$$U_B(z)\frac{\partial \Delta\bar{u}}{\partial x} + \frac{\partial \Delta\bar{p}}{\partial x} \sim \frac{\partial \Delta\tau}{\partial z} \tag{10}$$

where $\Delta\bar{u}$, $\Delta\bar{p}$, $\Delta\tau$ are the perturbations in mean streamwise velocity, kinematic pressure, and shear stress induced by the hill and $U_B(z)$ denotes

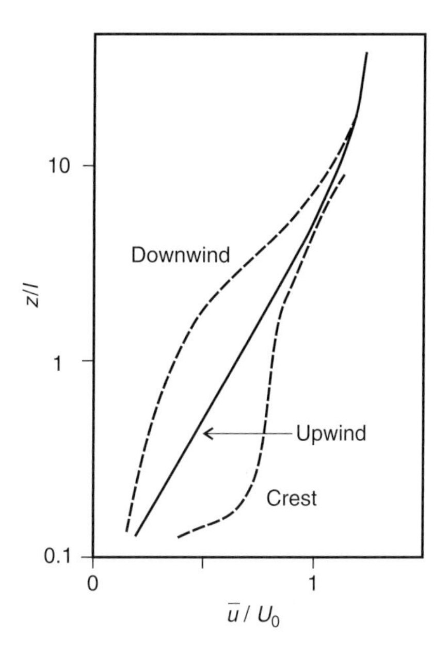

FIG. 12 Profiles of mean velocity observed upwind, on the crest, and in the wake region of a hill. The vertical scale is made dimensionless with the inner layer depth, *l*. Note the position of the maximum speed-up above the crest at $z \sim l/3$. (From Kaimal and Finnigan [1994], reprinted with permission from Oxford University Press © 1994.)

the undisturbed flow upwind of the hill or, in the case of a range of hills, the horizontally averaged wind. Well above the surface in the outer region, perturbations in stress gradient are small and advection and pressure gradient, the first two terms in Equation (10), are essentially in balance. Closer to the surface, an imbalance develops between these terms as the perturbation stress gradient grows. The inner layer height is defined as the level at which the left side of Equation (10) equals the right side (Hunt *et al.*, 1988; henceforth, HLR).

If the undisturbed upwind profile is taken as logarithmic, $U_B(z) = u^*/\kappa \ln(z/z_0)$, this definition for the inner layer depth leads to an implicit expression for l (HLR):

$$\frac{l}{L}\ln\left(\frac{l}{z_0}\right) = 2\kappa^2 \tag{11}$$

The pressure field that develops over the hill deflects the entire boundary flow over the obstacle. Its magnitude is determined, therefore, by the inertia of the faster-flowing air in the outer region and is also related to the steepness of the hill, so we expect:

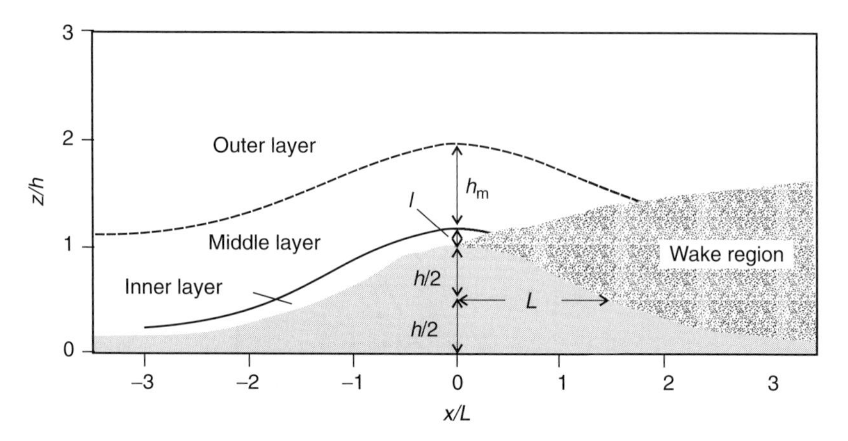

FIG. 13 The different regions of the flow over an isolated hill, comprising inner, middle, outer, and wake layers and their associated length scales. (From Kaimal and Finnigan [1994], reprinted with permission from Oxford University Press © 1994.)

$$\Delta\overline{p} \sim \frac{H}{L} U_0^2 \tag{12}$$

Scaling arguments (HLR) reveal that when the winds are strong so that the logarithmic law describes the wind profile to heights much larger than H, the appropriate definition of U_0 is

$$U_0 = U(h_m); \quad \frac{h_m}{L} \ln^{1/2}\left(\frac{h_m}{z_0}\right) \sim 1 \tag{13}$$

The middle layer height h_m divides the outer region into a middle layer between l and h_m, where shear in the approach flow exerts an important influence on the flow dynamics, and an upper layer, where shear in the approach flow is negligible. In contrast to the inner layer, the perturbations in the outer layer are controlled by inviscid flow dynamics, obeying the equations of rotational inviscid flow in the middle layer and irrotational or potential flow equations in the upper layer. For a hill with L = 200 m, u_* = 0.3 m s^{-1} and z_0 = 0.02 m, typical sizes of the key parameters are l = 10 m, h_m = 70 m, U_0 = 6 m s^{-1} and $U_B(l)$ = 4.5 m s^{-1}. Note that in the linear theory the vertical extent of the regions influenced by the hill depend only on the hill length, L. The hill height enters only through the influence of the steepness H/L on the pressure perturbation that drives all other changes in the flow field.

The pressure perturbation falls to a minimum at the hilltop and then rises again behind the hill. It varies in both the horizontal and vertical directions over distances of order L. Hence, because l is much smaller than L, the horizontal pressure gradient $\partial\Delta\overline{p}/\partial x$ is essentially constant with height through the inner layer. The scaling of the pressure perturbation field gives a strong clue as to why the velocity speed-up peaks in the inner layer. Referring again to Equation (10), in the outer layer the momentum balance is dominated by the pressure gradient and the advection so that $U_B(z)\,\Delta\bar{u}\,(x,z)/L \sim (H/L)\,U_0^2/L$. Within the inner layer, as the background flow $U_B(z)$ becomes much smaller than U_0, the velocity perturbation $\Delta\bar{u}$ must grow to compensate. Eventually, at the bottom of the inner layer the shear stress gradient dominates the momentum balance and reduces Δu so that the peak in speed-up is found at about $z \sim l/3$ (Fig. 12).

Before considering how the presence of a canopy modifies this picture, we must note that strong stable stratification can have a large influence on the perturbation pressure field and the subsequent velocity and stress perturbations. In particular, it moves the position of maximum velocity from the hilltop to the downwind slope (see Kaimal and Finnigan (1994) and references therein). This is important because the pressure field around large hills, defined as hills that occupy a large fraction of the boundary layer, is often dominated by the displacement of the stable troposphere above the boundary layer rather than processes within the boundary layer itself. A consequence of this can be severe "downslope" windstorms on the lee slopes of forested mountains. The front range of the Rocky Mountains, west of Denver, Colorado, has provided many destructive examples of this phenomenon (Duran, 1990).

Canopies on Hills

Even though so many of the forests and other canopies in which we are interested are on hills, we have very few comprehensive measurements of the way topography alters the homogeneous canopy turbulence described previously. Measurements in the field almost all come from single towers so that the streamwise development of the windfield is not recorded. The expense of multiple tower studies in forests is one reason for their rarity, but even in wind tunnels only a few published model studies are available. That of Finnigan and Brunet (1995) still provides the only comprehensive data set of within-canopy velocity statistics, although Ruck and Adams (1991) and Neff and Meroney (1998) have made measurements above the model canopy.

We can turn instead to theoretical models for insight, but even there few signposts are available. The numerical model study by Wilson *et al.* (1998) reproduced the data of Finnigan and Brunet (1995), but it did not make major strides in unraveling the basic physics. A recent analytical linear model of flow in a tall canopy on a low hill by Finnigan and Belcher (2004) has, however, provided some insights comparable to the linear analytic models of flow over rough hills that followed the pioneering work of Hunt *et al.* (1988). Like that of Hunt *et al.*, the analysis of Finnigan and Belcher divides the flow in the canopy and that in the free boundary layer above into a series of layers with essentially different dynamics. The dominant terms in the momentum balance in each layer are determined, and the complete solution for the flow field is achieved by asymptotically matching the solutions for the flow in each layer. The model applies in the limit of "tall" canopies, by which we mean that almost all the momentum is absorbed as aerodynamic drag on the foliage and not as shear stress on the underlying surface. We can see from Fig. 4B that this condition is satisfied in most continuous natural canopies.

Finnigan and Belcher calculate the perturbations to a background flow $U_B(z)$ that now consists of a logarithmic velocity profile above the canopy merging smoothly with an exponential profile in the canopy. These formulae are easier to state if we take the origin of coordinates at the canopy top rather than the ground surface so that:

$$\begin{aligned} U_B(z) &= \frac{u^*}{\kappa}\ln\left(\frac{z+d_t}{z_0}\right); z>0 \\ U_B(z) &= U_h e^{\frac{\beta z}{l}} \qquad ; z\leq 0 \end{aligned} \tag{14}$$

where $U_h = U_B(0)$ is the mean wind speed at the top of the canopy, u^* is the friction velocity, $l = 2\beta^3 L_c$ is the mixing length in the canopy $l = \kappa u^*(z + d_t)$ is the mixing length above the canopy and $\beta = u^*/U_h$ quantifies the momentum flux through the canopy. For closed uniform natural canopies, $\beta \approx 0.3$ (Raupach *et al.*,1996). Note that with the change of z origin we have also introduced the redefined displacement height d_t, where $d_t = h-d$ [compare with Equations (5) and (6)]. Matching both mean wind and shear stress at the canopy top also fixes the following relationships:

$$U_h = \frac{u^*}{\kappa}\ln\left(\frac{d_t}{z_0}\right); \quad d = l/\kappa; \quad z_0 = \frac{l}{\kappa}e^{-\kappa/\beta} \tag{15}$$

The essential features of canopy flow on low hills under neutral stratification can now be summarized as follows. Above the canopy, the flow behaves as described previously. The magnitude of the driving perturbation pressure gradient is essentially invariant with height through the canopy as it is through the inner layer. The canopy flow itself divides naturally into an upper and a lower layer. In the upper canopy layer, the dominant terms in the momentum balance are the streamwise pressure gradient induced by the hill, the shear stress divergence, and the aerodynamic drag on the foliage. In comparison to these three terms, advection is small. In the lower canopy layer, the shear stress divergence becomes small compared to the pressure gradient and the aerodynamic drag. The depths of the upper and lower canopy layers are defined in Finnigan and Belcher (2004).

In the air layers above the canopy, the velocity perturbation caused by the hill is in phase with (minus) the pressure perturbation so that the maximum velocity perturbation (speed-up) occurs over the hilltop:

$$\textit{Outer Layer: } \Delta\bar{u}(x,z) \propto -\sqrt{|\Delta\bar{p}|}\, Sgn(\Delta\bar{p}) \tag{16}$$

In the lower canopy layer, in contrast, the velocity perturbation is in phase with (minus) the *gradient* of the pressure perturbation, which attains its maximum value on the upwind slope of the hill and reverses on the lee side (Fig. 14):

$$\textit{Lower Canopy Layer: } \Delta\bar{u}(x) = -\sqrt{L_c|\partial\Delta\bar{p}/\partial x|}\, Sgn(\partial\Delta\bar{p}/\partial x) \tag{17}$$

Matching the solutions in the different layers tells us that the shear stress layer and upper canopy layer together form a region of adjustment across which the mean flow perturbations change from being in phase with (minus) the pressure well above the surface [Equation (16)] to being in

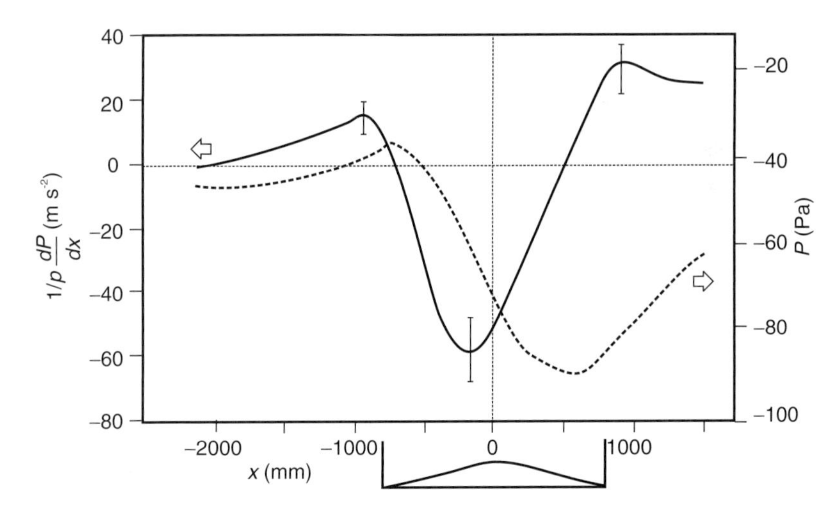

FIG. 14 The pressure perturbation and its gradient induced by a hill. Curves of static pressure measured along a streamline within the canopy (dashed line) and the kinematic static pressure gradient averaged over the height range 1.5 > z/h > 0. (Data from the wind tunnel model experiment of Finnigan and Brunet [1995]. Reprinted with permission from Cambridge University Press © 1995.)

phase with (minus) the pressure gradient deep in the canopy [Equation (17)]. This adjustment strongly modulates the shear across the canopy top. These features are clearly illustrated in Fig. 15, from Finnigan and Belcher (2004), where the perturbations in mean streamwise velocity at a series of stations across one of a range of sinusoidal ridges are plotted. Included for comparison are solutions for a rough surface with the same z_0 from the theory of Hunt *et al.* (1988). The extra turbulent mixing generated by the canopy reduces the sharp speed-up peak on the hillcrest predicted by Hunt *et al.* and moves it from around $z \approx h_i/3$ to $z \approx h_i$.

Fig. 16 shows consecutive vertical profiles of total mean velocity, $U_B(z) + \Delta\bar{u}(x,z)$, from the wind tunnel model study of Finnigan and Brunet (1995). Although this hill is too steep to satisfy the $H/L << 1$ limits of linear theory, upwind of the hillcrest we can still see the main features predicted by the Finnigan and Belcher model. The maximum velocity in the lower canopy occurs well before the crest and is falling by the hilltop. The difference between lower canopy and outer layer velocities is greatest at the hilltop and maximizes the canopy-top shear at that point, with consequences for the magnitude and scale of turbulence production. Conversely, the difference is at a minimum halfway up the hill, where the lower-canopy velocity is maximal but the outer layer flow has not yet increased much. This effect is so marked that the inflexion point in the velocity profile at the top of the canopy has disappeared. Note also that on this steep hill we observe a large separation bubble behind the hillcrest.

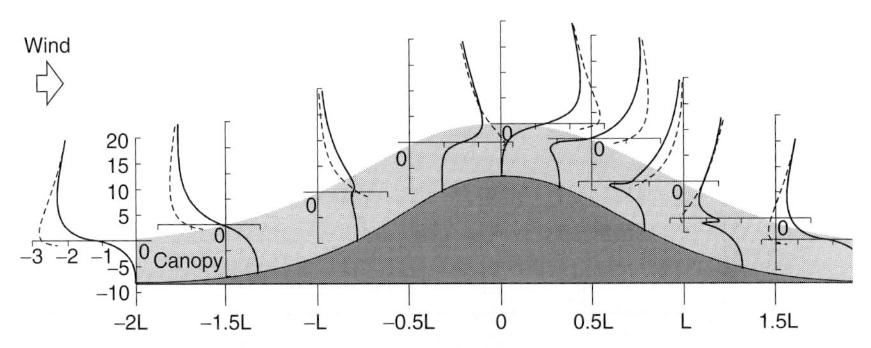

FIG. 15 Profiles of velocity perturbations caused by a low hill. Comparison of velocity perturbation $\Delta u/U_{SC}$ over a hill covered with a tall canopy with the no-canopy Hunt *et al.* (1988) solution (dotted line). Note the Hunt *et al.* solution is only valid to $z = -d + z_0$. Profiles are plotted at a series of X/L values between $X/L = -2$ (upwind trough) and $X/L = 2$ (downwind trough). The units of Z are meters and the vertical range is $2l_i > Z > L_C$. (From Finnigan and Belcher [2004], reprinted with permission from the *Quarterly Journal of the Royal Meteorological Society*, Volume 130, © 2004.)

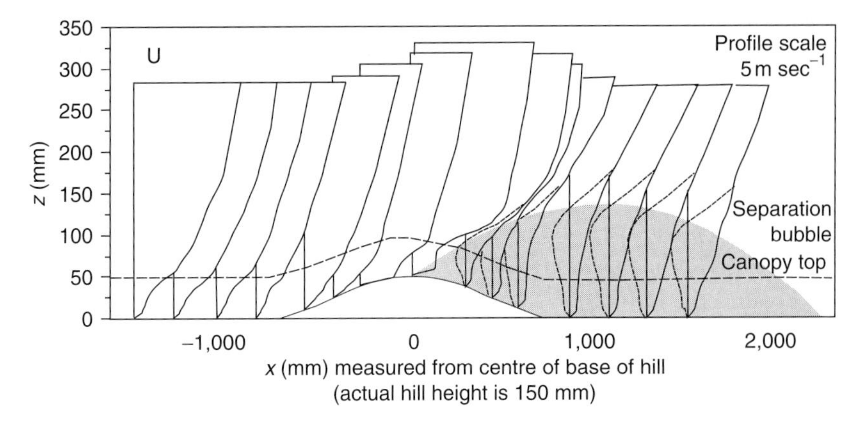

FIG. 16 Consecutive profiles of $\bar{u}$ (z) at consecutive x locations over a wind tunnel model of a tall canopy on a hill. (From Finnigan and Brunet [1995], reprinted with permission from Cambridge University Press © 1995.)

The vertical profiles of streamwise velocity variance, $\overline{u'^2}$, from the same experiment of Finnigan and Brunet (1995) are shown in Fig. 17. Two features are noteworthy, particularly so since they affect fluctuating forces on the plants, a point I return to later in this chapter. The first thing to note is that the peak in $\overline{u'^2}$ that is seen at the canopy top upwind of the hill is substantially reduced halfway up the hill, the position where the shear across the upper canopy and inner layers is reduced by the phase difference in the velocity perturbations above and below the canopy, as seen in Figs.

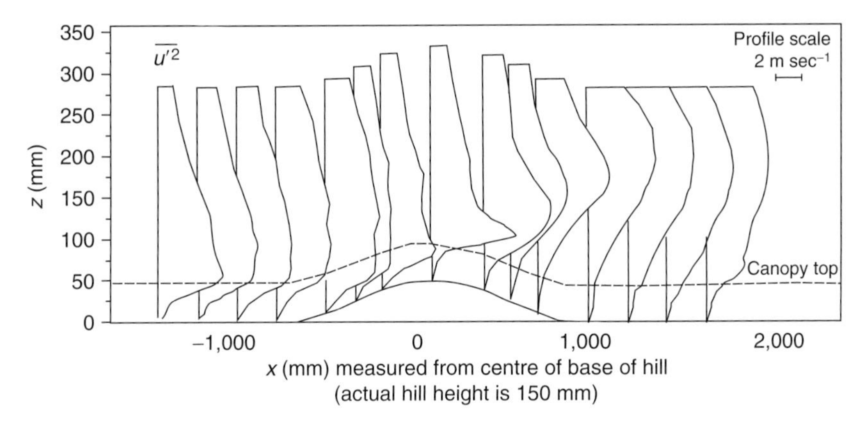

FIG. 17 Consecutive profiles of $\overline{u'^2}$ (z) at consecutive x locations over a wind tunnel model of a tall canopy on a hill. (From Finnigan and Brunet [1995], reprinted with permission from Cambridge University Press © 1995.)

15 and 16. The second is the strong peak in $\overline{u'^2}$ at the top of the canopy over the hilltop. This peak coincides with the strong shear region of Figs. 15 and 16. Behind the hill the peak in $\overline{u'^2}$ follows the separated shear layer that bounds the separation bubble.

A series of consequences flows from these dynamics. First is a marked increase in the aerodynamic drag even on low hills covered with tall canopies because the negative velocity perturbation behind the crest displaces streamlines away from the surface, increasing the asymmetry of flow around the hill and thereby the pressure drag. This increased drag of the hills on the airflow can affect predictions of wind speeds at larger scale around a region of forested hills, and many meteorological models in current use have poor parameterizations of this effect. Equally important is the interaction of the negative velocity perturbation and the mean flow behind the hillcrest. Positive mean flow behind the hillcrest is maintained by turbulent transport of momentum from the faster-moving fluid above the vegetation, but this mechanism is ineffective deep in a canopy. The pressure field, in contrast, passes essentially unimpeded through the canopy; thus, the pressure gradient that is acting to decelerate the flow can easily overwhelm the turbulent transport. The result is reversal of the total velocity and flow separation. Hence, even on low hills, if the canopy is deep enough we can expect reversed flow and separation near the ground.

Finally, I need to point out that both the wind tunnel model and theoretical studies I have quoted were performed with the wind perpendicular to two-dimensional ridges. This configuration maximizes the speed-up and associated distortions of the wind and scalar fields. Isolated axisymmetric hills are likely to exhibit effects that are similar in kind but smaller in degree than ridges. Conversely, saddles between two isolated hills can exhibit even larger effects than ridges when the wind blows across the saddle.

Before moving on to consider situations with smaller scale forcing, I remind the reader that most of the features of the perturbed flow in a canopy on a hill should be observed in any large-scale pressure field, whether caused by a hill or not. Large-scale contrasts in surface energy balance are a ubiquitous source of such pressure fields, and with appropriate redefinition of coordinate frames, we can expect canopy flow dynamics similar to those described previously in such cases.

Forest Edges and Clearings

There has been a series of field studies of this situation, but in Fig. 18 we again have recourse to a wind tunnel study (Raupach *et al.*, 1987) because

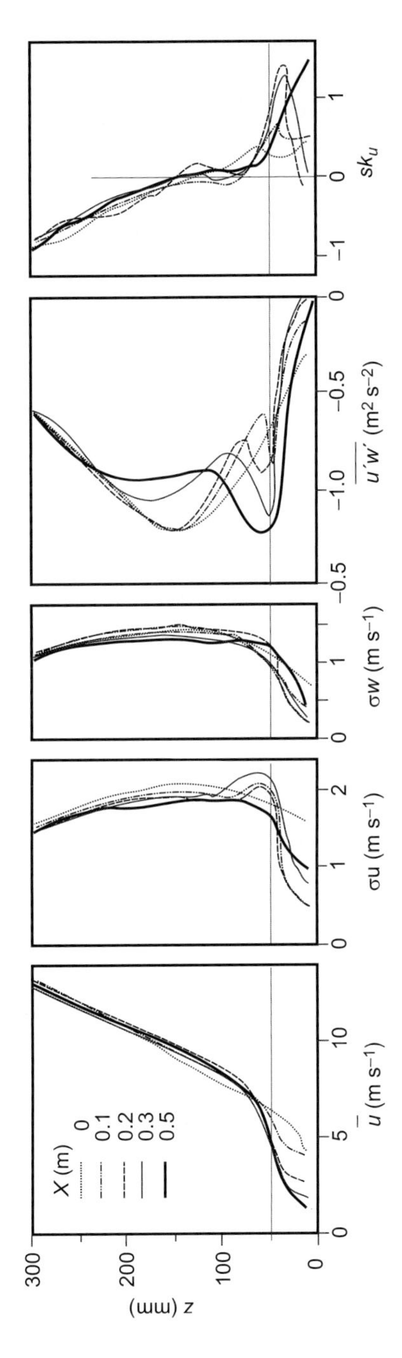

FIG. 18 Consecutive vertical profiles of velocity statistics at and behind a forest edge. The wind tunnel model canopy is 0.05 m high so profiles are at $x/h = 0, 2, 4, 6$, and 10. (From Raupach *et al.* [1987], reprinted with permission from CSIRO Marine and Atmospheric Research.)

of the far greater data density it affords. These results from a simulation of flow into a forest comprise a set of profiles of first, second, and third moments of velocity: $\bar{u}(z)$, $\sigma_u(z)\left(=\sqrt{\overline{u'^2}}\right)$, $\sigma_w(z)\left(=\sqrt{\overline{w'^2}}\right)$, $\overline{u'w'}$, $Sk_u\left(=\overline{u'^3}/\sigma_u^{3/2}\right)$. They show several characteristic features. The flow near the surface decelerates and streamlines are displaced upward as the flow encounters the resistance offered by the aerodynamic drag of the canopy. Very close to the surface in the canopy we see deceleration for a distance $x > 10h$. In contrast, just above the canopy top we see the flow accelerate. The accelerated flow above the top of the forest produces the characteristic inflexion point velocity profile that we expect at the top of a canopy (see previous section on characteristics of turbulent flow in and above plant canopies), but the depth of this inflected shear layer and the scale of the energetic large eddies that develop within it are initially much smaller than the canopy height, h.

As we progress downwind, this shear layer deepens, eventually reaching its equilibrium width as the decelerated region of the within-canopy flow grows. The large eddies that are generated in this shear layer penetrate progressively deeper into the canopy as the shear layer grows with x, transferring momentum deeper into the foliage, and eventually, at downstream distances greater than ~$10h$, lead to an acceleration of the flow lower in the canopy. The result is a reconvergence of the streamlines that were originally displaced upward (Morse *et al.*, 2002). Since this process is governed by the canopy aerodynamic resistance and the adjustment of the within-canopy flow, we can expect it to scale on L_C rather than h so long as the ratio h/L_C is sufficiently small that almost all the streamwise momentum is absorbed as foliage drag rather than as shear stress on the surface beneath the trees. A value of $h/L_C \geq 1$ is sufficient to ensure this. Both wind tunnel and field experiments suggest that the maximum streamline displacement and, therefore, the minimum velocity at mid canopy height ($z/h \sim 0.5$) occur in the region $10 > x/L_C > 5$. These features are also in broad agreement with the recent analytical theory of Belcher *et al.* (2003), although that theory, which parameterizes momentum transfer through a mixing length, cannot reproduce flow features that depend on the character of the large turbulent eddies, as I discuss next.

Equilibrium flow within the canopy is almost re-established by $x/L_C \sim 10$, but turbulence statistics just above the canopy continue to develop in the growing internal boundary layer (Morse *et al.*, 2002). Peaks in streamwise and vertical velocity variance and in shearing stress are observed in the region $10 > x/L_C > 5$, suggesting that the turbulent eddies attain maximum energy when the inflected shear layer at the canopy top has reached an intermediate depth, sufficient for the inviscid instability process by which these large eddies gain energy from the mean flow to be well established but before the mechanisms that dissipate the energy in the coherent eddies have developed fully. The velocity moment in which this is most marked is the streamwise velocity skewness, Sk_u, which in the region $6 > x/L_C > 4$ becomes two or three times larger than its canopy-top equilibrium value (compare Fig. 4I and Fig. 18). Such behavior is consonant with what is known of the development of plane-mixing layers (see preceding section on characteristics of turbulent flow in and above plant canopies).

For details of the structure of the wind blowing from a continuous canopy into a clearing, I again refer to a wind tunnel study by Chen *et al.* (1995). References to several other studies, both in wind tunnels and the field, are quoted there. The wind structure they observed was quite similar to what is found behind thin porous windbreaks. A sheltered region just behind the forest edge extends for around 3–5 h, at which point the inflection in the velocity profile at $z = h$ has disappeared and a boundary layer profile is re-established at the ground, even though the overall mean velocity levels are much lower than further downwind. Over this "quiet zone," mean windspeeds rise; the fastest increase is at $z = h$, while the turbulent intensities, $\sigma_u/\bar{u}$ and $\sigma_w/\bar{u}$, rise below $z = h$ and fall at and above $z = h$.

Behind the quiet zone, the effect of the upwind canopy is felt through a wake zone extending to at least 20 h. Mean velocities are suppressed in this region and turbulence levels enhanced compared to far downwind. Total readjustment of the flow to the downwind surface can require distances of as much as 50 h. The speed of adjustment is controlled by the roughness of the downwind surface: the rougher the surface, the more rapidly the influence of the upwind canopy is lost (Kaimal and Finnigan, 1994).

IMPLICATIONS OF THIS VELOCITY STRUCTURE FOR CANOPY DISTURBANCE

Windthrow and Crop Lodging

This subject is treated in much more detail in Chapter 4. Here I simply make the connection between the properties of the turbulent wind field in canopies and the key factors contributing to windthrow. Considerable work has been done in the last two decades on the detailed mechanisms by which individual trees or swathes of trees or crop plants are flattened by the wind. This research has treated the plants as aeroelastic cantilevers and utilized the considerable mathematical apparatus developed by aeronautical engineers to quantify their interaction with the turbulent canopy wind (e.g., Finnigan and Mulhearn, 1978; Baker, 1995; Marshall *et al.*, 2002). Aeroelastic canopy models representing cereal crops (Finnigan and Mulhearn, 1978; Brunet *et al.*, 1994) and conifer forests (Stacey *et al.*, 1994) have been constructed and studied and the results compared with simultaneous measurements of bending and turbulence in real canopies. The several individual mechanisms leading to failure of stems, trunks, root plates, and the soil matrix have also been studied intensively (Coutts and Grace, 1995). We can conclude several things from this body of research.

The first is that the key parameter determining whether a tree or plant will blow down is the maximum bending moment at the plant's root. The root bending moment is defined as:

$$M_R = \int_0^h z\, F(z)\, dz \qquad (18)$$

where $F(z)$ is the aerodynamic force on the plant at height z and h, the plant or canopy height. The rapid absorption of momentum by the canopy is demonstrated by the shear stress profiles of Fig. 4C and results in the characteristic exponential velocity profile seen in Fig. 4A. In horizontally homogeneous flow, the time-averaged aerodynamic force on the plants at a height z is simply $F(z) = \partial \langle \rho \overline{u'w'} \rangle / \partial z$, the vertical derivative of the shear stress, so that we see that aerodynamic force is concentrated in the upper canopy, maximizing the root bending moment. This is in contrast to the forces on an isolated tree, which for the same vertically integrated force will experience a much smaller root bending moment than its counterpart in a forest.

Two modes of failure are commonly seen in windthrow of plants: stem breakage and root plate failure. In tree stands, static failure by either mechanism is rare because the average wind speeds required to generate time-averaged root bending moments large enough to exceed the failure threshold are very large. Typical calculations show that steady wind speeds greater than 20 m s^{-1} would be required for static failure of trees (Baker, 1995). Cereal crops, in contrast, could fail at wind speeds about half this, particularly when rain adds weight to stalks displaced from the vertical (Baker, 1995). As we have seen in the earlier sections, however, the canopy wind is not steady and peak loads much higher than the average are readily generated. Furthermore, plants are dynamic elastic objects that interact with the wind as forced resonant cantilevers. The peak loads experienced by the plants are, therefore, a function of the turbulent wind, which determines the spectrum of forcing, and the elastic nature and resonant properties of the individual plants, including the damping of the motion of individual plants. In canopies, this is determined primarily by the mechanical interference of adjacent plants and the interaction of the roots and the soil matrix rather than by internal damping in the flexing stems (Finnigan and Mulhearn, 1978; Baker, 1995; Marshall *et al.*, 2002).

When the aeroelastic nature of the plants is taken into account in this way, it is clear that it is possible to exceed the threshold for failure much more easily than the static load criterion would suggest. Marshall *et al.* (2002) have shown that if the peak root bending moment is used as a trigger for conditional sampling of the windfield, then wind patterns that correspond to the characteristic ejection-sweep sequence of the large canopy eddies are obtained. In other words, the peak overturning moments on the plants are associated directly with the coherent canopy eddies discussed previously. The intermittent loading by these eddies, however, is not the only mechanism causing the root bending moment to exceed the failure threshold.

Baker (1995) has suggested that the best way to view forcing by the turbulent wind is as a sequence of intermittent step changes in root bending moment, each triggered by the arrival of a canopy gust or large eddy. After being bent over by the gust, the plant executes several damped oscillations at its natural frequency until the next gust arrives. If these oscillations are of large enough amplitude, then the cohesion between soil and root plate can be reduced (especially in very wet soils) and the root bending moment required for failure reduced so that a later gust may cause the plant to fall. Gardiner (1994), in contrast, using measurements in a spruce plantation, has demonstrated an alternative mechanism whereby resonant interaction between the wind and plants can cause the failure threshold to be exceeded. A wind tunnel model study supported his conclusions. As noted previously, we expect the canopy eddies generated by the mixing layer instability process to come in groups of two, three, or four as the canopy-top instability is triggered by a large-scale gust from the boundary layer above so that resonant interaction is clearly possible. Indeed, Finnigan (1979b) speculated that the most spectacular cases of honami[3] occur when this gust-arrival frequency coincides with the natural waving frequency of the plants.

In summary, we see that windthrow is caused by the intermittent penetration of canopy eddies, interacting with the resonant swaying of plants. The largest overturning forces are caused by the intermittent canopy eddies generated by the mixing layer instability, and we expect these to arrive in groups

[3] Honami is defined as the characteristic coherent waving of cereal crops and grasses that can be observed on windy days.

leading to the possibility of resonant amplification of the peak loads. These eddies are of relatively small, cross-stream extent, as we have seen (Fig. 7), so that it is unlikely that wide cross-wind swaths of trees will be laid low by this mechanism, although entanglement of branches may spread the effect of failure. The rapid absorption of momentum in the upper canopy amplifies the bending moment relative to isolated trees, and this is as true for the impact of the large eddies as for the mean wind (Marshall *et al.*, 2002 and Fig. 4C).

Rapid momentum absorption in the upper part of horizontally homogeneous canopies (Fig. 4A) is echoed by the peak in $\overline{u'^2}$ near the top of the foliage (Fig. 4C) so that both the mean and the fluctuating contributions to the root bending moment are enhanced for plants in canopies compared to the same plants in isolation. We can see from Figs. 16 and 17 that this picture is altered for canopies on hills. On the hilltop, not only are the mean wind and $\overline{u'^2}$ increased relative to values on flat ground, but their profiles are more sharply peaked around the canopy top, ensuring that for the same total force on the plant the mean and fluctuating root bending moments would be substantially increased. Halfway up the hill, in contrast, the profiles of $\overline{u}$ and $\overline{u'^2}$ are both less peaked than on the flat, leading to reduced root M_R, other factors being equal. Hence, trees near the hilltop will suffer greater static and fluctuating bending moments than trees in the valley, and those on the upwind slope will have the smallest M_R. Of course, this does not mean that trees on the hilltop are the first to blow over, as they will have grown in this environment, but it does indicate where in the landscape any given threshold M_R will be exceeded first as overall wind speeds rise.

Spread of Spores and Pathogens

Long-distance dispersal of seeds, spores, and insects is central to species expansion following ecological disturbance. Recent work by Nathan *et al.* (2002) has shown that the particular nature of the turbulent wind structure determines whether a species will spread over significant distances in a canopy. The key to long-distance dispersal is ejection of the seed or spore from within the canopy to the ASL above. Objects diffusing within the canopy are typically intercepted by the ground or foliage elements within a distance of order h. Nathan *et al.* (2002) found that h was the modal distance of within-canopy travel for a variety of seeds in a deciduous forest. In contrast, seeds ejected from the canopy traveled tens to hundreds of canopy heights downwind (hundreds of meters to several kilometers). The most energetic ejection events are probably associated with the outer edges of the turbulent large structures generated by the mixing layer instability. To eject seeds or spores from the canopy, these ejection events must have vertical

velocities greater than the terminal velocity of the particles. According to the calculations of Nathan *et al.*, (2002), the probability of such events ranges between 1% and 5% for seeds ranging from 5 to 50 mg in weight.

No measurements are currently available to indicate how these probabilities are modified by topography or homogeneity. Nevertheless, we can infer from published and unpublished data from the wind tunnel experiment of Finnigan and Brunet (1995) that ejection from a canopy on a hill is most likely to occur halfway up the windward slope, where the inflection in the velocity profile is weakest, and that it is in this region also that spores or seeds from the lower canopy space have the greatest chance of being fully ejected from the canopy. Conversely, spores, seeds, or pathogens being intercepted by canopy plants are more likely to be transported all the way to the ground by turbulence in this region, while on the hilltop they will be intercepted by the overstory.

Fire

At first thought it seems reasonable to imagine that the character of the turbulent canopy airflow should strongly influence forest fires. However, a revealing similarity analysis of bushfire plumes by Raupach (1990) has shown that this can be true only for fires of very low intensity. By representing a fire front as a line source of buoyancy, Raupach showed how to relate the heat released by the fire in W/m to the characteristic vertical velocity in the ensuing buoyant plume. Classifying fires in terms of the heat released, he associated updraft velocities with fire strengths from a small burn at 10^5 W m^{-1} to an inferno at 10^8 W m^{-1} (see Table 2).

We can compare the updraft velocities, w_B in Table 2, with the mean and turbulent components of the horizontal wind speed at the canopy top to estimate the dynamic significance of the wind patterns generated by the canopy relative to those produced by the fire front. First, we note even moderate burns of 10^6 W m^{-1} produce updraft velocities around 5 m s^{-1}. Next we can calculate that to generate a velocity $\bar{u}(h)$ of 5 m s^{-1} over a forest canopy of moderate density (leaf area index ~ 4, say) requires windspeeds at $z = 2h$

TABLE 2 Updraft Velocity Generated by Wildfires of Specified Heat Release

Type of Fire	Heat Release H (W m^{-1})	Updraft Velocity w_B (m s^{-1})
Small burn	10^5	2.4
	10^6	5.0
	10^7	10.9
Inferno	10^8	23.4

of 10 to 15 m/s. Finally, we recall from Fig. 4 that in the roughness sublayer the streamwise velocity variance $\sigma_u \leq \bar{u}$. Hence, we see that only in very strong winds are the mean wind fields and turbulent velocities generated by the canopy comparable to the buoyant updrafts generated by fires of even moderate intensity.

Another measure of the windfield distortion calculated by Raupach (1990) was the effect of the buoyant plume on the horizontal wind. He showed by a mass conservation argument that the difference in horizontal mean velocity across the plume could be estimated as $\bar{u}_{in} - \bar{u}_{out} \approx 0.2\,w_B$. This implies that the streamwise velocity difference across the fire front is small for small to moderate fires ($H \leq 10^6$ W/m) but ranges up to about 5 m/s for intense fires with H ~ 10^8 W/m. In other words, the action of the fire line as a windbreak is weak except for very strong fires.

It is clear that the particular structure of the canopy windfield is likely to play only a minor role in determining fire dynamics, except in the case of rather cool burns confined to the under story. While this is itself a case of practical importance, because it corresponds to fuel reduction burns that are a part of fire management regimes, wildfires that cause major ecological disturbance probably have windfields totally dominated by the intense buoyant updrafts accompanying the fire front itself and the more diffuse buoyant plume generated over the heated burned region behind the front. Katabatic winds produced in this way typically result in fires burning fiercely to ridge tops on the windward slopes of hills and can result in near-sterilization of hill crests, with profound consequences for the structure of the reemerging canopy.

SUMMARY

I have covered a great deal of ground in this chapter, so it might be useful to recall the approach taken. The chapter began by looking at the structure of the atmospheric boundary layer over land from a traditional micrometeorological point of view. It is worth noting that this point of view tends to be biased toward midlatitude conditions when it isn't raining. Not only do most meteorological instruments (rain gauges excepted) fail to work then, but there are theoretical grounds for believing that turbulent structure changes qualitatively when it is raining. This said, I followed the traditional division of the boundary layer into the mixed or convective boundary layer by day with the surface layer beneath and a shallower stable boundary layer at night.

Discussing the daytime surface layer allowed an introduction to Monin-Obukhov similarity theory, and the departures from this theory close to the canopy pointed to the distinctive nature of turbulence there. I began a closer dissection of canopy flows with a brief history of the development of ideas and followed this with a set of diagrams illustrating typical common features of turbulent and mean flow in uniform canopies. In particular, we saw how turbulence moments and integral scales departed from surface layer norms. Moving on to the cause of these differences, I showed that the flow structure in the canopy was dominated by large coherent eddies that owed their origin and behavior to a hydrodynamic instability of the mean flow that mimics a plane-mixing layer rather than a conventional boundary layer.

All these ideas had been developed for the simplest case of uniform canopies on flat ground, so next I considered heterogeneity in the form of clearings, forest edges, and canopies on hills. From a fluid mechanical point of view, the distinction between these situations is determined by the scale of the heterogeneous forcing rather than by the canopy response, which operates through the same mechanism in both cases. Canopies on hills are subject to forcing by a perturbation pressure field varying on the scale of the hill, a scale much larger than the canopy height. A series of consequences follows from this including, at the local scale, strong perturbations to the mean and turbulent flow fields in and just above the canopy and, at the global scale, phenomena such as increased topographic drag over ranges of hills. The effects of flow across a forest edge are qualitatively different. The pressure field that develops and drives the perturbations varies on the scale of the canopy height and the development of a velocity profile of mixing layer type in a streamwise distance of around 5 to 10 canopy heights plays a critical role in the adjustment of turbulence structure from field to forest.

Finally, I considered the implications of uniform and distorted flows for three kinds of disturbance: windthrow, seed or spore dispersal, and fire. The root bending moment necessary to blow a tree over is difficult to achieve with the drag force developed by the mean wind but is much more easily exceeded if we consider the resonant interaction of flexible plants and the turbulent wind. Seed and spore dispersal, which can govern the spread of weeds, pathogens, or even insect infestations, depends on the ejection of these vectors from the canopy. This, in turn, is directly dependent on the turbulence structure and is probably strongly modulated by topography. In the last example of the implications of canopy wind structure, I analyzed its impact on fire. I compared typical canopy mean and turbulent velocities with the updraft velocities developed around a fire front and also looked at the ability of the fire front to perturb the canopy flow.

The conclusion was that in all but the least intense burns the detailed structure of the canopy wind plays little part in controlling the fire. Rather, the windfield is dominated by the buoyancy field of the fire itself.

REFERENCES

Allen, L. H. (1968). Turbulence and wind speed spectra within a Japanese larch plantation. *J Appl Meteorol* 7, 73–78.

Amiro, B. D. (1990). Drag coefficients and turbulence spectra within three boreal forest canopies. *Bound Lay Meteorol* 52, 227–246.

Baker, C. J. (1995). The development of a theoretical model for the windthrow of plants. *J Theor Biol* 175, 355–372.

Baldocchi, D. D., and Meyers, T. P. (1988a). A spectral and lag-correlation analysis of turbulence in a deciduous forest canopy. *Bound Lay Meteorol* 45, 31–58.

Baldocchi, D. D., and Meyers, T. P. (1988b). Turbulence structure in a deciduous forest. *Bound Lay Meteorol* 43, 345–364.

Belcher, S. E., Jerram, N., and Hunt, J. C. R. (2003). Adjustment of a turbulent boundary layer to a canopy of roughness elements. *J Fluid Mech* 488, 369–398.

Bohm, M., Finnigan, J. J., and Raupach, M. R. (2000). Dispersive fluxes and canopy flows: just how important are they? In *Proceedings of the 24th Conference on Agricultural and Forest Meteorology, Davis, California, 14–18 August 2000*, pp. 106–107. American Meteorological Society, Boston, MA.

Brunet, Y., Finnigan, J. J., and Raupach, M. R. (1994). A wind tunnel study of air flow in waving wheat: single-point velocity statistics. *Bound Lay Meteorol* 70, 95–132.

Chen, J. M., Black, T. A., Novak, M. D., and Adams, R. S. (1995). A wind tunnel study of turbulent airflow in forest clearcuts. In *Wind and Trees* (Coutts, M. P., and Grace J., Eds), pp. 71–87. Cambridge University Press, Cambridge, UK.

Collineau, S., and Brunet, Y. (1993). Detection of turbulent coherent motions in a forest canopy, Part II: Timescales and conditional averages. *Bound Lay Meteorol* 66, 49–73.

Coutts, M. P., and Grace, J. (Eds.) (1995). *Wind and Trees*. Cambridge University Press, Cambridge, UK.

Denmead, O. T., and Bradley, E. F. (1987). On scalar transport in plant canopies. *Irrigation Sci* 8, 131–149.

Duran, D. R. (1990). Mountain waves and downslope winds. In *Atmospheric Processes over Complex Terrain* (Blumen, W., Ed.), pp. 59–81. American Meteorological Society, Boston, MA.

Finnigan, J. J. (1979a). Turbulence in waving wheat. I. Mean statistics and honami. *Bound Lay Meteorol* 16, 181–211.

Finnigan, J. J. (1979b). Turbulence in waving wheat. II. Structure of momentum–transfer. *Bound Lay Meteorol* 16, 213–236.

Finnigan, J. J. (2000). Turbulence in plant canopies. *Annu Rev Fluid Mech* 32, 519–571.

Finnigan, J. J., and Belcher, S. E. (2004). Flow over a hill covered with a plant canopy. Part 1 analytical model for the flow. *Q J R Meteorol Soc* 130, 1–29.

Finnigan, J. J., and Brunet, Y. (1995). Turbulent airflow in forests on flat and hilly terrain. In *Wind and Trees* (Coutts, M. P., and Grace, J., Eds.), pp 3–40. Cambridge University Press, Cambridge, UK.

Finnigan, J. J., and Mulhearn, P. J. (1978). Modelling waving crops in a wind tunnel. *Bound Lay Meteorol* 14, 253–277.

Finnigan, J. J., and Shaw, R. H. (2000). Turbulence in waving wheat: an empirical orthogonal function analysis of the large-eddy motion. *Bound Lay Meteorol* 96, 211–255.

Finnigan, J. J., Einaudi, F., and Fua, D. (1984). The interaction between an internal gravity wave and turbulence in the stably-stratified nocturnal boundary layer. *J Atmos Sci* 41, 2409–2436.

Fleagle, R. G., and Businger, J. A. (1963). *An Introduction to Atmospheric Physics*. Academic Press, New York.

Gao, W., Shaw, R. H., and Paw U, K. T. (1989). Observation of organized structure in turbulent-flow within and above a forest canopy. *Bound Lay Meteorol* 47, 349–377.

Gardiner, B. A. (1994). Wind and wind forces in a plantation spruce forest. *Bound Lay Meteorol* 67, 161–186.

Garratt, J. R. (1992). *The Atmospheric Boundary Layer*. Cambridge University Press, Cambridge, UK.

Hunt, J. C. R., Leibovich, S., and Richards, K. J. (1988). Turbulent shear flow over low hills. *Q J R Meteorol Soc* 114, 1435–1470.

Kaimal, J. C., and Finnigan, J. J. (1994). *Atmospheric Boundary Layer Flows: Their Structure and Measurement*. Oxford University Press, New York.

Lu, C. H., and Fitzjarrald, D. R. (1994). Seasonal and diurnal variations of coherent structures over a deciduous forest. *Bound Lay Meteorol* 69, 43–69.

Lumley, J. L. (1967). The structure of inhomogeneous turbulent flows. In *Atmospheric Turbulence and Radio Wave Propagation* (Yaglom, A. C., and Tatarsky, V. I., Eds.), pp. 166–xxx. Nauka, Moscow

Lumley, J. L. (1981). Coherent structures in turbulence. In *Transition and Turbulence* (Meyer, R. E., Ed.), pp. 215–241. Academic Press, New York.

Marshall, B. J., Wood, C. J., Gardiner, B. A., and Belcher, R. E. (2002). Conditional sampling of forest canopy gusts. *Bound Lay Meteorol* 102, 225–251.

Moin, P., and Moser, R. D. (1989). Characteristic-eddy decomposition of turbulence in a channel. *J Fluid Mech* 200, 471–509.

Morse, A. P., Gardiner, B. A., and Marshall, B. J. (2002). Mechanisms controlling turbulence development across a forest edge. *Bound Lay Meteorol* 103, 227–251.

Nathan, R., Katul, G. G., Horn, H. S., Thomas, S. N., Oren, R., Avissar, R., and Pacala, S. W. (2002). Mechanisms of long-distance dispersal of seeds by winds. *Nature* 418, 409–413.

Neff, D. E., and Meroney, R. N. (1998). Wind-tunnel modeling of hill and vegetation influence on wind-power availability. *J Wind Eng Industrial Aerodynam* 74–6, 335–343.

Novak, M. D., Warland, J. S., Orchansky, A. L., Ketler, R., and Green, S. R. (2000). Comparison between wind tunnel and field measurements of turbulent flow. Part I: Uniformly thinned forests. *Bound Lay Meteorol* 95, 457–495.

Panofsky, H. A., and Dutton, J. A. (1984). *Atmospheric Turbulence: Models and Methods for Engineering Applications*. Wiley-Interscience, John Wiley and Sons, New York.

Raupach, M. R. (1990). Similarity analysis of the interaction of bushfire plumes with ambient winds. *Math Comput Model* 13, 113–121.

Raupach, M. R., Antonia, R. A., and Rajagopalan, S. (1991). Rough-wall turbulent boundary layers. *Appl Mech Rev* 44, 1–25.

Raupach, M. R., Bradley, E. F., and Ghadiri, H. (1987). *Wind Tunnel Investigation into the Aerodynamic Effect of Forest Clearing on the Nesting of Abbot's Booby on Christmas Island*. CSIRO Centre for Environmental Mechanics Technical Report T12, CSIRO Marine and Atmospheric Research, Canberra, Australia.

Raupach, M. R., Coppin, P. A., and Legg, B. J. (1986). Experiments on scalar dispersion within a model plant canopy. Part I: The turbulence structure. *Bound Lay Meteorol* 35, 21–52.

Raupach, M. R., Finnigan, J. J., and Brunet, Y. (1996). Coherent eddies in vegetation canopies—the mixing layer analogy. *Bound Lay Meteorol* 78, 351–382.

Rogers, M. M., and Moser, R. D. (1992). The three-dimensional evolution of a plane mixing layer: the Kelvin-Helmholtz rollup. *J Fluid Mech* 243, 183–226.

Ruck, B., and Adams, E. (1991) Fluid mechanical aspects of the pollutant transport to coniferous trees. *Bound Lay Meteorol* 56, 163–195.

Seginer, I., Mulhearn, P. J., Bradley, E. F., and Finnigan, J. J. (1976). Turbulent flow in a model plant canopy. *Bound Lay Meteorol* 10, 423–453.

Shaw, R. H., and Patton, E. G. (2003). Canopy element influences on resolved- and subgrid-scale energy within a large-eddy simulation. *Agricult Forest Meteorol* 115, 5–17.

Shaw, R. H., Brunet,Y., Finnigan, J. J., and Raupach, M. R. (1995). A wind tunnel study of airflow in waving wheat: two point velocity statistics. *Bound Lay Meteorol* 76, 349–376.

Shaw, R. H., Finnigan, J. J., and Patton, E. G. (2004). An analysis of the structure of the large-eddy motion in a simulated vegetation canopy. In *Proceedings of the 26th Conference on Agricultural and Forest Meteorology, Vancouver, Canada, August 23–26, 2004*. American Meteorological Society, Boston, MA.

Shaw, R. H., Silversides, R. H., and Thurtell, G. W. (1974). Some observations of turbulence and turbulent transport within and above plant canopies. *Bound Lay Meteorol* 5, 429–449.

Shaw, R. H., Tavanger, J., and Ward, D. P. (1983). Structure of the Reynolds stress in a canopy layer. *J Climate Appl Meteorol* 22, 1922–1931.

Stacey, G. R., Belcher, R. E., and Wood, C. J. (1994). Wind flows and forces in a model spruce forest. *Bound Lay Meteorol* 69, 311–334.

Townsend, A. A. (1976). *The Structure of Turbulent Shear Flow*. Cambridge University Press, Cambridge, UK.

Wilson, J. D., Finnigan, J. J., and Raupach, M. R. (1998). A first-order closure for disturbed plant-canopy flows, and its application to winds in a plant canopy on a ridge. *Q J R Meteorol Soc* 124, 705–732.

Wilson, J. D., Ward, D. P., Thurtell, G. W., and Kidd, G. E. (1982). Statistics of atmospheric turbulence within and above a corn canopy. *Bound Lay Meteorol* 24, 495–519.

3

Microbursts and Macrobursts: Windstorms and Blowdowns

Mark R. Hjelmfelt
South Dakota School of Mines and Technology

INTRODUCTION

Winds causing plant disturbances can arise from a number of sources: (1) hurricanes; (2) downbursts and tornadoes from convective storms; (3) intense winds from winter cyclonic storms; and (4) topographically induced downslope windstorms. These can vary from marginal events (clearing forests of dead or weakened branches in the canopy) to extreme events (flattening all vegetation in the path of the winds and causing high mortality of many animal species). The size scale of these events and their effects may vary from tens of square meters to hundreds or thousands of square kilometers. While forest blowdowns represent the most notable of plant disturbances from windstorms, agricultural crops, especially corn (Fujita, 1979), and even prairie ecosystems can experience devastating ecological effects from extreme events.

Obviously the extent of plant disturbance depends on many factors. In addition to wind speed, gustiness, and duration of the winds, the simultaneous occurrence of large hail may greatly increase the damage. Besides the meteorological wind storm, the type, height, and condition of the plants, their location relative to open spaces, soil conditions, and other ecological parameters may strongly affect the type and degree of disturbance.

An example of a tornado blowdown event occurred in the Yellowstone National Park area in 1987 (Fujita, 1989). Intense winter cyclonic storms, especially when interacting with topography, can produce winds in excess of 50 $m s^{-1}$ (Weaver, 1999; Browning, 2004), and have resulted in many large blowdown events, including the October 1962 Oregon event (Lynott and Cramer, 1966) and the "Great Storm" of October 1987 in southeast England (Browning, 2004). In mountainous regions, flow over the mountain ridges under certain conditions may result in extreme winds on the downslope side (Lilly and Zipser, 1972; Lilly, 1978). Winds in excess of 60 $m s^{-1}$ have been recorded. A recent significant blowdown event from downslope winds occurred in the Routt National Forest, Colorado, in 1999 (Meyers *et al.*, 2003). Downbursts—a localized descending current of air (or downdraft) from a cumulus (convective) cloud that induces an outward burst of damaging wind on or near the ground (Fujita and Byers, 1977)—are the focus of this chapter. Other example references from forest and ecology literature are found in Table 1 in Chapter 4.

CONVECTIVE STORMS AND DOWNBURSTS

Convective storms produce downdrafts that descend to the surface and spread out, forming a gust front at the leading edge. The gust front is often important in maintaining the storm as it flows outward from the storm and forces the warm prestorm air upward to initiate new convection. However, the strong gusty winds may be sufficient to cause crop, tree, or structural damage or lead to aircraft accidents. Knupp and Cotton (1985) presented a general review of convective downdrafts. Downbursts and microbursts have been described by Fujita (1981; 1985) and others. Most recently, Wakimoto (2001) has provided an excellent comprehensive scientific review of convectively driven high-wind events.

The general topic of downbursts can be viewed as consisting of phenomena on two different scales:

1. More or less isolated incidences of individual pulses of strong downdraft and outflow. Depending on the size of the outflow, these may be

classified as *microbursts* (outflows for which the horizontal extent of the strong, damaging winds is less than 4 km) or *macrobursts* (if the outburst winds extend over distances greater than 4 km). Fig. 1A shows Fujita's (1978; 1985) conceptual model of a microburst downdraft. The descending current of air, which may be rotating, impacts the ground

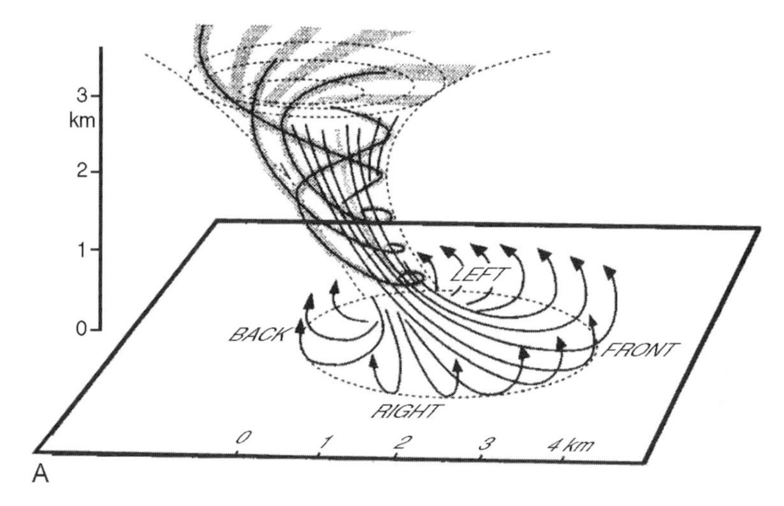

FIG. 1 The microburst: (**A**) Fujita's conceptual model of a microburst (Fujita and McCarthy, 1989); and (**B**) an aerial photograph of forest blowdown damage attributed to a rapidly traveling microburst (Fujita and Byres, 1977).

and spreads out, forming a sharp gust front and a curl of air (sometimes creating an actual horizontal "roll" vortex) on the leading edge. Fig. 1B shows an example of forest blowdown damage that Fujita attributed to a rapidly moving microburst. In this case we see a small area of downed trees, lying in a diverging-line pattern from the bottom to the top of the photo. Microbursts may be produced from thunderstorms that produce heavy rain (sometimes called "wet" microbursts) or even innocuous-appearing shallow, high-based convective clouds producing only light showers or virga ("dry" microbursts). Examples are shown in Fig. 2. The panels on the left are photographs of events; those on the right are the respective plan views of wind vectors analyzed from Doppler radar observations and radar reflectivity contours. Strong diverging flow is evident in both cases. The reflectivity for the "wet" microburst was greater than 55 dBz (upper right panel), whereas the reflectivity for the "dry" microburst was only 15 dBz (lower right panel).

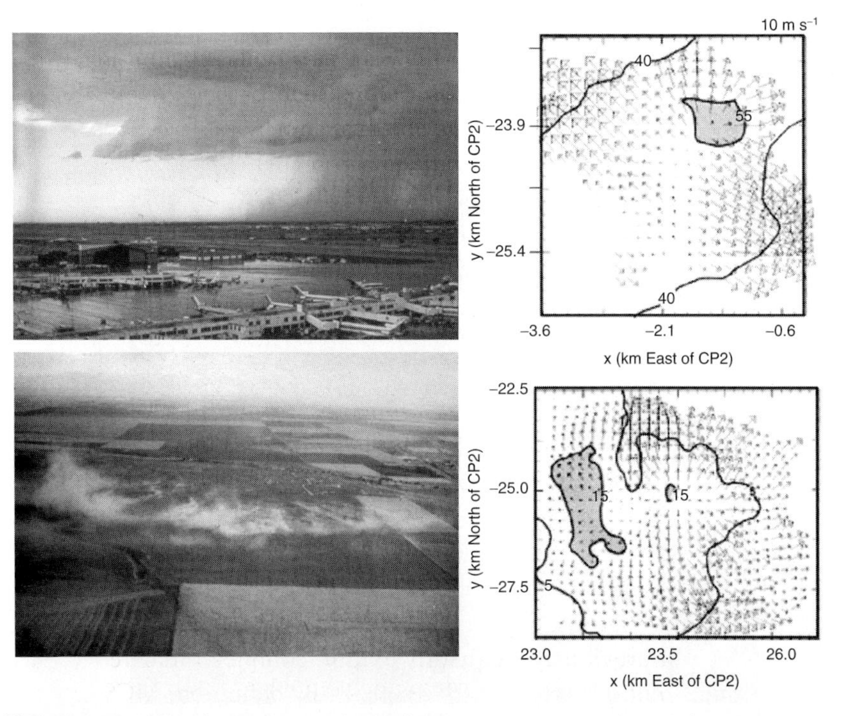

FIG. 2 Examples of microbursts observed in Joint Airport Weather Studies (**top**) "wet," with radar reflectivity greater than 55 dBz shaded, and (**bottom**) "dry," with reflectivity greater than 15 dBz shaded. Wind vectors drawn to scale shown at top right. (Photos courtesy of NCAR-ATD, from Hjelmfelt [1988].)

2. Longer-lived, organized, severe wind–producing systems may produce incidences of severe winds over a sizable area for an extended path of many tens of kilometers. These are typically associated with supercell thunderstorms and groups of thunderstorms known as mesoscale convective systems (MCSs), such as squall lines, and especially bow echo storms. [Houze (2004) presents a detailed technical review of MCSs.]

It is also useful to separate downbursts that are associated with strong, small-scale rotation (mesocyclones or mesoanticyclones, depending on the direction of rotation) in severe storms, which I refer to as *mesocyclone-related*, from other downbursts. In the Northern Hemisphere, the Coriolis force acting on the larger-scale environmental winds leads to conditions that favor mesocyclones.

The primary organized convective severe wind–producing systems (squall lines, bow echoes, and supercell thunderstorms) may be defined following Klimowski *et al.* (2003). Example radar images are shown in Fig. 3. Supercell thunderstorms are distinguished from nonsupercell storms by the presence of a rotating updraft or mesocyclone. The mesocyclone may be associated with several distinctive radar features (Forbes, 1981; Moller *et al.*, 1994), which include hook and pendant echoes and, in three dimensions, a bounded weak echo region containing the very strong updraft, which may exceed 50 m s^{-1} (Musil *et al.*, 1986). Supercells often exhibit motion that deviates significantly to the right or left of other nonsupercell storms (Bunkers *et al.*, 2000). Squall lines are lines of thunderstorms that maintain a definite linear organization for some period of time, perhaps 30 minutes or so, as in Bluestein and Jain (1985). Bow echoes can be defined following the original definitions of Fujita (1978); see Fig. 4. A bow echo is a bow- or crescent-shaped radar echo with a tight reflectivity gradient on the convex (leading) edge, consistent with storms that propagate along a strong outflow. The bow echo demonstrates linear organization, rather than a brief grouping of cells into an otherwise unassociated arc-like structure. A bowing segment embedded within a squall line may be identified as a line echo wave pattern (Nolan, 1959). Occasionally, storms are associated with multiple severe wind reports, do not demonstrate linear organization, and are not supercells; these we may call "irregular." The MCSs described previously frequently exhibit complex radar reflectivity patterns and evolutions (Houze, 1993; 2004). By definition, MCSs consist of several individual (thunderstorm) convective cells, or updrafts and downdrafts, and it is not uncommon for one severe wind–producing convective mode to evolve into another, as seen in radar imagery (e.g., squall

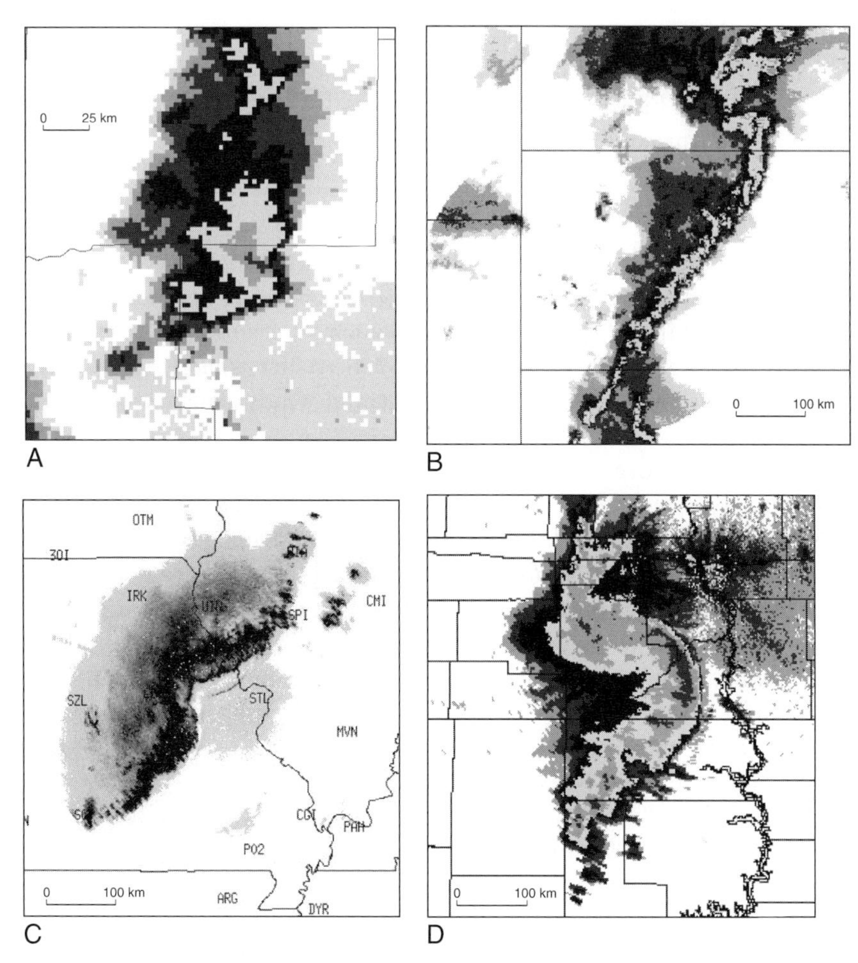

FIG. 3 Radar view of storm types producing organized downburst events: (**A**) supercell; (**B**) squall line; (**C**) squall line with a line echo wave pattern (LEWP); and (**D**) bow echo. Figures not all to same scale. (Photos courtesy of Matthew Bunkers, NWS, Rapid City, SD.)

line–bow echo or supercell–bow echo transitions). Also, one high wind–producing storm type may occur *within* another (bow echoes within larger-scale squall lines and high wind–producing supercells within squall lines). High-wind events frequently occur as a part of the evolution of an MCS, and tornadoes and downbursts may occur simultaneously from the same storm (as in Forbes and Wakimoto, 1983).

MCSs, such as squall lines and bow echo storms, may produce mesoscale downdrafts that cause widespread damage of varying intensity

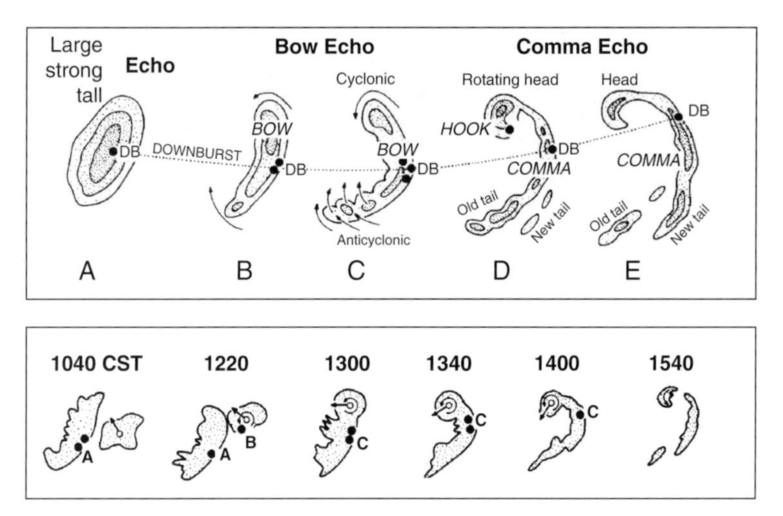

FIG. 4 Fujita's explanation of bow echo evolution, top, and a drawing of the echo evolution of the storm that caused the July 4, 1977 Wisconsin downburst shown in Fig. 5 (Fujita, 1978; courtesy of the Tetsuya "Ted" Fujita Collection, Southwest Collection/Special Collections Library, Texas Tech University, Lubbock, TX).

over hundreds of square kilometers as a result of clusters of downbursts. Much of the damage caused by nontornadic winds from convective storms is caused by these mesoscale downburst clusters. Fujita (1978) showed that even the largest severe wind systems exhibited embedded smaller-scale swaths of more intense damage. The July 4, 1977 bow echo storm (Fig. 4) produced severe winds for over 12 hours as it traveled more than 1300 km from Minnesota to eastern Ohio. Along this path were three definable areas of damage. The one shown in Fig. 5 was associated with peak winds in excess of 50 m s^{-1}, produced a damage swath over 260 km long and up to 27 km wide, and damaged more than 3400 km^2 of forest (Fujita, 1978; Canham and Loucks, 1984). Fig. 5 also shows small patches, covering several square kilometers, of very intense damage (dark shading) embedded within the large-scale outline of the overall damage area; this figure illustrates that the damage swath is not continuous—not every tree or other vegetation is damaged, but the wind is gusty and highly variable. Fujita and Wakimoto (1981) described such patterns in terms of five scales associated with downbursts. The largest scale, families of downburst clusters, may cover a path length extending over entire states. A downburst cluster, as in Fig. 5, may extend over 100 km in length, representing serial downbursts (identified by numbers in Fig. 5) from a storm system. The downburst itself

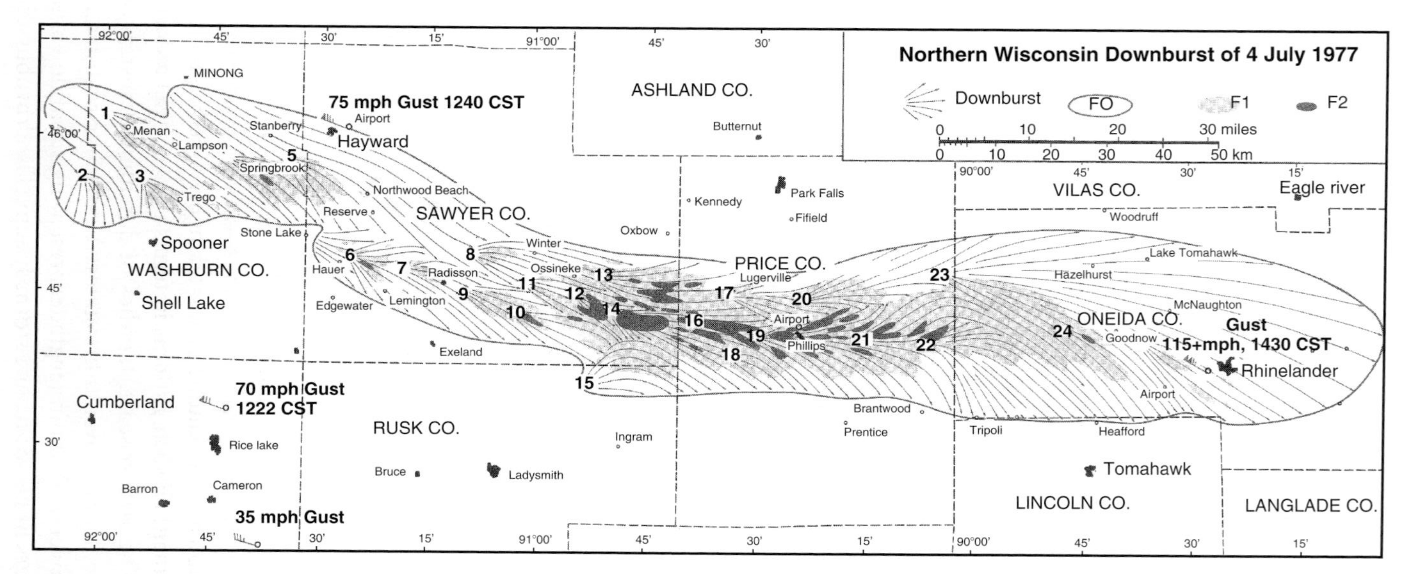

FIG. 5 Fujita's damage path analysis of the July 4, 1977 downburst "derecho" (Fujita, 1978; courtesy of the Tetsuya "Ted" Fujita Collection, Southwest Collection/Special Collections Library, Texas Tech University, Lubbock, TX). Intensity of damage indicated by shading according to Fujita's F scale (Fujita, 1981); see Table 1. This figure shows one downburst cluster, which was part of a family of downburst clusters produced by the bow echo illustrated in Fig. 4. The small numbers correspond to individual downbursts, which contain smaller-scale microbursts and yet even smaller burst swaths (see discussion in text for more details).

may cover several tens of kilometers, while embedded microbursts are restricted to a few kilometers in size and are associated with individual convective cell downdrafts. Within a microburst, small swaths of more intense damage may appear. These burst swaths have scales of a few hundred meters or less. Thus, even the largest downburst events have small-scale embedded microbursts that produce pockets of more severe disturbance.

The largest damage swaths are called *derechos*—a family of downburst clusters associated with an MCS (after Hinrichs, 1988). The term "derecho" is increasingly being reserved by forecasters to specific criteria, following Johns and Hirt (1987) and Coniglio and Stensrud (2004):

- An area of wind reports of convectively induced wind damage or gusts $>26\ \mathrm{m\,s^{-1}}$, with length > 400 km.
- The wind reports must have a chronological progression as a continuous swath or series of swaths.
- The associated convective weather system must have spatial and temporal continuity, and multiple damage swaths must be from the same system as indicated by available data.
- No breaks of more than 2 to 3 hours in time may occur between successive reports.

The intensity of the damage caused by windstorms and tornadoes is described by the F-scale, developed by Fujita (1971; 1981); see Table 1. This scale, ranging from 0 to 6, is uncalibrated, but an attempt was made to roughly relate the scale to wind speeds, as shown in Table 1. Official severe storm reports include estimates of damage severity according to this scale. An effort is underway to update this scale, especially the relation to wind speed, based on research advances over the past 30 years (McDonald, 2002).

In addition to wind measurements and damage reports, storm analyses, such as that shown in Fig. 5, rely heavily on surveys, especially careful aerial surveys and photography (e.g., Fig. 1 and examples shown later). Satellite imagery can also be used to determine the spatial extent of damage (Klimowski *et al.*, 1998; Bentley *et al.*, 2002).

Research associated with microbursts has been related primarily to aircraft safety and has typically emphasized smaller, isolated events. Research on the larger-scale downbursts and derechos has primarily analyzed events that cause major damage (Fujita, 1978; Klimowski *et al.*, 1998) and climatologies related to forecasting criteria. Existing research on severe wind–producing convective systems covers a wide variety of convective storm types (Klimowski *et al.*, 2003), including small microburst–producing storms (e.g., Fujita, 1981; Wakimoto, 1985), supercells (Brooks *et al.*,

TABLE 1 F-scale Damage Specification (Fujita, 1981)

(FO)	18–32 m s^{-1} (40–72 mph): Light damage Some damage to chimneys; break branches off trees; push over shallow-rooted trees; damage sign boards.
(Fl)	33–49 m s^{-1} (73–112 mph): Moderate damage The lower limit (73 mph) is the beginning of hurricane wind speed; peel surface off roofs; mobile homes pushed off foundations or over-turned; moving autos pushed off the roads.
(F2)	50–69 m s^{-1} (113–157 mph): Considerable damage Roofs torn off frame houses; mobile homes demolished; boxcars pushed over; large trees snapped or uprooted; light-object missiles generated.
(F3)	70–92 m s^{-1} (158–206 mph): Severe damage Roofs and some walls torn off well-constructed houses; trains overturned; most trees in forest uprooted; heavy cars lifted off ground and thrown.
(F4)	93–116 m s^{-1} (207–260 mph): Devastating damage Well-constructed houses leveled; structure with weak foundation blown off some distance; cars thrown and large missiles generated.
(F5)	117–142 m s^{-1} (261–318 mph): Incredible damage Strong frame houses lifted off foundations and carried considerable distance to disintegrate; automobile-sized missiles fly through the air in excess of 100 m; trees debarked; incredible phenomena will occur.
(F6-F12)	142 m s^{-1} to Mach I, the speed of sound The maximum wind speeds of tornadoes are not expected to reach the F6 wind speeds.

1994; Moller *et al.*, 1994), bow echoes (e.g., Przybylinski, 1995; Weisman, 2001), squall lines (e.g., Bluestein and Jain, 1985; Smull and Houze, 1985), and MCSs and mesoscale convective complexes (Maddox, 1980; Schmidt and Cotton, 1989). These convective modes cover a substantial range of time and space scales, spanning from less than half an hour and a few kilometers on the lower end to many hours and hundreds of kilometers on the higher end. Derechos and derecho-producing storms have been a focus of much recent study (e.g., Miller and Johns, 2000; Evans and Doswell, 2001; Coniglio *et al.*, 2004; Coniglio and Stensrud, 2004; and a recent major field study, Davis *et al.*, 2004).

VERTICAL EQUATION OF MOTION

The physics of the downdraft may be described by the equation of motion for vertical velocities. One form that illustrates the physical principles related to severe wind storms is given by Equation (1) (Wakimoto, 2001). In this form, the temperature is described by the virtual potential temperature, θ_v, which, as described in Chapter 2, is the temperature corrected for the natural expansion of the atmosphere as the pressure decreases with height and for the effect of water vapor on the buoyancy of the air. θ_v is defined as $\theta_v = T_v(p_0/p)^{.286}$, where T_v is the virtual temperature, given by $T_v = T - 0.61q$ (where T is temperature and q is the mixing ratio of water vapor), p is pressure, and p_0 is a reference pressure, taken as 1000 hPa. The virtual potential temperature is widely used in atmospheric science because it is conserved for adiabatic processes. Therefore,

$$\underset{1}{\frac{d\overline{w}}{dt}} = \underset{2}{-\frac{1}{\rho}\frac{\partial p'}{\partial z}} + g\left[\underset{3}{\frac{\theta_v'}{\theta_{vo}}} - \underset{4}{\frac{c_v}{c_p}\frac{p'}{p_o}} - \underset{5}{(r_c + r_r + r_i)}\right]$$

where $\overline{w}$ is the vertical velocity, g is gravity, and primed quantities represent the deviation between an air parcel and the environment at the same level.

The terms of this equation are:

1. The vertical acceleration
2. The vertical gradient of perturbation pressure
3. The thermal buoyancy term
4. The perturbation pressure buoyancy
5. The loading by the weight of condensed water substance

Equation (1) shows that, as long as the forcing terms on the right hand side are negative, there will be a downward acceleration to a parcel. Thus, for example, as long as the parcel is colder than its environment, it will continue to accelerate downward, even if no additional cooling occurs.

The first term on the right (term 2), the vertical gradient of perturbation pressure, has generally been found to be weak in ordinary cumulus convection (Schlesinger, 1980), and has not been found to be appreciable in most microbursts (Hjelmfelt, 1987; Lee *et al.*, 1992). However, in supercells (Newton and Newton, 1959; Klemp and Rotunno, 1983), bow echoes, and other mesocyclone situations, this term may be of primary importance (Carbone, 1983). By the cyclostrophic relation, a strong low-level mesocyclone (or mesoanticyclone) would have low pressure in the center. This low pressure near the surface would create a vertical perturbation pressure gradient, which would induce a downdraft. The pressure perturbation term may be an order of magnitude or more than loading and may dominate the downdraft forcing (Carbone, 1983; Wakimoto *et al.*, 1996). In supercells, such downdrafts exist as the "occlusion downdraft," discussed later in this chapter, and are also frequently observed in severe bow echo storms (Trapp and Weisman, 2003; Atkins *et al.*, 2004).

Thermal buoyancy, term 3, is the primary driving force for convection and represents the effects of temperature difference between a parcel and its environment. Colder air is denser and thus negatively buoyant. Water vapor is lighter than air; thus, a mixture of air and water vapor is less dense (more buoyant) than dry air, as a function of the mixing ratio of water vapor to air. This may be explained through use of the virtual (potential) temperature. Significant errors can result if the effects of the presence of water vapor are not included. Phase changes caused by transformations of the water substance in a parcel release or consuming considerable amounts of latent heat and can greatly affect the buoyancy. In many downburst situations this term dominates the vertical velocity forcing through the effects of sublimation/evaporation or melting and must be carefully treated in calculations (Saunders, 1957; Orville and Hubbard, 1973).

The perturbation pressure buoyancy, term 4, is the buoyancy effect of a parcel having a lower pressure than its surroundings. It has been shown to be generally weaker than buoyancy or pressure gradient effects (Schlesinger, 1980).

The total weight of any condensed phase water in a parcel will exert a downward force, term 5, according to the sum of the mixing ratios of cloud water and cloud ice, rainwater, snow, graupel, and hail. Most downdrafts reaching the ground are initiated by the weight of developing precipitation, and, in intense precipitation, this term may be sufficient by itself to produce strong outflows.

Mixing caused by turbulence will tend to slow any acceleration. Turbulent mixing with environmental air will also affect the buoyancy by reducing the temperature contrast with the environment through mixing, but may enhance evaporation (cooling) and result in changes in the total parcel air density and precipitation loading.

If any of these terms, or their combination, is strong enough, strong downdrafts and outflows may result. For example, strong gusty surface winds may occur from high-based clouds in dry environments, with little or no precipitation reaching the ground as a result of evaporation and sublimation of precipitation below cloud base. Heavy rain gushes may similarly be accompanied by strong outflows even in moist stable boundary layers as a result of the weight of the precipitation and melting within the cloud. Additionally, if the downdraft occurs in a rapidly moving storm or transports high momentum air downward from aloft, outflows may be further strengthened in this direction. This additive effect may lead to damaging winds along the resulting outflow axis, even if the downdraft outflow itself was not strong enough to produce damage.

CLIMATOLOGY

The most detailed specific high-wind climatologies from convective storms have been developed in the United States. Kelly *et al.* (1985) and Doswell and Bosart (2001) have described the climatology of severe convective wind events over the United States. Klimowski *et al.* (2003) provide details of the climatology over the northern High Plains. Several climatological studies have been made specifically of derecho events (e.g., Johns and Hirt, 1987; Bentley and Mote, 1998; Coniglio and Stensrud, 2004). Other information comes from field research programs and properly applies to only the specific small region (and time period) of the experiment. A critical problem with these climatologies is extreme population and station bias (Kelly *et al.*, 1985). Many reports are based on damage, which requires someone to observe the damage and report it to the right people so that it gets recorded in the system. If population is high and there are structures or particularly vulnerable crops (e.g., corn: Fujita, 1979) or trees, reports are much more likely, even for weak events. With an absence of observers, destruction will go unreported. Thus, over much of the western half of the United States and Canada, events are greatly under-reported.

Kelly *et al.* (1985) presented a climatology of nontornadic convective severe storm events. They found, as expected, that convective high-wind events occur primarily in the summer, with a June and July maximum over the United States, though in the southern states, events may occur even in the winter. There is a late afternoon peak, corresponding to the time of maximum instability in the lower atmosphere, with activity extending after midnight; more than 75% of the events occurred during the afternoon and evening hours. This is somewhat dependent on location, with the High Plains experiencing a stronger diurnal dependence and a clearer afternoon peak (Klimowski *et al.*, 2003). Nocturnal events are more common over the southeastern United States.

If we consider the geographical distribution, there is a difference between strong events (>25 m s^{-1}) and violent events (>32 m s^{-1}). Maps of the probability of strong and violent wind storms are presented by Doswell and Bosart (2001) and are available from the National Severe Storms Laboratory (2005). For strong events, as shown in Fig. 6A, maximum frequencies occur over a broad area extending from the southern Great Plains across the Midwest into Pennsylvania. Another area of higher frequency occurs across the southern states to South Carolina. Violent events are reported most frequently in a corridor that runs from north central Texas northward through eastern Nebraska and extending eastward across Iowa (Fig. 6B). Coastal regions generally show less frequent reports.

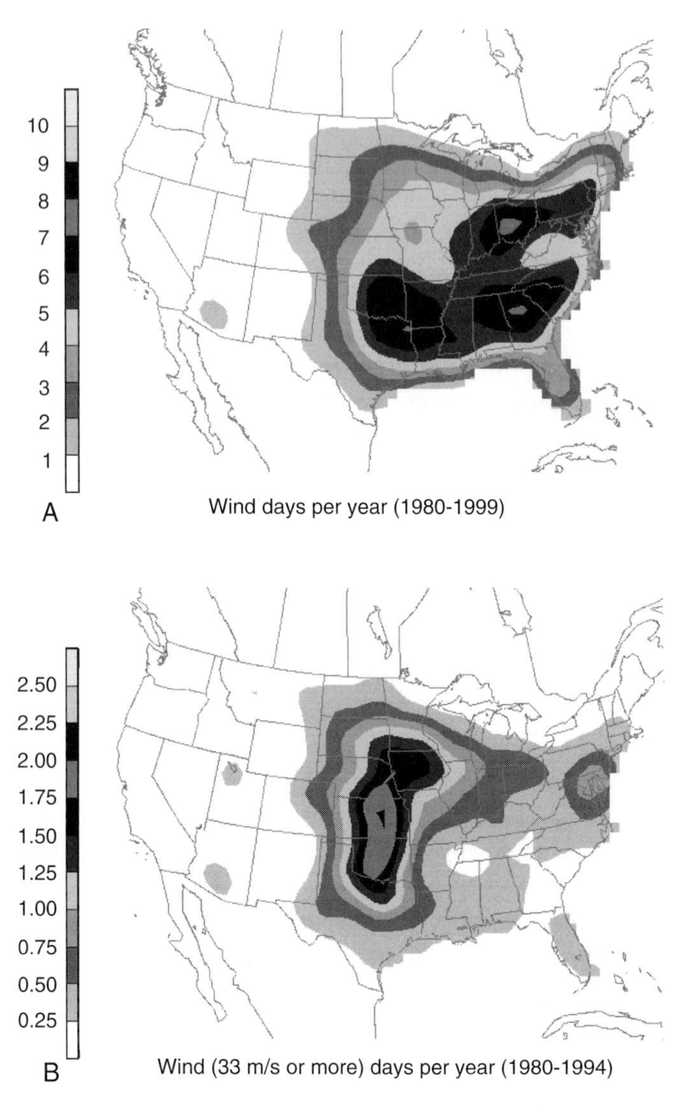

FIG. 6 Severe wind events over the United States: (**A**) frequency of severe (>25 m s^{-1}) wind days per 1600 km^{-2} y^{-1}; (**B**) winds greater than 32 m s^{-1}, days per 1600 km^{-2} y^{-1}. (From Doswell and Bosart, [2001].)

Our knowledge of isolated microburst climatology comes primarily from individual field experiments and includes many weak cases of marginal interest for plant disturbance. There is no good climatology on the frequency of occurrence across the country. Being very small-scale events, only a small fraction are reported and included in the statistics above, except in heavily populated areas. Microburst data from four field projects in different parts of the United States suggest that 60% to 80% of days with thunderstorms may have microbursts. During the Joint Airport Weather Studies (JAWS) project near Denver in 1982 (McCarthy *et al.*, 1982), 186 microbursts were observed in an 86-day period within an area about 100 km in diameter, although many of these were of marginal intensity and of little importance for plant disturbances. Relatively more nocturnal microbursts were observed in the southeastern United States. These projects all reported an approximate exponential decrease in occurrence with increasing maximum wind speed. Data from JAWS near Denver showed a distribution extending beyond 45 m s^{-1}. The maximum wind speed that has been observed for any isolated microburst was one that occurred at Andrews Air Force Base with a measured wind speed of more than 65 m s^{-1} (Fujita, 1985).

In their study of severe (>25 m s^{-1}) and damaging convective windstorms over the northern High Plains, Klimowski *et al.* (2003) found that about 29% of the official severe wind reports were from isolated microbursts and 65% were from organized downburst–producing systems (6% were unknown). On the average, each organized event resulted in three to four reports, so that 35% of the severe wind events were caused by organized systems and 60% were from isolated microbursts (though lack of reporting could have caused some cases to be misclassified as isolated and many isolated cases to be missed). Other convective storm and severe weather climatologies suggest that in less arid regions, such as the southeastern United States, the proportion of isolated events should be lower.

Coniglio and Stensrud (2004) examined the distribution of derechos occurring from 1986 to 2001 over the central and eastern United States (Fig. 7). For warm season (May to August) events, a broad area of maximum occurrence extended from eastern Oklahoma and Arkansas northeastward to Minnesota and Ohio. The strongest events were more prevalent across Minnesota and Iowa to Ohio. In the cold season, the maximum was centered over Mississippi and extended only weakly northward to Ohio. The distributions changed over time so that the areas of greatest occurrence were different for the first half of the period from the second half of the period. Some of this variation is expected with the small number

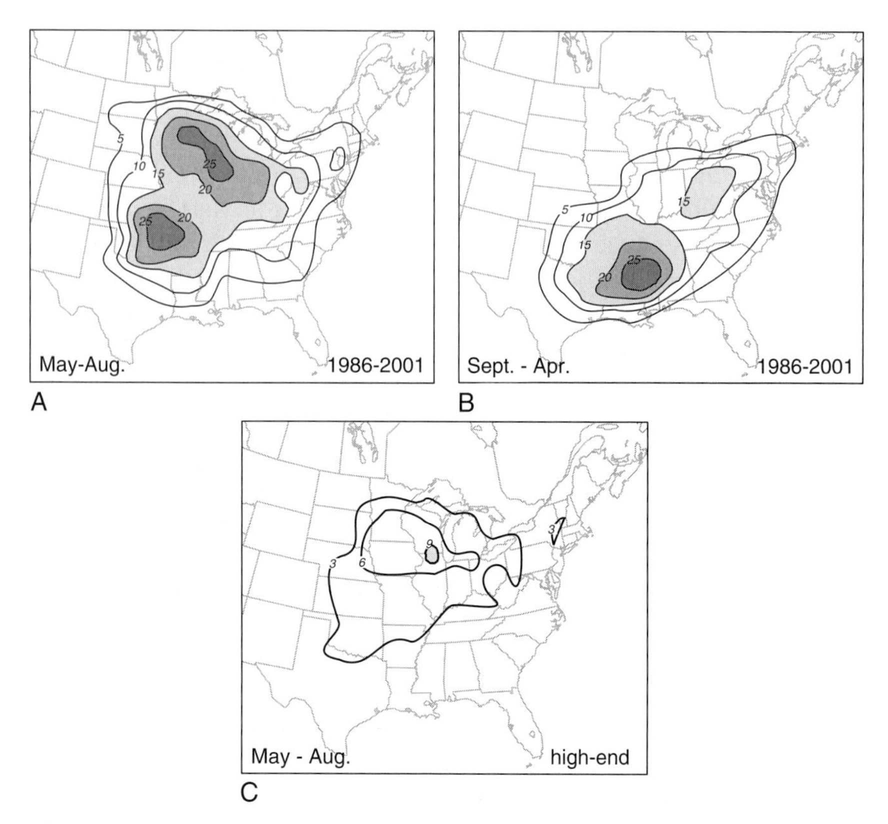

FIG. 7 Derechos over the United States. 1986–2001: (**A**) warm season (May–August); (**B**) cold season (September–April); (**C**) only the strongest (>36 m s^{-1}) warm season derechos. Contours are number of events in 200 km × 200 km grid cells during the period. (Adapted from Coniglio and Stensrud, [2004].)

of events, but some may be attributed to changes in storm tracks between these periods.

Information on the distribution and climatology of severe convective winds in other parts of the world is sparser, but aircraft incidents attributable to microbursts have been reported worldwide (Serafin *et al.,* 1999). Diurnal variations and maximum wind speed statistics for microbursts very similar to those reported in the United States have been obtained in Australia (Potts, 1991) and Japan (Ohno *et al.,* 1996). Downbursts have been observed in tropical regions, such as Brazil (Garstang *et al.*, 1998; Atlas *et al.*, 2004) and Singapore (Choi, 1999), the Hawaiian Islands (Businger *et al.*, 1998), and in monsoon regions, such as Bangladesh (Peterson and Dewan, 2002), as well as in Europe (Simeonov and

Georgiev, 2003; Gatzen, 2004). Geerts (2001) published a regional high-wind climatology for New South Wales, Australia. The global distribution of microbursts might be expected to be somewhat correlated with the global occurrence of thunderstorms and, therefore, of lightning. Fig. 8 depicts satellite-observed lightning occurrence for 1995–1996. In addition to the midlatitude continental areas, large areas of the tropics are covered. This is consistent with reports of blowdown events and microbursts and downbursts in such areas as the Amazon (Nelson *et al.*, 1994; Atlas *et al.*, 2004). The thermodynamic structure of the atmosphere over the tropical oceans may not permit the strong cooling necessary to drive strong downdrafts and downburst outflows, so the correlation may be small in these areas. A possible indication of the worldwide distribution of large, organized, severe convective wind storms and derechos can be obtained from the global distribution of large MCSs and mesoscale convective complexes, a very high percentage of which produce high winds (Fig. 9, from Fritsch and Forbes, 2001). On the basis of satellite observations, Laing and Fritsch (1997) found that these large convective storm complexes occur frequently in the central United States, the Amazon, and northern Argentina-Uruguay; on the African continent over sub-Saharan Africa and South Africa; and across India and parts of southeast Asia. Brooks *et al.* (2003) presented a map of the global distribution of favorable environmental sounding parameters for severe convective storms (Fig. 10). This figure suggests that conditions conducive to severe convective storms may exist in many areas. Notable areas include the eastern half of the United States, Argentina, sub-Saharan Africa and South Africa, northern India, Pakistan, Bangladesh, Australia, western Turkey, and parts of Europe, especially Spain. Annually, the distribution generally follows the sun, with summertime maximum in each hemisphere, but cold season derechos are observed in the southeastern United States (Burke and Schultz, 2004 [Fig. 7]). Note also that neither figure shows strong numbers in most of Europe, the western United States and Canada, or Russia.

DOWNDRAFTS, MESOCYCLONES, AND OUTFLOWS

Updrafts in clouds are always saturated, as required for instability. Downdrafts of any strength are often unsaturated, as the evaporation from the precipitation particles is unable to keep up with the requirements for saturation (Das and Subbarao, 1972). The thermodynamic structure of the air below cloud base and microphysical details (concentration, size, phase,

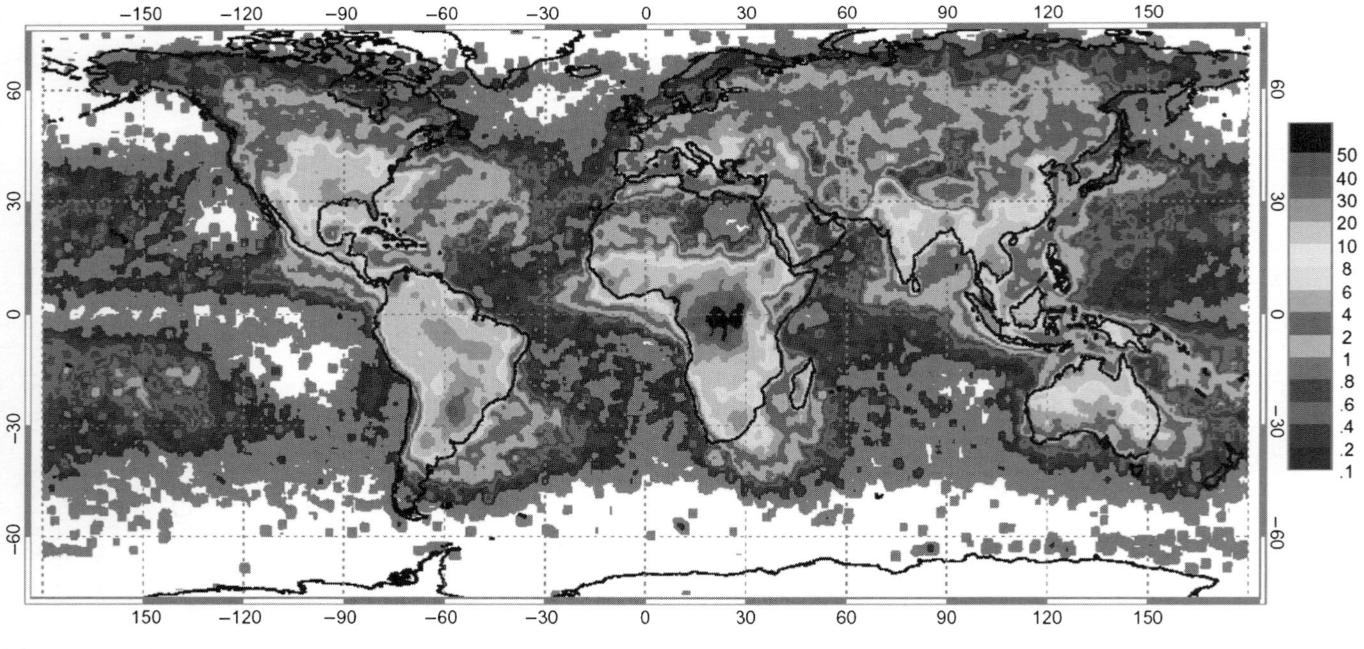

FIG. 8 Where lightning strikes—the global incidence of lightning September 1995–August 1996, as observed from space by the optical transient detector. Shading gives flashes km^{-2} y^{-1}. (NASA Image Exchange, image courtesy of NSSTC Lightning Team.)

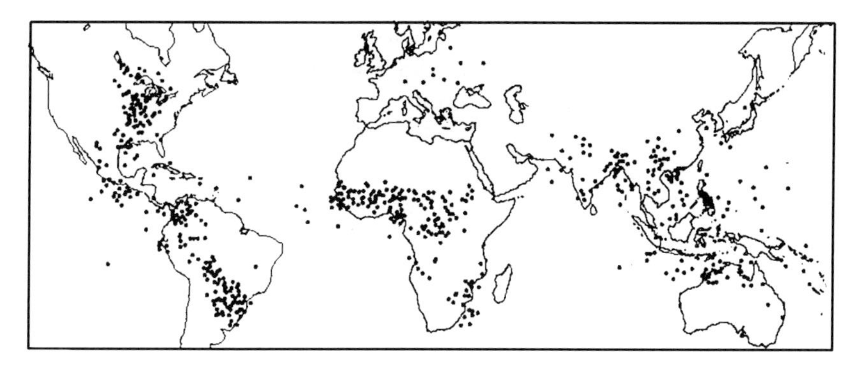

FIG. 9 Locations of mesoscale convective complexes based on 1- to 3-year samples of satellite imagery. (From Forbes and Fritsch [2001], based on the study of Laing and Fritsch [1997].)

and density of particles) are important and can greatly affect the potential strength of the downdraft and resulting outflow.

While many different types of downdrafts may occur in the upper levels of convective clouds, downdrafts that reach the surface usually start no greater than about 4 km above the ground. Newton (1966) and Betts and SilvaDias (1979) suggested that downdrafts reaching the surface probably do not originate much above the level of minimum of potential buoyancy, which is about 500 to 600 mb or 4 to 6 km. This is confirmed by many studies and is consistent with the analysis of forcing, as discussed previously in this chapter (Knupp and Cotton, 1982; 1985). One possible

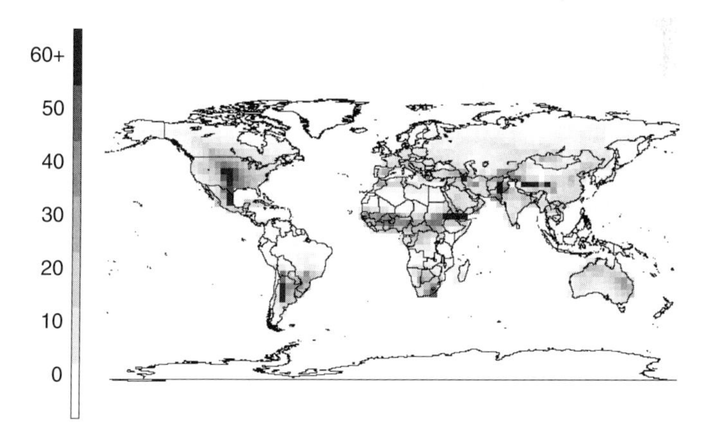

FIG. 10 Incidence of atmospheric sounding conditions favorable for severe weather in days per year (Brooks *et al.*, 2003). (With permission from Elsevier. Reprinted from Brooks *et al.* [2003], Atmospheric Research Vol. 67–68.)

exception, discussed below, is the occlusion downdraft associated with mesocyclones (Klemp and Rotunno, 1983; Wakimoto, 2001).

As described by Knupp and Cotton (1982) and Wakimoto (2001), the primary downdrafts reaching the surface from convective storms can be separated into the precipitation-driven convective downdrafts, dynamically driven occlusion or mesocyclone-related downdrafts, and the mesoscale-stratiform region wake-low downdrafts in MCSs.

The precipitation-driven downdraft originates within the storm as the weight of the precipitation causes it to begin to fall; the drag caused by this falling precipitation initiates a downdraft. The downdraft is further accelerated as melting, evaporation, or sublimation inhibits the compressional heating of the descending air. As long as the downdraft air is denser (colder) than the environmental air at the same level, it will continue to accelerate. It will not decelerate until it becomes less dense (warmer) than the environment or until it begins to spread out in response to the surface. Thus, the air may actually reach the surface warmer than the air at that level.

Supercells exhibit two distinct downdraft areas, the forward-flank downdraft and the rear-flank downdraft, and may have an additional "occlusion downdraft" (Lemon and Doswell, 1979; Klemp and Rotunno, 1983; Fig. 11). Intense downdrafts and severe winds have been frequently reported from the rear-flank and occlusion downdrafts (Hall and Brewer, 1959; Fujita, 1985; Wakimoto, 2001). The forward-flank downdraft represents a precipitation-driven downdraft. While important in storm dynamics, this downdraft is normally relatively benign in terms of severe surface winds. The rear-flank downdraft is the stronger downdraft in the storm. The origin of this air has been shown to be from the mid-troposphere, 3 to 5 km above the ground. The rear-flank downdraft is primarily precipitation driven, with contribution from the dynamic storm–environment interaction caused by the supercell's deviant motion (Lemon and Doswell, 1979; Rotunno and Klemp, 1982).

The occlusion downdraft is driven by the strong low-level rotation that forms at the occlusion of the rear-flank and forward-flank outflows. By the cyclostrophic relation, a strong low-level mesocyclone or mesoanticyclone would have low pressure in the center. This low pressure near the surface would create a vertical perturbation pressure gradient, which would induce a downdraft. In supercells these occlusion downdrafts are also intimately connected to tornado development through the mesocyclone and interaction of the outflow with the storm updraft (Klemp and Rotunno, 1983; Markowski, 2002).

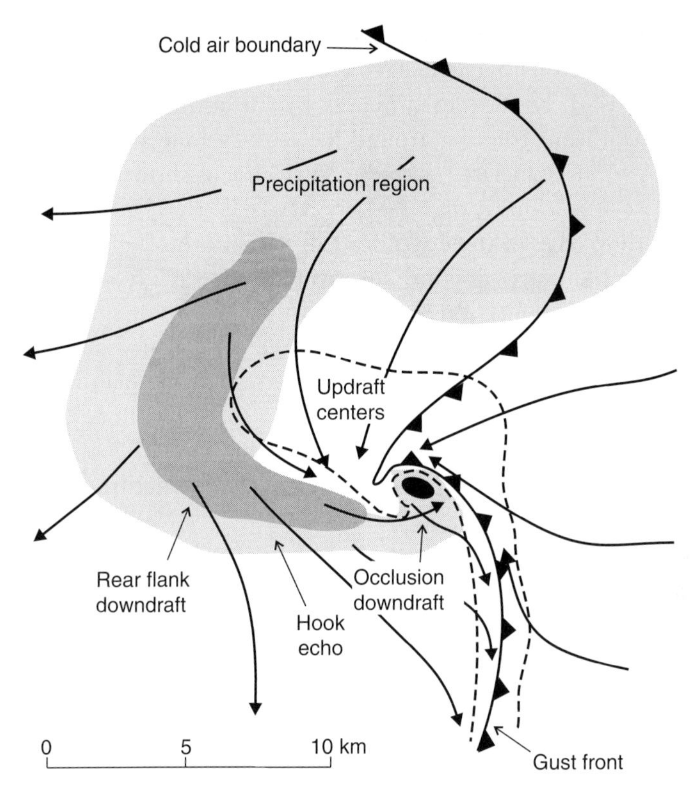

FIG. 11 Schematic of surface-level winds and precipitation patterns in a supercell storm. Shading denotes areas of precipitation and downdraft. (Adapted from Klemp and Rotunno [1983]; Weisman and Klemp [1986].)

Similar mesocyclone-related downdrafts also have been observed with cold fronts (Carbone, 1983) and along gust fronts (Wilson, 1986; Mueller and Carbone, 1987), as well as in supercells. Recent studies have shown that the most intense damaging winds from bow echo storms often are associated with low-level mesocyclones (Miller and Johns, 2000; Trapp and Weisman, 2003; Weisman and Trapp, 2003; Atkins *et al.*, 2004).

The mesocyclone rotation typically originates from tilting of the vorticity because of the increase in horizontal wind speed with height in the boundary layer or caused by the wrap-up of instabilities along the very strong horizontal gradients of temperature and winds at the leading edge of the storm outflow (i.e., Kelvin-Helmholtz instabilities). Many of the mesocyclones observed in bow echoes appear to be similar to the gust front and cold front occlusion downdraft cases; other mesocyclones may be deeper and may even become tornadic (Trapp and Weisman, 2003; Atkins *et al.*, 2004; 2005). Sometimes the severe surface winds from embedded mesocyclones in bow echoes can be explained merely by the addition of the tangential wind speed of the air rotating around the mesocyclone and the larger-scale straight-line winds associated with the bow echo, as shown in the top panel of Fig. 4 (also see Atkins *et al.*, 2004).

The leading edge of the downdraft outflow forms a gust front, which propagates as a density (gravity) current (Charba, 1974; Simpson, 1987). Observations and simulations have shown that the propagation speed of thunderstorm gust fronts can be described by gravity current theory (Wakimoto, 1982). These results also indicate that the maximum surface wind speed is about 1.5 times the propagation speed (Mahoney, 1988). The structure of the gust front was described by Charba (1974) and Wakimoto (1982). Along the gust front is a lobe and cleft structure (Mueller and Carbone, 1987; Weckwerth and Wakimoto, 1992), which can result in vortices sufficient to drive occlusion downdrafts (Mueller and Carbone, 1987). These small-scale perturbations in the outflow may be responsible for the finescale "microburst swaths" described by Fujita and Wakimoto (1981).

A schematic of the surface pressure field and surface winds beneath a squall line is shown in Fig. 12A. A vertical cross-section of a squall line with a trailing stratiform rain region is shown in Fig. 12B. The prominent mesohigh feature in the surface field can be identified with the rain-cooled stratiform rain downdrafts. The wake low is positioned toward the rear of the storm. The wake low is attributed to subsidence warming (Zipser, 1977; Johnson and Hamilton, 1988; Loehrer and Johnson, 1995) and is maximized at the back edge of the precipitation where evaporative cooling is insufficient to offset the warming caused by the subsidence (Fig. 13). This subsidence is also related to the descending rear-inflow jet shown in Fig. 12B. Loehrer and Johnson (1995) have shown that the pressure gradient between the mesohigh and wake low can be as large as 5 mb per 10 km. This can drive strong winds in excess of 25 m s^{-1} in a direction opposite that of the squall line motion. A discussion of damaging severe wind events involving wake lows was presented by Gaffin (1999); see section on large-scale systems later in this chapter.

MICROBURSTS

Several focused field programs were carried out to understand microbursts and to develop detection and warning capabilities. These programs focused on documenting the structure and lifecycle of the outflows and developing

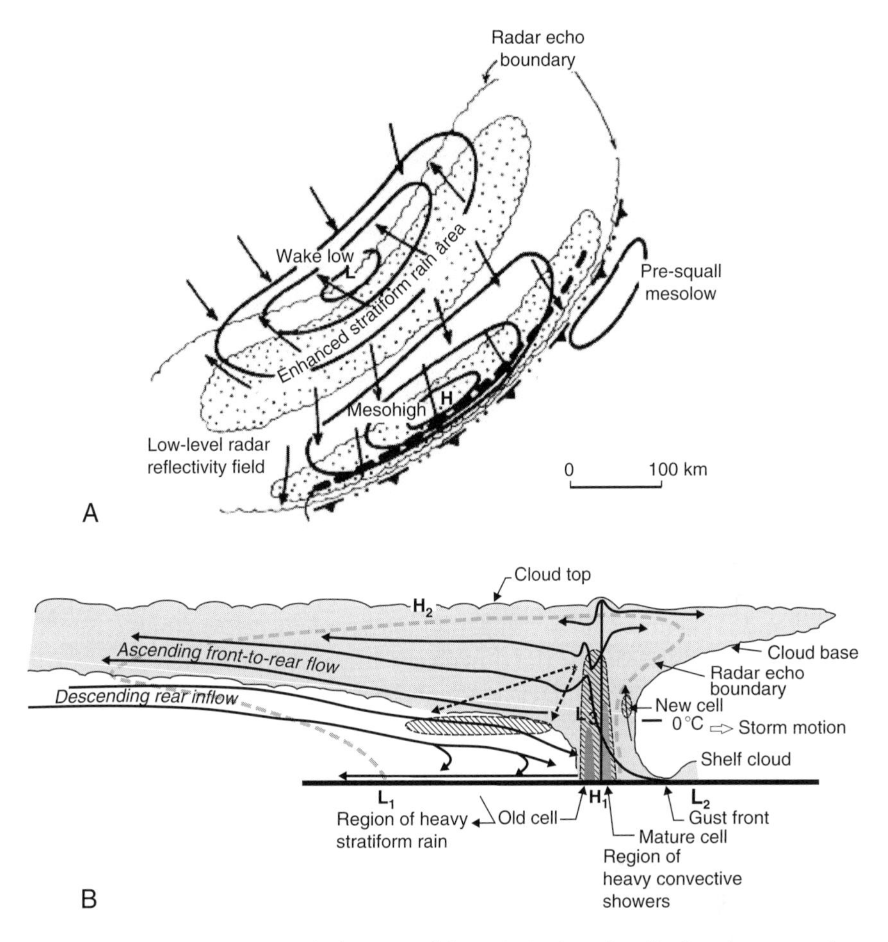

FIG. 12 Schematic diagrams of a large squall line: (**A**) Surface plot. Shading denotes moderate-heavy rain areas. Wake low and mesohigh pressure areas are contoured (from Johnson and Hamilton, [1988]). (**B**) Vertical cross-section. Shaded areas denote higher reflectivity associated with convective line and bright band. Ascending front-to-rear outflow in deep anvil and descending rear-to-front flow at midlevels are also shown (Houze *et al.* [1989]).

and testing radar and surface capabilities for detection and warning for airports. Fujita (1985) documented a total of 186 microbursts in the 3-month JAWS project near Denver (McCarthy *et al.*, 1982). Many of these were analyzed in detail and provide a wealth of knowledge about microbursts (Wilson *et al.*, 1984; Fujita, 1985; Hjelmfelt, 1988).

Wilson *et al.* (1984) demonstrated that the microbursts observed in JAWS were all associated with descending precipitation downdrafts.

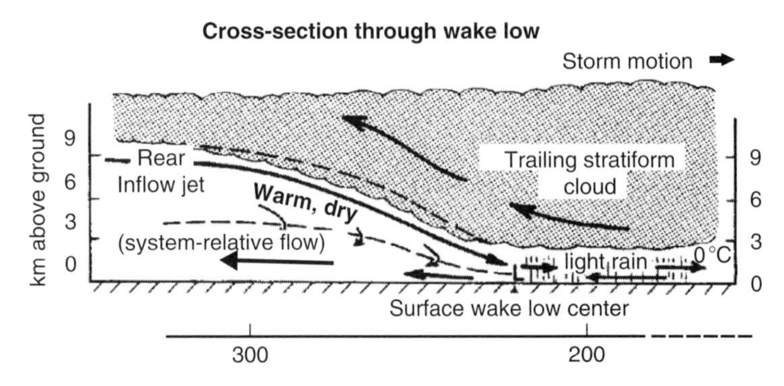

FIG. 13 Schematic vertical cross-section through the wake-low region, showing descending rear-to-front "jet" and strong surface outflow to rear of wake low center (Johnson and Hamilton [1988]).

Srivastava (1985; 1987) considered the physics of downdrafts in a simple one-dimensional time-dependent model and examined the roles of loading, evaporation, sublimation, melting, and entrainment below cloud base. He found that the intensity of the downdraft increased with increasing lapse rate of the environment, increasing rainwater content, increasing humidity below cloud base, and decreasing raindrop size. Ice melting was considerably more effective than rain evaporation in cooling air. Although the latent heat of fusion is much less than the latent heat of evaporation, melting occurs rapidly below the 0°C level. This cools the air and permits acceleration all the way to the surface. Evaporation takes longer and may not be complete before the ground is reached. Hjelmfelt (1987) and Wakimoto *et al.* (1994) found that the model could explain observed cases of both "dry" and "wet" microbursts.

Numerical cloud models have also been used to study the processes leading to microburst formation. For example, Orville *et al.* (1990) simulated a heavy-precipitation microburst from Alabama and found that loading initiated the downdraft, melting of hail was important, but evaporation was the number one forcing mechanism. Trajectory analyses showed that the top of trajectories in the microburst were not much above the height of the freezing level, and most trajectories originated below 3 to 4 km (consistent with observations). Similar results were obtained in Proctor's (1988) simulation of the 1985 Dallas airline crash case. Evaporation provided the primary forcing, then cooling caused by the melting of hail and graupel. Details of outflow structure, including a "roll" vortex associated with the maximum winds, were obtained. Proctor (1989) performed numerous sensitivity experiments and found that intensity is sensitive to the following: (1) temperature and humidity stratification of the environment; (2) width of downdraft; (3) magnitude of precipitation loading; (4) type of precipitation; and (5) duration of precipitation. Snow was very effective in accelerating downdrafts. In addition to cloud dynamic models used to study microburst production and outflow processes, several empirical models of microburst winds have been developed for input to engineering studies for aircraft and structures; examples include those of Holmes and Oliver (2000) and Chen and Letchford (2004).

The lifecycle of a typical isolated microburst is illustrated in Fig. 14, showing a descending precipitation downdraft that spreads out as it reaches the surface, forming the microburst outflow, intensifying, and then either decaying or growing to a larger-scale, less intense downburst. The lifecycle is short, typically lasting only 10 to 20 minutes. From initial detection of an outflow, 50% reached maximum intensity within 5 minutes, and 90% within 10 minutes. At a given location on the ground, the microburst winds increase impulsively, representing an extreme gust condition that greatly enhances their destructive effect (Fig. 15). A visual indicator used by pilots for detecting the presence of microbursts is "trees flailing wildly" in response to these gusts. Comparisons of surface wind records for thunderstorm severe wind events versus nonthunderstorm severe winds in Singapore revealed that the thunderstorm-related events are characterized by much greater gustiness and more extreme gusts than those due to other causes (Choi, 1999).

Fujita (1985) presented schematics for microburst surface wind patterns for stationary and rapidly translating microbursts (Fig. 16). Such conceptual pictures aid in interpreting observed damage swaths. Hjelmfelt (1988) described the structure of isolated microburst outflows observed in JAWS in detail. The height of the maximum winds was found to be about 80 m above the surface, but this is probably an overestimate because of

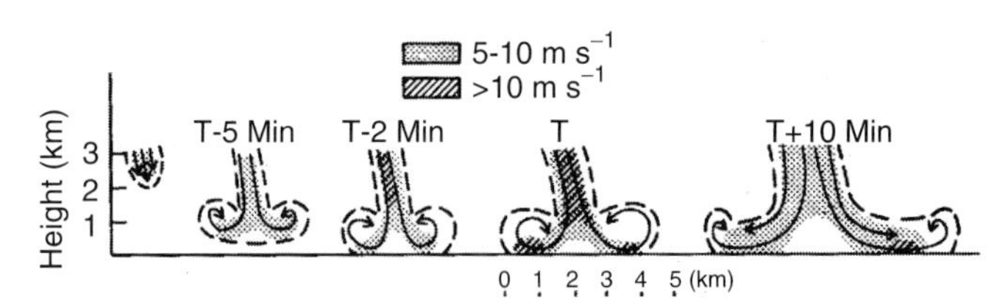

FIG. 14 Schematic of typical Joint Airport Weather Studies microburst lifecycle, time in minutes relative to microburst touchdown (Wilson *et al.* [1984]).

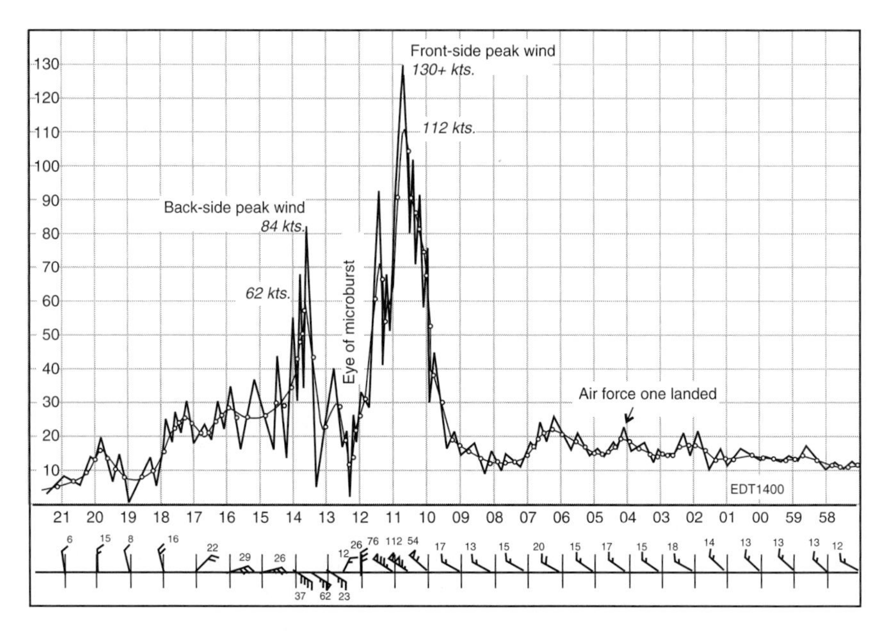

FIG. 15 Anemometer wind trace showing measured surface winds for microburst crossing station at Andrews Air Force Base, August 1, 1983. 130 kts = 67 m s^{-1}. (From Wakimoto, 2001, after Fujita, 1985; courtesy of the Tetsuya "Ted" Fujita Collection, Southwest Collection/Special Collections Library, Texas Tech University, Lubbock, TX.)

lack of resolution in the data. The actual height above the surface was probably lower, perhaps on the order of 40 to 60 m, as has been estimated for tornadoes (Golden, 1999). The maximum near-surface winds were associated with the curl or "roll" near the leading edge of the outflow. As indicated in the schematics in Fig. 16, a microburst in a strong mean flow, or a rapidly moving microburst, has an outflow that appears in only one

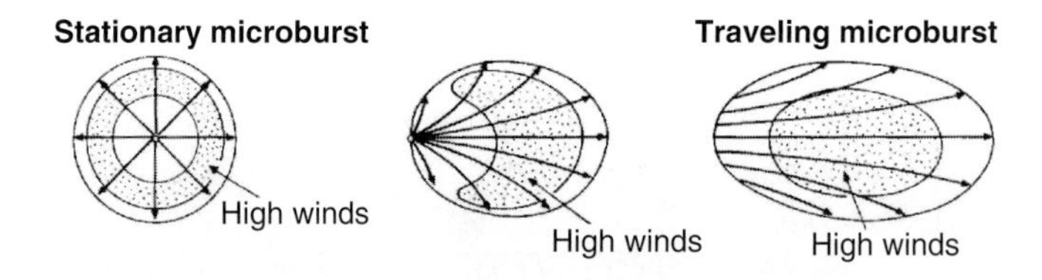

FIG. 16 Schematic of microburst outflows affected by storm movement. Stationary microburst with radial outflow, moderate speed, and high-speed traveling microburst (Fujita, 1985; courtesy of the Tetsuya "Ted" Fujita Collection, Southwest Collection/Special Collections Library, Texas Tech University, Lubbock, TX).

direction (Fig. 17). Such a wind pattern might be expected to produce a blowdown pattern similar to that shown in Fig. 1. The structure of microburst outflows observed in JAWS is summarized in Fig. 18.

Roberts and Wilson (1989) examined observed radar features associated with microburst production. They found that they could track descending cores of reflectivity, increasing radial convergence into the cloud near the level of the start of the descending core (in response to the accelerating downdraft), weak-echo notches in the reflectivity pattern (caused by evaporation in response to entrainment of environmental air), and rotation (which could often be explained in terms of conservation of vorticity of the air converging into the downdraft).

Downburst-producing storms can be detected by radar, where radar echo features such as bow echoes, line echo wave patterns, and hook echoes can be used to identify severe storms. Doppler radar permits measurement of the wind components toward and away from the radar and can be used to estimate the actual outburst wind speeds if the storms are within about 100 km of the radar (so that the radar beam is not too high off the ground). In fact, weather network radars, such as the WSR-88D in the United States, have automated algorithms for downburst detection (Smith *et al.*, 2004). Several research-computerized systems exist and have been tested in operational settings, as at the 2000 Sydney, Australia Olympic Games (Joe *et al.*, 2004). Detection of severe convective storms likely to produce damaging downburst winds is also possible from satellites (Pryor and Ellrod, 2004).

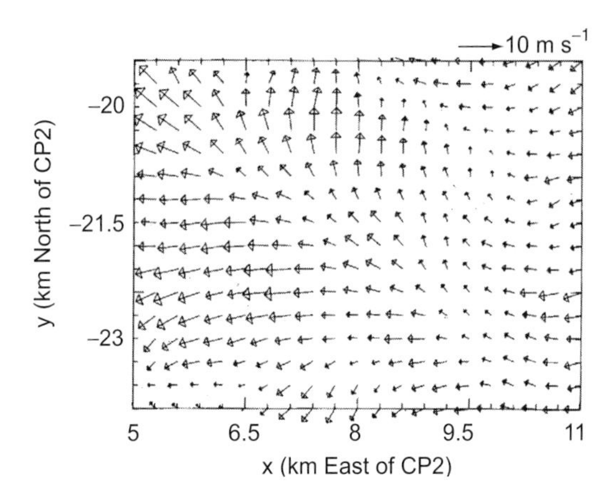

FIG. 17 Horizontal winds at 100 m above ground for a microburst manifested as a strongly diverging directional flow (Hjelmfelt [1988]).

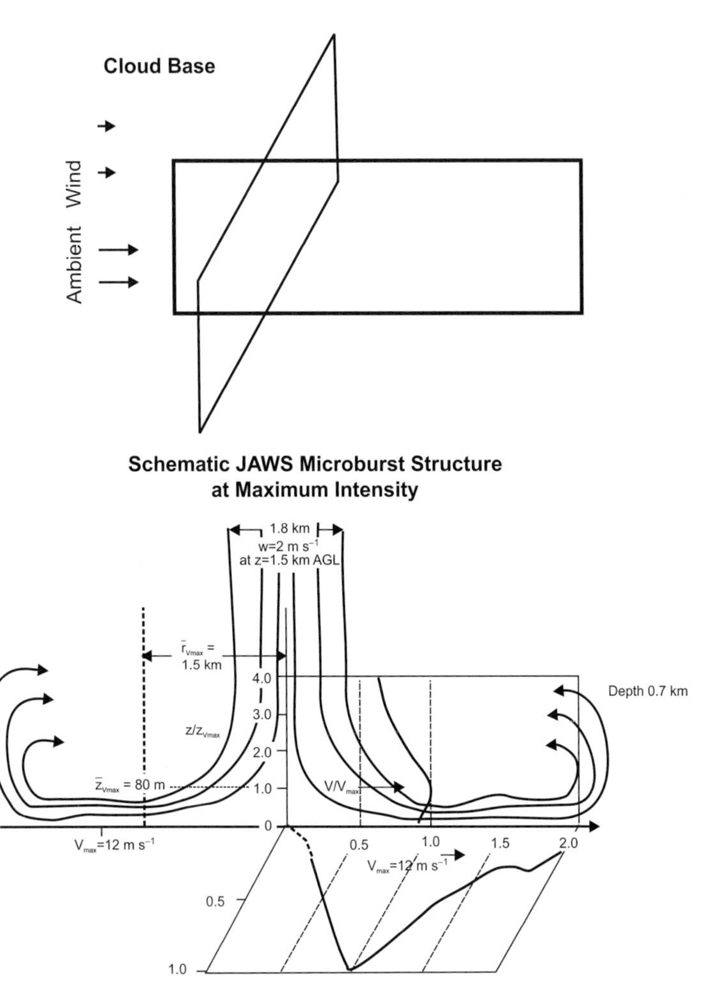

FIG. 18 Summary of microburst observations showing typical dimensions of microbursts observed in Denver area (Hjelmfelt [1988]).

Forecasting methods for microbursts rely primarily on analysis of atmospheric soundings. For low-precipitation, "dry" microbursts (Krumm, 1954; Caracena *et al.*, 1983; Wakimoto, 1985), the typical profile indicates a deep convective boundary layer with a high cloud base, perhaps as much as 3 km or so above the ground and often at a temperature near 0°C. Sufficient moisture exists above cloud base to permit precipitation to develop. For heavy-precipitation microbursts in a moist environment (Atkins and Wakimoto, 1991), the sounding profile emphasizes a layer of dry, potentially cool air above cloud base with much warmer moist air near the surface, so that a parcel above cloud base will be colder, accounting for the effects of latent heating, if it descends to the surface. Operational forecasting criteria have been developed (Proctor, 1989; McCann, 1994) and modified by Geerts (2001). Such methods can be adapted for use with satellite-derived atmospheric soundings (Ellrod *et al.*, 2000; Pryor and Ellrod, 2004). Operational nowcasting methods using radar observations have been implemented (Wolfson *et al.*, 1994), based on the radar signatures identified by Roberts and Wilson (1989) as discussed previously. Cloud models also have been applied in a forecast mode, with some success, to predict convective cloud development and potential for microbursts (Tuttle *et al.*, 1989; Kopp and Orville, 1994; and Elmore *et al.*, 2002a; 2002b).

LARGE-SCALE SYSTEMS

Mesoscale downbursts can occur from a variety of storm types (Klimowski *et al.*, 2003), including bow echoes (Fujita, 1978; Przybylinski, 1995; Weisman, 2001), squall lines, line echo wave patterns (Nolan, 1959), supercells (Moller *et al.*, 1994), and organized clusters of thunderstorms. These storms are complex and undergo complex lifecycles, often involving interactions with other storms and boundaries (Finley *et al.*, 2001; Klimowski *et al.*, 2003; 2004; Burke and Schultz, 2004).

Synoptic conditions favorable to large-scale derecho events were given by Johns and Hirt (1987) and Coniglio *et al.* (2004); see Fig. 19. A key feature

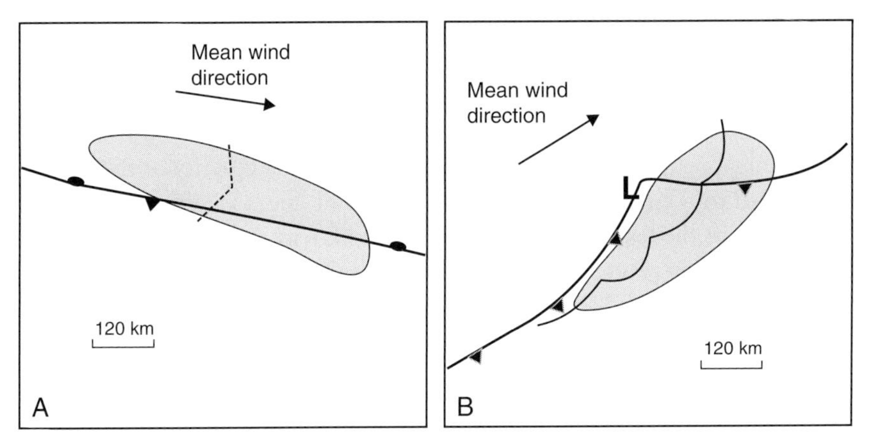

FIG. 19 Schematic of synoptic situations conducive to derecho events: (**A**) warm season event with weak large-scale dynamics; (**B**) dynamically active situation (from Johns and Hirt [1987]).

is that the mean wind tends to be along a front rather than across a front. The pattern in Fig. 19A tends to produce relatively short lines and bows. Typically, instabilities are very large, with relatively small turning of the wind with height. The pattern in Fig. 19B is most frequent in the southeastern and eastern United States and is more common in spring and fall than mid- to late summer. Longer squall lines, with serial production of downbursts, are common in this situation. Coniglio *et al.* (2004) also found that some derechos occur in less dynamic, zonal flow situations. Further discussion of mesoscale downburst environments can be found in Pryzbylinski (1995), Evans and Doswell (2001), and Klimowski *et al.* (2003). A consistent feature is relatively strong shear in the horizontal wind, especially from the surface to 3 km aloft. Recent observations and modeling studies have emphasized the importance of deeper tropospheric shear (Lafore and Moncrieff, 1989; Coniglio and Stensrud, 2001), though storm-relative winds aloft often tend to be less than typical for tornado situations.

Further understanding of the large mesoscale downbursts and derechos from MCS storms requires additional explanation of the structure of these storms. A prototype for these storms may be taken as the large squall line with an extensive trailing stratiform rain region, shown in Fig. 12. Intense convective downdrafts occur near the leading edge, driven by precipitation loading and melting and evaporation. The trailing, weaker, stratiform precipitation is marked by a radar "bright band" just below the 0°C temperature level caused by melting of snowflake aggregates to form rain. This is accompanied by a layer of subsiding air, directed in a storm-relative rear-to-front flow. Above is a layer of growing snowflakes in a weakly ascending storm-relative front-to-rear flow. The mesoscale downdrafts within the stratiform region are driven by evaporative cooling and melting. In some cases, a substantial rear inflow develops, which can exceed 10 m s^{-1} relative flow (Smull and Houze, 1987). The descent of this rear inflow to the surface may produce broad areas of high winds as the velocity of the inflow jet is added to the translation speed of the squall line. If the inflow jet continues aloft to the leading edge of the line, it may interact with the convective downdraft to produce extreme winds (Wakimoto, 2001) as seen in bow echo–type storms (Fig. 20).

As strong winds develop, the parent storm (often a squall line or sometimes a supercell) may develop a characteristic shape with a protruding segment in the location of the strongest winds (a bow echo or line echo wave pattern). Note that the bow echo is a response to the strong winds and is not a precursor to them. As the strong winds develop, cyclonic and anticyclonic circulations are noted on the flanks of the bow. At later stages, the

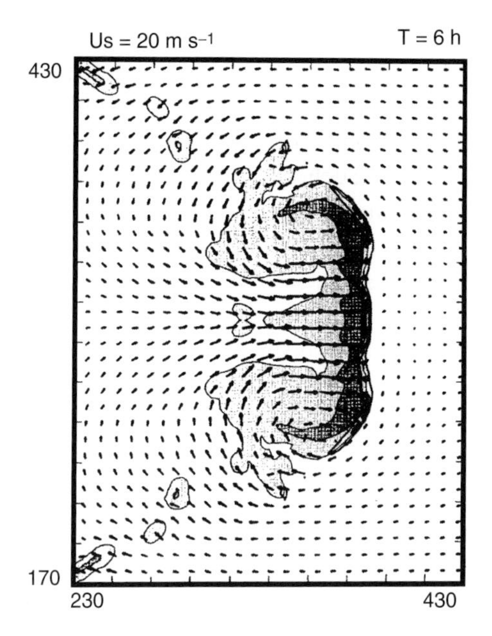

FIG. 20 Numerical simulation of a bow echo storm showing the system-relative low level wind field (vectors), rainwater mixing ratio (shaded) and updrafts (contours) after 6 hours of simulation. A vector length equal to the spacing between vectors (8 km) represents a system-relative wind of 20 m s^{-1}. (After Weisman and Davis [1998].)

cyclonic vortex tends to dominate in the Northern Hemisphere, and frequently the system evolves into a comma shape with a cyclonically rotating head in its decaying stage, as shown in Fig. 4. Most bow echo storms produce severe winds, and most, if not all, derechos are caused by bow echoes. These storms are believed to be responsible for the majority of the largest forest blowdowns.

The origin of the midlevel circulations at either end of the bow ("bookend vortices") can be described in terms of the growth of an updraft in an environment in which the low-level winds vary strongly with height (Klemp, 1987; Lee *et al.,* 1992; Weisman, 1993). As the horizontal vorticity associated with this flow is tilted upward and enhanced by stretching in the updraft, it produces cyclonic and anticyclonic vortices on either side of the updraft. The development of precipitation deforms the pattern. This process is fundamental to the development of supercell storms and may lead to storm splitting, with either a left-moving or right-moving rotating storm dominant (Klemp, 1987). In the case of the bow echo storm, the system does not split. The circulations act together to enhance the rear inflow, as is suggested by the pattern of wind vectors in Fig. 20, and may enhance it by as much as 30% to 50%. Resultant surface outflows may exceed 30 m s^{-1}. Klimowski *et al.* (2004) have shown that often this development may occur very rapidly when storm interactions, mergers, and interactions with surface boundaries (such as outflows from other storms) occur. Observations have shown that within the large-scale swaths of damage, there are smaller-scale embedded areas of more intense destruction associated with individual convective-scale downdrafts and more intense winds. Recent research has indicated that the most intense winds within MCS storms are often associated with small-scale mesocyclones and embedded supercells (Miller and Johns, 2000; Trapp and Weisman, 2003; Atkins *et al.* 2004; 2005). Study of these embedded mesocyclone–related severe winds is very active.

Klimowski *et al.* (2003) found that over the northern High Plains of the United States, more than 85% of all bow echo storms and about 50% of observed squall lines resulted in severe wind reports. These two storm types represented over 60% of the organized severe wind events and accounted for nearly 50% of all reports. In the northern High Plains these storms produced more than three reports of severe weather per event and lasted for about 3.5 hours. In other, less arid, parts of the country, large MCS storms last much longer, as much as 6 to 12 hours (Geerts, 1998; Parker and Johnson, 2000). The shorter lifetimes of the storms in the northern High Plains are also illustrated by the fact that only 3% of the mesoscale severe wind events met the derecho criteria.

Three example cases are given below to illustrate. One is a dramatic bow echo case from July 4–5, 1999. The second case is a large-scale wake-low event from April 11, 1995. In contrast, the third case is a severe supercell wind and hail case that occurred during the evening hours of July 5, 1996.

The first example of a dramatic blowdown event is the Boundary Waters–Canadian Derecho event of July 4–5, 1999. This event extended for a distance of over 2000 km, from North Dakota and Minnesota, across southern Ontario, Quebec, and northern New England, to the Atlantic Ocean, with winds reaching 45 ms^{-1} (Fig. 12A). The derecho was caused by a long-lived bow echo storm (Fig. 21B). Note on this bow echo the notches of weak echo on the back side. These indicate strong evaporation driving the strong surface winds. The storm movement was tracked at an average speed of about 25 ms^{-1} over most of its lifetime.

In the Superior National Forest in Minnesota, including the Boundary Waters region, over 1500 km^2 of forest were damaged (Fig. 21C). Examples of the damage are shown in Fig. 22. Note that the heaviest blowdown areas

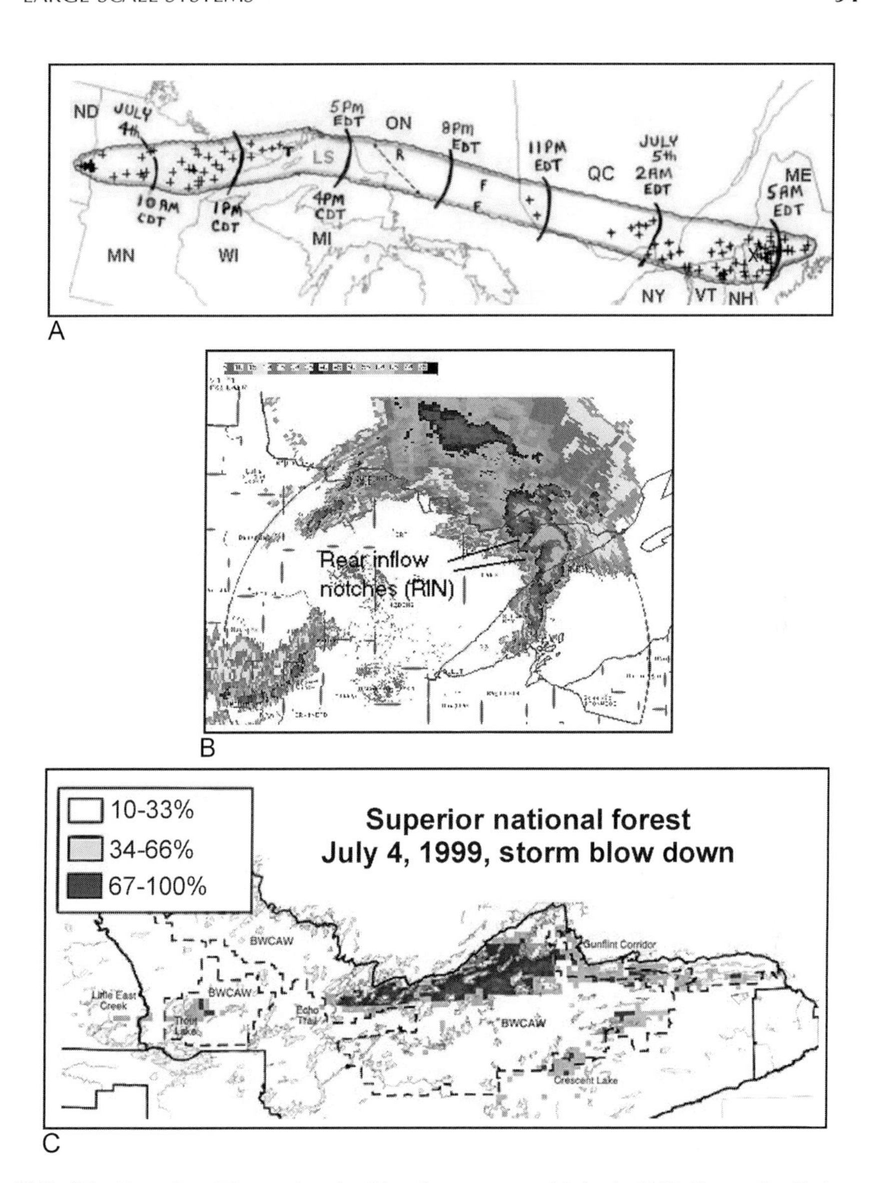

FIG. 21 Boundary Waters derecho blowdown event of July 4, 1999 (Storm Prediction Center, 2005): (**A**) map of geographical extent of storm damage from North Dakota to the Atlantic Ocean; (**B**) radar reflectivity display of the bow echo storm that caused the derecho (courtesy of Peter Parke and Norvan J. Larson, NWS Duluth); (**C**) survey of damage in the Superior National Forest of Minnesota. Shading indicates percentage of forest disturbed (adapted from map by Craig Perrault, Minnesota Department of Natural Resources).

FIG. 22 Photos of the July 4, 1999 blowdown: (**A**) an aerial view of one of the large areas of devastation; (**B**) a close-up view of the damage; (**C**) embedded areas of microburst blowdown within the larger damage; (**D**) a surface view. (Photographs courtesy of Craig Perrault, Minnesota Department of Natural Resources.)

are only a small part of thc broader swath of less severe damage, and even in the areas of heavy blowdown, there are patches of standing trees. This is consistent with the 1977 example described by Fujita and the different scales of motion associated with downbursts. Similar scenes could be reproduced in Canada and New England.

Severe winds produced by the squall line wake-low, shown in Fig. 13, have been discussed by Gaffin (1999). For example, during the April 11, 1995 event, surface observations at Little Rock, Arkansas, and Tupelo, Mississippi, showed that the maximum winds occurred just before passage of the pressure minimum (Fig. 23A). The winds were from the east-southeast, while the storm was moving to the east-northeast. Thus, the winds were nearly opposite to storm motion. The radar cross-section near the time of the peak winds in Mississippi is shown in Fig. 23B. The cross-section shows a structure very similar to that indicated in the schematic of Johnson and Hamilton (1988) shown in Fig. 13. Garstang *et al.* (1998) implicated such mesoscale downbursts from MCS storms as the cause of the large blowdowns in the Amazon rainforest documented by Nelson *et al.* (1994).

In the Klimowski *et al.* (2003) study, supercell storms accounted for a smaller fraction of events and reports; however, they produced some of the most devastating destruction of any storms. This is in part because of extreme

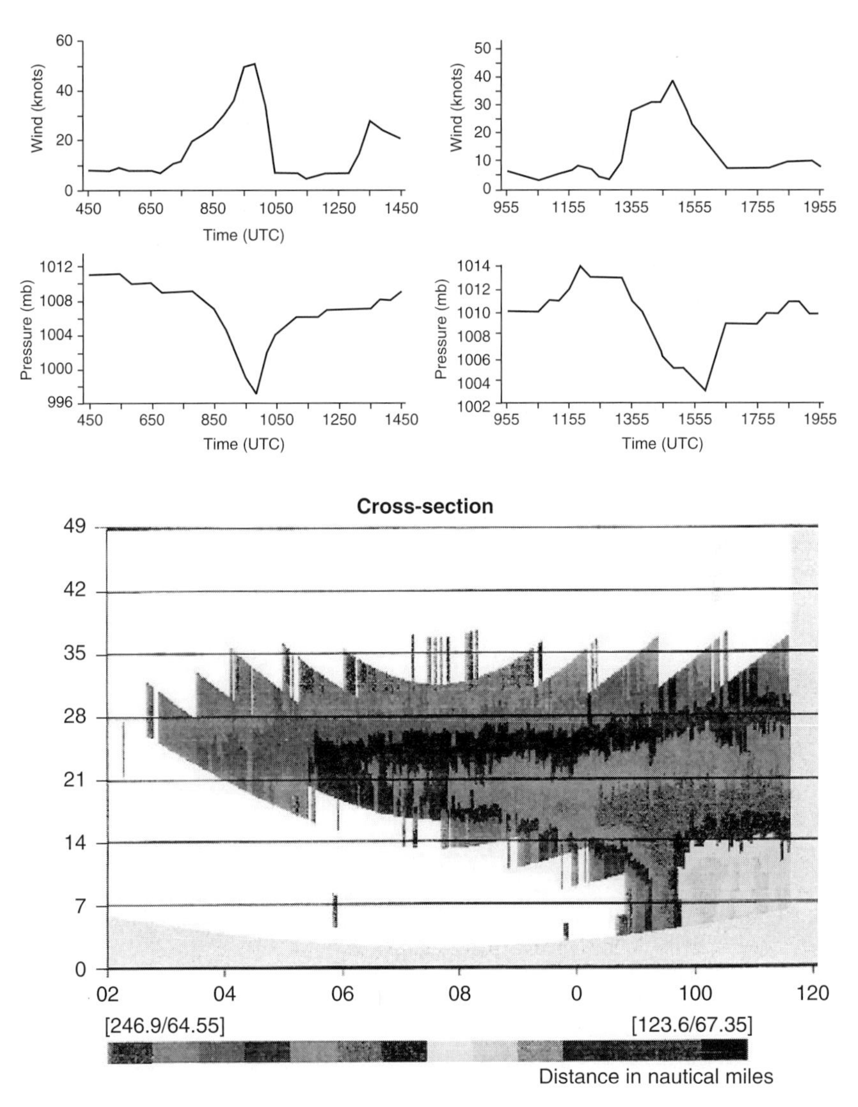

FIG. 23 Observations of April 11, 1995 wake low downburst event: (**A**) wind and pressure traces for Little Rock, AR; (**B**) wind and pressure traces for Tupelo, MS; (**C**) radar reflectivity cross-section of storm in Mississippi. (From Gaffin, 1999.)

winds, but especially because the severe winds are frequently associated with large hail. An example is the case analyzed by Klimowski *et al.* (1998) (Fig. 24). This case, which occurred over the South Dakota prairie, included winds in excess of 50 m s^{-1} and hailstones greater than 5 cm in diameter. The most severe damage extended along a 120-km long by 7- to

FIG. 24 Severe wind and hail event from a supercell storm (Klimowski *et al.* [1998]): (**A** and **B**) photographs of resulting plant disturbance. (**C**) Bare branches stripped of leaves and twigs. (**D**) Damage path, bold contour shows region of intense damage; shaded region is area of visible damage from geostationary satellite, 2 weeks after event; dots indicate locations of official severe wind reports.

11-km wide path. The impact of the winds and hail not only destroyed standing plants but also in some areas disturbed the soil such that roots were exposed. Where trees existed, in addition to blowdowns, in many areas the leaves were stripped off and branches less than 1 cm in diameter were snapped off. In many areas, standing crops were reduced to bare soil. Most insects and birds were killed in the path, with some species of birds not seen in the center of the path for several years afterward. Significant losses of animal life and livestock were also reported.

SUMMARY

Downbursts that produce blowdown events occur over a variety of scales, from microbursts on scales less than a few kilometers to derechos with length scales of hundreds of kilometers and damage swaths covering thousands of square kilometers. The winds and resulting plant disturbances are not uniform in these events but are characterized by extreme gustiness and large spatial variability, with small patches of intense damage embedded in larger areas of less severe damage.

Downbursts are caused by convective storms and are physically related to precipitation processes within these storms. The simplest case of isolated microburst is driven by the weight of precipitation causing a downdraft and evaporation and melting cooling of the descending air, which further accelerates the downdraft. As the descending current impinges on the ground, an outburst of diverging winds is produced.

Larger-scale downbursts are caused by supercell and mesoscale convective storms, notably squall lines and bow echoes. Downbursts associated with supercells are frequently associated with large hail, increasing the damaging impact. The bowing shape of the bow echo storms occurs in response to the strong wind circulations occurring in the storm; thus, most bow echoes are associated with damaging winds. Most of the largest damage swaths are associated with bow echoes. The most intense winds associated with these larger downburst-producing storms appear to be associated with small-scale low-level rotation centers (mesocyclones or mesoanticyclones).

The actual extent of plant disturbances caused by windstorms depends on many other factors than just the meteorology. Some of these include the type of plants or trees, their exposure to the wind, the height and shape of the tree, the size (strength) of the tree, the nature and extent of the root system, the health of the tree, and the wetness of the soil (e.g., King, 1986; Foster, 1988; Webb, 1989; Foster and Boose, 1992). These topics are discussed in other chapters in this book.

REFERENCES

Atkins, N. T., and Wakimoto, R. M. (1991). Wet microburst activity over the southeastern United States. *Weather Forecast* 6, 470–482.

Atkins, N. T., Arnott, J. M., Przybylinski, R. W., Wolf, R. A., and Ketcham, B. D. (2004). Vortex structure and evolution within bow echoes. Part I: single-Doppler and damage analysis of the 29 June 1998 derecho. *Mon Weather Rev* 132, 2224–2242.

Atkins, N. T., Bouchard, C. S., Przybylinski, R. W., Trapp, R. J., and Schmocker, G. (2005). Damaging surface wind mechanisms within the 10 June 2003 Saint Louis bow echo during BAMEX. *Mon Weather Rev* 133, 2275–2296.

Atlas, D., Ulbrich, C. W., and Williams, C. R. (2004). Physical origin of a wet microburst: observations and theory. *J Atmos Sci* 61, 1186–1195.

Bentley, M. L., and Mote, T. L. (1998). A climatology of derecho-producing mesoscale convective systems in the central and eastern United States, 1986-95. Part I: Temporal and spatial distribution. *B Am Meteorol Soc* 79, 2527–2540.

Bentley, M. L., Mote, T. L., and Thebpanya, P. (2002). Using Landsat to identify thunderstorm damage in agricultural regions. *B Am Meteorol Soc* 83, 363–376.

Betts, A. K., and SilvaDias, M. F. (1979). Unsaturated downdraft thermodynamics in cumulonimbus. *J Atmos Sci 26,* 1061–1071.

Bluestein, H. B., and Jain, M. H. (1985). Formation of mesoscale lines of precipitation: severe squall lines in Oklahoma during the spring. *J Atmos Sci* 42, 1711–1732.

Brooks, H. E., Doswell, C. A., and Cooper, J. (1994). On the environments of tornadic and nontornadic mesocyclones. *Weather Forecast* 9, 606–618.

Brooks, H. E., Lee, J. W., and Craven, J. P. (2003). The spatial distribution of severe thunderstorm and tornado environments from global reanalysis data. *Atmos Res* 67–68, 73–94.

Browning, K. A. (2004). The sting at the end of the tail: damaging winds associated with extratropical cyclones. *Q J R Meteor Soc* 130, 375–390.

Bunkers, M. J., Klimowski, B. A., Zeitler, J. W., Thompson, R. L., and Weisman, M. L. (2000). Predicting supercell motion using a new hodograph technique. *Weather Forecast* 15, 61–79.

Burke, P. C., and Schultz, D. M. (2004). A 4-yr climatology of cold-season bow echoes over the continental United States. *Weather Forecast* 19, 1061–1074.

Businger, S., Birchard, Jr., T., Kodama, K., Jendrowski, P. A., and Wang, J.-J. (1998). A bow echo and severe weather associated with a Kona low in Hawaii. *Weather Forecast* 13, 576–591.

Canham, C. D., and Loucks, O. L. (1984). Catastrophic windthrow in the presettlement forests of Wisconsin. *Ecology* 65, 803–809.

Caracena, F., McCarthy, J., and Flueck, J. A. (1983). Forecasting the likelihood of microbursts along the Front Range of Colorado. In: *Preprints, 13th Conference on Severe Local Storms: October 17-20, 1983, Tulsa, Okla.* pp. 261–264. American Meteorological Society, Boston.

Carbone, R. (1983). A severe frontal rainband. Part II: Tornado parent vortex circulation. *J Atmos Sci* 40, 2639–2654.

Charba, J. (1974). Application of a gravity current model to analysis of squall line gust front. *Mon Weather Rev* 102, 140–156.

Chen, L., and Letchford, C. W. (2004). A deterministic-stochastic hybrid model of downbursts and its impact on a cantilevered structure. *Eng Struct* 26, 619–629.

Choi, E. C. C. (1999). Extreme wind characteristics over Singapore—an area in the equatorial belt. *J Wind Eng Ind Aerodynam* 83, 61–69.

Coniglio, M. C., and Stensrud, D. J. (2001). Simulation of a progressive derecho using composite initial conditions. *Mon Weather Rev* 129, 1593–1616.

Coniglio, M. C., and Stensrud, D. J. (2004). Interpreting the climatology of derechos. *Weather Forecast* 19, 595–605.

Coniglio, M. C., Stensrud, D. J., and Richman, M. B. (2004). An observational study of derecho-producing convective systems. *Weather Forecast* 19, 320–337.

Das, P., and Subbarao, M. C. (1972). The unsaturated downdraught. *Indian J Meteor Geophys* 23, 135–144.

Davis, C., Atkins, N., Bartels, D., Bosart, L., Coniglio, M., Bryan, G., Cotton, W., Dowell, D., Jewett, B., Johns, R., Jorgensen, D., Knievel, J., Knupp, K., Lee, W.-C., McFarquhar, G., Moore, J., Przybylinski, R., Rauber, R., Smull, B., Trapp, R., Trier, S., Wakimoto, R., Weisman, M., and Ziegler, C. (2004). The bow echo and MCV experiment: observations and opportunities. *B Am Meteorol Soc* 85, 1075–1093.

Doswell, C. A., and Bosart, L. F. (2001). Extratropical synoptic-scale processes and severe convection. In: *Severe Convective Storms*. (Doswell, C. A., Ed.) pp. 27–69. Meteorological Monographs, v. 28, no. 50, American Meteorological Society, Boston.

Ellrod, G. P., Nelson III, J. P., Witiw, M. R., Bottos, L., and Roeder, W. P. (2000). Experimental GOES sounder products for the assessment of downburst potential. *Weather Forecast* 15, 527–542.

Elmore, K. L., Stensrud, D. J., and Crawford, K. C. (2002a). Explicit cloud-scale models for operational forecasts: a note of caution. *Weather Forecast*. 17, 873–884.

Elmore, K. L., Stensrud, D. J., and Crawford, K. C. (2002b). Ensemble cloud model applications to forecasting thunderstorms. *J Appl Meteorol* 41, 363–383.

Evans, J. S., and Doswell, III C. A. (2001). Examination of derecho environments using proximity soundings. *Weather Forecast* 16, 329–342.

Finley, C. A., Cotton, W. R., and Pielke, Sr., R. A. (2001). Numerical simulation of tornadogenesis in a high-precipitation supercell. Part I: Storm evolution and transition into a bow echo. *J Atmos Sci* 38, 1597–1629.

Forbes, G. S. (1981). On the reliability of hook echoes as tornado indicators. *Mon Weather Rev* 109, 1457–1466.

Forbes, G. S., and Wakimoto, R. M. (1983). A concentrated outbreak of tornadoes, downbursts, and microbursts and implications regarding vortex classification. *Mon Weather Rev* 111, 220–235.

Foster, D. R. (1988). Species and stand response to catastrophic wind in central New England, USA. *J Ecol* 76, 135–151.

Foster, D. R., and Boose, E. R. (1992). Patterns of forest damage resulting from catastrophic wind in central New England, USA. *J Ecol* 80, 79–98.

Fritsch, J. M., and Forbes, G. S. (2001). Mesoscale convective systems. In: *Severe Convective Storms*. (Doswell, C. A., Ed.), pp. 27–69. Meteorological Monographs, v. 28, no. 50, American Meteorological Society, Boston.

Fujita, T. T. (1971). Proposed Characterization of Tornadoes and Hurricanes by Area and Intensity. SMRP Research Paper No. 91, University of Chicago, Chicago. [Available from NTIS, US Dept of Commerce, 5285 Port Royal Rd., Springfield VA, 22161]

Fujita, T. T. (1978). Manual of Downburst Identification for Project NIMROD. SMRP Research Paper No. 156, University of Chicago, Chicago. [NTIS PB-2860481.]

Fujita, T. T. (1979). Objective, operation, and results of project NIMROD. In: *Preprints, 11th Conference on Severe Local Storms of the American Meteorological Society, October 2-5, 1979, Kansas City, Missouri*. pp. 259–266. American Meteorological Society, Boston.

Fujita, T. T. (1981). Tornadoes and downbursts in the context of generalized planetary scales. *J Atmos Sci* 38, 1511–1534.

Fujita, T. T. (1985). *The Downburst*. SMRP Research Paper No. 210, University of Chicago, Chicago. [NTIS PB-148880.]

Fujita, T. T. (1989). The Teton-Yellowstone tornado of 21 July 1987. *Mon Weather Rev* 117, 1913–1940.

Fujita, T. T., and Byers, H. R. (1977). Spearhead echo and downbursts in the crash of an airliner. *Mon Weather Rev* 105, 129–146.

Fujita, T. T., and Wakimoto, R. M. (1981). Five scales of airflow associated with a series of downbursts of 16 July 1980. *Mon Weather Rev* 109, 1438–1456.

Gaffin, D. M. (1999). Wake low severe wind events in the Mississippi River Valley: a case study of two contrasting events. *Weather Forecast* 14, 581–605.

Garstang, M., White, S., Shugart, H. H., and Halverson, J. (1998). Convective cloud downdrafts as the cause of large blowdowns in the Amazon rainforest. *Meteorol Atmos Phys* 67, 199–212.

Gatzen, C. (2004). A derecho in Europe: Berlin, 10 July 2002. *Weather Forecast* 19, 639–645.

Geerts, B. (1998). Mesoscale convective systems in the southeast United States during 1994-95: a survey. *Weather Forecast* 13, 860–869.
Geerts, B. (2001). Estimating downburst-related maximum surface wind speeds by means of proximity soundings in New South Wales, Australia. *Weather Forecast* 16, 261–269.
Golden, J. H. (1999). Tornadoes. In: *Storms*. Vol. II (Pielke Sr., R. A., and Pielke Jr., R. A., Eds.), pp. 103–132. Routledge, New York.
Hall, F., and Brewer, R. D. (1959). A sequence of tornado damage patterns. *Mon Weather Rev* 87, 207–216.
Hinrichs, G. (1988). Tornadoes and derechos. *Am Meteorol J* 5, 306–317, 341–349.
Hjelmfelt, M. R. (1987). The microburst of 22 June 1982 in JAWS. *J Atmos Sci* 44, 1646–1665.
Hjelmfelt, M. R. (1988). Structure and life cycle of microburst outflows observed in Colorado. *J Appl Meteorol* 27, 900–927.
Holmes, J. D., and Oliver, S. E. (2000). Empirical model of a downburst. *Eng Struct* 22, 1167–1172.
Houze, Jr., R. A. (1993). *Cloud Dynamics*. Academic Press, New York.
Houze, Jr., R. A. (2004). Mesoscale convective systems. *Rev Geophys* 42(4), Art. No. RG4003, Dec. 31, 2004.
Houze, Jr., R. A., Biggerstaff, M. I., Rutledge, S. A., and Smull, B. F. (1989). Interpretation of Doppler weather radar displays of midlatitude mesoscale convective systems. *B Am Meteorol Soc* 70, 608–619.
Joe, P., Burgess, D., Potts, R., Keenan, T., Stumpf, G., and Treloar, A. (2004). The S2K severe weather detection algorithms and their performance. *Weather Forecast* 19, 43–63.
Johns, R. H., and Hirt, W. D. (1987). Derechos: widespread convectively induced windstorms. *Weather Forecast* 2, 32–49.
Johnson, R. H., and Hamilton, P. J. (1988). The relationship of surface pressure features to the precipitation and air flow structure of an intense midlatitude squall line. *Mon Weather Rev* 116, 1444–1472.
Kelly, D. L., Schaefer, J. T., and Doswell, III C. A. (1985). Climatology of nontornadic severe thunderstorm events in the United States. *Mon Weather Rev* 113, 1997–2014.
King, D. A. (1986). Tree form, height growth, and susceptibility to wind damage in *Acer saccharum*. *Ecology* 67, 980–990.
Klemp, J. B. (1987). Dynamics of tornadic thunderstorms. *Annu Rev Fluid Mech* 19, 369–402.
Klemp, J. B., and Rotunno, R. (1983). A study of the tornadic region within a supercell thunderstorm. *J Atmos Sci* 40, 359–377.
Klimowski, B. A., Bunkers, M. J., Hjelmfelt, M. R., and Covert, J. (2003). Severe convective windstorms over the northern high plains of the United States. *Weather Forecast* 18, 502–519.
Klimowski, B. A., Hjelmfelt, M. R., Bunkers, M., Sedlacek, D., and Johnson, L. R. (1998). Hailstorm damage observed from the GOES-8 satellite: The 05-06 July Butte-Meade Storm. *Mon Weather Rev* 126, 831–834.
Klimowski, B. A., Hjelmfelt, M. R., and Bunkers, M. J. (2004). The early evolution of bow echoes. *Weather Forecast* 19, 727–734.
Knupp, K. R., and Cotton, W. R. (1982). Characteristics of downdrafts and turbulence within thunderstorms. In: *Conference on Cloud Physics: November 15–18, 1982, Chicago, Illinois [Proceedings]*, pp. 539–542. American Meteorological Society, Boston.
Knupp, K. R., and Cotton, W. R. (1985). Convective cloud downdraft structure: an interpretive survey. *Rev Geophys* 23, 183–215.
Kopp, F. J., and Orville, H. D. (1994). The use of a two-dimensional, time-dependent cloud model to predict convective and stratiform clouds and precipitation. *Weather Forecast* 9, 62–77.

Krumm, W. R. (1954). On the cause of downdrafts from dry thunderstorms over the plateau area of the United States. *B Am Meteorol Soc* 35, 122–125.

LaFore, J.-P., and Moncrieff, M. W. (1989). A numerical investigation of the organization and interaction of the convective and stratiform regions of tropical squall lines. *J Atmos Sci* 46, 521–544.

Laing, A. G., and Fritsch, J. M. (1997). The global population of mesoscale convective complexes. *Q J R Meteor Soc* 123, 389–405.

Lee, W.-C., Wakimoto, R. M., and Carbone, R. E. (1992). The evolution and structure of a "bow-echo-microburst" event. Part II: The bow echo. *Mon Weather Rev* 120, 2211–2225.

Lemon, L. R., and Doswell, C. A. (1979). Severe thunderstorm evolution and mesocyclone structure as related to tornadogenesis. *Mon Weather Rev* 107, 1184–1197.

Lilly, D. K. (1978). A severe downslope windstorm and aircraft turbulence event induced by a mountain wave. *J Atmos Sci* 35, 59–77.

Lilly, D. K., and Zipser, E. J. (1972). The Front Range windstorm of 11 January 1972—a meteorological narrative. *Weatherwise* 25, 56–63.

Loehrer, S. M., and Johnson, R. H. (1995). Surface pressure and precipitation life cycle characteristics of PRE-STORM mesoscale convective systems. *Mon Weather Rev* 123, 600–621.

Lynott, R. E., and Cramer, O. P. (1966). Detailed analysis of the 1962 Columbus Day windstorm in Oregon and Washington. *Mon Weather Rev* 94, 105–117.

Maddox, R. A. (1980). Mesoscale convective complexes. *B Am Meteorol Soc* 61, 1374–1387.

Mahoney, W. P. (1988). Gust front characteristics and the kinematics associated with interacting thunderstorm outflows. *Mon Weather Rev* 116, 1474–1491.

Markowski, P. M. (2002). Hook echoes and rear flank downdrafts: a review. *Mon Weather Rev* 130, 852–876.

McCann, D. W. (1994). WINDEX—a new index for forecasting microburst potential. *Weather Forecast* 9, 532–541.

McCarthy, J., Wilson, J. W., and Fujita, T. T. (1982). The joint airport weather studies project. *B Am Meteorol Soc* 63, 15–22.

McDonald, J. R. (2002). Development of an enhanced Fujita scale for estimating tornado *intensity.* In: *Preprints, 21st Conference on Severe Local Storms, 12-16 August, 2002, San Antonio, Texas,* pp. 174–177. American Meteorological Society, Boston.

Meyers, M. P., Snook, J. S., Wesley, D. A., and Poulos, G. S. (2003). A Rocky Mountain storm. Part II: The forest blowdown over the west slope of the northern Colorado Mountains—observations, analysis, and modeling. *Weather Forecast* 18, 662–674.

Miller, D. L., and Johns, R. H. (2000). A detailed look at extreme wind damage in derecho events. In: *Preprints, 20th Conference on Severe Local Storms, 11-14 September, 2000, Orlando, Florida,* pp. 52–55. American Meteorological Society, Boston.

Moller, A. R., Doswell, III C. A. , Foster, M. P., and Woodall, G. R. (1994). The operational recognition of supercell thunderstorm environments and storm structures. *Weather Forecast* 9, 327–347.

Mueller, C. K., and Carbone, R. E. (1987). Dynamics of a thunderstorm outflow. *J Atmos Sci* 44, 1879–1998.

Musil, D. J., Heymsfield, A. H., and Smith, P. L. (1986). Microphysical characteristics of a well-developed weak echo region in a High Plains supercell thunderstorm. *J Clim Appl Meteorol* 25, 1037–1051.

NASA (2005). Optical Transient Detector. NASA Global Hydrology and Climate Center, Marshall Space Flight Center. Available at http://thunder.nsstc.nasa.gov/otd/. Accessed September 7, 2006

Nelson, B. W., Kapos, V., Adams, J. B., Oliveira, W. J., and Braun, O. P. G. (1994). Forest disturbance by large blowdowns in the Brazilian Amazon. *Ecology* 75, 853–858.

Newton, C. W. (1966). Circulations in large sheared cumulonimbus. *Tellus* 18, 699–713.

Newton, C. W., and Newton, H. R. (1959). Dynamical interaction between large convective clouds and environment with vertical shear. *J Meteorol* 16, 483–496.

Nolan, R. H. (1959). A radar pattern associated with tornadoes. *B Am Meteorol Soc* 40, 277–279.

Ohno, H., Suzuki O., and Kusunoki, K. (1996). Climatology of downburst occurrence in Japan. In: *Preprints, 18th Conference on Severe Local Storms: February 19-23, 1996, San Francisco, CA*, pp. 87–90. American Meteorological Society, Boston.

Orville, H. D., and Hubbard, K. G. (1973). On the freezing of liquid water in a cloud. *J Appl Meteorol* 12, 671–676.

Orville, H. D., Todey, D. P., Farley, R. D., and Kopp, F. J. (1990). More on the microphysics and dynamics of microbursts. In: *Preprints, 1990 Conference on Cloud Physics: July 23-27, 1990, San Francisco, Calif*, pp. 582–588. American Meteorological Society, Boston.

Parker, M. D., and Johnson, R. H. (2000). Organizational modes of midlatitude mesoscale convective systems. *Mon Weather Rev* 128, 3413–3436.

Peterson, R. E., and Dewan, A. M. (2002). Damaging Nor'westers in Bangladesh. In: *Preprints, 21st Conference on Severe Local Storms, 12-16 August, 2002, San Antonio, Texas*, pp. 389–392. American Meteorological Society, Boston.

Potts, R. J. (1991). Microburst observations in tropical Australia. In: *Preprints, 25th International Conference on Radar Meteorology, June 24-28, Paris, France*, pp. J67–J72. American Meteorological Society, Boston.

Proctor, F. H. (1988). Numerical simulations of an isolated microburst. Part I: Dynamics and structure. *J Atmos Sci* 45, 313–360.

Proctor, F. H. (1989). Numerical simulations of an isolated microburst. Part II: Sensitivity experiments. *J Atmos Sci* 46, 2143–2165.

Pryor, K. L., and Ellrod, G. P. (2004). Recent improvements to the GOES microburst products. *Weather Forecast* 19, 582–594.

Przybylinski, R. W. (1995). The bow echo: observations, numerical simulations, and severe weather detection methods. *Weather Forecast* 10, 203–218.

Roberts, R. D., and Wilson, J. W. (1989). A proposed microburst nowcasting procedure using single-Doppler radar. *J Appl Meteorol* 28, 285–303.

Rotunno, R., and Klemp, J. B. (1982). The influence of the shear-induced pressure gradient on thunderstorm motion. *Mon Weather Rev* 110, 136–151.

Saunders, P. M. (1957). The thermodynamics of saturated air: a contribution to classical theory. *Q J R Meteor Soc* 83, 342–350.

Schlesinger, R. E. (1980). A three-dimensional numerical model of an isolated thunderstorm. Part II: Dynamics of updraft splitting and mesovortex couplet evolution. *J Atmos Sci* 37, 395–420.

Schmidt, J. M., and Cotton, W. R. (1989). A high plains squall line associated with severe surface winds. *J Atmos Sci* 46, 281–302.

Serafin, R. J., Wilson, J. W., McCarthy, J., and Fujita, T. T. (1999). Progress in understanding windshear and implications for aviation. In: *Storms*. Vol. II (Pielke, Sr., R. A., and Pielke, Jr., R. A., Eds.), pp. 237–251. Routledge, New York.

Simeonov, P., and Georgiev, C. G. (2003). Severe wind/hail storms over Bulgaria in 1999-2001 period: synoptic and meso-scale factors for generation. *Atmos Res* 67–68, 629–643.

Simpson, J. E. (1987). *Gravity Currents: in the Environment and the Laboratory*. Ellis Horwood Limited, Chichester, UK.

Smith, T. M., Elmore, K. L., and Dulin, S. A. (2004). A damaging downburst prediction and detection algorithm for the WSR-88D. *Weather Forecast* 19, 240–250.

Smull, B. F., and Houze Jr. R. A. (1985). A midlatitude squall line with a trailing region of stratiform rain: radar and satellite observations. *Mon Weather Rev* 113, 117–133.

Smull, B. F., and Houze Jr. R. A. (1987). Rear inflow in squall lines with trailing stratiform precipitation. *Mon Weather Rev* 115, 2869–2889.

Srivastava, R. C. (1985). A simple model of evaporatively driven downdraft: application to microburst downdraft. *J Atmos Sci* 42, 1004–1023.
Srivastava, R. C. (1987). A model of intense downdrafts driven by the melting and evaporation of precipitation. *J Atmos Sci* 44, 1752–1773.
Trapp, R. J., and Weisman, M. L. (2003). Low-level mesovortices within squall lines and bow echoes. Part II: Their genesis and implications. *Mon Weather Rev* 131, 2804–2823.
Tuttle, J. D., Bringi, V. N., Orville, H. D., and Kopp, F. J. (1989). Multiparameter radar study of a microburst: comparison with model results. *J Atmos Sci* 46, 601–620.
Wakimoto, R. M. (1982). The life cycle of thunderstorm gust fronts as viewed with Doppler radar and rawinsonde data. *Mon Weather Rev* 110, 1060–1082.
Wakimoto, R. M. (1985). Forecasting dry microburst activity over the high plains. *Mon Weather Rev* 113, 1131–1143.
Wakimoto, R. M. (2001). Convectively driven high wind events. In: *Severe Convective Storms.* (Doswell, C. A., Ed.), pp. 255–298. Meteorological Monographs v. 28, no. 50, American Meteorological Society, Boston.
Wakimoto, R. M., Kessinger, C. J., and Kingsmill, D. E. (1994). Kinematic, thermodynamic, and visual structure of low-reflectivity microbursts. *Mon Weather Rev* 122, 72–92.
Wakimoto, R. M., Lee, W.-C., Bluestein, H. B., Liu, C.-H., and Hildebrand, P. H. (1996). ELDORA observations during VORTEX 95. *B Am Meteor Soc* 77, 1465–1481.
Weaver, J. F. (1999). Windstorms associated with extratropical cyclones. In: *Storms.* Vol. I (Pielke, R. A. Sr., and Pielke, R. A. Jr., Eds.), pp.449–459. Routledge, New York.
Webb, S. L. (1989). Contrasting windstorm consequences in two forests, Itasca State Park, Minnesota. *Ecology* 70, 1167–1180.
Weckwerth, T. M., and Wakimoto, R. M. (1992). The initiation and organization of convective cells atop a cold-air outflow boundary. *Mon Weather Rev* 120, 2169–2187.
Weisman, M. L. (1993). The genesis of severe, long-lived bow echoes. *J Atmos Sc,* 50, 645–670.
Weisman, M. L. (2001). Bow echoes: a tribute to T. T. Fujita. *B Am Meteor Soc* 82, 97–116.
Weisman, M. L., and Davis, C. (1998). Mechanisms for the generation of mesoscale vortices within quasi-linear convective systems. *J Atmos Sci* 45, 1990–2013.
Weisman, M. L., and Klemp, J. B. (1986). Characteristics of convective storms. In: *Mesoscale Meteorology and Forecasting.* (Ray, P. S., Ed.), pp. 331–358. American Meteorological Society, Boston.
Weisman, M. L., and Trapp, R. J. (2003). Low-level mesovortices within squall lines and bow echoes. Part I: overview and dependence on environmental shear. *Mon Weather Rev* 131, 2779–2803.
Wilson, J. W. (1986). Tornadogenesis by nonprecipitation induced wind shear lines. *Mon Weather Rev* 114, 270–284.
Wilson, J. W., Roberts, R. D., Kessinger, C., and McCarthy, J. (1984). Microburst wind structure and evaluation of Doppler radar for airport wind shear detection. *J Clim Appl Meteorol* 23, 898–915.
Wolfson, M. M., Delanoy, R. L., Forman, B. E., Hallowell, R. G., Pawlak, M. L., and Smith, P. D. (1994). Automated microburst wind-shear prediction. *Lincoln Lab J* 7, 399–426.
Zipser, E. J. (1977). Mesoscale and convective-scale downdrafts as distinct components of squall-line circulation. *Mon Weather Rev* 105, 1568–1589.

ADDITIONAL REFERENCE

Storm Prediction Center (2005). About Derechos. Available at www.spc.noaa.gov/misc/AbtDerechos/derechofacts.

4

Understanding How the Interaction of Wind and Trees Results in Windthrow, Stem Breakage, and Canopy Gap Formation

Christopher P. Quine and Barry A. Gardiner
Forestry Commission Research Agency

INTRODUCTION

The interaction of wind and trees can result in substantial changes to forest structure (Fig. 1) and is of interest to many forest ecologists, but the complexity of the relationship has confounded many studies. The result of the interaction can take many forms across a range of scales, may have chronic

FIG. 1 Windthrow in boreal forest of Charlevoix, Quebec, eastern Canada, from strong winds of November 7, 1994.

(acclimative) and acute (failure) components, and can be exacerbated by other conditions, such as wet snowfall and salt deposition. Leaves may be *abraded*, causing subsequent desiccation; young trees may *socket* (i.e., become loosened around the root collar by swaying) and in extreme cases *topple* because of inadequate rooting; leaders, branches, and crowns may *break*; older trees may be *windthrown* when stem and root plate overturn (Fig. 2) or may experience *windsnap* when the stem fails above ground level (Fig. 3). However, many studies commence after the interaction is

FIG. 2 Windthrown Sitka spruce (*Picea sitchensis*) tree, Birkley Wood, north England. Shallowness of rooting was caused by high winter water tables.

FIG. 3 Wind snapped tree, Tarn Hows, English Lake District. Root anchorage has been good and stem failure has resulted from loading caused by strong winds.

complete. Such postevent investigation frequently seeks simple relationships and frequently yields disappointing results. There is no escaping the fact that the interaction is complex, and an understanding of the process and response requires the integration of multiple disciplines, such as soil science, physiology, ecology, mechanics, meteorology, and climatology (Fig. 4). This chapter seeks to demonstrate that such integration, although difficult, is possible. A key point that we seek to emphasize is that the interaction between wind speed and tree response is a nonlinear relationship, and this has implications for both ecological understanding and tree/forest management; for example, small increments in wind speed can result in substantial differences in mode and scale of wind disturbance. Tree response may occur over a variety of time scales, with some being rapid (deflection and streamlining caused by acute loading) and some lagged (acclimation to chronic wind exposure). An improved understanding of the process gives insights into ecological processes and consequences.

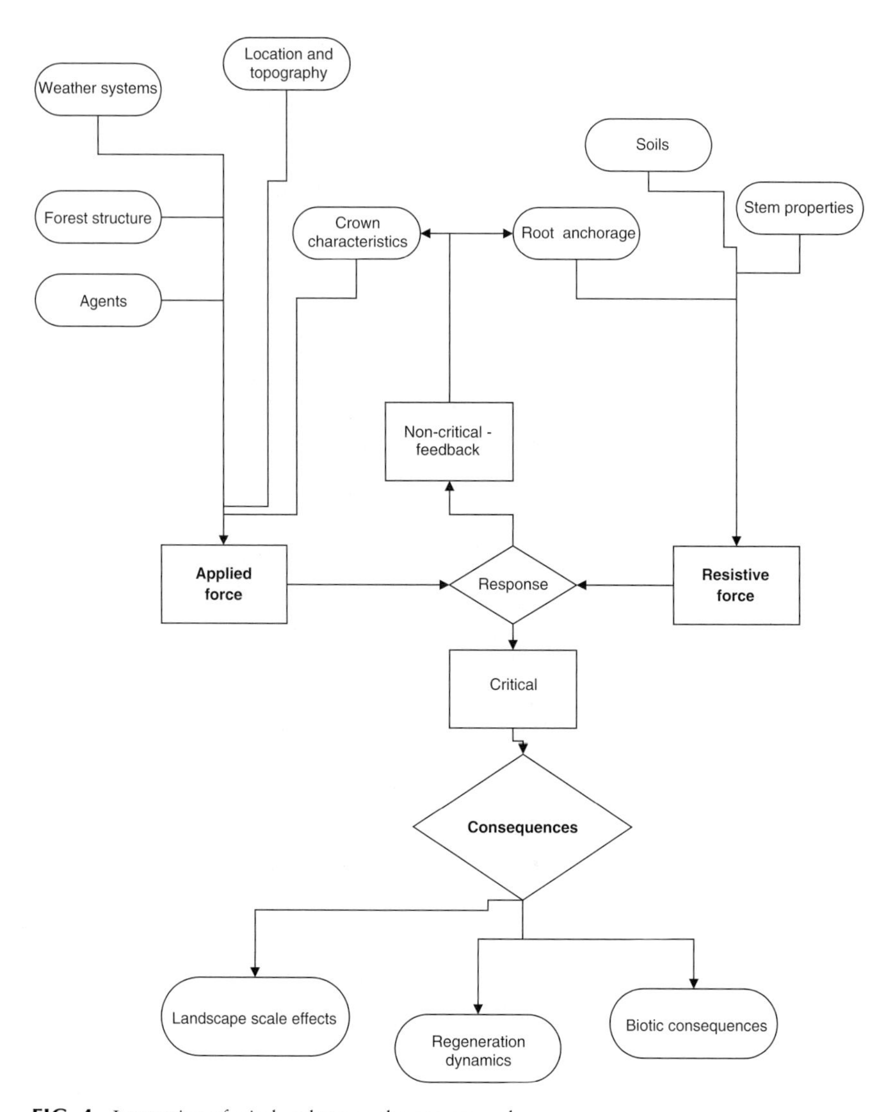

FIG. 4 Interaction of wind and trees—the process and response.

We write from a background in wind risk modeling of managed forests in a temperate maritime climate. Our understanding of forest–wind interaction has been aided by the limited variability in tree-stand characteristics because of the activities of forest management, the relative frequency of disturbance, and the relatively comprehensive data available. Nevertheless, the insights of the process and response are of wider applicability, and should complement and inform further ecological studies, which have been dominated by descriptive–empirical approaches, often relating to discrete events causing substantial change.

In this chapter we first look at the theoretical background to wind damage and methods for calculating the risk of windthrow and windsnap. We then explore the factors affecting this risk at different scales from tree to landscape, and the direct and indirect consequences of the interaction between strong winds and trees.

THEORETICAL CORE

To calculate the risk of wind damage to trees, it is necessary first to account for all the key components controlling the risk to individual trees (Fig. 5).

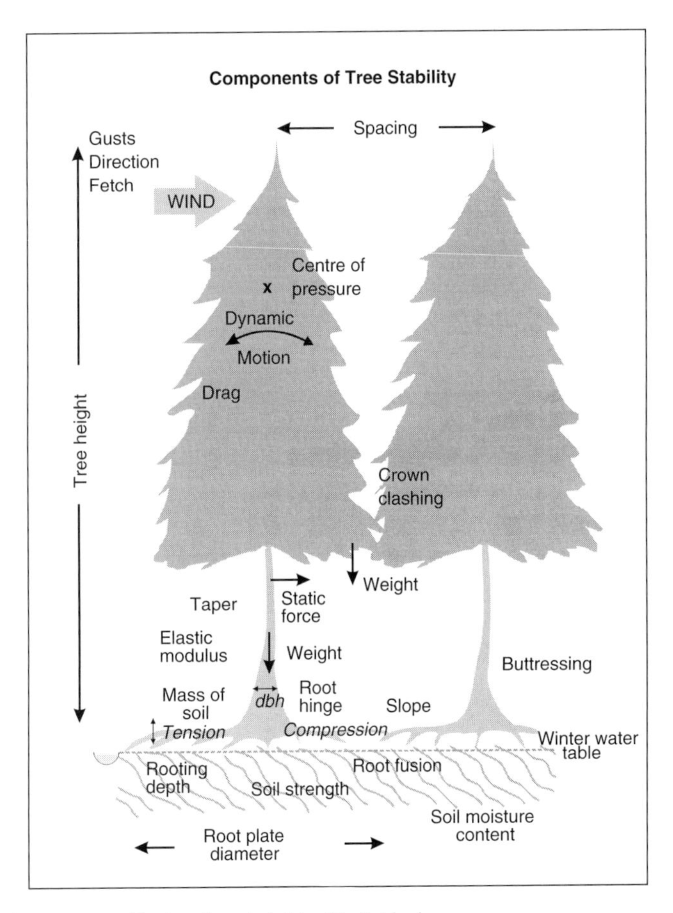

FIG. 5 Components affecting the wind risk of individual trees.

This risk can then be scaled up to the stand/patch, forest, landscape, and even national scale if enough basic information on these factors is available.

The approach adopted was to treat the tree as a cantilever beam set in the ground and subjected to a fluctuating load because of the wind. The load is first calculated for a steady wind, and the gustiness of the wind is then taken into account. The steady wind loading is calculated by determining the drag of the forest on the atmosphere and apportioning this drag force to the individual trees at a specific height in their canopy called the zero-plane displacement height. The method is outlined below and given in detail in Gardiner *et al.* (2000). A glossary of terms is provided in the Appendix.

Wind Loading

The wind speed (u) over a forest canopy is given by a logarithmic profile of the form:

$$u(z) = \frac{u_*}{k} \ln\left(\frac{z-d}{z_0}\right) \tag{1}$$

where z is height above the surface, u_* is the friction velocity, k is von Karman's constant (≈ 0.4), d is the zero-plane displacement, and z_0 is the aerodynamic roughness. The friction velocity is defined by:

$$\tau = -\rho u_*^2 \tag{2}$$

where τ is the shear stress on the surface (drag/force per unit area) and ρ is the air density. If the average spacing between trees is D (m), then τD^2 is the average force acting on each tree, assuming trees are evenly spaced. This force acts at the height of the zero-plane displacement (d) (Thom, 1971) giving the mean bending moment (BM_{mean}, Nm) at any height (z) on the stem as $(d\text{-}z)\tau D^2$. This moment is that caused by the mean wind only, and we account for the gusty nature of the wind by multiplying by a gust factor (G). By substituting for τ using Equations (1) and (2), we obtain a general expression for the maximum bending moment ($BM_{max,}$) at any point on the tree stem as a function of the canopy top wind speed (u_h):

$$BM_{\max}(z) = f_{CW}(d-z)\rho G\left(\frac{Du_h k}{\ln\left(\frac{h-d}{z_0}\right)}\right)^2 \tag{3}$$

where h is tree height and f_{CW} is a factor to account for the additional moment provided by the overhanging displaced mass of the canopy.

The gust factor was empirically established from wind tunnel studies (Gardiner *et al.*, 1997).

$$G = \frac{(2.7193(D/h) - 0.061) + (-1.273(D/h) + 0.9701)(1.1127(D/h) + 0.0311)^{x/h}}{(0.68(D/h) - 0.0385) + (-0.68(D/h) + 0.4785)(1.7239(D/h) + 0.0316)^{x/h}} \quad (4)$$

where x is the distance from the forest edge and $0.075 < D/h < 0.45$.

Resistance to Breakage

Assuming that wind-induced stress in the outer fibers of the tree stem is constant (Morgan and Cannell, 1994), the stress needs only to be calculated at breast height (z = 1.3 m). The stem will break when the stress exceeds the modulus of rupture (MOR) and the critical bending moment ($BMcrit_{break}$) at which this occurs is given (Jones, 1983) by the expression:

$$BMcrit_{break} = \frac{\pi}{32} f_{knot} MOR \times dbh^3 \quad (5)$$

where dbh is the stem diameter at breast height and f_{knot} is a factor to account for the presence of knots and other defects, which reduces the modulus of rupture from the values obtained for clear-wood samples (Lavers, 1969). When the bending moments calculated in Equation (3) (with z = 1.3) and Equation (5) are equal, the canopy top wind speed is sufficient to break the tree, and we can predict the critical wind speed for stem breakage (uh_{break}) as follows:

$$uh_{break} = \frac{1}{kD}\left[\frac{\pi \times MOR \times dbh^3}{32\rho G(d - 1.3)}\right]^{\frac{1}{2}}\left[\frac{f_{knot}}{f_{CW}}\right]^{\frac{1}{2}} \ln\left(\frac{h-d}{z_0}\right) \quad (6)$$

Resistance to Overturning

The resistance to overturning is calculated from tree pulling experiments (e.g., Fraser and Gardiner, 1967; Ray and Nicoll, 1998; Moore, 2000; Peltola *et al.*, 2000; Nicoll *et al.*, 2006) rather than trying to construct theoretical models of root anchorage. Trees are pulled over by a winch attached at approximately half tree height, and the force required to uproot the tree, together with detailed measurements of the physical characteristics of the tree, is measured. The best predictor of the maximum base bending moment recorded ($BMcrit_{over}$) has consistently been found to be stem weight or a surrogate for stem weight (SW), such as $h \times dbh^2$ (Cucchi and Bert, 2003; Cucchi *et al.*, 2005), giving a relationship of the form:

$$BMcrit_{over} = C_{reg}\, SW \tag{7}$$

where C_{reg} is a regression constant and the regression is forced through zero.

When the bending moments calculated in Equation (3) (with $z = 0$) and Equation (7) are identical, the wind speed is sufficient to overturn the tree and we can predict the critical wind speed for overturning (uh_{over}) as follows:

$$uh_{over} = \frac{1}{kD}\left[\frac{C_{reg}\, SW}{\rho G d}\right]^{\frac{1}{2}}\left[\frac{1}{f_{CW}}\right]^{\frac{1}{2}} \ln\left(\frac{h-d}{z_0}\right) \tag{8}$$

APPLIED FORCE

Windthrow or stem breakage occurs to a single tree in an instant, but the sequence leading to this event may unfold over a much longer period and be the result of a cascade of processes over greater spatial scales. This section deals with the processes that influence the force experienced by a tree at each scale from the pressure systems that generate the wind to the influence of the tree itself. The resistance of the tree to this wind loading is covered in the subsequent section.

Weather Systems and Their Influence on Regional Differences in Wind Climate

Wind arises from the existence of a pressure gradient across a volume of atmosphere and the consequent flow of air from high-pressure to low-pressure areas (Stull, 1988). The flow of air is deflected by the rotation of the earth, and the resultant geostrophic wind blows parallel to notional lines of equal pressure (isobars) around the areas of high and low pressure. Where the gradient of pressure is intense, there is an additional rotational element known as vorticity, which deflects the geostrophic wind. The geostrophic wind is slowed by friction at the Earth's surface, but the effect is less significant when wind speeds are high.

Five types of strong winds have been identified (ESDU, 1987; Hunt, 1995; Wyatt, 1995): extra tropical cyclones, tropical cyclones (hurricanes), thunderstorms, tornadoes, and orographically induced winds (such as föhn, and other downslope winds). Each mechanism can deliver wind speeds sufficient to overturn or snap trees. However, the mechanism influences the spatial extent, duration, and severity of the strongest winds, and recurrence at any one location (Table 1). In general, in temperate maritime regions, such as Great Britain, extra-tropical storms dominate. Such storms are frequent and may cause wind speeds close to critical to be experienced a number of times during the life of a tree. There may well be a relationship between chronic and catastrophic disturbance and between mean and extreme processes. In locations where strong winds are rare—and delivered by spatially discrete mechanisms such as tornadoes—the repeatability may be very low and beyond the life span of trees. In areas such as the subtropics, there may be mixed wind climates. Thus, the eastern United States experiences hurricanes, tornadoes, and extreme thunderstorms. Such mixed climates present problems in assessing recurrence and identifying the dominant mode of disturbance, particularly given the relatively short measurement history.

The frequency of strong winds varies markedly with regional location relative to the main storm tracks. For example, strong winds are much more prevalent in Great Britain than in central Europe, most of Scandinavia, and central North America. Wind speeds that are experienced every year in the British uplands on sites such as Eskdalemuir (Quine, 1995) are experienced only every few decades in Ontario (Smith *et al.*, 1987) and Alberta (Flesch and Wilson, 1993). Comparable frequencies of strong winds are found in the extreme coastal fringe of the Pacific Northwest (especially southeast Alaska) (Quine *et al.*, 1999) and in Tierra del Fuego. There may be a substantial decline in wind speed with distance from coast at the regional and continental scale.

Temporal Variability in the Occurrence of Strong Winds

There is substantial temporal variability in the wind experienced at a location, spanning seconds (gusts), hours (diurnal flows), days (movement of fronts), and greater (seasonal weather patterns, decadal shifts in location of polar front affecting depression tracks). Regional patterns of strong winds can be identified, but the precise track of the individual intense storms is very influential in determining the affected area. This variation cannot be anticipated, and so the climatology of strong winds depends on probabilistic statements of frequency and magnitude. The Weibull distribution is commonly used to describe the parent wind distribution and the extreme value distribution to represent the strong wind climatology (Cook, 1985). The understanding of the long-scale temporal variability is limited by the

TABLE 1 Main Climatological Mechanisms Providing Strong Winds*

Storm Mechanism	Spatial and Temporal Scale	Characteristics	Sources—Examples of Related Forest Damage
Extratropical cyclone	Scale of weather system: diameter up to 4000 km Duration: up to 3 days Main affected area may be several hundred kilometers in diameter Recurrence: may be once in 50–400 years at any one point; large population of storms useful for recurrence calculations	Often occur in families—fatiguing Winter: wet soils, and sometimes wet snow May contain squall lines and other intense gust fronts embedded within them	Western Europe—Great Britain (Andersen, 1954; Quine, 1991); Germany (Mulder *et al.*, 1973); Denmark (Jakobsen, 1986; Wolf *et al.*, 2004); France (Pontailler *et al.*, 1997; Doll, 2000); Ukrainian Carpathians (Lavnyy and Lässig, 2003); Poland (Mikulowski and Gil, 2003); Switzerland (Schönenberger *et al.*, 2003) Pacific Northwest of America (Harris and Farr, 1974; Russell, 1982; Taylor, 1990; Kramer *et al.*, 2001) South Island of New Zealand (Jane, 1986) Eastern Canada (Ruel and Benoit, 1999) South America—southern tip of Argentina, Chile (Rebertus *et al.*, 1997) Exacerbated by snow/ice, especially in maritime areas (Bruederle and Stearns, 1985; Guild, 1986; Boerner *et al.*, 1988)
Thunderstorm	Scale: downbursts—tens of kilometers Duration: up to 30 minutes at any one spot Recurrence: calculation is problematic	Location and direction may be very different from prevailing wind and dependent upon properties of the convective cell Summer—dry soils; hail stones (trees in leaf)	Eastern Canada/USA—Great Lakes of Canada/USA (Flannigan *et al.*, 1989; Frelich and Lorimer, 1991; Frelich, 2002); Wisconsin (Dunn *et al.*, 1982; Canham and Loucks, 1984); Virginia (Orwig and Abrams, 1995); New York (Kearsley and Jackson, 1997) Europe: Southern Germany, Bavaria (Fischer, 1992); Slovakia (Pavlik and Pavlik, 2003) Australia (Sheehan *et al.*, 1982) Russia—Urals (Lässig and Mocalov, 2000)

Tornado	Scale: up to 2 km wide, 400 km long Duration: seconds–minutes Recurrence: at a single location 1 in 1500 years (Oklahoma), >1 in 10000 years (western Europe)	Location and direction may be insensitive to topography Extreme gradient and extreme exceedance of critical wind speed Torsional effects	Eastern USA (Glitzenstein and Harcombe, 1988, Peterson and Pickett, 1991; 1995; Peterson, 2000); Texas (Liu *et al.*, 1997); Georgia (Harrington and Bluhm, 2001); Minnesota (Dyer and Baird, 1997) Eastern Canada—Ontario (Harrington and Newark, 1986); Quebec (Lehmann and Laflamme, 1975) Russia (Lyakhov, 1987; Syrjanen *et al.*, 1994)
Tropical cyclone	Scale: up to 600 km in diameter, main affected area approximately 200 km in diameter Duration: 1–2 days Recurrence: calculated at the regional scale—1 in 8 to 1 in 14 for southeastern USA	Extreme rainfall (e.g., 35 cm in 24 hours) Steep gradient Concentric wind field—range of wind directions associated with storm passage	Caribbean (Tanner *et al.*, 1991); Puerto Rico (Basnet *et al.*, 1992) Eastern USA (Foster, 1988; Boose *et al.*, 1994; Haymond *et al.*, 1996; Clinton and Baker, 2000) Japan (Nakashizuka and Iida, 1995; Ida and Nakagashi, 1998; Ishizuka *et al.*, 1998; Peters, 1998; Chiba, 2000) North Island of New Zealand (Shaw, 1983) Northern Australia (Unwin *et al.*, 1988)
Orographic and other local winds	Scale: kilometers Duration: hours–days	Repeatable but discontinuous Discrete boundaries and locations due to topography and wavelength	Great Britain—Sheffield, (Aanenson, 1965). South Island of New Zealand—Canterbury Plains (Somerville, 1989) USA—Colorado (Baker *et al.*, 2002, Kulakowski and Veblen, 2002)

*Documented examples of forest damage associated with different storm types. Examples from natural and managed forests combined.

short history of quantitative measurements of wind speed. For example, in Great Britain surface pressure measurements began in a systematic way only in 1880, and in 1919 only 14 stations were recording wind speed (Hunt, 1995). There are additional problems with homogeneity of record and, in particular, with changes to instruments, location and exposure, and method of data summary (Palutikof *et al.*, 1997; Robinson, 1999). Furthermore, the potential for climate change affecting storminess (frequency, location, and intensity) and the potential for century-long changes in storm frequency (Lamb, 1991) add to the difficulties.

Strong winds are infrequent. Rarity is often expressed as return period; this does not imply equal temporal spacing but the average interval between winds of particular magnitudes. The stochastic nature of storm occurrence means that the degree of wind damage can vary significantly from year to year. A succession of quiet winters (or decades) can be followed by a sequence of winters (or decades) with very damaging storms. The occurrence of a particularly severe storm in one year does not modify the chance of the same intensity of storm occurring in the following year.

Some mechanisms for strong winds have large spatial extent and recur quite often; this can have implications for expansion of windthrown gaps. However, not all strong winds have high repeatability. In the case of tornadoes, the discrete and rare nature of severe wind speeds makes further enlargement by a repeat episode highly unlikely; even in the most prone areas, the recurrence is of the order of once in 1500 years.

Note that this variability is only one element affecting future frequency of disturbance. Modeling this requires not only a probabilistic treatment of wind climate (with appropriate assumptions over stationarity) but also treatment of the changing vulnerability of the forest as individual stands age and regenerate and the forest mosaic ebbs and flows. Climate change predictions concerning the frequency and location of strong winds have been tentative, but some authors suggest an increase in intensity of events and regional shifts in typical storm tracks. There are clear implications for the assessment of recurrence—and the degree to which experience (as captured in empirical/statistical studies of historical damage) is a guide to the future. A process-response approach allows predictions to be made provided there are sound estimates of future wind climates.

Spatial Variability Related to Topography

Strength of wind in a single event is influenced by location relative to the storm centre (Fig. 6). For example, in hurricanes there is a relatively symmetric

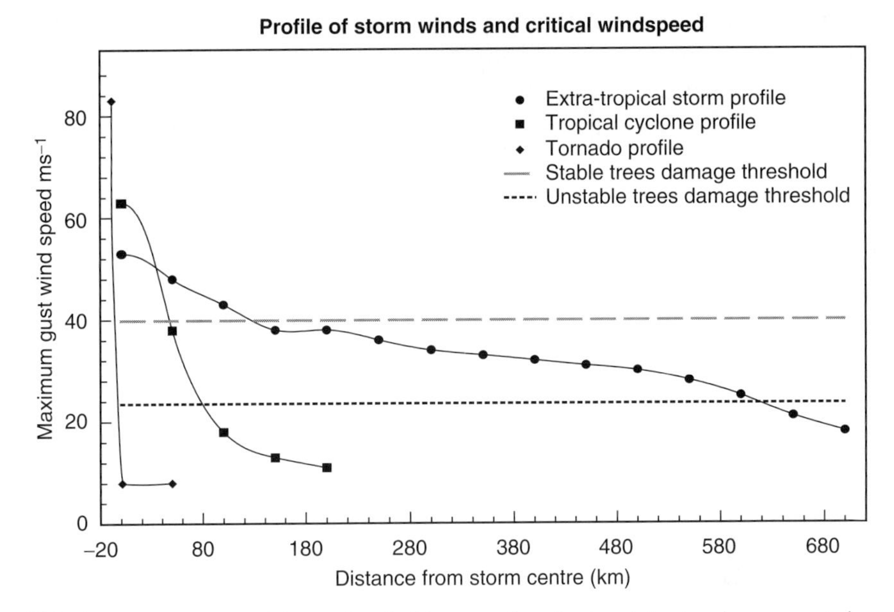

FIG. 6 Contrast in potential area disturbed because of mechanism delivering the strong winds, for stable and unstable trees.

distribution of strong winds concentrically from the eye of the storm, and there may be a rapid gradient in wind speed out from the eye wall. In many extratropical cyclones the pressure field is skewed, resulting in very different wind fields on opposite sides of the center of the depression. In the northern hemisphere, circulation around a depression is counterclockwise and the strongest winds are commonly found to the west and south of the center of the low pressure.

Strength of wind is also strongly influenced by topography (see Chapter 2). Wind speed frequently increases with elevation and may double over a range of a few hundred meters. Aspect of hill slopes and orientation of valleys with respect to the prevailing wind direction can have a considerable influence on the magnitude of wind speed experienced at a location. In many storms, the highest wind speeds are found on ridge tops and on gentle slopes facing the wind. In complex terrain, the topography can alter the wind speed and direction significantly with wind accelerating over hilltops and being funneled in valleys (Fig. 7).

The mechanism delivering the strong winds can also influence the interaction with topography; in particular, lee slopes can be either the most or least sheltered location, depending upon temperature profiles aloft and

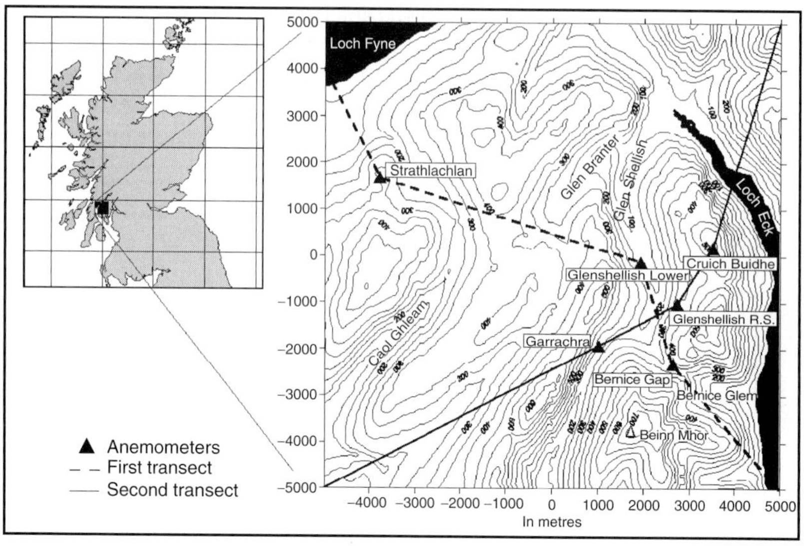

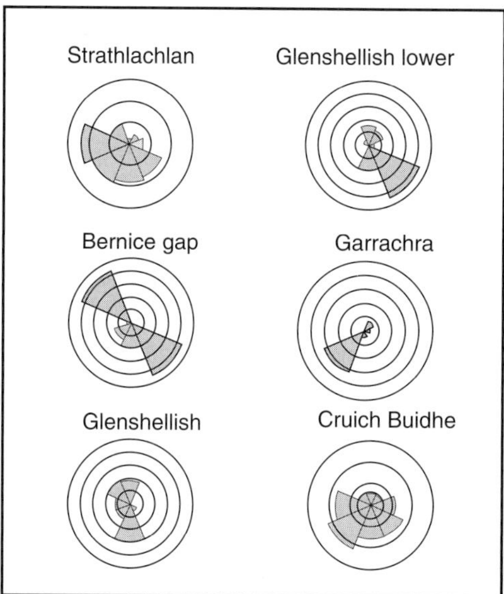

FIG. 7 Wind roses from simultaneous anemometer data in complex terrain in the Cowal district of west Scotland. Funneling caused by the orientation of valleys is readily apparent. (Reproduced courtesy of *Met Apps*, 6, pp. 329–342.)

the degree of flow separation. Under certain conditions (the presence of a strong temperature inversion close to a hilltop), air can be forced down the lee slope like water flowing over a weir, creating extremely strong downslope winds in an area that, under normal circumstances, might be reasonably sheltered (Barry, 1992). Examples include the "helm wind" in the

Eden valley, Cumbria; the "Bora wind" along the Adriatic coast of Yugoslavia; the föhn winds of the Alps; and the winds experienced on the Canterbury Plains of the South Island of New Zealand. Such orographic winds may affect particular topographic locations repeatedly (Troen and Petersen, 1989).

Surface roughness can add to the topographic effect. Trees on the windward lower slopes of a hill will tend to reduce the wind strength at the hilltop. Trees also induce the flow to separate from leeward slopes so that large areas of relatively calm air may be found behind a forested hill (see also Chapter 2).

Topography may also influence accompanying canopy loading, such as by determining the transition from rain to snow (Chapter 5). Note that topography may also indirectly affect disturbance through effects on tree growth and substrate type. Soils vary with slope position, and tree growth, stature, and root anchorage (i.e., contributing to resistive forces) will reflect these topographic influences. Species distribution may also be strongly related to such factors, confounding simple comparison of species stability.

A variety of methods have been employed to account for and represent the effects of topography (Hannah *et al.*, 1995). Direct measurements of wind speed, although desirable, have been limited by the number of instruments available and the practicalities of deploying them in difficult terrain where change of location by a few meters can significantly influence the results. Nevertheless, improvements in instrumentation have recently made this more feasible. An indirect form of wind speed measurement was provided by the attrition of cotton flags, with the advantage of cheap and easy installation (Miller *et al.*, 1987). These have been used in Great Britain to establish topographic relationships that can subsequently be applied through geographic information systems (Quine and White, 1994). A variety of indices of topographic representation have also been derived—by expert judgment, statistical analysis of wind measurement (e.g., correlations with upper air measurements), or disturbance patterns (e.g., shelter angle) (Boose *et al.*, 1994). Another method that appears attractive is the use of physical models in a wind tunnel (Ruel *et al.*, 1998), though with substantial costs in model construction and wind tunnel operation. Finally, with the advent of powerful computers, there has been a growing use of numerical models, operating at a wide variety of scales and with a particular set of assumptions (Ruel *et al.*, 1997; Suárez *et al.*, 1999). However, representation of complex terrain and the extreme surface roughness generated by presence of forests remains a challenge (see Chapter 2).

Spatial Variability Related to Nature of Ground Surface

The strength and turbulent structure of the wind are further modified by the roughness of the ground surface. Thus, forest structure—including the mosaic formed by open ground, water bodies, pre-existing gaps, and stands of different height/density—will influence the speed and turbulence of the air flow upwind of the object of study. Rough surfaces may reduce mean wind speed but induce turbulence and increase the gustiness of the wind. The flow arriving at a particular stand will be influenced by upwind shelter, such as extensive mature forest, and the profile of the stand edge.

What constitutes an edge? In aerodynamic terms, edges are formed by a discontinuity in the surface—for example, a step change in height of the structural elements. In forests, relatively permanent edges are found at the margins of landscape types (lakes, rivers, bogs, bare rock, frost hollows) and may develop a streamlined profile that reflects the strength of the wind (Fig. 8); new edges may be formed by disturbance (e.g., gap/patch formation, riparian processes, harvesting). The presence of an edge directly influences the airflow and thereby affects the wind speeds experienced within a stand to the lee. The airflow is further modified by the stand itself.

Wind Distribution within a Forest Stand

The structure of a forest stand is important in determining the wind loading on individual trees within the stand. There have been suggestions that

FIG. 8 An adapted edge formed by persistent exposure to strong winds. The canopy has adopted a streamlined profile. Lynsey Lee plantation, Shetland Isles.

multistoried "natural" forests with a large range of tree sizes have less severe wind shear at canopy top than forests with uniform canopies and that this might lead to reduced gustiness and damage (Finnigan and Brunet, 1995). In fact, the opposite seems to be the case; Gardiner *et al.* (2005) found that the tallest trees (22.5 m) in a mixed-height model forest (mean height = 15 m) had 7 times the mean wind loading and 15 times the extreme wind loading compared to the mean trees; such dominant trees are often greater in diameter and less slender and so have greater resistive moments than co-dominants (see further discussion later in this chapter). In more even-sized forests, the dominant trees will have increased wind loading because of their height and canopy size relative to the mean trees. Even for trees that are extremely similar in size, wind loading can vary widely because of variations in the shelter and damping provided by the neighbors (Gardiner, 1995; Gardiner *et al.*, 1997). Smaller trees gain substantial shelter from taller trees (Gardiner *et al.*, 2005); however, they may become extremely vulnerable if the taller trees are felled (by wind or harvesting), in part because of their lack of mechanical adaptation to the new, more extreme wind climate.

In addition to the variation in wind loading because of stature, loading can vary because of the relative position of a tree within a stand and, in particular, its proximity to stand edges. The highest wind loading is on trees at an exposed edge, and the wind loading declines back into the forest (Gardiner *et al.*, 1997); the most rapid decline occurs in the densest forests. The change in the mean and extreme wind loading back from an edge is not consistent, and at a distance of 4 to 5 tree heights the extreme wind loading starts to increase again (Morse *et al.*, 2002). This is due to the time it takes for canopy scale eddies to develop over the new surface and explains the observation that damage often appears to occur a few tree heights back from established edges (Somerville, 1989). Trees at or close to newly created edges (because of natural causes or human activity) will be more vulnerable than those close to established edges because of lack of adaptation (acclimation) to their new wind environment.

Gaps within the stand, such as those created by the wind, will lead to increased wind loading on the remaining trees and consequent expansion of the gap. This is well documented over medium to long periods (Quine, 2001; Quine, 2003a) but not over the period of an individual storm. The changes in vulnerability of individual trees during a storm because of the loss of neighboring trees require further research.

Seasonal changes tend to have more impact in broadleaved forests than in coniferous forests. There has been much debate about the impact of leaf loss during the winter, but it appears that for dense forests the loss of leaves leads to increased wind penetration into the canopy and increased loading even though the "sail area" is reduced (Sigmon *et al.*, 1984; Dolman, 1986) and drag is less (Vollsinger *et al.*, 2005). In mixed conifer/deciduous forests, the loss of leaves from the deciduous component may increase drag on the conifers during winter. For widely spaced broad-leaved trees, the loss of leaves during the winter will reduce the wind loading.

Other agents, such as rain and snow, can increase the loading on trees within the forest and exacerbate the effect of wind loading. Although substantial weights of water can be intercepted within canopies (up to 10 kg on a single tree canopy), the water will have been blown or evaporated off by the time wind speeds reach high enough values to cause damage. Increases in air density caused by the presence of rain are too small to be of any importance. Snow, however, can accumulate in enough quantities (and often be asymmetrically distributed in the crown), particularly under certain meteorological conditions (Nykänen *et al.*, 1997). When the snow is wet, it can directly damage trees or reduce the wind speed at which damage occurs (Petty and Worrell, 1981).

Influence of Tree Crown

The crown shape and density, together with the wind profile through the canopy, determine the total force of the wind (F) that acts on trees (Jones, 1983):

$$F = \int_0^h \frac{1}{2} C_d \rho u(z)^2 A(z)\, dz \qquad (9)$$

where $A(z)$ is the area of the canopy presented to the airflow and C_d is the drag coefficient. Both of these are functions of tree species and the structure of the forest. The drag coefficient is very variable between species (Mayhead, 1973; Rudnicki *et al.*, 2004) so that flexible species, such as western hemlock [*Tsuga heterophylla* (Raf.) Sarg.], have much lower drag coefficients than stiffer branched species, such as Sitka spruce [*Picea sitchensis* (Bong.) Carr.]. Unfortunately, there are few data other than for coniferous species; a recent exception is the work by Vollsinger *et al.* (2005). As the wind speed increases, the effective area [$A(z)$] exposed to the wind decreases as the branches streamline so that the effective area can decrease by a factor of more than 2 as the wind speed increases from 10 to 30 $m\,s^{-1}$.

Certain species, such as *Pinus* spp., are particularly susceptible to damage due to snow accumulation. Other species (e.g., *Picea engelmannii* Parry ex Engelm.) are adapted through evolution to locations with high snow, and their branches are designed to shed snow. Even within the same species, canopy form can vary widely among different provenances, depending on whether they are adapted to high snowfall (Büsgen *et al.*, 1929) (Fig. 9).

Influence of Tree Bending and Dynamics

The shape of a close grown conifer tree stem is reasonably well represented by the equation (Gardiner, 1989):

$$d_x = d_b\,(x/h)^{0.6} \tag{10}$$

where d_b is the tree diameter at the tree base, x is the distance from the top of the tree, and h is the tree height. Then the deflection of a tree

FIG. 9 A narrow crowned form of Norway spruce (*Picea abies*), evolved to prevent excessive loading of the canopy through snow.

under a steady load can be calculated (Gardiner, 1989) by an equation of the form:

$$\frac{y}{h} = \frac{Fh^2}{EI_h}\left[2.5\left(\frac{x}{h}\right) - 4.17\left(\frac{x}{h}\right)^{0.6} - 0.71r\left(\frac{x}{h}\right) - 1.79r\left(\frac{x}{h}\right)^{-0.4} + 1.67 + 2.5r\right] \quad (11)$$

where F is the force of the wind, E is the Young's modulus, r is the ratio of the distance from the top of the tree to the point at which the force acts to the total tree height, and I_b is the area moment of inertia at the base of the tree ($\pi d_b^4/64$). For a wind load it can be assumed (Thom, 1971) that the wind force acts at the zero-plane displacement height (d). For forests d is found to be close to $0.8h$ (Raupach, 1992) and, therefore, a reasonable estimate for r is given by $r = (h - 0.8h)/h = 0.2$. A similar approach can be adopted for any taper function used to describe the stem shape [see Gardiner (1992) and Petty and Swain (1985) for more details].

However, for the fluctuating wind conditions in a storm, we have to account for the dynamic behavior of trees. Gardiner (1992) reviewed the understanding of the dynamical behavior of trees under wind loading and showed how, to a first approximation, a tree can be treated as a simple cantilever beam. The mass of the tree is modeled as a single equivalent lumped mass on a massless beam, and energy losses in the stem, branch clashing, and the drag of the canopy are represented as a damping coefficient:

$$m_e\frac{d^2y}{dt} + c\frac{dy}{dt} + ky = F_0e^{i\omega t} \quad (12)$$

where y is the displacement, m_e is the equivalent mass of the tree, c is a damping coefficient, k is a spring constant, F_0 is the amplitude of the wind force, ω is the natural frequency of the wind gusts, and t is time. Moore and Maguire (2004; 2005) have recently reviewed the data on tree swaying behavior and compared measurements against this simple model as well as cataloguing measurements of tree damping.

Work by Kerzenmacher and Gardiner (1998) has shown that the above simple representation of a tree is unable to reproduce measured tree movement without recourse to artificial adjustment to the stiffness of each stem section because of the complex interaction of the branches and the stem, which do not allow the branches to be simply treated as masses attached to the stem. Rather, the branches are dynamically coupled to the stem in a way that leads to high levels of damping. Without damping, oscillations caused by the wind would build up until the tree broke or fell over. Damping reduces the amplitude of these oscillations. Some of the damping is caused by hysteresis in the stem bending with the production of heat, some is caused by crown clashing, but the majority is caused by turbulence generated by the tips of the branches as they move. What is fascinating is that the tree appears to be designed to dissipate energy in a particularly efficient manner. The major branches have natural frequencies close to the first harmonic of the whole tree frequency, and the small branches have natural frequencies close to the first harmonic of the large branches (Gardiner, 1995). Therefore, the transfer of energy from the movement of the whole tree through to turbulence at the branch tip is very well coupled and efficient. Recent work by Moore (2002) has used finite element modeling to simulate the dynamic behavior of the whole tree, including branches, and this approach offers the possibility of much better understanding of the coupling of the wind and tree movement. Damping ratios were higher when the branches were treated as coupled harmonic oscillators than if they were simply treated as discrete masses.

RESISTIVE FORCE

The resistance of a tree to wind loading is through a variety of processes that are covered in this section.

Stem

As we saw previously, if the wind loading on the canopy is such that the stress becomes greater than the modulus of rupture of the wood, the stem will break. Trees adopt a number of mechanisms to maximize the strength of the wood and the resistance to breakage with the minimum amount of material; this is shown by the fact that the strength-to-weight ratio of wood is comparable to that of any human-made material (Gordon, 1991).

Wood fibers are weaker in compression, and trees tend to fail first on the side of the stem under compression. However, to minimize this imbalance, trees produce growth stresses (Fournier *et al.*, 1994) that result in the outside of the tree being in tension in the longitudinal direction (up and down the stem) and the inside of the tree in compression. Thus, if a notch is cut in the outside of the stem, the tensile stresses act as though to pull the notch apart. These tensile stresses counteract the compressive stresses imposed by the wind on the leeward side of the stem; lee-side stresses on the tree are less than those on the windward side, which is under additional tension because of the wind. Wood tends to be weaker in compression than in tension, which helps to reduce the risk of the stem breaking on the side in compression. Sometimes, it is possible to see when a tree has been close to failing by the presence of compression bulges on the outside of the stem made up of fibers that have failed by local buckling.

The process that produces longitudinal tensile stresses in the outer part of the tree also produces compressive stresses in the tangential direction. These compressive stresses help to stop wood splitting in a radial direction. At the same time, the parenchyma rays that run radially hold the wood together and help to prevent delamination between the different growth layers (Burgert *et al.*, 1999).

Weakness will occur in tree stems because of the presence of any kind of defects. These may be caused by branch knots or rot. Knots cause weakness because the wood fibers in the stem are forced around knots and there are localized stress concentrations. If the knots are large or there is a concentration of knots, such as at a whorl of branches, the weakness can be significant and reduce the wood strength by up to 50%. This can often lead to the tops of trees being broken by the wind or heavy snow.

Rot from fungal attack is associated with a wound in the stem or infection through the roots. Stem damage can occur by breakage in the crown as above, rubbing by a neighboring tree, or herbivore damage. Different types of rot attack different parts of the wood structure, but the result is always a severe reduction in strength (Mattheck and Breloer, 1995). However, it is possible to have a considerable amount of rot in a stem before the reduction in strength of the stem is significant (Mattheck and Breloer, 1995). For example, a tree that is 70% rotten in cross-sectional area on the inside still has 76% of the strength of a solid tree, and a tree that is 50% rotten will still have 91% of the strength. Stem damage and decay on the periphery of a stem have a much greater effect on bending resistance than does heart rot.

Mechanical adaptation, or acclimation, of the tree stem to wind loading has been observed in many circumstances (Mattheck, 1991; Telewski, 1995). The cells of wood subjected to flexure or loading are found to be shorter and to have thicker walls. The result is that, in areas of the tree stem where stresses are greatest, there is increased growth and the tree "adapts" its shape to try to keep the stress as even as possible and to reduce the chance of breakage. This produces idealized shapes for the connection between stem and roots, stem and branches, and within forks. In tropical trees, this mechanical adaptation can result in immense buttresses, which act as guys from the stem to the shallow root systems typically found in tropical soils (Crook *et al.*, 1997). In gymnosperms, buttresses are constructed where the stem is in compression and tend to be thicker and less dramatic. In all parts of the tree subjected to loading, the structure is developed to avoid high stresses (Mattheck and Breloer, 1995; Telewski, 1995).

Within the cells that make up the wood, the key components in determining the stiffness of the material are the cellulose microfibrils and, in particular, the angle of these microfibrils. Within the juvenile core of trees (i.e., wood laid down in the crown at the top of the tree), the microfibril angle is large; this makes the wood more flexible and appears to help reduce the risk of breakage at the tip of the tree where wind loads are highest. This increase in microfibril angle is also observed in trees subjected to increased flexing or wind loading (Meguro and Miyawaki, 1994; Zipse *et al.*, 1998) and allows the tree greater flexibility and an ability to withstand higher strains.

Roots

The resistance of shallow-rooted trees to uprooting has been shown by Coutts (1986) to be caused by four major components: the tensile strength of the windward roots, the weight of the root/soil plate, the stiffness of the root hinge on the leeward side of the tree, and the shearing strength of the soil. Generally, the hinge strength and soil shearing strength are much smaller than the other two components. Thus, a moment equation for an individual tree on the flat can be written according to Achim *et al.* (2003), following earlier work by Blackwell *et al.* (1990), as:

$$M_{wind} + M_{SW} = M_{rw} + M_{spring} \qquad (13)$$

where $M_{wind} = F\,(h_f \cos\alpha + l_h \sin\alpha)$ is the moment at the tree base, $M_{SW} = m_s g\,(l_h \cos\alpha - h_c \sin\alpha)$ is the moment caused by the overhanging stem weight, $M_{rw} = m_r g l_r \cos\alpha/2$ is the moment caused by the displaced root/soil plate weight, and $M_{spring} = k l_r^2 \sin\alpha$ is the resistance of windward roots in tension. In addition, α is the angle of deflection of the root plate, h_f is the height at which the wind acts, h_c is the height of the shoot (stem and crown) centre of gravity, l_h is the length of the root hinge, l_r is the length of the root plate, m_r is the mass of the root/soil plate, and m_s is the shoot weight. If the moments on the left side of Equation (13) are greater than those on the right side, then the tree will continue to be displaced until an equilibrium is reached or the roots/root-soil plate fail when they pass their elastic limit and the tree falls over.

As discussed previously and given in Equation (7), the resistance of trees to overturning can be readily predicted from the stem weight if the data from tree-pulling experiments are separated by soil type (Moore, 2000; Peltola *et al.*, 2000; Meunier *et al.*, 2002; Nicoll *et al.*, in press). The stem weight provides a measure of the below ground biomass (and associated soil mass) and, consequently, the resistance of the root system to overturning.

Experimental work has found that trees on slopes appear on average to be as well anchored as trees on the flat. However, trees are more vulnerable to being windthrown downslope and more resistant to being thrown upslope, and this appears to be an effect of the architecture of the roots (Nicoll *et al.*, 2005). This is possibly a mechanical adaptation to the stronger winds, which tend to blow upslope and the shelter effect on the lee-side of a hill.

Differences in root architecture and shape are particularly obvious in trees subjected to a strong prevailing wind (Nicoll and Ray, 1996). Close to the stem on the leeward-side, the roots of conifers tend to develop a shape like a T-beam, which is effectively an extension to the stem buttress. Further away from the stem, and particularly on the windward side, the roots may form shapes similar to engineering I-beams, which place more material further from the neutral axis and are particularly suited to resisting flexing (Fig. 10). This produces stiffer root systems, which reduce the chance of the soil shearing and separating from the roots under wind loading. Root rigidity may be further increased by root fusing wherever there is contact (Mattheck and Breloer, 1995). Crook and Ennos (1997) found that the presence of the tap-root in tap-rooted systems was an important contribution to the overall stability of such trees.

Mechanical adaptation can also take the form of biomass reallocation. In shallow root systems, additional biomass is allocated from the stem to the root system in order to develop lateral growth in an attempt to compensate for a lack of vertical depth (Nicoll and Ray, 1996) (Figs. 11 and 12). As a consequence of a change in wind loading on trees after opening up of

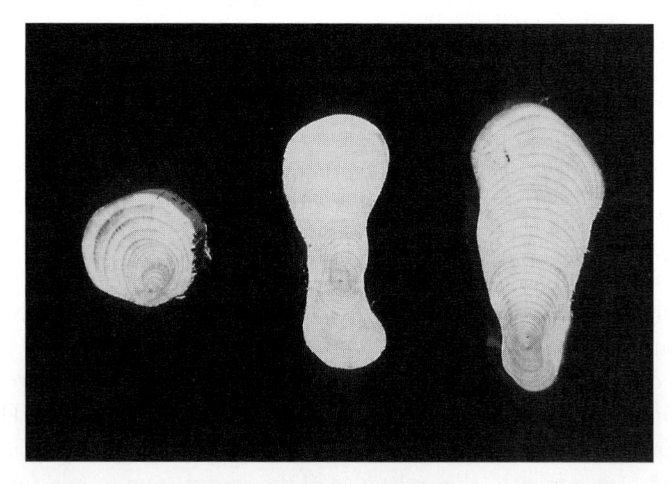

FIG. 10 Examples of the adaptive growth of roots in response to mechanical stresses showing uneven growth leading to roots resembling I beams and T beams.

FIG. 11 Shallow root system of Sitka spruce (*Picea sitchensis*) on a seasonally waterlogged gley soil. Note the development of lateral structural roots as the tree attempts to compensate mechanically for lack of rooting depth; Kershope Forest, north England.

FIG. 12 Deep root system of lodgepole pine (*Pinus contorta*) on a deep peat soil. Lodgepole pine roots can cope with a greater degree of anaerobic conditions than Sitka spruce, enabling it to root more deeply on wet soils; Inchnacardoch Forest, west Scotland.

the canopy, the immediate mechanical response is a strengthening of the root system in an attempt to re-establish root anchorage; this is followed by adjustment to the stem shape a number of years later (Urban *et al.*, 1994). Such adjustments are influenced by competitive status (Mitchell, 2000) and therefore by thinning operations (Ruel *et al.*, 2003). Ideally, from a mechanical point of view, the strength of the stem and root systems would be identical at all points; however, this may not be the case because of restrictions imposed by soil compaction, water table depth, presence of stones, and other factors (Crook and Ennos, 1997). When roots do meet obstacles, such as stones, they may be deflected but will re-establish their original direction once past the obstacle. At points of contact, the root will grow to increase the contact zone in order to minimize contact stresses, which could lead to breakage (Mattheck and Breloer, 1995).

DIRECT CONSEQUENCES

Consequences at Wind Speeds Below Critical (Applied < Resistive)

When the applied force does not exceed the resistive force, the critical wind speed is not reached. Nevertheless, some changes of significance may occur. For example, there may be substantial branch, stem, and root movement. Browning of leaves (desiccation) may occur, particularly where the wind is accompanied by substantial aerosol deposition, such as marine salt or pollutants, or contains other abrasive material, such as sand particles. The clashing of crowns (Rudnicki *et al.*, 2003) may result in the loss of buds and tips of branches, leading to crown shyness. Wind abrasion of terminal shoots can affect subsequent form and growth rate (Wierman and Oliver, 1979; Larson, 1992). If the wind loading becomes too severe, needles and branches will be shed in some trees to reduce the chance of the whole tree being destroyed (Hedden *et al.*, 1995). Branches appear to have a lower safety factor (critical bending moment/bending moment caused by self-weight) than stems in order to act as a "weak link" under extreme loading (Spatz and Bruechert, 2000).

We noted previously that persistent stem and root movement through the action of the wind may stimulate mechanical adaptation or acclimation—including stem thickening, buttressing, and adoption of I-beam or T-beam cross-section by roots (Ruel *et al.*, 2003) (Fig. 10). In some cases, the root-soil interface may shear, without subsequent overturning; this will allow the tree to rock and "pump," contributing to fatiguing of root anchorage. The plastic limit of stem strength may be exceeded and the stem may remain bowed and likely to snap in future events. Compression fractures may form in some bent stems, stimulating the development of reaction wood and callus swelling (Arnold, 2003); such reaction wood may occur repetitively and has been observed as far back as 1930 in trees felled in 1999. Rocking of trees on stony soils can lead to root abrasion (Rizzo and Harrington, 1988).

Consequences of Exceedance of Critical Wind Speed (Applied > Resistive)

Wind throw, or wind snap, occurs in the instant when the applied force within a gust overcomes the anchorage of a tree, or the strength of its stem. Failure of the stem occurs through a variety of modes, and height of failure varies substantially, reflecting points of particular structural weakness (e.g., branch nodes, wounds, rot) or points of localized stress as a consequence of the sway/vertical wind profile. Failure of root anchorage results in overturning. In shallow-rooted trees, this appears to follow movement of the soil/rootplate mass, shear of the rootplate–soil interface, snapping (or extraction of upwind tension roots), and folding of downwind roots under compression (Coutts, 1983; 1986). Deeply rooted trees and broadleaves, with a sharp divide between rootplate and surrounding soil space, may fail through rotation around a downwind point rather than folding (Fig. 13).

FIG. 13 Rotational overturning of broadleaved tree caused by a lack of strong lateral roots; Floors Estate, south Scotland, after storm of January 1990.

The surface area (and mass of soil) disturbed can be substantial. Overturning in sandy soils and highly saturated soils may result in less soil mass being moved.

The proportion of trees damaged by wind throw or wind snap varies according to many site and stand characteristics. When rooting is relatively unconstrained and the stand has self-thinned, there may be roughly equal numbers of overturned and snapped trees—reflecting the adaptive growth of the tree balancing the strength of the various structural elements. Breakage is more common on freely draining soils with deep rooting where the ground is frozen (Peltola *et al.*, 2000), where stem rot is prevalent (Fig. 14) (Barden, 1981), where stem density is high and taper is low, and where additional loading because of snow or ice occurs; breakage may also predominate where there has been a sudden step change in wind speed caused by explosive phenomena, such as volcanic eruptions or certain types of fire storm. Overturning is common on soils where vertical root development is limited, where root rots are present, and where the soil is saturated (e.g., due to heavy rainfall associated with the strong wind event). For example, more than 90% of Sitka spruce growing on wet soils in the British uplands (e.g., Birkley Wood) were overturned (Quine, 2001) whereas at Cascade Head, Oregon, 84% of gaps were caused by stem snap, 12% to wind throw, and 4% to standing death (Taylor, 1990).

Once trees begin to fall or break, the risk to nearby trees, particularly those downwind, will increase either directly, by their being hit by the

FIG. 14 Root rot in Scots pine (*Pinus sylvestris*) leading to stem failure at the root collar.

falling stems, or indirectly, through the creation of gaps that allow increased wind penetration into the canopy. Falling trees can cause canopy loss and stem breakage to neighbors downwind.

Edges that have been long established may represent relatively stable features with relatively stable trees (Fig. 15), whereas trees on recently formed edges will be relatively vulnerable and be more susceptible to disturbance (Fig. 16). This reflects the adaptation of trees to mechanical stimuli, so resistance of trees may differ markedly on a long-established and a new edge, even though aerodynamically they experience similar applied forces. The difference can be such that edge trees can be the sole remaining trees left—or the only trees disturbed!

It has often been assumed that species will exert a significant effect on the likelihood or actual occurrence of windthrow or windsnap. Differences in crown form, root morphology, and stem strength (e.g., wood density) are frequently mentioned. However, rather less attention is typically paid to the potential association between species and site type (with effect of soil strength and wetness), species and stand structure (including effects of species assemblages on canopy form and surface roughness), species and topographic location (affecting strength of wind and thus applied force), or a combination of any of these and others (e.g., frequency of occurrence in a given landscape). As a consequence, many studies are unconvincing, and conflicting statements on the relative stability of different species are not

FIG. 15 A stable edge—long-established edge trees remaining standing after catastrophic damage (including to interior stand trees) in storm of October 1987, Rendlesham Forest, east England.

FIG. 16 An unstable edge—recently exposed edge trees that have since been windthrown (when interior stand trees remained undamaged); Kielder Forest, north England.

hard to find (Meunier *et al.*, 2002, Achim *et al.*, 2005). Rarely are sample sizes sufficient or conditions controlled or characterized enough to identify true species effects.

Dose–Response Relationships

Analysis of risk often considers the relationship between the "dose" of a potentially damaging agent and the response of the affected population. The severity of disturbance will depend upon the vulnerability of the population of trees and the intensity of the event. Wind strength varies substantially over space and time. The resultant extent of disturbance may vary from a few trees (small gaps) (Fig. 17) to patches of several hundred hectares. In particularly extreme cases, a large proportion of the mature forest within a region may be disturbed; studies of catastrophic events in planted forests have shown that 4% to 30% of the mature forest area may be blown over (Holtam, 1971; Grayson, 1989; Quine, 1991) and similar proportions of natural forests have also been observed.

As indicated above, numerous factors interact to determine the critical wind speed. There are very few documented cases of within stand wind speed measurements during a disturbance event. A study in a pine forest in East Anglia recorded gust wind speeds at the canopy top of 17.5 m s^{-1} (28 m s^{-1} at 31 m above ground, 16 m above canopy) on a day when the

FIG. 17 Small gap creation within large areas of Sitka spruce of similar age, possibly reflecting particular site conditions; Glentrool Forest, southwest Scotland.

surrounding stand was damaged (Oliver and Mayhead, 1974); inspection of Meteorological Office records for the day suggests maximum gust speeds of 30 to 32 m s^{-1} at open field stations in the region. Only rarely is it possible to measure repeat disturbance, but Peterson (2000) recorded the passage of two tornadoes: one (F1 tornado) causing 1.5 ha of forest destruction, the second (of F4 magnitude) causing 386 ha of forest destruction.

However, at large spatial scales, it is possible to identify broad dose–response relationships. Studies in the boreal forest have linked treefall with hourly mean wind speeds in excess of 14 m s^{-1} (approximate gust 22 m s^{-1}; with a frequency of 1.4/year) (Liu and Hytteborn, 1991), or hourly mean in excess of 11 m s^{-1} (approximate gust 18 m s^{-1}) (Jonsson and Dynesius, 1993). Gusts of 26 to 32 m s^{-1}were sufficient for small-scale disturbance of temperate deciduous forest and 32 to 39 m s^{-1} for catastrophic events, which occurred approximately three times per century (Pontailler *et al.*, 1997). In Great Britain, the winds recorded at conventional Meteorological Office sites during catastrophic storms in the 20th century have been reviewed (Quine, 1991). In broad terms these indicate that gusts of 35 m s^{-1} experienced at low-lying sites are associated with low percentages of damage (i.e., less than 5% of vulnerable growing stock). Gusts greater than 40 m s^{-1} result in widespread damage (10%–30%); damage is rarely noted when gusts less than 30 m s^{-1} are recorded (Quine, 1991; Mason and Quine, 1995). In studies of gap formation and expansion in planted spruce forests, gust wind speeds were higher (27–33 m s^{-1}) during periods in which new gap formation was observed than periods in which only gap expansion (20–24 m s^{-1}) or no change in gaps (19–22 m s^{-1}) was observed (Quine, 2003).

Nature and Distribution of Disturbance

The scale, nature, and distribution of damage reflect the interaction of the applied and resistive forces, and the spatial coincidence in the variability of these two forces.

At the stand scale, the pattern of disturbance may reflect change in rooting conditions, localized changes in growth rate, edge effects (Fig. 18), rooting conditions (Fig. 19), stand vertical structure, or location of the most damaging gust.

At the forest scale, the pattern of disturbance may be strongly influenced by topography (Fig. 20), pre-existing mosaic (distribution of mature stands), time since past disturbance, and location within the storm field. In managed forests, the location of forest harvesting (clear-cut edges) and other operations (such as thinning) can be influential.

At the regional scale, the pattern of disturbance may reflect topography, forest types (and interaction with other land uses), and location relative to passage of the strong winds (Fig. 21).

FIG. 18 Complex pattern of windthrow reflecting species differences, aspect effects, and stability of long-established edge trees; storm of January 1968, west Scotland.

FIG. 19 Substantial area of snapped trees caused by good root anchorage; Dartmoor, southwest England, following storm of January 1990.

FIG. 20 Topographically determined damage in GlenBranter Forest, west Scotland, in storm of January 1968; highest wind speeds have occurred where interfluve has constricted flow along the valley.

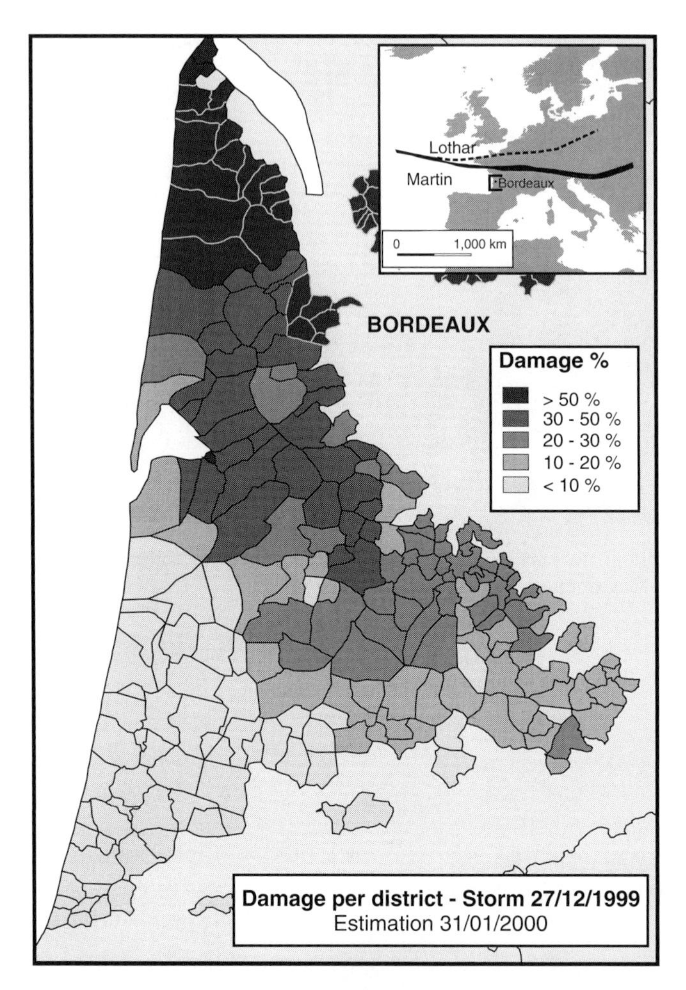

FIG. 21 Map of southwest France showing windthrow in storm Martin of December 1999; the regional pattern of damage reflects the large-scale wind field and location relative to passage of center of the storm; note that the storm Lothar, which passed further north, did not affect the area to the same degree. The forest area has little topography and extensive stands of Maritime pine (*Pinus pinaster*). (Reproduced with permission from *Ann Forest Sci*, 2003, vol 60, pp. 209–226.)

SUBSEQUENT IMPACT OF WINDTHROW, STEM BREAKAGE, AND GAP/PATCH FORMATION

The impact, beyond the immediate disturbance event, will be considered at three scales: the tree, the gap or patch, and the landscape.

Tree Scale Impacts

Individual trees may survive overturning or snap, or may die. Some species appear better able to cope with such disturbance and have developed adaptive mechanisms. A variety of vegetative propagation methods may be stimulated. These include sprouting from snapped and fallen trees, layering from branches of fallen trees, and upturning of original canopy branches. Sprouting is a mechanism that has been frequently observed in tropical forests, for example, after hurricane disturbance. In temperate forests, the evidence is conflicting. Peterson found sprouting unimportant in his study of mixed forests in the eastern United States (Peterson and Pickett, 1991). However, Rebertus and Veblen (1993a) found sprouting important in temperate *Nothofagus* forests in Chile, and Ohkubu *et al.* (1996) identified sprouting from stools important in Japanese beech. Sprouting may be from buds that are at rest, inhibited, or quiescent (Morey, 1973). Sprouting has been observed on 61% to 82% of trees in tropical forests hit by hurricanes (Bellingham *et al.*, 1994; Negrelle, 1995) but on 17% to 25% in temperate forests (Peterson and Pickett, 1991; Bellingham *et al.*, 1996). Some conifers appear to sprout as a form of canopy rebuilding following disturbance (Ishii and Ford, 2001; Ishii and McDowell, 2002; Deal *et al.*, 2003; Quine, 2004). Many species may produce stem sprouts, layered branches, and new adventitious roots following overturning.

Gap/Patch Scale Impacts

Gap formation brings about changes in environmental conditions at the forest floor. Removal of overstory trees leads to increased light levels—and may also affect, either directly or indirectly, other aspects of the microclimate, such as temperature and moisture availability. The exposure of mineral soil may provide a suitable germination site (Hutnik, 1952), and herbaceous vegetation may be suppressed by fallen trees (Everham and Brokaw, 1996). As a result, wind-formed gaps may fill by seedling growth (advanced or germinating after gap formation), by vegetative regrowth of fallen or snapped trees, or by lateral crown development from edge trees.

However, herbivores may also be attracted to open areas (Messier *et al.*, 1999), and a competitive ground flora may also develop, leading to a failure of gap-filling. The relative importance of these mechanisms has been found to vary between forest types because of species ecophysiology, shade tolerance, canopy cover, spacing, and gap size (Coates and Burton, 1997).

The rapidity of response to the gap formation will depend upon the presence of advanced regeneration, or a seedbank, and the frequency of seed production from neighboring canopy trees. The process of windthrow may create particularly suitable seedbeds through soil and vegetation disturbance, resulting in pit and mound microtopography; in some areas of primeval forest, up to 25% of ground surface may be formed of these features (Ulanova, 2000). In time, fallen trees may provide regeneration microsites through the supply of rotting wood (Harmon, 1987; 1989; Harmon and Franklin, 1989). However, shading by rootplates and fallen stems may be detrimental, and some of the substrates may be unstable (e.g., sloughing of old bark from fallen stems).

Gap expansion may be important. Where repeat disturbance occurs (e.g., through temperate storms), many gaps are found to be the result of multiple episodes. For example, 65% of gaps in a boreal forest reflected multiple episodes (Liu and Hytteborn, 1991), 75% in a temperate coniferous forest (Taylor, 1990), and 53% in a temperate deciduous forest (Rebertus and Veblen, 1993a). In a subalpine coniferous forest, half of all gaps expanded within a single year (Perkins *et al.*, 1992). Gap expansion may have particular implications. Observations have suggested that gap expansion can lead to release of regeneration that has formed on the margin of gaps (Lees, 1969; McNeill and Thompson, 1982). Several episodes of gap expansion may be required for the regeneration to reach the canopy (Liu and Hytteborn, 1991). However, a continuous input of dead wood and fallen trees may smother emergent regeneration (Everham and Brokaw, 1996). Expansion may lead to a spread in age or size of regeneration across a gap—for example, as a form of wave (Perkins *et al.*, 1992; Rebertus and Veblen, 1993b). The directional fall of trees, leading to an abrupt edge of standing trees and a "ramp" of fallen trees, may lead to markedly different conditions across the gap (Jackson *et al.*, 2000).

Landscape Scale Impacts

At the landscape (or larger) scale there are a number of consequences. The direct consequence is upon the mosaic of age and stand types represented in the landscape. Repeat disturbance will probably be found in some locations. For example, in temperate maritime regions there is regular stress on the forest, and wind is likely to be a major disturbance agent both as chronic disturbance and occasional acute stand-replacing disturbance. Evidence from documentary sources confirm that wind-induced change to forest structure occurs somewhere in Great Britain at a frequency of approximately once every 3 years. Substantial change occurs at a regional scale once every 10 to 50 years. In *Nothofagus* forests in Chile and spruce-hemlock forests in southeast Alaska, there appear to be wind-created mosaics in extensive areas of natural forest (Rebertus *et al.*, 1997; Kramer *et al.*, 2001). On the most exposed sites, a shifting mosaic with repetitive, but variable, disturbance may be found.

The proportion of stands in the different stages and structure types will reflect the frequency of disturbance, and this is likely to depend upon the topographic and site type variability within the landscape. A study of forests on Kuiu island, southeast Alaska, illustrates the range of stand types that can result (Nowacki and Kramer, 1998). Approximately 35% of the forested landscape reflected sheltered conditions conducive to old-growth development, 15% was very wind prone and dominated by single-cohort stands (with individual patches of 400 ha), and 50% was intermediate and susceptible to occasional extreme events. A study of Glen Affric, west Scotland, using the ForestGALES model to calculate the probability of damage to Scots pine and then inferring resultant stand types, proposed a similar assembly of stand types within a landscape. The model interpretations suggested that 25% of the area may be of the single cohort type, 18% may form "old growth," 42% may be intermediate, and the remaining 15% comprise tree-line woodland or open ground (Quine, 2003b) (Fig. 22). It is likely that landscapes with substantial topographic shelter may contain relatively high proportions of stands with mature trees and fine-scale disturbance may be predominant. In contrast, exposed sites with little shelter, and with uniformly poor soils, may be dominated by young stands resulting from frequent, large-scale disturbance by wind.

There are also important indirect consequences for water quality, site nutrition, and soil development that are due to the soil turnover (and litter fall) and carbon dynamics. The incidence of other types of disturbance may be affected. For example, windthrown trees can contribute to increased fuel loading, leading to enhanced potential for fire disturbance (Myers and van Lear, 1998). Similarly, the volume of downed timber may provide breeding material for bark beetles that then spread to surrounding stands and cause direct effects there (Wichmann and Ravn, 2001).

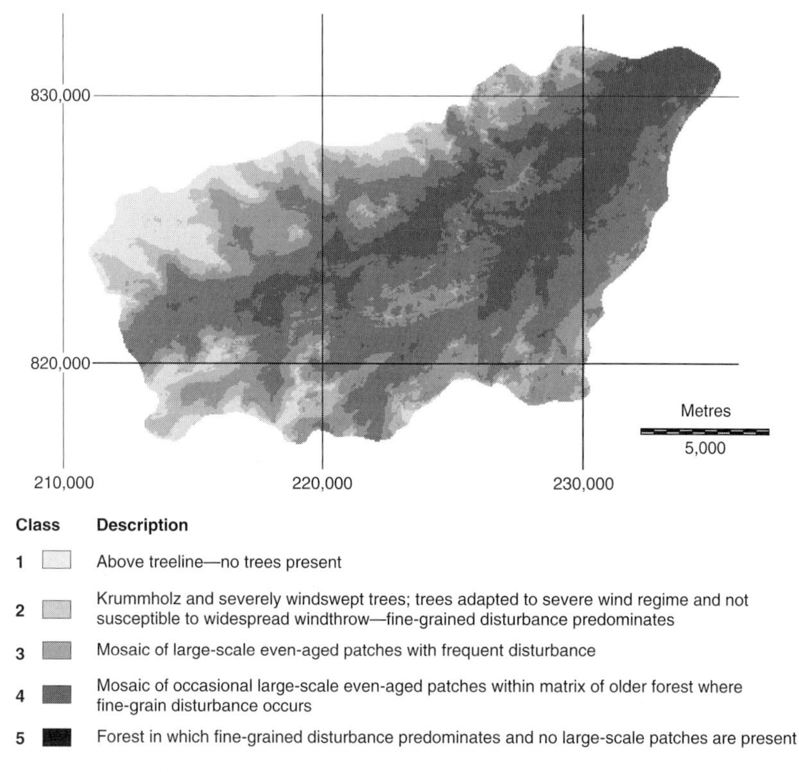

FIG. 22 Modeled disturbance classes, using ForestGALES wind risk model, for Glen Affric semi-natural pine forest in northwest Scotland.

Feedbacks to Likelihood and Character of Subsequent Disturbance

Several potential feedback routes may affect the process–response relationship.

Loss of branches and crown may make trees less vulnerable through reduction in applied force. However, damage to tree stems or roots may make trees more vulnerable through reducing resistive force. There is ample evidence that trees respond to mechanical stress (Telewski, 1995). Stem and root adaptation to loading occurs in response to tree-swaying and is in some way proportional to such stimulus. The magnitude of the difference between the mean and the extreme wind conditions may be influential in determining the degree of mechanical adaptation.

Changes to exposure of trees/stands as a consequence of surrounding damage may change the magnitude of the applied force. Openings in the canopy allow the penetration of strong winds and generate turbulence (Gardiner, 1995), leading to increased crown damage, rocking of trees, and fatiguing of the root/soil interface. Studies with model trees in a wind tunnel have shown that loading of trees on the downwind edge of a gap is substantially greater than that on the upwind edge, which does not differ from the mid forest values. Loading increases very rapidly with increasing size of opening up to a gap diameter of twice the height of the edge trees. Trees on the downwind side of gaps of one tree height in diameter have double the loading of trees in mid forest (Stacey *et al.*, 1994). Response to changing conditions may take some time, such as 8 to 10 years for root recovery.

Change to the forest mosaic (patch formation) will increase the number of unadapted trees that are exposed. Altered exposure may stimulate the biomechanical response of trees. Response to a reduction in stem numbers by thinning (Mitchell, 2000), windthrow, or edge creation has been shown to be slow and involve an initial increased allocation to root growth, followed by stem and crown adaptation (Valinger, 1992; Urban *et al.*, 1994; Lundqvist and Valinger, 1996).

The degree to which these feedback loops are significant or visible or manifest will depend upon the frequency of strong winds (versus any recovery period or time for a new cohort to become vulnerable). Thus, the impact of a storm sequence will depend upon relative rates of removal of vulnerable trees through disturbance and enhanced vulnerability of remaining trees through exposure. Past event history will play a role in determining particular disturbance in an event. Time since last exceedance may have a profound effect on the impact of a particular storm (particularly given that vulnerability is age sensitive)—akin to the "winter harvest" of human populations.

SUMMARY AND CONCLUSIONS

Process and Response

It seems likely that the wind as a disturbance agent has been underemphasized in past research. Strong winds are likely to be highly influential in shaping the structure of forests in maritime temperate (Rebertus *et al.*, 1997; Nowacki and Kramer, 1998) and some tropical regions. For example, in Great Britain there is a substantial documentary record of instances of damage at frequent occasions over the past 300 years and evidence from as early as the 13th century (Linnard, 1982), and in paleoecological material such as stems and stumps found in peat bogs and those exposed in coastal areas (Allen, 1992a; b). In other areas, the frequency may be less (and hence have gone underrecognized), but the individual impact can be substantial—for example, the storm "Lothar" disturbed 195 million m^3 of timber trees in western Europe!

Work on the ecology of forests subjected to wind damage has tended to rely on postdamage analysis to draw conclusions (e.g., Wolf *et al.*, 2004). Although valuable, such an approach relies on a primarily empirical or statistical approach that can only hypothesize on mechanisms of damage. It does not provide any understanding of the processes involved in wind damage. In other words, it is difficult to transfer knowledge gained in one set of circumstances to another situation. Indeed, studies tend to lump together and draw comparisons between descriptions of disturbance relating to very different mechanisms—for example, failing to distinguish storm events where the lee slope effect was likely to be present from those where it was extremely unlikely.

We have attempted to outline the processes and mechanisms of wind and tree interaction that can be used to understand (and model) the likelihood of wind disturbance (windthrow, stem breakage, and canopy gap formation). Knowledge of the cascade of factors that influence the applied and resistive forces and combine to determine whether the critical wind speed is exceeded provides a sound basis for understanding the process and the response.

There has been surprisingly little crossover between studies of wind disturbance ecology and wind risk management—and there is considerable scope for progress. A number of models have been developed to provide tools for managers to predict the risk of wind damage to their forests based on the understanding of tree biomechanics discussed in the preceding sections (Peltola *et al.*, 1999; Gardiner *et al.*, 2000; Gardiner and Quine, 2000). These models can be used to make forecasts of the probable future forest structure (Quine *et al.*, 2002; Hope, 2003). This is possible only by a blending of the separate disciplines of biomechanics, climatology, and ecology. The goal should now be to extend the experience gained in understanding the dynamics of uniform forests to more complex forest structures (whether natural forests or those derived through management).

Further Challenges to Understanding Wind and Tree Process/Response Interactions

Several factors provide a challenge to incorporate, and indeed may frustrate, attempts to explain (or model) damage. For example:

- It has been suggested that the most extreme gusts relate to convective activity, even in extra-tropical cyclones, and involve bundles of fast-moving air coming from aloft; if this occurs, relationships with topography and stand characteristics are likely to be confounded.

- There are often very strong gradients in wind speed at the margins of major storms such that the relationship between disturbance and site/stand characteristics and topography varies. Similarly, the character of snow deposition (wet/dry snow or rainfall) changes with location. The network of surface observations is rarely sufficiently comprehensive to capture this variability.
- There may be many influential factors that are hard to measure before or after any event, including localized differences in soil conditions, extent of rot, time since the wind speed was last exceeded, and even some influential factors that have yet to be identified.
- The infrequency of damage events and the age sensitivity of vulnerability mean that the compilation of a time series of comparable events (e.g., required to determine dose–response relationships) is difficult. Subsequent events operate on a surviving cohort that will differ in many ways from the initial population. Damage propagation in multistoried stands is a particularly challenging topic.

These topics, and many more components identified in the body of the chapter, provide an exciting challenge for future research into our understanding of the interaction of wind and trees. Considerable progress has been made in the past 20 years, and our understanding of the process–response relationship already provides a basis that could be better exploited in ecological studies related to wind disturbance.

ACKNOWLEDGMENTS

We wish to acknowledge a wide network of past and present colleagues and collaborators who have assisted and challenged us over the past 20 years. In particular, it is appropriate to thank Calvin Booth, Andy Neustein, and Mike Coutts, who provided encouragement and wisdom for our initial work. The chapter benefited from the comments of Jean-Claude Ruel, John Moore, Steve Mitchell, and Alexis Achim, and we also thank the editors for their forebearance.

REFERENCES

Aanenson, C. J. M. (1965). *Gales in Yorkshire in February 1962*. Meteorological Office Geophysical Memoirs No. 108 (Vol. 14, No. 3). H.M. Stationery Office, London.

Achim, A., Nicoll, B., Mochan, S., and Gardiner, B. A. (2003). Wind stability of trees on slopes. In: Wind Effects on Trees: Proceedings of the International Conference, University of

Karlsruhe, Germany, September 16-18, 2003. (Ruck, B., Kottmeier, C., Mattheck, C., Quine, C. P., and Wilhelm, G., Eds.), pp. 231–237. University of Karlsruhe, Karlsruhe, Germany.

Achim, A., Ruel, J.-C., Gardiner, B. A., Laflamme, G., and Meunier, S. (2005). Modelling the vulnerability of balsam fir forests to wind damage. *Forest Ecol Manag* 204, 35–50.

Allen, J. R. L. (1992a). Strong winds and windblown trees: a paleoclimatic indicator. *Weather* 47, 410–416.

Allen, J. R. L. (1992b). Trees and their response to wind: mid Flandrian strong winds, Severn Estuary and inner Bristol Channel, southwest Britain. *Phil Trans Roy Soc Lond Series B* 338, 335–364.

Andersen, K. F. (1954). Gales and gale damage to forests, with special reference to the effects of the storm of 31st January 1953, in the north east of Scotland. *Forestry* 27, 97–121.

Arnold, M. (2003). Compression failures in wind-damaged spruce trees. In: Wind Effects on Trees: Proceedings of the International Conference, University of Karlsruhe, Germany, September 16-18, 2003. (Ruck, B., Kottmeier, C., Mattheck, C., Quine, C. P., and Wilhelm, G., Eds.), pp. 253–260. University of Karlsruhe, Karlsruhe, Germany.

Baker, W. L., Flaherty, P. H., Lindemann, J. D., Veblen, T. T., Eisenhart, K. S., and Kulakowski, D. W. (2002). Effect of vegetation on the impact of a severe blowdown in the southern Rocky Mountains, USA. *Forest Ecol Manag* 168, 63–75.

Barden, L. S. (1981). Forest development in canopy gaps of a diverse hardwood forest of the southern Appalachians. *Oikos* 37, 205–209.

Barry, R. G. (1992). *Mountain Weather and Climate*. Routledge, London.

Basnet, K., Likens, G. E., Scatena, F. N., and Lugo, A. E. (1992). Hurricane Hugo: damage to a tropical rain forest in Puerto Rico. *Oecologia* 8, 47–55.

Bellingham, P. J., Kohyama, T., and Aiba, S. (1996). The effects of a typhoon on Japanese warm temperate rainforests. *Ecol Res* 11, 229–247.

Bellingham, P. J., Tanner, E. V. J., and Healey, J. R. (1994). Sprouting of trees in Jamaican montane forests, after a hurricane. *J Ecol* 82, 747–758.

Blackwell, P. G., Rennolls, K., and Coutts, M. P. (1990). A root anchorage model for shallowly rooted Sitka spruce. *Forestry* 63, 73–91.

Boerner, R. E. J., Runge, S. D., Cho, D., and Kooser, J. G. (1988). Localized ice storm damage in an Appalachian Plateau watershed. *Am Midl Nat* 119, 199–208.

Boose, E. R., Foster, D. R., and Fluet, M. (1994). Hurricane impacts to tropical and temperate forest landscapes. *Ecol Monogr* 64, 369–400.

Bruederle, L. P., and Stearns, F. W. (1985). Ice storm damage to a southern Wisconsin mesic forest. *B Torrey Bot Club* 112, 167–175.

Burgert, I., Bernasconi, A., and Eckstein, D. (1999). Evidence for the strength function of rays in living trees. *Holz Roh-Werkstoff* 57, 397–399.

Büsgen, M., Munch, E., and Thomson, T. (1929). *The Structure and Life of Forest Trees*. Chapman and Hall, London.

Canham, C. D., and Loucks, O. L. (1984). Catastrophic windthrow in the presettlement forests of Wisconsin. *Ecology* 65, 803–809.

Chiba, Y. (2000). Modelling stem breakage caused by typhoons in plantation *Cryptomeria japonica* forests. *Forest Ecol Manag* 135, 123–131.

Clinton, B. D., and Baker, C. R. (2000). Catastrophic windthrow in southern Appalachians: characteristics of pits and mounds and initial vegetation responses. *Forest Ecol Manag* 126, 51–60.

Coates, K. D., and Burton, P. J. (1997). A gap-based approach for development of silvicultural systems to address ecosystem management. *Forest Ecol Manag* 99, 337–354.

Cook, N. J. (1985). *The Designer's Guide to Wind Loading of Building Structures. Part 1: Background, Damage Survey, Wind Data and Structural Classification*. Butterworths, London.

Coutts, M. P. (1983). Root architecture and tree stability. *Plant Soil* 71, 171–188.

Coutts, M. P. (1986). Components of tree stability in Sitka spruce on peaty gley soil. *Forestry* 59, 173–197.

Crook, M. J., and Ennos, A. R. (1997). Scaling of anchorage in the tap rooted tree *Mallotus wrayi*. In: *Plant Biomechanics 1997: Conference Proceedings* (Jeronimidis, G., and Vincent, J. F., Eds.), pp. 31–36. Centre for Biomimetics, University of Reading, Reading, UK.

Crook, M. J., Ennos, A. R., and Banks, J. R. (1997). The function of buttress roots: a comparative study of the anchorage systems of buttressed (*Aglaia* and *Nephelium ramboutan* species) and non-buttressed (*Malloyus wrayi*) tropical trees. *J Exp Bot* 48, 1703–1716.

Cucchi, V., and Bert, D. (2003). Wind-firmness in *Pinus pinaster* Ait. stands in southwest France: influence of stand density, fertilisation and breeding in two experimental stands damaged during the 1999 storm. *Ann Forest Sci* 60, 209–226.

Cucchi, V., Meredieu, C., Stokes, A., de Coligny, F., Suarez, J., and Gardiner, B. A. (2005). Modelling the windthrow risk for simulated stands of Maritime pine (*Pinus pinaster* Ait.). *Forest Ecol Manag* 213, 184–196.

Deal, R. L., Barbour, R. J., McClellan, M. H., and Parry, D. L. (2003). Development of epicormic sprouts in Sitka spruce following thinning and pruning in south-east Alaska. *Forestry* 76, 401–412.

Doll, D. (2000). Les ouragans de Noël: premiers bilans. *La Forêt Privée* 251, 23–40.

Dolman, A. J. (1986). Estimates of roughness length and zero plane displacement for a foliated and non-foliated canopy. *Agr Forest Meteorol* 36, 241–248.

Dunn, C. P., Guntenspergen, G. R., and Dorney, J. R. (1982). Catastrophic wind disturbance in an old growth hemlock-hardwood forest, Wisconsin. *Can J Bot* 61, 211–217.

Dyer, J. M., and Baird, P. R. (1997). Wind disturbance in remnant forest stands along the prairie-forest ecotone, Minnesota, USA. *Plant Ecol* 129, 121–134.

ESDU. (1987). *World-wide Extreme Wind Speeds. Part 1: Origins and Methods of Analysis.* Data Item 87034. ESDU International, London.

Everham III, E. M., and Brokaw, N. V. L. (1996). Forest damage and recovery from catastrophic wind. *Bot Rev* 62, 113–185.

Finnigan, J. J., and Brunet, Y. (1995). Turbulent airflow in forest on flat and hilly terrain. In: *Wind and Trees.* (Coutts, M. P. and Grace, J., Eds.) pp. 3–40. Cambridge University Press, Cambridge, UK.

Fischer, A. (1992). Long-term vegetation development in Bavarian Mountain forest ecosystems following natural destruction. *Vegetatio* 103, 93–104.

Flannigan, M. D., Lynham, T. J., and Ward, P. C. (1989). An extensive blowdown occurrence in northwestern Ontario. In: 10th Conference on Fire and Forest Meteorology. (MacIver, D. C., Auld, H., and Whitewood, R., Eds.), pp. 65–71. Forestry Canada, Ottawa, ON.

Flesch, T. K., and Wilson, J. D. (1993). *Extreme Value Analysis of Wind Gusts in Alberta.* Project report FO42-91/107-1993E. Forestry Canada and Alberta Land and Forest Services, Edmonton, AB.

Foster, D. R. (1988). Species and stand response to catastrophic wind in central New England USA. *J Ecol* 76, 135–152.

Fournier, M., Bailleres, H., and Chanson, B. (1994). Tree biomechanics: growth, cumulative prestresses and reorientations. *Biomimetics* 2, 229–252.

Fraser, A. I., and Gardiner, J. B. H. (1967). *Rooting and Tree Stability in Sitka Spruce.* Forestry Commission Bulletin 40. H.M. Stationery Office, London.

Frelich, L. E. (2002). *Forest Dynamics and Disturbance Regimes: Studies from Temperate Evergreen-Deciduous Forests.* Cambridge University Press, Cambridge, UK.

Frelich, L. E., and Lorimer, C. G. (1991). Natural disturbance regimes in the hemlock-hardwood forests of the Upper Great Lakes region. *Ecol Monogr* 61, 145–164.

Gardiner, B. A. (1989). *Mechanical Characteristics of Sitka Spruce.* Occasional Paper 24. Forestry Commission, Edinburgh, Scotland.

Gardiner, B. A. (1992). Mathematical modelling of the static and dynamic characteristics of plantation trees. In: *Mathematical Modelling of Forest Ecosystems: Proceedings of a Workshop Organized by Forstliche Versuchsanstalt Rheinland-Pfalz and Zentrum für Praktische Mathematik, Lambrecht/Pfalz, May 27-31, 1991.* (Franke, J., and Roeder, A., Eds.), pp. 40–61. J. D. Sauerländer's Verlag, Frankfurt, Germany.

Gardiner, B. A. (1995). The interactions of wind and tree movement in forest canopies. In: *Wind and Trees.* (Coutts, M. P. and Grace, J., Eds.), pp. 41–59. Cambridge University Press, Cambridge, UK.

Gardiner, B. A., and Quine, C. P. (2000). Management of forests to reduce the risk of abiotic damage—a review with particular reference to the effects of strong winds. *Forest Ecol Manag* 135, 261–277.

Gardiner, B. A., Marshall, B. J., Achim, A., Belcher, R. E., and Wood, C. J. (2005). The stability of different silvicultural systems: a wind tunnel investigation. *Forestry* 78, 471–484.

Gardiner, B. A., Peltola, H., and Kellomäki, S. (2000). Comparison of two models for predicting the critical wind speeds required to damage coniferous trees. *Ecol Model* 129, 1–23.

Gardiner, B. A., Stacey, G. R., Belcher, R. E., and Wood, C. J. (1997). Field and wind-tunnel assessment of the implications of respacing and thinning on tree stability. *Forestry* 70, 233–252.

Glitzenstein, J. S., and Harcombe, P. A. (1988). Effects of the December 1983 tornado on forest vegetation of the Big Thicket, southeast Texas, USA. *Forest Ecol Manag* 25, 269–290.

Gordon, J. E. (1991). *Structures, or, Why Things Don't Fall Down.* Penguin, London.

Grayson, A. J. (1989). *The 1987 Storm: Impact and Responses.* Forestry Commission Bulletin 87. H. M. Stationery Office, London.

Guild, D. W. (1986). Snow damage in plantation forests in southern New Zealand. *N Z J Forest* 31, 9–14.

Hannah, P., Palutikof, J. P., and Quine, C. P. (1995). Predicting windspeeds for forest areas in complex terrain. In: *Wind and Trees* (Coutts, M. P. and Grace, J., Eds.), pp. 113–129. Cambridge University Press, Cambridge, UK.

Harmon, M. E. (1987). The influence of litter and humus accumulations and canopy openness on *Picea sitchensis* (Bong.) Carr. and *Tsuga heterophylla* (Raf.) Sarg. seedlings growing on logs. *Can J Forest Res* 17, 1475–1479.

Harmon, M. E. (1989). Retention of needles and seeds on logs in *Picea sitchensis-Tsuga heterophylla* forests of coastal Oregon and Washington. *Can J Bot* 67, 1833–1837.

Harmon, M. E., and Franklin, J. F. (1989). Tree seedlings on logs in *Picea-Tsuga* forests of Oregon and Washington. *Ecology* 70, 48–59.

Harrington, J. B., and Newark, M. J. (1986). The interaction of a tornado with rough terrain. *Weather* 41, 310–318.

Harrington, T. B., and Bluhm, A. A. (2001). Tree regeneration responses to microsite characteristics following a severe tornado in the Georgia Piedmont, USA. *Forest Ecol Manag* 140, 265–275.

Harris, A. S., and Farr, W. A. (1974). *The Forest Ecosystem of Southeast Alaska: 7. Forest Ecology and Timber Management.* General Technical Report PNW-25. USDA Forest Service, Pacific Northwest Research Station, Portland, OR.

Haymond, J. L., Hook, D. D., and Harms, W. R. (1996). *Hurricane Hugo: South Carolina Forest Land Research and Management Related to the Storm.* General Technical Report SRS-5. USDA Forest Service, Southern Research Station, Asheville, NC.

Hedden, R. L., Fredericksen, T. S., and Williams, S. A. (1995). Modelling the effect of crown shedding and streamlining on the survival of loblolly pine exposed to acute wind. *Can J Forest Res* 25, 704–712.

Holtam, B. W. (Ed.). (1971). *Windblow of Scottish Forests in January 1968.* Forestry Commission Bulletin 45. H. M. Stationery Office, London.

Hope, J. E. (2003). Dynamic modelling of a forest ecosystem. In: *The Potential of Applied Landscape Ecology to Forest Design Planning*. (Bell, S., Ed.), pp. 71–86. Forestry Commission, Edinburgh, Scotland.

Hunt, J. C. R. (1995). The contribution of meteorological science to wind hazard mitigation. In: *Windstorm: Coming to Terms with Man's Worst Natural Hazard*. pp. 9–17. The Royal Academy of Engineering, London.

Hutnik, R. J. (1952). Reproduction on windfalls in a northern hardwood stand. *J Forestry* 50, 693–694.

Ida, H., and Nakagashi, N. (1998). A large gap formation in a beech forest on Mt. Garyu in southwestern Japan by Typhoon 9119. *J Sustain Manag* 6, 237–250.

Ishii, H., and Ford, E. D. (2001). The role of epicormic shoot production in maintaining foliage in old *Pseudotsuga menziesii* (Douglas-fir) trees. *Can J Bot* 79, 251–264.

Ishii, H., and McDowell, N. (2002). Age-related development of crown structure in coastal Douglas-fir trees. *Forest Ecol Manag* 169, 257–270.

Ishizuka, M., Toyooka, H., Osawa, A., Kushima, H., Kanazawa, Y., and Sato, A. (1998). Secondary succession following catastrophic windthrow in a boreal forest in Hokkaido, Japan: the timing of tree establishment. *J Sustain Manag* 6, 367.

Jackson, R. G., Foody, G. M., and Quine, C. P. (2000). Characterising windthrown gaps from fine spatial resolution remotely sensed data. *Forest Ecol Manag* 135, 253–260.

Jakobsen, B. (1986). [Abiotic injuries in Danish forests in the period 1890-1984.] *Forstlige Forsogsvaesen: Danmark* 40, 213–224.

Jane, G. T. (1986). Wind damage as an ecological process in mountain beech forests of Canterbury, New Zealand. New Zeal. *J Ecol* 9, 25–39.

Jones, H. G. (1983). *Plants and Microclimate: A Quantitative Approach to Environmental Plant Physiology*. Cambridge University Press, Cambridge, UK.

Jonsson, B. G., and Dynesius, M. (1993). Uprooting in boreal spruce forests: long-term variation in disturbance rate. *Can J Forest Res* 23, 2383–2388.

Kearsley, J. B., and Jackson, S. T. (1997). History of a *Pinus strobus*-dominated stand in northern New York. *J Veg Sci* 8, 425–436.

Kerzenmacher, T., and Gardiner, B. A. (1998). A mathematical model to describe the dynamic response of a spruce tree to the wind within a forest. *Trees* 12, 385–394.

Kramer, M. G., Hansen, A. J., Taper, M. L., and Kissinger, E. J. (2001). Abiotic controls on long-term windthrow disturbance and temperate rain forest dynamics in southeast Alaska. *Ecology* 82, 2749–2768.

Kulakowski, D. W., and Veblen, T. T. (2002). Influences of fire history and topography on the pattern of a severe wind blowdown in a Colorado subalpine forest. *J Ecol* 90, 806–819.

Lamb, H. H. (1991). *Historic Storms of the North Sea, British Isles and Northwest Europe*. Cambridge University Press, Cambridge, UK.

Larson, B. C. (1992). Pathways of development in mixed-species stands. In: *The Ecology and Silviculture of Mixed-Species Forests* (Kelty, M. J., Larson, B. C., and Oliver, C. D., Eds.), pp. 3–10. Kluwer, Dordrecht, Netherlands.

Lässig R., and Mocalov, S. A. (2000). Frequency and characteristics of severe storms in the Urals and their influence on the development, structure and management of the boreal forests. *Forest Ecol Manag* 135, 179–194.

Lavers, G. M. (1969). *The Strength Properties of Timbers*. Forest Products Research Bulletin 50, Second Edition. H. M. Stationery Office, London.

Lavnyy, V., and Lässig, R. (2003). Extent of storms in the Ukrainian Carpathians. In: *Wind Effects on Trees: Proceedings of the International Conference, University of Karlsruhe, Germany, September 16-18, 2003*. (Ruck, B., Kottmeier, C., Mattheck, C., Quine, C. P., and Wilhelm, G., Eds.), pp. 341–348. University of Karlsruhe, Karlsruhe, Germany.

Lees, J. C. (1969). *Spruce Regeneration on Wet Forest Sites*. PhD dissertation. University of Edinburgh, Edinburgh, Scotland.

Lehmann, A., and Laflamme, J. (1975). [Tornados in Quebec: research made from the study of fallen wood areas.] Les tornades au Québec: recherche a partir de Chablis. *Rev G Montr* 29, 357–366.

Linnard, W. (1982). *Welsh Woods and Forests: History and Utilization*. National Museum of Wales, Cardiff, UK.

Liu, C., Glitzenstein, J. S., Harcombe, P. A., and Knox, R. G. (1997). Tornado and fire effects on tree species composition in a savanna in the Big Thicket National Preserve, southeast Texas, USA. *Forest Ecol Manag* 91, 279–289.

Liu, Q., and Hytteborn, H. (1991). Gap structure, disturbance and regeneration in a primeval *Picea abies* forest. *J Veg Sci* 2, 391–402.

Lundqvist, L., and Valinger, E. (1996). Stem diameter growth of Scots pine trees after increased mechanical load in the crown during dormancy and (or) growth. *Ann Bot* 77, 59–62.

Lyakhov, M. Y. (1987). Tornadoes in the midland belt of Russia. *Sov Geogr* 1986, 562–570.

Mason, W. L., and Quine, C. P. (1995). Silvicultural possibilities for increasing structural diversity in British spruce forests: the case of Kielder forest. *Forest Ecol Manag* 79, 13–28.

Mattheck, C. (1991). *Trees: The Mechanical Design*. Springer-Verlag, Munich, Germany.

Mattheck, C., and Breloer, H. (1995). *The Body Language of Trees: A Handbook of Failure Analysis*. H. M. Stationery Office, London.

Mayhead, G. J. (1973). Some drag coefficients for British forest tree derived from wind tunnel studies. *Agr Meteorol* 12, 123–130.

McNeill, J. D., and Thompson, D. A. (1982). Natural regeneration of Sitka spruce in the Forest of Ae. *Scot Forest* 36, 269–282.

Meguro, S., and Miyawaki, A. (1994). A study of the relationship between mechanical characteristics and coastal vegetation among several broadleaf trees in Miura Peninsula in Japan. *Vegetatio* 112, 101–111.

Messier, C., Zasada, J., and Greene, D. (1999). Workshop on functional aspects of regeneration of the boreal forest in the context of sustainable forest management: preamble. *Can J Forest Res* 29, 791–795.

Meunier, S., Ruel, J.-C., and Achim, A. (2002). Résistance comparée de l'épinette blanche et du sapin baumier au renversement. *Can J Forest Res* 32, 642–652.

Mikulowski, M., and Gil, W. (2003). Wind-induced damage to Polish forests and the methods of mitigating its effect. In: *Wind Effects on Trees: Proceedings of the International Conference, University of Karlsruhe, Germany, September 16-18, 2003*. (Ruck, B., Kottmeier, C., Mattheck, C., Quine, C.P., and Wilhelm, G., Eds.), pp. 349–356. University of Karlsruhe, Karlsruhe, Germany.

Miller, K. F., Quine, C. P., and Hunt, J. (1987). The assessment of wind exposure for forestry in upland Britain. *Forestry* 60, 179–192.

Mitchell, S. J. (2000). Stem growth responses in Douglas-fir and Sitka spruce following thinning: implications for assessing wind-firmness. *Forest Ecol Manag* 135, 105–114.

Moore, J. R. (2000). Differences in maximum resistive bending moments of *Pinus radiata* trees grown on a range of soil types. *Forest Ecol Manag* 135:63–71.

Moore, J. R. (2002). *Mechanical Behavior of Coniferous Trees Subjected to Wind Loading*. Ph.D. dissertation. Oregon State University, Corvallis, OR.

Moore, J. R., and Maguire, D. A. (2004). Natural sway frequencies and damping ratios of trees: concepts, review and synthesis of previous studies. *Trees-Struct Funct* 18, 195–203.

Moore, J. R., and Maguire, D. A. (2005). Natural sway frequencies and damping ratios of trees: influence of crown structure. *Trees-Struct Funct* 19, 363–373.

Morey, P. R. (1973). *How Trees Grow*. Edward Arnold, London.

Morgan, J., and Cannell, M. G. R. (1994). Shape of tree stems—a re-examination of the uniform stress hypothesis. *Tree Physiol* 14, 49–62.

Morse, A. P., Gardiner, B. A., and Marshall, B. J. (2002). Mechanisms controlling turbulence development across a forest edge. *Bound Lay Meteorol* 103, 227–251.

Mulder, D., Delorme, A., Wujaciak, R., Haberle, S., and Waldschmidt, M. (1973). The Guttingen Conference of forest storm damage February 1973. *Forestarchiv* 44, 41–75.

Myers, R. K., and van Lear, D. H. (1998). Hurricane-fire interactions in coastal forests of the south: a review and hypothesis. *Forest Ecol Manag* 103, 265–276.

Nakashizuka, T., and Iida, S. (1995). Composition, dynamics and disturbance regime of temperate deciduous forests in monsoon Asia. *Vegetatio* 121, 23–30.

Negrelle, R. R. B. (1995). Sprouting after uprooting of canopy trees in the Atlantic rain forest of Brazil. *Biotropica* 27, 448–454.

Nicoll, B. C., and Ray, D. (1996). Adaptive growth of tree root systems in response to wind action and site conditions. *Tree Physiol* 16, 891–898.

Nicoll, B. C., Achim, A., Mochan, S., and Gardiner, B. A. (2005). Does steep terrain influence tree stability? A field investigation. *Can J Forest Res* 35, 2360–2367.

Nicoll, B. C., Gardiner, B. A., Rayner, B., and Peace, A. J. (2006). Anchorage of coniferous trees in relation to species, soil type and rooting depth. *Can J Forest Res* 36, 1871–1883.

Nowacki, G. J., and Kramer, M. G. (1998). *The Effects of Wind Disturbance on Temperate Rain Forest Structure and Dynamics of Southeast Alaska.* General Technical Report PNW-GTR-421. USDA Forest Service, Pacific Northwest Research Station, Portland, OR.

Nykänen, M.-L., Peltola, H., Quine, C. P., Kellomäki, S., and Broadgate, M. (1997). Factors affecting snow damage of trees with particular reference to European conditions. *Silva Fenn* 31, 192–213.

Ohkubo, T., Tanimoto, T., and Peters, R. (1996). Response of Japanese beech (*Fagus japonica* Maxim.) sprouts to canopy gaps. *Vegetatio* 124, 1–8.

Oliver, H. R., and Mayhead, G. J. (1974). Wind measurements in a pine forest during a destructive gale. *Forestry* 47, 185–195.

Orwig, D. A., and Abrams, M. D. (1995). Dendroecological and ecophysiological analysis of gap environments in mixed-oak understories of northern Virginia. *Funct Ecol* 9, 799–806.

Palutikof, J., Holt, T., and Skellern, A. (1997). Wind: resource and hazard. In: *Climate of the British Isles: Present, Past and Future.* (Hulme, M., and Barrow, E., Eds.), pp. 220–242. Routledge, London.

Pavlik, M., and Pavlik, S. (2003). Gale disasters in Slovakia: consequences and management implications. In: *Wind Effects on Trees: Proceedings of the International Conference, University of Karlsruhe, Germany, September 16-18, 2003.* (Ruck, B., Kottmeier, C., Mattheck, C., Quine, C. P., and Wilhelm, G., Eds.), pp. 199–206. University of Karlsruhe, Karlsruhe, Germany.

Peltola, H., Kellomäki, S., Hassinen, A., and Granander, M. (2000). Mechanical stability of Scots pine, Norway spruce and birch: an analysis of tree-pulling experiments in Finland. *Forest Ecol Manag* 135, 143–153.

Peltola, H., Kellomäki, S., Väisänen, H., and Ikonen, V.-P. (1999). A mechanistic model for assessing the risk of wind and snow damage to single trees and stands of Scots pine, Norway spruce, and birch. *Can J Forest Res* 29, 647–661.

Perkins, T. D., Klein, R. M., Badger, G. J., and Easter, M. J. (1992). Spruce-fir decline and gap dynamics on Camels Hump, Vermont. *Can J Forest Res* 22, 413–422.

Peters, R. (1998). Dynamics of two Japanese *Fagus* forests: the relation between stem growth and local wind climates. *J Sustain Manag* 6, 267–280.

Peterson, C. J. (2000). Damage and recovery of tree species after two different tornadoes in the same old growth forest: a comparison of infrequent wind disturbances. *Forest Ecol Manag* 135, 237–252.

Peterson, C. J., and Pickett, S. T. A. (1991). Treefall and resprouting following catastrophic windthrow in an old-growth hemlock-hardwoods forest. *Forest Ecol Manag* 42, 205–217.

Peterson, C. J., and Pickett, S. T. A. (1995). Forest reorganization: a case study in an old-growth forest catastrophic blowdown. *Ecology* 76, 763–774.

Petty, J. A., and Swain, C. (1985). Factors influencing stem breakage of conifers in high winds. *Forestry* 58, 75–84.

Petty, J. A., and Worrell, R. (1981). Stability of coniferous tree stems in relation to damage by snow. *Forestry* 54, 115–128.

Pontailler, J.-Y., Faille, A., and Lemée, G. (1997). Storms drive successional dynamics in natural forests: a case study in Fontainebleau forest (France). *Forest Ecol Manag* 98, 1–15.

Quine, C. P. (1991). Recent storm damage to trees and woodlands in southern Britain. In: *Research for Practical Arboriculture.* (Hodge, S. J., Ed.), pp. 83–94. Forestry Commission Bulletin 93. H. M. Stationery Office, London.

Quine, C. P. (1995). Assessing the risk of wind damage to forests: practice and pitfalls. In: *Wind and Trees.* (Coutts, M. P., and Grace, J., Eds.), pp. 379–403. Cambridge University Press, Cambridge, UK.

Quine, C. P. (2001). *The Role of Wind in the Ecology and Naturalisation of Sitka Spruce in Upland Britain.* PhD dissertation. University of Edinburgh, Edinburgh, Scotland.

Quine, C. P. (2003a). Wind-driven gap formation and expansion in spruce forests of upland Britain. In: *Wind Effects on Trees: Proceedings of the International Conference, University of Karlsruhe, Germany, September 16-18, 2003.* (Ruck, B., Kottmeier, C., Mattheck, C., Quine, C. P. and Wilhelm, G., Eds.), pp. 101–108. University of Karlsruhe, Karlsruhe, Germany.

Quine, C. P. (2003b). Wind as a forest disturbance agent—and its implications for forest landscape patterns at Glen Affric. In: *The Potential of Applied Landscape Ecology to Forest Design Planning.* (Bell, S., Ed.), pp. 55–62. Forestry Commission, Edinburgh, Scotland.

Quine, C. P. (2004). Development of epicormic sprouts on Sitka spruce stems in response to windthrown gap formation. *Forestry* 77, 225–233.

Quine, C. P., and White, I. M. S. (1994). Using the relationship between rate of tatter and topographic variables to predict site windiness in upland Britain. *Forestry* 67, 245–256.

Quine, C. P., Humphrey, J. W., and Ferris, R. (1999). Should the wind disturbance patterns observed in natural forests be mimicked in planted forests in the British uplands? *Forestry* 72, 337–358.

Quine, C. P., Humphrey, J. W., Purdy, K., and Ray, D. (2002). An approach to predicting the potential forest composition and disturbance regime for a highly modified landscape: a pilot study of Strathdon in the Scottish Highlands. *Silva Fenn* 36, 233–247.

Raupach, M. R. (1992). Drag and drag partitioning on rough surfaces. *Bound Lay Meteorol* 60, 375–395.

Ray, D., and Nicoll, B. C. (1998). The effect of soil water-table depth on root-plate development and stability of Sitka spruce. *Forestry* 71, 169–182.

Rebertus, A. J., and Veblen, T. T. (1993a). Structure and treefall gap dynamics of old-growth *Nothofagus* forests on Tierra del Fuego, Argentina. *J Veg Sci* 4, 641–653.

Rebertus, A. J., and Veblen, T. T. (1993b). Partial wave formation in old-growth *Nothofagus* forests on Tierra del Fuego, Argentina. *B Torrey Bot Club* 120, 461–470.

Rebertus, A. J., Kitzberger, T., Veblen, T. T., and Roovers, L. M. (1997). Blowdown history and landscape patterns in the Andes of Tierra del Fuego, Argentina. *Ecology* 78, 678–692.

Rizzo, D. M., and Harrington, T. C. (1988). Root movement and root damage of red spruce and balsam fir on subalpine sites in the White Mountains, New Hampshire. *Can J Forest Res* 18, 991–1001.

Robinson, M. (1999). The consistency of long-term climate datasets: two UK examples of the need for caution. *Weather* 54, 2–9.

Rudnicki, M., Lieffers, V. J., and Silins, U. (2003). Stand structure governs the crown collisions of lodgepole pine. *Can J Forest Res* 33, 1238–1244.

Rudnicki, M., Mitchell, S. J., and Novak, M. D. (2004). Wind tunnel measurements of crown streamlining and drag relationships for three conifer species. *Can J Forest Res* 34, 666–676.
Ruel, J.-C., and Benoit, R. (1999). Analyse du chablis du 7 novembre 1994 dans les régions de Charlevoix et de la Gaspésie, Québec, Canada. *Forest Chron* 75, 293–301.
Ruel, J.-C., Larouche, C., and Achim, A. (2003). Changes in root morphology after precommercial thinning in balsam fir stands. *Can J Forest Res* 33, 2452–2459.
Ruel, J.-C., Pin, D., and Cooper, K. (1998). Effect of topography on wind behaviour in a complex terrain. *Forestry* 71, 261–265.
Ruel, J.-C., Pin, D., Spacek, L., Cooper, K., and Benoit, R. (1997). The estimation of wind exposure for windthrow hazard rating: comparison between Strongblow, MC2, Topex and a wind tunnel study. *Forestry* 70, 253–266.
Russell, K. (1982). *Deterioration of Blowdown Timber on the Olympic Peninsula from the Lincoln Day Storm*. Department of Natural Resources Note 36. State of Washington, Olympia, WA.
Schönenberger, W., Fischer, A., and Innes, J. L. (2003). Vivian's legacy in Switzerland—impact of windthrow on forest dynamics. *Forest Snow Landscape Res* 77, 1–224.
Shaw, W. B. (1983). Tropical cyclones: determinants of pattern and structure in New Zealand indigenous forests. *Pac Sci* 37, 405–414.
Sheehan, P. G., Lavery, P. B., and Walsh, B. M. (1982). Thinning and salvage strategies in plantations prone to storm damage: case study of radiata pine plantations in Owens Valley, Victoria. New Zeal. *J Forest Sci* 12, 269–280.
Sigmon, J. T., Knoerr, K. R., and Shaughnessy, E. J. (1984). Leaf emergence and flow-through effects on mean windspeed profiles and microscale pressure fluctuations in a deciduous forest. *Agr Forest Meteorol.* 31, 329–337.
Smith, V. G., Watts, M., and James, D. F. (1987). Mechanical stability of black spruce in the clay belt region of northern Ontario. *Can J Forest Res* 17, 1080–1089.
Somerville, A. (1989). Tree wind stability and forest management practices. In: *Workshop on Wind Damage in New Zealand Exotic Forests: Proceedings*. (Somerville, A., and Wakelin, S., Eds.), pp. 38–58. Forest Research Institute Bulletin 146. Ministry of Forestry, Rotorua, New Zealand.
Spatz, H.-C., and Bruechert, F. (2000). Basic biomechanics of self-supporting plants: wind loads and gravitational loads on a Norway spruce tree. *Forest Ecol Manag* 135, 33–44.
Stacey, G. R., Belcher, R. E., Wood, C. J., and Gardiner, B. A. (1994). Wind and wind forces in a model spruce forest. *Bound Lay Meteorol* 69, 311–334.
Stull, R. B. (1988). *An Introduction to Boundary Layer Meteorology*. Kluwer Academic Publishers, Boston, MA.
Suárez, J. C., Gardiner, B. A., and Quine, C. P. (1999). A comparison of three methods for predicting wind speeds in complex forested terrain. *Meteorol Appl* 6, 1–14.
Syrjanen, K., Kalliola, R., Puolasmaa, A., and Mattson, J. (1994). Landscape structure and forest dynamics in subcontinental Russian European taiga. *Ann Zool Fenn* 31, 19–34.
Tanner, E. V. J., Kapos, V., and Healey, J. R. (1991). Hurricane effects on forest ecosystems in the Caribbean. *Biotropica* 23, 513–521.
Taylor, A. H. (1990). Disturbance and persistence of Sitka spruce (*Picea sitchensis* (Bong.) Carr.) in coastal forests of the Pacific Northwest, North America. *J Biogeogr* 17, 47–58.
Telewski, F. W. (1995). Wind-induced physiological and developmental responses in trees. In: *Wind and Trees*. (Coutts, M. P., and Grace, J., Eds.), pp. 237–263. Cambridge University Press, Cambridge, UK.
Thom, A. S. (1971). Momentum absorption by vegetation. *Q J R Meteor Soc* 97, 414–428.
Troen, I., and Petersen, E. L. (1989). *European Wind Atlas*. Published for the Commission of the European Communities, Directorate-General for Science, Research, and Development, Brussels, Belgium by Risø National Laboratory, Roskilde, Denmark.

Ulanova, N. G. (2000). The effects of windthrow on forests at different spatial scales: a review. *Forest Ecol Manag* 135, 155–167.

Unwin, G. L., Appelgate, G. B., Stocker, G. C., and Nicholson, D. I. (1988). Initial effects of Tropical Cyclone "Winifred" on forests in north Queensland. *P Ecol Soc Aust* 15, 283–296.

Urban, S. T., Lieffers, V. J., and Macdonald, S. E. (1994). Release in radial growth in the trunk and structural roots of white spruce as measured by dendrochronology. *Can J Forest Res* 24, 1550–1556.

Valinger, E. (1992). Effects of wind sway on stem form and crown development of Scots pine (*Pinus sylvestris* L.). *Aust Forest* 55, 15–22.

Vollsinger, S., Mitchell, S. J., Byrne, K. E., Novak, M. D., and Rudnicki, M. (2005). Wind tunnel measurements of crown streamlining and drag relationships for several hardwood species. *Can. J. Forest Res.* 35, 1238–1249.

Wichmann, L., and Ravn, H. P. (2001). The spread of *Ips typographus* (L.) (Coleoptera, Scolytidae) attacks following heavy windthrow in Denmark, analysed using GIS. *Forest Ecol Manag* 148, 31–39.

Wierman, C. A., and Oliver, C. D. (1979). Crown stratification by species in even-aged mixed stands of Douglas fir/western hemlock. *Can J Forest Res* 9, 1–9.

Wolf, A., Møller, P. F., Bradshaw, R. H. W., and Bigler, J. (2004). Storm damage and long-term mortality in a semi-natural, temperate deciduous forest. *Forest Ecol Manag* 188, 197–210.

Wyatt, T. A. (1995). Risk analysis: tropical and temperate climates contrasted. In: *Windstorm: Coming to Terms with Man's Worst Natural Hazard*. pp. 27–33. The Royal Academy of Engineering, London.

Zipse, A., Mattheck, C., Gräbe, D., and Gardiner, B. A. (1998). The effect of wind on the mechanical properties of wood in Scottish beech trees. *Arbor J* 22, 247–257.

APPENDIX 1

GLOSSARY AND DEFINITIONS

Aerodynamic roughness (z_0)
The height at which the wind speed is taken as zero and a measure of the effectiveness of the surface in absorbing momentum.

Anchorage
The complex of mechanisms by which the root system and soil resist the wind forces on the stem and crown.

Bending moment
The torque (force × distance) exerted by the wind on a tree. The bending moment is largest at the base of the tree.

Center of pressure
The average position in the crown of the tree where the total force of the wind can be said to act.

Coherent gusts
Organized rotational motions in the air (= vortices).

Critical wind speed
The wind speed at which damage to a tree from either breakage or overturning occurs.

Damping
The processes by which oscillations are reduced in size and tend to stop. Damping includes canopy clashing, canopy drag through the air, and frictional movement of stem fibers.

Displacement height (d)
The effective ground surface over vegetation to airflow.

Drag area
The surface area of the tree (canopy and stem) presented to the wind. Drag area can be reduced as windspeed increases due to streamlining of the tree.

Drag coefficient
A dimensionless number reflecting the proportion of the cross-sectional area of a body that contributes to its drag.

Drag force
The force on the tree caused by the pressure exerted by the wind on the crown (= wind-loading).

Frequency of oscillation
The number of sway cycles of the tree per second.

Fulcrum
The position on the lee side of the tree where the root system pivots when the tree is bent by the wind (= hinge).

Gust

A rapid increase in windspeed over a short period of time (seconds rather than minutes).

Gustiness (G)

A measure of the ratio of the extreme to mean wind loading on trees.

Hinge

See Fulcrum.

Height/diameter ratio

A measure of the taper of a tree. Often taken to indicate stability with tapers less than 80 regarded as signifying tree stability.

Leaf area

The total area provided by the foliage of a tree and likely to provide drag to the wind.

Leeward

The side of the tree/stand facing away from the wind.

Lever

The distance between the point of action of a force and the fulcrum.

Modulus of elasticity

A measure of the stiffness of a material; the rate of change of stress with a change in strain. In wood, modulus of elasticity depends on the direction of interest because wood is anisotropic.

Modulus of rupture

The stress in the outer fibers of the tree at which the fibers rupture and the tree fails.

Overturning moment

The force on the tree multiplied by the distance from where the force acts (the center of pressure) to the fulcrum, plus the additional moment due to the weight of the overhanging crown.

Return period

The average number of years between storms of sufficient intensity to damage the tree in question.

Risk (for a tree)

The probability in a particular year of the threshold wind speed being exceeded (see Vulnerability).

Root anchorage

The combination of root and soil that anchors a tree. The resistance to uprooting is a combination of the tension in the windward roots, the resistance of the root/soil plate to bending, the weight of the root plate, and the strength of the soil at the base of the root plate.

Root architecture
The appearance and structure of the root system, particularly the number and arrangement in three dimensions of the thickest roots.

Snap
Failure of a tree when the outer fibers break and the stem of the tree breaks. Snap is usually a combination of wood fiber breakage and delamination.

Strain
The ratio of extension to length when a material is put under load.

Stress
The ratio of force to cross-sectional area when a material is put under load.

Temperature inversion
A zone or layer in the atmosphere where the temperature increases with height. In windy conditions the air beneath the inversion may be confined and thus accelerated over mountains.

Turbulence
The random variations in wind speed and direction.

Vortices
See Coherent gusts.

Vulnerability (of a tree)
The threshold wind speed required to blow over a particular tree on a particular site.

Wind loading
See Drag Force.

Windthrow
The overturning of a tree when the wind induced moments at the base of the tree are greater than the resistance offered by the root system. On sites with poor rooting, windthrow is the normal mode of failure but on deeper rooted sites both windthrow and snap will occur.

Windward
The side of the tree facing towards the wind.

5

Meteorological Conditions Associated with Ice Storm Damage to Forests

Kaz Higuchi and Amir Shabbar

Environment Canada

INTRODUCTION

During a typical winter, freezing rain can occur several times, usually associated with synoptic weather systems, over the northeastern United States and southwestern Canada. Impact of a freezing rain event depends on its intensity and duration. Over the past several decades, a countable number of major freezing events have struck the region; the most damaging by far in recent years was the large ice storm that occurred in early January 1998.

For a period lasting from around January 5 to 9, 1998, a combination of various meteorological factors and conditions produced a series of synoptic weather events that produced freezing rain and drizzle (inclusively defined as freezing precipitation), snow, or ice pellets that fell over a region stretching from southeastern Ontario to southwestern Quebec, New Brunswick, and Nova Scotia in Canada, as well as northern New York and the New England states in the United States. Most locations experienced large amounts of freezing precipitation, with northern parts receiving snow and ice pellets. Some regions received more than 100 mm in accumulated precipitation that came in two major synoptic systems. Because of its severity in terms of economic and ecosystem damage, the entire 7-day meteorological event has been popularly called "Ice Storm '98."

Ice Storm '98 has been classified as one of the worst weather-related disasters in Canada. Large accumulations of ice on trees caused branches and small trunks to break, resulting in tree mortality in some cases (see Chapter 6 by Greene *et al.*). DeGaetano (2000) reported that damage to forest land and harvestable timber stands was "considerable." According to the U.S. Forest Service, more than 3 billion board feet of lumber were lost because of the ice storm. How a major ice storm event affects trees depends greatly on its intensity, duration, and wind speeds, as well as on the type of icing associated with the storm. As explained in Chapter 6, glaze ice and rime ice pose the greatest potential for bole and branch damage.

With rapidly changing climatic regimes induced by the increasing CO_2 concentration in the atmosphere, a possible increase in potential ecosystem disturbance from a series of such freezing rain events under warming climatic conditions needs to be examined. To proceed with such an examination, we first need to understand the meteorological processes and mechanisms that produce atmospheric conditions most favorable for formation of a severe freezing rain event. In this chapter, we describe the salient features of such processes, mechanisms, and conditions, and where they are likely to occur, thereby providing the current scientific understanding and account of the phenomenon of freezing rain.

SYNOPTIC CONDITIONS FOR FREEZING RAIN

An event that can be classified as an ice storm is usually associated with a relatively long-lasting freezing rain over a certain time period. "Freezing precipitation" is a general term that includes both freezing rain and freezing drizzle. By definition, freezing rain is composed of supercooled liquid water drops with diameter 0.5 mm or greater that freeze on contact with a surface with a temperature less than 0°C. Freezing drizzle behaves similarly but is composed mostly of drops with diameter less than 0.5 mm (Atmospheric Environment Service, 1977). Although freezing drizzle is an important component of freezing precipitation, an "ice storm" (such as Ice Storm '98) is characterized mainly by freezing rain. Therefore, we will confine our discussion to synoptic processes and conditions favorable for development of freezing rain. [For those readers who are interested in finding out the details of freezing drizzle, we recommend Huffman and Norman (1988) and Isaac *et al.* (1996).] An ice storm can be heuristically defined as a synoptic event in which one receives a lot of freezing rain. There are currently no rules that define "a lot of freezing rain."

The synoptic condition under which freezing rain can occur is now relatively well understood (Stewart, 1985; Stewart and King, 1990; Cyzs *et al.*, 1996; Zerr, 1997). A classic model of freezing rain requires a melting layer at the top of an inversion layer in the lower atmosphere, with a temperature greater than 0°C, sandwiched between layers at top and bottom with temperatures less than 0°C (Fig. 1A). Frozen precipitation (usually in the form of snow) falls through the melting layer and turns into supercooled liquid drops. These drops then refreeze upon contact with the ground or any other surface objects with temperature less than 0°C. It is important that the bottom layer, below the melting layer, be of just the right thickness. If it is too thick, then the liquid drops will refreeze (and become snow or ice) before impacting the ground. Fig. 1 also shows other vertical temperature profiles that could result in ice pellets (usually called sleet in the United States) or just snow. In the case of snow (Fig. 1B), there is no melting layer, and frozen precipitation does not encounter melting conditions during its fall to the ground. In the case of ice pellets (Fig. 1C), the bottom layer is thick enough that liquid drops from the melting layer above have a chance to refreeze before landing on the ground. Therefore, the type of precipitation observed on the ground depends sensitively on the vertical atmospheric temperature profile (Bourgouin, 2000; Lackmann *et al.*, 2002).

The type of precipitation one observes on the ground also depends on the size and density of snowflakes that fall through the melting layer. For example, if the snowflakes are large and do not melt completely as they pass through the warm layer, it is likely that much of the precipitation will remain in the form of snow. The airflow pattern associated with a synoptic weather system forces precipitation particles to take various paths (and thus through various temperature zones) before reaching the ground, and the complexity of the paths and accompanying particle–particle interactions, as

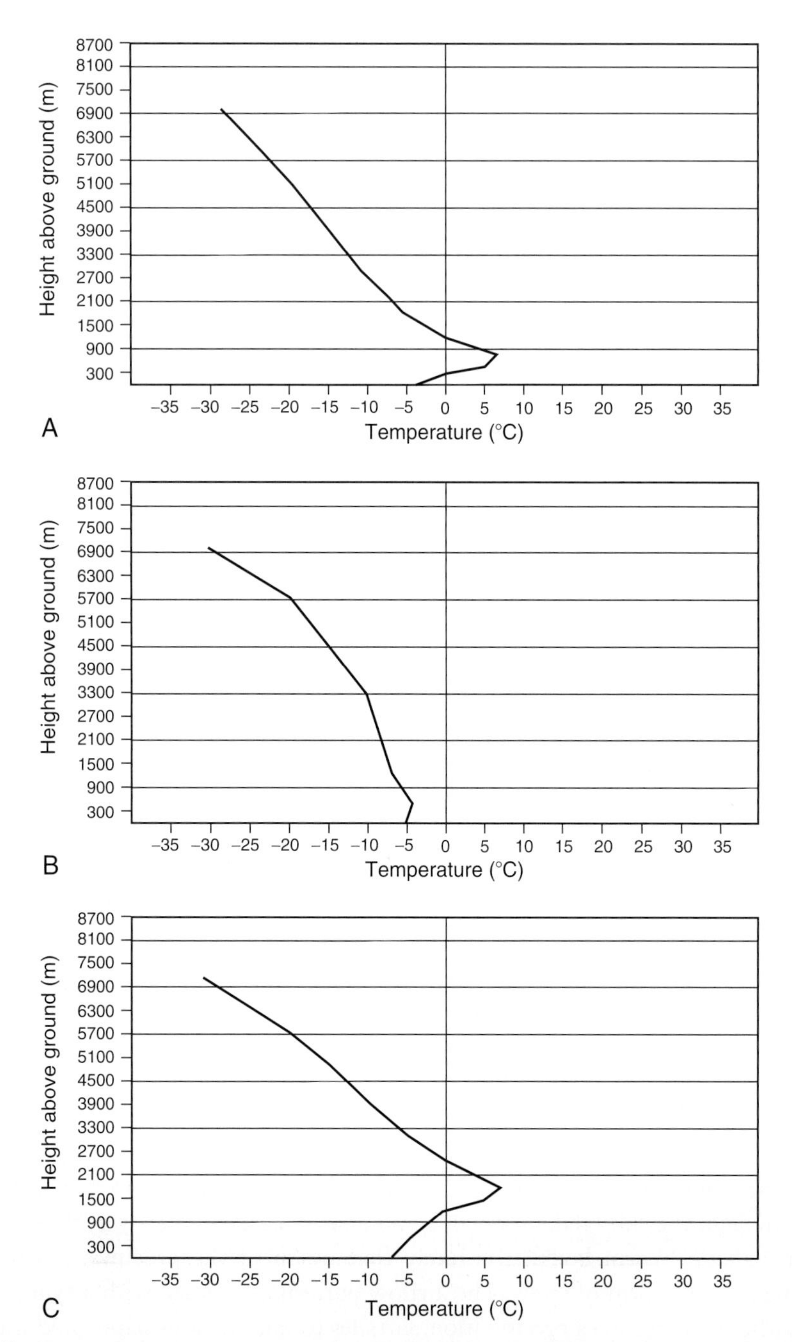

FIG. 1 A typical vertical temperature profile for formation of: (**A**) freezing rain, (**B**) snow formation, and (**C**) ice pellets. (From Milton and Bourque, [1999].)

well as particle–environment interactions, make the type of precipitation reaching the ground quite variable over short time and space scales.

Various synoptic-scale mechanisms can produce different vertical temperature structures. Some of the main mechanisms include evaporation, melting, vertical motion, and thermal advection. For the formation of synoptic conditions most favorable for development of freezing rain, thermal advection is the main process (Cortinas, 2000). In a winter synoptic system, snow is the most likely precipitation form, at least in the areas that were affected by Ice Storm '98. When this system is accompanied by an intrusion of warm air (with temperatures greater than 0°C) from the south in the lower troposphere, the likelihood of snow becoming freezing rain increases. Therefore, a freezing rain event is usually associated with movement, and evolution, of a synoptic frontal system. A simplified schematic representation showing the likely location of freezing rain formation in a frontal cross section is shown in Fig. 2.

It should be noted, however, that small changes in the vertical temperature structure induced by thermodynamics of melting and freezing can result in different precipitation types and amount (Lackmann *et al.*, 2002). For example, in a freezing rain situation, latent heat is released when the super-

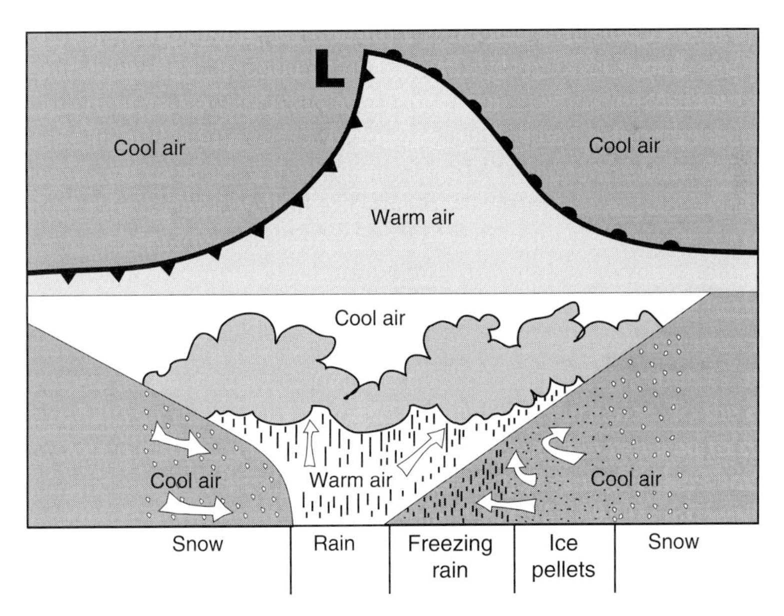

FIG. 2 A simple schematic cross-section of a frontal system showing likely locations of various precipitation types. (Modified from a figure in Heidorn [2006] with permission from the author.)

cooled liquid raindrops impact the ground (or any surface objects) and freeze. The release of latent heat warms the shallow layer (temperature less than 0°C) above the ground. In many cases, the amount of heat released is enough to warm the layer temperature to 0°C or slightly above, resulting in precipitation change from freezing rain to liquid rain (Stewart, 1985). Unless the situation is overridden by strong synoptic and/or topographic conditions (such as strong, cold, dry air advection in the lowest tropospheric layer, or adiabatic cooling caused by upslope flow), this is one of the reasons that many freezing rain events are relatively short-lived. In his study of freezing rain climatology in the Great Lakes region, Cortinas (2000) found that continuous freezing rain normally lasts less than an hour and only about 7% of freezing rain events lasted longer than 5 hours. In this way, freezing rain is a self-limiting process.

The meteorological conditions for freezing rain are usually not long lasting. The location and strength of the thermal advection need to be just right. A freezing rain event usually covers a line or an area with a spatial dimension on an order of a few hundred kilometers, associated mostly with the warm front of a synoptic system, and lasts for only a few hours (Stewart and King, 1990; Cortinas, 2000). Comparatively, a total of 80 hours or more of freezing rain or drizzle was reported in some areas during Ice Storm '98 (Milton and Bourque, 1999).

A typical synoptic surface pressure configuration for a significant freezing rain over the areas affected by Ice Storm '98 is given by Koolwine (1975), Stewart and King (1987), Cantin and Bachand (1990), and Cortinas (2000). (For convenience, we will hereafter refer to those areas as the IS98 region.) It is characterized by a high-pressure system over northern Quebec advecting cold northeasterly air. The high-pressure system becomes situated in juxtaposition with a low-pressure system that moves into the IS98 region from the southwest, forcing southerly intrusion of warm, moist air into the lower tropospheric layer centered on 850 hPa (Fig. 3). Cantin and Bachand (1990) found that a vertical temperature structure associated with this kind of synoptic pattern is characterized by a warm layer (with a temperature greater than 0°C) centered around 850 hPa, with 700- to 850-hPa thickness greater than 1540 m and 850- to 1000-hPa thickness somewhere between 1290 to 1310 m. The thickness of the warm air must be sufficient to melt the snow that falls from the below-zero layer above. The layer that is sandwiched between the warm layer and the ground must be below zero, with a thickness of around 500 m or less. This will ensure that the supercooled liquid drops will not freeze while falling through the bottom layer. Ice Storm '98 and other major freezing rain events experienced in

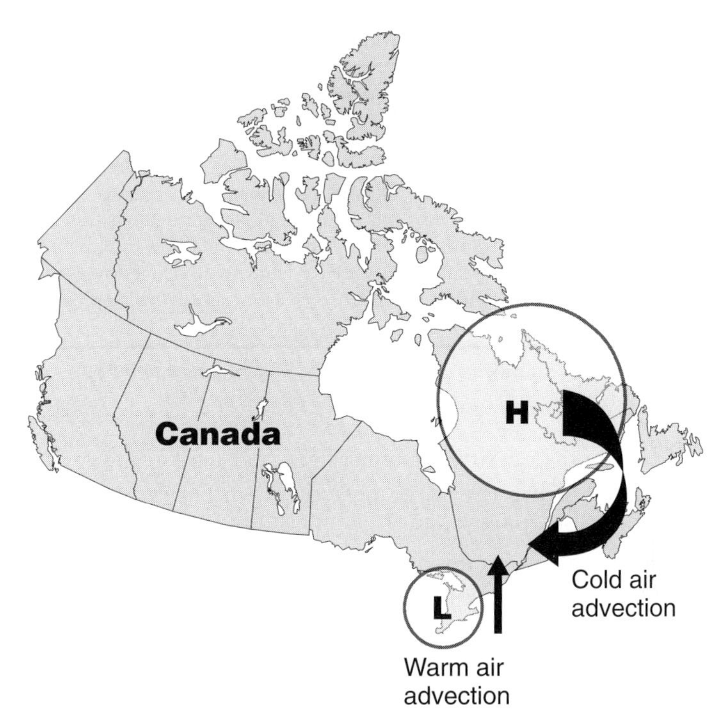

FIG. 3 A simplified schematic showing a typical synoptic configuration for freezing rain formation over southern Ontario, southwestern Quebec, and northeastern United States. Intrusion of warm air caused by northeastward moving low system takes place in the lower layers of the troposphere.

the recent past over the IS98 region satisfy these meteorological conditions (Rauber *et al.*, 1994; Keeter *et al.*, 1995; Milton and Bourque, 1999).

Table 1 lists five major freezing rain events over the IS98 region since 1960 (Milton and Bourque, 1999). We find that each event (not counting Ice Storm '98) is typically characterized by a similar synoptic evolutionary pattern. (The synoptic evolution of Ice Storm '98 is described separately later in this chapter.) To bring out the climatology of freezing rain synoptic system evolution, we perform a composite analysis (i.e., take the average) of the 1000-hPa height and 850- to 925-hPa moisture flux/divergence fields associated with each of the major freezing rain events.

Fig. 4 shows composite daily time frames of the 1000-hPa height field from day 1 to day 7. Day 5 corresponds to the time of maximum freezing rain over the IS98 region. From day 1 to day 4, we can see the development of a high-pressure system over northern Quebec that is connected to the

TABLE 1 Five Major Freezing Precipitation Events over the IS98 Region Since 1960

Time	Cumulative Amount
February 23–25, 1961	Ottawa: 32 mm Montreal: 26 mm
March 22–23, 1972	15–40 mm in Lower Laurentians Mix of rain and freezing rain in Montreal in western Quebec
December 12–14, 1983	30–>50 mm in Montreal region and Lower Laurentians
January 4–5, 1997	20–40 mm, mostly over Laurentians-Three Rivers regions
January 5–10, 1998	70– >100 mm over some areas, mostly from the second synoptic weather system

Amount is given in water equivalent. (Data from Milton and Bourque [1999].)

broad band of high-pressure systems stretched across the mid-latitude North Atlantic. During the same period, an intensifying low-pressure system approaches the IS98 region from the southwest. On day 5, the low-pressure system reaches its maximum intensity as it arrives over the IS98 region, with the Quebec high to the north. The spatial pattern of the 1000-hPa height field over eastern Canada and the northeastern United States is identical to the pattern identified by Koolwine (1975) for a major freezing rain occurrence. The Quebec high advects cold air near the surface towards the IS98 region, while the low advects warm, moist air northeastward over the surface cold air. After day 5, the synoptic flow configuration for freezing rain disappears as the low-pressure system moves northeastward and weakens, while distorting the high-pressure system over northern Quebec.

Associated evolution of the 850- to 925-hPa moisture transport and the 850-hPa moisture flux divergence fields from day 1 to day 7 are shown in Fig. 5. The moisture transport and flux divergence fields are calculated, respectively, as follows:

$$\frac{1}{g}\int_{925}^{850} q\vec{V}dp \quad \text{and} \quad \nabla \cdot q\vec{V},$$

where g is the gravitational acceleration (= 9.8 m s^{-2}), q is the moisture mixing ratio, $\vec{V}$ is the wind vector, and p is the pressure. On day 3 we see a distinct area of moisture flux convergence over the mid-southern United States. Moisture transport indicates the Gulf of Mexico to be the main moisture source. As the low-pressure system intensifies as it moves towards the IS98 region, the moisture flux convergence ahead of the system also intensifies and reaches a maximum on day 5. By this time, the main moisture source has shifted to the subtropical western North Atlantic Ocean. After day 5, atmospheric flow and moisture flux convergence have changed

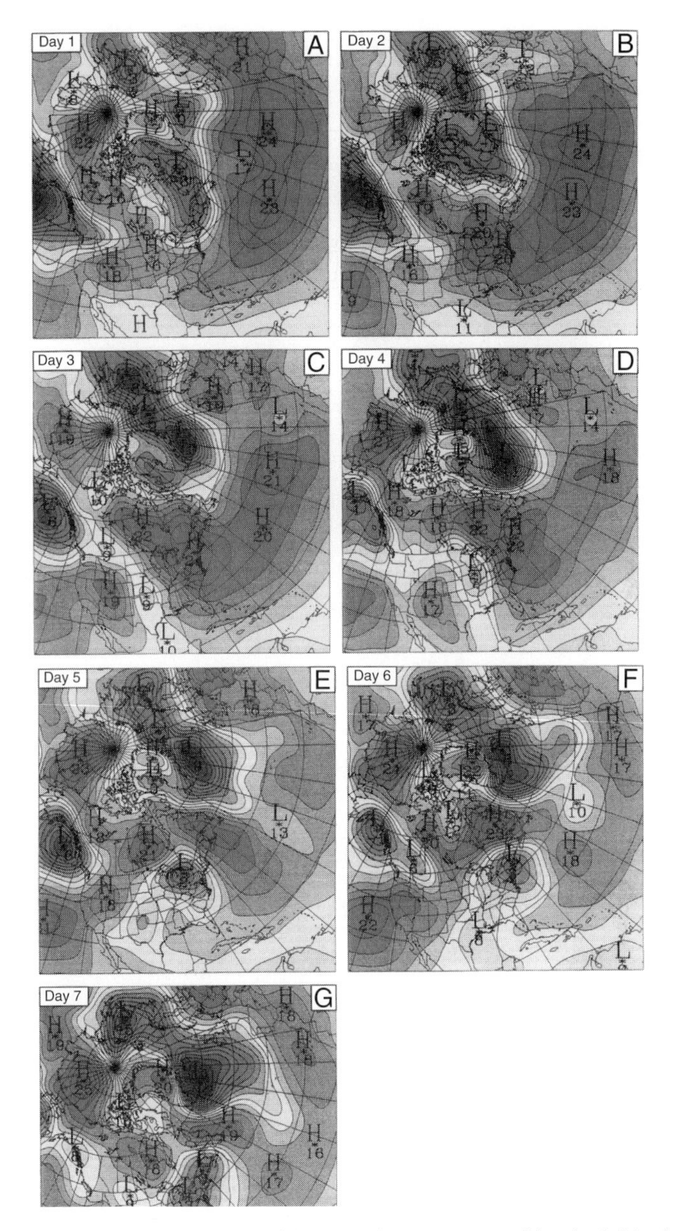

FIG. 4 Composite daily time series of the 1000-hPa geopotential height field, obtained by averaging the field associated with previous major freezing rain events. Day 5 corresponds to the peak activity (in terms of moisture flux convergence over the IS98 region) of a freezing rain event. Contour interval is 2 decameters (dams). (Adapted from Higuchi *et al.*, [2000].)

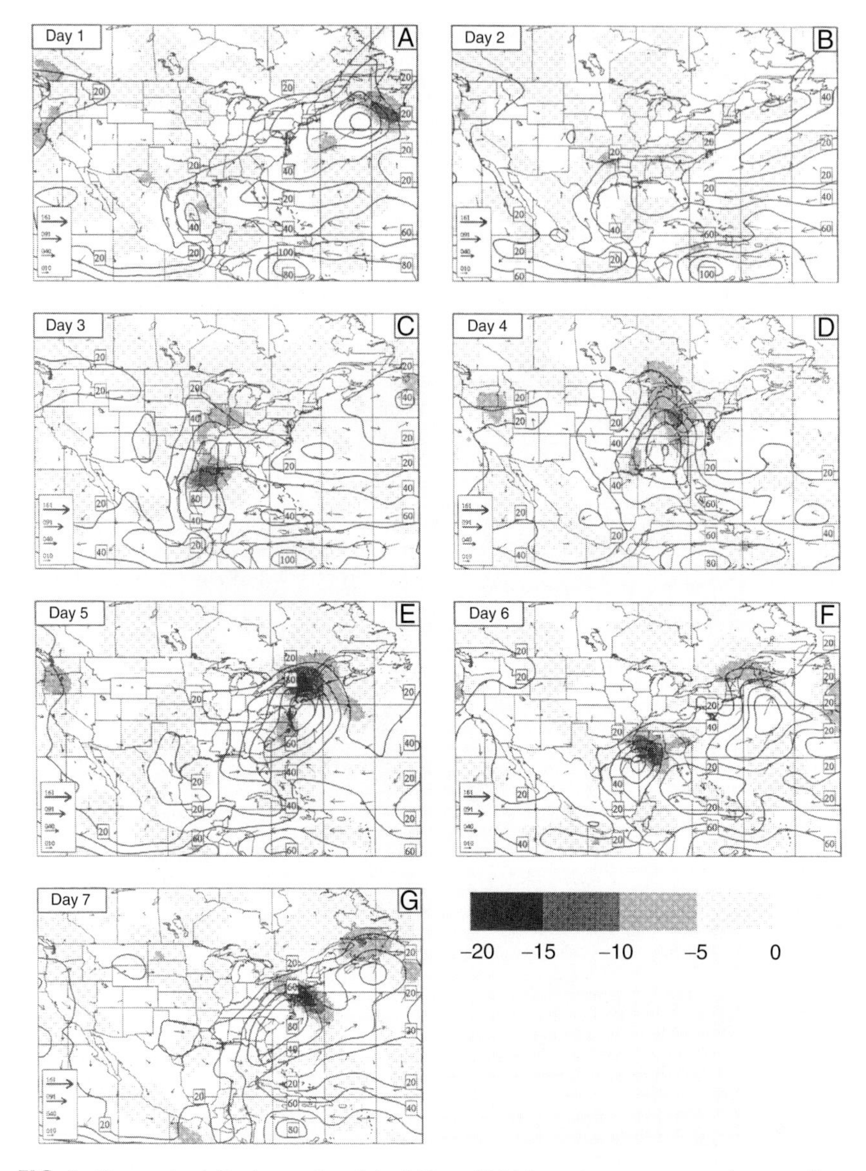

FIG. 5 Composite daily time series of the 850- to 925-hPa moisture transport (contour lines, kg m^{-2} s^{-1}) and the 850-hPa moisture flux convergence (shaded regions, 10^{-8} s^{-1}) associated with time frames shown in Fig. 4. Arrows indicate magnitude and direction of the moisture transport. (Adapted from Higuchi *et al.* [2000].)

so that the meteorological conditions favorable for freezing rain no longer exist over the IS98 region. Typically, a major freezing event involves only one synoptic system that usually lasts for no more than 3 days and drops about 20 to 50 mm of cumulative precipitation.

CLIMATOLOGY OF FREEZING RAIN IN CANADA

The earliest published climatology of freezing precipitation in Canada was by McKay and Thompson (1969) for 1957 to 1966. With more observations and better and faster analysis techniques, the earlier work was recently updated and extended by Stuart and Isaac (1999) for 1961 to 1990. It is interesting to note some differences, but the overall pattern of spatial distribution of mean annual hours of freezing precipitation obtained by Stuart and Isaac (1999) is very similar to that obtained by McKay and Thompson (1969) more than 30 years ago. A spatial distribution of mean annual freezing rain hours for Canada from Stuart and Isaac (1999) is reproduced here as Fig. 6. Typically, eastern Canada (southern

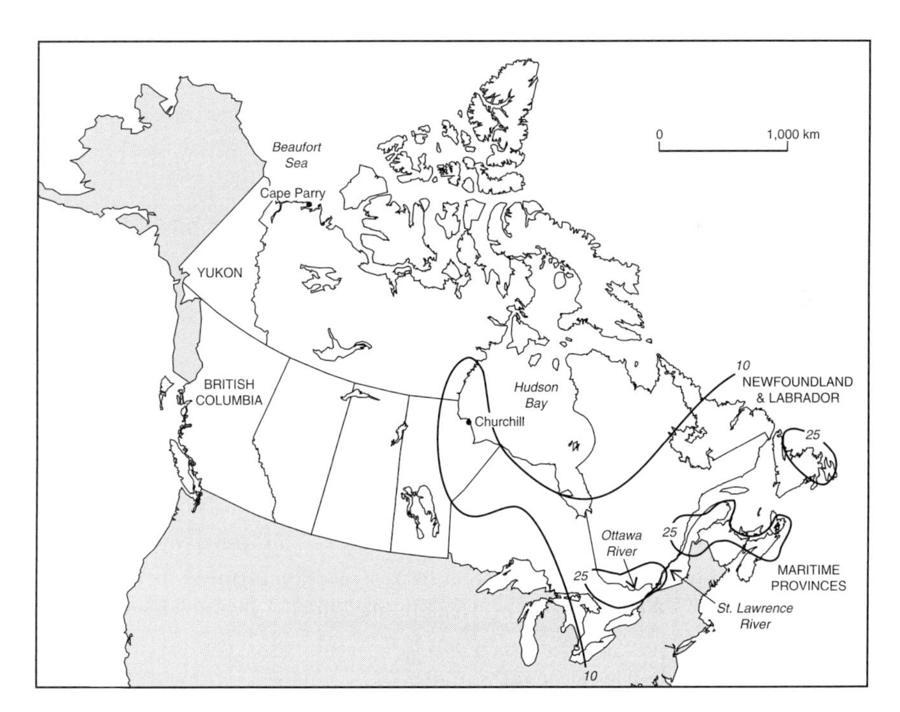

FIG. 6 Spatial distribution of climatological annual freezing rain hours in Canada. (From Stuart and Isaac [1999], reproduced here with authors' permission.)

Ontario and Quebec, Maritime and Atlantic provinces) experiences 10 to 25 hours of accumulated freezing rain per year. The southeastern Ontario, southwestern Quebec (including the St. Lawrence Valley), Maritime Provinces, and Newfoundland regions show an average of 25 hours or more of freezing rain in a year. This regional distribution reflects quite well the favorable location of the IS98 region to experience the appropriate meteorological conditions described previously for freezing rain formation. It also reflects the importance of local topographical effect on the synoptic conditions, as is the case along the St. Lawrence Valley. Such topographical features can channel the cold air at low levels from the high-pressure system.

Stuart and Isaac (1999) found that, depending on the season, freezing precipitation could occur anywhere in Canada. During the summer, freezing precipitation is confined to the far north, with a maximum frequency of occurrence no greater than 1%. Much of the freezing precipitation is in the form of freezing drizzle. As the season changes into autumn, the region where freezing precipitation could occur extends southward and covers most of Canada. In the winter (December, January, and February), as the northern part of the country becomes too cold, the region of freezing precipitation becomes confined to the southern, and in particular to the eastern, half of the country. The frequency of occurrence of freezing rain and drizzle over southern and southwestern Ontario and St. Lawrence River Valley regions in the winter season is about 1% to 2%. By April, freezing precipitation in Ontario and Quebec is very rare.

Using meteorological observations obtained from the National Center for Environmental Prediction (NCEP) in the United States, Cortinas (2000) carried out a 15-year (1976 to 1990) climatology of freezing rain events in the Great Lakes and surrounding regions of North America. In this region, freezing rain is usually short-lived and occurs near sunrise mainly between December and March, consistent with the results obtained by Stuart and Isaac (1999). The frequency of freezing rain increases from west to east, and seems to be related to the climatological paths of synoptic storms, the proximity of the Atlantic Ocean, and topographical features (i.e., valleys). The proximity of the eastern regions to the Atlantic Ocean is an important factor that provides a source of large amounts of moisture. The intensity of a freezing rain event depends on the amount of moisture available to a cyclonic storm, and when it is combined with long duration, freezing rain can cause significant damage to trees (see Chapter 6).

METEOROLOGICAL EVOLUTION OF ICE STORM '98

Because of its unprecedented impact, we briefly describe the meteorological evolution that led to the development of the ice storm event in early January 1998. Ice Storm '98 was the most extreme freezing rain event ever recorded in recent history, and it showed the potential destructive power of a prolonged freezing rain event, both in terms of human suffering and ecological damage. More detailed synoptic discussions of Ice Storm '98 are found in Milton and Bourque (1999), Higuchi *et al.* (2000), Gyakum and Roebber (2001), and Roebber and Gyakum (2003). Unlike the previous major freezing events, Ice Storm '98 was composed of two successive synoptic disturbances instead of just one. The two systems dropped, in some areas, a cumulative amount of more than 100 mm water equivalent of freezing precipitation. Individually, the two synoptic weather systems were not significantly more extreme in freezing rain precipitation than other major freezing rain events in the recent past. Combined, however, they produced the necessary intensity and duration to cause major tree damage, as described in Chapter 6.

Brief Description of Synoptic Evolution

Fig. 7 shows a series of 1200 UTC snapshots of the 1000-hPa height field from January 3 to 10. By January 4, the high-pressure ridge observed 24 hours previously over western Canada had joined the mid-latitude North Atlantic high-pressure system to form a high-pressure belt stretching from the Yukon to the southeastern United States. As already indicated, the high-pressure region over northern Quebec is an important component in creating a meteorological condition for freezing rain. This synoptic configuration remained quasi-stationary until its decay around January 10. By January 5, a trough protruded into southern Ontario and Quebec from the midwestern United States, allowing the first wave of northward flux of warm air and moisture into the IS98 area. A cross-sectional analysis along the transect extending from the Laurentian Mountains (north of Montreal) southward to New York shows a clear intrusion of a "warm tongue" (a warm layer with temperatures greater than 0°C) between the 850- and 930-hPa levels on January 5. By 0000 UTC, January 7 (not shown), the trough weakened as the high-pressure system observed at 1200 UTC, January 6, established itself firmly over northern Quebec, ending the first wave associated with the ice storm.

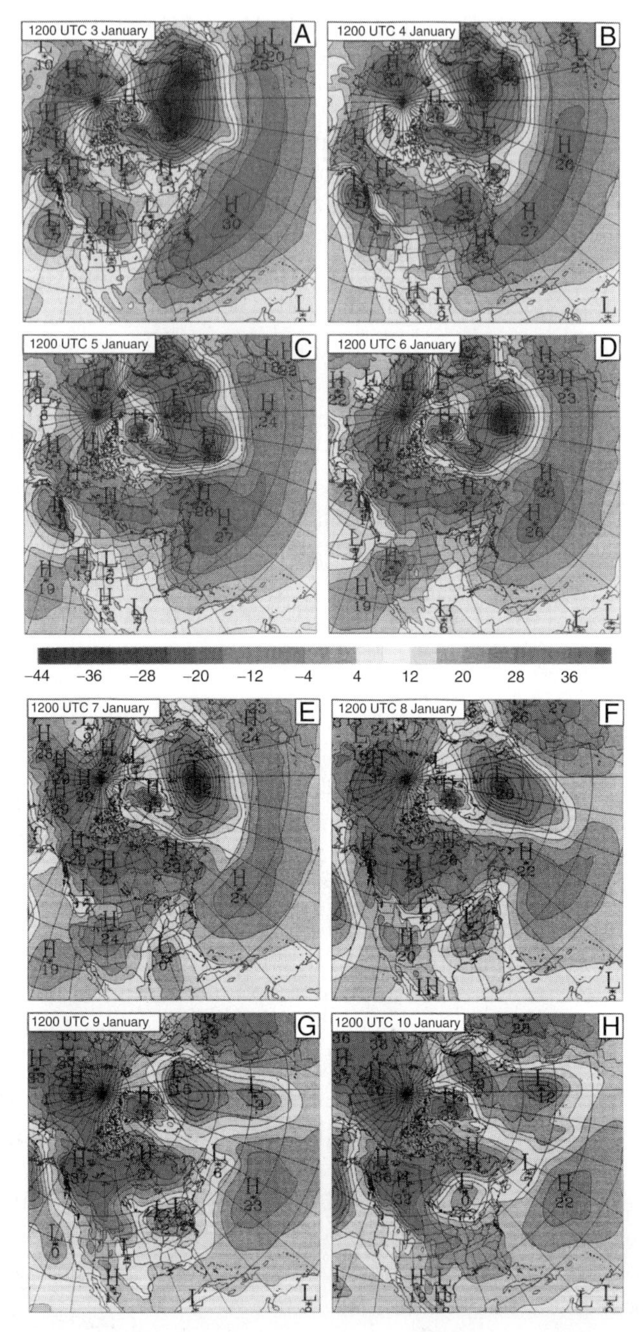

FIG. 7 Daily evolution of the 1000-hPa geopotential height field during Ice Storm '98, beginning 12:00 UTC, January 3, 1998. (Adapted from Higuchi *et al.* [2000].)

The second wave of moisture and warm air flux into the IS98 region started with a developing low-pressure system in the southeastern United States, with a trough extending into the IS98 area (Fig. 7E, January 7). Over the next 3 days, this low-pressure system slowly moved northward while the northward-extending trough shifted slowly eastward. The zonal high-pressure belt over the mid-latitude North Atlantic split into two cells, with the western cell persisting to give a strong southwesterly flow over the eastern seaboard. By January 10 (Fig. 7H), the low center moved to just south of Hudson Bay, with a trough extending eastward. Much of the freezing rain experienced during Ice Storm '98 came from the second wave.

The areas that experienced more than 100 mm in cumulative precipitation were in northwestern New York, around Montreal, and along the St. Lawrence Valley. The first indication of freezing precipitation in Ottawa and Montreal that was associated with the first wave came on January 5. About 10 mm fell in Montreal on January 5 and 6; Ottawa received almost 25 mm on January 6. After a break on January 7, the second wave of freezing precipitation started on January 8. It was more extensive in amount and duration. The precipitation peaked around 38 mm water equivalent at Montreal on January 10, while nearly 20 and 25 mm water equivalent fell in Ottawa on January 8 and 9, respectively.

Role of North Atlantic Oscillation and El Niño–Southern Oscillation

One of the main contributing meteorological factors that transformed what could have been just another major freezing rain event into an ice storm was the "coincidental" occurrences of El Niño and a relatively intense positive phase of the North Atlantic Oscillation (NAO). The NAO is one of the major low-frequency variability modes observed in the atmosphere. It describes the see-saw movement of atmospheric mass between the high-latitude (near Greenland and Iceland) and the mid-latitude (between 35° and 40°N) North Atlantic Ocean. In the positive phase of the NAO, there is an anomalous deficiency in atmospheric mass over the Greenland-Iceland region and an anomalous surplus of atmospheric mass over the mid-latitude North Atlantic Ocean. The opposite occurs in the negative phase of the NAO. The NAO teleconnection pattern is a dominant mode of atmospheric variability in the North Atlantic sector, accounting for more than 30% of the hemispheric variability in the surface pressure (Cayan, 1992). The NAO low-frequency variability mode is most pronounced during the winter. The NAO index that is designed to capture the salient feature of the

NAO variability is usually defined as the difference in the normalized sea level pressure between the Azores and Iceland (Walker and Bliss, 1932; Rogers, 1984). [A recent monograph by Hurrell *et al.* (2003) thoroughly reviews the NAO.] Indeed, whereas the previous major freezing events occurred in non–El Niño winters, Ice Storm '98 took place during one of the most intensive El Niño–Southern Oscillation (ENSO) winters in recent times. Barsugli *et al.* (1999) described the relationship between El Niño and Ice Storm '98.

A typical moderate to strong ENSO winter is usually characterized by a distinctive positive phase Pacific/North American (PNA) low-frequency variability mode pattern. This teleconnection pattern has a low-pressure anomaly over the Gulf of Alaska area, a high-pressure anomaly over western Canada, and a low-pressure anomaly over the southeastern United States (Horel and Wallace, 1981; Wallace and Gutzler, 1981; Barnston and Livezey, 1987). This spatial distribution in the geopotential height anomaly is caused by a standing Rossby wave that carries energy northward from the ENSO region in the tropical Pacific; the wave bends eastward with increasing Coriolis force in high latitudes (Hoskins and Karoly, 1981). A climatological 500-hPa composite geopotential anomaly height field for December/January of an El Niño winter is shown in Fig. 8A. The anomaly field is characterized by an identifiable positive phase of the PNA. There is no noticeable difference in the height anomaly field over the northern regions of the North Atlantic between the El Niño and the non–El Niño December/January period.

The atmospheric circulation anomaly pattern observed during the 1997–1998 winter was associated with a very strong El Niño event, with magnitude similar to the one observed during the 1982–1983 winter. An excessive amount of moisture flux into the United States caused severe flooding in northern California and in central and eastern regions of the southern United States. The subtropical jet stream went through northern California and flowed zonally eastward across the southern United States. The difference in the Northern Hemisphere 500-hPa height anomaly between the December 1997–January 1998 field and the typical December/January El Niño winter field is shown in Fig. 8B. We note that the overall spatial pattern of the 1997–1998 El Niño is similar to the climatological El Niño pattern, except for the intensity. Over eastern Ontario and southwestern Quebec, the positive height anomaly was much more intense, "pushing" the low anomaly center over the southeastern United States further south, and making the trough more zonal than in a typical El Niño winter in the southern United States. Another notable difference can

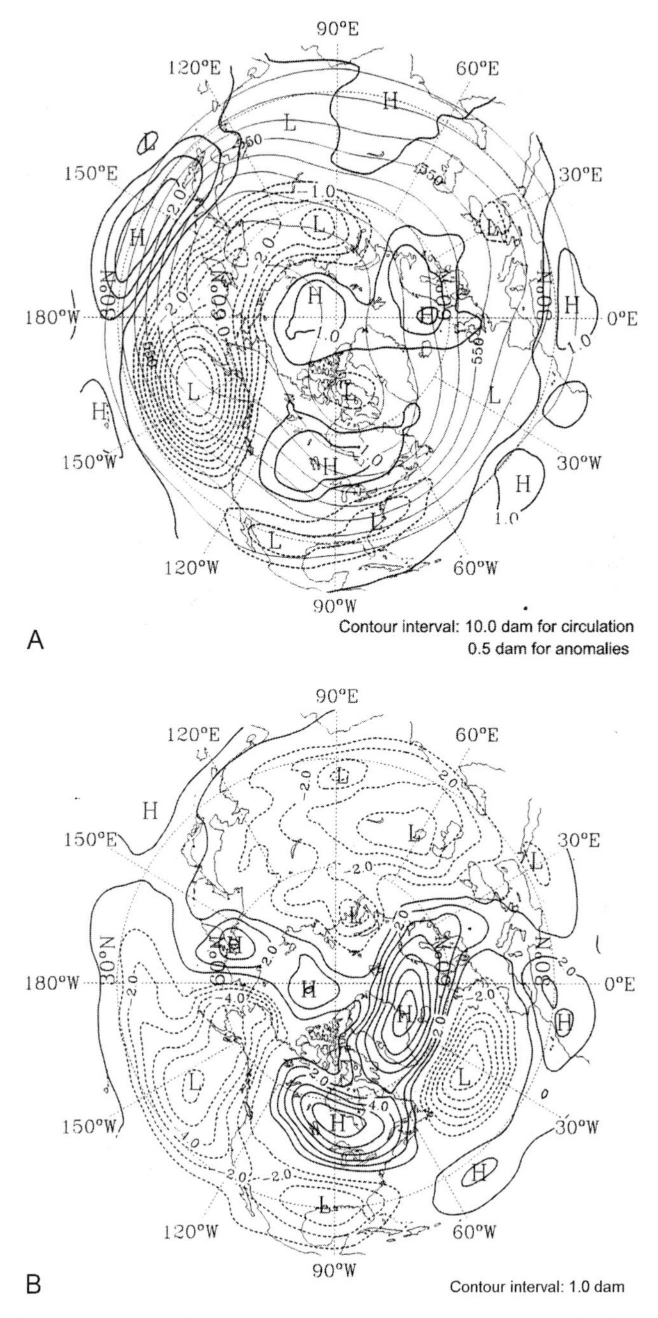

FIG. 8 (**A**) 500-hPa winter climatological and anomaly geopotential height fields associated with moderate to strong El Niño. (Contour interval: 10 dams for the height and 0.5 dam for the anomaly.) (**B**) Difference in the 500-hPa winter height field associated with the 1997–1998 El Niño and the climatological El Niño. (Adapted from Higuchi *et al.*, [2000].)

be seen over the North Atlantic where, compared to a typical El Niño winter, a very distinctive negative phase of the NAO is observed. Had the NAO index remained negative during the first week in January 1998, the flow of synoptic weather systems would probably have remained in the southern United States. However, beginning late December 1997, the NAO started to evolve into a positive phase pattern, with its index reaching a maximum greater than 3 on January 3. This was a relatively intense positive NAO pattern, because the composite NAO index value associated with the previous major freezing rain events is about 1.5. The NAO index stayed positive during much of the time period associated with Ice Storm '98. The positive phase of the NAO changed the flow pattern of the upper air, blocking the west-to-east flow of the synoptic systems that existed before the first week of January 1998 and causing them to move northeastward from the southern United States. Whereas this also characterizes the basic climatology of the synoptic pattern associated with a major freezing rain event, the occurrence of the positive NAO phase within the temporal context of the 1997–1998 El Niño intensified the freezing rain (transforming it into an ice storm) event because (1) the subtropical jet stream was farther south; and (2) the upper air trough over the eastern United States was deeper, giving rise to enhanced vorticity generation. Initial moisture flux from the Gulf of Mexico and then from the western tropical Atlantic Ocean created the increased moisture flux and convergence over the IS98 region.

Besides blocking the eastward movement of the low-pressure synoptic systems across the southern United States and thus forcing them to move toward the IS98 area, the positive NAO also linked with the high-pressure system over northern Quebec. This could have contributed to the persistence of the Quebec high-pressure system, thus prolonging the freezing precipitation during the ice storm.

Fig. 9A shows the position of the 250-hPa jet stream over North America associated with day 5 of the composite case shown in Fig. 4; this can be compared to the position of the jet stream at the same pressure level on January 8, 1998 (Fig. 9B). During Ice Storm '98, the jet stream flowed over the Gulf of Mexico before it was forced to veer northward toward the IS98 region because of the blocking action of the high-pressure system over the mid-latitude North Atlantic associated with the southern cell of the bipolar structure of the positive phase of the NAO. The jet stream of the composite case has a shallower trough, in contrast to the accentuated trough over the central United States on January 8, 1998 (Fig. 9B). The flow of the jet stream during Ice Storm '98 diverted moisture and warm air from the Gulf of Mexico and the western tropical Atlantic Ocean toward the IS98 region.

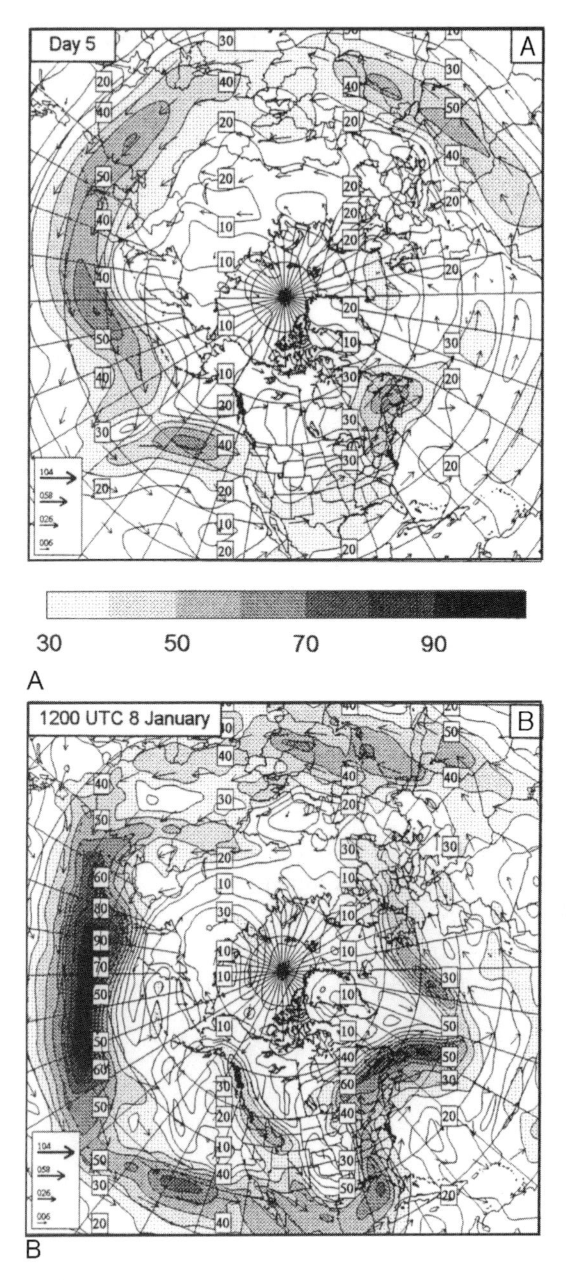

FIG. 9 Winds (m s^{-1}) at 250 hPa, showing jet streams at (**A**) day 5 of the composite average and (**B**) 1200 UTC, January 8, 1998. (Adapted from Higuchi *et al.* [2000].)

POSSIBLE CHANGES IN ICE STORM FREQUENCY UNDER A WARMING CLIMATE

As with many other issues related to climatic change, it is only natural to ask, How will the meteorological conditions and processes conducive to ice storm formation change over the northeastern region of North America under a warming climate? Will the frequency, intensity, and duration of ice storm events increase or decrease as climate becomes warmer? These are all interesting and valid questions, not only from the perspective of climate change science but also from the viewpoint of impact on economic and societal structures. However, these are also very difficult questions to answer. But perhaps it is possible to obtain an educated guess as to what could happen over the next several decades.

Numerical experiments with general circulation models (GCMs) of the climate tell us that when atmospheric CO_2 is increased to 600 to 700 ppm (parts per million), the atmosphere will probably become warmer, with the globally averaged annual surface temperatures rising by 2 to 4°C (IPCC, 2001). The atmospheric CO_2 concentration is now about 370 ppm, with an average increase of about 1.4 ppm/y. For the sake of argument, let us assume for the moment that the CO_2-induced climatic warming predicted by the GCMs is correct. What impact will this have on the meteorological conditions favoring the formation of an ice storm? To provide some sort of guidance about the answer, we focus on the surface temperature. (Surface temperature is one of the predictable variables that climatologists have most confidence in; there is less confidence in the prediction of thickness change.) We noted earlier that for a freezing rain event to occur, we need to have a surface temperature below 0°C.

Since there is less confidence in the details of the GCM climate change predictions at regional and short-time scales, we will focus on the seasonally averaged values of the surface temperature change produced by the Canadian GCM (McFarlane *et al.*, 1992). Fig. 10 shows the difference in the winter (December, January, February) surface temperature between the $1 \times CO_2$ (330 ppm) and the $2 \times CO_2$ (660 ppm) simulations for Canada and the northern United States. Under the $2 \times CO_2$ climate regime, the model predicts an increase of 7 to 10°C in the IS98 region. If this is true, and it is a big IF, then the likelihood of surface temperatures falling below zero in winter in the IS98 region is decreased, resulting in fewer hours of freezing rain. Analogous to the currently observed northward movement of areas of freezing rain during summer, it is possible that the freezing rain region in winter is "pushed" further north under the $2 \times CO_2$ warming

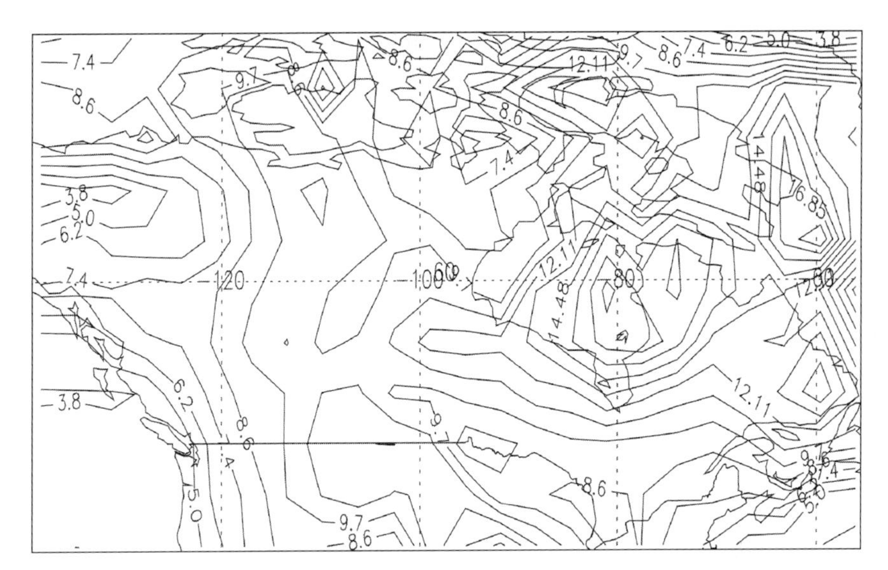

FIG. 10 Difference in the surface temperature (Kelvin) between the $1 \times CO_2$ and the $2 \times CO_2$ simulations obtained by the Canadian general circulation model.

climate. But with a significant temperature increase indicated over Hudson Bay, the region of winter freezing rain would probably be located further north, in the subarctic region.

SUMMARY

For a freezing rain to occur, certain meteorological conditions have to exist, brought about by a unique combination of processes related to vertical differential advection of air masses with very different temperature regimes. These meteorological conditions usually occur in front of an advancing warm front. The differential advection produces a vertical temperature profile that is characterized by a warm layer (with temperatures greater than 0°C) centered about 850 hPa, with the 700- to 850-hPa geopotential height thickness greater than 1540 m. The warm layer must be thick enough to melt the snow that falls from the layer above with temperatures less than 0°C. The cold layer (with temperatures below zero) between the warm layer and the surface should be less than 500 m to ensure that the supercooled liquid drops falling out of the warm layer will not freeze until they hit the ground or any surface object.

The vertical temperature profile for freezing rain occurrence can be brought about by a certain synoptic weather configuration that is relatively common over eastern Canada and the northeastern United States during winter. This configuration is characterized by a high-pressure system over northern Quebec that advects cold northeasterly air. When a low-pressure system moves in from the southwest, guided by the upper jet stream, a layer of warm air centered on 850 hPa is advected into the cold air mass ahead of the warm front.

What transforms an "ordinary" freezing rain event into an ice storm is its duration and precipitation amount. Typically, a freezing rain event covers a line or an area with a spatial dimension on the order of few hundred kilometers ahead of a warm front and lasts for only a few hours. On the basis of past observations, a severe freezing rain event associated with any one particular synoptic system can last for a few days and can drop something on the order of 20 to 30 mm water equivalent of freezing precipitation. The first severe freezing rain in recent history to be classified as an ice storm was Ice Storm '98, which lasted for almost a week and dropped more than double the usual amount of freezing precipitation associated with a "typical" severe freezing rain. Our analysis of the Ice Storm '98 event shows that the coincidental occurrence of a strong ENSO event during the 1997–1998 winter caused the severity of the freezing rain by producing meteorological conditions favoring cyclogenesis over the southeastern United States, as well as "easy" access to moisture from the Gulf of Mexico and the western subtropical Atlantic Ocean.

As the climate warms, various climate models indicate an increase in moderate to strong ENSO events. Were it not for the fact that these models also show a dramatic increase in the winter surface temperature over the region that was affected by Ice Storm '98, we would speculate that we will have more ice storms similar to Ice Storm '98. However, a 5° to 10°C increase in the surface temperature during the winter season under $2 \times CO_2$ condition will decrease the number of days with temperatures below zero, one of the meteorological criteria for producing freezing rain. Climate model predictions seem to indicate a likely decrease in the number of severe freezing rain events, at least over the IS98 region.

REFERENCES

Atmospheric Environment Service. (1977). *Manual of Surface Weather Observations*, seventh edition. Atmospheric Environment Service, Toronto.

Barnston, A. G., and Livezey, R. E. (1987). Classification, seasonality and persistence of low-frequency atmospheric circulation patterns. *Mon Weather Rev* 115, 1083–1126.

Barsugli, J. J., Whitaker, J. S., Loughe, A. F., Sardeshmukh, P. D., and Toth, A. (1999). The effect of the 1997/98 El Niño on individual large-scale weather events. *B Am Meteorol Soc* 80, 1399–1411.

Bourgouin, P. (2000). A method to determine precipitation types. *Weather Forecast* 15, 583–592.

Cantin, A., and Bachand, D. (1990). Synoptic pattern recognition and partial thickness techniques as a tool for precipitation types forecasting associated with a winter storm. In: *Proceeding of the 3rd Workshop on Operational Meteorology*, pp. 424–432. Canadian Meteorological and Oceanographic Society, Montreal.

Cayan, D. R. (1992). Latent and sensible heat flux anomalies over the northern oceans: the connection to monthly atmospheric circulation. *J Climate* 5, 354–369.

Cortinas Jr., J. (2000). A climatology of freezing rain in the Great Lakes region of North America. *Mon Weather Rev* 128, 3574–3588.

Czys, R. R., Scott, R. W., Tang, K. C., Przybylinski, R. W., and Sabones, M. E. (1996). A physically based, nondimensional parameter for discriminating between locations of freezing rain and ice pellets. *Weather Forecast* 11, 591–598.

DeGaetano, A. T. (2000). Climatic perspective and impacts of the 1998 northern New York and New England Ice Storm. *B Am Meteorol Soc* 81, 237–254.

Gyakum, J. R., and Roebber, P. J. (2001). The 1998 ice storm—analysis of a planetary-scale event. *J Climate* 129, 2983–2997.

Heidorn, K. C. (2006). *And Now . . . The Weather.* Fifth House Publishers, Calgary, AB.

Higuchi, K., Yuen, C. W., and Shabbar, A. (2000). Ice Storm '98 in southcentral Canada and northeastern United States: a climatological perspective. *Theor Appl Climatol* 66, 61–79.

Horel, J. D., and Wallace, J. M. (1981). Planetary scale atmospheric phenomena associated with the Southern Oscillation. *Mon Weather Rev* 109, 813–829.

Hoskins, B. J., and Karoly, D. (1981). The steady linear response of a spherical atmosphere to thermal and orographic forcing. *J Atmos Sci* 38, 1179–1196.

Huffman, G. J., and Norman, G. A. (1988). The supercooled warm rain process and the specification of freezing precipitation. *Mon Weather Rev* 116, 2172–2182.

Hurrell, J. W., Kushnir,Y., Ottersen, G., and Visbeck, M. (Eds). (2003). The North Atlantic Oscillation: climate significance and environmental impact. *Geophys Mon Series* 134.

IPCC. (2001). *Climate Change 2001: The Scientific Basis* (Houghton, J. T., Ding, Y., Griggs, D. J., Noguer, M., van der Linden, P. J., Dai, X., Maskell, K., and Johnson, C. A., Eds). Cambridge University Press, Cambridge.

Isaac, G. A., Korolev, A., Strapp, J. W., Cober, S. G., Tremblay, A., and Stuart, R. A. (1996). Freezing drizzle formation mechanisms. In: *Proceedings of the 12th International Conference on Clouds and Precipitation, 19–23 August 1996, Zurich, Switzerland*, pp. 11–14. International Association of Meteorology and Atmospheric Sciences, Zurich.

Keeter, K. K., Businger, S., Lee, L. G., and Waldstreicher, J. S. (1995). Winter weather forecasting throughout the eastern United States. Part III: The effects of topography and the variability of winter weather in the Carolinas and Virginia. *Weather Forecast* 10, 42–60.

Koolwine, T. (1975). *Freezing Rain.* MSc thesis, University of Toronto, Toronto.

Lackmann, G. M., Keeter, K., Lee, L. G., and Ek, M. B. (2002). Model representation of freezing and melting precipitation: implications for winter weather forecasting. *Weather Forecast* 17, 1016–1033.

McFarlane, N., Boer, G. J., Blanchet, J.-P., and Lazare, M. (1992). The Canadian Climate Centre second-generation general circulation model and its equilibrium climate. *J Climate* 5, 1013–1044.

McKay, G. A., and Thompson, H. A. (1969). Estimating the hazard of ice accretion in Canada from climatological data. *J Appl Meteorol* 8, 927–935.

Milton, J., and Bourque, A. (1999). *A Climatological Account of the January 1998 Ice Storm in Quebec*. Scientific Report CES-Q99-01, Environment Canada, Quebec Region.

Rauber, R. M., Ramamurthy, M. K., and Tokay, A. (1994). Synoptic and mesoscale structure of a severe freezing rain event: The St. Valentine's Day ice storm. *Weather Forecast* 9, 183–208.

Roebber, P. J., and Gyakum, J. R. (2003). Orographic influences on the mesoscale structure of the 1998 ice storm. *J Climate* 131, 27–50.

Rogers, J. C. (1984). The association between the North Atlantic Oscillation and the Southern Oscillation in the Northern Hemisphere. *Mon Weather Rev* 112, 1999–2015.

Stewart, R. E. (1985). Precipitation types in winter storms. *Pure Appl Geophys* 123, 597–609.

Stewart, R., and King, P. (1987). Freezing precipitation in winter storms. *Mon Weather Rev* 115, 1270–1279.

Stewart, R., and King, P. (1990). Precipitation type transition regions in winter storms over southern Ontario. *J Geophys Res* 95, 22355–22368.

Stuart, R. A., and Isaac, G. A. (1999). Freezing precipitation in Canada. *Atmos-Ocean* 37, 87–102.

Walker, J. M., and Bliss, E. W. (1932). World weather V. *Mem R Meteorol Soc* 4, 53–84.

Wallace, J. M., and Gutzler, D. S. (1981) Teleconnection in the geopotential height field during the Northern Hemisphere winter. *Mon Weather Rev* 109, 784–812.

Zerr, R. J. (1997). Freezing rain: an observational and theoretical study. *J Appl Meteorol* 36, 1647–1661.

6

The Effect of Icing Events on the Death and Regeneration of North American Trees

David F. Greene

Department of Geography, Concordia University

Kathleen F. Jones

Cold Regions Research & Engineering Laboratory, U.S. Army Corps of Engineers

Olga J. Proulx

Department of Geography, Concordia University

INTRODUCTION

Literature dating back to the early 20th century reveals a long-standing interest, almost entirely by foresters in eastern North America, on the effect of ice deposited on trees (Abell, 1934; Whitney and Johnson, 1984; Bruederle and Stearns, 1985; Belanger *et al.*, 1996). Although ice events also occur in the boreal forest and west of the prairies, little has been written about the effect of freezing precipitation on forests in this area. The map in Chapter 5, Fig. 6, showing the annual average number of hours of freezing rain, provides only limited information on the probability of damaging freezing rain (and drizzle) storms. The term "freezing precipitation" refers to both freezing rain, with drop diameters of about 1 mm, and freezing drizzle, with drop diameters of about 0.1 mm. As shown by the maps of icing severity for a 50-year return time in reports by the Canadian Standards Association (CSA, 2001), Jones *et al.* (2002), and the American Society of Civil Engineers (ASCE, 2005), freezing precipitation storms are particularly severe in a swath extending northeasterly from Oklahoma, Arkansas, and Missouri, across southwest Ontario and adjacent portions of the northeastern states, and into the Atlantic provinces of Canada. South, west, and north of this belt, freezing precipitation events tend to be less severe, but storms that damage trees and power lines occur even in Texas and the southeastern states and are relatively frequent in the Piedmont of the Carolinas. In the northwestern United States, freezing precipitation tends to occur with high winds in the Columbia River Gorge, and significant events with lower wind speeds occur in the Willamette Valley of Oregon and the Fraser Valley area in Washington and British Columbia. Elsewhere west of the Rockies, damaging freezing precipitation storms are quite rare.

Early studies of ice events (Von Schrenk, 1900; Harshberger, 1904; Illick, 1916; Rhoades, 1918; Rogers, 1922; 1923; 1924) were descriptive, telling us that an event had occurred and affected a smaller or greater area of forest. The more recent literature (Cannell and Morgan, 1989; Nicholas and Zedaker, 1989; Seischab *et al.*, 1993; Jones *et al.*, 2001; Proulx and Greene, 2001; Rhoads *et al.*, 2002; Yorks and Adams, 2003) adopts a more analytical approach to freezing precipitation events and forest damage. While research interests and methods vary widely, studies have centered on species-specific responses (based on post-event damage assessments) and have often been presented as an ordinal scale comparison of species' susceptibility to damage (e.g., low, moderate, and high, or some similar system) (Abell, 1934; Downs, 1938; Kienholz, 1941; Carvell *et al.*, 1957; Lemon, 1961; Goebel and Deitschman, 1967; Whitney and Johnson, 1984; Bruederle and Stearns, 1985; Boerner *et al.*, 1988; Hauer *et al.*, 1993; Rebertus *et al.*, 1997; Boulet, 1998). This more recent literature often leaves the impression that *everything* matters: accumulated ice thickness, wind velocity, topography, tree size (bole diameter and major branch diameter and length), and species' "intrinsic" resistance to injury, exemplified by attributes such as branching architecture and wood density, and has led to little agreement on species-specific or size-specific susceptibility. A process-oriented approach could not merely provide useful data sets but, more important, unambiguously specify what *should* be measured in future investigations.

There has been another major problem in almost all of these studies: aside from icing events in urban settings, it is difficult—but, as we will argue below, not impossible—to measure ice accretion in a stand because: (1) the ice is often melting within 24 hours of the damage; (2) some roads are closed during the period that the ice remains on branches; and (3) we normally cannot extrapolate from ice thickness measurements elsewhere because icing severity varies significantly within the affected region and there may well be no ice or much more ice at the nearest reporting station. Thus, few studies of damage to trees have estimated ice thickness. Further, in post hoc, interspecific comparisons within a single study, it is implicitly assumed that ambient ice thickness was similar for all species. But this may not be true when many transects are used, especially in the mountains, where small differences in altitude can lead to great differences in ice thickness and, of course, species tend to be arrayed along contours.

The climatology and meteorology of icing events were reviewed in Chapter 5. Individual ice events vary considerably in terms of the type of icing that occurs, as well as the intensity (ice thickness), persistence, and wind speeds during the disturbance. In addition to snow storms, the two types of icing with the greatest potential for bole or branch damage are glaze ice and rime ice. Glaze ice is usually clear, smooth, and hard; contains some air bubbles; and has a density of approximately 900 $kg\,m^{-3}$. Rime ice is opaque, with individually frozen cloud or fog droplets separated by pockets of air and, therefore, of a density about 30% to 90% that of glaze ice. Hoarfrost, a low-adhesion, feathery deposit of ice crystals formed from water vapor, may also build up on surfaces but is unlikely to cause branch damage because the total added mass is quite small (McKay and Thompson, 1969). We discuss damage associated with glaze ice from freezing precipitation storms.

Meanwhile, the consequences of major icing episodes for stand competitive dynamics will depend greatly on how often such events recur (Fig. 1). The measure of interest is the return time, which is defined as the inverse of the annual occurrence rate. For example, an ice thickness with a 50-year return time has a 1/50 or 0.02 probability of being exceeded in any

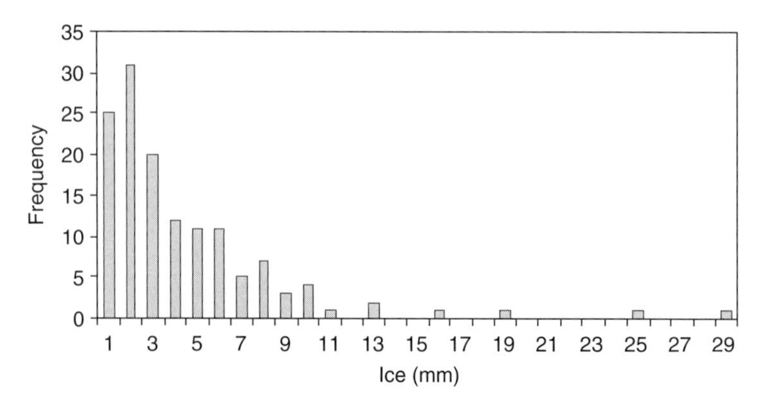

FIG. 1 Ice thickness (added cylinder diameter = $2t_{max}$) at St. Faustin (Quebec) as measured on horizontal wooden 0.025-m dowels for the period 1977–1997. This data set represents 136 events in 21 years. (Data courtesy of Hydro-Quebec.)

year, and a probability of 0.33 of being exceeded in any 20-year period. These measures for return time have often been reported informally as *regional* return times for major events. Thus, we have a report of a return time of 20 to 100 years for "major" storms of unspecified intensity in northern hardwood forests of eastern North America (Melancon and Lechowicz, 1987) and approximately 20 years for northern Missouri forests (Rebertus *et al.*, 1997). A regional return time is not useful for understanding the risk posed by recurrent icing events of specific magnitude for a randomly chosen tree. These foregoing quantifications cannot even suggest that icing disturbance in a particular region is more or less recurrent, at a point, than other forest disturbances, such as stand-replacement fires (Lorimer and Frelich, 1994) or perhaps catastrophic blowdowns (Lorimer, 1977; Canham and Loucks, 1984; Frelich and Lorimer, 1991). There are, however, point estimates for ice thickness for Canada (CSA, 2001) and the United States (ACSE, 2005) for use in the design of ice-sensitive structures, including power transmission lines and tall towers. These maps, showing equivalent radial ice thickness with concurrent gust speeds for a 50-year return time (with factors for adjusting the mapped values to other return times), have not been used in the ecological literature until now.

Our primary objective in this chapter is to introduce a biomechanical interpretation of branch breakage as a function of ice thickness, wind speed, and branch length. The exercise is perhaps too simple, but it will serve to express the interaction of these three fundamental factors. We then discuss other factors that affect the likelihood that any given branch in a forest canopy will break in an ice storm. We finish with a review of the empirical literature on tree damage caused by ice.

THE BIOMECHANICS OF BRANCH BREAKAGE DURING ICE EVENTS WITH AND WITHOUT WIND

Ice Accretion on Branches

As discussed in Chapter 5, freezing rain occurs when a warm air mass with a temperature aloft greater than 0°C slides above a colder air mass with a ground-level temperature below freezing. A snowflake is formed above the warm air layer, melting as it falls through the warm air and becoming supercooled during the descent through the cold air layer at the surface. Eventually it may strike a horizontal surface (flat roof, sidewalk, horizontal branch) that has had time to adjust to the freezing temperatures of the cold air mass. Assuming that it is cold and windy enough that the intercepted drops freeze and if the precipitation is not intercepted by nearby taller elements, the thickness of the ice, t_s (m), on a horizontal surface, is

$$t_s = 1.1\ t_p \tag{1}$$

where t_p is the depth (m) of the accumulated water and 1.1 is the ratio of the density of water to glaze ice.

The literature on the icing of cables, towers, and other structures uses the *equivalent radial ice thickness* (t) as the standard measure of ice accretion in freezing rain. This is the radius of ice added to the cylinder radius that would occur if the ice were distributed uniformly around the cylinder circumference. Jones (1998) developed a simple calculation for the equivalent radial thickness of ice accumulating on a horizontal cylinder, oriented with its axis perpendicular to the wind direction, intercepting rain or drizzle drops. Assume that the ice freezes uniformly thickly around the cylinder, which might occur when the branch or wire or cable twists in response to an asymmetric accretion (because of wind-blown precipitation or when the impinging water flows on the surface briefly before freezing). Then, in the absence of wind, the equivalent radial ice thickness is

$$t = t_s/\pi = 0.35t_p \tag{2}$$

Note that this radius of ice is independent of the original cylinder diameter, and thus we expect the ratio of ice thickness to branch diameter to be very large near the branch tip but much more reduced axially. By adding the effect of wind to this idealized accretion process, and simplifying the relationship between the precipitation rate and the liquid water content of the air, then

$$t = (t_s / \pi)\left[1 + (\bar{u}/5)^2\right]^{0.5} = 0.35 t_p\left[1 + (\bar{u}/5)^2\right]^{0.5} \tag{3}$$

where $\bar{u}$ is the local mean wind speed in m s^{-1} during freezing rain. This is the "back-of-the-envelope" version of the Simple model of Jones (1998), in which the factor 5 is the fall speed of the rain drops in m s^{-1}. The complete Simple model uses the time-varying precipitation rate and wind speed, and the fall speed of the drops is a function of the precipitation rate. The added ice mass ($m_i(x)$) per unit length at x along the branch (i.e., the added area multiplied by the density of ice, p_i) is

$$m_i(x) = p_i \, \pi \, (t^2 + d(x)t) \tag{4}$$

where $d(x)$ is the wood diameter at distance x distal from the branch tip. The ratio of ice mass to wood mass (assuming 800 kg m^{-3} for wood density) is shown as a function of branch diameter in Fig. 2. For this figure, the assumed wind speed is 5 m s^{-1}, and the total ice thickness on a horizontal surface, t_s, is 0.09 m (an extremely large value), giving a radial ice thickness from Equation (3) of 0.04 m (a bit less than the maximum reported during the 1998 event discussed in Chapter 5). Near the branch tip at a diameter of 0.002 m, the ice/wood mass ratio is about 1900, while near the base of a 3-m long branch at a branch diameter of 0.053 m, the ratio is about 6. Thus, the lack of dependence of the radial ice thickness on diameter leads to a much greater relative increase in the vertical load at the branch tip than at the base. However, the mass of ice per unit length [Equation (4)] near the tip of the branch is smaller than near the base, 4.9 kg m^{-1} compared to 10.7 kg m^{-1}. Note that the independence of radial ice thickness and diameter is not unique to this Simple model; in other ice accretion models for freezing rain in which the accreted ice thickness is allowed to vary with the branch diameter and a heat balance calculation is used to determine the fraction of the impinging water that freezes, it still remains essentially constant (Jones, 1998).

Now we need to embed this accretion model into the calculation of the stress on a branch. The accumulated ice can increase the resultant force on a branch element in two ways. First, the added area increases the drag exerted by the wind and second, the added mass increases the weight on the branch. We deal first with the drag.

Effect of Wind on the Ice-laden Branch

The wind's drag force per unit length ($D(x)$) on the shoot element at distance x distal from the branch tip is

$$D(x) = (p_a/2)\ \mathrm{d}(x)\ u^2\ C_D \tag{5}$$

where p_a is the density of air (1.2 kg m^{-3}), u is the ambient wind speed (m s^{-1}) near the element of interest (typically increasing with height above ground), $d(x)$ is the diameter at x (m), and C_D is the drag coefficient (which we can assume to be 1.0 at the Reynolds numbers of interest).

What wind speeds might be expected while the branches are ice-covered? Imagine a 25-m tall hardwood forest with all the branches located in the upper 50% of the forest and with crown midpoints located at 18.75 m. In *open* areas, the expected 3-s gust speed following major icing events while the air temperature remains below freezing and the accumulated ice remains on branches is 20 m s^{-1} in much of the United States (ASCE 2005). Speeds are lower in the southern states, where the air tends to warm more quickly after freezing precipitation ends (so there is less opportunity for very strong winds to occur concurrently with ice-covered branches), and higher in the north-central states, where winters tend to be cold. How strong will the winds be within an adjacent forest near the open area where

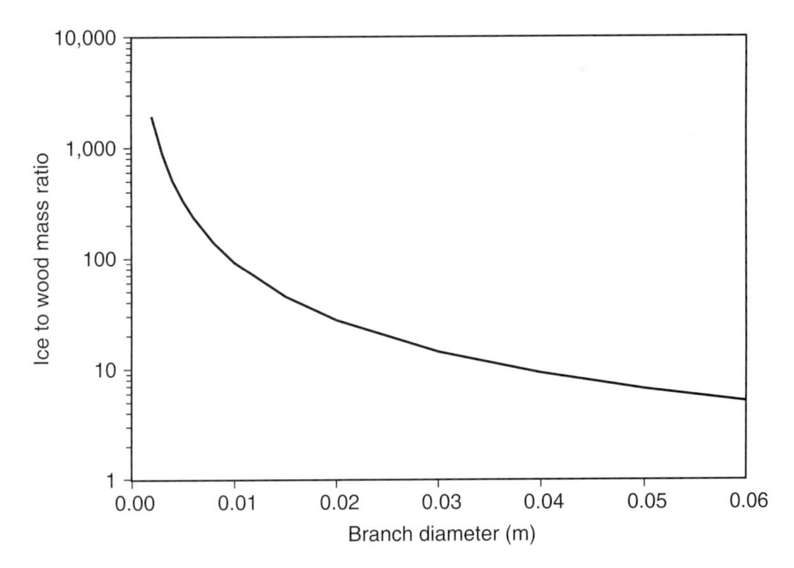

FIG. 2 The ratio of ice mass to wood mass as a function of position (diameter) on a branch for an equivalent radial ice thickness (t) of 0.04 m.

measurements were made? There will, of course, be a steep gradient in the wind speed profile just above and within the crown space (Greene and Johnson, 1996). Assuming a 20 m s^{-1}, 3-s gust at 10 m in the open, we can estimate the *fastest speed* at 18.75 m in a *leafless* forest using the protocol of Greene and Johnson (1996): $u_c = 0.47\ u_{r10}$ where u_{r10} is the reference speed at 10 m height in an open area and u_c is the horizontal speed midway in the crown. Thus, the gust speed expected to occur at the midcrown height while ice remains on the branches following a major storm is about 10 m s^{-1}.

There is a further consideration regarding the relationship between branch length and wind speed: given ramification, longer branches tend to be located lower within the crown and therefore experience slower wind speeds than branches higher in the crown. This complexity is ignored in the following calculations, and a constant wind speed is assumed over the length of the branch.

We initially idealize a branch as a cantilever lacking both side shoots and leaves with diameter increasing basally from the branch tip toward the bole. We need to allometrically relate distance from the branch tip (x) to the diameter ($d(x)$) at that point. Niklas (2000) has calculated the relationship for the eastern hardwood species *Prunus serotina* Ehrh. (black cherry) as:

$$\log x = 1.16 + 0.166 \log \mathrm{d}(x) - 0.29 (\log \mathrm{d}(x))^2 \qquad (6)$$

where the logarithm is base 10. [Note that Niklas (2000) erroneously writes 1.66 rather than 0.166 (as above) in both his figure caption and in his text.] In what follows we will use the allometry of this cherry as typical of eastern hardwoods. For branches up to 3 m long, Equation (6) is fit very well by the linear relationship

$$x = ad(x) + b \qquad (6a)$$

where $a = 59.26$ and $b = -0.0956$ m. This linear form will be used in calculating the stress at the base of the branch.

The bending moment M (defined as a force multiplied by a length) on the entire branch is the integral of the moments from the elemental sections dx from the tip to the base:

$$M = \int_0^{x_T} \frac{1}{2} C_D p_a u^2 [d(x) + 2t] (x_T - x)\, dx \qquad (7)$$

where x_T is the total length of the branch. Using the linear variation of the diameter of the branch along its length results in

$$M = \frac{1}{2} C_D p_a u^2 x_T^2 \left(\frac{x_T}{6a} - \frac{b}{2a} + t \right) \tag{7a}$$

Finally, the bending stress (a force per area: a pressure), σ, at the branch base is given by

$$\sigma = Md_T/(2I) \tag{8}$$

where $d_T = (x_T - b)/a$ is the basal diameter and I is the moment of inertia ($I = \pi(d_T/2)^4/4$). The stress at which the branch breaks (the breaking stress) can be found in published compendia for green (recently cut, not dried) wood specimens. These values range from about 30 to 90 MN m^{-2} for North American trees; in what follows we will use 60 MN m^{-2} as a typical value of the breaking stress. Of course, any prior damage to the branch acts as a stress riser, so a branch with a cavity, or canker, or any other sort of damage will break at a stress lower than the nominal breaking stress for that kind of tree (Kane *et al.*, 2001).

The result of this calculation of ice-augmented drag on a 3-m-long branch is shown in Fig. 3 for the range of radial ice thicknesses (0.01–0.05 m) that encompasses the deposition experienced from north to south across

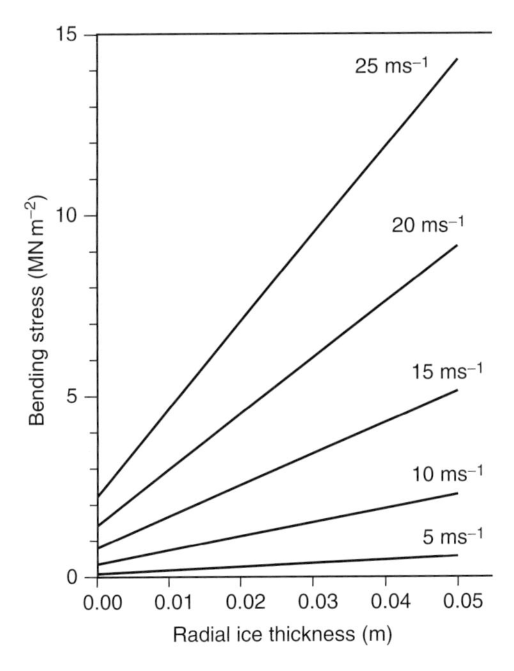

FIG. 3 The effect of wind-induced drag: stress vs equivalent radial ice thickness (t) where there has been no interception by higher branch elements for a 3-m long horizontal branch. The velocities (m s^{-1}) denote the ambient wind speed.

southern Quebec during the 1998 icing event. Given the dependence of drag on the square of the wind speed, it is not surprising that the bending stress is more sensitive to wind speed than to ice thickness, which is a linear term in Equation (7a).

The effect of branch length on the bending stress is shown in Fig. 4 for a wind speed of 10 m s^{-1} for leafless branches and across the range of radial ice thicknesses of interest. The stress peaks at a branch length of about 0.2 m. The magnitude and location of the peak are determined by the parameters *a* and *b* (Equation (6a)), where *b*/*a* is the minimum branch diameter (1.6 mm in this case). If *b* is doubled, the peak stress decreases by about half and occurs at a branch length of 0.4 m. If the rate of increase of branch diameter with length (the parameter *a*) is greater than in Equation (6a), the peak stress increases, but the peak still occurs at a branch length of 0.2 m. At a wind speed of 10 m s^{-1}, the stress from wind drag on this nonramified cantilever is significantly less than the breaking stress, even at the peak.

We should perhaps not take too seriously the mode in Fig. 4 because this calculation has ignored the probably much less compact shape of the actual ice accretion and has not included shoot ramification; the

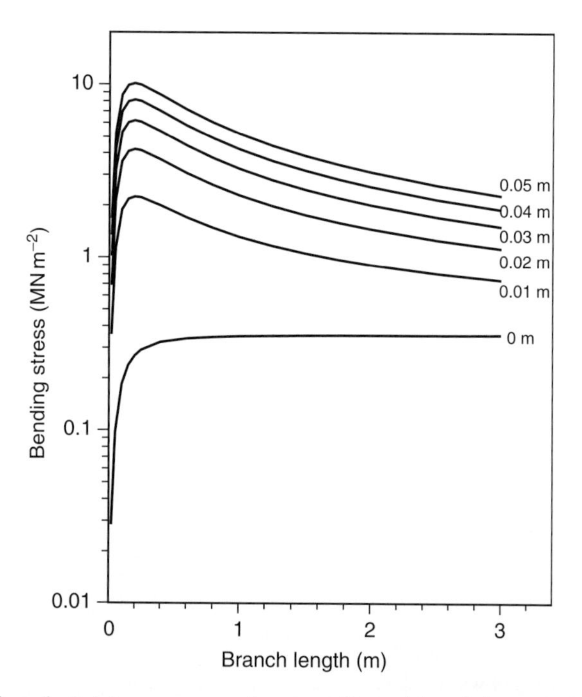

FIG. 4 The effect of wind drag: stress as a function of branch length for equivalent radial ice thicknesses from 0 to 0.05 m with an ambient wind speed of 10 m s^{-1}.

ice-enhanced drag will be greater for entire branch units than pictured here, and the bias is especially great for longer branches, which will have more side shoots than shorter branches. This added complexity is discussed further later in this chapter.

Effect of Gravity on the Ice-laden Branch

Of course, the additional weight of the ice applies a significant load to the branch, whether or not the wind is blowing. The drag in Equation (7) is replaced by the weight (total mass multiplied by the gravitational acceleration), and so the bending moment M_g is

$$M_g = \int_0^{x_T} \pi p_i g [d(x)t + t^2](x_T - x)\,dx \qquad (9)$$

where g is the gravitational acceleration (9.8 m s^{-2}). This equation uses the ice mass per unit length from Equation (4). The bending stress is, as in Equation (8), equal to $M_g\, d_T/(2I)$, and the diameter variation is given by Equation (6a). Integrating Equation (9) gives

$$M_g = 0.5\pi p_i g x_T^2 t\left(\frac{x_T}{3a} - \frac{b}{a} + t\right) \qquad (9a)$$

The bending stress from the ice weight is plotted in Fig. 5. In the absence of ice (t_s = 0), the stress rises with branch length but is, as one would expect, far below the breaking stress for healthy branches.

As for wind drag, the situation is quite different as ice is added: stress peaks at a branch length of about 0.2 m, and the location and magnitude of the peak depend, as with wind drag, on the minimum branch diameter and the rate of increase of diameter with distance from the tip. Interestingly, the stress from branch weight, both with and without ice, is much greater than that from the wind drag: an icing event with a radial ice thickness greater than 0.04 m should cause the loss of branches less than 0.5 m long even in the absence of wind. By contrast, the 10 m s^{-1} gust speeds at midcanopy height that are expected after an icing event could not break healthy branches if wind drag were the only stress. The effects of ramification on gravity-induced stress are examined in the following section.

Complications

This modeling is an initial foray; much that could make it more realistic and applicable to individual trees or forest stands is still missing. In this section we discuss (1) branch ramification, (2) bole snapping from wind drag

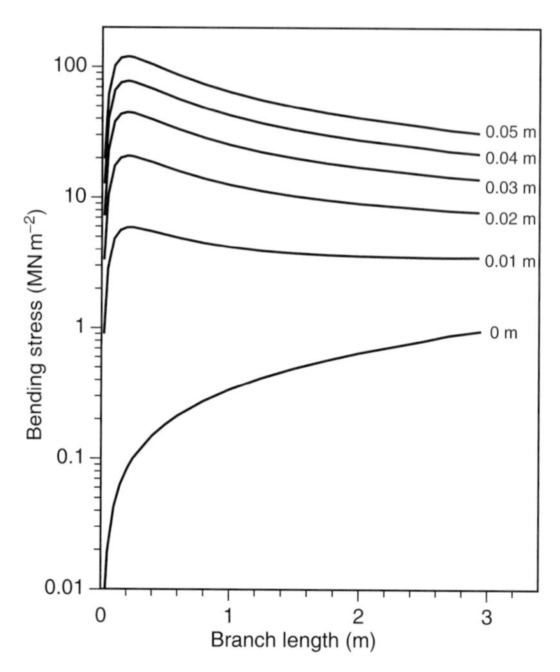

FIG. 5 The effect of gravity: stress as a function of branch length for equivalent radial ice thicknesses from 0 to 0.05 m.

on an entire tree, (3) branch deflection and orientation, (4) ice persistence, (5) canopy position, and (6) breaking strength.

Individual branches need to be treated more realistically as a ramified set of shoots possessing a varied set of branching angles and, in the case of non-*Larix* conifers, leaves attached. (Occasionally, of course, icing events occur when hardwoods or *Larix* still have leaves attached, particularly in regions where cold winters are not the norm.) To examine the effect of ramification on ice-augmented stresses from both wind and gravity, we cut a healthy, 2.76-m long black cherry branch from a 15-m tall tree at Mt. St. Hilaire, Quebec. Ignoring leaves, we measured the length and branching point of every shoot, working outward apically from the base. The total length of these side shoots was 30.7 m, increasing the length of the entire unit by an order of magnitude. The actual branch geometry was then idealized to simplify the calculation of moments and to examine the effects of branching angle and length of side shoots. All shoots arising from the main branch were assumed to be at the same constant angle to the main branch, and each subshoot was assumed to be parallel to these main shoots. The bending stress at the base of the ramified branch (as well as that of the simple cantilever we assumed earlier in this chapter) is shown in Fig. 6A (wind drag) and 6B (gravity) for angles of 90° (main shoots horizontal and perpendicular to the main branch), 45°, and 0° (main shoots pointing in the same direction as the main branch). Also shown is the bending stress for a pruned branch, with all except the side shoots arising directly from the main branch removed—this pruning reduces the total length of shoots from 30.7 to 10.6 m.

For our expected 10 m s^{-1} maximum gust at midcanopy, the wind load on this branch unit still does not cause a bending stress exceeding the average breaking stress of 60 MN m^{-2}, even for a 0.05-m radial ice thickness. However, the addition of side shoots also increases the bending stress from the gravity load so that this ramified branch could be broken by the 10 m s^{-1} gust if the radial ice thickness exceeded 0.015 m. This threshold ice thickness is roughly correct for the ramified branch for all assumed branching architectures. The removal of the higher-order shoots reduces the bending stress from either wind load or gravity load by about 60% for this range of ice thicknesses. For this pruned branch, the required radial ice thickness to cause failure at a gust speed of 10 m s^{-1} doubles to about 0.03 m. While these calculations ignore the real variations in branching architecture that occur, they show that the ice on side shoots in a branch unit contributes significantly to the bending stress at the base of the branch, and that the total length of the side shoots is more important than the branching architecture. Note that these results are for this particular 2.76-m branch unit and provide

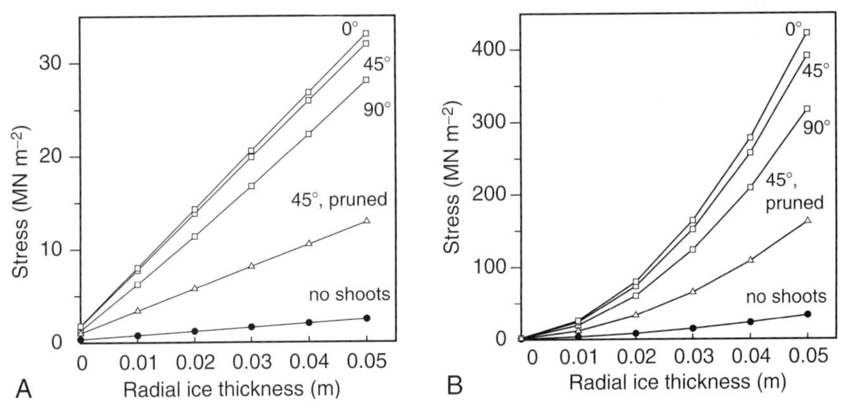

FIG. 6 Bending stress on a ramified 2.76-m branch compared to a branch with no side-shoots, showing the variation of stress with branching architecture and pruning. (**A**) Wind load (with an ambient speed of 10 m s^{-1}; (**B**) gravity load.

only one example of the increase in bending stress that might be expected for a ramified branch compared to a nonramified branch.

The augmentation of the wind-induced bending moment due to ice on an entire tree (i.e., bole snapping) is worth examining as well. A rough calculation can be done by using the 8.5-m tall ash (*Fraxinus*) in Jones (1999). If we assume that the 1538 m of total branch length on the ash are uniformly distributed over the canopy from 3 to 8.5 m, then the moment from the wind drag on the tree can be estimated from $L(d)$, which is the cumulative distribution of branch length with diameter less than d

$$L(d) = L_T e^{-(d_0/d)^2} \tag{10}$$

where the total branch length is L_T = 1538 m and a characteristic branch diameter is d_0 = 0.0041 m for this tree. For this cumulative distribution, the median branch diameter (half of the total branch length has a larger diameter) is $-d_0/\ln(0.5^{1/2})$ = 0.012 m. At the base of the stem, the bending stress (σ_{tree}) is

$$\sigma_{tree} = 0.5 C_D p_a u^2 H L_T (2t + d_0\sqrt{\pi}) / \left(0.25\pi\,(d_{tree}/2)^3\right) \tag{11}$$

where H = 5.75 m is the distance from the base of the stem, with diameter d_{tree} = 0.28 m, to the midcanopy height. For this tree, the value of $d_0\sqrt{\pi}$ is 0.007 m, so a radial ice thickness of only 0.01 m would more than triple the bending stress in the stem from the wind drag. A radial ice thickness of 0.03 m increases the bending stress by an order of magnitude. Note that these increases in stress are similar in magnitude to what we calculated for 1-m long unramified branches (Fig. 4). The moment in the stem is high compared to the moment in either the unramified branches in Fig. 4 or the ramified cherry branch in Fig. 6A because the wind is acting on all 1538 m of branches. However, because of the large stem diameter compared to the branch diameter, the stress in the stem exceeds the breaking stress only at high wind speeds. With no ice, the bending stress at the base of the stem exceeds the average breaking stress at a wind speed of 58 m s^{-1}. At radial ice thicknesses of 0.01 and 0.03 m, the wind speed necessary to exceed this breaking stress is reduced to 30 and 19 m s^{-1}, respectively.

We have ignored the branch deflection due to the ice loading and the wind drag, which will decrease the stress at the base of the branch (Bisshopp and Drucker, 1945; Leister and Kemper, 1973). Another factor—not dealt with here—is the relative orientation of the branch to the wind direction and to the vertical. For simplicity, we have assumed a horizontal branch oriented perpendicular to the wind. But of course branches with orientations oblique to the wind and to gravity will be the norm: wind and ice loads on those branches will be smaller than shown here. At the extreme, the wind load on a branch parallel to the wind direction will be from skin friction only, which is very small compared to the form drag, and the additional bending load due to ice on a branch (or, more cogently, a bole) that is close to vertical will be almost zero. However, the orientation of the branch may change in the course of an ice storm as the branch unit or tree sags, bends, or leans, changing the wind and gravity environment of individual branches. This latter point re-emphasizes the preliminary nature of our modeling effort.

As mentioned earlier, the concurrent gust speeds in the ASCE map of 50-year ice thicknesses generally increase from regions that tend not to have cold winters to regions where long cold winters are the norm, reflecting the expected increase in persistence of the accreted ice. This regional map does not show, however, local areas where temperatures remain cold longer than in the surrounding region perhaps because of higher terrain; nor does it show local areas where wind speeds tend to be higher than at the airport weather stations. A serious analysis of ice-induced damage in a particular region will need to deal with the local residency period; that is, how long the accumulated ice remains on the trees. Clearly, the chance of wind-related tree damage increases with the length of time the ice persists (Elfashny *et al.*, 1996; Jones *et al.*, 2002). Often, ice accretion is followed by above-freezing temperatures as the cold air layer at the surface is eroded away by the advancing warm air mass, but ice can remain long after an icing event (Schaub, 1996). In Quebec, Elfashny *et al.* (1996) reported that ice residence time for most events was limited to between 1 and 2 days. Nonetheless, at a few stations in Quebec, especially at high latitudes, residency periods over a month long have been recorded. As pointed out in Chapter 5, the 1998 event that devastated the forests of southeastern Canada and the northeastern United States was unusual because it included two periods of precipitation over a 5-day period with the initially accreted ice not melted before precipitation was renewed.

We have ignored the dependence of ice accretion on canopy position, using the expected gust speed at midcanopy in calculating the bending stress from the wind load. However, less ice is expected to accrete on lower branches and more on branches near the top of the canopy, both because higher branches in the canopy intercept some of the precipitation and because wind speeds decrease from the top to the bottom of the canopy. For the typical 5 m s^{-1} wind speeds in open areas during freezing rain storms, the ice thickness gradient in the canopy is small because of the low wind speeds, ranging from 3.8 m s^{-1} at the top of the canopy to

1.4 m s^{-1} at the bottom. Using this wind gradient from Greene and Johnson (1996) results in radial ice thicknesses for t_p = 0.03 m varying from 0.011 to 0.013 m from the bottom to the top of the canopy. However, the prior interception of precipitation by the upper canopy will steepen this gradient. The median interception fractions for deciduous and coniferous forests for liquid precipitation are 13% and 22%, respectively (Dunne and Leopold, 1978). The interception fraction tends to be higher in light-precipitation (the norm for freezing precipitation events) than in heavy-precipitation events, and tends to be lower in the dormant season than in the growing season in deciduous forests. During freezing precipitation, the interception fraction will be greater than suggested by rain studies because of the near-zero stemflow and throughfall by leaf drip as the intercepted water freezes to the branches. Assuming 20% interception with the wind speed and precipitation depth given above, the ice thickness at the base of the canopy is decreased to 0.009 m. The positive feedback of the accreting ice causing the canopy to become denser preferentially near the top will further increase the vertical gradient in ice thickness over the course of the storm. This is compounded by the expected vertical variation in gust speeds following the freezing precipitation event. The 10 m s^{-1} gust speed at midcanopy is consistent with winds of 15 m s^{-1} at the top of the canopy and 6 m s^{-1} at the bottom of the canopy. Thus, we expect greater stresses from both ice and wind for branches that are higher in the canopy.

Finally, we have ignored the variation in breaking stress of wood among different tree species, in the same species grown in different climates, and in different parts of the stem and canopy of an individual tree (Kane *et al.*, 2001). For branches with a low breaking stress of 30 MN•m^{-2}, Fig. 5 shows that unramified branches less than about 0.8 m long would break with a radial ice thickness of 0.03 m, while an ice thickness of over 0.05 m would be required to break a branch in that same length range with a breaking stress as high as 90 MNm^{-2}.

Predictions Based on This Biomechanical Exercise

Our first prediction from the forgoing modeling is that, assuming there is enough ice to make the event worth categorizing as a serious icing phenomenon, the stress caused by ice is far more important than that caused by wind. Nonetheless, the drag of the wind can be the hair that broke the camel's back. Toward the end of the 1998 event in the wooded area on Mount Royal in Montreal before the ice began to melt, one of us (Greene) noted qualitatively, yet unambiguously, that gusts of wind (during a day with very average wind speeds according to measurements at the nearby airport) were immediately followed by the "popping" sound of branches breaking and then the crash of the branches as they hit the ice-covered ground. One almost never heard branches breaking when there was no local gust simultaneously felt on one's face. There was no precipitation at the time of these observations; presumably, those branches sufficiently loaded with ice to break in relatively calm air had already fallen earlier in the week.

Oddly, the only quantitative information indicating that the wind matters in the amount of breakage during an icing event is that a few reports show that trees snapped and bent in the direction of the prevailing wind (with the prevailing wind taken from the nearest reporting station). Proulx and Greene (2001) found that the orientation of damaged stems was the same as that of the local wind on flat ground during the 1998 event in southern Quebec. This observation is similar to that of Nicholas and Zedaker (1989), who studied two severe icing events where the observed maximum icing diameter (*not* radial ice thickness) ranged from 0.07 to 0.10 m: they reported that generally, within a stand, tree stems and crowns were all snapped in a similar direction. We can imagine, then, that as ice accumulates on a branch, there comes a point where it can be broken by a small gust before it can be broken by the slower process of additional ice accretion. Thus, branches will lie on the ground primarily in the direction of the prevailing wind, and can be used as an ex post facto indicator of the wind azimuth during the event. [We must, however, be careful about using the orientation of snapped or bent boles as evidence for the effect of wind because, as reported by Proulx and Greene (2001), trees with asymmetrical crowns (typical of steep slopes and forest edges) will bend or snap in the direction of the center of gravity (i.e., downslope or toward the clearing) regardless of the prevailing wind—at least when the wind is modest, as in the 1998 event. That is, one should use only flat areas to show, indirectly, the prevailing wind.]

A second prediction is that in a typical icing event (and the subsequent period before the ice melts) assuming gusts of 10 m s^{-1} in the midcanopy of a leafless hardwood forest, breakage of a ramified branch about 3 m long would occur with an equivalent radial ice thickness (t) on the order of 0.015 m (from Figs. 5 and 6). In the absence of wind, the required radial ice thickness is only marginally smaller. One of us (Jones) has noted that significant tree damage begins at about 0.005 m radial ice thickness, although most shoots are simply sagging near their apical ends in response to the added weight. Proulx and Greene (2002) showed that breakage beyond the "background level" of crown loss from other previous minor disturbances began to be noticeable in northern hardwood stands only at ice diameters d_{max} on the order of 0.015 m. Branches with defects (see below) will have even lower minimum icing levels for breakage. Indeed, Lemon (1961) argued that breakage of branches with defects typically becomes noticeable with an added ice diameter of 0.006 m, and very common among such branches with an ice diameter of 0.0125 m. The ice diameter measurements $d_{\max}$ used by Lemon (1961) and Proulx and Greene (2002) are at least twice and perhaps significantly greater than twice the equivalent radial ice thickness: $t \leq d_{\max}/2$. These authors' d_{max} values of 0.015, 0.006, and 0.0125 m are consistent with equivalent radial ice thicknesses less than 0.007, 0.003, and 0.006 m, respectively. Thus, these studies are very roughly consonant: significant damage begins at a t value of about 0.003 to 0.005 m, with large numbers of (especially) defective branches falling if t reaches about 0.006 or 0.007 m.

As for healthy branches, Proulx and Greene (2002) observed that about 50% of the branches were removed (permanent bending and snapping are included in this tally) with icing diameters on the order of 0.045 m. Lemon (1961) observed that damage to healthy branches and snapping of small trees became widespread when the added ice diameter exceeded about 0.024 m. This range (t less than roughly 0.012 to 0.022) includes our expectation for breakage of a ramified 3 m long branch at $t = 0.015$ m (Fig. 6).

A third (and very obvious) prediction is that hollowed branches and those with stress risers should be the first to break. As the proportion of the central portion of a branch rots (and thus effectively hollows it out), the bending stress increases. Hollowness by itself only slightly increases the stress if the hollow is centered (Kane *et al.*, 2001): e.g., if half the diameter is rotted wood, the stress is only about 6% higher. Much more important are cracks and cavities in the wood near the outside of the shoot (caused by frost, lightning, beetle galleries, pitch pockets at old injury sites) that amplify stress and lead to crack propagation and, thus, eventually breakage. Structural failure of branches or boles has often been linked to the degree of rot and thus, very generally, to tree age. Empirical evidence and anecdotes for enhanced breakage caused by rot and stress risers are numerous (e.g., Seischab *et al.*, 1993; Bruederle and Stearns, 1985; Rhoads *et al.*, 2002). Additionally, the resulting new wounds from breakage significantly weaken the individual tree and greatly increase the risk of insect and fungal attacks after the icing event (Campbell, 1937; Melancon and Lechowicz, 1987; Rebertus *et al.*, 1997).

A fourth prediction involves higher versus lower branches. We set the height for evaluation of the wind speed at 50% of crown length for a

25-m tall forest, pointing out the steep gradient from the top to the bottom of the canopy. Given that branch length increases with increasing distance from the top of the canopy, the short branches are mainly in the upper crown and are subjected to greater drag during an ice storm and more added ice weight. Thus, we can make a simple prediction: upper branches are much more likely to be broken than lower branches because they experience greater wind speeds while ice remains on the branches (causing larger wind drag) and they intercept some of the freezing precipitation so that less water is available to freeze on the lower branches. For our assumed relationship between diameter and length, the highest stress at the base of unramified branches was for branches about 0.2 m long. These short branches will also be ramified but with many fewer side shoots than our sample 2.76-m long branch. The probably greater radial ice thicknesses (along with the stress peak) for short branches should more than compensate for the greater ramification of the more lightly iced longer branches lower in the canopy. We can find no quantification of this in the literature, although it seems to us qualitatively to be correct on the basis of observations by two of us (Greene and Proulx) during modest icing levels. Note, however, how messy testing this prediction really will be: falling upper branches can add to the moment on lower branches with which they collide, thus decreasing the expected vertical gradient in branch loss.

A fifth prediction is that interspecific differences in wood density (characteristic breaking stress of the unseasoned wood) ought to be important. As mentioned above, the range of breaking stress values for unseasoned (green) wood is about 3-fold for most North American tree species. However, within a diverse eastern hardwood stand we might expect only a 2-fold range in breaking stress from 40 to 80 MN m^{-2}. A radial ice thickness of 0.03 m or more would cause inflexible unramified branches with a breaking stress of 40 MN m^{-2} to fail. However, only at a radial ice thickness of 0.05 m or more would branches with a breaking stress of 80 MN m^{-2} fail. This effect of species-specific breaking stress may be difficult to recognize in the field because of other factors that influence the ability of the tree species in that hardwood stand to withstand ice loads, including the variation in ramification and branch angle (branching architecture), differences in branch flexibility, characteristic differences in the position of that species in the canopy (this is not a successional argument: on average, shade-intolerant species will place their crowns higher in the canopy layer than will shade-tolerant species because the former cannot be present in the lower canopy stratum), differences in the age of the various tree species (and thus time for the development of stress risers caused by fungal rot or insect-induced injury or previous exposure to damaging storms). Further, falling branches breaking lower branches and variations in breaking stress in different parts of the stem and canopy of individual trees will, of course, make it even more difficult to quantify the effect of wood density.

ICE MEASUREMENTS IN THE FIELD

The foregoing analysis is useful only if it can be tested with field data. Clearly, we need studies that relate the consequences of an icing event (e.g., branch loss, subsequent mortality) to the ice thickness. Previous attempts at measurement of ice thicknesses prior to melting have been based on a few estimates made on fallen branches at one or two sites, or on interpolation of meteorological station data (e.g., Proulx and Greene, 2002). While it is difficult to make field measurements, it can be done, and we turn to field problems here.

Given the typically short period of ice persistence, it is crucial that the investigator drive to the site as soon as it becomes clear that a major event is occurring nearby. Some roads are usually closed because of fallen branches, boles, and power lines, but often a 4-wheel drive vehicle can get around these obstacles.

Having arrived at a site of interest, one measures the ice on standing and fallen branches. The ice-covered branches are readily cut with a saw, and the mean radial ice thickness for a branch segment can be calculated [solving Equation (4) for t] given the measured masses (ice plus wood minus wood alone), branch length, and branch diameter. The ice-covered branch samples are stored in zipper-lock plastic bags in a cooler and processed at the end of the day. The cooler keeps the ice at least partially frozen until the end of the day, and the sealed plastic bags (double bags, preferably) retain the melted water for that sample. (The goal at the end of the day is to find a place to stay where the power is on so that the samples collected that day can be processed!) The samples and bare branches are weighed on an electronic balance to 0.1 g, the branch length measured to 1 mm, and the diameters of the branch segment measured at a few locations to 0.1 mm, from which the average diameter is determined for use in calculating the equivalent radial ice thickness. To provide a context for the ice thickness data, it is useful to record the GPS coordinates of the site (later one can determine elevation, slope angle, and other factors from digital elevation models), measure the wind speed and direction and temperature, record the height above ground and orientation of the sample, note the severity of damage, and photograph the site and samples.

The collection of ice data during and following a storm can be dangerous. The dangers include but are not limited to one-car and multi-car accidents, live power lines on the ground, falling trees and branches, slippery surfaces underfoot, and cold and wet conditions. (It becomes clearer why we have so little data on this disturbance type.)

In many cases, however, the investigator cannot reach the site until after the ice has melted. Using standard weather station data, rough stand-scale estimates of ice thickness can sometimes still be made. If the reporting station is near the site and at a similar altitude, then one can use the hourly measurements of the amount of freezing rain in the precipitation gauge. To calculate the radial ice equivalent (t) using Equation (3), the mean wind speed at the reporting station can be used, although this must be corrected for canopy position [e.g., the protocol in Greene and Johnson (1996)]. However, if the site of interest is at a different altitude from the reporting station (e.g., a mountain slope vs. an airport in the valley bottom) or some distance away, then the icing thickness cannot be reliably estimated; we are dealing with a wedge of cold air lying at an angle under warm air so, even if the temperatures at both the site and the station are below freezing, the type of precipitation will depend on the thickness of the wedge of cold air at each location. If the cold air layer is too thick or the warm air layer above it is thin, the precipitation will be ice pellets rather than freezing rain. Where there is no upper warm air layer, the precipitation will be snow. Furthermore, the amount of precipitation and the wind speed may differ significantly between the station and the site, both of which control the amount of ice that can accrete.

A final issue is whether we can experimentally test loading equations or the relationship between ice accumulation and branch diameter in the field. Lanctot *et al.* (1960) sprayed water on horizontal cylinders to simulate freezing rain; however, this experiment was done in the absence of wind to obtain even exposure of the cylinders to the spray. While it would be interesting to use a spray rig to create artificial freezing rain on tree branches, care must be taken in simulating natural icing conditions with appropriate drop sizes, precipitation rates, air temperatures, and wind speeds. Furthermore, the drops must be created far enough above the branch that they have time to supercool before impact.

A REVIEW OF THE LITERATURE ON TREE DAMAGE CAUSED BY ICING EVENTS

Proulx and Greene (2001) argued that the *severity* of the damage will be primarily controlled by ice thickness while the *type* of damage would be controlled by the size of the tree. By contrast, other authors (e.g., Lemon, 1961;

Bruederle and Stearns, 1985; Cannell and Morgan, 1989; Hauer *et al.*, 1993; Seischab *et al.*, 1993; Sampson and Wurtz, 1994) have claimed nontrivial roles for species-specific characteristics, such as branching architecture and wood density, and spatial effects, such as proximity to an edge or slope angle.

Types of Damage in Relation to Tree Size

Broadly, there are three types of significant ice-induced damage to trees: permanent bole bending, bole failure (snapping), and major branch loss. The biomechanical analysis above was for the latter category but is also applicable to bole failure. Permanent bending ranges from minor deviations from the perpendicular position to the extreme case in which a tree bole is arched over until its crown is resting on the ground. The extreme case is of more interest to us because stems bent to the ground for a long period almost never, it would appear, regain their place in the canopy (Carvell *et al.*, 1957; Lemon, 1961), although we can find no quantitative study of this. Tree injury by stem bending ranges from relatively inconsequential (temporary) to catastrophic (permanent). Smaller-diameter canopy individuals are more likely to bend permanently in response to ice-loading (Downs, 1938; Spaulding and Bratton, 1946; Shepard, 1975; Sisinni *et al.*, 1995). One study found that for three northern hardwood species, permanent bending was almost entirely limited to stems with diameter at breast height (dbh) less than 0.18 m (Proulx and Greene, 2001) and was the predominant mode of failure for stems with dbh less than about 0.1 m. Also, these authors found that dbh was a more crucial determinant of bending than was species identity, accounting for four times more of the explained variation. This is not to deny that some species with characteristically fine branches (e.g., birch) tend to bend more than others (Lemon, 1961) but merely to relegate species differences to the status of minor factors.

Many authors have speculated that an individual stem's position in the canopy influences its susceptibility to icing injury (e.g., Bruederle and Stearns, 1985), assuming that understory stems are less susceptible to damage (Downs, 1938; Boerner *et al.*, 1988; Seischab *et al.*, 1993; Hauer *et al.*, 1994) because most of the ice is intercepted by the top of the canopy. While this ought to be true (as we argued above), we point out that there is no convincing evidence that it is indeed correct. Perhaps a comparison of the breakage among boles and branches in the subcanopy of an intact forest versus equal-sized stems in adjacent small forest gaps would be the best way to tackle this issue: differences in wind speed and ice deposition would be minimized while the contrast in interception would be strong.

Snapping refers to trees whose trunks have been completely severed or remain attached by only a small amount of bark or wood. Proulx and Greene (2001) found that boles of intermediate diameter (approximately 0.1 to 0.2 m dbh) were most likely to fail. We should note that unlike the damage during catastrophic wind events, uprooting is not common for storms in the higher latitudes (e.g., southern Canada and the northern tier of states in the United States) that occur in mid- to late winter because the ground is, except in rare instances, frozen at the time of the icing event. Further south, however, and in that subset of major icing events occurring near the beginning of the winter season, uprooting during severe icing events can be common.

The bigger trees, whose basal diameters preclude bending or snapping, tend to suffer severe branch loss given high-magnitude ice loading (Hauer *et al.*, 1993; Sisinni *et al.*, 1995; Proulx and Greene, 2001; Rhoads *et al.*, 2002). The intuitive explanation (from Proulx and Greene, 2001) for the forgoing size-related damage under a severe loading is as follows. Very small boles (<0.01 m dbh) sometimes bend dramatically but are sufficiently resilient that they bend back after the load is released. Boles with a dbh of 0.01 to 0.1 m will bend without snapping but often cannot bend back if the load has been maintained for too long a period. It is noteworthy that their small low-order branches do bend back after the ice melts [a qualitative observation by two of us (Greene and Proulx)] while the trunk remains strongly curved. Larger boles (0.1–0.2 m dbh) have high-order branches that bend without breaking, but the accumulated load is too great and the bole itself breaks. [Whether wind is crucial for snapping (and less important for branch loss or permanent bending) has never been discussed in the literature.] Finally, the larger hardwood boles (>0.2 m dbh) have high-order branches, at a variety of angles but not departing too far from being parallel to the forest floor (the angle where weight will be most effective) that break, thus reducing the load on the bole. Older stands composed of such large trees tend to look, after a major icing event, like World War I battlefields, the individual boles standing with their appended large-diameter branch stubs, a meter or so in length, each revealing cream-colored wood.

Other Factors Influencing Icing Damage to Trees

Aside from tree size, susceptibility to icing damage is influenced by other factors that are typically defined qualitatively: branching architecture, species, crown asymmetry, and position of the tree in the canopy (Lemon,

1961; Bruederle and Stearns, 1985; Cannell and Morgan, 1989; Seischab *et al.*, 1993; Rebertus *et al.*, 1997; Proulx and Greene, 2001).

Branching Architecture

Strong horizontal branching and a large surface area associated with many fine branches increase exposed surface area and ought to increase susceptibility to icing injury (Lemon, 1961; Cannell and Morgan, 1989). Such assertions have been quite informal in the literature, but it might be useful to look at the extremes. Do conifers (ignoring *Larix*), possessing much larger amount of winter surface area (leaves), tend to experience more damage? The discussion is complicated by the fact that non-*Pinus* conifers such as *Tsuga* or *Abies* tend to have thin, short branches (increasingly short with height), while *Pinus* is more like hardwoods in possessing longer, thicker branches. A final complication is the lower wind speeds (at any height) within a stand rich in conifers: the same needles that gather ice will also reduce the ambient wind speed. (Our argument here is for the augmentation of ice by the wind rather than the drag it induces.) Amid this confounding of our intuition, the empirical literature provides no guidance—at least at first glance. While several studies of mixed stands have found needle-leaved trees more vulnerable to icing injury (Illick, 1916; Lemon, 1961; Whitney and Johnson, 1984; Bruederle and Stearns, 1985; Boerner *et al.*, 1988; Warrillow and Mou, 1999; Smith, 2000), other research has claimed that conifer species are less susceptible to icing damage than hardwoods (Rogers, 1924; Carvell *et al.*, 1957; Hauer *et al.*, 1993, Amateis and Burkhart, 1996; Irland, 1998; Manion *et al.*, 2001). Note, however, that if we remove *Pinus* from the analysis, the preponderance of the forgoing evidence is that non-*Pinus* conifers are much less damaged than are hardwoods. Nonetheless, the reader is reminded that in none of these studies is there any surety that ambient icing precipitation and species identity are independent when the study area is sufficiently large to include significant differences in ice thickness. In mountains in particular, both species' abundance and ice thickness can change markedly across a few hundred meters perpendicular to the contours.

Species Identity

Lemon (1961) first proposed the idea that species-specific wood density or (a measure it is proportional to) breaking stress ought to be important determinants of the probability of damage or death. It is no coincidence that these two parameters, the most easily obtained quantitative measures of interest, have been discussed so much in the literature. Nonetheless,

Lemon (1961) found no significant relationship between either measure and his qualitative damage categories. Since then, his conclusion that there was no relationship between wood strength and susceptibility to icing injury has been supported by others (Deuber, 1941; Carvell *et al.*, 1957, Hauer *et al.*, 1994), although Jones *et al.* (2001) suggested that greater wood strength may help protect against secondary damage inflicted by overstory stem failure or falling branches. To provide an example with a common species from eastern forests, three studies of three different icing events resulted in very different conclusions for red maple (*Acer rubrum* L.). On a relative scale in diverse forests, we find that it is one of the most resistant species (Whitney and Johnson, 1984), moderately susceptible (Seischab *et al.*, 1993), or one of the least resistant (Siccama *et al.*, 1976). Clearly, as we have stressed above, many other confounding factors can play a role. But it is disheartening to note that the only good generalization we can make from the voluminous ice damage literature is that non-*Pinus* conifers are damaged more than hardwoods.

Crown Asymmetry (Proximity to Edge or Slope Angle)

It has been asserted that the smaller, more narrow, crowns of interior forest trees are less likely to be damaged than the wider crowns of edge trees (Hauer *et al.*, 1993). Crown asymmetry at stand perimeters or large gap margins appears to predispose edge trees to bending damage and, of course, edge trees should experience higher wind speeds (and therefore greater ice deposition) on the nonwindward side of a clearing (Illick, 1916; Seischab *et al.*, 1993; Hauer *et al.*, 1994; Sampson and Wurtz, 1994). Similarly, crown asymmetry of trees located on slopes predisposes them to greater damage because of unbalanced ice-loading (Bruederle and Stearns, 1985; Seischab *et al.*, 1993). Proulx and Greene (2001) found that edge trees on flat ground bent away from the hardwood forest irrespective of the prevailing wind or direction that the edge faced. Trees on slopes also bent downslope, regardless of the prevailing wind or azimuth. Nonetheless, Proulx and Greene found that edge trees were not significantly more likely to be in the heavily damaged category than interior trees when the ice load was extreme (1998: near the Quebec–New York border; modest ambient wind speeds) and with heavy damage including catastrophic branch loss and snapping as well as permanent bending. We should not exaggerate the increase in wind speed on nonwindward versus windward clearing edges in clearings (say, 200 m on a side). For such openings in forests or in agricultural terrain *in full leaf*, the wind has little time to accelerate across the clearing (less than 2-fold) by the time it is within 1 tree height of the leeward edge (Greene and Johnson, 1996). Further, as the wind approaches within 1 tree height of the leeward edge, it decelerates by about 50%. (Unfortunately, we know of no comparable data for leafless forests, but the point is that the extra ice due to enhanced wind speeds on nonwindward edges will not be nearly as strong an effect as one might have initially supposed. If the wind at midcrown height in a leafless forest far from any edge is about 45% of the speed at some reference height, clearly the speed at a nonwindward clearing edge cannot be even 2-fold greater.)

Position of the Tree in the Canopy

In addition to direct damage (wind-augmented or not) by ice accumulation, understory trees are, at the same time, prone to injury resulting from indirect damage as snapped boles or branches fall from the canopy. Duguay *et al.* (2001) reported that only 22% of the shorter understory saplings avoided being pinned or crushed by falling branches at St. Hilaire, Quebec, during and after the 1998 ice storm. Likewise, one presumes (there are no empirical reports) that many of the lower branches on canopy trees are pruned by the additional load imposed by falling branches from higher in the crown.

THE POPULATION CONSEQUENCES OF MAJOR ICE EVENTS

By definition, any disturbance has a direct effect on population dynamics because it represents the death of a reproductive mature tree. But, as was made clear by the gap dynamics studies of the 1970s and 1980s (Pickett and White, 1985), disturbances are also sources of "births" (broadly defined): postdisturbance asexual and sexual recruitment, as well as release among the advance regeneration (e.g., Oliver and Stephens, 1977). While gap creation and the pulse of woody debris following severe events have numerous effects on animal communities (e.g., Faccio, 2003) and hydrological systems (Kraft *et al.*, 2002), our concern here is with tree populations. Arched trees and snapped trees can be considered as deaths. Stems bent dramatically toward the ground never regain their original form (Rhoades, 1918; Carvell *et al.*, 1957; Lemon, 1961), while snapped trees have lost their crowns completely. They may still be alive, resprouting vigorously from the root collar or, for a few species, from root suckers. But they will have to rebuild their trunks and crowns *de novo*.

For hardwood trees suffering major branch loss, Shortle and Smith (1998) suggested the following rules for predicting long-term impacts: (1) greater than 75% live crown loss: survival rate very poor, with surviving individuals likely to become heavily infected by fungi; (2) 50% to 75% crown loss: many trees likely to persist, with some loss of vigor owing to internal infections and growth suppression; and (3) less than 50% crown loss: high survivorship, with some individuals likely to experience only delayed growth. To a remarkable degree, this simple expectation has been borne out by permanent plot studies in sugar maple stands in southern Quebec (Boulet, 1998).

Immediately following a severe icing event, hardwoods with catastrophic branch loss produce small numbers of adventitious shoots along the trunk and major branches, but these shoots become less numerous and vigorous each year, the annual diameter increment narrowing. Meanwhile, fungal infection proceeds rapidly from each exposed branch stub, and insect predation becomes chronic (Campbell, 1937; Melancon and Lechowicz, 1987; Rebertus *et al.*, 1997). An additional cause of stress is the immediate exposure of stems to the heat and drying effect of the sun (Spaulding and Bratton, 1946). Death for these trees is a slow process, occurring within 3 to 10 years after the event.

When viewed years later, ice-induced canopy gaps have a random, patchy distribution (Bruederle and Stearns, 1985) that may be influenced by local variation in precipitation and wind during the event. Unfortunately, there are no estimates of local variations in radial ice thickness at the scale of, say, 20 to 200 m. Within these gaps, as discussed below, advance regeneration and asexual recruits from "dead" canopy trees will, one presumes, seize the vacant space. Between the gaps, where canopy trees were heavily damaged but nonetheless recovered, shade-tolerant, surviving subcanopy stems can be expected to show temporary release as in the classical gap dynamics argument (e.g., Frelich and Lorimer, 1991).

Following icing events, three methods of recruitment occur: release of advance regeneration and postdisturbance sexual and asexual regeneration. Perhaps the most crucial effect of a severe icing event on the subsequent regeneration is the dramatic increase of light at lower canopy levels and at the forest floor. Not surprisingly, it appears that the least damaged portion of the advance regeneration, the subcanopy stems not pinned by falling debris, dominate the race to form the new canopy (e.g., Whitney and Johnson, 1984). But this generalization is true, of course, only for stands with appreciable amounts of predisturbance subcanopy stems. Any disturbance regime, such as ice, wind, or insects, that leaves the lowest stratum of the canopy relatively unaffected will eventually have a low abundance of shade-intolerant species among both the canopy trees and the subcanopy stems; these disturbances promote regeneration by advance regeneration, which is necessarily moderately to very shade-tolerant. For example, in the old-growth hardwood forest of St. Hilaire (Quebec) following the December 1983 ice storm, there was a modest pulse of new recruits, but most of the regeneration was composed of advance regeneration, and these in turn supplied the individuals that are now the tallest members of the emerging new canopy (Fig. 7).

Why is there so little contribution made by the postdisturbance sexual recruitment? There is abundant light and, for the small-seeded species at least, seedbeds are much improved by the dramatic reduction in leaf-fall that persists for the first few years. A permanent plot study at St. Hilaire, begun the year before the 1998 icing event and following a large seed crop (1996) by many species, has revealed that there has been almost no sexual reproduction since 1998. Fig. 8 shows eastern hemlock [*Tsuga canadensis* (L.) Carr.] as an example. The 2000 *Tsuga* cone crop (leading to the 2001 seedling cohort) was a mast event across southern Quebec except in the areas devastated by the 1998 ice storm. For equivalent tree size (basal area), it was visually estimated by using binoculars that the cone crop was eight times smaller for *Tsuga* at the badly damaged St. Hilaire site than at undamaged *Tsuga* stands located 100 km to the east in the foothills. This is hardly surprising—the living St. Hilaire *Tsuga* canopy trees had lost about half their crowns. Similarly, the 2000 cone crop for *Pinus strobus* L. (white pine) was six times smaller at St. Hilaire than the crop for equal-sized trees outside the ice storm area. Thus, a severely damaged tree with a reduced crown cannot produce large seed crops. The literature emphasis on "stress

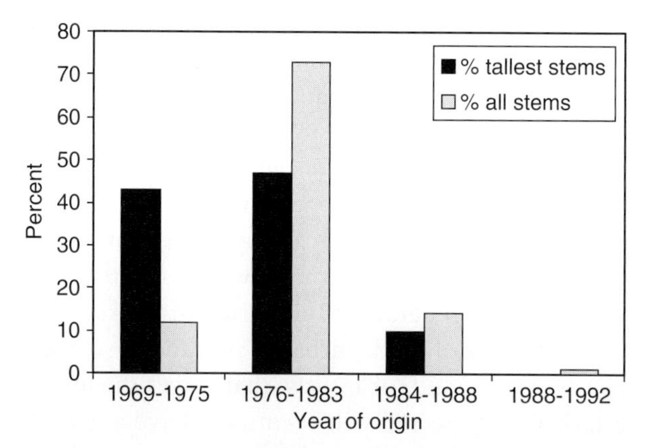

FIG. 7 Age structure of stems (almost entirely *Fagus grandifolium* and *Acer saccharum*) sampled in 1992 in a large gap created by the December 1983 ice storm at St. Hilaire, Quebec.

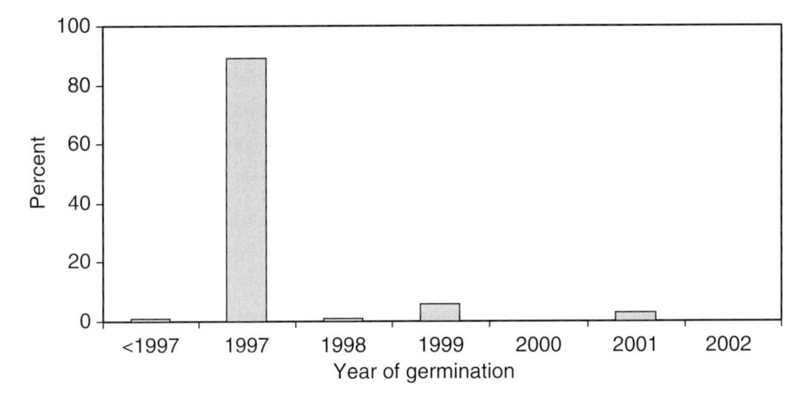

FIG. 8 Age structure of *Tsuga canadensis* derived from permanent plots in a stand devastated by the 1998 (January) ice storm at St. Hilaire, Quebec.

crops" (i.e., unusually large crops occurring in response to damage such as girdling or severe droughts) makes little sense to the present authors for severe icing events or any other natural disturbance.

Another way to regenerate following icing disturbance is asexually from basal sprouts or root suckers. Fig. 9 shows a permanent plot example for suckering in *Populus grandidentatum* Michx.; the results are similar to those for other *Populus* species following fire or clearcutting (Greene and Johnson, 1999). The sudden reduction of auxins transported from the crown following the disturbance ends the dormancy of subaerial buds. The rapid proliferation of fast-growing asexual stems leads to a quick onset of

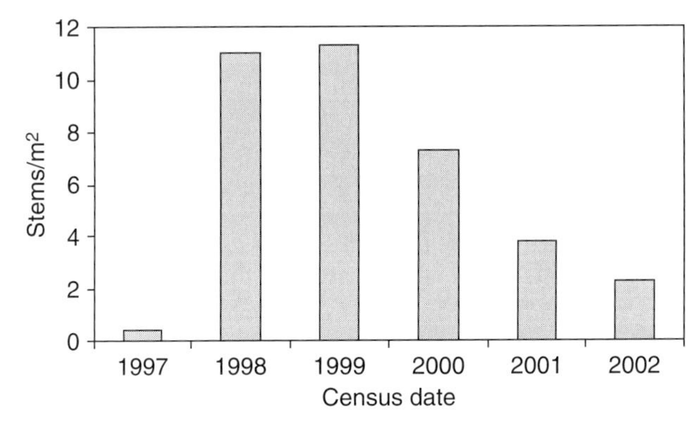

FIG. 9 Number of asexual stems of *Populus grandidentatum* in a permanent plot at St. Hilaire (Quebec) as effected by the January 1998 ice storm. "Births" were mainly in 1998; self-thinning began in 2000.

self-thinning as these stems compete for both light and a shrinking starch reserve in the communal root system. Likewise, many hardwood species such as yellow poplar (*Liriodendron tulipifera* L.) and red maple commonly regenerate well from basal sprouts (Whitney and Johnson, 1984) and then rapidly thin. Whether or not sprouts or suckers seize the opened canopy space will depend on whether taller nearby advance regeneration has survived. This competition with the surviving advance regeneration is especially important for shade-intolerant species such as *Populus*.

Whether or not icing events are important for tree population dynamics in a region depends on the return time for severe events relative to the return time for other types of disturbance. Because of the small bending stresses from wind drag over the range of likely gust speeds, severe ice storms can be quantified by the radial ice thickness alone. Given the 75% crown loss rule for mortality in hardwood species, Proulx and Greene (2001) showed that the probability of a canopy tree dying was about 0.1 for a maximum radial ice thickness (t_{max}) of 0.025 m, 0.25 for t_{max} of 0.03 m, and about half the trees in a stand should die when t_{max} = 0.045 m. (Note: these ice thicknesses are from the Hydro Quebec Passive Ice Meters from the measurement of the *maximum* iced diameter of 0.025-m diameter horizontal dowels. The relationship between t_{max} and t depends on the shape of the ice accretion, with t_{max} often significantly greater than t.) In Quebec, near the Canada–United States border, we expect t to be about 0.025 m for a 50-year return time and about 0.028 for a 100-year return time (CSA, 2001). Thus, we expect that icing events will be recurrent, devastating events in the maple- and beech-dominated forests that typify this region (Melancon and Lechowicz, 1987; Dugauy *et al.*, 2001; Proulx and Greene, 2001). For a canopy tree of some shade-tolerant species, much of its life (typically 150 years in total: references in Proulx and Greene, 2001) will have been spent in the understory where, as discussed above, it would have been relatively impervious to ice storm damage because of its small pliable stem. Released by the death of an overtopping tree, the residency of this shade-tolerant individual in the canopy stratum after a brief period of rapid height growth would be, one speculates, quite short because of subsequent icing events.

By contrast, to the north of this area in the eastern boreal forest, the 50-year return time radial ice thickness is much smaller (CSA, 2001). This region has a fire regime with a mean return time (at least for the last few centuries) of about 150 years. Thus, given a cumulative exponential distribution with constant hazard of burning, the probability of a stand burning within 50 years is 0.28. In short, icing events ought to be a minor player in the eastern boreal forest because, simply, other events (fire and, for conifer species, spruce budworm epidemics) are far more common. As for the western North American boreal forest, the radial ice thickness for a 50-year return time is even smaller than in the east, while the mean fire interval is shorter (i.e., ice events are far too rare to have a major role in shaping the canopy tree mortality regime).

REFERENCES

Abell, C. A. (1934). Influence of glaze storms upon hardwood forests in the Southern Appalachians. *J Forest* 32, 35–37.

Amateis, R. L., and Burkhart, H. E. (1996). Impact of heavy glaze in a loblolly pine spacing trial. *South J Appl For* 20, 151–155.

ASCE. (2005). *Minimum Design Loads for Buildings and Other Structures, Standard 7.* American Society of Civil Engineers, Washington, DC.

Belanger, R. P., Godbee, J. F., Anderson, R. L., and Paul, J. T. (1996). Ice damage in thinned and nonthinned loblolly pine plantations infected with fusiform rust. *South J Appl For* 20, 136–140.

Bisshopp, K. E., and Drucker, D. C. (1945). Large deflection of cantilever beams. *Q Appl Math* 3, 272–275.

Boerner, R. E. J., Runge, S. D., Cho, D-S., and Kooser, J. G. (1988). Localized ice storm damage in an Appalachian plateau watershed. *Am Midl Nat* 119, 199–208.

Boulet, B. (1998). *Management of Ice Storm Damaged Stands.* Ministère des Ressources Naturelles, Charlesbourg, QC.

Bruederle, L. P., and Stearns, F. W. (1985). Ice storm damage to a southern Wisconsin mesic forest. *B Torrey Bot Club* 112, 167–175.

Campbell, W. A. (1937). Decay hazard resulting from ice damage to northern hardwoods. *J Forest* 35, 1156–1158.

Canadian Standards Association (CSA). (2001). *CSA Standard C22.3 No. 1-01 Overhead Systems.* Canadian Standards Association, Toronto, ON.

Canham, C. D., and Loucks, O. L. (1984). Catastrophic windthrow in the presettlement forest of Wisconsin. *Ecology* 65, 803–809.

Cannell, M. G. R., and Morgan, J. (1989). Branch breakage under snow and ice loads. *Tree Physiol* 5, 307–317.

Carvell, K. L., Tryon, E. H., and True, R. P. (1957). Effects of glaze on the development of Appalachian hardwoods. *J Forest* 55, 130–132.

Deuber, C. G. (1940). The glaze storm of 1940. *Am Forests* 46, 210.

Downs, A. A. (1938). Glaze damage in the birch-beech-maple-hemlock type of Pennsylvania and New York. *J Forest* 36, 63–70.

Duguay, S. M., Arii, K., Hooper, M., and Lechowicz, M. J. (2001). Ice storm damage and early recovery in an old-growth forest. *Environ Monit Assess* 67, 97–108.

Dunne, T., and Leopold, L. (1978). *Water in Environmental Planning.* W. H. Freeman and Co., New York.

Elfashny, K., Chouinard, L. E., and Laflamme, J. (1996). Estimation of combined wind and ice loads on telecommunication towers in Québec. Phase 1: Modeling of the ice and wind observations. In: *Seventh International Workshop on Atmospheric Icing of Structures: Proceedings* (Laflamme, J., and Farzaneh, M., Eds.), pp.137–141. Université du Québec à Chicoutimi, Chicoutimi, QC.

Faccio, S. D. (2003). Effects of ice storm-created gaps in forest breeding bird communities in central Vermont. *Forest Ecol Manag* 186, 133–145.

Frelich, L. E., and Lorimer, C. G. (1991). Natural disturbance regimes in hemlock-hardwood forests of the Upper Great Lakes Region. *Ecol Monogr* 61, 159–162.

Goebel, C. J., and Deitschman, G. H. (1967). Ice storm damage to planted conifers in Iowa. *J Forest* 65, 496–497.

Greene, D. F., and Johnson, E. A. (1996). Wind dispersal of seeds from a forest into a clearing. *Ecology* 77, 595–609.

Greene, D. F., and Johnson, E. A. (1999). Modeling the recruitment of *Populus tremuloides*, *Pinus banksiana*, and *Picea mariana* following fire in the mixedwood boreal forest of central Saskatchewan. *Can J Forest Res* 29, 462–473.

Harshberger, J. W. (1904). The relation of ice storms to trees. *Contributions from the Botanical Laboratory of the University of Pennsylvania* 2, 345–349.

Hauer, R. J., Hruska, M. C., and Dawson, J. O. (1994) *Trees and Ice Storms: The Development of Ice Storm-resistant Urban Tree Populations*. Special Pub. 94-1, Department of Forestry, University of Illinois at Urbana-Champaign, Urbana, IL.

Hauer, R. J., Wang, W., and Dawson, J. O. (1993). Ice storm damage to urban trees. *J Arboriculture* 19, 187–193.

Illick, J. S. (1916). A destructive snow and ice storm. *Forest Leaves* 15, 103–107.

Irland, L. C. (1998). Ice storm 1998 and the forests of the northeast: a preliminary assessment. *J Forest* 96, 32–40.

Jones, J., Pither, J., DeBruyn, R. D., and Robertson, R. J. (2001). Modeling ice storm damage to a mature, mixed-species hardwood forest in eastern Ontario. *Ecoscience* 8, 513–521.

Jones, K. F. (1998). A simple model for freezing rain loads. *Atmos Res* 46: 87–97.

Jones, K. F. (1999). Ice storms, trees and power lines. In: *Cold Regions Engineering: Putting Research into Practice: Proceedings of the Tenth International Conference*. (Zufelt, J. E., Ed.), pp. 757–767. American Society of Civil Engineers, Reston, VA.

Jones, K. F., Thorkildson, R., and Lott, N. (2002). The development of the map of extreme ice loads for ASCE Manual 74. In: *Electrical Transmission in a New Age: Proceedings of the Conference, September 9–September 12, 2002, Omaha, Nebraska*. (Jackman, D. E., Ed.), pp. 9–31. American Society of Civil Engineers, Reston, VA. Also published at ftp://ftp.ncdc.noaa.gov/pub/data/techrpts/tr200201/tr2002-01.pdf as "The Development of a U.S. Climatology of Extreme Ice Loads." Technical Report 2002-01, U.S. Dept. of Commerce, NOAA/NESDIS, National Climate Data Center, Asheville, NC.

Kane, B., Ryan, D., and Bloniarz, D. V. (2001). Comparing formulae that assess strength loss due to decay in trees. *J Aboriculture* 27, 78–87.

Kienholz, R. (1941). Jack pine in Connecticut damaged by sleet storm. *J Forest* 39, 874–875.

Kraft, C. E., Schneider, R. L., and Warren, D. R. (2002). Ice storm impacts on woody debris and debris dam formation in northeastern U.S. streams. *Can J Fish Aquat Sci* 59, 1677–1684.

Lanctot, E. K., Peterson, E. L., House, H. E., and Zobel, E. S. (1960). Ice build-up on conductors of different diameters. *Transactions of the AIEE* 46, 1610–1615.

Leister, A. T., and Kemper, J. D. (1973). Analysis of stress distribution in the sapling tree trunk. *J Am Soc Hort Sci* 98, 164–170.

Lemon, P. C. (1961). Forest ecology of ice storms. *B Torrey Bot Club* 88, 21–29.

Lorimer, C. G. (1977). The presettlement forest and natural disturbance cycle of northeastern Maine. *Ecology* 58, 139–148.

Lorimer, C. D., and Frelich, L. E. (1994). Natural disturbance regimes in old-growth northern hardwoods. *J Forest* 92, 33–38.

Manion, P. D., Griffin, D. H., and Rubin, B. J. (2001). Ice damage impacts on the health of the northern New York State forest. *Forest Chron* 77, 619–625.

McKay, G. A., and Thompson, H. A. (1969). Estimating the hazard of ice accretion in Canada from climatological data. *J Appl Meteorol* 8, 927–935.

Melancon, S., and Lechowicz, M. J. (1987). Differences in the damage caused by glaze ice on codominant *Acer saccharum* and *Fagus grandifolia*. *Can J Bot* 65, 1157–1159.

Nicholas, N. S., and Zedaker, S. M. (1989). Ice damage in spruce-fir forests of the Black Mountains, North Carolina. *Can J Forest Res* 19, 1487–1491.

Niklas, K. J. (2000). Computing factors of safety against wind-induced tree stem damage. *J Exp Bot* 51: 797–806.

Oliver, C. D., and Stephens, E. P. (1977). Reconstruction of a mixed species forest in central New England. *Ecology* 58, 562–572.

Pickett, S. T. A., and White, P. S. (1985). *The Ecology of Natural Disturbance and Patch Dynamics*. Academic Press, Orlando, FL.

Proulx, O. J., and Greene, D. F. (2001). The relationship between ice thickness and northern hardwood tree damage during ice storms. *Can J Forest Res* 31, 1758–1767.

Rebertus, A. J., Shifley, S. R., Richards, R. H., and Roovers, L. M. (1997). Ice storm damage to an old-growth oak-hickory forest in Missouri. *Am Midl Nat* 137, 48–61.

Rhoades, V. (1918). Ice storms in the Southern Appalachians. *Mon Weather Rev* 46, 373–374.

Rhoads, A. G., Hamburg, S. P., Fahey, T. J., Siccama, T. G., Hane, E. N., Battles, J., Cogbill, C., Randall, J., and Wilson, G. (2002). Effects of an intense ice storm on the structure of a northern hardwood forest. *Can J Forest Res* 32, 1763–1775.

Rogers, W. E. (1922). Ice storms and trees. *Torreya* 22, 61–63.

Rogers, W. E. (1923). Resistance of trees to ice-storm injury. *Torreya* 23, 95–99.

Rogers, W. E. (1924). Trees in a glaze storm. *Tycos* 14, 4–8.

Sampson, G. R., and Wurtz, T. L. (1994). Record interior Alaska snowfall effect on tree breakage. *North J Appl For* 11, 138–140.

Schaub, W. R. (1996). Methods to estimate ice accumulations on surface structures. In: *Seventh International Workshop on Atmospheric Icing of Structures: Proceedings*. (Laflamme, J., and Farzaneh, M., Eds.), pp.183–188. Université du Québec à Chicoutimi, Chicoutimi, QC.

Seischab F. K., Bernard, J. M., and Eberle, M. D. (1993). Glaze storm damage to western New York forest communities. *B Torrey Bot Club* 120, 64–72.

Shepard, R. K. (1975). Ice storm damage to loblolly pine in northern Louisiana. *J Forest* 73, 420–423.

Shortle, W. C., and Smith, K. T. (1998). Ice Storm 1998. Northeastern Forest Experiment Station Information Sheet # 1, March 3. USDA Forest Service, Durham, NH.

Siccama, T. G., G. Weir, and Wallace, K. (1976). Ice damage in a mixed hardwood forest in Connecticut in relation to *Vitis* infestation. *B Torrey Bot Club* 103, 180–183.

Sisinni, S. M., Zipperer, W. C., and Pleninger, A. C. (1995). Impacts from a major ice storm: street-tree damage in Rochester, New York. *J Arboriculture* 21, 156–167.

Smith, W. H. (2000). Ice and forest health. *North J Appl For* 17, 16–19.

Spaulding, P., and Bratton, A. W. (1946). Decay following glaze storm damage in woodlands of central New York. *J Forest* 44, 515–519.

Von Schrenk, H. (1900). A severe sleet-storm. *Trans Acad Sci St. Louis* 10, 143–160.

Warrillow, M., and Mou, P. (1999). Ice storm damage to forest tree species in the ridge and valley region of southwestern Virginia. *J Torrey Bot Soc* 126, 147–158.

Whitney H. E., and Johnson, W. C. (1984). Ice storms and forest succession in southern Virginia. *B Torrey Bot Club* 111, 429–437.

Yorks, T. E., and Adams, K. B. (2003). Restoration cutting as a management tool for regenerating *Pinus banksiana* after ice storm damage. *Forest Ecol Manag* 177, 85–94.

7

Disturbance Processes and Dynamics in Coastal Dunes

Patrick A. Hesp
Louisiana State University

M. Luisa Martínez
Instituto de Ecología, A.C.

INTRODUCTION

Coastal dunes occur throughout the world and thus in a diversity of climatic regimes (Martínez and Psuty, 2004). However, independently of the geographical location, they all share a set of environmental characteristics (wind, sand deposition and erosion, substrate mobility, salt exposure, flooding, drought, and nutrient deficiency) that greatly affect seed germination, seedling establishment, and adult performance (Moreno-Casasola, 1986; Ehrenfeld, 1990; Hesp, 1990; 1991; Maun, 1998; 2004; Ripley and Pammenter, 2004). Repeatedly, and for a wide variety of locations ranging from tropical to subarctic latitudes, it has been demonstrated that the predominant factors that have a significant impact on the dynamics of coastal dune vegetation can be divided into two groups: (1) environmental gradients, and (2) recurring disturbances. Typically, salinity, substrate mobility, radiation, and nutrient contents vary following a gradient. Salinity, near-surface wind speed, radiation, and substrate mobility decrease inland, while nutrient and biotic pressures increase (McLachlan, 1991; Martínez and Psuty, 2004). In turn, disturbance events occur mostly through wave scarping, water intrusion, substrate erosion, and burial by sand (or snow) and may be gradual or abrupt, and are commonly spatially and temporally variable. The intensity of these events will depend on the orientation of the coast, wind speed, time of year (or season), storm frequency and intensity, tide regime, surfzone-beach type, and perhaps mean sediment size. Environmental gradients and disturbance factors and/or events are often considered mutually exclusive. However, recent perspectives state that disturbances overlay environmental gradients (Stallins and Parker, 2003), and their impact is, therefore, influenced by such gradients (e.g., Forman and Godron, 1986; Odum *et al.*, 1987). In this sense, the relative importance of each depends on the spatial and temporal scale at which each occurs (Peet, 1992).

In this chapter, we examine the types of disturbances and impacts in each of the major coastal dune types. We attempt to provide a sense of the range of different disturbance factors in each of the dune or dunefield types and, where possible, an indication of the scale and magnitude of those disturbances. A disturbance is herein regarded as any relatively discrete event in time that disrupts ecosystem, community, or population structure and that changes resources, substrate availability or condition, or the physical environment (Pickett and White, 1985). Space limitations restrict us from examining disturbance processes associated with animal and human behavior.

DUNE TYPES AND DISTURBANCE TYPES AND PROCESSES

A variety of classifications of coastal dune types exist. However, we follow a simple classification that recognizes four main types: foredunes, blowouts, parabolic dunes, and transgressive dunefields and sheets. The following examines each of these in turn and, in addition, briefly discusses barrier islands dominated by hurricane and storm events and overwash sheets and fans.

Foredunes

Foredunes are shore-parallel, convex, symmetric to asymmetric, foremost dune ridges formed on the backshore by aeolian (wind blown) sand accretion within vegetation. Incipient foredunes (sometimes known as embryo dunes) are newly forming types typically developing in "pioneer" vegetation, while established foredunes typically develop from incipient foredunes and are older and more ecologically and morphologically complex (Cowles, 1898; Salisbury, 1952; Olson, 1958; Ranwell, 1972; Hellemaa, 1998; Hequette and Ruz, 1991; Hesp, 1988a; 1991; 2000; 2002).

Because foredunes are sited on the backshore, they can be affected by a significant number of types and magnitudes of natural disturbances, including swash inundation, wave erosion (scarping and/or overwash), salt spray, sand deposition and erosion, wind abrasion and erosion, drought, predation or disease, and fire. The first four are most apparent on foredunes, while the other disturbance types may occur on all dune types regardless of proximity to the sea.

Incipient Foredunes

Incipient foredunes form on the backshore by: (1) the germination of seeds that have been swash deposited or wind blown to a position typically above the high spring tide line; (2) plants growing seawards by rhizome and stolon development from a landward source region (often an older established foredune); or (3) a combination of these two. They may form on a backshore devoid of wrack or one extensively covered in wrack or flotsam (e.g., seaweed, driftwood piles). Incipient foredunes may be present for only a season if the colonizing plants are annuals, or for only a short period of a year or two if the beach and foredune are in a long-term erosional state. The degree of incipient foredune development also depends on the number of plants that propagate on the backshore, the colonizing species, and its habit. If few seeds are wind blown or swash is transported to the backshore, or if only a few plants survive, an irregular incipient zone of nebkhas and shadow dunes results (Ranwell, 1972). A nebkha is a discrete dune formed around an individual plant (Cooke *et al.*, 1993). If the new incipient vegetation zone is laterally continuous, then an extensive ramp, terrace, or ridge incipient dune may develop (Hesp, 1990).

The survival of seedlings on the backshore above the high tide line and the development of an initial incipient foredune depend on many conditions, but particularly on the presence of minimal wind deflation or erosion, aeolian sand deposition at a scale commensurate with the upward growth ability of the plant (Maun, 1998; Martínez and Moreno-Casasola, 1996), some moisture availability, and little or no swash inundation. Any nominal increase in these factors amounts to a disturbance factor for new seedlings and commonly results in the demise of the incipient vegetation zone. Thus, the development of a new vegetation zone and incipient foredune on the backshore may be entirely dependent on those factors being optimized.

Many dune species cope with sand inundation (e.g., *Ammophila* or *Spinifex* species; Maun and Lapierre, 1984; Maze and Whalley, 1992), and a few can cope with some seawater immersion. For example, *Spinifex sericeus* R. Br. and *S. longifolia* in Australia, *Cakile maritima* Scop. and *Salsola kali* L. in temperate latitudes, and *Sporobolus virginicus* (L.) Kunth and *Ipomoea pes-caprae* (L.) R. Br. in the tropics respond well to sand deposition (up to some maximum level as with many other dune species; Maun, 1998) and can actually cope with one or two minor seawater swash events (Hesp, 1983). All species, however, have a maximum tolerance to sand burial, after which they usually die (Maun, 1998). Occasional burial versus extreme burial may elicit quite different responses in different species of plants (Kent *et al.*, 2001).

Fig. 1 illustrates a long-term survey of incipient foredunes formed at Dark Point, Fens Embayment, Myall Lakes National Park, New South Wales, Australia, over 14 years. After a major storm erosion event in 1974, the beach slowly recovered and an incipient foredune was formed by approximately 1976. This foredune continued to build until around 1983, when it was gradually replaced by the formation of a new incipient foredune to

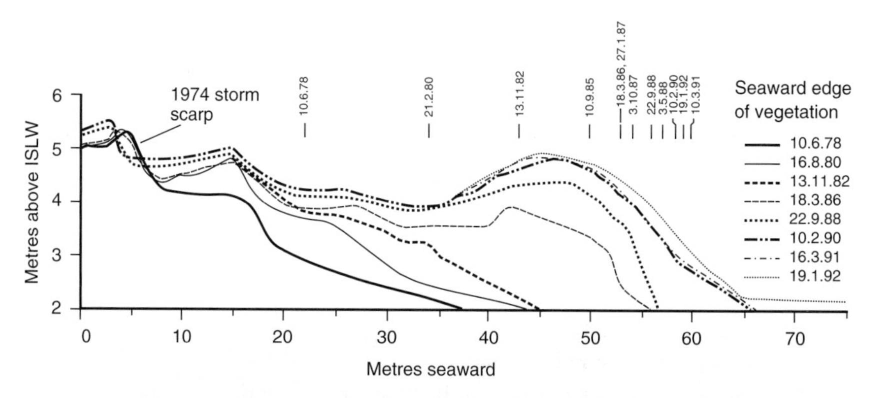

FIG. 1 Topographic surveys of incipient foredunes developing at Dark Point, Fens Embayment, Myall Lakes National Park, New South Wales, Australia, from 1978 to 1992. The arrows and dates indicate the seaward edge of vegetation at the time of each survey. Over time, two incipient foredunes have formed, and the delivery rate, volume, and spatial distribution of aeolian sand deposition (the major disturbance factor here) have changed significantly over time.

seaward (the 1986 dune). This dune eventually grew upward and outward, largely isolating the first dune from any further deposition by 1988. The dune development here illustrates how, over time, the magnitude and rate of disturbance change over a relatively short time scale and spatial distance. As plants propagated seaward on the first dune, aeolian deposition became more and more restricted to the lower seaward slope (compare the 1980 profile with the 1982 profile). This process continued while the beach prograded, isolating the dominant species, *Spinifex sericeus*, from sand deposition in the 0- to 30-m zone and leading to *Spinifex* decay and invasion by other species (Hesp, 1990). Once beach progradation had largely ceased (by around 1990), aeolian accretion diminished, as did seaward and vertical growth. In the first case, the major disturbance factor, aeolian transport and deposition, was reduced because of seaward growth and isolation of the landward portion from deposition. In the second case, aeolian deposition was reduced by a reduction in nearshore/beach sediment supply. Dune development and the seaward extension of the plant boundary were rapid in the early stages after storm recovery and slowed considerably in the latter stages (see the arrows indicating the edge of vegetation/backshore boundary). Carter and Wilson (1990) describe a similar sequence, showing that the seaward-most accreting incipient foredune had sand deposition rates of 0.3 to 0.4 m a^{-1} and was dominated by *Ammophila arenaria* (L.) Link. Landward ridges had deposition rates of less than 0.05 m a^{-1} and within 2 to 3 years following a decline in sediment supply displayed many more species.

The occurrence or level of localized disturbance may influence the species present and incipient foredune development. On north Padre Island (United States), for instance, three dominant species are present: *Ipomoea pes-caprae, Uniola paniculata* L., and *Panicum amarum* Ell. In some places on the seaward face of the foredune one dominates, while in other places the other dominates, and these may occur within 30 to 50 m of each other. It is unclear whether the dominance of one species or other is caused by pure chance or natural patchiness; it may also be that the defining factor producing the longshore variation in species dominance is localized storm swash events, which, for example, occur with differential swash uprush on the beach or, more commonly, where large-scale surfzone rips locally erode several to tens of meters of beach and dune front (cf. Psuty, 1990).

Established Foredunes

Established foredunes may vary in size from a meter or two in height to more than 20 m in some circumstances. They may be covered by a few species, sometimes with a very dominant species, or may display a relatively large number of species depending on geobotanical region (Morrison and Yarranton, 1974; Moreno-Casasola, 1988), climate, and beach or barrier history and stability (eroding, stable or prograding) and exposure, among other factors (Hesp, 1988a; 1991). They may be initiated and dominated by plants ranging from very low prostrate species (e.g., *Ipomoea*), short to tall grasses and sedges (e.g., *Spinifex, Uniola, Panicum, Ammophila* spp.), to tall shrubs and trees (e.g., *Atriplex, Populus* spp.).

Aeolian Sediment Transport The rate of aeolian sand supply and deposition and the degree of plant burial, as noted above, are strong factors in determining and/or altering plant density and composition (Salisbury, 1952; Marshall, 1965; Moreno-Casasola, 1986; Maun and Perumal, 1999; Martínez *et al.*, 2001). While an incipient foredune may be present, this does not necessarily negate sediment transport to the established foredune, because strong winds, a low- to moderate-density vegetation cover on the incipient dune, and sand bypassing of the incipient dune during early winter conditions can all lead to sediment being supplied to the established foredune (Hesp, 1988a; Davidson-Arnott and Law, 1990; 1996; Arens, 1996).

The rate of aeolian sand supply, and therefore the magnitude of this primary disturbance, may be critical in determining the species present and the species richness. For example, in southern Brazil, *Panicum racemosum* (P. Beauv.) Sprengel dominates foredunes as a virtual monospecific stand on dunes that are receiving a significant sediment supply, while a variety of species (e.g., *Blutaparon, Senecio, Hydrocotyle, Spartina,* and *Panicum* spp.) is more common where foredunes are receiving a lower sediment supply. Rates of sediment supply may vary considerably from beach to beach, from season to season, and from year to year, and depend on a variety of factors (Sarre, 1989; Davidson-Arnott and Law, 1990; Arens, 1997). During active periods of beach deflation, Carter and Rihan (1978) and Carter and Wilson (1990) observed rates of 15,000 to 19,000 m^3 of sediment moved by the wind across an 80-m shore normal transect in 1 week in northern Ireland. Typical sediment supply rates to a foredune averaged between 8 to 12 $m^3 m$ $shoreline^{-1} a^{-1}$. Transport onshore in low–wind-energy environments may be exponentially less (Nordstrom and Jackson, 1994). Offshore winds can also be important in redistributing lee slope and crestal sand back onto the foredune stoss (seaward) slope (Gares and Nordstrom, 1995; Psuty, 2005). Sediment supply may vary spatially and especially seasonally in arctic and subarctic regions and areas receiving seasonal snow, with a partial to complete shutdown of sand transport when snow is falling or present (Ruz and Allard, 1994a; 1994b). Snow deposition obviously also represents a significant disturbance in these environments (Saunders and Davidson-Arnott, 1991).

Blum and Jones (1985) examined disturbance to foredunes from recreational pressures. In essence, this disturbance may mimic natural disturbance where natural, very stable, well-vegetated foredunes are akin to foredunes with minimal human disturbance, and highly unstable natural foredunes are akin to foredunes with high pedestrian pressures (Fig. 2). They demonstrated that the vegetation density was highest, diversity was greatest, and zonation patterns were minimal on the least disturbed foredunes. Similar effects have been noted for restored dunes on nourished beaches (e.g., Freestone and Nordstrom, 2001). These findings can probably be applied to foredunes exposed to disturbance pressures of an intermediate intensity and frequency (e.g., intermediate levels of sand movement; cf. Godfrey and Godfrey, 1973).

Foredunes range in morphology and morpho-ecological state from very stable, well-vegetated types to highly unstable, poorly vegetated types (Hesp, 1988a; Fig. 2). They may slowly evolve through a series of types becoming progressively more erosional or jump from one type to another, skipping

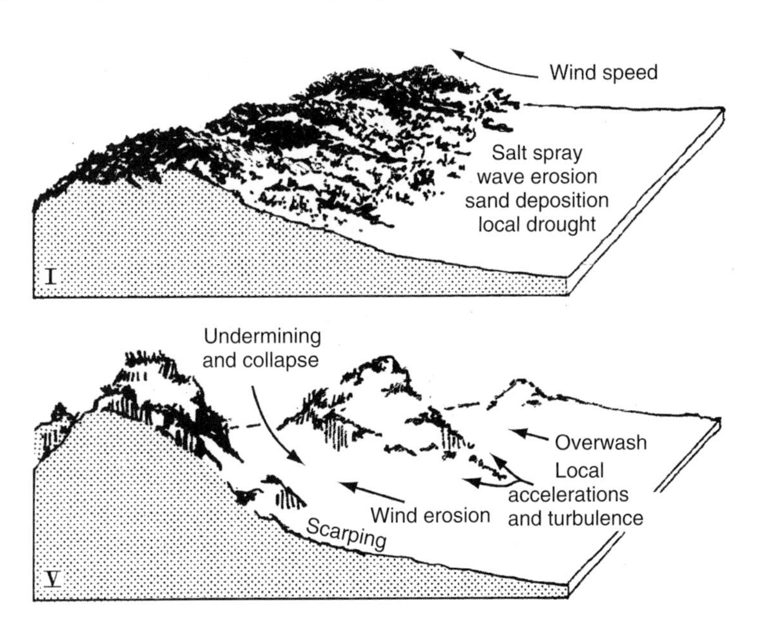

FIG. 2 The two extremes of established foredune types and their typical disturbance factors. The foredunes display a morpho-ecological variability ranging from type I (high species cover and presence, simple morphology, high stability, and relatively lower disturbance levels) to type V (low species cover and presence, high instability, and relatively higher disturbance levels).

some stages (Hesp, 2002; Sherman and Bauer, 1993). The latter occurs following catastrophic storm scarping or overwash events. Inter-foredune disturbance levels will be significantly different between types (e.g., compare type 1 with type 5 in Fig. 2). Sand erosion, deposition, and burial are far greater and occur at a variety of scales (small patches to large blowouts and sand sheets) and across the whole foredune on type 5 dunes compared to type 1 dunes. Salt spray will be intercepted less and travel further on type 5 dunes. Species distribution and richness can therefore be significantly different between the foredune types even when they may exist on the same beach. These differences can have significant downstream effects such that, years after both foredune types noted above (and illustrated in Fig. 2) have become relict and fully vegetated (usually by the development of a new foredune to seawards), differences in plant species composition may still be present because of their previously different morpho-ecological states and histories.

Within an individual embayment there can be considerable spatial variation in the magnitude of disturbance. Many zeta-form or parabolic-shaped embayments display a considerable change in orientation along their length—from low-energy to high-energy waves, from one breaking wave and minimum salt spray (reflective beach segment) to many breaking waves and maximum salt spray aerosols (dissipative beach segment), from low to high sediment supply, from narrow steep beaches with minimum aeolian sand transport to wide flat beaches with maximum aeolian sand transport, and from low to high wind exposure (Short and Hesp, 1982; Hesp, 1988b). For example, Ingleses Beach, Florianopolis, Brazil, displays such an alongshore variation in shoreline orientation. Foredunes are codominated by *Panicum racemosa, Ipomoea* spp., and *Senecio crassifolia* in the southern protected, least exposed end of the beach, where little aeolian sand movement takes place. The foredune lee slope is dominated by *Dalbergia ecastophylla* L. Taub. This species is absent once exposure increases and aeolian sand transport increases. *Panicum* alone dominates the foredune stoss slope and crest where sand transport and exposure are at a maximum in the northern end of the embayment (cf. Santos *et al.*, 1996).

The stability of the beach system also plays a role in determining disturbance frequency and magnitude. For instance, reflective beach types tend to be more stable over tens of years compared to dissipative beaches, and this leads to more stable foredunes and greater species richness on reflective beaches (Hesp, 1991).

Climate can play a role. Beaches in the wet and/or humid tropics tend to be more stable and have significantly less aeolian sand transport, less salt spray, and higher humidity than beaches in temperate regions. This leads to less disturbance on wet/humid tropical coasts. The dry tropics (or tropical regions with a substantial dry season) tend to have higher sediment availability, or the climate means that plant growth is less, sand transport is greater, and stability is less, resulting in greater disturbance (Hesp, 2004).

Beach and Foredune Erosion Foredunes are commonly part of the dynamic swept prism. During storms, sediments making up the backshore and foredune may be wave eroded and transported into the surfzone to maintain beach and surfzone profile equilibrium. Thus, depending on storm magnitude, foredunes may experience slight, moderate, or severe scarping and disturbance, and the level of disturbance can have a profound effect on the nature and rate of subsequent foredune recovery (Saunders and Davidson-Arnott, 1991). Because foredunes range in height from 1 to approximately 20 m, these disturbance levels are rather arbitrarily defined in the following way: slight disturbance or scarping—waves may wash up and onto the first few meters of the foredune, possibly leading to dieback or death of some plants because of salinity intolerance, or less than 10% to 20% removal of the dune volume (i.e., small scale scarping [or cliffing] and limited horizontal erosion of the dune); moderate disturbance or scarping—20% to 40% volumetric loss (i.e., moderate to significant relative vertical scarping, some localized overwash, and relative horizontal loss of the foredune); and severe disturbance or scarping—greater than 40% to 50% volumetric loss (i.e., significant relative vertical scarping and/or significant relative horizontal foredune erosion), or a major overwash event.

Slight disturbance and/or scarping may result in the death of leading edge plants if they have suffered swash or wave inundation (or slight overwash), but this is unlikely to significantly alter the species composition. Regular swash inundation or minor scarping is common on many stable to prograding beaches, especially during high spring tides and storms. This disturbance, while not significant in erosional terms, may still dictate which pioneer species survive following colonization since species intolerant to swash inundation will be unable to survive.

Moderate scarping or overwash may result in the removal of the entire incipient vegetation zone and possibly portions of the landward established foredune. Dieback of plants on the scarp frontline may occur and plant cover will be reduced as plants respond negatively to increased impact from salt spray, sand transport and deposition, root disturbance caused by slumping and a reduction in local stability (tensional cracking), and wind abrasion (Fig. 3). Other new species may possibly occupy the spaces created, although it is likely that little species composition change will occur, because most plants able to survive in the affected zone will

A
B
C

already be present. However, when foredune scarping is intense, inland vegetation will be closer to the beach, resulting in temporarily higher species diversity on the foredune (García-Mora *et al.*, 2001; Martínez *et al.*, 2006). Certainly, with recurring disturbances and wave overwash, these inland species will eventually become locally extinct. Vegetated slump blocks (Carter *et al.*, 1990) commonly slide, slump, or fall from the scarp crest. Plants with deep, fibrous, and/or significant root development may survive slumping or sliding and continue to grow in their new geographic positions on or at the base of the scarp. This results in a rearrangement of the species' geographic positions relative to prescarp conditions.

Severe scarping (or overwash) may result in the entire removal of the incipient and landward zones and, depending on the width of foredune species zonation, the lee slope (or landward) dominant species may be impacted to various degrees (Saunders and Davidson-Arnott, 1991). Blowouts (erosional saucers, bowls, and troughs) may evolve and foredune degradation may continue for some time. Following very severe storms in New South Wales, Australia, in 1974, several large foredunes suffered severe scarping and erosion. Erosion occurred landward into the lee slope tall shrubs–woodland zone in some cases. Slow, large-scale dieback of those species took place. Grass, sedge, and herb species colonized the base of the scarp, and sometimes the scarp crest, while a suite of shrubs only very slowly re-established in the former dieback woodland zone but over a significant period of time (several years). It is likely that some species never returned and species composition was significantly altered, if not permanently, at least for a long time.

On moderately (and severely) scarped dunes, the locus of maximum sand deposition will move landward as sediment, which would normally be trapped in the incipient foredune and seaward slope of the established foredune, is transported further across the established foredune and through low points (Hesp, 1988a). This is often because the surviving

FIG. 3 (**A**) An 8-m high foredune near Dongara, Western Australia, on February 9, 1983. The dune is dominated by *Spinifex longifolia* (stoss face) and *Atriplex* (crest). (**B**) The same foredune during a major storm disturbance event on July 9, 1983. Half of the dune was removed during the 2-day storm. (**C**) The same foredune location on February 14, 1984. All the vegetation is dead, slump blocks litter the stoss face, and incipient blowouts are beginning to form on the crest.

species present a low-density cover and are not conducive to extensive regeneration (Saunders and Davidson-Arnott, 1991). On long-term eroding coasts, slow scarp or dune crestline retreat occurs (McCann and Byrne, 1994), and the species zonation and richness will be continuously impacted and permanently affected.

Recolonization by pioneers is, in part, determined by the regenerative ability and strategies of the local "pioneer plants." For example, the seeds of *Cakile* species colonize new beaches or backshores by being able to float and survive in seawater, while *Ammophila* rhizomes colonize new areas by surviving breakage and seawater transport (Maun, 1984). Pioneer plants with stoloniferous rather than rhizomatous or seed establishment growth habits will have an advantage, because they are able to grow down the scarp face even before full scarp filling via sand accretion has taken place.

Quite large foredunes (approximately 8 m high) typically occur along the Dongara region, west Australian coast north of Perth, and front large parabolic dune systems (see below). One question applicable to many similar coastal sites is, "How do the parabolic dunes form on a coast with a large foredune?" In 1983, a severe storm caused massive erosion of the Dongara region foredunes and more than 50% of the foredune was removed in a single day, exposing the landward heath species to the full force of wind and salt spray. Large-scale dieback occurred (Fig. 3A and B) and the foredune began to be locally eroded and destroyed by wind events, leading to the development of blowouts. Over time, some of the blowouts probably evolve into parabolics. This severe disturbance resulted in short-term absolute removal of the most seaward species, longer-term removal of the "intermediate species," and dieback of much of the landward (lee slope) zone. Significant alongshore variation in dune morphology, species composition, and zonation resulted as blowouts alternated with re-establishing foredune segments. Might large storms of the magnitude seen at this location permanently alter the species composition and zonation patterns of the foredune? Certainly such disturbance leads to increased patchiness, spatial variations in species density and presence, and possibly removal of certain species from the foredune for a period of time, or permanently.

The storm history, magnitude, and return period could be critical for many foredunes. Major damage (and permanent or long-term vegetation change) occurs where a site is affected by several large storms occurring at high frequency rather than a few storms many years apart. The storm magnitude and occurrence may be the most critical elements in terms of site flexibility, resistance to change, and biodiversity (Hayden *et al.*, 1995).

Wind Speed and Turbulence The average or prevailing wind energy is important not only at a regional level for a given beach or group of beaches but also at a local level. Wind speed and turbulence will vary depending on plant canopy species' presence, structure, height, density, and distribution (see Chapter 2). The degree of wind flow speedup and therefore wind velocity at different points across a foredune will vary according to the height and morphology of the foredune as well as the canopy structure. Increasing dune height leads to increasing wind speedup over a dune, such that dune crest velocities can be greatly in excess of the dune toe (Hesp *et al.*, 2005a). Speedup leads to greater volumes of sand reaching higher elevations on a dune and salt spray aerosols carried further across dunes. These wind flow disturbances lead to leaf burn, wind pruning, stunting of plant growth, elimination of certain species once a certain dune height is reached, and temporal changes in plant species cover as dune height changes.

Overall, the pattern and horizontal distance (zonation) occupied by foredune species, and plant species richness on a foredune, are probably related to the number of wave scarping, overwash, and swash inundation events (Oertel and Larsen, 1976; Saunders and Davidson-Arnott, 1991; Cordazzo, 1999), as well as the volume of sand supply (related to beach type; Short and Hesp, 1982; Hesp, 1988a; 1988b), beach fetch and water levels (Davidson-Arnott and Law, 1990; 1996; Saunders and Davidson-Arnott, 1991), exposure, dune height and size, differences in salinity, flooding and moisture content (Olff *et al.*, 1993), the amount and residence time of snow accumulation (Ruz and Allard, 1994b), other biogeographic factors (e.g., climate and latitude; Moreno-Casasola, 1988; Hesp, 1991), and beach and barrier state (prograding, stable, or retrograding).

Blowouts and Parabolic Dunes

Blowouts are cup-, saucer-, bowl-, and trough-shaped depressions formed by wind erosion of a sandy substrate. They typically comprise a deflation basin, steep erosional side walls, and a downwind depositional lobe. Parabolic dunes are U- or V-shaped dunes characterized by short to elongate trailing ridges that terminate downwind in U- or V-shaped depositional lobes. Deflation plains, basins, and slacks occupy the area between the trailing ridges. Parabolic dunes are similar to blowouts in that they advance downwind and to a lesser degree expand laterally, often completely covering and destroying the vegetation and older dunes as they advance. The principal difference between blowouts and parabolic dunes is that parabolic dunes are typified by trailing ridges while blowouts generally do not exhibit trailing ridges (Hesp, 1999).

The major disturbances in such dune types are sand deposition, sand erosion, undermining, scarping and cliff collapse, substrate changes, water erosion (sheet wash and rilling/gullying), increasing temporal and spatial variation in exposure to solar radiation (shade versus sun), and locally increased wind flows and turbulence (Fig. 4).

The most common disturbance in blowouts and parabolic dunes is wind erosion (Jungerius and van der Meulen, 1989). In order to form, the foredune or dune is eroded by the wind and a deflation hollow or trough is formed. The eroded sediment is transported some distance downwind, and a small depositional lobe or sand sheet is formed. The degree of lateral and horizontal wind erosion will determine the magnitude of disturbance, so some blowout deflation hollows, for example, can display a quite reasonable cover of plants, particularly where deflation is minimal (e.g., in some shallow saucer blowouts and low-wind-energy environments), while others will be completely bare.

Within a blowout, wind speeds are locally accelerated and, within trough blowouts, jets (localized high speed flows) and pronounced turbulence (e.g., corkscrew vortices) may occur up the deflation floor, above the bed, and along the erosional walls (Hesp and Hyde, 1996).

The extent of horizontal wind deflation is restricted by the presence of the groundwater and overlying capillary fringe, the seasonally lowest water table, a lag (concentration of shell, pebble, wood, and other materials), or

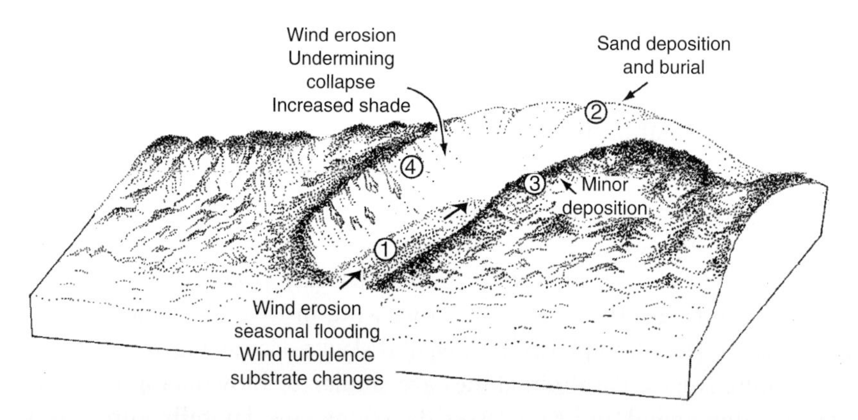

FIG. 4 A trough blowout and the kinds of disturbances present in different parts of the dune. Similar disturbances are found in parabolic dunes but typically at a larger scale of operation.

an older soil, palaeosol, carbonate surface, rock, or wind-resistant surface. The plant community that subsequently evolves in the deflation hollow or plain will be determined not only by the degree of wind erosion but also by the final surface that appears once deflation ceases.

High-speed wind flows and jets up the blowout deflation floor cause undermining and erosion of the adjacent lateral slopes (Gares, 1992; Pluis, 1992; Fraser *et al.*, 1998; Hesp, 2002). Slumping, avalanche, and collapse of the slopes occur and are a major disturbance to plants on the slopes or along the slope or erosional wall crest. Aeolian transport of sediment along the walls also increases this disturbance process and often leads to plant root exposure and dieback of the vegetation growing near the wall crests.

The obvious, and usually the other most significant, disturbance is sand deposition on and over the depositional lobe. The vegetation upon which the downwind migrating depositional lobe of a blowout (or a parabolic dune) advances is inundated with sand and often completely smothered and killed. Plant response to such moderate to large-scale disturbance varies according to location and proximity to the advancing parabolic lobe, volume and thickness of the lobe, speed of downwind migration, plant adaptation strategies (particularly to sand burial), and geobotanical province. For example, in eastern, southern, and western Australian parabolic dunes, some *Acacia* species respond positively to plant burial, sending out adventitious roots and coping with low to moderate burial. If the parabolic lobe is migrating rapidly downwind, then a formerly buried *Acacia* tree may be re-exposed (exhumed) on the upwind side, continue to survive, and regrow, acting both as an "island refuge" and seed and plant parent source for recolonization of the deflation plains. Significant patchiness may be introduced by this process (Fig. 5).

Similar to transgressive dunefields (below), the movement of blowout and parabolic depositional lobes results in the translocation or appearance of "pioneer" species far downwind of the beach and normal foredune location. The outside, lower slopes are often colonized by adventitious, pioneer, stoloniferous species that may grow up and over the lobe (e.g., *Spinifex* spp. in Australia and New Zealand; either *Ammophila* or *Panicum* spp. in the United States, Canada, and Brazil; *Ipomoea* and *Canavalia* spp. in the subtropics and tropics). Once stabilized, the lobe may be colonized by transitional species quite different from the transitional species found on the foredune. If the downwind movement of the lobe is slow, and the lobe is not higher than the plant or tree tops, some tree species will survive burial, grow new roots further up the trunks, and emerge intact out onto the

FIG. 5 An active parabolic dune migrating over relict, vegetated parabolic dunes near Dongara, Western Australia. The large-scale disturbance created by dune migration leads to almost complete destruction of the downwind heath community, *Acacia* species colonization of the trailing ridges, some survival of individual *Acacia*s on the parabolic lobe, and a completely new suite of plant species colonizing the newly formed deflation plain (just visible at the bottom of the photograph).

windward side of the lobe. If movement is rapid or the lobe is higher than most plants, then plant growth is impossible or "pioneer" species dominate.

In moderate- to high-energy wind environments, fine to very fine sand may be transported by suspension significant distances downwind into older, more stable canopies and dunes (Carter *et al.*, 1990). This fine suspended sand "rain" may act as a fine scale disturbance, aiding soil accession and adding nutrients and potentially seed material.

In higher latitudes, the solar azimuth is important in determining the degree of shade or sun on slopes, particularly seasonally (Oke, 1978), and this may affect plant growth rates and species colonization or presence. For example, in New Zealand at 40° south on the west coast, the northern slopes of deep trough blowouts are permanently in the shade for the entire winter. The northern (south facing) slopes almost never dry out; slope angles are typically maintained at 70° to 90° and are largely bare. Opposite slopes receive solar radiation and dry out, thus allowing slumping to occur and lower slopes to be formed (<30°). These slopes are more typically vegetated to various degrees. In summer, both slopes receive solar radiation but the northern slopes dry out, collapse, and slump, and are highly unstable, thus restricting plant colonization or growth.

Blowouts and parabolics create patchy disturbances in foredunes or other dunes allowing gap specialists to colonize and proliferate in disturbed areas. Lee (1995) found that even after blowouts were colonized by pioneer or original predisturbance species, greater densities of the gap specialist(s) could be detected years after the blowout revegetated. It is therefore important to recognize that where blowouts or parabolic dunes are common, they may have been formed individually or in phases over significant temporal and spatial scales (Thompson, 1983), and that moderate to high spatial variability in species composition and richness will be possible.

In the case of parabolic dunes, trailing ridges are formed by plants trapping and also colonizing sand on the outside margins of the depositional lobe (Fig. 5). It is common around the world for just one or a very few species to begin the trailing arm colonization process, mainly because they can cope with relatively high sand burial rates and minimum nutrients. Pioneer species normally found on the backshore may begin the process, but shrubs and trees may also be involved. In Australia, for example, as noted above, certain *Acacia* species are primarily responsible for preferentially growing onto or colonizing the outside margins of the trailing arms.

As the parabolic lobe moves downwind, the windward (upwind) edge of the lobe is deflated down to some base level determined by the water table, pebble or shell lag, older soils, palaeosols, bedrock, or a calcrete/cemented horizon. A bare deflation surface (basin or plain) is produced by this large-scale disturbance process (Fig. 5). A variety of plants then colonize the upwind, erosional, typically unvegetated, deflation basin or plain. This colonization process may be relatively random, especially where the plains are eroded down to a base level determined by the fresh water table. Ponds and wetlands formed in the deflation plains and basins provide a major source of freshwater in the dunefields for animals and birds, and these are important vectors for the transfer, introduction, or dispersion of seeds. Collapse and slumping of the inside margins of trailing arms also lead to the introduction of plants that would not normally be present (or would be introduced much later) onto lower slopes and deflation floors. In some cases, however, plants may colonize the plain in a well-defined sequence (Hesp and Pelham, 1984).

The coastal plain between Dongara (300 km north of Perth, Western Australia) and Cliff Head, Western Australia, is dominated by an extensive system of long-walled parabolic dunes. These dunes are typically initiated at the coast by localized foredune destruction (Fig. 4) and subsequent formation of blowouts, which then elongate downwind and evolve into parabolic dunes under the influence of strong south–southwest onshore winds. In this highly calcareous environment (Beard, 1976), the deflation plains are eroded down to a calcrete surface, which is extensive in the region. This surface is then colonized by vegetation in a relatively well-defined sequence (Hesp and Pelham, 1984).

Hesp and Pelham (1984) examined the vegetation in six parabolic dune deflation plains (dunes 3, 6, 8, 10, 11, and 12) of various ages and extents. Because all parabolic dunes are initiated at the coast the greater the distance the depositional lobe is downwind from the coast the older the dune and the longer the deflation plain. Analysis of aerial photographs of different years (1969, 1983) allowed determination of rates of dune migration and vegetation colonization. The pattern of colonization along the deflation basins of the dunes is generally consistent from dune to dune. Minor variations in the structure of the various zones reflect the dynamic processes operating within each dune and patchiness associated with survival or regrowth of individual *Acacia* shrubs and trees as the parabolic depositional lobe has passed.

The rate of downwind migration of each dune (i.e., the rate and magnitude of disturbance) and dune age (Table 1) appear to have the greatest influence on the vegetation zonation. Table 1 compares three dune deflation plains, 6 (the youngest) to 12 (the oldest). The pioneer zone is the most extensive in the youngest and rapidly moving parabolic dune (i.e., number 6), and the total number of species present in the deflation basins is low (9 species). The intermediate-age parabolic dune (i.e., number 3) displays a limited *Angianthus* zone (the pioneer in this environment) and large *Scaevola* zone. Parabolic dune 12 is the oldest dune and has not advanced since 1969 (at least). It completely lacks the *Angianthus*- and *Scaevola*-dominated zones but has 19 species present. Table 1 also shows that the total percentage cover of the pioneer and early intermediate stages tends to decrease and that of the older zones increases as the rate of dune movement decreases. Soil analyses show that this process is independent of edaphic influences (cf. Chadwick and Dalke, 1965).

This study indicates that it is the dynamic processes occurring within each dune (with the primary factors being rate of downwind advance by the dune sheet and deflation basin, and dune age) that result in changes in the plant communities.

TABLE 1 Morphometric and Vegetation Patterns in Three Representative Parabolic Dunes in the Dongara – Cliff Head Region, Western Australia*

		Rate Downwind Movement ($m \bullet yr^{-1}$)		Extent (m), Mean % Cover, Point of Maximum Cover of Vegetation Zones (m)			
Parabolic Dune	Distance Inland (m) Relative Age	Slipface Advance	Deflation Plain Formation	Ang.	Scaev.	Cas.	Mel.
6	500 Very young	15	10	0–400 16.5% 154	200–500 11% 460	Not Present –	 –
3	2500 Intermediate	5	8.5	0–160 12% 25	35–800 8% 225	80–915 13% 330	650+ 25% 800
12	4500 Old	None	None	–	Some Spyridium present	110–1050 25% 150+	400+ 34% 660+

*As the dunes get older or the rate of downwind migration slows, the presence of dominant species changes. Ang: *Angianthus cunninghamii* pioneer zone; Scaev: *Scaevola crassifolia* intermediate zone; Cas: *Casuarina lehmanniana* zone; Mel: *Melaleuca spp*. zone.

Barrier Island Dunes, Overwash Plains, Nebkha Fields, and Erosion Rates

The rate at which barrier islands are eroding is related to the frequency of storm/hurricane overwash (Leatherman, 1979), sea level rise, and sediment loss. Barriers experiencing low overwash frequencies will usually show richer vegetation communities and topographically more stable sites, while barriers experiencing frequent overwash are characterized by sparse vegetation (e.g. Hosier and Cleary, 1977; Cleary and Hosier, 1979; Roman and Nordstrom, 1988; Ehrenfeld, 1990; Hayden *et al.*, 1995). On many long-term eroding barrier islands there is significant spatial topographic variability and, particularly, longshore variation in vegetation cover and species richness. Ritchie and Penland's (1990) model of dune and overwash terrains on a Louisiana barrier island as modified by Hesp and Short (1999) defines five terrain types: (1) washover sheet (large-scale bare sheets); (2) washover terrace; (3) discontinuous foredune and overwash fan; (4) continuous foredune and scattered nebkha fields; and (5) continuous foredune and foredune plain. The magnitude of disturbance (particularly overwash) determines the landform type and may also drive species dominance on these Louisiana (and other) terrains. On average, *Spartina patens* (Aiton) Muhl. dominates the low, regularly overwashed, washover sheets and terraces because of its ability to rapidly colonize recently overwashed, shell-dominated areas (cf. Stallins and Parker, 2003). In contrast, *Panicum* species dominate the foredune terrains and areas that are more stable and receiving a sand supply (Ritchie and Penland, 1990). On many barrier islands that are only slowly eroding, landward dunes and associated landforms tend to have more diverse and richer plant communities than areas regularly overwashed (Roman and Nordstrom, 1988).

Scattered nebkha [equivalent to Hayden *et al.*'s (1995) "pimples"] are the only dune types possible on many long-term eroding barrier islands, particularly those that are regularly impacted by storms and hurricanes. Even defining a clear incipient foredune zone may be difficult. Here regular storm wave overwash can result in foredunes being unable to build or removal of foredunes, large-scale damage and removal of landward dunes, random isolation of dune segments which become remnant knobs, and the scattering of seed and plant material. Regrowth of plants can be largely random and/or dictated to various degrees by the extent of overwash—whether it is restricted to local patches (fans) or extensive (sheets and terraces; Ritchie and Penland, 1990). Hosier and Cleary (1977) provide an elegant cyclic model of dune response and associated vegetation development on a barrier island undergoing various patterns of washover. Stallins and Parker (2003) note that an infrequently overwashed barrier displayed greater spatial compositional variability compared to a barrier that was frequently overwashed.

The disturbance "type" may also vary temporally and result in differential responses. During Hurricane Ivan in 2004, large areas of the Santa Rosa (Florida Gulf coast) barrier were overwashed and marine overwash sediments were greater than a meter thick in places. Overwash terraces and fans smothered vegetation, while storm surge waters eroded the margins of larger dune segments, large nebkha, and remnant knobs (Fig. 6). In the following months, the overwash sediments were deflated by the wind and deposited on the downwind margins of the barrier, forming new, discontinuous precipitation ridges, hummocky dunes, and nebkha. Some dunes and vegetated areas were smothered by sand inundation. New blowouts formed in large nebkha, and the margins of remnant knobs became increasingly unstable and erosional.

FIG. 6 A portion of the Santa Rosa barrier island on the Gulf of Mexico coast, Florida, on September 24, 2004, following storm overwash by Hurricane Ivan. The storm surge and hurricane overwash destroyed a low foredune and produced a large overwash terrace across half of the barrier. Larger dunes and dune remnants were eroded on the margins. The surge flows were redirected and concentrated around these dunes, producing overwash channels and fans. The landward margin shrub and forest vegetation survived to various degrees, but is now (2006) subject to aeolian deposition and blowout development as the sediments making up the overwash terrace and fans are redistributed by wind. Some of the overwash channels have become wet slacks, and scattered nebkha are forming across the formerly bare terrace. (Photograph courtesy of Dr. Gregory Stone.)

Transgressive Dune Sheets and Dunefields

Transgressive sheets and dunefields are relatively large aeolian sand deposits formed by the downwind, oblique, and/or alongshore migration of sand. Sand sheets or sandy plains are relatively featureless undulating to flat forms. Dunefields and machair contain dunes of various types on the surface (e.g., blowouts, parabolic dunes, transverse, oblique, barchan, dome, linear, and star dunes). The sheets and dunefields may be quite small (approximately 10,000 m^2) to very extensive (many square kilometers), completely unvegetated, partially vegetated, or fully vegetated (relict; Hunter *et al.*, 1983; Hesp and Thom, 1990; Hesp *et al.*, 2005b). They may contain small to extensive deflation basins and plains, lakes, wetlands and slacks, nebkha fields, trailing ridges, precipitation ridges, bush pockets, remnant knobs, parabolic dunes, and gegenwalle ridges (Ranwell, 1972; Martinho *et al.*, 2006).

Dunefield Migration

The development and downwind movement of a transgressive dunefield or sheet can result in wholescale obliteration and massive disturbance of the downwind terrain being transgressed in some cases, and little vegetation may survive (Fig. 7). In these cases, such large-scale disturbance may later result in very large-scale physical and biological change. In contrast, within foredunes, blowouts, and probably many parabolics, the disturbances are more spatially limited and patch dynamics may be more common (van der Maarel, 1993; Lee, 1995). Seeliger *et al.* (2000) and Seeliger (2003) document the human-induced lowering of the water table, death of the dominant pioneer foredune species (*Panicum racemosum*), collapse of the foredune system, and subsequent formation of a transgressive dunefield. The inland movement eliminated the vegetation and fauna of downwind hummock dunes, dry and wet slacks, and marshes (Seeliger *et al.*, 2000).

As dunefields develop, a precipitation ridge is formed on the downwind edge. Cooper (1958, p. 55) defined a precipitation ridge as a "ridge due to precipitation of sand at a forest edge," but the term has been generally extended to include sand precipitation into any vegetation type or community. In essence, sand rains (precipitates) down onto the adjacent vegetation by grainfall, avalanche, saltation, and suspension, forming a steep slope, often but not always lying at the angle of repose. While it is often assumed that precipitation ridges must be large (e.g., Buffault, 1942; Bressolier *et al.*, 1990), they can be quite small (a meter or less) where a low sand sheet or plain is migrating into vegetation. Where an active dunefield is

FIG. 7 An extensive transgressive dunefield migrating alongshore at Van Stadens/Gamtoos, south coast, South Africa. Downwind vegetation is almost entirely smothered and destroyed by this large-scale disturbance. No foredune is present because of the alongshore dune migration. Bush pockets occur in the interdune depressions and flats (left side) and over time may expand (right side) and result in vegetative colonization of the dunefield. If the sand supply diminishes or ceases, small to extensive deflation plains may develop upwind, resulting in a suite of new disturbances, a variety of new landforms, and significant vegetation patchiness.

migrating transversely onshore there may be only one precipitation ridge bounding the landward margin. However, in dunefields that are crossing headlands or migrating alongshore to obliquely onshore, precipitation ridges may occur on both sides of the dunefield as well as on the downwind margin. In some of the alongshore migrating dunefields north of Veracruz, Mexico, the seaward-most precipitation ridge is formed along the backshore and the seaward margin functions as the foredune as well.

The level of disturbance depends very much on the rate and volume of sand migration. Dunefield and dune sheet migration may lead to a range of levels of disturbance from: (1) slow sand accretion, especially in partially vegetated sand sheets; (2) small- to large-scale localized inundation and burial, particularly where individual dunes are migrating downwind; to (3) massive inundation, burial, and destruction, particularly within the dunefields where sand supply is high, where dunes are large, and along the downwind and lateral margins where precipitation ridges are forming (Cooper, 1967; Hunter *et al.*, 1983; Thom, 1984; Orme and Tchakerian, 1986; Borowka, 1990; Hesp and Thom, 1990; Orme, 1990; Thom *et al.*, 1992).

Dune migration often occurs in phases or episodes and, as Ranwell (1972) notes, there is periodic re-exposure of bare sand at intervals in the mobile part of the system. There is also a tendency for cyclic alternation of dune and slack at any particular point within it. So there is often an extended period where deflation plains, basins, local slacks, and water bodies/wetlands develop in the wake of the dune transgression (Paul, 1944; Willis *et al.*, 1959; Garcia Novo, 1997). This allows a variety of species to colonize the dunefield, especially once the seasonal water table is reached via deflation or a limiting element is reached (e.g., calcrete layer, palaeosol, shell lag). Then, after some time, these sites may again be inundated by a new phase of dune advance.

Deflation Basins, Plains and Slacks

In transgressive dunefields and sheets, another major disturbance is deflation and wind erosion, leading to a range of effects, including (1) slight to extensive surface erosion (deflation) and (2) creation of interdune depressions and slacks.

Deflation plains and basins or slacks are formed by wind deflation of the surface. They can form nearly half of the dune system, according to Ranwell (1972), and, in some cases where the dunefield is in an advanced evolutionary stage, may dominate the landform units present. Hesp and Thom (1990) demonstrate that there may be a variety of types and sizes from small deflation basins and ponds to very large and very extensive deflation plains. There may be multiple suites of them separated by dune phases (e.g., the Donana system; Garcia Novo, 1979, 1997; Garcia Novo *et al.*, 2004). Ranwell (1972) defined three types of slack: semi-aquatic slacks (watertable <0.5 m below the surface), wet slacks (water table never below 1 m), and dry slacks (water table between 1 and 2 m below the surface at all seasons). On some occasions the slacks may be saline.

The development of deflation basins and plains (or slacks) typically provides two major disturbances. The first is clearly wind erosion of the substrate, undermining plants and exposing root systems, with attendant downwind deposition of sediment in other communities. New niches develop and new plant communities evolve as the water table is approached; for example, sedges may gradually replace grasses (Jones and Etherington, 1971). The second major disturbance factor is water, particularly flooding. Once established, wet slacks probably suffer the most disturbance, because they may be affected by drought during summer or periods of low rainfall and low water table, and by flooding during periods of high and sustained water table (Ranwell, 1959; 1960). Plants respond in a highly variable manner to flooding and intricate patterns of species distribution may be produced (Crawford, 1972). In some cases, following long periods of abnormally low water tables, deflation leads to the creation of small- to large-scale and relatively permanent ponds and lakes (e.g., Lubke and Avis, 1988; Crawford *et al.*, 1997; Garcia-Novo, 1997; Hesp, 2004; Lubke, 2004).

Deflation plains may also be later modified and disturbed to various degrees by subsequent sand inundation or by the secondary development of erosional dunes, such as blowouts and parabolics. The prototype golf bunker—the saucer blowout, for example—is very characteristic of machair terrains (Ritchie, 1976; Bassett and Curtis, 1985). The Joaquina dunefield in Florianopolis, Brazil, displays a large deflation plain currently undergoing significant modification by blowout and parabolic dune development (Bigarella, 1975). Regional climate change may switch a dunefield from a largely erosional phase to a stabilization phase if the rainfall increases.

Sand Accretion/Erosion and Development of Nebkha, Trailing Ridges, and Gegenwalle Ridges

As dunes migrate downwind, a variety of other dune types may develop within and on the margins of deflation plains, sand sheets, and dunes. Trailing ridges are formed where the outside margins of transverse dunes are colonized by vegetation in similarity to trailing ridges formed during the migration of parabolic dunes (Hesp, 2004; Martinho *et al.*, 2006). Gegenwalle ridges (Paul, 1944)—literally counter ridges [also the "contradunas" of Allier *et al.* (1974)]—may be formed in many ways, but the classic type develops where sand is blown off dunes by offshore winds back upwind into vegetation colonizing the deflation plains, basins, or slacks. In both cases of trailing and gegenwalle ridges, initial development results from disturbance where sand depositing within vegetation generally elicits a plant growth and community change response.

Individual and fields of nebkha [equivalent to Ranwell's (1972) "hummocks" or "hedgehogs"] may develop almost anywhere within a dunefield and may act as one of the important sites from which plants spread and whereby dunefields eventually become vegetated. As nebkha trap sand and become larger, they impose more and more into the boundary layer and may eventually evolve from largely depositional forms to erosional forms. In addition, surface deflation of the surrounding nonvegetated surface can indirectly increase the height of the nebkha and disturb the margins of the dunes. Disturbance processes include undermining, cliffing, slope retreat, and the creation of erosional nebkha, turrets, and remnant knobs.

In general and throughout the dunefields, wind flow, local wind decelerations, accelerations, and jets, and increased turbulence are common disturbances. Sheet wash and gullying can also be locally significant (Bakker *et al.*, 1990). In addition, many of the disturbances typical of the backshore and foredune, such as high surface heating, drought, salt spray, and lack of nutrients, occur in transgressive dunefields.

Overall, there are a wide variety of locations, habitats, and niches in the dunefields and dune sheets where different species or plant functional types may colonize and/or revegetate the environment (McLachlan *et al.*, 1987; Lubke and Avis, 1988; Hesp and Thom, 1990; Hayden *et al.*, 1995; García-Mora *et al.*, 2001; Martínez *et al.*, 2001; Grootjans *et al.*, 2004; Martinho *et al.*, 2006). Many of these niches, habitats, and local environments also display fascinating interrelationships between geomorphological and ecological processes and faunal dynamics (McLachlan *et al.*, 1987; Hesp and McLachlan, 2000). As Garcia-Novo *et al.* (2004) note, the character of these dunefield environments is one of fragmentation, where terrestrial habitats fluctuate in sediment transport, water availability, surface stability, and temperature. The range of disturbances can be significant (from inundation and burial to erosion and flooding) and vary temporally at various scales.

CONCLUSION

The four main coastal dune types—foredunes, blowouts, parabolic dunes, and transgressive sheets and dunefields—may experience significantly differing levels of disturbance. Foredunes, coastal barriers subjected to regular overwash events, and large-scale active (mobile) transgressive dunefields may experience the highest levels of disturbance. Blowouts and parabolic dunes are strongly associated with patch dynamics and gap specialists or sand accretion opportunists.

A host of questions arise from this brief overview of disturbance processes in dunes. We perhaps understand the types of disturbance, but there appears to be little research by geomorphologists on determining the levels of disturbance in dunes and the range of those disturbances. For example, what actually constitutes a mild or slight versus a severe disturbance for deflation, flooding, and sand inundation? Botanists, biologists, and ecologists have long studied the response of various plant species to sand burial, nutrient supply, water relations, and other factors, but there is relatively little work linking physical rates of dunefield and dunefield landform unit changes with the botanical research. For example, while there are some data on rates of parabolic and transgressive dune advance or migration, our guess is that there are few data on the relationships between dune migration rates (different rates equal different levels of disturbance) and woodland or forest response. How do plants respond and what community changes take place where the temporal and spatial variability of disturbance and disturbance type is significant?

In general, much more work along the lines of that conducted by Hayden *et al.* (1995) needs to be done on the interrelationships between geomorphological and hydrological processes, disturbance types and events, and plant and vegetation responses in order to obtain an improved understanding of disturbance and coastal ecosystem response. In particular, transgressive dunefields are largely ignored in current geological, geomorphological, and ecological research and literature, yet provide many opportunities for fundamental and exciting research.

ACKNOWLEDGMENTS

Thanks to Jean Ellis for helping P. H. with parts of this research, Sergio Dillenburg for wonderful trips to Brazilian dunes, our respective departments for support, Mary-Lee Eggart for cartography, and several anonymous referees.

REFERENCES

Allier, C., Gonzalez Bernaldez, F., and Ramirez Diaz, L. (1974). Reserva Biológica de Doñana. Ecological Map. Estacion Biologica de Doñana. C.S.I.C., Sevilla, Spain.

Arens, S. M. (1996). Patterns of sand transport on vegetated foredunes. *Geomorphology* 17, 339–350.

Arens, S. M. (1997). Transport rates and volume changes in a coastal foredune on a Dutch Wadden Island. *J Coastal Conserv* 3, 49–56.

Bakker, T., Jungerius, P. D., and Klijn, J. A., Eds. (1990). *Dunes of the European Coasts: Geomorphology - Hydrology - Soils.* Catena Supplement 18, Catena, Cremlingen-Destedt, W. Germany.

Bassett, J. A., and Curtis, T. G. F. (1985). The nature and occurrence of sand dune machair in Ireland. *Proc R Irish Acad* 85B, 1–20.

Beard, J. S. (1976). *The Vegetation of the Dongara Area, Western Australia.* Vegmap Publications, Perth, Australia.

Bigarella, J. J. (1975). Lagoa dunefield, Santa Catarina, Brazil—a model of aeolian and pluvial activity. *B Para Geosci* 33, 133–167.

Blum, M., and Jones, J. R. (1985). Variation in vegetation density and foredune complexity at North Padre Island, Texas. *Tex J Sci* 37, 63–73.

Borowka, R. K. (1990). The Holocene development and present morphology of the Leba dunes, Baltic coast of Poland. In: *Coastal Dunes: Form and Process.* (Nordstrom, K. F., Psuty, N. P., and Carter, R. W. G., Eds.), pp. 289–314. Wiley, Chichester, UK.

Bressolier, C., Froidefond, J. M., and Thomas, Y.-F. (1990). Chronology of coastal dunes in the south-west of France. In: *Dunes of the European Coasts: Geomorphology - Hydrology -*

Soils. (Bakker, T., Jungerius, P. D., and Klijn, J. A., Eds.), pp. 101–107. Catena Supplement 18, Catena, Cremlingen-Destedt, W. Germany.

Buffault, P. (1942). *Histoire des Dunes Maritimes de la Gascogne*. Editions Delmas, Bordeaux, France.

Carter, R. W. G., Hesp, P. A., and Nordstrom, K. (1990). Geomorphology of erosional dune landscapes. In: *Coastal Dunes: Form and Process* (Nordstrom, K. F., Psuty, N. P., and Carter, R. W. G., Eds.), pp. 217–250. Wiley, Chichester, UK.

Carter, R. W. G., and Rihan, C. L. (1978). Shell and pebble pavements on beaches: examples from the north coast of Ireland. *Catena* 5, 365–374.

Carter, R. W. G., and Wilson, P. (1990). The geomorphological, ecological and pedological development of coastal foredunes at Magilligan Point, Northern Ireland. In: *Coastal Dunes: Form and Process*. (Nordstrom, K. F., Psuty, N. P., and Carter, R. W. G., Eds.), pp. 130–157. Wiley, Chichester, UK.

Chadwick, H. W., and Dalke, P. D. (1965). Plant succession on dune sand in Fremont County, Idaho. *Ecology* 46, 765–780.

Cleary, W. J., and Hosier, P. J. (1979). Geomorphology, washover history, and inlet zonation: Cape Lookout, NC, to Bird Island, NC. In: *Barrier Islands from the Gulf of Saint Lawrence to the Gulf of Mexico* (Leatherman, S. P., Ed.), pp. 237–271. Academic Press, New York.

Cooke, R. U, Warren, A., and Goudie, A. (1993). *Desert Geomorphology*. UCL Press, London.

Cooper, W. S. (1958). *Coastal Sand Dunes of Oregon and Washington*. Memoir 72 Geological Society of America, New York.

Cooper, W. S. (1967). *Coastal Dunes of California*. Memoir 104, Geological Society of America, Boulder, CO.

Cordazzo, C. V. (1999). Effects of salinity on seed germination, seedling growth and survival of *Spartina ciliata* Brong. *Acta Bot Bras* 13, 317–322.

Cowles, H. C. (1898). The ecological relations of the vegetation on the sand dunes of Lake Michigan. *Bot Gaz* 27, 97–117.

Crawford, R. M. M. (1972). Some metabolic aspects of ecology. *T Edin Bot Soc* 41, 309–316.

Crawford, R. M. M., Studer-Ehrensberger, K., and Studer, C. (1997). Flood-induced change on a dune slack observed over 24 years. In: *The Ecology and Conservation of European Dunes*. (Garcia-Novo, F., Crawford, R. M. M., and Diaz Barradas, M. C., Eds.), pp. 27–40. Serie Ciencias núm. 38, Universidad de Sevilla, Sevilla, Spain.

Davidson-Arnott, R. G. D., and Law, M. D. (1990). Seasonal patterns and controls on sediment supply to coastal foredunes, Long Point, Lake Erie. In: *Coastal Dunes: Form and Process*. (Nordstrom, K. F., Psuty, N. P., and Carter, R. W. G., Eds.), pp. 177–200. Wiley, Chichester, UK.

Davidson-Arnott, R. G. D., and Law, M. D. (1996). Measurement and prediction of long-term sediment supply to coastal foredunes. *J Coastal Res* 12, 654–663.

Ehrenfeld, J. G. (1990). Dynamics and processes of barrier island vegetation. *Rev Aquat Sci* 2, 437–480.

Forman, R. T. T., and Godron, M. (1986). *Landscape Ecology*. Wiley and Sons, New York.

Fraser, G. S., Bennet, S. W., Olyphant, G. A., Bauch, N. J., Ferguson, V., Gellasch, C. A., Millard, C. L., Mueller, B., O'Malley, P. J., Way, N., and Woodfield, M. C. (1998). Windflow circulation patterns in a coastal dune blowout, south coast of Lake Michigan. *J Coastal Res* 14, 451–460.

Freestone, A. L., and Nordstrom, K. F. (2001). Early development of vegetation in restored dune plant microhabitats on a nourished beach at Ocean City, New Jersey. *J Coastal Conserv* 7, 105–116.

García-Mora, M. R., Gallego-Fernández, J. B., Williams, A. T., and García Novo, F. (2001). A coastal dune vulnerability classification: a case study of the SW Iberian Peninsula. *J Coastal Res* 17, 802–811.

Garcia-Novo, F. (1979). The ecology of vegetation of the dunes in Doñana National Park (South-west Spain). In: *Ecological Processes in Coastal Environments.* (Jefferies, R. L., and Davy, A. J., Eds.), pp. 571–592. Blackwell Science, London.

Garcia-Novo, F. (1997). The ecosystems of Doñana National Park. In: *The Ecology and Conservation of European Dunes.* (Garcia-Novo, F., Crawford, R. M. M., and Diaz Barradas, M. C., Eds.), pp. 97–116. Serie Ciencias núm. 38, Universidad de Sevilla, Sevilla, Spain.

Garcia-Novo, F., Diaz Barradas, M. C., Zunzunegui, M., Garcia-Mora, R., and Gallego Fernandez, J. B. (2004). Plant functional types in coastal dune habitats. In: *Coastal Dunes: Ecology and Conservation.* (Martínez, M. L., and Psuty, N. P., Eds.), pp. 155–169. Springer, Berlin.

Gares, P. A. (1992). Topographic changes associated with coastal dune blowouts at Island Beach State Park, New Jersey. *Earth Surf Proc Land* 17, 589–604.

Gares, P. A., and Nordstrom, K. F. (1995). A cyclic model of foredune blowout evolution for a leeward coast: Island Beach, NJ. *Ann Assoc Am Geogr* 85, 1–20.

Godfrey, P. J., and Godfrey, M. M. (1973). A comparison of ecosystems and geomorphic interaction between altered and unaltered barrier island systems in North Carolina. In: *Coastal Geomorphology*. (Coates, D. F., Ed.), pp. 239–258. Publications in Geomorphology, State University of New York, Binghamton, NY.

Grootjans, A. P., Adema, E. B., Bekker, R. M., and Lammerts, E. J. (2004). Why young coastal dune slacks sustain a high biodiversity. In: *Coastal Dunes: Ecology and Conservation* (Martínez, M. L., and Psuty, N. P., Eds.), pp. 85–101. Springer, Berlin.

Hayden, B. P., Santos, M. C. F. V., Shao, G., and Kochel, R. C. (1995). Geomorphological controls on coastal vegetation at the Virginia Coast Reserve. *Geomorphology* 13, 283–300.

Hellemaa, P. (1998). The development of coastal dunes and their vegetation in Finland. *Fennia* 176, 111–221.

Hequette, A., and Ruz, M-H. (1991). Spit and barrier island migration in the southeastern Canadian Beaufort Sea. *J Coastal Res* 7, 677–698.

Hesp, P. A. (1983). Morphodynamics of incipient foredunes in N.S.W., Australia. In: *Eolian Sediments and Processes* (Brookfield, M. E., and Ahlbrandt, T. S., Eds.), pp. 325–342. Elsevier, Amsterdam.

Hesp, P. A. (1988a). Foredune morphology, dynamics and structures. *J Sediment Geol Special Issue: Aeolian Sediments* 55, 17–41.

Hesp, P. A. (1988b). Surfzone, beach and foredune interactions on the Australian south east coast. *J Coastal Res Special Issue* 3, 15–25.

Hesp, P. A. (1990). A review of biological and geomorphological processes involved in the initiation and development of incipient foredunes. *Proc R Soc Edin B-BI* 96, 181–202.

Hesp, P. A. (1991). Ecological processes and plant adaptations on coastal dunes. *J Arid Environ* 21, 165–191.

Hesp, P. A. (1999). The beach backshore and beyond. In: *Handbook of Beach and Shoreface Morphodynamics* (Short, A. D., Ed.), pp. 145–170. John Wiley, New York.

Hesp, P. A. (2000). *Coastal Dunes.* Forest Research (Rotorua) and NZ Coastal Dune Vegetation Network (CDVN), Rotorua, NZ.

Hesp, P. A. (2002). Foredunes and blowouts: initiation, geomorphology and dynamics. *Geomorphology* 48, 245–268.

Hesp, P. A. (2004). Coastal dunes in the tropics and temperate regions: Location, formation, morphology and vegetation processes. In: *Coastal Dunes: Ecology and Conservation* (Martínez, M. L., and Psuty, N. P., Eds.), pp. 29–49. Springer, Berlin.

Hesp, P. A., and Hyde, R. (1996). Flow dynamics and geomorphology of a trough blowout. *Sedimentology* 43, 505–525.

Hesp, P. A., and McLachlan, A. (2000). Morphology, dynamics, ecology and fauna of *Arctotheca populifolia* and *Gazania rigens* nebkha dunes. *J Arid Environ* 44, 155–172.

Hesp, P. A., and Pelham, A. F. (1984). Vegetation succession in parabolic dune deflation basins, Dongara to Cliff Head, W.A. In: *MEDECOS IV: Proceedings of the 4th International Conference on Mediterranean Ecosystems held at Perth, Western Australia, August 13–17, 1984* (Dell, B., Ed.). Botany Department, University of Western Australia, Nedlands, W.A., Australia.

Hesp, P. A., and Short, A. D. (1999). Barrier morphodynamics. In: *Handbook of Beach and Shoreface Morphodynamics* (Short, A. D., Ed.), pp. 307–333. John Wiley, New York.

Hesp, P. A., and Thom, B. G. (1990). Geomorphology and evolution of transgressive dunefields. In: *Coastal Dunes: Ecology and Conservation* (Martínez, M. L., and Psuty, N. P., Eds.), pp. 253–288. Springer, Berlin.

Hesp, P. A., Walker, I., Davidson-Arnott, R., and Ollerhead, J. (2005a). Flow dynamics over a vegetated foredune at Prince Edward Island, Canada. *Geomorphology* 65, 71–84.

Hesp, P. A., Rebellio Dillenburg, S., Guimaraes Barboza, E., Tomazelli, L., Ayup Zouain, R. N., Luciana Slomp Esteves, L., Sambaqui Gruber, N. L., Toldo, E. E. Jr., and Cavalcanti de Albuquerque Tabajara, L. L. (2005b). Beach ridges, foredunes or transgressive dunefields? Definitions and initiation, and an examination of the Torres to Tramandaí Barrier System. *Anais da Academia Brasileira de Ciencias (Annals of the Brazilian Academy of Sciences)* 77, 493–508.

Hosier, P. E., and Cleary, W. J. (1977). Cyclic geomorphic patterns of washover on a barrier island in southeastern North Carolina. *Environ Geol* 2, 23–31.

Hunter, R. E., Richmond, B. R., and Alpha, T. R. (1983). Storm-controlled oblique dunes of the Oregon coast. *B Geol Soc Am* 94, 1450–1465.

Jones, R., and Etherington, J. R. (1971). Comparative studies of plant growth and distribution in relation to water-logging IV. The growth of dune and slack plants. *J Ecol* 59, 793–801.

Jungerius, P. D., and van der Meulen, F. (1989). The development of dune blowouts, as measured with erosion pins and sequential air photos. *Catena* 16, 369–376.

Kent, M., Owen, N. W., Dale, P., Newnham, R. M., and Giles, T. M. (2001). Studies of vegetation burial: a focus for biogeography and biogeomorphology? *Prog Phys Geog* 25, 455–482.

Leatherman, S. P. (1979). Migration of Assateague Island, Maryland, by inlet and overwash processes. *Geology* 7, 104–107.

Lee, P. C. (1995). The effect of gap dynamics on the size and spatial structure of *Solidago sempervirens* on primary coastal dunes. *J Veg Sci* 6, 837–846.

Lubke, R. (2004). Vegetation dynamics and succession on sand dunes of the eastern coasts of Africa. In: *Coastal Dunes: Ecology and Conservation* (Martínez, M. L., and Psuty, N. P., Eds.), pp. 67–84. Springer, Berlin.

Lubke, R., and Avis, A. M. (1988). Succession on the coastal dunes and dune slacks at Kleinemonde, Eastern Cape, South Africa. *Monogr Syst Bot Missouri Bot Gard* 25, 599–622.

Marshall, J. K. (1965). *Corynephorus canescens* (L.) P. Beauv. as a model for the *Ammophila* problem. *J Ecol* 53, 447–463.

Martínez, M. L., Gallego-Fernández, J. B., García-Franco, J.G., Moctezuma, C., and Jiménez, C. D. (2006). Coastal dune vulnerability along the Gulf of Mexico. *Environ Conserv* 33, 109–117.

Martínez, M. L., and Moreno-Casasola, P. (1996). Effects of burial by sand on seedling growth and survival in six tropical sand dune species from the Gulf of Mexico. *J Coastal Res* 12, 406–419.

Martínez, M. L., and Psuty, N., Eds. (2004). *Coastal Dunes: Ecology and Conservation.* Springer, Berlin.

Martínez, M. L., Vázquez, G., and Sánchez-Colón, S. (2001). Spatial and temporal variability during primary succession on tropical coastal sand dunes. *J Veg Sci* 12, 361–372.

Martinho, C. T., Giannini, P. C. F., Sawakuchi, A. O., and Hesp, P. A. (2006). Morphological and depositional facies of transgressive dunefields in the Imbituba-Jaguaruna region, Santa Catarina State, southern Brazil. *J Coastal Res*. In press.
Maun, M. A. (1984). Colonizing ability of *Ammophila breviligulata* through vegetative regeneration. *J Ecol* 72, 565–574.
Maun, M. A. (1998). Adaptations of plants to burial in coastal sand dunes. *Can J Bot* 76, 713–738.
Maun, M. A. (2004). Burial of plants as a selective force in sand dunes. In: *Coastal Dunes: Ecology and Conservation* (Martínez, M. L., and Psuty, N. P., Eds.), pp. 119–135. Springer, Berlin.
Maun, M. A., and Lapierre, J. (1984). Effects of burial by sand on *Ammophila breviligulata. Am J Bot* 73, 450–455.
Maun, M. A., and Perumal, J. (1999). Zonation of vegetation on lacustrine coastal dunes: effects of burial by sand. *Ecol Lett* 2, 14–18.
Maze, K., and Whalley, R. (1992). Effects of salt spray and sand burial on *Spinifex sericeus* R. Br. (Poaceae). *Aust J Ecol* 17, 9–19.
McCann, S. B., and Byrne, M. L. (1994). Dune morphology and the evolution of Sable Island, Nova Scotia in historic times. *Phys Geogr* 15, 342–357.
McLachlan, A. (1991). Ecology of coastal dune fauna. *J Arid Environ* 21, 229–244.
McLachlan, A., Ascaray, C., and du Toit, P. (1987). Sand movement, vegetation succession and biomass spectrum in a coastal dune slack in Algoa Bay, South Africa. *J Arid Environ* 12, 9–25.
Moreno-Casasola, P. (1986). Sand movement as a factor in the distribution of plant communities in a coastal dune system. *Vegetatio* 65, 67–76.
Moreno-Casasola, P. (1988). Patterns of plant species distribution on coastal dunes along the Gulf of Mexico. *J Biogeogr* 15, 787–806.
Morrison, R. G., and Yarranton, G. A. (1974). Vegetational heterogeneity during a primary sand dune succession. *Can J Bot* 52, 397–410.
Nordstrom, K. F., and Jackson, N. L. (1994). Aeolian processes and dune fields in estuaries. *Phys Geogr* 15, 358–371.
Odum, W. E., Smith III, T. J., and Dolan, R. (1987). Suppression of natural disturbance: Long-term ecological change on the Outer Banks of North Carolina. In: *Landscape Heterogeneity and Disturbance* (Turner, M. G., Ed.), pp. 123–135. Springer-Verlag, New York.
Oertel, G. F., and Larsen, M. (1976). Developmental sequences in Georgia coastal dunes and distribution of dune plants. *B Georgia Acad Sci* 34, 35–48.
Oke, T. R. (1978). *Boundary Layer Climates*. Methuen and Company, London.
Olff, H., Huisman, J., and Van Tooren, B. F. (1993). Species dynamics and nutrient accumulation during early primary succession in coastal sand dunes. *J Ecol* 81, 693–706.
Olson, J. S. (1958). Lake Michigan dune development: 2. Plants as agents and tools in geomorphology. *J Geol* 66, 345–351.
Orme, A. R. (1990). The instability of Holocene coastal dunes: the case of the Morro dunes, California. In: *Coastal Dunes: Form and Process* (Nordstrom, K. F., Psuty, N. P., and Carter, R. W. G., Eds.), pp. 315–336. Wiley, Chichester, UK.
Orme, A. R., and Tchakerian, V. P. (1986). Quaternary dunes of the Pacific coast of the Californias. In: *Aeolian Geomorphology* (Nickling, W. G., Ed.), pp. 149–175. Allen and Unwin, London.
Paul, K. H. (1944). Morphologie und vegetation der Kurischen Nehrung. *Acta Nova Leopoldina Carol NF* 13, 217–378.
Peet, R. K. (1992). Regeneration dynamics. In: *Plant Succession: Theory and Prediction* (Glenn-Lewin, D. C., Peet, R. K., and Veblen, T. T., Eds.), pp. 152–176. Chapman and Hall, London.

Pickett, S. T. A., and White, P. S. (1985). *The Ecology of Natural Disturbance and Patch Dynamics*. Academic Press, Orlando, FL.

Pluis, J. L. A. (1992). Relationships between deflation and near surface wind velocity in a coastal dune blowout. *Earth Surf Proc Land* 17, 663–673.

Psuty, N. P. (1990). Foredune mobility and stability, Fire Island, New York. In: *Coastal Dunes: Form and Process* (Nordstrom, K. F., Psuty, N. P., and Carter, R. W. G., Eds.), pp. 159–176. Wiley, Chichester, UK.

Psuty, N. P. (2005). Coastal foredune development under a diurnal wind regime, Paracas, Peru. *J Coastal Res* 42, 68–73.

Ranwell, D. S. (1959). Newborough Warren, Anglesey. I. The dune system and dune slack habitat. *J Ecol* 47, 571–601.

Ranwell, D. S. (1960). Newborough Warren, Anglesey. II. Plant associes ands succession cycles of the sand dune and dune slack vegetation. *J Ecol* 48, 117–141.

Ranwell, D. S. (1972). *Ecology of Salt Marshes and Sand Dunes*. Chapman and Hall, London.

Ripley, B. S., and Pammenter, N. W. (2004). Physiological characteristics of coastal dune pioneer species from the eastern Cape, South Africa, in relation to stress and disturbance. In: *Coastal Dunes: Ecology and Conservation* (Martínez, M. L., and Psuty, N. P., Eds.), pp. 137–153. Springer, Berlin.

Ritchie, W. (1976). The meaning and definition of machair. *T Edin Bot Soc* 42, 431–440.

Ritchie, W., and Penland, S. (1990). Aeolian sand bodies of the south Louisiana coast. In: *Coastal Dunes: Form and Process* (Nordstrom, K. F., Psuty, N. P., and Carter, R. W. G., Eds.), pp. 105–128. Wiley, Chichester, UK.

Roman, C. T., and Nordstrom, K. F. (1988). The effect of erosion rate on vegetation patterns of an east coast barrier island. *Estuar Coast Shelf Sci* 26, 233–242.

Ruz, M.-H., and Allard, M. (1994a). Foredune development along a subarctic emerging copastline, eastern Hudson Bay, Canada. *Mar Geol* 117, 57–74.

Ruz, M.-H., and Allard, M. (1994b). Coastal dune development in cold-climate environments. *Phys Geogr* 15, 372–380.

Salisbury, E. J. (1952). *Downs and Dunes: Their Plant Life and Environment*. G. Bell and Sons, London.

Santos, C. R., Castellani, T. T., Horn Filho, N. O. (1996). "Pioneer" vegetation dynamics at the beach and fore dunes in Joaquina Beach, Santa Caterina Island, Brazil. *Ann Brazilian Acad Sci* 68, 495–508.

Sarre, R. (1989). The morphological significance of vegetation and relief on coastal foredune processes. *Z Geomorph N F* 73, 17–31.

Saunders, K. E., and Davidson-Arnott, R. G. D. (1991). Coastal dune response to natural disturbances. In: *Proceedings Canadian Symposium on Coastal Sand Dunes 1990*, pp. 321–346. National Research Council of Canada, Ottawa, ON.

Seeliger, U. (2003). Response of southern Brazilian coastal foredunes to natural and human-induced disturbance. *J Coastal Res* 35, 51–55.

Seeliger, U., Cordazzo, C. V., Oliveira, C. P. L., and Seeliger, M. (2000). Long-term changes of coastal foredunes in the southwest Atlantic. *J Coastal Res* 16, 1068–1072.

Sherman, D., and Bauer, B. (1993). Dynamics of beach-dune systems. *Prog Phys Geogr* 17, 413–447.

Short, A. D., and Hesp, P. A. (1982). Wave, beach and dune interactions in south eastern Australia. *Mar Geol* 48, 259–284.

Stallins, J. A., and Parker, A. J. (2003). The influence of complex systems interactions on barrier island dune vegetation pattern and process. *Ann Assoc Am Geogr* 93, 13–29.

Thom, B. G. (1984). Transgressive and regressive stratigraphies of coastal sand barriers in eastern Australia. *Mar Geol* 56, 137–158.

Thom, B. G., Shepherd, M. J., Ly, C., Roy, P., Bowman, G. M., and Hesp, P. A. (1992). *Coastal Geomorphology and Quaternary Geology of the Port Stephens-Myall Lakes Area.*

Dept. of Biogeography and Geomorphology, Australian National University Monograph No. 6. ANU Tech P.L., Canberra, Australia.
Thompson, C. H. (1983). Development and weathering of large parabolic dune systems along the subtropical coast of eastern Australia. *Z Geomorph Suppl Bd* 45, 205–225.
van der Maarel, E. (1993). Some remarks on disturbance and its relations to diversity and stability. *J Veg Sci* 4, 733–736.
Willis, A. J., Folkes, B. F., Hope-Simpson, J. F., and Yemm, E. W. (1959). Braunton Barrows: the dune system and its vegetation. Parts I and II. *J Ecol* 47, 1–24 and 249–288.

8

Coastal Dune Succession and the Reality of Dune Processes

Kiyoko Miyanishi
University of Guelph

Edward A. Johnson
University of Calgary

INTRODUCTION

Coastal dunes were one of the first ecological systems studied in some detail (e.g., Cowles, 1899; 1901), and observation of the distribution of vegetation on coastal dunes gave rise to the idea of succession, defined as the sequential replacement of dominant species over time in the absence of disturbance (Clements, 1916). The early studies of coastal dune vegetation, particularly of the Indiana Dunes of Lake Michigan (e.g., Cowles, 1899; Fuller, 1912; Downing, 1922), were some of the first to apply the chronosequence method based on the ergodic hypothesis of substituting spatial sequences of vegetation for the temporal sequences of vegetation development. This method was subsequently used for the study of other types of succession, such as on glacial till following glacial retreat (Cooper, 1923; 1931; 1939; Crocker and Major, 1955) and on abandoned agricultural land (Billings, 1938; 1941). It is still used extensively today to study succession (e.g., Thuille and Schulze, 2006; Merila *et al.*, 2006). However, as discussed in Chapter 1 and in more detail later in this chapter, the chronosequence-based successional sequences encounter problems as soon as we begin to examine the assumptions of this space-for-time substitution. For example, for primary succession on the Indiana Dunes (Fig. 1), cottonwoods (*Populus deltoides* Bartr. ex Marsh.) are typically shown growing on the second dune ridge behind the foredunes that are dominated by the pioneer dune grass *Ammophila breviligulata* Fernald, leading to the chronosequence interpretation that *P. deltoides* is a seral species that establishes *after* the dune has been built and stabilized by *A. breviligulata* (e.g., McNaughton and Wolf, 1973). However, several studies (Fuller, 1912; Downing, 1922; Olson, 1958a; Poulson, 1999) have noted that the extremely small-seeded *P. deltoides* establishes only on moist germination beds, such as low pannes, swales, or recently in-filled runnels with surfaces close to the water table (Fig. 2); the margins of beaches or blowout ponds; or depressions on the beach, such as those caused by wheel ruts. No studies have shown evidence of successful establishment of these tree seedlings on the crests or slopes of dunes previously built up by the establishment of grasses. Furthermore, well-established seedlings of *P. deltoides* have been found to be very tolerant of sand deposition because of their rapid vertical growth and adventitious roots (Fuller, 1912; Poulson, 1999). Thus, these

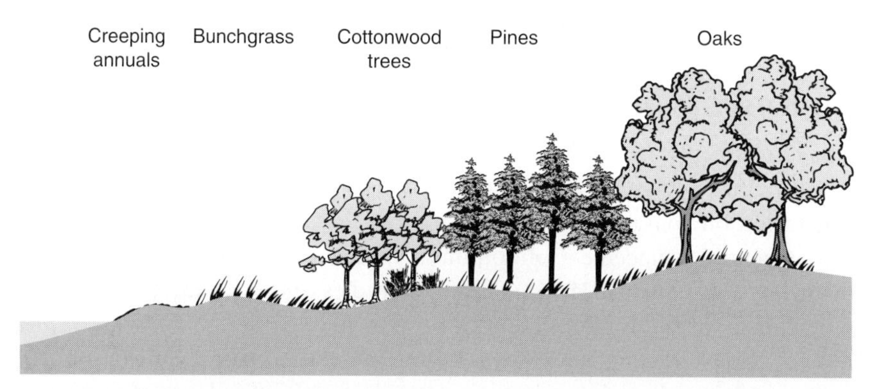

FIG. 1 Traditional dune successional sequence of plant communities attributed to Cowles (1898). (From *General Ecology*, 1st edition, by McNaughton and Wolf © 1973. Reprinted with permission of Brooks/Cole, a division of Thomson Learning: www.thomsonrights.com.)

FIG. 2 Establishment of cottonwoods (*Populus deltoides*) at Long Point on Lake Erie, Ontario, Canada. (**A**) Runnel developed during longshore sandwave migration by the onshore migration, emergence, and welding of a nearshore sand bar. (**B**) *Populus* seedlings 1 year after recession of the runnel. (**C**) Established *Populus* saplings in a former runnel. (Photographs by R. G. D. Davidson-Arnott.)

trees, once established, can survive subsequent establishment of *A. breviligulata* and the growth of a dune. This would explain how mature *P. deltoides* can be found growing on the dunes despite their inability to establish on such dry sites.

This example clearly illustrates how the chronosequence-based spatial pattern of vegetation can lead to an incorrect interpretation of the temporal sequence of plant species establishment. It also shows the importance of the contemporary understanding of aeolian sand transport processes, together with the ecological requirements of plant species for germination, establishment, growth, and survival. In fact, sand erosion and deposition have been widely recognized as a principal cause of plant species distribution in the coastal dune environment (Martin, 1959; Ayyad, 1973; Moreno-Casasola, 1986; Maun and Perumal, 1999; Poulson, 1999; Maun, 2004). Marsh (1987) proposed the relationship between sand mobility and dune species distribution illustrated in Fig. 3.

Therefore, rather than invoking the developmental argument of dune succession based on flawed chronosequence evidence, in this chapter we will use the concept of sediment budgets and the processes of coastal and aeolian sediment transport on beaches and dunes to largely explain the spatial and temporal distribution of coastal dune vegetation. However, before presenting this alternative explanation of dune vegetation dynamics and patterns, we first look at the traditional hypothesis of dune plant succession and review the major problems with this hypothesis. We then provide some basic explanation of sand budgets and transport on beaches and foredunes and show how an understanding of sand movement can be used to explain the observed distribution and abundance of dune vegetation. While the previous chapter (Chapter 7) and this chapter provide an introduction to some of the relevant geomorphic literature on coastal processes, we would encourage anyone interested in dune vegetation dynamics to use this as a mere starting point for any substantial understanding of sediment transport and coastal processes.

TRADITIONAL DUNE SUCCESSION HYPOTHESIS

The standard textbook diagram of dune succession on the Indiana dunes along the southern shores of Lake Michigan (Fig. 1) shows a linear sequence from bare sand to dune grasses (primarily *Ammophila breviligulata*), then cottonwood (*Populus deltoides*), pines (*Pinus strobus* L. and *P. banksiana* Lamb.), and black oak (*Quercus velutina* Lam.). Each of these dominant species is accompanied by its associated subordinate species

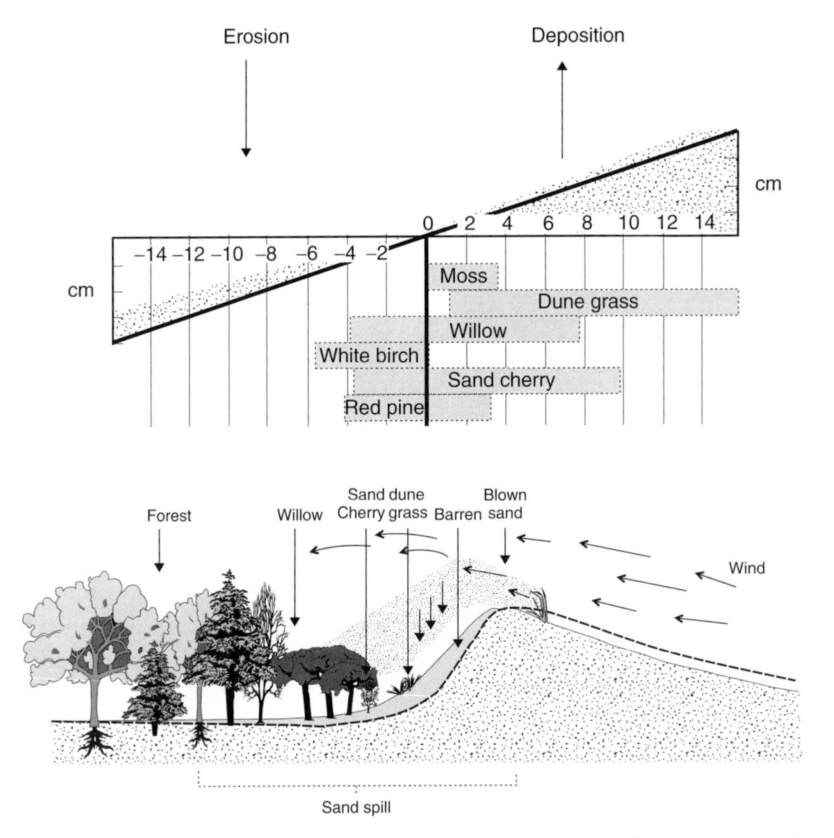

FIG. 3 Ranges of tolerance to sand erosion and deposition for six dune species and their corresponding distribution on the lee slope of an active sand dune in the Great Lakes region of the United States. (From Marsh [1987], redrawn with permission from W.M. Marsh.)

(various graminoids, forbs, and shrubs). At times the sequence that has been proposed includes a stage of annuals (e.g., *Cakile edentula* (Bigel.) Hook.) preceding the dune grass stage (McNaughton and Wolf, 1973) or a climax beech-maple (*Fagus grandifolia* Ehrh.-*Acer saccharum* Marsh.) community following the black oak stage (Cowles, 1901; Clements, 1916). This vegetation sequence was derived from observations by Cowles (1899), as well as other early ecologists (e.g., Fuller, 1912; Downing, 1922), of the spatial distributions of these dominants on dune sequences along transects moving inland from the lake edge. This spatial pattern was interpreted as representing the temporal sequence of vegetation development that occurs on each dune, based on the reasonable assumption that the dune nearest the lakeshore is the youngest and the dune furthest inland is the oldest. This assumption was based on evidence of the long-term drop in water level of Lake Michigan during the Holocene. Thus, the hypothesis was that (1) the oldest dunes currently covered by a black oak–dominated community had undergone temporal changes in vegetation cover that essentially followed the observed spatial sequence and (2) all dunes in this area would trace this same sequence of communities in their development. Similar sequences but with different species [e.g., *Ammophila arenaria* (L.) Link instead of *A. breviligulata* (Hewett, 1970), *Cakile maritima* Scop. instead of *C. edentula* (Barbour, 1972)] have been postulated for marine and coastal dunes in various locations around the world.

The explanation that was given for the succession of plant communities on sand dunes (as well as on other substrates) was Clements' (1916) theory of habitat alteration and facilitation, in which each dominant species alters the habitat in a way that facilitates the establishment of the next-dominant species in the sequence and inhibits or prevents its own regeneration. The "evidence" for this theory provided by various studies (e.g., Morrison, 1973) typically involved showing significant differences in soil properties (e.g., pH, organic matter content, field capacity, cation exchange capacity, and nitrogen) along the spatial dune sequence. The soil properties found associated with each dominant were interpreted as an indication of the soil conditions necessary for the establishment of each dominant species. [Note that Colinvaux (1993) used Crocker and Major's (1955) similar chronosequence study of soil properties as solid evidence for the facilitation theory of plant succession on glacial till, although this conclusion was subsequently shown by Chapin *et al.* (1994) and Fastie (1995) to be flawed.]

According to Clements' theory, the endpoint of succession was the climax community, which was stable since the dominant of this community was believed to alter the habitat such that it allowed no further invasions. Thus, in the absence of any major disturbance that might destroy the climax dominant, there would be no further change in species composition. According to Clements (1928), the species composition of the climax community was determined by the regional climate alone since any other factors (such as the characteristics of the initial substrate) would be altered sufficiently by the seral dominants to become irrelevant. For the Indiana dunes along Lake Michigan, the climatic climax was presumed to be mesic beech-maple forest. There was some subsequent disagreement on whether the final stage in the sequence was a black oak or beech-maple community, perhaps because of the lack of widespread beech-maple communities growing on sandy substrate, regardless of the distance from the lake or the presumed age of the dunes. In fact, Olson (1958a) argued that substrate acidification resulting from carbonate leaching and the tolerance of black oaks and intolerance of beech and sugar maple to these acidic soil conditions can explain why the black oak–covered dunes would not be expected to be replaced by a beech-maple forest, even on the oldest dunes.

Regardless of what the final stage of this traditional dune succession was presumed to be, the sequence of stages leading up to and including the black oak community (Fig. 1) was generally accepted and widely presented in North American ecology textbooks for many years (e.g., McNaughton and Wolf, 1973). As Colinvaux (1993) noted: "Cowles' hypothesis of dune succession dominated ecology teaching in America for 60 years." Furthermore, Clements' explanation of such succession also dominated; according to Burrows (1990), while "[m]uch of his [Clements'] terminology was never generally accepted and many of his ideas have been abandoned by most vegetation ecologists in recent times . . . the basic concept of sequential development of vegetation on bare surfaces (first a colonizing phase, followed by immature ('seral') phases and culminating in a mature and stable ('climax') phase) is firmly embedded in the literature of vegetation ecology and in the minds of many plant ecologists." This was still reflected in textbooks in the last decade of the 20th century (e.g., Scott, 1991; Bradshaw and Weaver, 1995; de Blij and Muller, 1996; Christopherson, 1998), with Clements' theory even upgraded by Castillon (1992) to the "law of succession."

PROBLEMS WITH THE DUNE SUCCESSION HYPOTHESIS

Chronosequence Assumptions

The first problem with the traditional dune succession sequence of plant communities is that it is based on the results of chronosequence studies. This approach has many critical assumptions that must be valid in order to infer that the observed spatial sequence of plant communities represents the temporal sequence that each of the dunes has followed and will follow. The basic assumption is that the dune sites along the spatial sequence differ only in age and that, regardless of the spatial location, they have all traced exactly the same sequence of species and environment. However, in order to conclude that all dunes have followed and will follow the same developmental sequence, one must assume that (1) the conditions that influence successful establishment of the dominant species of each seral community have been the same and are determined only by the previous dominant species and (2) the establishment of species is not significantly influenced by temporal variation in environmental factors, such as temperature or precipitation. In other words, the climate at the time of dune initiation or any other seral stage plays little or no role in selecting which species can successfully establish; all that matters is the seral stage of dune development.

There is little, if any, evidence for the validity of either of these two assumptions. First, there is now abundant and conclusive evidence that significant climatic variation occurs at all time scales and that environmental conditions, such as temperature and precipitation, are therefore not constant over the time span represented by the ages of the Indiana dunes or any dune system (i.e., centuries to millennia). Furthermore, such climatic variation is not primarily controlled by the vegetation. Second, plant history studies generally show that the most vulnerable stages in a plant's life are seed germination and seedling establishment (e.g., Laing, 1958; Sharitz and McCormick, 1973; Houle, 1995; Clark *et al.*, 1999; Rey and Alcántera, 2000; also see Grubb, 1977; Harper, 1977; Grime, 1979), and most mortality in dune plant populations occurs within the first year (Lichter, 2000) or even the first few days or weeks (Maun, 1994). Many, if not most, plant species have a relatively narrow window of conditions that allow successful establishment and survival of seedlings. As a result, many plant species show episodic establishment in response to environmental variability; for example, establishment of the dune grass, *Ammophila arenaria* (L.) Link (Huiskes, 1977), and the pines, *Pinus strobus* and *P. resinosa* Ait. (Lichter, 2000), is episodic because of high desiccation mortality of seedlings. Thus, plant species may encounter variable establishment success on dunes at different times and locations, depending on the prevailing environmental conditions and regardless of the age of the dune. The environment encountered by so-called later successional species as their propagules arrive is not determined only by the dominant species that have established before them.

An additional assumption of the chronosequence-based succession argument is that propagules of the species appropriate for each stage are always available (i.e., that establishment of species is limited by environmental conditions and not by species availability). However, numerous studies have concluded that distance to seed source and propagule availability are significant limiting factors in species recruitment on a wide range of substrates, including sand dunes (Olson, 1958a; Saunders and Davidson-Arnott, 1991; Lichter, 2000), glacial till (Chapin *et al.*, 1994; Fastie, 1995), alluvium (Walker and Chapin, 1986; Walker *et al.*, 1986), and volcanic deposits (Wood and del Moral, 1987; del Moral and Bliss, 1993). The importance of species availability was a key component of Gleason's (1939) individualistic view of community development and composition. This individualistic view was supported by Raup (1975), who observed that because of the stochastic nature of such species availability, no two communities in similar environments have identical species composition.

Data Collection Techniques

A second problem is that the traditional dune succession sequence was originally based on casual observations rather than systematic data collection. In subsequent years, it has often been obtained by averaging relative abundance data from several transects across a dune system. The pattern obtained by such averaging does not necessarily represent the sequence found for any individual transect. However, if the sequence of dune communities illustrated by Cowles (1899) actually represented a temporal sequence, one should expect most, if not all, transects to show this consistent pattern. Johnson and Muller (1992) and Stallins and Parker (2003) did not find consistent sequences of similar vegetation across transects on a marine coastline. Stallins and Parker's explanation for this inconsistency was that the vegetation was responding not only to environmental gradients along transects but also to disturbances that did not occur simultaneously along the coast. Similarly, when Olson (1958a) arranged plant communities on the basis of the radiocarbon ages of the dunes on which they were found, he found much variation in the dominant species for any particular age or stage of dune development, depending on the particular location and disturbance history of the site (Fig. 4). Olson noted that the usual ordering of dune communities into a temporal sequence was a fiction, derived from patching together what appears to be the pattern for the region; he dubbed such a sequence a "synthetic succession" since communities at one end are not actually likely to be transformed into those at the other end.

Species Replacement

A third problem concerns the position in the dune succession sequence of annuals and cottonwoods (*Populus deltoides*) on the Indiana dunes (Fig. 1). McNaughton and Wolf (1973) described how the "grasses replace the annuals and then in about 20 years they are invaded by cottonwood tree seedlings." However, as we will explain here, grasses do not replace

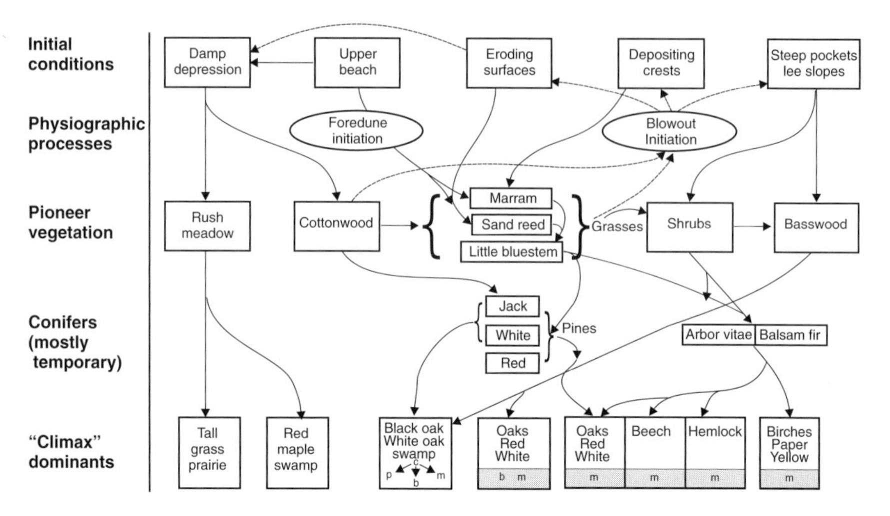

FIG. 4 Plant communities found on dunes of various ages and explained by topographic location (e.g., protected lee slopes and wet depressions) and disturbance history. (From Olson [1958a], redrawn with permission from The University of Chicago Press.)

annuals and cottonwoods do not invade areas previously colonized by grasses. Annuals (e.g., *Cakile edentula*) are primarily found on bare sandy areas at the driftline and midbeach, which are kept bare in the fall and early spring months by high waves (Payne and Maun, 1981). Their habitat is largely segregated from the upper beach and dune habitat of the perennial grasses (e.g., *Ammophila breviligulata*) by seasonal wave disturbance. Any *A. breviligulata* that may advance by rhizomatous growth during the summer months into the driftline beach habitat of *C. edentula* are subsequently destroyed by high waves. Furthermore, while some individual annual plants may be found within the grass populations on the upper beach and dunes, this is not their normal habitat; seeds of *C. edentula* are primarily dispersed by the action of high waves that cast the fruits onto the driftline during autumn and winter (Payne and Maun, 1981). Similarly, Barbour (1972) noted that *Cakile maritima* is not found in the coastal dunes of California because the seeds are not carried inland beyond the reach of the water.

As previously discussed in this chapter, the dunes on which cottonwoods are found would have been initiated by the establishment of *Ammophila breviligulata* after the cottonwood trees had already successfully established on a moist surface close to the water table. Similarly, studies of dunes in northern Canada concluded that small seeded woody plants, such as *Salix* and *Betula,* could not germinate on dunes since their seedlings were known only from moist to wet dune slacks (e.g., Raup and Argus, 1982), although *Salix* also appears to establish from branches/cuttings that wash up onto the driftline (A. M. Maun, personal communication). If a dune invades an area where these woody species have already established, these shrubs and trees may be found occupying the lee slope of the dune because they can withstand and thrive with sand burial (Fuller, 1912; Olson, 1958b; Raup and Argus, 1982; Poulson, 1999). This explanation of how nonxerophytic species establish and are maintained on dunes would help to explain the observation by McLeod and Murphy (1977) that, although the Lake Michigan sand dune vegetation has been described as xerophytic, it includes many species and genera that are not typically considered xerophytic and are primarily distributed along waterways (e.g., *Cornus stolonifera* Michx., *Populus deltoides, Salix glaucophylloides* Fern.).

Soil Development and Colonization Constraints

A fourth problem with the traditional dune succession hypothesis is the facilitation argument that mid- and late-successional species can establish only after habitat alteration (primarily soil development and nitrogen fixation) by early successional species (Clements, 1916). However, several studies have not found soil development or soil nutrient status to be a significant factor in explaining species distribution on dunes (Chadwick and Dalke, 1965; Lichter, 2000). Furthermore, through seed addition and seedling transplant studies, Lichter (2000) showed that tree species traditionally considered to be mid- to late-successional on dunes (e.g., *Juniperus communis* L., *Pinus strobus*, *Quercus rubra* L.) were capable of germinating and establishing seedlings on young dunes (30 years old) still dominated by *Ammophila breviligulata* where soil development would have been minimal. Olson (1958a) also found that the tree species *Tilia americana* L. can establish on steep lee slopes and protected pockets of young dunes along with grasses and shrubs. Furthermore, where *T. americana* has established, other hardwood trees can also establish without the dune going through any pine or oak stages. Thus, abundant evidence suggests that most, if not all, of the dominant species on dune systems can establish without the changes in pH or the addition of organic matter or nitrogen brought about by plants considered to be pioneers or early seral species. In fact, Baldwin and Maun (1983) found no significant differences in organic matter content within the top 5 cm of soil taken from the high beach, first dune ridge, slack, and second dune ridge on the Lake Huron sand dunes in

Ontario, Canada. Similarly, Poulson (1999) found that depth of A-horizon, amount of humus, cation exchange capacity, and levels of potassium and phosphorus on the Miller dunes were not statistically different than in a beech-sugar maple old-growth forest on sandy soil at Warren Woods in Michigan. Thus, Lichter (2000) concluded that, on his study dunes, successful establishment of "later successional" species depends not on facilitation by previous dominant species but on interacting constraints of chance seed dispersal, stochastic weather conditions, and fluctuating populations of rodent seed predators.

Physical and Biotic Disturbances

Probably the most important problem or limitation with the traditional view of dune succession is the lack of incorporating disturbance as an influential factor in the establishment and survival of plants on all of the dunes, regardless of their position relative to the shoreline and regardless of their age. For example, the common explanation that was given for the role of the dune grasses, such as *Ammophila breviligulata* or *A. arenaria*, was that of dune stabilization. As explained in one textbook (Strahler and Strahler, 1978), "the shoots of dune grass act to form a baffle that suppresses movement of sand, and thus the dune becomes more stable. With increasing stabilization, plants that are adapted to the dry, extreme environment but cannot withstand much burial begin to colonize the dune." The assumption here appears to be that sand accumulation on the dune is somehow reduced by the establishment of the sand-binding dune grass. However, while the dune grasses do certainly act in reducing the windflow (e.g., Arens *et al.*, 2001), thus trapping sand and resulting in vertical growth of the dune, they cannot stop the *continued* influx of sand from the beach. Neither can the establishment of trees stop such influx, as indicated by the sand burial of some established forests on dunes, resulting in reduced growth or even death of the trees (Wolfe, 1932; Kumler, 1969; Marin and Filion, 1992).

Established dunes may experience repeated episodes of sand deposition. For example, studies of the large (>30 m) perched dunes on the eastern shores of Lake Superior and Lake Michigan have found buried soil horizons beneath pine forests, indicating repeated episodes of dune-building over the past several millennia linked to changing lake levels and changing sand supply (Anderton and Loope, 1995; Loope and Arbogast, 2000). As a result of these repeated burials, the surfaces of these dunes are actually much younger than expected. In fact, buried organic horizons in dunes indicating episodic aeolian sand movement and dune building are not limited to dunes along the Great Lakes but have been found elsewhere (e.g., Farrow, 1919; Filion, 1984). The implication of these buried organic horizons in dunes is that stability of the dune surface is determined not just by vegetation cover but also by sand supply (this is discussed in more detail in the later section on sand budgets). Furthermore, sand movement is not the only disturbance affecting dune vegetation; other disturbances include windstorms, fire, grazing, insect outbreaks, and trampling. Thus, the plant community patterns observed on any dune system at any given time reflect not only environmental gradients of abiotic and biotic factors but also disturbance history (Downing, 1922; Olson, 1958a; Hobbs and Grace, 1981; Reice, 1994; Saunders and Davidson-Arnott, 1991; Stallins and Parker, 2003). Chapter 7 provides further details on the coastal processes and disturbances prevalent in coastal dune environments.

PROCESS-RESPONSE ALTERNATIVE TO TRADITIONAL SUCCESSION HYPOTHESIS

While most ecology textbook authors have indicated an awareness of some of the above-mentioned issues and arguments against the traditional dune successional sequence, probably a major reason that this interpretation of dune vegetation persists has been the lack of an alternative conceptual framework to replace it. We propose a framework based on a process-response approach that links plant population processes of birth and death to the principal physical processes operating.

Because of the low field capacity and limited capillarity of sand (especially in medium to coarse sand) and the resultant rapid drying of sandy surfaces, as well as the wind exposure of most coastal environments, the principal disturbance of coastal dunes is aeolian sand transport, which both creates and modifies the dunes. Although coastal dune ecosystems are also influenced by natural disturbances and processes other than sand transport (e.g., overwash, salt spray, herbivory), we would argue that the movement of sand is the principal filter for species selection on the backbeach, foredunes, and second dune ridge. Therefore, this section will briefly explain some background on budgets and transport of sand and then show how these can be used to explain the distribution of dune vegetation. Processes and disturbances, such as competition, grazing, and fire, that become more important on the stabilized dunes with more complete vegetation cover will not be discussed in this chapter.

Sand Transport and Sand Budgets

"Dunes are rarely stable landforms . . . and are liable to migration (by aeolian transport of sand-grains from the windward to the leeward slopes) and severe erosion" (Small and Witherick, 1986). Thus, the essential feature of most sand dune systems is the movement of sand by wind and a key element of such sand movement is the sand budget (i.e., the availability of sand for transport). As noted by Psuty (1988), "All variations of beach/dune forms are tied to sediment availability, the relative input to the dunes and to the beach, and whether these inputs are decreasing or increasing over some duration." Thus, Fig. 5 presents Psuty's ideas on morphologic development of the beach-dune system under varying combinations of sand budgets for the two components.

When sediment (sand) inputs exceed outputs, the system is described as having a positive budget; a negative budget indicates the reverse (i.e., outputs exceed inputs). Each component of the coastal beach/dune system (beach, foredunes, and so forth) has its sediment budget and can react in relatively short time frames to imbalances between inputs and outputs, producing changes in the coastal topography. For example, coastal dunes can grow in elevation by as much as 58 to 90 cm in a year (Ranwell, 1958; Tyndall *et al.*, 1986; Costa *et al.*, 1991) or can deflate by 60 to 75 cm in a blowout over 3 to 14 days (Gares, 1992). Note that within a beach-dune system that may have an overall positive sediment budget, there may be localized areas of deflation/erosion.

For coastal dunes to be in a stable state (neither growing nor deflating), three conditions are required: (1) that the dune be far enough from the shoreline to be unaffected by erosive wave action; (2) that the dune have complete vegetation cover such that the surface sand is not exposed to erosive wind action; and (3) that there be no external source of sand for deposition.

The first condition is obviously affected by beach width; factors and processes influencing spatial and temporal variation in beach width are discussed later. However, in general, embryo dunes and first dune ridges on beaches are subject to erosion during storm surges or longer-term cycles of high water (Saunders and Davidson-Arnott, 1991) and are therefore rarely stable while the dunes landward of and protected by the foredunes are more likely to be stable. If close to the water, even high-cliffed dunes along the coastline are vulnerable to undercutting by wave action, subsequently exposing the cliff sand to transport by wind (Anderton and Loope, 1995; Arbogast *et al.*, 2002).

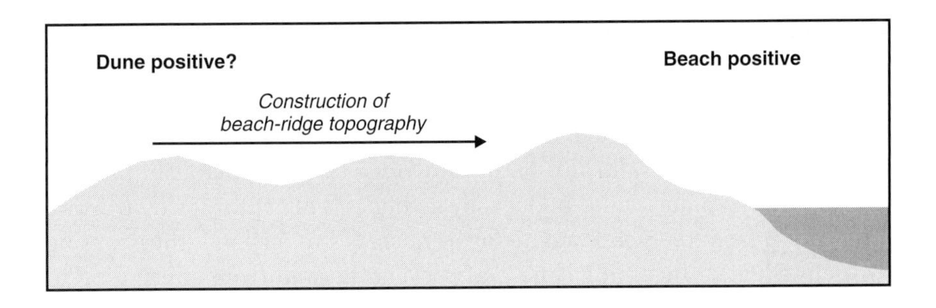

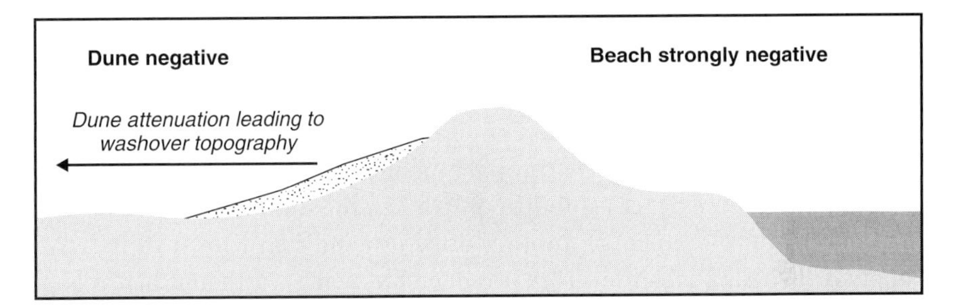

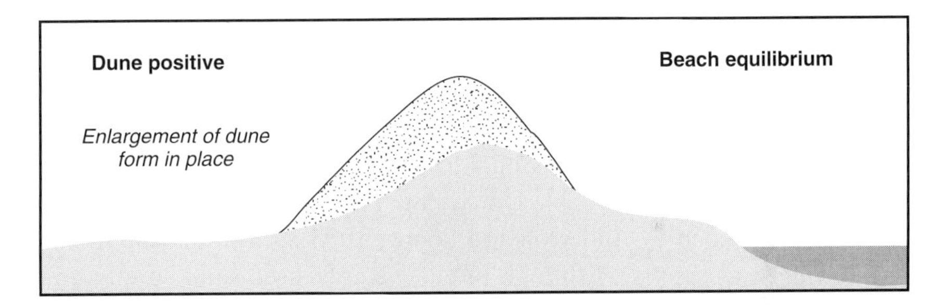

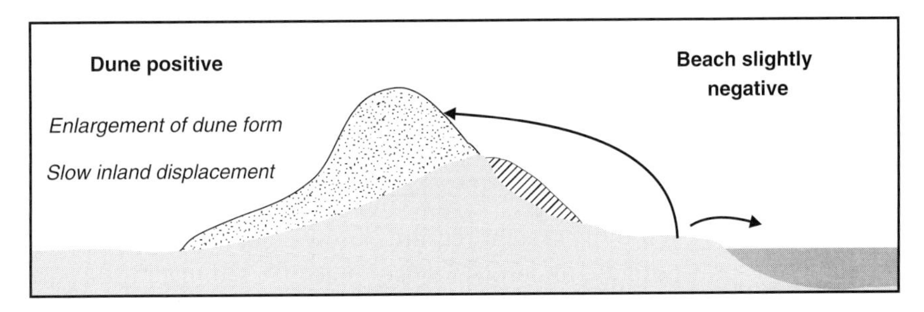

FIG. 5 Development of various beach and dune morphologies under various sediment budget combinations. (From Psuty [1988], redrawn with permission from Coastal Education and Research Foundation, Inc.)

Concerning the second condition, parts of dunes that lack complete plant cover (either through initial patchiness of plant establishment or subsequent disturbance such as grazing, fire, or trampling) have sand exposed to aeolian transport, particularly when the sand is dry and wind speeds are sufficient to initiate saltation (e.g., blowouts). Plants obviously play an important role in dune sand budgets by acting as sediment traps (extracting momentum from the wind and acting as obstacles to airflow), thus causing the deposition of incoming sand and preventing sand from continuing to move downwind (Lancaster and Baas, 1998; Arens *et al.*, 2001). Plant cover also inhibits loss of sand from surfaces on which they occur by protecting the sand grains from the action of wind. However, plant cover of a dune *by itself* cannot stabilize a dune surface since it cannot control the input of sand from upwind sources. Thus, establishment of *Ammophila* may facilitate dune growth by trapping incoming sand, but the dunes *stop* growing and their surfaces stabilize only when the sand supply is removed (e.g., by the development of another dune upwind that robs it of its sand supply or by rising water levels that reduce the beach width and hence area of sand exposed to wind). Growth of a stabilized dune may be reinitiated if sand again becomes available (e.g., by destruction of the upwind dune).

Finally, if the availability of sand from external source areas increases (e.g., when falling lake levels expose a greater area of dry beach sand), then even previously stable, completely vegetated dunes downwind from this increased sand supply may become buried in sand. For example, buried soils in the perched Grand Sable Dunes near the east end of Lake Superior have been explained by Anderton and Loope (1995) as a result of episodic changes in sand supply. Although in general increased sand supply and foredune progradation occur during falling lake levels (Olson, 1958c), in the case of the perched Grand Sable Dunes, Anderton and Loope argued that sand supply increased during periods of *rising* lake levels. At such times, the bluffs were destabilized through basal erosion, producing actively eroding slopes that provided a fresh source of sediments for aeolian transport to these perched dunes.

Determination of sand budgets (i.e., changes in mass flux of sand) over the surface of a beach-dune system requires knowledge of the process by which sand is transported. Numerous models of aeolian sediment transport have been proposed (Bagnold, 1936; Kawamura, 1951; Zingg, 1953; Owen, 1964; Kadib, 1965; Hsu, 1971; Lettau and Lettau, 1978; Nakashima, 1979; Horikawa *et al.*, 1984, Anderson and Haff, 1991; McEwan and Willetts, 1993; Spies and McEwan, 2000). In these mechanistic models of saltation, sediment transport is a function of wind speed (or shear velocity), air density, mineral density, and sediment size (Bauer and Davidson-Arnott, 2002).

Most of these models consider transport under the near-ideal conditions of a uniform wind field, dry uniform sand particles, and a horizontal, flat, unobstructed surface, even though field conditions are rarely, if ever, ideal (Namikas and Sherman, 1998). The presence of moisture in the sand or a sloping surface (upward or downward) would have a significant effect on the transport rate. Namikas and Sherman listed several moisture correction models (Belly, 1964; Kawata and Tsuchiya, 1976; Hotta *et al.*, 1984; Gregory and Darwish, 1990) and slope correction models (Bagnold, 1956; Allen, 1982; Dyer, 1986; Hardisty and Whitehouse, 1988; Zeman and Jensen, 1988) that account for these factors. They then presented a computer program (AEOLUS II) that allows iterative selection of combinations of appropriate models to simulate aeolian transport on a complex surface with varying grain size, slope, and moisture content. Since the sand budgets of the various components of the dune-beach system are necessarily connected as sand is transported across the system, AEOLUS II calculates different transport rates and resulting mass flux within each spatially explicit component or "bin."

Furthermore, wind flow is strongly affected by topography (Namikas and Sherman, 1998; van Boxel *et al.*, 1999; van Dijk *et al.*, 1999), creating a positive feedback situation. For example, studies of wind flow in blowouts show that once a blowout is initiated, the structure of the blowout influences subsequent wind flow patterns (Landsberg and Riley, 1943; Olson, 1958b; Hesp, 1996; Hesp and Hyde, 1996; Pringle *et al.*, 1999; Hesp, 2002). Any wind flow approaching a trough blowout from a seaward or alongshore direction in an arc of at least 180° to the entrance orientation is directed into the blowout. Thus, the direction that sand is transported within the blowout is determined by the topography created by the blowout. Besides changes in wind direction, the morphology of the blowout creates jets in the deflation basin and depositional lobe, flow deceleration and expansion up the depositional lobe, further jets over the crest of the depositional lobe, and finally flow separation and corkscrew vortices over the crests of the erosional walls (Hesp and Hyde, 1996). AEOLUS II attempts to account for some of the feedback effect of topography on wind flow by iteratively adjusting the topography as a result of the mass flux at each time step (Namikas and Sherman, 1998).

The output of this model gives simulated topographic change on a three-dimensional landscape (Fig. 6). While this model is the only three-dimensional spatially explicit distributed model of sand transport and

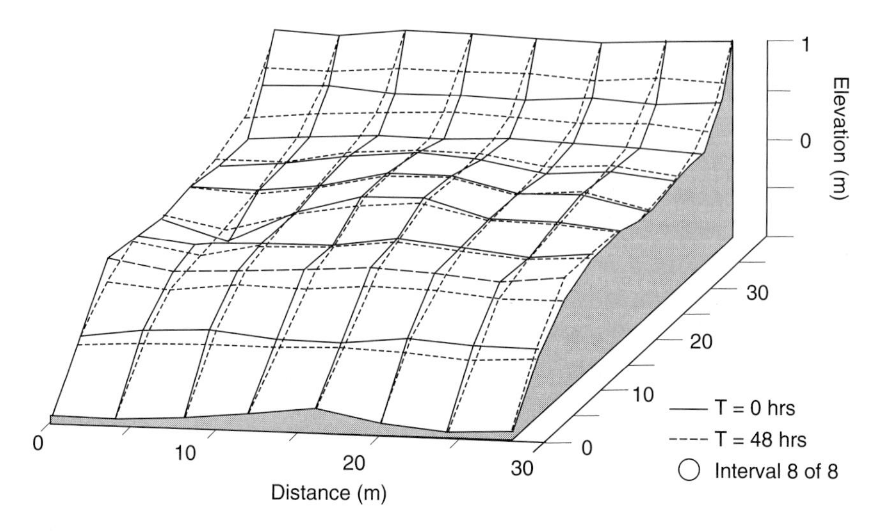

FIG. 6 Example of simulated three-dimensional topographic change on a beach-dune surface resulting from output of AEOLUS II. (From Namikas and Sherman [1998], redrawn with permission from Elsevier.)

topographic change that we could find in the literature, it has the major limitation of applying only to unvegetated surfaces and does not allow for changes in windflow and deposition due to the presence of plants.

An alternative approach that does include vegetation is the model by van Dijk *et al.* (1999) that uses published deterministic and empirical relationships describing the influence of meteorological conditions, topography, sediment characteristics, and vegetation to determine sediment transport. Their model produced two-dimensional simulated dunes whose morphology is consistent with those of natural dunes, both unvegetated and vegetated. As indicated in Fig. 7, model simulations showed that the height of vegetation influenced the morphology of the developing dune and empirical studies appear to support these model results (Olson, 1958b; Oertel and Larsen, 1976). However, this influence of vegetation height on dune morphology may be a temporary depositional phenomenon and may not be detectable as the dune evolves over several decades.

Arens *et al.* (2001) used the model by van Dijk *et al.* (1999) to investigate the influence of experimentally planted reed stem density on sediment transport and on the resulting dune forms and found that vegetation density also influenced the shape of the dune, both in the empirical and simulated results (Fig. 8). The highest stem density produced a steeper dune because most of the sand was trapped by the first row of planted stems.

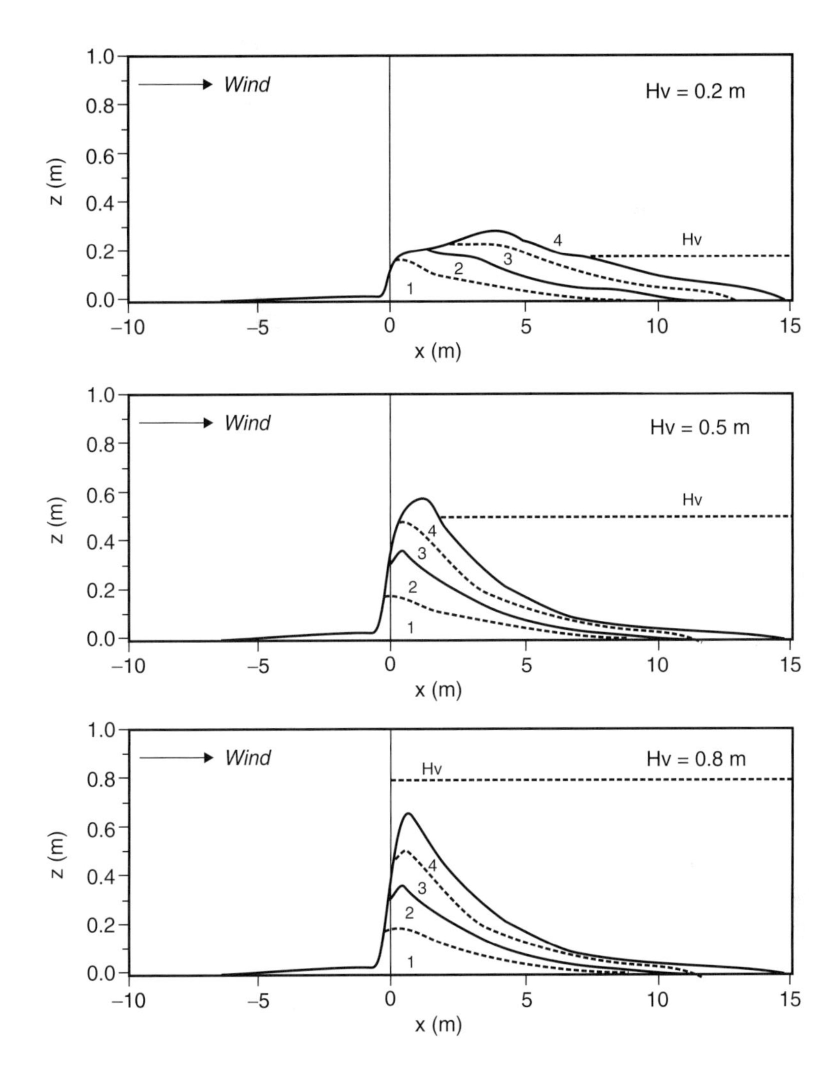

FIG. 7 Simulated two-dimensional dune development on an initially flat partly vegetated surface during 4 days of a constant wind (0.4 m s^{-1}) for three different vegetation heights (0.2 m, 0.5 m, and 0.8 m). Vegetation is present downwind (i.e., to the right) of $x = 0$. (From van Dijk *et al.* [1999], redrawn with permission from Wiley.)

The lowest stem density produced a smoother wider dune since sand was deposited more evenly throughout the planted area. Despite these differences in sand transport and deposition, the total amount of sand that accumulated was not affected by stem density. This indicates that the mass flux was limited by the amount of sediment transported to the planted area from the upwind beach.

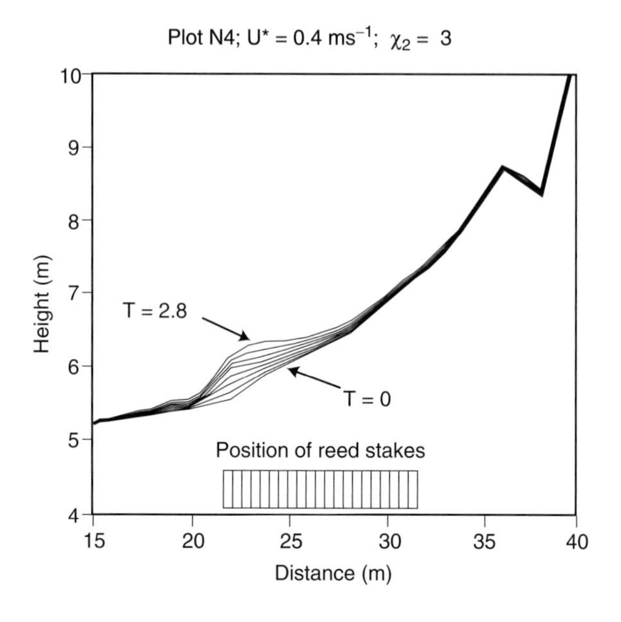

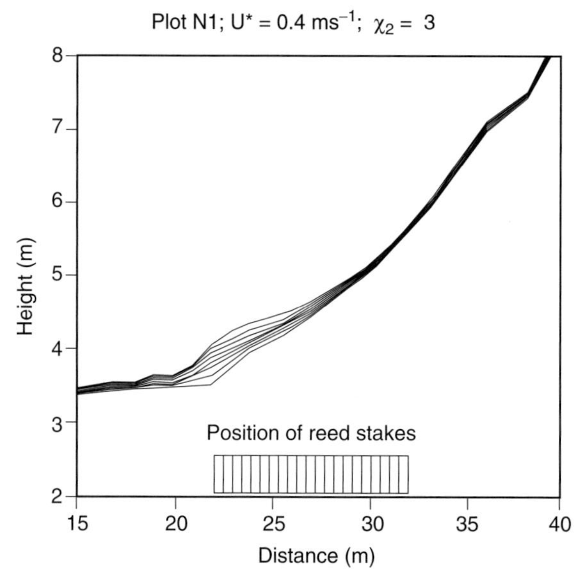

FIG. 8 Simulated two-dimensional dune development resulting from a high-density (plot N4) and a low-density (plot N1) of reed stems planted in rows on the windward bare slope of a foredune. (From Arens *et al.* [2001], redrawn with permission from Wiley.)

Thus, what is required here is a model to quantify the rate of sediment supply from the beach to the first zone of net deposition (typically where vegetation acts as an obstacle to windflow and extracts momentum from the wind and saltating sand). Bauer and Davidson-Arnott (2002) developed a model of sand transport processes operating on the beach to predict sand supply to the dunes. A problem they addressed was that none of the equilibrium sediment transport models account for the fact that sediment transport may be less than the maximum for any given windspeed and set of sand characteristics because of the fetch effect. As the wind begins to entrain sand and cause saltation of individual grains, there is a cascading effect of the impact of saltating grains dislodging more grains from the surface. As a result, the number of saltating particles increases exponentially and asymptotically downwind to a limiting maximum (saturation) condition (Chepil, 1957; Anderson *et al.*, 1991; Gillette *et al.*, 1996). Whether the sand transport rate is at equilibrium upon reaching the vegetation depends on the beach geometry, angle of the wind approach relative to shorenormal, and any "end effects" (since the beach is not infinitely long and would have limiting lateral barriers to alongshore transport with an oblique wind). These barriers may include both natural and artificial features, such as rivers, vegetated zones, and seawalls. Bauer and Davidson-Arnott's model could be useful in explaining some of the alongshore differences in dune formation.

While Bauer and Davidson-Arnott's (2002) model provides a more realistic approach to determining sand supply from the beach to the dunes, there is still the issue of sand supply to the beach. In van Dijk *et al.*'s (1999) model of sand transport and dune formation, the beach lost sand but had no mechanism for replenishment. As a result, the researchers had to artificially replenish the beach with sand in order to allow the model to continue. Certainly, if sand supply to the beach is a limiting factor as on eroding shorelines, it would be necessary to incorporate a model of sediment transport from longshore drift cells and wave action that would determine sand supply to the beach. Thus, Bauer and Davidson-Arnott suggested coupling their model to a littoral sediment budget model (e.g., Bowen and Inman, 1966) for a more comprehensive model of beach–dune interaction.

Beaches may vary in their sand budgets over varying temporal scales because of short-term changes in water levels (e.g., storm surges), cyclic changes in water level (e.g., seasonal as well as quasi-periodic lake level fluctuations; see Figure 3 in Davidson-Arnott and van Heyningen, 2003), and long-term trends in water levels (e.g., sea level change resulting from global warming or cooling). As water levels rise, beach widths decrease and vice versa. Beach width acts as a major control on sediment supply to the foredunes as well as on the extent of dune erosion during storms (Davidson-Arnott, 1988). As previously discussed, these changes in beach width result in increased availability of sand for aeolian transport and dune formation/growth or scarping/erosion of established dunes.

Besides changes in water levels, beach width and onshore sediment movement are also affected by the alongshore migration of sandwaves (where they occur) on a temporal scale of years to decades (Stewart and Davidson-Arnott, 1988). Stewart and Davidson-Arnott describe sandwaves as local beach protuberances with lengths of 500 to 2500 m, maximum widths of 50 to 90 m, and lifespans of 10 years or more, which migrate alongshore at 150 to 300 m y^{-1}. These sandwaves migrate through the onshore migration and attachment of a nearshore bar, after which the bar becomes emergent, thus extending the width of the beach. As a result, the beaches are very narrow between the sandwaves (Davidson-Arnott, 1988). Because they influence onshore sediment transport and beach width, these migrating sandwaves affect erosion and accretion of dunes (Davidson-Arnott and Stewart, 1987), particularly during storms that occur during high water levels (Davidson-Arnott and Stewart, 1987; Saunders and Davidson-Arnott, 1991). Ruessink and Jeuken (2002) also found evidence of migrating sand units that traveled along sections of the coast of Holland. As with the sandwaves along the Lake Erie coast in Canada studied by Davidson-Arnott and coworkers, these sandwaves in Holland were found to play an important role in foredune dynamics through temporal changes in beach width and resulting sand budgets.

Coupling Sand Budgets and Vegetation Distribution

Although past studies examining the distribution of coastal dune vegetation have cited influential factors such as desiccation (De Jong and Klinkhamer, 1988), nutrient shortage, and salt spray or soil salinity (Barbour *et al.*, 1976; Barbour and De Jong, 1977; Barbour, 1978; Barbour *et al.*, 1987; Olff *et al.*, 1993), as well as herbivory and seed predation (Boyd, 1988a; 1988b; 1991; Klinkhamer *et al.*, 1988), sand burial has generally been considered the most important factor (Ranwell, 1958; Hewett, 1970; van der Valk, 1974; Moreno-Casasola, 1986; Harris and Davy, 1987; Sykes and Wilson, 1990; Zhang and Maun, 1990a; Martínez and Moreno-Casasola, 1996; Maun and Perumal, 1999; Kent *et al.*, 2001; Maun, 2004). Since aeolian sand transport is the principal factor creating and modifying the dune surface on which plants are establishing and growing, it acts as a primary filter in determining which plants can successfully establish and survive.

Species differ in their tolerance of sand burial (positive sand budget) or erosion (negative sand budget) (Fig. 2). Numerous studies have investigated the responses of various species to varying rates and amounts of sand burial (Maun and Riach, 1981; Maun, 1985; Maun and Lapierre, 1984; 1986; Harris and Davy, 1987; Sykes and Wilson, 1990; Zhang and Maun, 1990a; 1990b; Yuan *et al.*, 1993; Maun, 1996; Chen and Maun, 1999; Martínez and Maun, 1999; Maun and Perumal, 1999). This variation results in selection for those species that tolerate the particular sand budget of a surface (Moreno-Casasola, 1986; Maun and Perumal, 1999; Dech and Maun, 2005), thus influencing or determining species composition of dune plant communities (Owen *et al.*, 2004; Kent *et al.*, 2005). For example, Tyndall *et al.* (1986) proposed that the restriction of species distributions on the beach and foredunes of the North Carolina Outer Banks could be explained by the differential ability of species to establish seedlings from buried seeds. Of six species occupying the foredunes, *Cakile edentula*, which was able to emerge from the greatest depths of experimental sand burial, was the only species occurring on the open beach. More recently, Dech and Maun (2005) showed clearly that the distribution and abundance of both woody and herbaceous species on the Lake Huron dunes in Ontario followed a gradient of sand burial, independent of any other environmental factors. Furthermore, species' tolerance of varying rates of sand deposition or erosion may change through their life cycle. Thus, while the extremely small seedlings of cottonwood (*Populus deltoides*) cannot survive sand burial, once well-established on a stable deflation panne with roots below the water table, the saplings can tolerate a considerable positive sand budget (Downing, 1922).

The first important implication of recognizing the key role that the sand budget plays in vegetation dynamics and vegetation patterns on dunes is the necessity to understand the spatial pattern of sand budgets resulting from aeolian sediment transport processes. As the sediment transport models mentioned in the previous section indicate, the beach-dune environment (and particularly the beach and foredune components) has a very dynamic surface over relatively short time periods. Thus, a change in the wind speed at the surface (e.g., a decrease resulting from planting sand reeds at any density) on the bare windward slope of a foredune resulted in a change from a negative sand budget to a positive one and vertical growth of the surface (Arens *et al.*, 2001). Changes in any of the factors that influence and determine aeolian sediment transport, such as wind speed, sand moisture content, or slope, have impacts on the transport rate and resulting mass flux. Furthermore, such transport produces topographic changes that, in turn, affect windflow and result in a feedback effect on subsequent transport and topographic changes. Once we understand the dynamic nature of the surface of the beach-dune environment, we can begin to relate this to the tolerances of various species for the various sand budget environments they would encounter. These conditions determine the subset of species that can establish in different parts of the beach-dune system (e.g., on the crests, lee slopes, and bases of foredunes).

The second implication is the recognition of the temporal instability of the sand budgets at varying time scales. In other words, the sand budget of any surface in the coastal environment can, and does, switch between positive, negative, or zero over time. At Great South Beach, New York, Clark (1986) found evidence of sand deposition sufficient to destroy all vegetation occurring at irregular intervals but at least once every 100 years over the past 340 years. This means that the particular species that have an advantage in terms of improved establishment and decreased mortality in response to the sand budget conditions will also change over time. For example, an area that was deflated to the water table following a period of negative sand budget may allow establishment of species that require wet surfaces for seed germination and seedling survival (e.g., *Populus deltoides*). Other species with similar requirements for establishment may also invade. However, if the sand budget changed subsequently to a positive one as a sand supply became available and the established plants began to act as sand traps, many of these species may become buried and die while *P. deltoides*, which can tolerate and continue to grow vertically when subjected to sand deposition, will continue to flourish. Other species that can establish and survive with a positive sand budget may then cohabit the area with *P. deltoides*. Thus, the composition of the community would change as the changing conditions affect species differentially and individualistically (Gleason, 1939).

Furthermore, because different areas of coastal environments will have different histories of changing sand budgets with different subsets of species available to them as the sand budgets changed, they would not be expected to follow any deterministic temporal pattern of community composition change. In other words, dunes that may be a similar distance from the shoreline and may be considered to have originated from the same dune-building event at some time in the past would not be expected to have the identical species composition of plant community, as Olson's (1958a) study showed (see Fig. 3). A good example of this is the beach topography resulting from the passage of an alongshore sandwave as described earlier. The attachment and emergence of a nearshore bar during sandwave migration can result in the formation of a runnel between the emergent bar and the beach (Davidson-Arnott, 1988; Stewart and Davidson-Arnott, 1988). The subsequently infilled runnel can provide the seedbed conditions for successful establishment of *Populus deltoides*. If the beach continues to prograde and the area is invaded by dune-building grasses such as *Ammophila breviligulata*, the foredune community that develops will include *P. deltoides*. On the other hand, a foredune that did not originate in this manner and did not involve successful *P. deltoides* establishment before establishment by *A. breviligulata* would not have *P. deltoides* as a component of the plant community.

CONCLUSION

Sand dune succession is not a fact or a phenomenon but a *hypothesis* of vegetation development on dunes that is based on chronosequence studies. A chronosequence is a series of spatially distinct sites of varying ages that is assumed to represent a temporal sequence of vegetation. Thus, the chronosequence assumes that the spatially distinct sites are identical in all aspects other than age and that all sites in the spatial sequence have followed or will follow the same temporal sequence. The validity of this assumption has not proven to be valid for many reasons. Empirical evidence has shown that sites are rarely, if ever, stable (i.e., constant climate and disturbance free) over the time period represented by the age of the oldest site. Because of a constantly changing climate, sites of different ages would have initiated under very different climatic conditions that would affect seedling establishment. As a result of other factors, such as changing water levels and migration of alongshore sandwaves, the history of foredune development would vary both spatially and temporally, affecting the sequence of species establishment. Also, because of disturbance regimes changing over time, the sites would have been subject not only to different types of disturbances but also to different disturbance regimes. In other words, not all sites would share a common disturbance history. Finally, the ubiquity of disturbance regardless of the age of the dune substrate would mean that the vegetation in all sites in the sequence, not just the youngest site, would be expected to have arisen from some disturbance.

In fact, the traditional hypothesis of dune succession (and of plant succession in general) assumes the development of vegetation in the *absence* of disturbance. That most coastal dune communities are subject to different disturbances (e.g., sand movement, overwash/inundation, hurricanes, fires, trampling, overgrazing) at rates frequent enough to play a significant role in the establishment and survival of plant species is undeniable and therefore, this point alone should be sufficient argument against the stability of vegetation on even "fixed" dunes and their development to a stable climax community. As noted by Carter (1990), "[m]any dunes systems are a mosaic of stable and unstable surfaces"; they do not consist simply of a growing dune inland of which are dunes stabilized for increasing lengths of time.

Whether the vegetation reflects the dominant influence of sand movement or one of the other disturbances, such as fire, depends on the relative frequency of these disturbances. For example, on the embryo and foredunes where vegetation is often too sparse to propagate fires and where sand transport is significant enough to be measurable at short time scales (<1 year), the vegetation would obviously reflect species' tolerances to sand erosion/deposition rather than to fire. On the other hand, on dunes further inland where the sand budget is neither significantly positive nor negative and where continuous ground cover by vegetation allows fires to spread, the vegetation would more likely reflect species' tolerances to the fire regime (frequency and intensity).

Much of the dune succession studies have focused on vegetation development on the first few dune ridges, where the stability of the surface is an issue. Despite this, there is a surprising lack of recognition of the geomorphological literature on sediment transport in coastal environments that provides an understanding of the temporal and spatial dynamics of the dune surfaces upon which vegetation must establish. Lacking such understanding, the succession argument (and particularly the facilitation explanation for community replacement) assumes that it is the establishment of the "pioneer" dune grasses that stabilizes the sand surface, thus allowing seral species that are intolerant of sand deposition to establish and replace the grasses that not only tolerate but require continued sand deposition for survival. While plants do influence the rates of sediment mobilization and deposition through their influence on windflow, they cannot prevent sand movement which depends on the availability of sand for transport (i.e., the sand budget).

Sand budgets are important in determining whether the surface of a particular site is accreting or deflating. The changes in surface morphology resulting from changes in sand availability and the aeolian processes of sand transport have been modeled relatively successfully for both bare and vegetated surfaces (e.g., van Dijk *et al.*, 1999; Arens *et al.*, 2001). Furthermore, such models can be used to explain and predict dune growth or deflation under changing conditions. The next step would be to couple these sediment transport and sand budget models with the ecological tolerances of species for varying rates of sand deposition or erosion in order to explain the distribution of species on actively accreting or deflating surfaces.

However, other factors are also recognized as playing a role in successful plant establishment and thus explaining species distribution in coastal dune environments. For short-lived species, the current environmental conditions of a particular site (e.g., the current sand budget, temperature and precipitation, distribution of seed predators and herbivores) may best explain the distribution of species. For old, long-lived species within the community such as trees, it would be necessary to know the developmental history of the site since the time of establishment of the population in order to account for its presence. Thus, Clark (1986) concluded that "[p]resent distributions [of the barrier-beach vegetation] are largely the product of historical events acting together with existing conditions."

The above proposed framework points out the importance of integrating the work of coastal and aeolian geomorphologists with that of physiological and population ecologists to develop a better and more realistic understanding of the spatial and temporal distribution of plant species populations in coastal dune environments (e.g., Hayden *et al.*, 1995). Without such integration, ecologists tend either to correlate current environmental conditions with population or community distribution patterns, assuming some kind of equilibrium state between the two without considering the dynamic nature of the dune environment, or to invoke a "synthetic successional" argument (sensu Olson 1958a) based on an invalid chronosequence.

ACKNOWLEDGMENTS

We thank M. Anwar Maun and Martin Kent for their comments and helpful suggestions on an earlier draft of the manuscript. We are especially grateful to Robin Davidson-Arnott, who critically read more than one draft of this chapter, provided many useful suggestions, and helped us to develop a better understanding of coastal geomorphic processes.

REFERENCES

Allen, J. R. L. (1982). Simple models for the shape and symmetry of tidal sand waves: (1) statically stable equilibrium forms. *Mar Geol* 48, 31–49.

Anderson, R. A., and Haff, P. K. (1991). Wind modification and bed response during saltation of sand in air. *Acta Mech* Suppl 1, 21–52.

Anderson, R. A., Sorenson, M., and Willetts, B. B. (1991). A review of recent progress in our understanding of aeolian sediment transport. *Acta Mech* Suppl 1, 1–19.

Anderton, J. B., and Loope, W. L. (1995). Buried soils in a perched dunefield as indicators of late Holocene lake-level change in the Lake Superior Basin. *Quatern Res* 44, 190–199.

Arbogast, A. F., Hansen, E. C., and Van Oort, M. D. (2002). Reconstructing the geomorphic evolution of large coastal dunes along the southeastern shore of Lake Michigan. *Geomorphology* 46, 241–255.

Arens, S. M., Baas, A. C. W., Van Boxel, J. H., and Kalkman, C. (2001). Influence of reed stem density on foredune development. *Earth Surf Process Landforms* 26, 1161–1176.

Ayyad, M. A. (1973). Vegetation and environment of the western Mediterranean coastal land of Egypt: 1. The habitat of sand dunes. *J Ecol* 61, 509–523.

Bagnold, R. A. (1936). The movement of desert sand. *Proc Royal Soc Lond, Series A* 157, 594–620.

Bagnold, R. A. (1956). Flow of cohesionless grains in fluids. *Phil Trans Roy Soc Lond Series A* 249, 235–297.

Baldwin, K. A., and Maun, M. A. (1983). Microenvironment of Lake Huron sand dunes. *Can. J Bot* 61, 241–255.

Barbour, M. G. (1972). Seedling establishment of *Cakile maritima* at Bodega Head, California. *B Torrey Bot Club* 99, 11–16.

Barbour, M. G. (1978). Salt spray as a microenvironmental factor in the distribution of beach plants at Point Reyes, California. *Oecologia* 32, 213–224.

Barbour, M. G., and De Jong, T. M. (1977). Response of west coast beach taxa to salt spray, seawater inundation and soil salinity. *B Torrey Bot Club* 104, 29–34.

Barbour, M. G., De Jong, T. M., and Johnson, A. F. (1976). Synecology of beach vegetation along the Pacific coast of United States of America: a first approximation. *J Biogeogr* 3, 55–69.

Barbour, M. G., Rejmanék, M., Johnson, A. F., and Pavlik, B. M. (1987). Beach vegetation and plant distribution patterns along the northern Gulf of Mexico. *Phytocoenologia* 15, 201–233.

Bauer, B. O., and Davidson-Arnott, R. G. D. (2002). A general framework for modeling sediment supply to coastal dunes including wind angle, beach geometry, and fetch effects. *Geomorphology* 49, 89–108.

Belly, P.-Y. (1964). *Sand Movement by Wind.* U.S. Army Corps of Engineers Tech. Memo. No. 1. Coastal Engineering Research Center, Washington, DC.

Billings, W. D. (1938). The structure and development of old field shortleaf pine stands and certain associated physical properties of the soil. *Ecol Monogr* 8, 437–500.

Billings, W. D. (1941). Quantitative correlations between vegetational changes and soil development. *Ecology* 22, 448–456.

Bowen, A. J., and Inman, D. L. (1966). *Budget of Littoral Sediments in the Vicinity of Point Arguello, California.* US Army Corps of Engineers Tech. Memo. No. 19. Coastal Engineering Research Center, Washington, DC.

Boyd, R. S. (1988a). Microdistribution of the beach plant *Cakile maritima* (Brassicaceae) as influenced by a rodent herbivore. *Am J Bot* 75, 1540–1548.

Boyd, R. S. (1988b). Herbivory and species replacement in the west coast searockets (*Cakile*, Brassicaceae). *Am Midl Nat* 119, 304–317.

Boyd, R. S. (1991). Population biology of west coast *Cakile maritima*: effects of habitat and predation by *Peromyscus maniculatus*. *Can J Bot* 69, 2620–2630.

Bradshaw, M., and Weaver, R. (1995). *Foundations of Physical Geography.* Wm. C. Brown Publishers, Dubuque, IA.

Burrows, C. J. (1990). *Processes of Vegetation Change.* Unwin Hyman, London.

Carter, R. W. G. (1990). The geomorphology of coastal dunes in Ireland. In: *Dunes of the European Coasts* (Bakker, T. W., Jungerius, P. D., and Klijn, J. A., Eds.), pp. 31–40. Catena Supplement 18, Catena Verlag, Cremlingen-Destedt, Germany.

Castillon, D A. (1992). *Conservation of Natural Resources.* Wm. C. Brown Publishers, Dubuque, IA.
Chadwick, H. W., and Dalke, P. D. (1965). Plant succession on dune sands in Fremont County, Idaho. *Ecology* 46, 765–780.
Chapin, F. S., Walker, L. R., Fastie, C. L., and Sharman, L. C. (1994). Mechanisms of primary succession following deglaciation at Glacier Bay, Alaska. *Ecol Monogr* 64, 149–175.
Chen, H., and Maun, M. A. (1999). Effects of sand burial depth on seed germination and seedling emergence of *Cirsium pitcheri. Plant Ecol* 140, 53–60.
Chepil, W. S. (1957). *Width of Field Strips to Control Wind Erosion.* Agricultural Experiment Station Bulletin 92. Kansas State College of Agriculture and Applied Science, Manhattan, KS.
Christopherson, R. W. (1998). *Elemental Geosystems.* 2nd edition. Prentice Hall, Upper Saddle River, NJ.
Clark, J. S. (1986). Dynamism in the barrier-beach vegetation of Great South Beach, New York. *Ecol Monogr* 56, 97–126.
Clark, J. S., Beckage, B., Camill, P., Cleveland, B., HilleRisLambers, J., Lichter, J., McLachlan, J., Mohan, J., and Wyckoff, P. (1999). Interpreting recruitment limitation in forests. *Am J Bot* 86, 1–16.
Clements, F. E. (1916). *Plant Succession: Analysis of the Development of Vegetation.* Carnegie Institute of Washington Publication 242, Washington, DC.
Clements, F. E. (1928). *Plant Succession and Indicators.* Hafner Publishing Co., New York.
Colinvaux, P. (1993). *Ecology 2.* John Wiley and Sons, New York.
Cooper, W. S. (1923). The recent ecological history of Glacier Bay, Alaska: Permanent quadrats at Glacier Bay: an initial report upon a long-period study. *Ecology* 4, 355–365.
Cooper, W. S. (1931). A third expedition to Glacier Bay, Alaska. *Ecology* 12, 61–95.
Cooper, W. S. (1939). A fourth expedition to Glacier Bay, Alaska. *Ecology* 20, 130–159.
Costa, C. S. B., Seeliger, U., and Cordazzo, C. V. (1991). Leaf demography and decline of *Panicum racemosum* populations in coastal foredunes of southern Brazil. *Can J Bot* 69, 1593–1599.
Cowles, H. C. (1899). The ecological relations of the vegetation on the sand dunes of Lake Michigan. *Bot Gaz* 27, 95–117, 167–202, 281–308, 361–391.
Cowles, H. C. (1901). The physiographic ecology of Chicago and vicinity. *Bot Gaz* 31, 73–108, 145–182.
Crocker, R. L., and Major, J. (1955). Soil development in relation to vegetation and surface age at Glacier Bay, Alaska. *J Ecol* 43, 427–448.
Davidson-Arnott, R. G. D. (1988). Temporal and spatial controls on beach/dune interaction, Long Point, Lake Erie. *J Coast Res* Special Issue No. 3, 131–136.
Davidson-Arnott, R. G. D., and Stewart, C. J. (1987). The effects of longshore sand waves on dune erosion and accretion, Long Point, Ontario. In: *Proceedings of the Canadian Coastal Conference* (Ouellet, Y., Ed.), pp. 131–144. National Research Council Canada, Associate Committee for Research on Shoreline Erosion and Sedimentation, Ottawa, ON.
Davidson-Arnott, R. G. D., and van Heyningen, A. G. (2003). Migration and sedimentology of longshore sandwaves, Long Point, Lake Erie, Canada. *Sedimentology* 50, 1123–1137.
De Blij, H. J., and Muller, P. O. (1996). *Physical Geography of the Global Environment.* John Wiley & Sons, New York.
Dech, J. P., and Maun, M. A. (2005). Zonation of vegetation along a burial gradient on the leeward slopes of Lake Huron sand dunes. *Can J Bot* 83, 227–236.
de Jong, T. D., and Klinkhamer, P. G. L. (1988). Seedling establishment of the biennials *Cirsium vulgare* and *Cynoglossum officinale* in a sand-dune area: the importance of water for differential survival and growth. *J Ecol* 76, 393–402.
del Moral, R., and Bliss, L. C. (1993). Mechanisms of primary succession—insights resulting from the eruption of Mount St-Helens. *Adv Ecol Res* 24, 1–66.

Downing, E. R. (1922). *A Naturalist in the Great Lakes Region*. University of Chicago Press, Chicago, IL.

Dyer, K. R. (1986). *Coastal and Estuarine Sediment Dynamics*. Wiley, New York.

Farrow, E. P. (1919). On the ecology of the vegetation of Breckland. *J Ecol* 7, 55–64.

Fastie, C. L. (1995). Causes and ecosystem consequences of multiple pathways of primary succession at Glacier Bay, Alaska. *Ecology* 76, 1899–1916.

Filion, L. (1984). A relationship between dunes, fire and climate recorded in the Holocene deposits of Quebec. *Nature* 309, 543–546.

Fuller, G. D. (1912). The cottonwood dune association. *Trans Ill Acad Sci* 5, 137–143.

Gares, P. A. (1992). Topographic changes associated with coastal dune blowouts at Island Beach State Park, New Jersey. *Earth Surf Process Landforms* 17, 589–604.

Gillette, D. A., Herbert, G., Stockton, P. H., and Owen, P. R. (1996). Causes of the fetch effect in wind erosion. *Earth Surf Proc Landforms* 21, 641–659.

Gleason, H. A. (1939). The individualistic concept of the plant association. *Am Midl Nat* 21, 92–110.

Gregory, J. M., and Darwish, M. M. (1990). Threshold friction velocity prediction considering water content. In: *Our Biosphere, Our Responsibility: the* 1990 *International Winter Meeting of the American Society of Agricultural Engineers, December* 18-21, 1990, Chicago, Illinois, paper No. 902562. American Society of Agricultural Engineers, St. Joseph, MI.

Grime, J. P. (1979). *Plant Strategies and Vegetation Processes*. John Wiley, New York.

Grubb, P. J. (1977). The maintenance of species richness in plant communities: the importance of the regeneration niche. *Biol Rev* 52, 107–145.

Hardisty, J., and Whitehouse, R. J. S. (1988). Evidence for a new sand transport process from experiments on the Sahara. *Nature* 332, 532–534.

Harper, J. L. (1977). *Population Biology of Plants*. Academic Press, London.

Harris, D., and Davy, A. J. (1987). Seedling growth in *Elymus farctus* after episodes of burial with sand. *Ann Bot* 60, 587–593.

Hayden, B. P., Santos, M. C. F. V., Shao, G., and Kochel, R. C. (1995). Geomorphological controls on coastal vegetation at the Virginia Coast Reserve. *Geomorphology* 13, 283–300.

Hesp, P. A. (1996). Flow dynamics in a trough blowout. *Bound-Layer Meteorol* 77, 305–330.

Hesp, P. A. (2002). Foredunes and blowouts: initiation, geomorphology and dynamics. *Geomorphology* 48, 245–268.

Hesp, P. A., and Hyde, R. (1996). Flow dynamics and geomorphology of a trough blowout. *Sedimentology* 43, 505–525.

Hewett, D. G. (1970). The colonization of sand dunes after stabilization with marram grass (*Ammophila arenaria*). *J Ecol* 58, 653–668.

Hobbs, R. J., and J. Grace. (1981). A study of pattern and process in coastal vegetation using principal components analysis. *Vegetatio* 44, 137–153.

Horikawa, K., Hotta, S., Kubota, S., and Katori, S. (1984). Field measurement of blown sand transport rate by trench trap. *Coast Eng Japan* 27, 214–232.

Hotta, S., Kubota, S., Katori, S., and Horikawa, K. (1984). Sand transport by wind on a wet sand surface. In: *Nineteenth Coastal Engineering Conference: Proceedings of the International Conference, September 3-7, 1984, Houston, Texas*. (Edge, B. L., Ed.), pp. 1263–1281. American Society of Civil Engineers, New York.

Houle, G. (1995). Environmental filters and seedling recruitment on a coastal dune in subarctic Quebec (Canada). *Can J Bot* 74, 1507–1513.

Hsu, S. A. (1971). Wind stress criteria in eolian sand transport. *J Geophys Res* 76, 8684–8686.

Huiskes, A. H. L. (1977). The natural establishment of *Ammophila arenaria* from seed. *Oikos* 29, 133–136.

Johnson, A. F., and Muller, J. W. (1992). *An Assessment of Florida's Remaining Coastal Upland Natural Communities: Panhandle Florida.* Unpublished report, Florida Natural Areas Inventory, Tallahassee, FL.

Kadib, A. A. (1965). *A Function for Sand Movement by Wind.* Hydraulics Engineering Laboratory Report HEL-2-12. University of California, Berkeley, CA.

Kawamura, R. (1951). *Study of Sand Movement by Wind.* Hydraulics Engineering Laboratory Report HEL-2-8. University of California, Berkeley, CA.

Kawata, Y., and Tsuchiya, Y. (1976). Influence of water content on the threshold of sand movement and the rate of sand transport in blown sand. In: *Proceedings of the Japanese Society of Civil Engineering 249"* pp. 95–100. (In Japanese)

Kent, M., Owen, N. W., Dale, P., Newnham, R. M., and Giles, T. M. (2001). Studies of vegetation burial: a focus for biogeography and biogeomorpology. *Prog Phys Geog* 25, 455–482.

Kent, M., Owen, N. W., and Dale, M. P. (2005). Photosynthetic responses of plant communities to sand burial on the machair dune systems of the Outer Hebrides, Scotland. *Ann Bot* 95, 869–877.

Klinkhamer, P. G. L., de Jong, T. J., and van der Meijden, E. (1988). Production, dispersal and predation of seeds in the biennial *Cirsium vulgare. J Ecol* 76, 403–414.

Kumler, M. L. (1969). Plant succession on the sand dunes of the Oregon coast. *Ecology* 50, 695–704.

Laing, C. C. (1958). Studies in the ecology of *Ammophila breviligulata* 1. Seedling survival and its relation to population increase and dispersal. *Bot Gaz* 119, 208–216.

Lancaster, N., and Baas, A. (1998). Influence of vegetation cover on sand transport by wind: field studies at Owens Lake, California. *Earth Surf Process Landforms* 23, 69–82.

Landsberg, H., and Riley, N. A. (1943). Wind influences on the transportation of sand over a Michigan sand dune. In: *Proceedings of the Second Hydraulics Conference.* (Rouse, H., and Howe, J. W., Eds.), pp. 342–352. University of Iowa, Iowa City, IA.

Lettau, K., and Lettau, H. (1978). Experimental and micrometeorological field studies of dune migration. In: *Exploring the World's Driest Climate.* (Lettau, H., and Lettau, K., Eds.), pp. 110–147. IES Report 101, Center for Climatic Research, Institute for Environmental Studies, University of Wisconsin, Madison, WI.

Lichter, J. (2000). Colonization constraints during primary succession on coastal Lake Michigan sand dunes. *J Ecol* 88, 825–839.

Loope, W. L., and Arbogast, A. F. (2000). Dominance of an ~150-year cycle of sand-supply change in late Holocene dune-building along the eastern shore of Lake Michigan. *Quatern Res* 54, 414–422.

Marin, P., and Filion, L. (1992). Recent dynamics of subarctic dunes as determined by tree-ring analysis of white spruce, Hudson Bay, Québec. *Quatern Res* 38, 316–330.

Marsh, W. M. (1987). *Earthscape: A Physical Geography.* John Wiley and Sons, Inc., New York.

Martin, W. E. (1959). The vegetation of Island Beach State Park, New Jersey. *J Ecol* 29, 1–46.

Martínez, M. L., and Maun, M. A. (1999). Responses of dune mosses to experimental burial by sand under natural and greenhouse conditions. *Plant Ecol* 145, 209–219.

Martínez, M. L., and Moreno-Casasola, P. (1996). Effects of burial by sand on seedling growth and survival in six tropical sand dune species from the Gulf of Mexico. *J Coastal Res* 12, 406–419.

Maun, M. A. (1985). Population biology of *Ammophila breviligulata* and *Calamovilfa longifolia* on Lake Huron sand dunes. I. Habitat, growth form, reproduction and establishment. *Can J Bot* 63, 113–124.

Maun, M. A. (1994). Adaptations enhancing survival and establishment of seedlings on coastal dune systems. *Vegetatio* 111, 59–70.

Maun, M. A. (1996). The effects of burial by sand on survival and growth of *Calamovilfa longifolia*. *Ecoscience* 3, 93–100.

Maun, M. A. (2004). Burial of plants as a selective force in sand dunes. In: *Coastal Dunes: Ecology and Conservation*. (Martínez, M. L., and Psuty, N. P., Eds.), pp. 119–135. Springer-Verlag, Berlin.

Maun, M. A., and Lapierre J. (1984). The effects of burial by sand on *Ammophila breviligulata*. *J Ecol* 72, 827–839.

Maun, M. A., and Lapierre, J. (1986). Effects of burial by sand on seed germination and seedling emergence of four dune species. *Am J Bot* 73, 450–455.

Maun, M. A., and Perumal, J. (1999). Zonation of vegetation on lacustrine coastal dunes: effects of burial by sand. *Ecol Lett* 2, 14–18.

Maun, M. A., and Riach, S. (1981). Morphology of caryopses, seedlings and seedling emergence of the grass *Calamovilfa longifolia* from various depths in sand. *Oecologia* 49, 137–142.

McEwan, I. K., and Willetts, B. B. (1993). Adaption of the near surface wind to the development of sand transport. *J Fluid Mech* 252, 99–115.

McLeod, K. W., and Murphy, P. G. (1977). Establishment of *Ptelea trifoliata* on Lake Michigan sand dunes. *Am Midl Nat* 97, 350–362.

McNaughton, S. J., and Wolf. L. L. (1973). *General Ecology*. Holt, Rinehart and Winston, New York.

Merila, P., Galand, P. E., Fritze, H., Tuittila, E. S., Kukko-oja, K., Laine, J., and Yrjala, K. (2006). Methanogen communities along a primary succession transect of mire ecosystems. *FEMS Microbiol Ecol* 55, 221–229.

Moreno-Casasola, P. (1986). Sand movement as a factor in the distribution of plant communities in a coastal dune system. *Vegetatio* 65, 67–76.

Morrison, R. G. (1973). *Primary succession on sand dunes at Grand Bend, Ontario*. Ph.D. thesis, University of Toronto, Toronto, ON.

Nakashima, Y. (1979). A fundamental study on the blown sand control. *Bull Kyushu University Forestry* 51, 125–183. (In Japanese)

Namikas, S. L., and Sherman, D. J. (1998). AEOLUS II: an interactive program for the simulation of aeolian sedimentation. *Geomorphology* 22, 135–149.

Oertel, G. F., and Larsen, M. (1976). Developmental sequences in Georgia coastal dunes and distribution of dune plants. *B Georgia Acad Sci* 34, 35–48.

Olff, H., Huisman, J., and Van Tooren, B. F. (1993). Species dynamics and nutrient accumulation during early primary succession in coastal sand dunes. *J Ecol* 81, 693–706.

Olson, J. S. (1958a). Rates of succession and soil changes on southern Lake Michigan sand dunes. *Bot Gaz* 119, 125–170.

Olson, J. S. (1958b). Lake Michigan dune development. 2. Plants as agents and tools in geomorphology. *J Geol* 66, 45–351.

Olson, J. S. (1958c). Lake Michigan dune development. 3. Lake level, beach and dune oscillation. *J Geol* 66, 473–483.

Owen, N. W., Kent, M., and Dale, M. P. (2004). Plant species and community responses to sand burial on the machair of the Outer Hebrides, Scotland. *J Veg Sci* 15, 669–678.

Owen, P. R. (1964). Saltation of uniform grains in air. *Fluid Mech* 20, 225–242.

Payne, A. M., and Maun, M. A. (1981). Dispersal and floating ability of dimorphic fruit segments of *Cakile edentula* var. *lacustris*. *Can J Bot* 59, 2595–2602.

Poulson, T. L. (1999). Autogenic, allogenic, and individualistic mechanisms of dune succession at Miller, Indiana. *Nat Areas J* 19, 172–176.

Pringle, A., Hesp, P., and Brough, G. (1999). Wind flow and topographic steering within a trough blowout at Tangimoana, New Zealand. In: 20th *New Zealand Geography Conference*. (Roche, M., McKenna, M., and Hesp, P., Eds.), pp. 270–272. New Zealand Geographical Society, Hamilton, New Zealand.

Psuty, N. P. (1988). Sediment budget and dune/beach interaction. *J Coast Res* Special Issue No. 3, 1–4.

Ranwell, D. S. (1958). Movement of vegetated sand dunes at Newborough Warren, Anglesey. *J Ecol* 46, 83–100.

Raup, H. M. (1975). Species versatility in shore habitats. *J Arnold Arboretum* 56, 126–163.

Raup, H. M., and Argus, G. W. (1982). *The Lake Athabasca Sand Dunes of Northern Saskatchewan and Alberta, Canada. 1. The Land and Vegetation*. Publications in Botany No. 12. National Museums of Canada, Ottawa, ON.

Reice, S. R. (1994). Nonequilibrium determinants of biological community structure—biological communities are always recovering from the last disturbance—disturbance and heterogeneity, not equilibrium generate biodiversity. *Am Sci* 82, 424–435.

Rey, P. J., and Alcántara, J. M. (2000). Recruitment dynamics of a fleshy-fruited plant (*Olea europaea*): connecting patterns of seed dispersal to seedling establishment. *J Ecol* 88, 622–633.

Ruessink, B. G., and Jeuken, M. C. J. L. (2002). Dunefoot dynamics along the Dutch coast. *Earth Surf Process Landforms* 27, 1043–1056.

Saunders, K. E., and Davidson-Arnott, R. G. D. (1991). Coastal dune response to natural disturbances. In: *Proceedings of the Canadian Symposium on Coastal Sand Dunes, September 1990, Guelph, Ontario*, pp. 321–346. National Research Council of Canada, Ottawa, ON.

Scott, R. C. (1991). *Essentials of Physical Geography*. West Publishing Company, St. Paul, MN.

Sharitz, R. R., and McCormick, J. F. (1973). Population dynamics of two competing annual plant species. *Ecology* 54, 723–740.

Small, R. J., and Witherick, M. E. (1986). *A Modern Dictionary of Geography*. E. Arnold, London.

Spies, P.-J., and McEwan, I. K. (2000). Equilibration of saltation. *Earth Surf Process Landforms* 25, 437–453.

Stallins, J. A., and Parker, A. J. (2003). The influence of complex systems interactions on barrier island dune vegetation pattern and process. *Ann Assoc Am Geogr* 93, 13–29.

Stewart, C. J., and Davidson-Arnott, R. G. D. (1988). Morphology, formation and migration of longshore sandwaves; Long Point, Lake Erie, Canada. *Mar Geol* 81, 63–77.

Strahler, A. N., and Strahler, A. H. (1978). *Modern Physical Geography*. Wiley, New York.

Sykes, M. T., and Wilson, J. B. (1990). An experimental investigation into the response of New Zealand sand dune species to different depths of burial by sand. *Acta Bot Neerl* 39, 171–181.

Thuille, A., and Schulze, E. D. (2006). Carbon dynamics in successional and afforested spruce stands in Thuringia and the Alps. *Global Change Biology* 12, 325–342

Tyndall, R. W., Teramura, A. H., Mulchi, C. L., and Douglass, L. W. (1986). Seed burial effect on species presence along a mid-Atlantic beach. *Can J Bot* 64, 2168–2170.

van Boxel, J. H., Arens, S. M., and van Dijk, P. M. (1999). Aeolian processes across transverse dunes. I: Modelling the air flow. *Earth Surf Process Landforms* 24, 255–270.

van der Valk, A. G. (1974). Environmental factors controlling the distribution of forbs on coastal foredunes in Cape Hatteras National Seashore. *Can J Bot* 52, 1057–1073.

van Dijk, P. M., Arens, S. M., and van Boxel, J. H. (1999). Aeolian processes across transverse dunes. II: Modelling the sediment transport and profile development. *Earth Surf Process Landforms* 24, 319–333.

Walker, L. R., and Chapin, F. S. (1986). Physiological controls over seedling growth in primary succession on an Alaskan floodplain. *Ecology* 67, 1508–1523.

Walker, L. R., Zasada, J. C., and Chapin, F. S. (1986). The role of life-history processes in primary succession on an Alaskan floodplain. *Ecology* 67, 1243–1253.

Wolfe, F. (1932). Annual rings of *Thuja occidentalis* in relation to climatic conditions and movement of sand. *Bot Gaz* 93, 328–335.

Wood, D. M., and del Moral, R. (1987). Mechanisms of early primary succession in subalpine habitats on Mount–St-Helens. *Ecology* 68, 780–790.

Yuan, T., Maun, M. A., and Hopkins, W. G. (1993). Effects of sand accretion on photosynthesis, leaf-water potential and morphology of two dune grasses. *Funct Ecol* 7, 676–682.

Zeman, O., and Jensen, N. O. (1988). *Progress Report on Modelling Permanent Form Sand Dunes*. Riso National Laboratory, RISO-2738, Roskilde, Denmark.

Zhang, J., and Maun, M. A. (1990a). Sand burial effects on seed germination, seedling emergence and establishment of *Panicum virgatum. Holarctic Ecol* 13, 56–61.

Zhang, J., and Maun, M. A. (1990b). Effects of sand burial on seed germination, seedling emergence, survival, and growth of *Agropyron psammophilum. Can J Bot* 68, 304–310.

Zingg, A. W. (1953). Wind tunnel studies of the movement of sedimentary material. In: *Proceedings of the Fifth Hydraulics Conference*. (McNown, J. S., and Boyer, M. C., Eds.), pp. 111–135. State University of Iowa, Iowa City, IA.

9

Fluvial Geomorphic Disturbances and Life History Traits of Riparian Tree Species

Futoshi Nakamura
Hokkaido University

Satomi Inahara
Niiza, Saitama

INTRODUCTION

The mechanism for maintaining species coexistence and richness in a forest is one of the central issues in plant community ecology. Early works identified two important concepts to explain forest structures and compositions: niche partitioning along environmental gradients (Whittaker, 1956; 1965) and the regeneration niche (Grubb, 1977). According to the first concept of niche partitioning, tree species are arranged along complex gradients of elevation, light, and edaphic conditions and form stand patches and canopy layers, thus partitioning a given habitat among the species present in the forest. The second concept of regeneration niche focuses on niche partitioning in reproduction, seed dispersal, seedling establishment, and further development of the immature plant. According to Grubb (1977), such partitioning can explain the indefinite persistence of many coexisting species. In temperate mixed deciduous forest, Nakashizuka (2001) argued that life history traits and their variations generate niche partitioning and tradeoffs, promoting tree coexistence in a community with heterogeneous abiotic and biotic conditions.

In riparian zones, where episodic disturbances provide spatial and temporal heterogeneity, life history traits and the availability of regeneration habitat associated with disturbance regimes may be important. Located at transition zones between terrestrial and aquatic ecosystems, riparian zones preserve a dynamic state in their interaction with streams (Swanson *et al.*, 1982). Energy and material flows can be concentrated during fluvial and geomorphic disturbances, which occur more frequently in riparian zones than in adjacent hillslope ecosystems, and thus their influence on forest communities is likely to be more significant in riparian zones (Junk *et al.*, 1989). Riparian forest communities survive such intensive and frequent disturbance regimes through physiological and structural adaptations, while utilizing the materials trapped by these disturbance events in a complex of river geomorphic surfaces (i.e., secondary channel, point-bar, backswamps) as habitat for their growth and maintenance.

Previous studies have consistently demonstrated that riparian forest structure is tightly linked to spatial and temporal patterns of fluvial and geomorphic disturbances (Niiyama, 1987; 1989; Ishikawa, 1988; Hughes, 1990, Aruga *et al.*, 1996; Nakamura *et al.*, 1997; Robertson and Augspurger, 1999; Nakamura and Shin, 2001; Shin and Nakamura, 2005). Intensive disturbance provides exposed sediment for seedling establishment and alters the edaphic environment (e.g., soil texture, moisture, organic content, and nutrients) of the floodplains. Inundation frequencies and durations, which differ at elevations from the low-lying active channels to the upper floodplain terraces, form the elevation gradient of soil conditions, resulting in the species distribution along the elevation gradient. This species distribution on the elevation gradient can be complicated by multiple disturbance regimes, with each different disturbance frequency and turnover time often forming a mosaic pattern of vegetation patches (Suzuki *et al.*, 2002; Shin and Nakamura, 2005). Timing, duration, and frequencies of disturbance influence the processes of erosion and deposition that promote the formation of channel bars and islands as regeneration habitats (Cordes *et al.*, 1997; Dykaar and Wingington, 2000). Accordingly, many studies have reported the alteration of forest communities in flow-regulated rivers, the lack of light-demanding early colonizers leading to dominance by shade-tolerant species, grasses, shrubs, and exotic species (Auble *et al.*, 1994; Johnson, 1998; Merritt and Cooper, 2000). Thus, temporal and spatial patterns of disturbance regimes and species responses to these patterns are the key to understanding the mechanisms of riparian forest maintenance.

The disturbance regime in riparian zones changes from headwaters to low-gradient alluvial rivers (Nakamura *et al.*, 2000; Nakamura and Swanson, 2003). A dominant disturbance in headwater streams is mass movement, such as landslide and debris flow, characterized by its intensiveness but rare occurrence. In downstream braided reaches, seasonal flood disturbances generate more frequent high flows, which form extensively developed geomorphic surfaces on floodplains. Low-gradient alluvial rivers in developing peat marshes rarely experience disturbances that result in landform alteration, although prolonged inundation in the marsh can cause a great deal of physiological stress to trees. Thus, a variety of disturbance types, frequencies, and intensities can be found at a catchment scale and may have diverse ecological consequences for riparian communities. Yet, many of the past studies have focused mainly on seasonal or relatively frequent flood disturbances in large rivers and on the responses of one group of species, the Salicaceae (e.g., Rood and Heinze-Milne, 1989; Begg *et al.*, 1998; Friedman and Lee, 2002). The regeneration of Salicaceae forests has been threatened in flow-regulated rivers by the lack of flow regimes that promote their seedling establishment (Auble *et al.*, 1994; Johnson, 1998; Merritt and Cooper, 2000). Life history traits characteristic of these short-lived Salicaceae species and of flood disturbances operating at the scale of seasonal to decades have now been reasonably understood; however, such understanding falls short of allowing any generalizations on the importance and role of geomorphic disturbances at the catchment scale. There may be a range of life history strategies to accommodate a variety of disturbance regimes, such as those with long recurrence intervals up to hundreds or thousands of years. In addition, Nakashizuka (2001) asserts that variation in life history strategies among different life stages of a species as well as among different species is important in explaining the relationships between environmental heterogeneity and the organization of temperate mixed deciduous forest communities.

The objective of this chapter is to describe the relationship between disturbance regimes, ranging from montane headwater streams to low-gradient meandering rivers, and life history traits of the dominant tree species from riparian forests in the Asia Monsoon Belt of Japan as an example. We define "regeneration habitat" as the light and edaphic conditions or opportunities suitable for successful seedling establishment and maturation of the young trees to allow seed production. Because disturbances operate in time and space, the availability of regeneration habitat also varies in time and space. Riparian trees are specialized to disturbance regimes characteristic of their given geomorphic settings by developing life history strategies that are adapted to the fluctuating availability of regeneration sites. Reproductive traits, such as fruiting, modes of seed dispersal, and their timing, can be constrained by the times and locations of regeneration habitats (Clark, 1991; Strykstra *et al.*, 2002). Structural adaptation and asexual regeneration, which can maintain the stages of post-seedling establishment, may be crucial during a period when regeneration habitat is not available. Thus, we propose that predictability in the occurrence of regeneration sites [referred to as "reliability" by Strykstra *et al.* (2002)], both in time and space, is the link between life history traits and disturbance regimes at the catchment scale. We begin with a geomorphic classification of riparian zones and describe the disturbance regimes dominant in each class. The availability of regeneration habitat determined by the disturbance regimes is then characterized in temporal and spatial dimensions. We discuss how riparian trees are adapted to the temporally and spatially dynamic regeneration habitat, with a particular emphasis on life history strategies of dominant species.

GEOMORPHIC CLASSIFICATION OF RIPARIAN ZONES AND DISTURBANCE REGIMES IN A CATCHMENT

For many ecological phenomena, especially disturbance in stream networks, it is important to place the analysis in a geographic context. River landscape conditions can be classified into three dominant morphologies: headwater streams, braided rivers on alluvial fans, and low-gradient meandering rivers (Nakamura and Swanson, 2003). Headwater streams are boulder-dominated and are characterized by their high-gradient and very narrow valley floors. Braided rivers, generally observed on alluvial fans, have a relatively steep gravel bed and multiple channels. Low-gradient, meandering rivers are typical of sand-bed streams. In many regions of the world, such as central North America and central Eurasia, a transitional system exists between braided gravel-bed rivers and meandering sand-bed rivers that is characterized by meandering reaches with moderate gradients. Some studies that we cite in our discussion of life history strategies may not fit within our classification because this transitional system is not usually developed in Japanese river landscapes.

Headwater Streams (0- to ~3rd-Order Streams)

A disturbance regime dominant in headwater streams is mass movement, such as landslide, debris flow, and earthflow, most of which are landform shifting caused by gravity (Fig. 1). Mass movement is characterized by a deep layer of massive soil moving at high velocities but at extremely low frequency on the order of every 100 to 1000 years; therefore, it is unpredictable. During the period of mass movement events, tree-fall and windthrow gaps are created more frequently. This type of gap formation may be localized, such as an individual tree-fall gap, or may be extended to a relatively large area by windthrow.

The depth of the soil layer in deep-seated landslides, initiated by episodic events such as an earthquake in a certain geological setting, can reach from a few to dozens of meters. These deep soil layers can move with

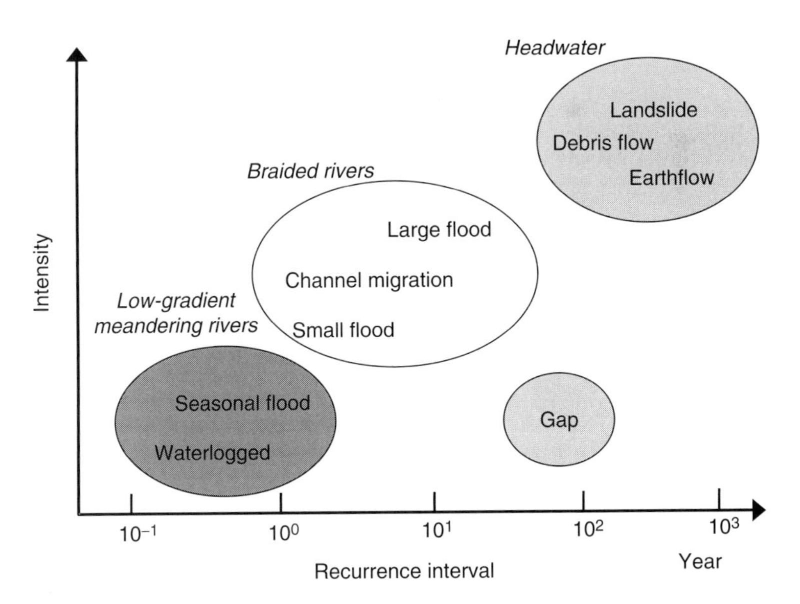

FIG. 1 Dominant disturbance regimes in riparian zones arranged along a temporal scale. Intensity indicates the level of physical disturbance that alters geomorphic surfaces.

trees that remain standing on the layers. The trees suffer from root cutting and may topple at slope failures and around landslide scars. The surviving individuals often exhibit architectural deformity and growth reduction, which can be visible as eccentric growth of their annual rings. This type of mass movement is definitely high in magnitude and detrimental to riparian trees. While its occurrence is extremely episodic, the locations are geologically limited.

Shallow landslides may be more important for forest composition and structure because they occur more frequently and can extend over a large area. They occur in headwater basins after heavy rainfall events and generate soil mass movement of layers that range from tens of centimeters up to two meters in depth. Depending on geological and geomorphological settings, a landslide can occur at a particular site from once every few decades (Shimokawa, 1984; Yanai and Usui, 1989) to centuries (Nakamura *et al.*, 1995). Shallow landslides occurring in headwater reaches often develop into debris flows (Nakamura *et al.*, 2002), cascading their influences through a stream network to the lower reaches. Consequently, a main tributary in a small basin with multiple landslide-prone headwaters would experience a debris flow disturbance more frequently, at a time scale of several decades. Shallow landslides typically form bedrock hollows (Tsukamoto, 1973; Dietrich and Dunne, 1978) and steep, convergent slopes (>30°) created by channel incision, then followed by being filled with colluvium (Hatano, 1974). Most of the established trees on the erosion surfaces of mass movements fall down, leaving few surviving individuals. However, where the energy of mass movement is small and the landslide deposits are shallow, trees may persist by forming adventitious roots and stems to anchor themselves on the deposits. Debris flows may stop at river confluences and at unconstrained reaches that extend over valley floors in mountainous regions. Mass movement deposits composing a river bottom substrate are large, reaching boulder sizes. Large logs also provide important regeneration sites for riparian tree species. In these depositional zones, undulating landforms are created to provide heterogeneous habitats.

Braided Rivers on Alluvial Fan (4th- to ~5th-Order Streams)

The most common type of disturbance in braided rivers is flooding (Fig. 1). In Asian monsoon areas, heavy seasonal rainfalls cause flooding in both the monsoons (June to early July) and typhoons (August to early October). In the northern part of Japan, snowmelt high flow is also predominant, occurring every year in spring (April and May). The magnitude and frequencies of typhoon- and monsoon-induced flooding show greater year-to-year variation than snowmelt high flows.

Valley floor landforms in alluvial fan systems are typically unconstrained, with relatively steep bed gradients of braided rivers. Multiple active channels migrate laterally, forming abandoned secondary channels. Bed materials, which consist of pebble- and gravel-size particles, are typically low in moisture content. Flood frequency is in the order of years near active channels and ten to a hundred years for more stable floodplains (Nakamura, 1990; Nakamura *et al.*, 1995). However, lateral channel migration can alter stable floodplains, changing them to highly unstable areas prone to intensive and frequent flood disturbances.

Flood disturbances in braided rivers influence riparian trees to varying degrees of severity. Some of the trees toppled during a flood may be completely uprooted and die, while others may survive by resprouting from their buried shoots and stems (Everitt, 1968; Karrenberg *et al.*, 2002). With less intensive flooding, trees that are left standing may be killed because of the anoxic conditions under sediment depositions, while others may survive by forming adventitious roots (Nakamura and Kikuchi, 1996). Disturbance intensity decreases along an elevation gradient from the active channel to higher terraces in a floodplain and is determined primarily by inundation frequency (Nakamura *et al.*, 1997). Flooding rarely reaches the high elevations, and, even if it does, removal of canopy trees is rare. Only understory vegetation on the high elevation surfaces is affected, resulting in the dominance of large canopy trees there. In contrast, geomorphic surfaces at low elevations near active channels are frequently subject to flood disturbance. On these low surfaces, fully mature trees are not established, while seedlings and saplings of light-demanding species are common.

When flooding completely eliminates the ground surface in a braided reach, new floodplain surfaces are formed during periods of water recession. Sediment and nutrients transported from upstream accumulate to form complex geomorphic surfaces that have a variety of substrate properties and nutrient conditions. Flood disturbances can occur at various frequencies and magnitudes, resulting in further topographically and edaphically diverse soil conditions (Wolman and Miller, 1960; Dollar, 2000; Shin and Nakamura, 2005). These new geomorphic surfaces are typical recruitment sites for wind-dispersed, light-demanding species, but may not be suitable as regeneration habitat where the seedlings can survive to reproductive maturity.

Low-Gradient Meandering Rivers (6th- or Higher-Order Streams)

Geomorphic disturbances in low-gradient alluvial rivers are less extreme in magnitude than those in headwater and braided rivers. The meandering process in alluvial rivers creates unique landforms over geologic time. Natural levees adjacent to the active channels are meander belts. Point bars, the areas of fresh, sandy sediment on the inside of the meander bends, are formed one after another to create ridge and swale topography of meander scrolls. Progressively older plant communities develop on the scrolls at increasing distance from the active channel. In back swamps behind the natural levees, a layer of peat soil develops in which water tables are constantly high (Nakamura *et al.*, 2002).

Seasonal flooding is the major disturbance in low-gradient meandering rivers (Fig. 1). However, compared to braided rivers, both runoff-rise and recession in meandering systems are slower with a less sharp peak. This is caused in part by the low gradient of the meandering rivers but also because of the ability for water retention in back swamps. Storm waters are intercepted by the sponge-like peat soil in the swamps, which have extremely low hydraulic conductivities, and released gradually to the stream channels.

Within a low-gradient meandering river, the influences of flood disturbances are different between natural levee belts and back swamps. In natural levee areas that include meander scrolls, the concave side of bends is continuously eroded by the concentration of stream flows and the occurrence of secondary flows. Trees established on the meander scroll areas, therefore, gradually collapse. The convex sides of the bends are depositional zones, which develop point bars available for seedling establishment. In the back swamps behind the levees, flooding that destroys trees is unlikely to occur. Most of the flooding is not accompanied by sediment movement in the back swamps. Instead of physical damage by erosion forces, flooding in the back swamps has a physiological impact caused by prolonged inundation. The dominant substrate in the back swamps is not sandy soil, as in point bar areas, but leaf litter from reeds and sedges or peat soil. Thus, because of the unique hydrology and substrate properties, the regeneration process in back swamps is expected to be very different from that on natural levees.

DISTURBANCE, RELIABILITY OF REGENERATION HABITAT, AND LIFE HISTORY OF DOMINANT TREE SPECIES

To evaluate the reliability of regeneration habitat, this section characterizes the occurrence of regeneration sites in terms of temporal frequency and spatial abundance and regularity. The reliability of regeneration habitat in time is evaluated as high when regeneration sites are formed regularly. We assumed that regularity of disturbances was closely related to their recurrence intervals (Fig. 1). For example, deep-seated landslides rarely occur and vary widely in their recurrence intervals. Alternatively, snowmelt flood in braided rivers recurs more frequently and is seasonal. Lower frequency represents less predictability, and vice versa. The reliability of regeneration habitat in space is high where regeneration sites are abundant and distributed regularly over the space. The evaluation is arbitrary, based on comparison among different disturbance regimes.

Headwater Streams

In Japan, landslides occur mostly in 0-order streams or on steep slopes (> 30°) underlain by granite or tertiary geology. Mass movement can create a large-scale exposure of sediment as a potential regeneration site. However, because it is highly episodic and has a wide range of recurrence intervals (from a few decades to several thousand years), a long life span is one of the most important strategies. We hereafter refer to species that are particularly adapted to mass movement as mass movement adapted (MMA) species.

One MMA species comprising a cool temperate deciduous forest in Japan is *Cercidiphyllum japonicum* Sieb. & Zucc. ex J. Hoffman & H. Schult. This long-lived species has a very distinctive stand architecture, forming an enormous canopy with numerous shoots (Fig. 2). Canopy trees are distributed sporadically on landslide deposits and lower parts of V-shaped valleys. Abundant small seeds (3 mm long) are produced every year without a masting year and are dispersed over a large area by wind (Kubo *et al.*, 2000; Sakio *et al.*, 2002). However, very few juvenile trees are found, suggesting low reliability of regeneration habitat for *C. japonicum*. The small seeds are incapable of germinating under leaf litter and are also susceptible to desiccation in sites that are too large and open and in sandy soils with variable moisture contents (Seiwa and Kikuzawa, 1996; Kubo *et al.*, 2000). While small seeds can easily attach themselves to steep slopes and logs and germinate on the exposed fine mineral soil without leaf litter, heavy rainfall and sediment erosion may wash away or bury the small seeds and seedlings on such unstable substrates. Thus, given these severe microsite constraints on seedling recruitment, the abundance of regeneration sites suitable for *C. japonicum* is limited and randomly

FIG. 2 A mature stand of *Cercidiphyllum japonicum* forming multiple trunks.

distributed (Table 1). The occurrence of regeneration sites in time is also unreliable because of episodic disturbances in headwater streams. The annually constant seed production deals with low reliability of regeneration habitat in time, and fecundity and dispersal capability increase the chances of arriving at spatially unreliable regeneration habitat for *C. japonicum*. Moreover, clonal sprouting is a crucial strategy for sustaining a long life span. An individual tree of *C. japonicum* forms numerous basal sprouts around the main trunk and replaces the main trunk as it dies with surrounding shoots. This vegetative growth consequently maintains canopy trees for 500 to 1000 years. While sprouting appears to be encouraged by substrate instability (Kubo *et al.*, 2001), its ultimate result is longevity where reliability of regeneration habitat for *C. japonicum* is extremely low both in time and space.

Pterocarya rhoifolia Sieb. & Zucc. is another MMA species that synchronously regenerates with the recurrence of mass movements (Sakio *et al.*, 2002). Mature stands establish on large-scale debris flows and landslide deposits but can also be found in a wide range of landforms, from floodplains to the adjacent terraces and slopes (Sato, 1992; Ann and Oshima, 1996; Sakio *et al.*, 2002). Although vegetative growth and adventitious root formation have been reported (Kaneko, 1995), seedlings are the primary means of regeneration (Kaneko and Kawano, 2002). Seeds of *P. rhoifolia* are carried by wind and quickly colonize fresh exposed surfaces to form an even-aged stand (Oshima *et al.*, 1990; Sato, 1992; Kaneko and Kawano, 2002; Sakio *et al.*, 2002). As seen with *C. japonicum*, regeneration of *P. rhoifolia* relies on a low frequency of mass movement. However, its seeds are larger than those of *C. japonicum* and are less constrained in substrate requirements for germination and seedling establishment. Therefore, regeneration habitats suitable for *P. rhoifolia* should be much less limited in their spatial abundance and regularity, implying higher reliability in space. Larger seeds are more tolerant of leaf litter accumulation, thereby allowing seedling establishment on flat surfaces where the ground is less disturbed. Where young *P. rhoifolia* and *C. japonicum* co-occur, *P. rhoifolia* grows faster to reach the canopy (Sakio *et al.*, 2002). The life span of *P. rhoifolia* is around 100 years and appears to be timed to disturbance frequencies ranging from a few decades to 100 years. Without a disturbance within this time scale, stands of *P. rhoifolia* are eventually replaced by species having longer life spans, such as *Fraxinus platypoda* Oliver, *C. japonicum,* and *Aesculus turbinata* Blume (Sato, 1995; Sakio *et al.*, 2002).

Euptelea polyandra Sieb. & Zucc. can also be classified as a MMA species that establishes even-aged stands on sandy gravel deposits in landslide sites and talus cones (Sakai *et al.*, 1995). It has small seeds, although larger than those of *C. japonicum*, that are dispersed by wind over long distances (Sakio, 2002). While *C. japonicum* and *P. rhoifolia* are adapted to intensive, infrequent mass movements such as a deep-seated landslide, *E. polyandra* regeneration can cope with more frequent and less intensive disturbances, such as a shallow landslide. The occurrence of regeneration sites for *E. polyandra* is qualified as moderate in time and space, with its abundant specific locations (landslide deposits and talus cones). However, talus cones and steep slopes, in which mature stands typically occur, are subject to frequent ground disturbance (Sakai *et al.*, 1995). *E. polyandra* forms multiple shoots to cope with erosion, sediment burial, and soil creep on the unstable slopes. The species starts sprouting at relatively early stages of its life history and constantly forms dormant buds to respond quickly to the frequent soil disturbance. Sprouting certainly has an important role in repairing damage by disturbances. Roots also are well adapted to anchor on steep slopes, protecting the trees from uprooting. These structural adaptations consequently contribute to sustaining a life span while regeneration habitat is not available.

TABLE 1 Summary of Disturbance Regimes and Tree Responses in Three Geomorphic Types

			Regeneration Habitat					
				Time	Space		Life History Strategy	
Geomorphic Classification	Disturbance Regime	Dominant Tree Species	Landform and Microsite Requirement	Frequency	Abundance	Regularity	Reliability in Time	Reliability in Space
Headwater streams (0- to 3rd-order streams)								
		Cercidiphyllum japonicum	Landslide and debris flow deposits, fine mineral soil	Low	Low	Low	Sprouting, longevity, constant seed production	Abundant seed production, high dispersal capability
	Landslide, debris flow, earth flow	*Pterocarya rhoifolia*	Landslide, debris flow deposit	Low to moderate	High	Moderate	Synchronous reproduction, fast growth	Wind dispersal, large seed
		Euptelea polyandra	Talus cone, shallow landslide	Moderate	Moderate	Moderate	Sprouting, dormant bud	Wind dispersal
	Tree-fall/ windthrow gap	*Fraxinus platypoda*	Abandoned channel, landslide and debris flow deposit, canopy gap	Low to moderate	High and low	Moderate and low	Seedling bank	Generalist, wind dispersal, large seed
		Aesculus turbinata	Canopy gap, terrace and toe slope	Low	Low	Low	Seedling bank	Large seed, secondary dispersal (animal dispersal)
Braided rivers on alluvial fan (4th- to 5th-order streams)								
	Large flood	*Ulmus davidiana* var. *japonica*	Upper floodplain, sand and gravel deposit	Moderate	High	High	Synchronous reproduction, longevity, seed dormancy	Wind dispersal
		Fraxinus mandshurica var. *japonica*	Upper floodplain, sand and gravel deposit	Moderate	High	High	Synchronous reproduction, longevity, seed dormancy	Wind dispersal

	Small flood	Salicaceae	Point bar, sand and gravel deposit	High	Low	Low	Constant seed production, timed seed dispersal, bending stability, sprouting, adventitious rooting, fast growth, early maturation	High dispersal capability (water and wind), abundant seed production
Low-gradient meandering rivers (6th-order streams)								
	Flood	Salicaceae	Natural levee, sand and gravel deposit	High	High	High	Constant seed production, timed seed dispersal, bending stability, sprouting, adventitious rooting, fast growth, early maturation	High dispersal capability (water and wind), abundant seed production
	Waterlogged	*Alnus japonica*	Backwater swamp (tussok)	Low	Low	Low	Sprouting, enhanced stem growth, hypertrophied lenticel, adventitious rooting, nitrogen fixation	Water and wind dispersal
		Fraxinus mandshurica var. *japonica*	Backwater swamp (tussok)	Low	Low	Low	Sprouting, enhanced stem growth, hypertr ophied lenticel, adventitious rooting	Water and wind dispersal

Some species in headwater streams may depend on gap formation (treefall and windthrow). Although gap formation can be more frequent than mass movement in a forest (Fig. 1), its abundance is low with irregular distribution in space (Table 1). Canopy gaps are low in light intensity, requiring shade tolerance of species in their seed and seedling stages. High moisture conditions of forest soil and thick leaf litter in the gaps make seeds and seedlings vulnerable to disease and herbivory (Hoshizaki *et al.*, 1997). We refer to species that thrive in these conditions as gap phase (GP) species.

Aesculus turbinata is the most typical GP species in headwater riparian zones in Japan. This species preferentially occurs on stable surfaces of floodplain terraces and their adjacent slopes (Oshima *et al.*, 1990; Sato, 1995; Kaneko *et al.*, 1999). A prominent reproductive characteristic of *A. turbinata* is its large seed size (6.2 g dry weight) (Hoshizaki *et al.*, 1997). The heavy seeds fall under parent crowns and are secondarily dispersed by rodents. This secondary seed dispersal plays a vital role in avoiding light-limited seedling mortality under parent trees, as well as potentially increasing the chance of reaching a regeneration site (Hoshizaki *et al.*, 1999). A large seed reserve enables the seedlings to grow above the leaf litter, which is crucial for successful seedling establishment (Sato, 1995; Hoshizaki *et al.*, 1997; Kubo *et al.*, 2000; Nakashizuka, 2001). *A. turbinata* seedlings also are shade tolerant, growing in low light conditions (Ishida and Peters, 1998).

The last species representing headwater riparian zones is *Fraxinus platypoda*. The most prominent feature of this species is its generalist strategy in regeneration site requirements, exhibiting a characteristic of both MMA and GP species (Table 1). Similar to *E. polyandra*, a seed of *F. platypoda* is relatively large (30 mm in length) and capable of germinating on various substrates, ranging from sandy pebbles and logs to leaf litter (Kubo *et al.*, 2000). Seedling regeneration forming even-aged stands often occurs on sandy gravel deposits in abandoned channels that are protected from frequent flooding (Sakio, 1997). *F. platypoda* can live around 180 years and is adapted to large-scale disturbances, such as an earthquake-induced mass movement. Among the MMA species considered in this study, *F. platypoda* is the most shade tolerant and can form a seedling bank under a closed canopy. Seedlings grow into canopy trees when the forest canopy opens. *F. platypoda* shows strong dominance in headwater streams in central Japan, which may be a result of its generalist strategy between GP and MMA (Sakio, 1997, Sakio *et al.*, 2002).

Although we found many riparian forest studies of headwater streams in Japan, no comparable examples were found for other regions. Gom and Rood (1999) compared reproductive strategies of two *Populus* sections and their hybrid between headwaters and the lower Great Prairie regions in Alberta, Canada. Clonal sprouting was encouraged in the headwater reach, partly by erosion, sedimentation, ice scouring, and flow forces during high flows. By contrast, seedling reproduction appeared to be more common in meandering and braided rivers in the lower basins where frequent flood disturbances promote seedling regeneration. Flood disturbances in the headwater site seemingly do not eliminate the canopy trees, and, thus, regeneration sites rarely occur. As inferred for *C. japonicum*, Gom and Rood concluded that clonal growth was important for the maintenance of cottonwood forests where regeneration habitat was unreliable both in time and space.

In summary, the species that are common in Japanese headwater riparian zones can be largely grouped as MMA or GP species. High dispersal potential by wind or animals is a strategy adapted to increase the chances of arriving at regeneration habitat (Table 1). Generally, life spans are long, ranging from 100 to 500 years and reflecting the long interval between disturbances in headwater systems. The life span of *P. rhoifolia* appears to be synchronized with disturbance frequency, while this assumption does not appear to be applicable to *C. japonicum*. *C. japonicum* can reach the canopy only in sites that remain undisturbed beyond the life span of *P. rhoifolia*. Seed size is associated with microsite requirements (soil moisture, light availability, and substrate texture and stability). The ultimate function of sprouting is to sustain longevity on unstable substrates. For GP species, shade tolerance and the seedling bank strategy are vital. A facultative strategy between GP and MMA may have a role in adapting species to low reliability of regeneration sites in time and space.

Braided Rivers on Alluvial Fans

One disturbance typical of braided rivers is large-scale flooding, which can extend across the floodplains and form various geomorphic surfaces that promote riparian forest regeneration. Compared to headwater streams, flood disturbance frequently occurs more regularly in braided rivers because of a combination of seasonal snowmelt and more erratic Asian monsoon and typhoon high flows. It is important for species to increase their reproductive potential by timing periods of recruitment and maturation to the periodic floods. Because of the periodic creation of recruitment habitats, seedling regeneration with a short life span should be favored in braided rivers.

Depending on the elevation within a floodplain, the factors influencing reliability of regeneration habitat are different (Table 1). At high elevations reachable by relatively large floods, regeneration sites occur moderately often. Near active channels, the frequency of regeneration site occurrence is much higher because of shorter flood intervals. However, seedlings established in the active channel may be washed away by frequent and extensive channel migrations, thereby limiting the abundance and location of regeneration habitat. Therefore, seedling establishment in lower floodplains requires morphological tolerance to high flows; it is not useful to have strategies to escape in time, such as seedling bank formation, vegetative growth to reach maturation, or a long life span. Species adapted to lower floodplain environments may extend their roots deep into the sediment to protect them from physical damage or from being flushed, or may bend themselves with flexible stems. They can also form adventitious roots and shoots after being buried by sediment.

These life history strategies can be seen in most species in riparian zones of braided reaches in Japan. The Salicaceae is dominant in braided rivers in Japan and comprises four genera and 17 species (Niiyama, 2002). Each species segregates its habitat along a stream with some overlapping. Fig. 3 shows a Salicaceae-dominant forest on the low floodplain of a braided river in an alpine region of Japan, where frequent channel migration and multiple channel development occur. Salicaceae trees reach reproductive maturity quickly, within a period that coincides with flood recurrence in the order of years to a decade (from 7.8 to 13.9 years on average; Shin and Nakamura, unpublished data). Their extremely

FIG. 3 Riparian forests dominated by *Chosenia arbutifolia* in a braided river, Azusa River.

light seeds (38–600 mg for 1000 seeds) are produced prolifically every year and are easily dispersed over a long distance by wind (Satoh, 1955; Niiyama, 2002). Seed germination requires an exposed surface, with mineral soil consisting of sandy pebbles, which is formed extensively by flood disturbances. Seedling establishment requires abundant light and high soil moisture content; thus, leaf litter under forest canopies and herbaceous layers does not provide suitable regeneration habitat. Seeds are released during the period of the descending limb in a snowmelt flood, between April and September, with slightly different timing among the Salicaceae species (Niiyama, 1990; Nagasaka, 1996; Fig. 4). *Salix rorida* Lacksch. begins seed dispersal the earliest (mid-May) and *S. subfragilis* Anders. the latest (June to early July). These differences in dispersal times by genus are rather distinct. Seed dispersal of *Chosenia arbutifolia* (Pall.) A. Skv. and *Populus maximowiczii* Henry occurs from late June to early July, which is later than those of Salix spp.; that of *Toisusu urbaniana* (Seem.) Kimura is even later, from August to September. Species with seed dispersal in the early periods of a snowmelt flood may take advantage of less competition, but may also be flushed away during small

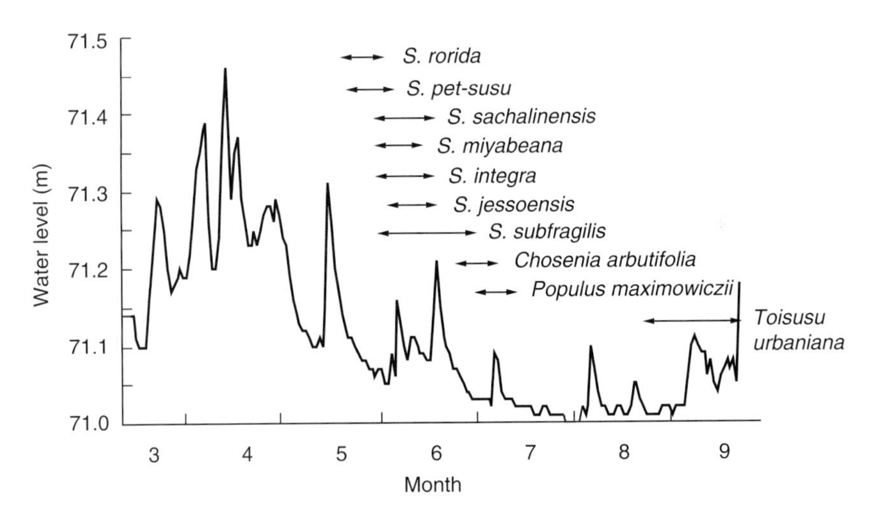

FIG. 4 Variation of water levels and seed dispersal periods of Salicaceae species in Hokkaido The water levels were measured at the Bannosawa gauging station in Bibai River, one of the tributaries of the Ishikari River in 1993. Original data on seed dispersal periods are from Niiyama (1990) and Nagasaka (1996), who monitored seed dispersal periods along the tributaries of the Ishikari River. (Adapted from Nakamura and Shin [2001]. Copyright 2001 American Geophysical Union. Modified by permission of American Geophysical Union.)

flood peaks following the major flooding (Fig. 4). In contrast, the seeds dispersed in the later periods may be protected from high flows but are subject to higher competition from early colonizers. Thus, there is a trade-off in the seed dispersal strategy associated with reliability of regeneration habitat.

The Salicaceae has extremely short seed longevity, and thus the period of seed dispersal is critical (Karrenberg *et al.*, 2002); it is known that spring floods promote its recruitment in Europe and North America (Fenner *et al.*, 1985; Rood and Mahoney, 1990; Nilsson *et al.*, 1991). For other species, Pettit and Froend (2001) reported seed dispersal adapting to flow dynamics in western Australia. They compared the reproductive characteristics of *Eucalyptus* and *Melaleuca* species in two rivers with contrasting flow dynamics (the Ord River and Blackwood River). Seed fall occurred in the periods of receding high flows in both rivers for both species. This timed seed fall can increase the chance of reaching regeneration habitat in time, as commonly seen in braided and meandering rivers.

However, a recent study in a regulated braided river, the Platte River in Colorado, demonstrated that small spikes of flows after cottonwood (*Populus*) germination greatly impacted their survival within a year (Johnson, 2000). Although seedling recruitment could occur extensively on alluvial surfaces after major floods, very few trees survived for one year. Regeneration sites occurred seasonally but with irregular and sparse distribution because of frequent high flows; these results are in agreement with the characteristics of regeneration habitat in low floodplains of Japanese braided rivers (Table 1).

On upper floodplains in the same braided rivers, regeneration habitats may be abundant but their occurrence is infrequent because of high elevations (Table 1). Dominant species are relatively shade tolerant, such as *Ulmus davidiana* var. *japonica* (Sarg. ex Rehder) Nakai and *Fraxinus mandshurica* Rupr. var. *japonica* Maxim. Seedlings of these species are less light demanding and can establish under canopy trees. However, in most cases, the opportunities for germination under canopy trees are limited because of shading, disease, herbivory, and, in some cases, desiccation above the leaf litter. Successful seedling establishment is more likely to occur on exposed mineral soil, thus relying on flood disturbances that can eliminate the forest canopy or, at least, the forest floor vegetation. As with the Salicaceae, time to reproductive maturity of these species seems to coincide with the longer-interval flood frequencies of the upper floodplains [e.g., 47 years for *F. mandshurica* var. *japonica* (Shin and Nakamura, unpublished data); Fig. 1]. Their seeds are heavier than those of the

Salicaceae and dispersed by wind to a limited distance from the parent trees. These species grow more slowly but live longer than the Salicaceae; for example, *U. davidiana* var. *japonica* can live for about 200 to 300 years (Kon and Okitsu, 1999).

Given less frequent recruitment opportunity on upper floodplains, a strategy to thrive within canopy stands may be important to some extent. It has been shown that *U. davidiana* var. *japonica* exhibits delayed germination (Seiwa, 1997). Its seeds are dispersed in summer and germinate in the same growing season. However, some of the seeds undergo dormancy during the winter and germinate early the following spring. This delayed germination takes advantage of the better light conditions before the leaves of the deciduous canopy trees are developed. It can also dilute the variation in seed production (mast seeding) and avoid sibling competition (Shibata and Nakashizuka, 1995).

Timed dispersal in the abovementioned Australian study by Pettit and Froend (2001) may be considered as another example of reducing seedling mortality under parent trees. Seasonal high flows in the Blackwood River were less intensive, and seed dispersal of *Eucalyptus rudis* Sm. and *Melaleuca rhaphiophylla* Schauer depended upon wind; seed density was a function of the distance from the parent trees. However, some portions of the seeds tended to be retained in the parent trees and released throughout the year. By contrast, wind-dispersed seeds of *E. camaldulensis* Dehn. and *M. leucadendra* (L.) L. in the Ord River, with its erratic and intensive flooding, appeared to be further transported by currents, presumably because of much greater flow forces (about 30 times greater average maximum discharge than the Blackwood River). Seed dispersal by river flows is less efficient in the Blackwood River, and, therefore, "delayed dispersal" by retention in the parent trees can contribute to avoidance of sibling competition and seedling mortality under canopy trees. This delayed dispersal may be an adaptation to temporally unreliable regeneration sites near parent trees, which is also assumed for "delayed germination" by *Ulmus davidiana* var. *japonica*.

In summary, both the Salicaceae and shade-tolerant species depend upon flood disturbances for their seedling establishment. Both groups exhibit a strategy to increase their chances of successful regeneration by synchronizing their maturation time to flood frequencies. Life span is longer for the shade-tolerant species than the Salicaceae, reflecting the difference in flood frequencies between the upper and lower floodplains. Furthermore, seed dispersal of the Salicaceae is timed to seasonal snowmelt flooding to increase their reproductive potential.

Low-Gradient Meandering Rivers

Flood disturbance in low-gradient meandering rivers is generally less destructive and more predictable in its spatial patterns. Intensive flood disturbances that cause large-scale landform alteration rarely occur. As described previously, two distinctive types of riparian habitats are formed: natural levees and backwater swamps. On natural levees, seedlings on the concave parts of a meander reach are subject to erosion and eventually being washed away. On the contrary, the convex sides develop point-bars and become regeneration sites because severe erosion is unlikely to occur there. Thus, the location of a regeneration site is predictable in natural levees, and abundance of regeneration sites is moderate. Overall, reliability of regeneration habitat in space is high in natural levees (Table 1). Seasonal high flows, which occur by snowmelt and monsoon rainfall, increase the temporal opportunity for successful regeneration in natural levees.

In contrast, backwater swamps behind the natural levees are less likely to support reliable regeneration habitat because of long-term inundation. Swamp habitats are characterized by high water tables synchronizing with adjacent river stages and by fine sediment deposition caused by overbank flooding. Vegetation in the backwater swamps, which typically is a cover of herbaceous species, such as sedges and reeds, is adapted to anoxic conditions. Few open sites are available for tree seedling establishment. Intensive high flows that can eliminate the herbaceous layer are rare. Accordingly, the occurrence of regeneration sites in backwater swamps is extremely rare in both time and space (Table 1). As for MMA species in headwater riparian forests, high dispersal ability and long life spans might be important.

Dominant species on natural levees are members of the Salicaceae, in particular, *Salix pet-susu* Kimura and *S. subfragilis* in cool temperate zones in Japan (Nakamura *et al.*, 2002). Seed dispersal of these two species is timed to water receding periods after snowmelt high flows, by which seedlings successfully establish on point bars. Abundant seeds are dispersed by stream flow and wind. At the margins between point bars and the stream current, water depth and velocity are greatly reduced, allowing the floating seeds to colonize on the point bars. Therefore, the Salicaceae seedlings are distributed in a belt on point bars along an active channel. This spatial pattern of cottonwood stands associated with lateral point-bar formation has also been described for meandering rivers in the United States (Everitt, 1968; Auble and Scott, 1998; Robertson and Augspurger, 1999; Merritt and Cooper, 2000). As periodical and moderate flooding grow convex point-bars, cottonwood seedlings become established, forming a predictable stand structure with increasing age with distance from the active channel.

One representative species in cool temperate backwater habitats in Japan is *Alnus japonica* (Thunb.) Steud. This species often grows in a dwarfed form with numerous shoots in sedge-reed communities (Fig. 5). The aboveground plant consists of a single shoot in the first generation that dies in a few decades. The remaining trunk again sprouts numerous shoots (Yamamoto, 2002) and continues increasing the number of shoots outward, forming an alder thicket in a dwarfed form (Yabe and Onimaru, 1997). Recent studies suggested that anoxic conditions and/or nutrient-poor conditions caused this vegetative growth (Yamamoto, 2002; Nakamura *et al.*, 2002). In such a stressful environment, plants could survive by keeping their aboveground biomass at a low level with a high turnover of repeating asexual regeneration (Yamamoto, 2002). This vegetative growth also complements poor seedling regeneration in backwater habitats where regeneration habitat is highly unreliable in both space and time. Although seed dispersal of *A. japonica* occurs by both wind and water, seedling establishment is difficult because of shading and constant inundation within a sedge-reed community. In deepwater swamps in the southeastern United States, viable bald cypress [*Taxodium distichum* (L.) L. C. Rich.] seeds are transported by water potentially for their buoyancy periods up to nearly 80 days and trapped by emergent substrates (Schneider and Sharitz, 1988). Seed germination requires occasional drawdowns, and otherwise floating

FIG. 5 Dwarfed form of *Alnus japonica* growing in a sedge-reed community.

vegetation mats can be important regeneration sites (Huffman and Lonard, 1983). Likewise, it has been proposed that floating peat mats provide regeneration sites for *A. japonica* (Yabe and Onimaru, 1997). However, successful seedling establishment on these mats has rarely been observed; thus, little is known about the regeneration mechanism of *A. japonica*. The species exhibits various structural adaptations to anoxic conditions of backwater swamps, including enhanced stem growth, development of hypertrophied lenticels, and adventitious root formation on submerged stems (Yamamoto *et al.*, 1995a, b). Hypertrophied lenticels and adventitious roots belowground can compensate for reduced oxygen supply and increase the efficiency of water absorption (Yamamoto *et al.*, 1995b). The ability for nitrogen fixation has also been known for this species.

Fraxinus mandshurica var. *japonica* often occurs with *A. japonica* in backwater swamps in cool temperate forests in Japan. Although the species exhibits similar traits, such as adventitious root formation and vegetative growth to overcome anoxic conditions (Yamamoto *et al.*, 1995a), and probably is adapted to unreliable regeneration habitat, the extent of its functional ability seems to be limited. For example, while *A. japonica* can continuously develop adventitious roots from spring through early fall, *F. mandshurica* var. *japonica* stops forming them in summer (Yamamoto, 2002). In addition, the larger seeds of *F. mandshurica* var. *japonica* may restrict dispersal potential by wind and water. Limited structural adaptation and ineffective seed dispersal may confine *F. mandshurica* var. *japonica* to being subdominant to *A. japonica* in backwater swamps.

In summary, the reliability of regeneration habitat is distinctly different between natural levees and backwater swamps. In natural levee habitats, regeneration sites (point-bars) are formed constantly in a predictable manner, in both time and space. The Salicaceae species dominate natural levee habitats using their strategies of fast growth, timed reproduction, and seed dispersal that utilize the directional flood disturbance in meandering alluvial rivers, as well as their structural adaptation to tolerate the high flows. *A. japonica* thrives in backwater swamps where sedge-reed communities are dominant. Structural adaptation and vegetative growth, particularly in a dwarfed form, enable the species to persist in waterlogged and oligotrophic wetlands where regeneration habitat reliability is extremely low, in both time and space.

CONCLUSION

The riparian forests in the Asia monsoon belt of Japan are subject to a variety of geomorphic and fluvial disturbances that can vary longitudinally at a catchment scale because of combined seasonal and extreme floods caused by snowmelt, heavy rainfalls initiated by monsoons and typhoons, and the high-gradient topography and extensive tectonic area. The temporal and spatial availability of regeneration sites can also vary depending upon the disturbance regime. Dominant species thus exhibit various life history strategies to cope with the temporal and spatial reliability of regeneration habitat, which is determined by disturbance regimes in a dynamic state. The specific strategies for each life history stage that were identified in this study are listed in Table 2. These strategies are found in three geomorphic types of habitat that are distinguished by their disturbance regime (Table 1). For all types of riparian systems, longevity and wind dispersal are the prominent strategies to cope with regeneration habitat that is unreliable in time and space. In habitats where reliability of regeneration habitat is extremely low in both time and space, vegetative growth is a vital strategy to complement seedling regeneration. Structural adaptation is common for

TABLE 2 Summary of Life History Strategies of Dominant Riparian Tree Species, Developed at Different Life Stages in Disturbance-Prone Japanese Rivers

Life History Stage	Strategies
Seed dispersal	Dispersal by wind over an extensive area Secondary dispersal by animals Seed provision on point bars by water Synchronization between the timings of seed dispersal and snowmelt flood
Germination, seedling, and sapling	Large seed Seed dormancy Fast growth Seedling/sapling bank Adventitious roots Bending ability Enhanced stem growth Hypertrophied lenticels
Reproductive mature	Synchronization between maturation time and disturbance frequency Longevity Sprouting Enhanced stem growth Hypertrophied lenticels Adventitious roots

most species. For some species, the periods of maturation and seed dispersal coincide with disturbance frequencies and timing to increase their reproductive potential both in time and space.

As predicted by Grubb's (1977) concept of regeneration niche, species colonize riparian zones by adapting to the reliability of regeneration habitat with different strategies at each developmental life stage. The variety of disturbance regimes promotes a diverse array of life history strategies that differ among species and life stages, thus contributing to species coexistence in riparian forests (Nakashizuka, 2001).

Focusing on life history strategies and the likelihood of regeneration habitat occurrence, we were able to describe the relationships between a number of species and disturbance regimes. While we included various disturbance regimes at a catchment scale, our review heavily relied upon examples of Japanese rivers. Some of these examples may be comparable to studies in other regions. However, much of the riparian studies in other regions, particularly in North America, lack variation in tree species and are limited to low-gradient alluvial rivers. Thus, roles of mass-movement–type disturbances have been poorly studied. Further efforts in these gaps are required to develop a generality of species–disturbance relationships in riparian zones.

ACKNOWLEDGMENTS

We wish to thank Drs. Hitoshi Sakio, Mahito Kamada, Tohru Nakashizuka, Gregor Auble, Edward Johnson, and Kiyoko Miyanishi for their critical and helpful comments on earlier drafts of this manuscript. This research was supported by Grants in Aid for Scientific Research (No. 13460061 and No. 14506039) from the Ministry of Education, Science and Culture, Japan, and a grant from the Water Resources Environment Technology Center.

REFERENCES

Ann, S. W., and Oshima, Y. (1996). Structure and regeneration of *Fraxinus spaethiana-Pterocarya rhoifolia* forests in unstable valleys in the Chichibu Mountains, central Japan. *Ecol Res* 11, 363–370.

Aruga, M., Nakamura, F., Kikuchi, S., and Yajima, T. (1996). Characteristics of floodplain forests and their site conditions in comparison to toeslope forests in the Tokachi River. *J Jpn Forest Soc* 78, 354–362 (in Japanese with English summary).

Auble, G. T., and Scott, M. L. (1998). Fluvial disturbance patches and cottonwood recruitment along the upper Missouri River, Montana. *Wetlands* 18, 546–556.

Auble, G. T., Friedman, J. M., and Scott, M. L. (1994). Relating riparian vegetation to present and future stream flows. *Ecol Appl* 4, 544–554.

Begg, C. S., Archibold, O. W., and Delanoy, L. (1998). Preliminary investigation into the effects of water-level control on seedling recruitment in riparian cottonwoods, *Populus deltoides*, on the South Saskatchewan River. *Can Field Nat* 112, 684–693.

Clark, J. S. (1991). Disturbance and tree life history on the shifting mosaic landscape. *Ecology* 72, 1102–1118.

Cordes, L. D., Hughes, F. M. R., and Getty, M. (1997). Factors affecting the regeneration and distribution of riparian woodlands along a northern prairie river: Red Deer River, Alberta, Canada. *J Biogeogr* 24, 675–695.

Dietrich, W., and Dunne, T. (1978). Sediment budget for a small catchment in mountainous terrain. *Z Geomorph N F* 29, 191–206.

Dollar, E. S. J. (2000). Fluvial geomorphology. *Prog Phys Geog* 24, 385–406.

Dykaar, B. B., and Wingington, Jr., P. J. (2000). Floodplain formation and cottonwood colonization patterns on the Willamette River, Oregon, USA. *Environ Manage* 25, 87–104.

Everitt, B. L. (1968). Use of the cottonwood in an investigation of recent history of a flood plain. *Am J Sci* 266, 417–439.

Fenner, P., Brady, W. W., and Patton, D. R. (1985). Effects of regulated water flows on regeneration of Fremont cottonwood. *J Range Manage* 38, 135–138.

Friedman, J. M., and Lee, V. J. (2002). Extreme floods, channel changes, and riparian forest along ephemeral streams. *Ecol Monogr* 72, 409–425.

Gom, L. A., and Rood, S. B. (1999). Patterns of clonal occurrence in a mature cottonwood grove along the Oldman River, Alberta. *Can J Bot* 77, 1095–1105.

Grubb, P. J. (1977). The maintenance of species-richness in plant communities: the importance of the regeneration niche. *Biol Rev* 52, 107–145.

Hatano, S. (1974). Landslide geomorphology (No.2). *Tsuchi-To-Kiso* 22, 85–93 (in Japanese).

Hoshizaki, K., Suzuki, W., and Nakashizuka, T. (1999). Evaluation of secondary dispersal in a large-seeded tree *Aesculus turbinata*: a test of directed dispersal. *Plant Ecol* 144, 167–176.

Hoshizaki, K., Suzuki, W., and Sasaki, S. (1997). Impacts of secondary seed dispersal and herbivory on seedling survival in *Aesculus turbinata*. *J Veg Sci* 8, 735–742.

Huffman, R. T., and Lonard, R. E. (1983). Successional patterns on floating vegetation mats in a southwestern Arkansas bald cypress swamp. *Castanea* 48, 73–78.

Hughes, F. M. R. (1990). The influence of flooding regimes on forest distribution and composition in the Tana River floodplain, Kenya. *J Appl Ecol* 27, 475–491.

Ishida, M., and Peters, R. (1998). Effects of potential PAR on shoot extension in juveniles of the main tree species in a Japanese temperate forest. *Ecol Res* 13, 171–182.

Ishikawa, S. (1988). Floodplain vegetation of the Ibi River in central Japan. Distribution behaviour and habitat conditions of the main species of the river bed vegetation developing on the alluvial fan. *Jpn J Ecol* 38, 73–84 (in Japanese with English summary).

Johnson, W. C. (1998). Adjustment of riparian vegetation to river regulation in the Great Plains, USA. *Wetlands* 18, 608–618.

Johnson, W. C. (2000). Tree recruitment and survival in rivers: influence of hydrological processes. *Hydrol Process* 14, 3051–3074.

Junk, W. J., Bayley, P. B., and Sparks, R. E. (1989). The flood pulse concept in river-floodplain systems. *Can Spec Publ Fisheries Aquatic Sci* 106, 110–127.

Kaneko, Y. (1995). Disturbance regimes of a mountainous riparian forest and effects of disturbance on tree population dynamics. *Jpn J Ecol* 45, 311–316 (in Japanese with English summary).

Kaneko, Y., and Kawano, S. (2002). Demography and matrix analysis on a natural *Pterocarya rhoifolia* population developed along a mountain stream. *J Plant Res* 115, 341–354.

Kaneko, Y., Takada, T., and Kawano, S. (1999). Population biology of *Aesculus turbinata* Blume: a demographic analysis using transition matrices on a natural population along a riparian environmental gradient. *Plant Species Biol* 14, 47–68.

Karrenberg, S. P., Edwards, P. J., and Kollman, J. (2002). The life history of Salicaceae living in the active zone of floodplains. *Freshwater Biol* 47, 733–748.

Kon, H., and Okitsu, S. (1999). Role of land-surface disturbance in regeneration of *Ulmus davidiana* var. *japonica* in a cool temperate deciduous forest on Mt. Asama, central Japan. *J Jpn Forest Soc* 81, 29–35 (in Japanese with English summary).

Kubo, M., Shimano, K., Sakio, H., and Ohno, K. (2000). Germination sites and establishment conditions of *Cercidiphyllum japonicum* seedlings in the riparian forest. *J Jpn Forest Sci* 82, 349–354 (in Japanese with English summary).

Kubo, M., Shimano, K., Sakio, H., and Ohno, K. (2001). Sprout trait of *Cercidiphyllum japonicum* based on the relationship between topographies and sprout structure. *J Jpn Forest Sci* 83, 271–278 (in Japanese with English summary).

Merritt, D. M., and Cooper, D. J. (2000). Riparian vegetation and channel change in response to river regulation: a comparative study of regulated and unregulated streams in the Green River Basin, USA. *Regul River* 16, 543–564.

Nagasaka, Y. (1996). Salicaceae species growing in riparian zone. *Koshunai-Kihou* 101, 12–17 (in Japanese).

Nakamura, F. (1990). Perspectives for the effects of geomorphological processes on forest ecosystems. *Biol Sci (Tokyo)* 42(2), 57–67 (in Japanese).

Nakamura, F., and Kikuchi, S. (1996). Some methodological developments in the analysis of sediment transport processes using age distribution of floodplain deposits. *Geomorphology* 16, 139–145.

Nakamura, F., and Shin, N. (2001). The downstream effects of dams on the regeneration of riparian tree species in northern Japan. In: *Geomorphic Processes and Riverine Habitat.* (Dorava, J. M., Ed.), pp. 173–181. American Geophysical Union, Washington, DC.

Nakamura, F., and Swanson, F. J. (2003). Dynamics of wood in rivers in the context of ecological disturbance. In: *The Ecology and Management of Wood in World Rivers.* (Gregory, S., Boyer, K. L., and Gurnell, A. M., Eds.), pp. 279–297. American Fisheries Society, Bethesda, MD.

Nakamura, F., Jitsu, M., Kameyama, S., and Mizugaki, S. (2002). Changes in riparian forests in the Kushiro Mire, Japan, associated with stream channelization. *River Res Appl* 18, 65–79.

Nakamura, F., Maita, H., and Araya, T. (1995). Sediment routing analyses based on chronological changes in hillslope and riverbed morphologies. *Earth Surf Proc Land* 20, 333–346.

Nakamura, F., Swanson, F. J., and Wondzell, S. M. (2000). Disturbance regimes of stream and riparian systems—a disturbance-cascade perspective. *Hydrol Process* 14, 2849–2860.

Nakamura, F., Yajima, T., and Kikuchi, S. (1997). Structure and composition of riparian forests with special reference to geomorphic site conditions along the Tokachi River, northern Japan. *Plant Ecol* 133, 209–219.

Nakashizuka, T. (2001). Species coexistence in temperate, mixed deciduous forests. *Trends Ecol Evol* 16, 205–210.

Niiyama, K. (1987). Distribution of Salicaceae species and soil texture of habitats along the Ishikari River. *Jpn J Ecol* 37, 163–174 (in Japanese with English summary).

Niiyama, K. (1989). Distribution of *Chosenia arbutifolia* and soil texture of habitats along the Satsunai River. *Jpn J Ecol* 39, 173–182 (in Japanese with English summary).

Niiyama, K. (1990). The role of seed dispersal and seedling traits in colonization and coexistence of *Salix* species in a seasonally flooded habitat. *Ecol Res* 5, 317–332.

Niiyama, K. (2002). Riparian forests. In: *Ecology of Riparian Forests.* (Sakio, H., and Yamamoto, F., Eds.), pp. 61–93. University of Tokyo Press, Tokyo (in Japanese).

Nilsson, C., Ekblad, A., Gardfjell, M., and Carlberg, B. (1991). Long-term effects of river regulation on river margin vegetation. *J Appl Ecol* 28, 963–987.

Oshima, Y., Yamanaka, N., Tamai, S., and Iwatsubo, G. (1990). A comparison of the distribution properties of two dominant species, *Aesculus turbinata, Pterocarya rhoifolia*, in the natural riparian forest of the Kyoto University forest in Ashiu. *B Kyoto Univ Forest* 62, 14–27 (in Japanese with English summary).

Pettit, N. E., and Froend, R. H. (2001). Variability in flood disturbance and the impact on riparian tree recruitment in two contrasting river systems. *Wetlands Ecol Manage* 9, 13–25.

Robertson, K. M., and Augspurger, C. K. (1999). Geomorphic processes and spatial patterns of primary forest succession on the Bogue Chitto River, USA. *J Ecol* 87, 1052–1063.

Rood, S. B., and Heinze-Milne, S. (1989). Abrupt downstream forest decline following river damming in southern Alberta. *Can J Bot* 67, 1744–1749.

Rood, S. B., and Mahoney, J. M. (1990). Collapse of riparian poplar forests downstream from dams in western prairies: probable causes and prospects for mitigation. *Environ Manage* 14, 451–464.

Sakai, A., Ohsawa, T., and Ohsawa, M. (1995). Adaptive significance of sprouting of *Euptelea polyandra*, a deciduous tree growing on steep slopes with shallow soil. *J Plant Res* 108, 377–386.

Sakio, H. (1997). Effects of natural disturbance on the regeneration of riparian forests in the Chichibu Mountains, central Japan. *Plant Ecol* 132, 181–195.

Sakio, H. (2002). Headwater riparian zone. In: *Ecology of Riparian Forests*. (Sakio, H., and Yamamoto, F., Eds.), pp. 21–60. University of Tokyo Press, Tokyo (in Japanese).

Sakio, H., Kubo, M., Shimano, K., and Ohno, K. (2002). Coexistence of three canopy tree species in a riparian forest in the Chichibu Mountains, central Japan. *Folia Geobotanica* 37, 45–61.

Sato, H. (1992). Regeneration traits of saplings of some species composing *Pterocarya rhoifolia* forest. *Jpn J Ecol* 42, 203–214 (in Japanese with English summary).

Sato, H. (1995). Studies on the dynamics of *Pterocarya rhoifolia* forest in southern Hokkaido. *B Hokkaido Forest Res Inst* 32, 55–96 (in Japanese with English summary).

Satoh, Y. (1955). Seed longevity of *Salix* species. *B Hokkaido Forest Res Inst* 17, 225–266 (in Japanese with English summary).

Schneider, R. L., and Sharitz, R. R. (1988). Hydrochory and regeneration in a bald cypress-water tupelo swamp forest. *Ecology* 69, 1055–1063.

Seiwa, K. (1997) Variable regeneration behaviour of *Ulmus davidiana* var. *japonica* in response to disturbance regime for risk spreading. *Seed Sci Res* 7, 195–207.

Seiwa, K., and Kikuzawa, K. (1996). Importance of seed size for the establishment of seedlings of five deciduous broad-leaved tree species. *Vegetatio* 12, 51–64.

Shibata, M., and Nakashizuka, T. (1995). Seed and seedling demography of four co-occurring *Caprinus* species in a temperate deciduous forest. *Ecology* 76, 1099–1108.

Shimokawa, E. (1984). A natural recovery process of vegetation on landslide scars and landslide periodicity in forested drainage basins. In: *Symposium on Effects of Forest Land Use on Erosion and Slope Stability*. (O'Loughlin, C. L., and Pearce, A. J., Eds.), pp. 99–107. Environment and Policy Institute, East-West Center, University of Hawaii, Honolulu, HI.

Shin, N., and Nakamura, F. (2005) Effects of fluvial geomorphology on riparian tree species in Rekifune River, northern Japan. *Plant Ecology* 178, 15–28.

Strykstra, J. R., Bekker, R. M., and Van Andel, J. (2002). Dispersal and life span spectra in plant communities: a key to safe site dynamics, species coexistence and conservation. *Ecography* 25, 145–160.

Suzuki, W., Osumi, K., Masaki, T., Takahashi, K., Daimaru, H., and Hoshizaki, K. (2002). Disturbance regimes and community structures of a riparian and an adjacent terrace stand in the Kanumazawa Riparian Research Forest, northern Japan. *Forest Ecol Manag* 157, 285–301.

Swanson, F. J., Gregory, S. V., Sedell, J. R., and Campbell, A. G. (1982). Land-water interactions: the riparian zone. In: *Analysis of Coniferous Forest Ecosystems in the Western United States*. (Edmonds, R.L., Ed.), pp. 267–291. Hutchinson Ross Publishing Co., Stroudsburg, PA.

Tsukamoto, Y. (1973). Study on the growth of stream channel. *Shin-Sabo* 87, 4–13 (in Japanese).

Whittaker, R. H. (1956). Vegetation of the Great Smoky Mountains. *Ecol Monogr* 26, 1–80.

Whittaker, R. H. (1965). Dominance and diversity in land plant communities. *Science* 147, 250–260.

Wolman, M. G., and Miller, J. P. (1960). Magnitude and frequency of forces in geomorphic processes. *J Geol* 68, 54–74.

Yabe, K., and Onimaru, K. (1997). Key variables controlling the vegetation of a cool-temperate mire in northern Japan. *J Veg Sci* 8, 29–36.

Yamamoto, F. (2002). Life history strategies of tree species in backwater swamps. In: *Ecology of Riparian Forests*. (Sakio. H., and Yamamoto, F., Eds.), pp. 139–167. University of Tokyo Press, Tokyo (in Japanese).

Yamamoto, F., Sakata, T., and Terazawa, K. (1995a). Growth, morphology, stem anatomy and ethylene production in flooded *Alnus japonica* seedlings. *IAWA J* 16, 47–59.

Yamamoto, F., Sakata, T., and Terazawa, K. (1995b). Physiological, morphological and anatomical responses of *Fraxinus mandshurica* seedlings to flooding. *Tree Physiol* 15, 713–719.

Yanai, S., and Usui, G. (1989). Measurement of the slope failure frequency on sediments with tephrochronological analysis. *Shin-Sabo* 163, 3–10 (in Japanese with English summary).

10

Water Level Changes in Ponds and Lakes: The Hydrological Processes

Masaki Hayashi

Department of Geology and Geophysics, University of Calgary

Garth van der Kamp

Environment Canada, National Hydrology Research Centre

INTRODUCTION

Lakes and ponds occur in a wide range of depths, sizes, and permanence—from deep lakes having a permanent body of surface water to shallow ponds having water for only a few weeks each year. These factors also vary within each lake or pond, resulting in diverse communities of aquatic plants growing in various patterns. Certain types of plants require relatively high water levels, while others cannot tolerate standing water. Therefore, water level change is considered a disturbance to many aquatic plants.

Dynamic changes in water level are controlled by the balance between inputs and outputs of water, which are in turn controlled by the hydrological processes. Many hydrological processes are sensitive to changes in climate. For example, during a prolonged drought, precipitation inputs generally decrease and evaporation outputs increase, resulting in a drawdown of lake level or even a complete drying out. Climate also affects the lake water balance by changing the amount of stream flow and groundwater flow into the lake, but the response of the hydrological processes to climate is complicated because of the complex interactions among climate, vegetation, soil, and groundwater. Such interactions are also strongly affected by land-cover change caused by natural (e.g., fire) or anthropogenic (e.g., agriculture) processes.

This chapter discusses the hydrological processes that control the water level change of surface water and shallow groundwater. The objective is to present essential hydrological principles and practices so that ecologists and hydrologists can engage in meaningful interdisciplinary research. We first present the concept of the water balance, followed by a discussion of individual components of the water balance. We also discuss the effects of climatic fluctuation and landuse change on the water balance by using examples from prairie wetland ecosystems, with particular emphasis on the ecohydrological linkage between water and riparian vegetation.

WATER BALANCE

Water Balance Equation

The water level in a lake is controlled by the balance between input and output:

$$Q_{in} - Q_{out} = A\frac{dh}{dt} \tag{1}$$

where Q_{in} [L^3 T^{-1}] is the sum of all water inputs, Q_{out} [L^3 T^{-1}] is the sum of all outputs, A [L^2] is the surface area of the lake, and dh/dt [L T^{-1}] is the rate of water-level (h) change. Note that A represents the water-covered area, which is normally dependent on h (the higher the water level, the greater the surface area). The water regime of a lake (i.e., how A and h change over time) is determined by the seasonal and interannual variability of $Q_{in} - Q_{out}$. Therefore, understanding the water regime requires some knowledge of the hydrological processes controlling Q_{in} and Q_{out}. The inputs include perennial and intermittent streams, groundwater inflow, direct precipitation onto the lake, diffuse runoff from the shoreline, and snowdrift. The outputs include streams, groundwater outflow, and evaporation and transpiration, commonly called evapotranspiration (Fig. 1). In addition to these natural processes, Q_{in} and Q_{out} may include artificial terms, such as water intake for irrigation or wastewater discharge.

Precipitation

Direct precipitation is an important component of the water balance. Depending on the atmospheric conditions, precipitation occurs as rain, snow, hail, or various other forms. Rainfall is relatively easily measured in principle and, hence, is usually the most accurately measured term in the water balance equation. However, precipitation may have significant spatial variability, even over a relatively small area. Therefore, accurate determination of event-by-event precipitation inputs requires multiple precipitation gauges distributed over the study area. Winter precipitation in cold regions mostly occurs as snow. Snowflakes are easily moved by wind, making it difficult to measure snowfall accurately, even with gauges equipped with windshields (Goodison *et al.*, 1998). Therefore, one should be aware of the uncertainty associated with measured winter precipitation, even though such uncertainty may be much smaller than the uncertainty regarding other terms in the water balance. Many hydrological studies only require periodic measurements of the amount of water equivalent in the snowpack. In these studies, the average snow water equivalent over a relatively large area is estimated from measurements of snow depth and density along representative survey lines.

Part of the precipitation is intercepted by leaves and stems of emergent plants and returned to the atmosphere by evaporation. This process can be important for the water balance of shallow marshes having a sizable area

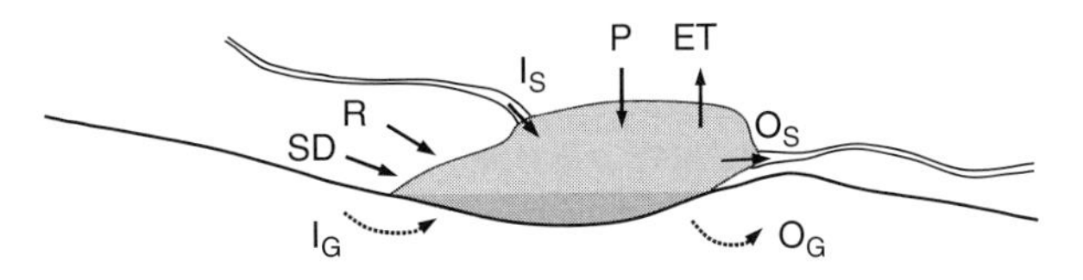

FIG. 1 Schematic diagram showing water-balance components: precipitation (P), evapotranspiration (ET), stream inflow (I_S), stream outflow (O_S), diffuse runoff (R), snow drift (SD), groundwater inflow (I_G), and groundwater outflow (O_G).

covered by plants. The amount of interception is controlled by the leaf area index, the total area of all plant leaves per unit area. Leaf area index is commonly estimated from the comparison of radiation measurements beneath and above the canopy because direct determination of the index is impractically labor intensive (Weiss *et al.*, 2004). Interception is also controlled by the shape and angle of leaves. Flat leaves oriented horizontally can intercept precipitation more effectively than narrow leaves oriented at steep angles (Crockford and Richardson, 2000).

Emergent vegetation also affects snow accumulation. Once deposited on the ground or frozen lake surface, the snowpack typically remains frozen for several days to months, allowing wind-driven snowdrifts to redistribute snow across the landscape (Pomeroy *et al.*, 1998). Tall grasses or shrubs effectively trap drifting snow and serve as a snow "sink." Where vegetation is nearly absent, such as on cultivated fields, snow tends to accumulate in topographic depressions and, upon melting, releases a major water input to upland ponds and wetlands that form in these depressions (van der Kamp *et al.*, 2003).

Evapotranspiration

Water is transferred from ponds and lakes to the atmosphere by direct evaporation from the water surface and transpiration by emergent plants. The two processes are driven by the same meteorological factors and are commonly lumped together as evapotranspiration (ET). The phase change of water from liquid to vapor is an essential part of ET. Therefore, ET can be understood in context of the energy exchange at the lake surface. However, the phase change needs to be accompanied by removal of vapor from the lake or leaf surface by the turbulent mixing of air. Therefore, ET can also be considered an aerodynamic mass-transfer process. Not surprisingly, most of the equations that best model ET over lakes and wetlands contain both energy and aerodynamic terms (Winter *et al.*, 1994; Rosenberry *et al.*, 2004). A comprehensive review of methods to estimate wetland ET is found in Drexler *et al.* (2004).

The primary source of energy for ET is radiation inputs. The wavelength of radiation is dependent on the temperature of the source. Solar radiation has much shorter wavelengths than the radiation emitted by the atmosphere and by the earth's surface. Therefore, solar radiation is commonly referred to as shortwave radiation, and the radiation emitted by lower-temperature surfaces is referred to as longwave radiation. The lake surface receives solar radiation and reflects part of it back. The lake surface also receives and emits longwave radiation. Water is an excellent absorber of shortwave radiation and allows only a small portion of incoming shortwave radiation to be reflected back. The ratio of reflected to incoming shortwave radiation (albedo) is typically about 0.03 to 0.05 for water surfaces. The albedo becomes much higher when the lake surface is partially covered by emergent plants or ice. The amount of longwave radiation emitted by a water surface ($Q_{LW\uparrow}$ [M T^{-3}]) is proportional to the fourth power of the temperature of the surface (T_s [K]):

$$Q_{LW\uparrow} = \varepsilon\sigma T_s^4 \tag{2}$$

where ε is the emissivity of the water surface (typically 0.95 to 0.98) and σ is the Stefan-Boltzmann constant (5.7×10^{-8} J m^{-2} s^{-1} K^{-4}). Therefore, outgoing longwave radiation is strongly dependent on the lake surface temperature. The net radiation (Q_n [M T^{-3}]) received by the surface is obtained by adding all incoming radiation and subtracting all outgoing radiation. Part of Q_n is stored in the water in an amount proportional to the total water volume times the change in average water temperature, while the rest is used to convert liquid water molecules to vapor and to warm the air in contact with the water.

The vapor-holding capacity of air increases with temperature (Lowe, 1977). A parcel of air saturated with water vapor has a relative humidity of 100%. A thin layer (<1 mm) of air immediately above the water is saturated with water vapor, and its temperature is equal to the water-surface temperature (T_s). Therefore, the partial pressure of water vapor (e_s) in this layer is uniquely determined by T_s. The rate of increase in saturation vapor pressure with respect to temperature has a special significance in evaporation and is represented by Δ [mb K^{-1}]. Suppose that the air temperature (T_a) at some distance, say 0.1 m, above the lake surface is lower than T_s, then mixing the warmer air, which is in contact with water, with the colder air above will result in the net upward transfer of energy. This process and the reverse, transfer of energy from warm air to cold water, is called sensible heat transfer, with a positive quantity (Q_h [M T^{-3}]) assigned for upward flux. Suppose now that the air at 0.1 m is relatively dry so that the vapor pressure (e_a) of this air is less than e_s; mixing more humid air, which is in contact with the water, with the less humid air above results in the net upward transfer of vapor. This process is called latent heat transfer because water vapor carries with it the latent heat of vaporization. A positive quantity (Q_e [M T^{-3}]) is assigned for upward flux so that the evaporation rate (E [L T^{-1}]) is directly proportional to Q_e:

$$E = Q_e/(\rho_w L_e) \tag{3}$$

where L_e is the latent heat of vaporization (2.47×10^6 J kg^{-1} at 293 K) and ρ_w is the density of water (1000 kg m^{-3}). A similar argument applies to the sensible and latent heat transfers between a leaf surface and ambient air. Since turbulent mixing is driven by the motion of air, it is reasonable to expect that Q_h and Q_e increase with wind speed. It is also reasonable to expect that Q_h and Q_e are dependent on the gradients of temperature and vapor pressure. Numerous meteorological studies have established empirical relations between mean horizontal wind speed (u [L T^{-1}]) and Q_h, and between u and Q_e:

$$Q_h = \rho_a C_p f(u)(T_s - T_a) \tag{4}$$

$$Q_e = \rho_a (0.622/p_b) L_e f(u)(e_s - e_a) \tag{5}$$

where ρ_a [M L^{-3}] is the density of air, C_p [L^2 T^{-2} K^{-1}] is the specific heat of air, and p_b [M L^{-1} T^{-2}] is barometric pressure. The constant 0.622 is the mass ratio between the molecules of dry air and water vapor (Arya, 1988). The wind function $f(u)$ has a dimension of [L T^{-1}] and represents the empirical relation between u and the degree of turbulent mixing. The most commonly used form of wind function is linear, such as:

$$f(u) = a + bu \tag{6}$$

where a [L T^{-1}] and b are empirical constants. The same wind function is commonly used for both sensible and latent heat transfers, assuming that the two processes are driven by the same mixing mechanism. If a and b are known, ET flux can be estimated from Equations (3) and (5) by using the observed u, e_s, and e_a. This is commonly called mass-transfer or Dalton-type formula in the literature (Bras, 1990).

In a well-developed boundary layer over a homogeneous surface, u is essentially zero at some short distance (z_0) above the land surface and increases with the logarithm of height (Arya, 1988). This distance is called the roughness length and has a strong influence on a and b in Equation (6). As the name suggests, z_0 is dependent on the physical roughness of the surface, such as that caused by waves or emergent plants. Dense growth of tall plants such as *Typha* or *Phragmites* can result in z_0 being substantially above the water surface. In addition to z_0, thermal stratification of the air can also have a major influence on a and b, especially when u is relatively small. The atmosphere above a lake surface is said to be stable when a cold layer of air is overlain by warm air because the upward motion of colder, denser air into warmer, lighter air is countered by gravity. In contrast, the atmosphere is unstable when a warm layer of air is overlain by cold air. Since a and b are dependent on so many variables, they need to be determined empirically for each site. Therefore, mass-transfer formulas are usually used in combination with energy balance equations described below.

The energy balance of a lake is given by:

$$Q_n - Q_g + Q_a - Q_w = Q_h + Q_e \tag{7}$$

where Q_n is net radiation averaged over the lake area, Q_a [M T^{-3}] is the net energy input by stream and groundwater divided by the lake area, Q_w [M T^{-3}] is the area-averaged rate of energy storage change in lake water, and Q_g [M T^{-3}] is the area-averaged energy flux into the lake bottom sediment. Q_a is usually a minor component for most lakes and ponds, but it may be important for systems with artificial inputs of warm water. In principle, Q_w is given by:

$$Q_w = \frac{\rho_w C_w}{A} \frac{d}{dt}(VT_w) \tag{8}$$

where ρ_w is the density and C_w [L^2 T^{-2} K^{-1}] is the specific heat of water, A is the lake area, V is the total volume of water, and T_w is average water temperature. Determining T_w may be difficult because it requires monitoring the depth-temperature profile at multiple locations (Stannard and Rosenberry, 1991). However, in some lakes, measuring the depth-temperature profile at only one location will result in very little loss of accuracy (Rosenberry *et al.*, 1993). The sediment heat flux (Q_g) may be estimated from temperature gradients and the thermal conductivity of sediments. Its magnitude is generally smaller than the magnitude of Q_w, except for very shallow ponds and wetlands.

The left-hand side of Equation (7) represents the available energy for turbulent heat transfer and can be lumped together into a single term Q_{h+e}:

$$Q_{h+e} = Q_n - Q_g + Q_a - Q_w \tag{9}$$

For large lakes with negligible "shoreline effects" (described below), Q_e can be calculated from Q_{h+e} and the Bowen ratio (β) given by:

$$\beta = Q_h / Q_e \tag{10}$$

It follows from Equations (4) and (5) that:

$$\beta = \gamma (T_s - T_a)/(e_s - e_a) \tag{11}$$

where γ is the psychrometric constant defined as

$$\gamma = C_p \, p_b/(0.622 \, L_e) \tag{12}$$

and takes a value of about 0.66 mb K^{-1} at T_a = 293 K and p_b = 1000 mb. From Equations (7), (9), and (10), Q_e is given by

$$Q_e = Q_{h+e}/(1 + \beta) \tag{13a}$$

The Bowen ratio energy balance method uses Equations (11) to (13a) to estimate ET flux (e.g., Burba *et al.*, 1999). Equation (13a) can be written in a different form called the Priestley and Taylor (1972) equation:

$$Q_e = Q_{h+e} \, \alpha\Delta/(\Delta + \gamma) \tag{13b}$$

where α is a dimensionless coefficient and Δ is the slope of the saturation vapor pressure-temperature curve. When data are averaged over a period longer than several days, many studies have found that $\alpha \cong 1.26$ for relatively large lakes and wetlands (e.g., Rosenberry *et al.*, 2004). Therefore, when detailed temperature and vapor pressure data are unavailable, Equation (13b) may be used to estimate Q_e.

For relatively small lakes and ponds having high shoreline-to-area ratios, the warm air coming from dry uplands (i.e., advection) may transfer sensible heat to the water, providing extra energy input in addition to radiation. In this case, Q_h is downward (negative) and Q_e is upward (positive), hence β is negative. Therefore, the actual ET flux can substantially exceed published values of potential evaporation that are based on evaporation measurements over large lakes, where radiation inputs dominate the energy balance. Some lakes and ponds are surrounded by shrubs or trees that grow in relatively moist environments, such as *Salix* or *Populus*. These riparian trees reduce shortwave radiation inputs, particularly during early spring and late fall, when the solar altitude is relatively low. More important, they block the advection of warm air from uplands and reduce the horizontal wind velocity over the pond, hence the turbulent mixing of air. In this case, actual ET from the pond can be substantially smaller than published values of potential evaporation (Hayashi *et al.*, 1998).

Radiation inputs have strong diurnal and seasonal variation. The variation of Q_{h+e} is strongly correlated with Q_n but may be modulated by the storage term (Q_w). In shallow ponds, where energy is stored and released on a daily cycle, ET flux remains positive during nighttime because of the energy released from storage. During daytime, on the other hand, a major portion of the incoming radiation is stored in the water, resulting in a relatively even distribution of ET flux between daytime and nighttime (Burba *et al.*, 1999). Evaporation flux in deep lakes has a lagged seasonal response, where the flux remains high into late fall, owing to the release of energy stored in the water (Blanken *et al.*, 2000). Transpiration by emergent plants requires transporting water from the root system to leaf surfaces, overcoming gravity and hydraulic resistance within the plant tissues. When the primary energy input is net radiation, total ET is limited by available energy incident upon a unit area regardless of the mass-transfer efficiency of leaves. In other words, transpiration mediated by emergent plants cannot be much higher than evaporation from open water surfaces. One could argue that plants with a large leaf area index can capture the sensible heat advected from uplands more effectively than smooth water surfaces. Few studies have compared the ET fluxes from open water and emergent vegetation, but it appears that average ET from vegetated water surfaces is somewhat smaller than evaporation from open water (Burba *et al.*, 1999). Plant leaves have a much smaller capacity to store energy compared to water. As a result, the radiation inputs during daytime are almost immediately released to the atmosphere at a relatively high Bowen ratio because the temperature of leaf surfaces increases much faster than that of water. Peacock and Hess (2004) reported daytime Bowen ratios in a range of 0.5 to 2.5 over a wetland covered by *Phragmites australis*. They noted that ET flux was about half of reference ET (i.e., potential ET) when leaves were dry, but was close to reference ET when leaves were wet from rain. However, transpiration by individual *Phragmites australis*, determined by the direct measurement of sap flow through stems, can be several times higher than potential ET (Moro *et al.*, 2004).

Evapotranspiration is a major component of the water balance, especially for closed lakes without outflow streams. Air temperature is sometimes used as an index of ET in water balance studies because it is correlated with net radiation. However, it is clear from the preceding discussion that ET is strongly influenced by wind, humidity, and energy storage in water, as well as the height and growth pattern of emergent and riparian plants. Therefore, a simple temperature-index method such as that given by Thornthwaite (1948) should be used with caution because it was originally intended as an index of climate, not as a predictive tool for daily or monthly ET (Dunne and Leopold, 1978). Models that use only solar radiation and temperature, however, have been shown to work surprisingly well in some lake and wetland settings (Winter *et al.*, 1994; Rosenberry *et al.*, 2004).

Groundwater Exchange

Lakes and ponds are almost always connected to groundwater (Fig. 1). Therefore, groundwater inputs and outputs always affect the water and dissolved mass balance of lakes and ponds (Winter *et al.*, 1998; Hayashi and Rosenberry, 2002). The magnitude of groundwater flux depends on the property of substrates and the gradient of hydraulic potential according to Darcy's law, as illustrated in an idealized example in which relatively permeable sand is underlain by impermeable bedrock (Fig. 2). The dashed line represents the water table, which, roughly speaking, is the boundary between saturated and unsaturated sand. The water level in monitoring wells indicates the hydraulic potential of groundwater, or hydraulic head, around the screen at the bottom of the well. The hydraulic head is higher at well A than at well B in Fig. 2, indicating lateral flow of groundwater towards the lake. The difference in head (i.e., potential drop) divided by the distance is the hydraulic gradient. According to Darcy's law:

$$Q = wdK\,(h_A - h_B)/\Delta x \tag{14}$$

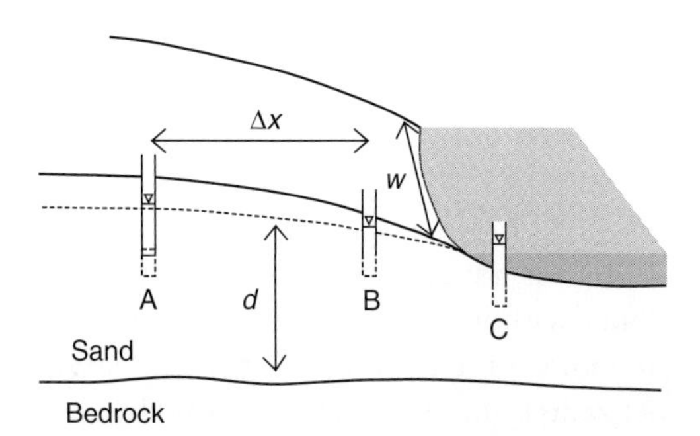

FIG. 2 Idealized example demonstrating Darcy's law. A, B, and C are groundwater monitoring wells having short screens at the bottom. Triangles indicate the water level.

where Q [$L^3 T^{-1}$] is the rate of groundwater flow into the lake, w [L] is the cross sectional width perpendicular to the flow direction, d [L] is the thickness of the saturated sediment, K [$L T^{-1}$] is the hydraulic conductivity of the saturated sediment, h [L] is the hydraulic head with subscripts indicating the location, and Δx [L] is the distance between the two points.

In more general form, Darcy's law is written as:

$$q = -K \, \Delta h/\Delta l \tag{15}$$

where q [$L T^{-1}$] is specific discharge given by $Q/(wd)$, and Δh is the change in head between two points separated by distance Δl in an arbitrary direction. If the head decreases ($\Delta h < 0$) from the first point to the second point, then groundwater is flowing from the first to the second point and q is positive. When the well screen is located below the water table, the hydraulic head around the screen may be different from the water table. In Fig. 2, the hydraulic head in the wells is higher than the water table, indicating that the vertical component of flow is upward. As a result, groundwater under the lakebed is flowing into the lake. Hydraulic conductivity of saturated sediments is dependent on the size and connectivity of pore spaces. Coarse materials such as sand and gravel generally have K ranging from 10^{-5} to 10^{-2} m s^{-1}, and fine materials such as unweathered clay commonly have K as low as 10^{-11} to 10^{-9} m s^{-1} (Freeze and Cherry, 1979, p. 29). The magnitude of hydraulic gradient below lakebeds is usually within a relatively narrow range between 0 and 0.2. Therefore, hydraulic conductivity is the decisive factor controlling groundwater flux.

The volume of water contained in a unit volume of sediment is called volumetric water content. Its value is roughly equal to porosity (i.e., sediment is nearly saturated) at the water table and gradually decreases from the water table to the land surface except during major infiltration events. Water in the zone above the water table, called the vadose zone, is suspended in pores by surface tension. The vadose zone immediately above the water table is under a tension-saturated condition, where sediments remain largely or entirely saturated. This zone is called the capillary fringe. Effects of surface tension are greater in smaller pores; hence, finer sediments have a greater ability to retain water and a thicker capillary fringe. Where the water table is near the ground surface, as is the case at the riparian fringe of lakes and ponds, the existence of a capillary fringe can cause the soil to be saturated right to the ground surface so that rain cannot infiltrate and runs off to the pond instead. The hydraulic conductance of pores dramatically decreases with pore radius, and only pores containing water contribute to the flow of water (Hillel, 1998). Therefore, hydraulic conductivity of sediments increases with water content as larger and larger pores become available for flow. Hydraulic conductivity of unsaturated sediments can be orders of magnitude smaller than that of saturated sediments, meaning that the majority of lateral groundwater flow occurs through the saturated sediments below the water table and within the capillary fringe. However, vertical flow of water in the vadose zone is important for evapotranspiration and infiltration processes.

Depth to the water table has some spatial variability within a given region, but its variability is generally much smaller than the variability of ground-surface elevation. Therefore, generally speaking, the water table is a subdued replica of the ground surface (Freeze and Cherry, 1979, p. 193). Lakes and ponds are usually located in topographic lows and hence receive groundwater discharge (Fig. 2). However, depending on the local topography and geology, a lake or pond can lose water to groundwater (i.e., groundwater recharge) or have separate areas of discharge and recharge (Born *et al.*, 1979; Winter, 1999). The direction of groundwater exchange may also change seasonally, depending on the hydraulic gradient between the pond and the underlying groundwater (Mills and Zwarich, 1985; Winter and Rosenberry, 1995).

Theoretical studies have shown that groundwater discharge flux under lakebeds is greatest near the shore and decreases toward the lake center (Pfannkuch and Winter, 1984). In an idealized cross-section of homogeneous sediment with the uniform input of groundwater from the left side (Fig. 3), the distribution of flow lines is much denser near the shore than the lake center. In this graphical representation, each flow "tube" between flow lines carries the same amount of water. Therefore, denser flow lines indicate higher groundwater input. Pfannkuch and Winter (1984) showed that the tendency of flow lines to be crowded near the shore is greater in

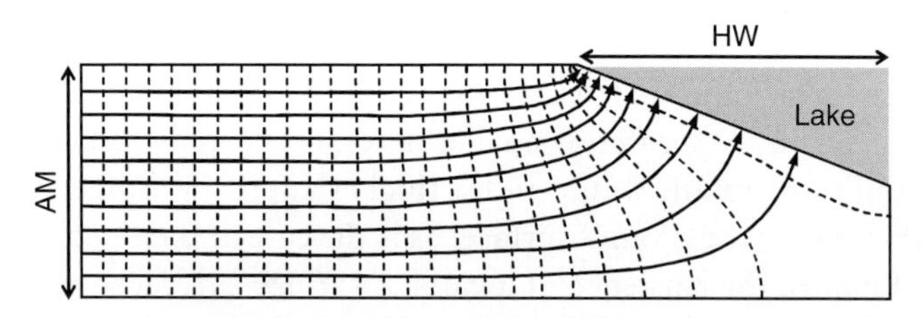

FIG. 3 Groundwater flow near a lake having half-width HW and the thickness of permeable material AM. Solid lines represent flow lines and dashed lines represent the contours of hydraulic head. (Modified after Pfannkuch and Winter [1984] with permission of Elsevier Science.)

lakes having greater width (HW) and smaller thickness of the flow domain (AM). Similar flow distribution has also been observed in numerous field studies (Shaw and Prepas, 1990). However, groundwater flux varies widely, even within small areas because heterogeneity in lakebed sediments causes hydraulic conductivity to vary substantially (Kishel and Gerla, 2002).

Geology has a major influence on groundwater exchange. Lakes located on relatively permeable materials, such as fluvial sands, tend to have high groundwater inputs. In a small (0.81 km^2) lake located on glacial outwash sediments in Wisconsin, Krabbenhoft *et al.* (1990) reported an average groundwater input of 290 $mm\,y^{-1}$, or 27% of total water input. Rosenberry *et al.* (2000) also studied a small lake (0.66 km^2) located on coarse glacial sediments in northern Minnesota and reported a total groundwater input of 57 $L\,s^{-1}$, or 2700 $mm\,y^{-1}$ averaged over the lake, of which about 30% was discharged at springs located on shore or near shore. These springs were relatively easily identified by the growth of marsh marigold (*Caltha palustris*). Lakes located on materials with low hydraulic conductivity, such as clay or shale, may have very little groundwater inflow or outflow so that the water balance of such lakes has no significant groundwater component.

Environmental tracer studies have shown that riparian trees may selectively use the deeper groundwater recharged under uplands, even when the shallower groundwater directly below the stream bed is readily available (Dawson and Ehleringer, 1991). Riparian trees and grasses not only intercept groundwater that would otherwise discharge into ponds but also induce drawdown of the water table around the pond perimeter (Rosenberry and Winter, 1997).

Water uptake by riparian vegetation has particularly pronounced effects on the water balance of relatively small ponds that have high ratios of perimeter (i.e., shoreline length) to area, P/A [L^{-1}]. Millar (1971) monitored the water level in 75 prairie wetlands in Saskatchewan, Canada, in 1963–1969 and determined the water-level recession rates during the periods between two consecutive measurements. Out of the 146 periods, varying in length from 1 to 4 weeks, 106 periods showed a significant correlation ($p < .01$) between the recession rate and P/A. Fig. 4 shows an example for the Saskatoon site, in which the recession rate increases linearly with P/A. Riparian trees encircling a pond block the flow of warm, dry air over the pond. Therefore, water uptake by riparian trees may be accompanied by suppression of ET from the pond. It is important to consider the complex ecohydrological linkages between ponds and riparian zones in evaluation of the water balance.

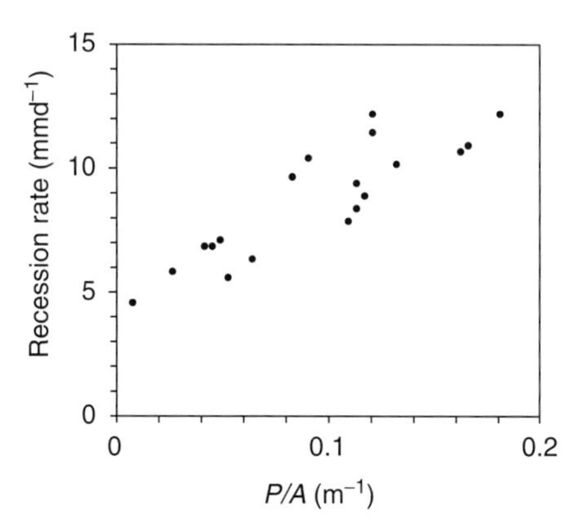

FIG. 4 Relation between perimeter-to-area ratio (*P/A*) and the water-level recession rate in 19 wetlands located near Saskatoon, Saskatchewan, Canada, measured in July 1969 (From Millar, [1971].)

Surface Water Input and Output

Some lakes and ponds have surface water inputs through confined stream channels, while others have diffuse inputs along the shoreline (Fig. 1). The amount of input is primarily controlled by the size of the contributing area (i.e., catchment) and the amount of runoff generated from a unit area. A catchment can be determined in principle by tracing drainage divides on a topographic map using computer programs (McMaster, 2002). The catchment delineated on published maps (1:50,000 is often a convenient scale) is called the gross catchment, while the actual area that drains into a lake is called the effective catchment. Gross catchments coincide with effective catchments reasonably well on mountainous terrain, where drainage divides are clearly represented on a map. On relatively flat terrain, however, effective catchments can be substantially smaller than gross catchments because subtle topography, too small to be shown on maps, may have major effects on drainage systems (Hubbard and Linder, 1986; Hayashi *et al.*, 2003). For example, the watershed of a lake may contain small wetlands and their associated subwatersheds that direct overland flow to the wetlands rather than to the nearby larger lake.

Flow rates in streams generally have a large degree of temporal variability. In unregulated streams without artificial reservoirs, the "baseflow" during dry periods is mostly sustained by groundwater discharge to the stream, while the high flow during storms or snowmelt consists of both groundwater and surface water inputs. Because surface water input is a major component of the water balance, lake water level is strongly influenced by the amount and timing of groundwater and surface water inputs to the streams feeding the lake. Older conceptual models, developed primarily for engineering purposes, assume that most runoff during storms occurs overland when the infiltration capacity of the surface soil is exceeded by rainfall intensity. Detailed field studies have shown, however, that such a simplistic model applies only to areas covered by an impervious surface such as bedrock, urban pavement, heavily compacted soils, or frozen ground (see Dunne and Leopold, 1978, for an excellent review). In most vegetated catchments, runoff occurs as a combination of surface and subsurface processes, and overland flow is normally limited to riparian zones and lake margins where the water table is close to the ground surface.

Recent studies indicate an important role of macropores, such as decayed root holes and animal burrows, in infiltration processes (Bodhinayake and Si, 2004). Pore diameters within the soil matrix are generally less than a few millimeters and groundwater flow velocity in the soil matrix is less than a few centimeters per hour. In contrast, macropores can have diameters as large as a few centimeters and, as a result, macropore flow velocity can be several orders of magnitude greater than soil matrix flow velocity. However, the soil needs to become saturated or near saturated in order to have water in macropores. Alteration of macropores by cultivation or compaction by animals or humans may have a major influence on infiltration and subsurface flow, hence on the input of water to streams and lakes.

The amount and timing of surface and subsurface runoff are strongly dependent on soil types and geology that control the infiltration rate of rain and snowmelt. They are also dependent on the soil moisture condition because wetter soils tend to generate more runoff than do drier soils that have a larger capacity to accommodate infiltration. Vegetation has a major influence on the soil water balance through rain and snow interception and transpiration. Therefore, major land-cover change caused by natural causes, such as fire or human activities, can result in a dramatic shift in the water regime of lakes and ponds receiving water from the affected area (van der Kamp *et al.*, 2003).

Surface water outflows from lakes and ponds are usually confined to one channel. The amount of outflow is determined by the width and depth of water, channel slope, and channel roughness, which is dependent on the type of channel material as well as presence of submergent and emergent vegetation. Flow velocity and channel cross-sectional area both increase with the depth of water, resulting in the strong dependence of volumetric flow rate on water depth. Water depth at the outlet increases with lake stage, implying that the rate of surface outflow is greater when the amount of water stored in the lake is greater. Therefore, a negative feedback mechanism exists in the lake, through which the water level is regulated around its average position.

Storage and Basin Morphometry

The volume (V) and area (A) of lakes are usually calculated from the depth of water (H) by using predetermined V-A-H relations because it is difficult to monitor V and A directly. Each lake has a unique V-A-H relation reflecting the complicated morphometry of the lake basin. However, it is useful to define simple mathematical forms of V-A-H that are applicable to many single-basin lakes for examining the effects of basin morphometry on water level fluctuation. For prairie wetlands forming in isolated basins in Saskatchewan and North Dakota, Hayashi and van der Kamp (2000) proposed the following:

$$A = s\left(\frac{H}{H_0}\right)^{2/p} \tag{16}$$

$$V = \frac{s}{(1 + 2/p)} \frac{H^{1 + 2/p}}{H_0^{2/p}} \tag{17}$$

where s [L^2] is a scaling factor, p is a dimensionless constant, and H_0 [L] is the unit length that keeps the dimension of the equations correct. For example, $H_0 = 1$ m is always used if the length is measured in meters. The two parameters, s and p, are specific to each basin, hence need to be determined by bathymetry survey. These equations have been successfully used for ephemeral forest pools in Massachusetts (Brooks and Hayashi, 2002).

The p takes a value of 1 for a hypothetical cone-shaped basin (Fig. 5A) and a value of 2 for a paraboloid basin, and increases with the curvature of the basin profile until it reaches ∞ for a cylinder. The A-h relation is linear for a paraboloid and nonlinear for other shapes (Fig. 5B). For hypothetical basins having a depth of 1 m and area of 1000 m^2, the volume corresponding to a given depth varies with p (Fig. 5B). Real basins have complex shapes, and p gives the characteristics of the basin in an average sense. Hayashi and van der Kamp (2000) reported $p = 1.5$ to 6.2 for 27 prairie wetlands, and Brooks and Hayashi (2002) reported $p = 0.6$ to 2.2 for 33 forest pools. Ponds with larger p have a more concave basin shape to store a larger amount of water for a given area; they also have smaller rates of

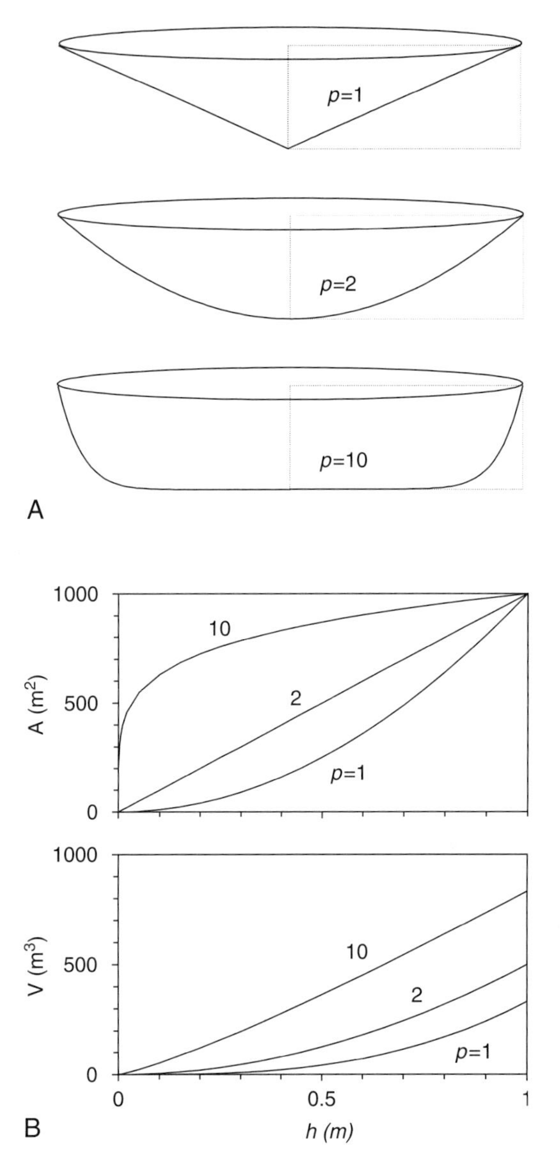

FIG. 5 Idealized symmetric basins (**A**) and their volume-area-depth relation (**B**).

area shrinkage for a given water level drop. These are important factors regarding the ecohydrological response of ponds to changes in inputs or outputs of water. For example, the zonation patterns of emergent vegetation are controlled by the changes in the depth of water and the location of the shoreline (van der Valk, 2000; also see Chapter 11).

Climate Effects

The average water level in ponds and lakes, and the water level changes around that average, reflect a dynamic balance among various water inputs and outputs (Fig. 1). The riparian vegetation around the water body responds to the degree of hydrological disturbance that is represented by the short-term and long-term changes of water level. In fact, one can often deduce a considerable amount of information about the water-level regime of a pond or lake by inspection of the vegetation in and around it.

The average water level is determined by landscape setting, hydrogeological setting, and the climate. In humid climates, where precipitation exceeds ET annually, most ponds and lakes have a surface outflow that controls the average water level and restricts water level changes to a fairly narrow range. High water levels caused by storms or snowmelt will tend to dissipate rather quickly. However, some ponds and lakes located on permeable materials are drained mainly by groundwater outflow. Since groundwater outflow is less sensitive to lake level fluctuations, changes in water inputs in these ponds and lakes lead to larger and longer-lasting changes of water level.

In semi-arid and arid climates, where ET exceeds precipitation on an annual basis, lakes and ponds can be hydrologically closed with no outflow. ET is then the dominant or only form of water output. Unlike water outflow by surface or groundwater, ET is not directly controlled by water level, although total lake ET is dependent on the lake area that expands and shrinks depending on water level. Therefore, large, closed water bodies may have only weak feedback mechanisms to stabilize the water level if the lake area does not change appreciably with changes of water level. Such lakes can be subject to large, multiyear water level changes. They may experience large sudden rises of water level as a result of high runoff, but the subsequent downward recession is slow, being limited by the difference between precipitation and evaporation. The water level regime of large closed lakes can be difficult to simulate or predict because relatively small changes in evaporation and precipitation, or of inputs such as changes of runoff due to landuse change, can lead to large cumulative changes of water level over the long term. As an example, Devil's Lake in North Dakota is a closed lake having a catchment of 8600 km^2 (Todhunter and Rundquist, 2004). The lake level increased by about 7 m between 1992 and 2001, most likely because of a shift of climate from dry to wet, although the effects of agricultural land conversion and drainage are also debated (Todhunter and Rundquist, 2004).

CASE STUDY: NORTHERN PRAIRIE WETLANDS

Overview of Prairie Wetlands

The northern prairie region of North America is characterized by undulating topography with numerous depressions. Prairie wetlands forming in these depressions are the focal point of aquatic ecosystems in the region. Dynamic water regimes of the prairie wetlands provide excellent case study examples demonstrating the effects of hydrological processes on the water levels. The studies presented below were conducted in St. Denis National Wildlife Area located 40 km east of Saskatoon, Saskatchewan, Canada. Over 100 wetlands and numerous smaller depressions occur on the 4 km^2 study site. The land use consists of the conventional rotation of cereal crops and oil seeds, tame grasses, and native prairie grasses.

Prairie wetlands usually have no permanent surface inflow or outflow streams and are regarded as hydrologically closed basins. The water balance is strongly influenced by the cold semi-arid climate, where annual evaporation from open water (600–800 mm) exceeds annual precipitation (300–400 mm). Winter snowfall accumulates up to 150 mm of water equivalent and the soil freezes to depths greater than 1 m (van der Kamp *et al.*, 2003). Glacial tills underlying the surface soils are rich in clay and have relatively low hydraulic conductivity, limiting the rates of groundwater flow (Hayashi *et al.*, 1998).

Major inputs of water are direct precipitation onto wetlands and the lateral transfer of snow-derived water from adjacent lands in the form of wind-driven snowdrift and snowmelt runoff over frozen ground. Summer runoff is rare because of the dry climate, and groundwater input is minor because of the low hydraulic conductivity of glacial till. Outputs are evapotranspiration and infiltration of surface water to groundwater since there is no outflow stream.

The water level is generally highest in early spring after snowmelt runoff is received and gradually declines in summer, primarily because of evaporation and infiltration exceeding the inputs of rain and occasional storm runoff (Winter, 1989). Many prairie wetlands hold surface water only for a few weeks to months. The duration of surface water, or hydroperiod, is a critical habitat parameter for waterfowl and other species dependent on water (Swanson and Duebbert, 1989). The water level change in a typical prairie wetland can be schematically represented by a triangle (Fig. 6) where the magnitude of spring rise and the slope of summer recession determine the duration of surface water. Therefore, the ecology of prairie wetlands is strongly dependent on the factors controlling spring rise and summer recession.

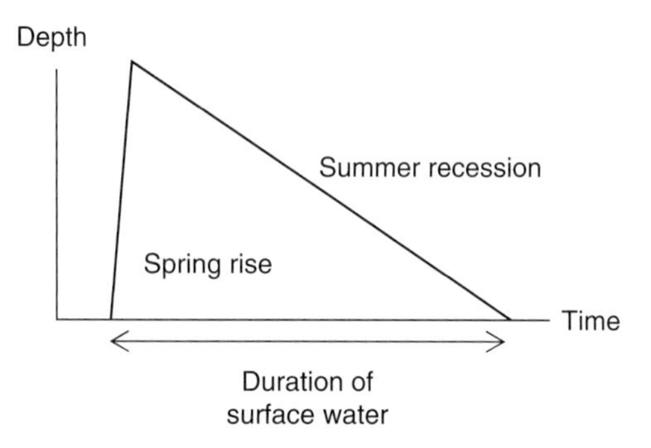

FIG. 6 Annual cycle of water level in typical prairie wetlands.

Effects of Upland Vegetation and Land Use

The magnitude of spring rise depends on the volume of snow drift and snowmelt runoff into the wetland. Snow drift is influenced by many factors, but microtopography and vegetative cover on the upland seem to be the most important (van der Kamp *et al.*, 2003). Tall stubble and perennial grasses retain snow on the upland, while resuspension and redistribution of snow is relatively unrestricted on cultivated fields with little stubble. Riparian trees around wetlands also function as effective snow accumulators (Hayashi *et al.*, 1998). In the absence of snow-trapping vegetation, snow accumulates preferentially in depressions.

The volume of snowmelt runoff is determined by the size of the catchment and the amount of runoff generated over a unit area. The latter depends on the amount of snow on the ground and the infiltration capacity of frozen soil, which in turn depends on soil water content, particle size, and structure (Gray *et al.*, 2001). Conditions favorable for large amounts of runoff are wet soil at the time of freeze up, high clay content, and the absence of soil macropores, such as decayed root holes and animal burrows (van der Kamp *et al.*, 2003). Macropores can easily be destroyed by cultivation. Therefore, development of macropores is enhanced under undisturbed vegetation compared to cultivated fields.

Long-term records of water level at Wetland 92 (Fig. 7A) and Wetland 109 (Fig. 7B) show that the two wetlands had similar water regimes until 1986. The two wetlands have similar size, and the uplands around these wetlands were subjected to dryland cultivation until 1983, when the uplands around 92 were converted to a dense nesting cover of *Bromus inermis* for the purpose of improving wildlife habitat. This dense nesting cover has not been disturbed by grazing, mowing, or burning (van der Kamp

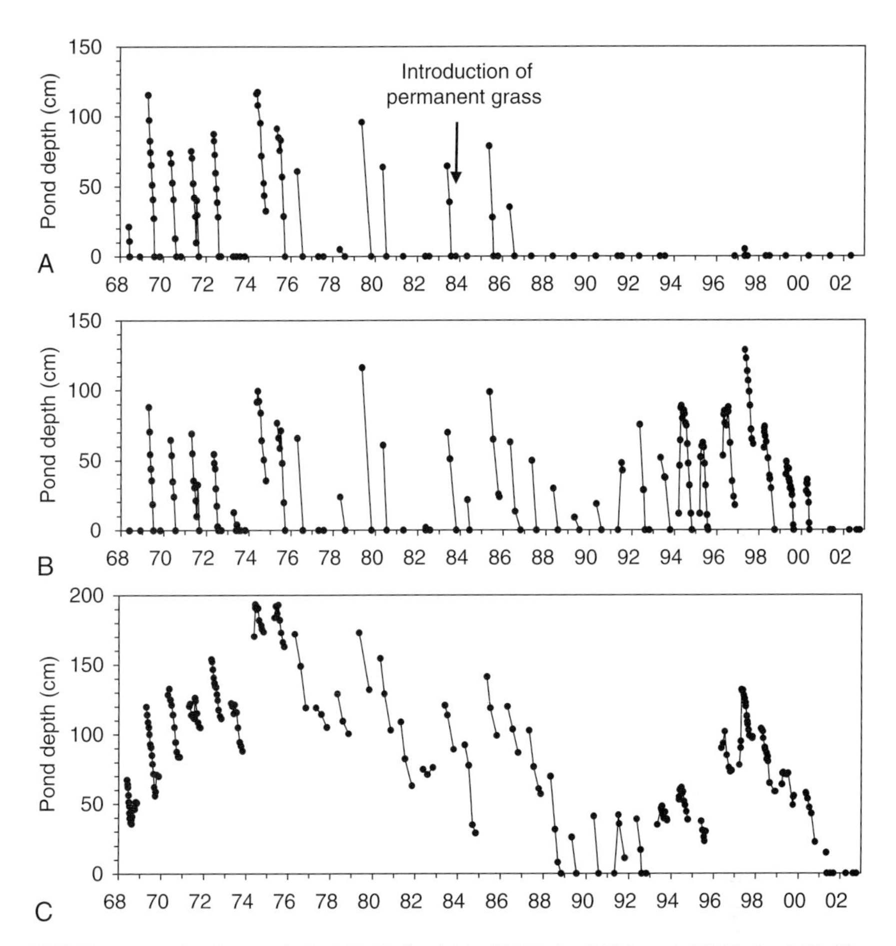

FIG. 7 Water depth records for (**A**) Wetland 92, (**B**) Wetland 109, and (**C**) Wetland 50. The uplands around Wetland 92 were converted to permanent grass in 1983.

et al., 2003). From 1987 onward, Wetland 92 dried out and has remained almost entirely dry ever since, while the water regime in Wetland 109 did not show a major change. Other small wetlands within the grassed area behaved similarly to Wetland 92 (van der Kamp *et al.*, 1999). Detailed studies of hydrological processes indicated that two factors operate together and lead to the dramatic shift in wetland water regime: (1) tall permanent grass cover is effective in trapping snow so that the wind-driven transport of snow into wetlands is reduced; and (2) the undisturbed grass cover leads to the development of a macropore network in the topsoil, which markedly increases the infiltrability of the soil, even when it is frozen.

The macropore network takes several years to develop after introduction of the grass, as indicated by the delayed response of water level in 1987 to the cultivated-to-grass conversion that took place in 1983.

Size and Permanence of Wetlands

In addition to the triangular annual cycles (Fig. 6), water levels in wetlands have interannual and interdecadal cycles related to the variability of climatic conditions. For example, relatively dry conditions in the late 1980s to early 1990s resulted in low water levels, which recovered during the relatively wet period of the mid 1990s (Fig. 7B). Winter and Rosenberry (1998) found similar patterns in the water levels of prairie wetlands in North Dakota and showed the correlation between water levels and a climatic index. Wetlands with seasonal ponds, such as Wetland 109, respond immediately to reduced water inputs during dry periods, resulting in a very short duration of surface water (Fig. 7B). In contrast, wetlands with more permanent ponds, such as Wetland 50, keep their surface water during multiple years of low water inputs (Fig. 7C).

The area of Wetland 109 is approximately 0.24 ha and its catchment is 2.4 ha (Hayashi *et al.*, 1998), meaning that the catchment-to-wetland area ratio is 10. The area of Wetland 50 is 3.4 ha and its catchment-to-wetland area ratio is 16 (Su *et al.*, 2000). Therefore, if the upland areas generate constant amounts of runoff on average, Wetland 50 receives 60% more spring runoff per unit area than Wetland 109. This partially explains why the former has a more permanent pond than the latter. The permanence of a pond is also related to its P/A. The rate of water level recession in summer is higher for those ponds with larger P/A than those with smaller P/A (Fig. 4). Therefore, smaller ponds in smaller wetlands have a larger P/A and tend to have shorter durations. Approximate P/A at peak water level is 0.02 m^{-1} for the pond in Wetland 50 and 0.07 m^{-1} for the pond in Wetland 109. Fig. 4 suggests that summer recession rates of the water level in Wetland 109 could be nearly 50% higher than that in Wetland 50, and the water level data in Fig. 7 show that this is indeed the case, with the water level recession in Wetland 109 always being greater than the corresponding recession in Wetland 50.

Many of the wetlands with permanent and semi-permanent ponds have high salinity, indicating the accumulation of dissolved salts transported to wetlands by groundwater and surface water. However, except for the shoreline-related seepage discussed below, the flow rate of groundwater is typically much less than 50 mm y^{-1} (van der Kamp and Hayashi, 1998), which is too low to have measurable effects on the water balance. Therefore, the groundwater input generally has little effect on the water balance. However, the groundwater input is important for the mass balance of dissolved chemical species, such as sulfates, because the concentration of these species in groundwater is several orders of magnitude higher than that in rain or snowmelt water.

Riparian Vegetation

The shoreline-related loss of water as indicated in Fig. 4 is linked to water uptake in the riparian zone around the wetland, where the water table is close to the ground surface, allowing plants to transpire at a rate approaching potential ET. Fig. 8 shows the elevation of the water surface in Wetland 109 and hydraulic head in two monitoring wells under the riparian zone.

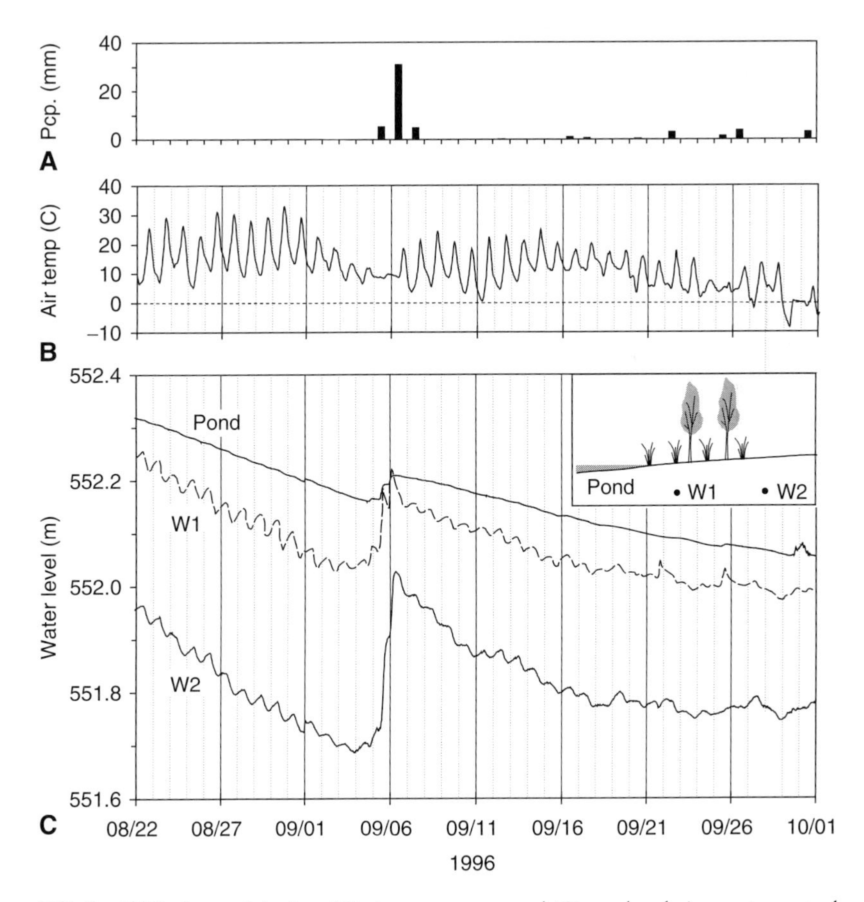

FIG. 8 (A) Daily precipitation, (B) air temperature, and (C) pond and piezometer water level at Wetland 109.

The relative position of the pond, monitoring wells, and riparian zone are schematically shown in the insert. The distance between W1 and W2 is approximately 7 m. The distance between the edge of the pond and W2 ranged between 5 and 13 m during this period. Hydraulic head decreased from the pond to W1, then to W2, indicating that the groundwater flow direction was away from the pond towards the edge of the wetland. Hydraulic head data show peaks around 10:00 to 11:00 and troughs around 19:00 to 22:00. It is clear that daytime drawdown of the groundwater level is caused by water uptake by plants. The nighttime recovery most likely represents the replenishment of water by the lateral flow of groundwater from the pond to the riparian zone. Parsons *et al.* (2004) reported that 60% to 70% of water loss from the pond in Wetland 109 during the summer months was caused by the groundwater flow induced by the riparian vegetation uptake, while the remainder was evapotranspiration within the pond area.

The clear diurnal pattern at W1 and W2 disappeared after September 17, and the slope of the pond water recession decreased, indicating that the water uptake by riparian plants had stopped. Air temperature dropped to 0.6°C on September 11 (Fig. 8B), suggesting a possibility of the first frost because the temperature of exposed leaves can be substantially lower than air temperature because of radiation cooling (Inouye, 2000). We did not document the phenology of trees, but it is likely that leaf senescence started sometime between September 11 and 17.

The water levels in the pond and wells rose rapidly on September 5 and 6 (Fig. 8C), responding to rainfall (Fig. 8A). The pond level rose by 43 mm, where total rainfall for this event was 36 mm, suggesting an input of 7 mm of "extra" water. This input was probably surface runoff generated in the riparian zone, where the water table is near the ground surface and the ground is almost entirely saturated because of a relatively thick capillary fringe in the clay-rich soil. The rainfall caused a rapid rise of the water table, leading to fully saturated conditions in the ground that restricted the infiltration of rainfall. Gerla (1992) and Winter and Rosenberry (1995) made similar observations at prairie wetlands in North Dakota.

CONCLUSIONS

Water-level changes in ponds and lakes occur as a result of the water input exceeding output or vice versa. Because inputs and outputs are controlled by hydrological processes, understanding the water-level changes and resulting ecological responses requires understanding the individual processes. It is particularly important to realize the intimate link between lakes and their catchments. Disturbance in the catchment, such as major land use change, can cause a dramatic change in hydrological processes, which ultimately affects the lake water level. This was clearly demonstrated in the case study of prairie wetlands, where grassing the uplands resulted in the drying out of wetlands. Climate changes also have major effects on water level. When the hydrological processes and their response to land use and climate are understood, it is reasonably straightforward to simulate the water level in a particular lake or pond using the water balance equation (e.g., Poiani *et al.*, 1996; Su *et al.*, 2000). However, hydrological processes are poorly understood in many cases, resulting in large uncertainty in model predictions. Ecohydrological linkage between plants and water presents a fruitful opportunity for collaboration between ecologists and hydrologists. The role of riparian trees in evapotranspiration and groundwater exchange, for example, is an important but relatively poorly understood process. It is hoped that collaborative research on ecohydrology will enable us to observe hydrological processes and ecological responses simultaneously and to develop coupled models for the prediction of ecosystem responses to land use and climate changes.

ACKNOWLEDGMENTS

Funds for the case studies were provided by Ducks Unlimited, Natural Sciences and Engineering Research Council of Canada, and Environment Canada. Constructive comments on earlier drafts by Don Rosenberry and Philip Gerla greatly improved the chapter.

APPENDIX: LIST OF SYMBOLS

a	constant in wind function (Equation 6) [L T^{-1}]
A	area of lake or pond [L^2]
b	constant in wind function (Equation 6)
C_p	specific heat of air [L^2 T^{-2} K^{-1}]
C_w	specific heat of water [L^2 T^{-2} K^{-1}]
d	thickness of saturated sediments [L]
e_a	vapor pressure of air [M L^{-1} T^{-2}]
e_s	vapor pressure at the water surface [M L^{-1} T^{-2}]
E	evaporation rate [L T^{-1}]
$f(u)$	wind function [L T^{-1}]
h	lake water level [L]

h_A	groundwater hydraulic head [L]
H	depth of lake or pond [L]
H_0	unit depth [L]
K	hydraulic conductivity [L T^{-1}]
L	length
L_e	latent heat of vaporization [L^2 T^{-2}]
M	mass
p	shape constant in Equation (16)
p_b	barometric pressure [M L^{-1} T^{-2}]
P	perimeter of lake or pond [L]
q	groundwater specific discharge [L T^{-1}]
Q	groundwater flow rate [L^3 T^{-1}]
Q_a	net energy input by stream and groundwater per area [M T^{-3}]
Q_e	latent heat transfer [M T^{-3}]
Q_g	energy flux into lake bottom sediment [M T^{-3}]
Q_h	sensible heat transfer [M T^{-3}]
Q_{h+e}	sum of latent and sensible heat [M T^{-3}]
Q_{in}	total water input [L^3 T^{-1}]
$Q_{LW\uparrow}$	outgoing long-wave radiation [M T^{-3}]
Q_n	net radiation [M T^{-3}]
Q_{out}	total water output [L^3 T^{-1}]
Q_w	rate of energy storage change per lake area [M T^{-3}]
s	scaling constant in Equation (16) [L^2]
T	time
T_a	air temperature [K]
T_w	average lake temperature [K]
T_s	temperature of surface [K]
u	wind speed [L T^{-1}]
V	lake or pond volume [L^3]
w	cross-sectional width of groundwater flow [L]
z_0	roughness length [L]
α	Priestley–Taylor coefficient
β	Bowen ratio
Δ	slope of vapor pressure-temperature curve [M L^{-1} T^{-2} K^{-1}]
ε	emissivity
γ	psychrometric constant [M L^{-1} T^{-2} K^{-1}]
ρ_a	density of air [M L^{-3}]
ρ_w	density of water [M L^{-3}]
σ	Stefan–Boltzmann constant [M T^{-3} K^{-4}]

REFERENCES

Arya, S. P. (1988). *Introduction to Micrometeorology*. Academic Press, San Diego.

Blanken, P. D., Rouse, W. R., Culf, A. D., Spence, C., Boudreau, L. D., Jasper, J. N., Kochtubajda, B., Schertzer, W. M., Marsh, P., and Verseghy, D. (2000). Eddy covariance measurements of evaporation from Great Slave Lake, Northwest Territories, Canada. *Water Resour Res* 36, 1069–1077.

Bodhinayake, W., and Si, B. C. (2004). Near-saturated surface soil hydraulic properties under different land uses in the St. Denis National Wildlife Area, Saskatchewan, Canada. *Hydrol Process* 18, 2835–2850.

Born, S. M., Smith, S. A., and Stephenson, D. A. (1979). Hydrogeology of glacial-terrain lakes, with management and planning applications. *J Hydrol* 43, 7–43.

Bras, R. L. (1990). *Hydrology: An Introduction to Hydrologic Science*. Addison-Wesley, Reading, MA.

Brooks, R. T., and Hayashi, M. (2002). Depth-area-volume and hydroperiod relationships of ephemeral (vernal) forest pools in southern New England. *Wetlands* 22, 247–255.

Burba, G. G., Verma, S. B., and Kim, J. (1999). A comparative study of surface energy fluxes of three communities (*Phragmites australis*, *Scirpus acutus*, and open water) in a prairie wetland ecosystem. *Wetlands* 19, 451–547.

Crockford, R. H., and Richardson, D. P. (2000). Partitioning of rainfall into throughfall, stemflow and interception: effect of forest type, ground cover and climate. *Hydrol Process* 14, 2903–2920.

Dawson, T. E., and Ehleringer, J. R. (1991). Streamside trees that do not use stream water. *Nature* 350, 335–337.

Drexler, J. Z., Snyder, R. L., Spano, D., and Paw U, K. T. (2004). A review of models and micrometeorological methods used to estimate wetland evapotranspiration. *Hydrol Process* 18, 2071–2101.

Dunne, T., and Leopold, L. B. (1978). *Water in Environmental Planning*. Freeman, San Francisco.

Freeze, R. A., and Cherry, J. A. (1979). *Groundwater*. Prentice-Hall, Englewood Cliffs, NJ.

Gerla, P. J. (1992). The relationship of water-table changes to the capillary fringe, evapotranspiration, and precipitation in intermittent wetlands. *Wetlands* 12, 91–98.

Goodison, B. E., Louie, P. Y. Y., and Yang, D. (1998). *WMO Solid Precipitation Measurement Intercomparison Final Report*. World Meteorological Organization Report No. 872, Geneva.

Gray, D. M., Toth, B., Zhao, L., Pomeroy, J. W., and Granger, R. J. (2001). Estimating areal snowmelt infiltration into frozen soils. *Hydrol Process* 15, 3095–3111.

Hayashi, M., and Rosenberry, D. O. (2002). Effects of ground water exchange on the hydrology and ecology of surface water. *Ground Water* 40, 309–316.

Hayashi, M., and van der Kamp, G. (2000). Simple equations to represent the volume-area-depth relations of shallow wetlands in small topographic depressions. *J Hydrol* 237, 74–85.

Hayashi, M., van der Kamp, G., and Rudolph, D. L. (1998). Water and solute transfer between a prairie wetland and adjacent uplands, 1. Water balance. *J Hydrol* 207, 42–55.

Hayashi, M., van der Kamp, G., and Schmidt, R. (2003). Focused infiltration of snowmelt water in partially frozen soil under small depressions. *J Hydrol* 270, 214–229.

Hillel, D. (1998). *Environmental Soil Physics*. Academic Press, San Diego.

Hubbard, D. E., and Linder, R. L. (1986). Spring runoff retention in prairie pothole wetlands. *J Soil Water Conserv* 41, 122–125.

Inouye, D. W. (2000). The ecological and evolutionary significance of frost in the context of climate change. *Ecol Lett* 3, 457–463.

Kishel, H. F., and Gerla, P. J. (2002). Characteristics of preferential flow and groundwater discharge to Shingobee Lake, Minnesota, USA. *Hydrol Process* 16, 1921–1934.

Krabbenhoft, D. P., Bowser, C. J., Anderson, M. P., and Valley, J. W. (1990). Estimating groundwater exchange with lakes, 1. The stable isotope mass balance method. *Water Resour Res* 26, 2445–2453.

Lowe, P. R. (1977). An approximating polynomial for the computation of saturation vapor pressure. *J Appl Meteorol* 16, 100–103.

McMaster, K. J. (2002). Effects of digital elevation model resolution on derived stream network positions. *Water Resour Res* 38(4), Art. No. 1042.

Millar, J. B. (1971). Shoreline-area ratio as a factor in rate of water loss from small sloughs. *J Hydrol* 14, 259–284.

Mills, J. G., and Zwarich, M. A. (1985). Transient groundwater flow surrounding a recharge slough in a till plain. *Can J Soil Sci* 66, 121–134.

Moro, M. J., Domingo, F., and López, G. (2004). Seasonal transpiration pattern of *Phragmites australis* in a wetland of semi-arid Spain. *Hydrol Process* 18, 213–227.

Parsons, D. F., Hayashi, M., and van der Kamp, G. (2004). Infiltration and solute transport under a seasonal wetland: bromide tracer experiments in Saskatoon, Canada. *Hydrol Process* 18, 2011–2027.

Peacock, C. E., and Hess, T. M. (2004). Estimating evapotranspiration from a reed bed using the Bowen ratio energy balance method. *Hydrol Process* 18, 247–260.

Pfannkuch, H. O., and Winter, T. C. (1984). Effect of anisotropy and groundwater system geometry on seepage through lakebeds. *J Hydrol* 75, 213–237.

Poiani, K. A., Johnson, W. C., Swanson, G. A., and Winter, T. C. (1996). Climate change and northern prairie wetlands: simulations of long-term dynamics. *Limnol Oceanogr* 41, 871–881.

Pomeroy, J. W., Gray, D. M., Shook, K. R., Toth, B., Essery, R. L. H., Pietroniro, A., and Hedstrom, N. (1998). An evaluation of snow processes for land surface modelling. *Hydrol Process* 12, 2339–2367.

Priestley, C. H. B., and Taylor, R. J. (1972). On the assessment of surface heat flux and evaporation using large-scale parameters. *Mon Weather Rev* 100, 81–92.

Rosenberry, D. O., and Winter, T. C. (1997). Dynamics of water-table fluctuations in an upland between two prairie-pothole wetlands in North Dakota. *J Hydrol* 191, 266–289.

Rosenberry, D. O., Stannard, D. I., Winter, T. C., and Martinez, M. L. (2004). Comparison of 13 equations for determining evapotranspiration from a prairie wetland, Cottonwood Lake area, North Dakota, USA. *Wetlands* 24, 483–497.

Rosenberry, D. O., Striegl, R. G., and Hudson, D. C. (2000). Plants as indicators of focused ground water discharge to a northern Minnesota lake. *Ground Water* 38, 296–303.

Rosenberry, D. O., Sturrock, A. M., and Winter, T. C. (1993). Evaluation of the energy-budget method of determining evaporation at Williams Lake, Minnesota, using alternative instrumentation and study approaches. *Water Resour Res* 29, 2473–2483.

Shaw, R. D., and Prepas, E. E. (1990). Groundwater-lake interactions: I. Accuracy of seepage meter estimates of lake seepage. *J Hydrol* 119, 105–120.

Stannard, D. I., and Rosenberry, D. O. (1991). A comparison of short-term measurements of lake evaporation using eddy correlation and energy budget methods. *J Hydrol* 122, 15–22.

Su, M., Stolte, W. J., and van der Kamp, G. (2000). Modelling Canadian prairie wetland hydrology using a semi-distributed streamflow model. *Hydrol Process* 14, 2405–2422.

Swanson, G. A., and Duebbert, H. F. (1989). Wetland habitats of waterfowl in the prairie pothole region. In: *Northern Prairie Wetlands*. (van der Valk, A., Ed.), pp. 228–267. Iowa State University Press, Ames, IA.

Thornthwaite, C. W. (1948). *Micrometeorology of the Surface Layer of the Atmosphere.* C. W. Thornthwaite Associates, Centerton, NJ.

Todhunter, P. E., and Rundquist, B. C. (2004). Terminal lake flooding and wetland expansion in Nelson County, North Dakota. *Phys Geogr* 25, 68–85.

van der Kamp, G., and Hayashi, M. (1998). The groundwater recharge function of small wetlands in the semi-arid Northern Prairies. *Great Plains Res* 8, 39–56.

van der Kamp, G., Hayashi, M., and Gallén, D. (2003). Comparing the hydrology of grassed and cultivated catchments in the semi-arid Canadian prairies. *Hydrol Process* 17, 559–575.

van der Kamp, G., Stolte, W. J., and Clark, R. G. (1999). Drying out of small prairie wetlands after conversion of their catchments from cultivation to permanent brome grass. *J Hydrol Sci* 44, 387–397.

van der Valk, A. G. (2000). Vegetation dynamics and models. In: *Prairie Wetland Ecology.* (Murkin, H. R., van der Valk, A. G., and Clark, W. R., Eds.), pp. 125–161. Iowa State University Press, Ames, IA.

Weiss, M., Baret, F., Smith, G. J., Jonckheere, I., and Coppin, P. (2004). Review of methods for in situ leaf area index (LAI) determination. Part II. Estimation of LAI, errors and sampling. *Agr Forest Meteorol* 121, 37–53.

Winter, T. C. (1989). Hydrologic studies of wetlands in the northern prairies. In: *Northern Prairie Wetlands.* (van der Valk, A., Ed.), pp. 17–54. Iowa State University Press, Ames, IA.

Winter, T. C. (1999). Relation of streams, lakes, and wetlands to groundwater flow systems. *Hydrogeol J* 7, 28–45.

Winter, T. C., Harvey, J. W., Franke, O. L., and Alley, W. M. (1998). *Ground Water and Surface Water: A Single Resource.* U.S. Geological Survey Circular 1139. U.S. Geological Survey, Denver, CO.

Winter, T. C., and Rosenberry, D. O. (1995). The interaction of ground water with prairie pothole wetlands in the Cottonwood Lake Area, east-central North Dakota, 1979-1990. *Wetlands* 15, 193–211.

Winter, T. C., and Rosenberry, D. O. (1998). Hydrology of prairie pothole wetlands during drought and deluge: a 17-year study of the Cottonwood Lake wetland complex in North Dakota in the perspective of longer term measured and proxy hydrological records. *Clim Change* 40, 189–209.

Winter, T. C., Rosenberry, D. O., and Sturrock. A. M. (1994). Evaluation of empirical methods of determining evaporation for a small lake in the north-central United States. *Water Resour Res* 31, 983–993.

11

Development of Post-Disturbance Vegetation in Prairie Wetlands

Arnold G. van der Valk
Iowa State University

INTRODUCTION

Why does a particular plant species or assemblage of plant species grow in a given place at a given time? In some cases, the answer to this question may be self-evident, for example, the species composition of a crop field in Iowa on July 1, 2005. In other cases, an answer may be beyond our grasp for the time being because of the complexity of the vegetation and absence of essential studies of its development, for example, a particular patch of lowland tropical rainforest in the Amazon basin on July 1, 2005. It has long been recognized (Cittadino, 1990) that plants are adapted to specific environmental conditions (e.g., soil moisture, soil chemistry, air temperature, and light levels) and that the distribution of plants varies along environmental gradients (McIntosh, 1967; Whittaker, 1967). Early on, consequently, plant ecologists believed that it should be possible to predict the distribution of species along environmental gradients on the basis of their physiological tolerances. The distribution of plant species along coenoclines (i.e., community gradients found along environmental gradients), however, is not necessarily solely a function of the physiological tolerances of adult plants. Plant species first have to reach a suitable site and become established for them to be found along a particular coenocline. In other words, seed dispersal patterns and seed germination requirements can also play a role in the distribution of plants along environmental gradients (van der Valk, 1992).

Plant ecologists have taken three different approaches to understanding and, ultimately, to predicting the distribution of species along coenoclines: (1) conducting observational studies of the distribution of species along environmental gradients; (2) doing experimental studies of coenocline development; and (3) developing mathematical models of coenocline development. Studies of coenocline development are still rare because it is often difficult to manipulate environmental conditions along coenoclines. Freshwater wetlands make an ideal system for studying coenoclines because their coenoclines are often very short and the main factor controlling the distribution of species along them is usually water depth (Spence, 1982). Water depths often change cyclically in wetlands (e.g., in North American prairie wetlands), and water depths can be manipulated in wetlands.

Prairie wetlands are found in the northcentral part of North America (van der Valk, 1989; 2005a). Major observational studies of the vegetation dynamics of prairie wetlands can be found in Walker (1959; 1965), Weller and Spatcher (1965), Millar (1973), Weller and Fredrickson (1974), van der Valk and Davis (1978), Shay and Shay (1986), Kantrud *et al.* (1989a; 1989b), Shay *et al.* (1999), Swanson *et al.* (2003), Euliss *et al.* (2004), and van der Valk (2005a). Reviews of observational studies are found in Kantrud *et al.* (1989a; 1989b) and Euliss *et al.* (2004). Information on water level fluctuations in prairie wetlands can be found in Euliss and Mushet (1996), LaBaugh *et al.* (1987; 1998), Woo and Rowsell (1993), Winter and Rosenberry (1995; 1998), Winter (1989; 2003), van der Kamp and Hayashi (1998), van der Kamp *et al.* (1999; 2003), Johnson *et al.* (2004), van der Valk (2005b), and Chapter 10 by Hayashi and van der Kamp.

In the prairie pothole region, changes in total annual precipitation result in oscillatory waterlevel fluctuations (van der Valk, 2005b). There are a variety of possible responses to these water level fluctuations along coenoclines in prairie wetlands (Euliss *et al.*, 2004; van der Valk, 2005b). The cyclical elimination and re-establishment of emergent vegetation caused by water levels described by Weller and Spatcher (1965) and van der Valk and Davis (1978) are features only of deeper and usually larger prairie wetlands. These vegetation cycles, locally called wet-dry cycles, typically have a period of 10 to 20 years. In these wetlands, high water levels periodically extirpate all or most of their emergent vegetation, and it becomes re-established during subsequent periods when standing water is absent or very shallow (drawdowns). In other words, because high water results in the destruction of most of their vegetation, these wetlands undergo periodic disturbances (*sensu* Grime, 1979). Unlike most other kinds of disturbances (e.g., fire), which are very intense, brief phenomena, periods of high water have to last for years for them to destroy the vegetation; even then, only a subset of the species, primarily emergent species, may be killed.

These periodic disturbances provide an opportunity to study how species become re-established along elevation gradients when water levels drop, that is, coenocline development. Prairie wetlands are a good model system for studying the development of post-disturbance coenoclines. One, they are herbaceous wetlands whose species reach maturity in 1 to 3 years. Two, the entire flora of these wetlands is present in their seed bank, which eliminates dispersal uncertainties. Three, because they contain a variety of emergent vegetation types along an elevation (water depth) gradient, it is feasible to study the development of an entire coenocline, not just one vegetation type. Four, it is possible to manipulate their water levels and to do controlled studies of the destruction and redevelopment of their coenoclines.

In this chapter, I review what is known about post-disturbance coenocline development in prairie wetlands. I emphasize identifying factors (e.g., seed germination, seedling survival, water depth tolerances) that control the establishment and distribution of species along an elevation gradient. Over the last 40 years, the effects of water level changes on the coenoclines in prairie wetlands have been investigated by using observational studies, experimental studies, and models. The results of experimental and modeling studies will be emphasized in this chapter, especially those from a 10-year experimental study of water level changes, the Marsh Ecology Research Program (MERP), carried out in an experimental complex in the Delta Marsh, Manitoba, Canada (Murkin *et al.*, 2000). After some brief background information about wet-dry cycles and MERP, I address three topics: (1) the development of a new coenocline when pre- and post-disturbance water levels are the same; (2) the development of new coenoclines when post-disturbance water levels differ from pre-disturbance water levels; and (3) the latest models of coenocline development.

WET-DRY CYCLES

The various stages during wet-dry cycles caused by water level fluctuations were first described by Weller and Spatcher (1965). They outlined a five-stage cycle: dry marsh (drawdown), dense marsh (mostly emergent cover), hemi-marsh (50% emergent, 50% open water), open marsh (more than 50% open water), and open water marsh (emergents largely or completely eliminated). The open water marsh persists until the next drought, which lowers water levels, eventually exposing the marsh sediments. This marks the start of the next dry marsh stage. Later studies of these wet-dry cycles by van der Valk and Davis (1976; 1978) established the central role of seed banks (viable seed in the sediments) for the persistence of plant species in these wetlands and the importance of the dry marsh stage for the re-establishment of emergent species. Fig. 1, which is taken from van der Valk and Davis (1978), outlines the vegetation cycle that occurs in deeper and larger prairie potholes [i.e., classes IV and V of Stewart and Kantrud (1971; 1972)], in which a significant range of water levels (ca. 1.5 to 2 m) can occur. In their version of the wet-dry cycle, van der Valk and Davis (1978) modified the cycle proposed by Weller and Spatcher (1965) by combining the hemi-marsh and open marsh stages, during both of which emergent

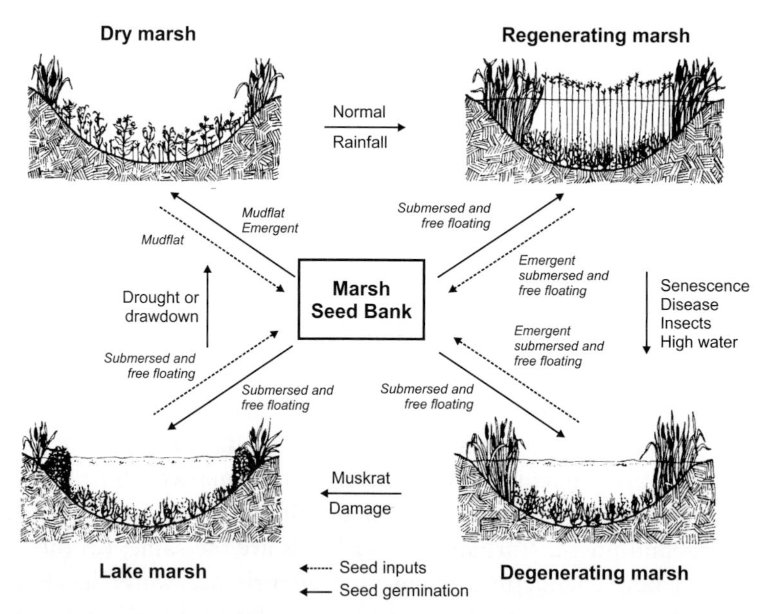

FIG. 1 Idealized wet-dry cycle in a semi-permanent prairie pothole. (From van der Valk and Davis [1978].)

vegetation is declining. The four stages in the van der Valk and Davis cycle (Fig. 1) are the dry marsh, regenerating marsh (= dense marsh), degenerating marsh (= hemi-marsh and open marsh), and lake marsh (= open water marsh). It is the re-development of the emergent sections of coenoclines initiated during the dry marsh stage and the fate of the species along them during the regenerating marsh stage that is the focus of this chapter.

MARSH ECOLOGY RESEARCH PROGRAM

Wet-dry cycles are not confined to palustrine wetlands but also occur in prairie lacustrine wetlands. Walker (1959; 1965) studied vegetation changes caused by water level fluctuations in the Delta Marsh at the south end of Lake Manitoba, Manitoba, Canada. The range of water levels observed in the Delta Marsh historically was about 1.5 m, which is similar to that observed in deeper palustrine prairie-pothole wetlands. In the early 1960s, water level control structures were built around Lake Manitoba to prevent flooding of farms and villages near the lake during wet years. This significantly reduced the range of water level fluctuations in the Delta Marsh and changed its flora and fauna; most notably it caused a large decline in the number of waterfowl nesting in the marsh (Batt, 2000). Concerns about the condition of the Delta Marsh because of the stabilization of its water levels prompted the organization of MERP.

As part of MERP, 10 experimental wetlands were constructed in 1979 by partitioning off a portion of Delta Marsh into 10 contiguous, rectangular cells, each between 5.5 and 7 ha in area (Fig. 2). Beginning in 1980, water levels were manipulated in these cells for the next 10 years to simulate a wet-dry cycle. There were three water level treatments (Fig. 3): normal, medium, and high. During the first year or two of MERP, water levels in the cells were maintained at the stabilized water level in the Delta Marsh (247.5 m); this is called the baseline period. (All water levels are elevations above mean sea level.) This was followed by 2 years of high water (248.5 m), the deep flooding period, and 1 or 2 years of low water (247.0 m), the drawdown period. During the drawdown years, the cells were free of standing water. Because the length of the drawdown had little impact on coenocline development, it is ignored in this chapter. During the last 5 years, the reflooding period, water levels were maintained at 247.5 m in the normal treatment, at 247.80 m in the medium, and at 248.1 m in the high treatment. There were four replicates of normal and three replicates of medium and high treatments. More detailed descriptions of MERP and the MERP cells can be found in Murkin *et al.* (2000).

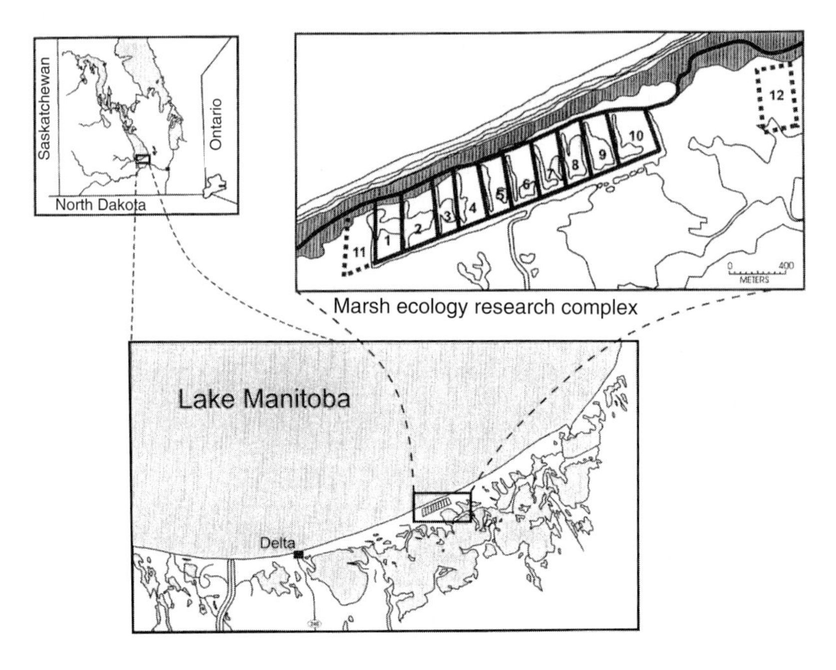

FIG. 2 Location of the Marsh Ecology Research Program study site in the Delta Marsh, Manitoba, Canada. (From Murkin *et al.* [2000].)

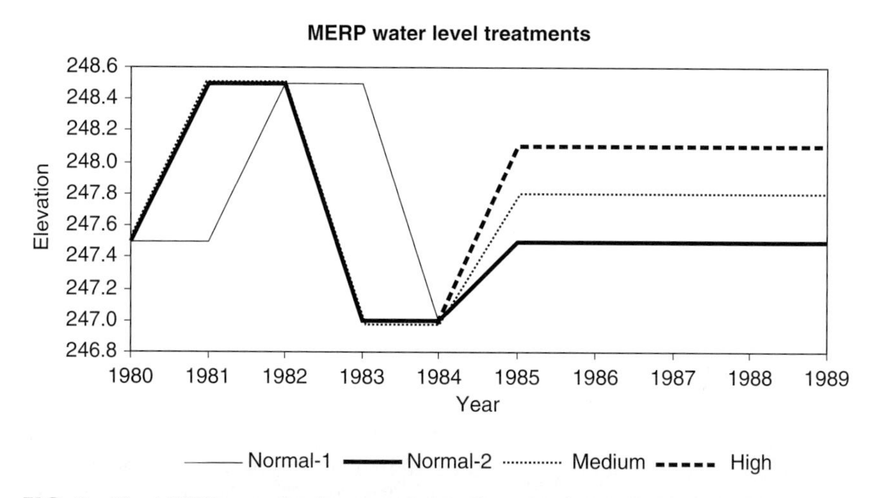

FIG. 3 The MERP water level treatments. In the normal-1 treatment, cells were drawn down (247 m) for only 1 year (1984), while in the normal-2, medium, and high treatments, cells were drawn down for 2 years (1983 and 1984). The baseline year when all cells were at 247.5 m was 1980. Each cell was deep flooded (248.5 m) for 2 years beginning in 1981 or 1982. The three reflooding treatments, normal (247.5 m), medium (247.8 m), and high (248.1 m), began in 1985 and ended in 1989.

The coenocline that was found in the cells at the beginning of MERP consisted primarily of four monodominant emergent zones and a submersed aquatic zone (Fig. 4). *Scolochloa festucacea* (Willd.) Link and *Phragmites australis* (Cav.) Trin. zones covered most of the upper sections of the gradient. The lower sections were dominated by *Typha glauca* Godr. and *Scirpus lacustris* L. spp. *glaucus* (Sm.) Hartm. The submersed zone at the lowest elevations was dominated by a mixture of *Potamogeton* spp., especially *P. pectinatus* L., *Utricularia vulgaris* L., *Myriophylum spicatum* L., *Najas flexilis* (Willd.) Rostk. & Schmidt, *Ceratophyllum demersum* L. Plant nomenclature is based on Scoggan (1978–1979).

Among the coenocline-related studies conducted as part of MERP were a detailed examination of the seed banks of the MERP complex before the start of water level treatments (Pederson, 1981; Pederson and van der Valk, 1985), field studies of recruitment during the drawdown years (van der Valk, 1986; Welling *et al.*, 1988a; 1988b), field studies of the fates of emergent species when the cells were flooded (Meredino and Smith, 1991; van der Valk, 1994; 2000; van der Valk *et al.*, 1994), experimental studies of recruitment from seed banks (van der Valk and Pederson, 1989; Meredino *et al.*, 1990; Meredino and Smith, 1991), seed germination studies of the dominant species (Galinato and van der Valk, 1986), and water-depth tolerance studies of the dominant emergent species (Squires and van der Valk, 1992). In addition, two new models were developed to predict the distribution of emergent species along coenoclines in the MERP cells (de Swart *et al.*, 1994; van der Valk, 2000; Seabloom *et al.*, 2001).

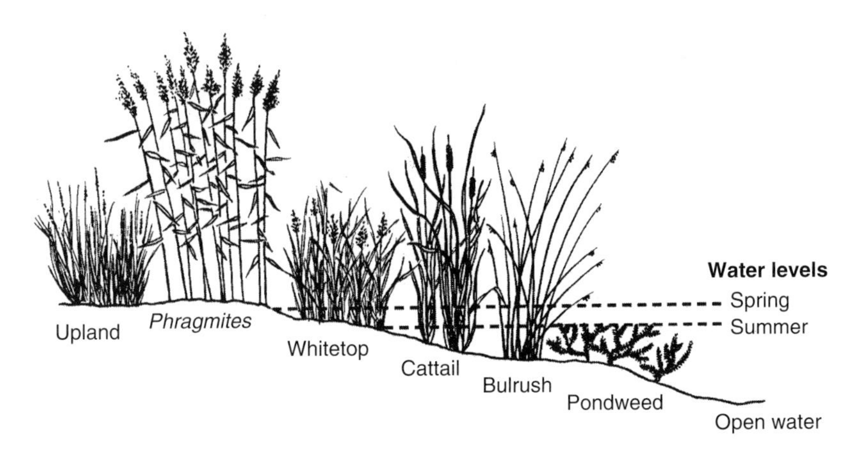

FIG. 4 Idealized coenocline in the Marsh Ecology Research Program cells in the Delta Marsh, Manitoba, Canada. (From Batt [2000].)

COENOCLINE DEVELOPMENT: SAME PRE- AND POST-DISTURBANCE WATER LEVELS

During the drawdown period (Fig. 3), new coenoclines began to develop in the MERP cells (Welling *et al.*, 1988a; 1988b; van der Valk and Welling, 1988). Several factors or some combination of them (Fig. 5) could be responsible for the final distribution of emergent species along these new coenoclines, including the following:

1. Composition of seed bank along the elevation gradient
2. Differential seed germination along the elevation gradient
3. Differential seedling survival along the elevation gradient during the drawdown
4. Differential adult survival along the elevation gradient after reflooding, and
5. Exploitative competition

The relative abundances (i.e., percentage of the total number of seeds, seedlings, or adults in a given elevation range out of the total number found over the entire elevation gradient) of *Scolochloa festucacea*, *Phragmites australis*, *Typha glauca*, and *Scirpus lacustris* as seeds, seedlings, and adults along coenoclines in the MERP cells are presented in Tables 1 to 4. Only data for the normal treatment in the reflooding years is presented in these tables.

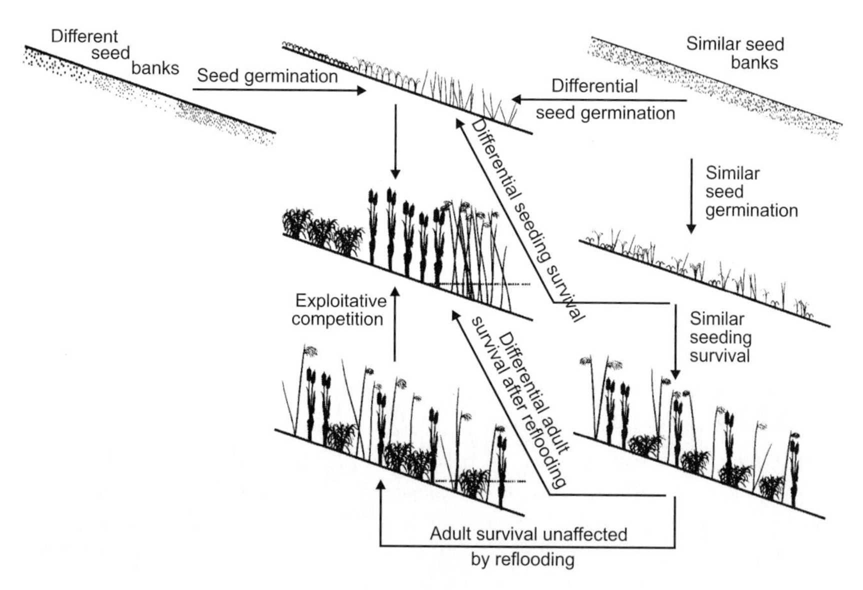

FIG. 5 The major factors that may affect the distribution of species along a post-disturbance coenocline. (From van der Valk and Welling [1988].)

The baseline data illustrate the pre-disturbance distribution of adults of these species along the elevation gradient and the seedbank data the pre-disturbance distribution of their seeds. The drawdown and reflooding data illustrate the post-disturbance distributions of their seedlings and adults.

Seeds of the four dominant emergents were found over a wider elevation range than adults of the same species. Seed densities were highest for most species at 247.5 m, the water's edge. Secondary dispersal by wind and water currents had a significant impact on the distribution of seeds of all species along the coenocline in the Delta Marsh (Pederson and van der Valk, 1985). During the baseline period, adult populations were most abundant at higher elevations (*Phragmites australis* and *Scolochloa festucacea*) or lower elevations (*Scirpus lacustris*) than their seeds in the seed bank. Because the pre-disturbance distribution of their seeds was over a wider range than the post-disturbance distribution of adult plants, seed distribution *per se* did not affect the final distribution of the dominant emergent species along post-disturbance coenoclines.

The elevations at which seedlings of emergent species were most abundant during the first year of the drawdown were either higher (e.g., *Scolochloa festucacea*) or lower (e.g., *Typha glauca*) than the elevations at which their seeds were most abundant. The largest discrepancy between seed and seedling distribution patterns was for *Phragmites australis*. The seeds of *P. australis* had a bimodal distribution, with maxima at 247.9 to 248.0 m and at 247.5 m. No seedlings of *Phragmites* were found during the drawdown years (or subsequently) at the higher elevations (Table 2). Seedling densities were much lower for *Scirpus lacustris*, *Scolochloa festucacea,* and *Typha glauca* during the second year of the drawdown (van der Valk and Welling, 1988). The net effect of this decline was that maximum seedling densities for most species were found at lower elevations during the second year of the drawdown. This shift resulted in *Phragmites australis* seedlings being most abundant at 247.4 m, an elevation at which very few adult plants were found in the pre-disturbance coenocline. Overall, the percentage overlap of adult distributions during the baseline years and seedlings during the second year of the drawdown were 59% for *Scolochloa*, 22% for *Phragmites*, 38% for *Typha*, and 67% for *Scirpus*. In other words, although all four dominant species were found along both the pre- and post-disturbance coenoclines, their initial post-disturbance distributions, especially those of *Phragmites* and *Typha*, were different from those on the pre-disturbance coenocline. This difference was primarily caused by seedling establishment (germination, survival) patterns.

TABLE 1 **Relative (%) Abundance of *Scolochloa festucacea* in the Marsh Ecology Research Program Cells During the Baseline Year, in the Seed Bank, During the First and Second Years of the Drawdown, and During the 5 Reflooding Years***

Elevation (m)	Baseline	Seedbank	Drawdown 1	Drawdown 2	Reflooding 1	Reflooding 2	Reflooding 3	Reflooding 4	Reflooding 5
248	0	8	1	4	0	0	0	0	0
247.9	2	3	0	1	0	0	0	0	0
247.8	14	5	15	3	0	1	5	9	5
247.7	**40**	13	**37**	**25**	5	**16**	**37**	**34**	**50**
247.6	**37**	**25**	19	**23**	22	**47**	**44**	**54**	**44**
247.5	8	**34**	**25**	**23**	**41**	15	9	3	1
247.4	0	10	3	19	**26**	13	4	0	0
247.3	0	0	0	0	2	2	1	0	0
247.2	0	0	0	3	2	0	0	0	0
247.1	0	0	0	0	3	6	0	0	0

*Relative abundance is based on areal cover during the baseline years and relative number of seeds, seedling, or shoots per unit area during all other years. The elevations with the highest relative abundances are given in bold. Only data for the cells in the normal treatment are presented for the reflooding years.

TABLE 2 **Relative (%) Abundance of *Phragmites australis* in the Marsh Ecology Research Program Cells During the Baseline Year, in the Seed Bank, During the First and Second Years of the Drawdown, and During the 5 Reflooding Years***

Elevation (m)	Baseline	Seedbank	Drawdown 1	Drawdown 2	Reflooding 1	Reflooding 2	Reflooding 3	Reflooding 4	Reflooding 5
248	16	**20**	0	0	0	0	0	0	0
247.9	**23**	**20**	0	0	0	0	0	0	0
247.8	**21**	9	12	0	0	0	0	0	0
247.7	16	8	8	20	0	0	8	11	19
247.6	14	11	12	1	0	0	2	1	2
247.5	6	**20**	23	3	38	**40**	**34**	**40**	**40**
247.4	2	6	**30**	**60**	**48**	**42**	**34**	**40**	33
247.3	0	2	15	13	1	5	11	8	4
247.2	0	1	0	3	12	14	11	1	2
247.1	0	3	0	0	0	0	0	0	0

*Relative abundance is based on areal cover during the baseline years and relative number of seeds, seedling, or shoots per unit area during all other years. The elevations with the highest relative abundances are given in bold. Only data for the cells in the normal treatment are presented for the reflooding years.

TABLE 3 Relative (%) Abundance of *Typha glauca* in the Marsh Ecology Research Program Cells During the Baseline Year, in the Seed Bank, During the First and Second Years of the Drawdown, and During the 5 Reflooding Years[*]

Elevation (m)	Baseline	Seedbank	Drawdown 1	Drawdown 2	Reflooding 1	Reflooding 2	Reflodding 3	Reflooding 4	Reflooding 5
248	0	4	0	0	0	0	0	0	0
247.9	0	3	0	0	0	0	0	0	0
247.8	1	8	0	0	0	0	0	0	0
247.7	7	6	2	1	0	1	1	2	2
247.6	19	17	0	2	0	0	8	6	5
247.5	**31**	**19**	9	6	6	6	23	19	23
247.4	**24**	**33**	7	8	14	9	7	10	13
247.3	14	8	**77**	13	**36**	**50**	**36**	**44**	**31**
247.2	3	1	4	**65**	**43**	**34**	**25**	**20**	**27**
247.1	0	2	0	5	0	0	0	0	0

[*]Relative abundance is based on areal cover during the baseline years and relative number of seeds, seedling, or shoots per unit area during all other years. The elevations with the highest relative abundances are given in bold. Only data for the cells in the normal treatment are presented for the reflooding years.

TABLE 4 Relative (%) Abundance of *Scirpus lacustris* in the Marsh Ecology Research Program Cells During the Baseline Year, in the Seed Bank, During the First and Second Years of the Drawdown, and During the 5 Reflooding Years*

Elevation (m)	Baseline	Seedbank	Drawdown 1	Drawdown 2	Reflooding 1	Reflooding 2	Reflooding 3	Reflooding 4	Reflooding 5
247.9	0	0	0	1	0	0	0	0	0
247.8	0	1	3	0	0	0	0	0	0
247.7	0	3	7	0	0	0	0	1	1
247.6	2	12	1	1	0	0	0	0	0
247.5	15	**48**	8	4	3	4	14	**28**	**38**
247.4	**34**	18	**66**	21	**34**	**27**	**28**	**50**	**50**
247.3	**32**	8	13	**24**	21	21	**25**	8	0
247.2	15	7	2	**43**	**36**	**38**	23	12	7
247.1	2	3	0	6	8	11	10	1	3

*Relative abundance is based on areal cover during the baseline years and relative number of seeds, seedling, or shoots per unit area during all other years. The elevations with the highest relative abundances are given in bold. Only data for the cells in the normal treatment are presented for the reflooding years.

Studies of the effect of environmental conditions on seed germination from the seed bank in the Delta Marsh (van der Valk and Pederson, 1989; Meredino *et al.*, 1990) and in other prairie wetlands (Seabloom *et al.*, 1998) have shown that seedling establishment is highly dependent on environmental conditions during drawdowns (Table 5), especially soil moisture, soil salinity, and soil temperature. Differences in soil moisture can alter not only seedling densities but also the species composition of the vegetation that develops during drawdowns (van der Valk and Pederson, 1989). Likewise, seedling survival is highly dependent on adequate soil moisture. The percentage germination of seeds of emergent species was not uniform over the entire elevation range over which they were found (van der Valk and Welling, 1988).

During the reflooding years in the normal treatment cells when these cells were flooded again to 247.5 m, the distribution of three species, *Scolochloa*, *Typha*, and *Scirpus*, gradually became more similar to their distribution along the pre-disturbance coenocline. By the fifth year of the reflooding period (Table 1), the percentage overlap in the distribution of *Scolochloa* with that during the baseline years was 89%. For *Typha* (Table 3) and *Scirpus* (Table 4), their overlap was 60%. The water depth tolerances of the adults eventually resulted in the post-disturbance coenocline coming to resemble the pre-disturbance coenocline more closely. For *Phragmites*, however, the overlap was only 27%. The post-disturbance coenocline also differs from the pre-disturbance coenocline in that *Typha* had its highest relative abundance at the lowest elevations, not *Scirpus lacustris*. Even after 7 years (Fig. 6), the distribution of emergent species along the post-disturbance coenocline was not identical to that along the pre-disturbance coenocline.

Why was the post-dispersal coenocline different from the pre-disturbance coenocline? Although all four emergent species became re-established along the post-dispersal coenoclines, they did not become re-established initially

TABLE 5 Number of Seedlings Recruited from Seed Bank Samples from the Delta Marsh in Drawdowns that Began in May, June, July, and August 1988*

Species	May	June	July	August
Scolochloa festucacea	10	39	58	0
Phragmites australis	16	21	11	0
Typha glauca	140	495	311	18
Scirpus lacustris	300	218	130	0

*Adapted from Meredino *et al.* (1990).

within the elevation ranges at which they occurred in the pre-disturbance coenoclines. After establishment, these species were not able to adjust their distributions. The only changes that occurred during the five reflooding years along the elevation gradient were that emergent species declined in abundance in areas too deep for them to persist and expanded clonally in areas with more optimal water depths. They never moved up- or down-slope into more

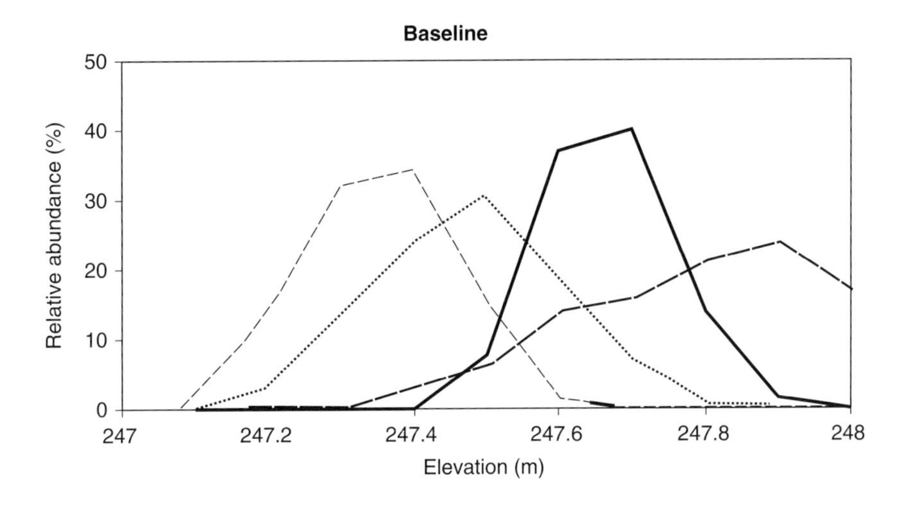

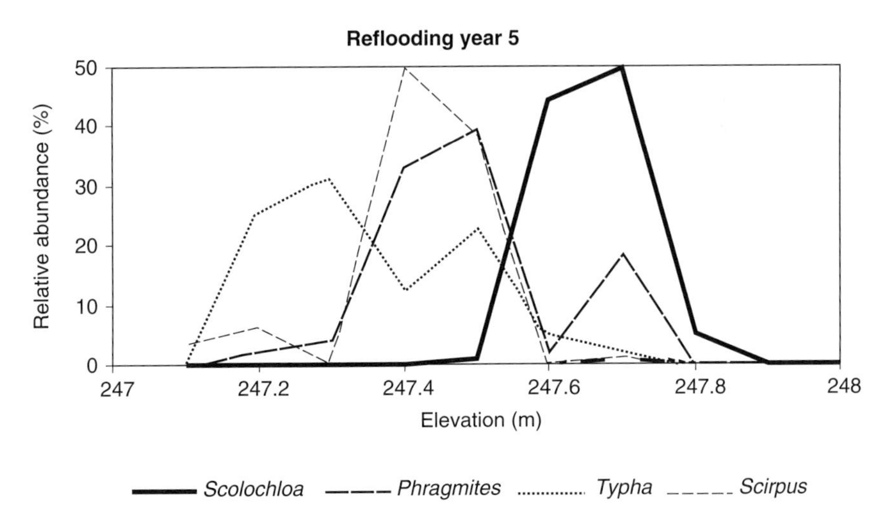

FIG. 6 Relative abundance of *Scolochloa festucea*, *Phragmites australis*, *Typha glauca*, and *Scirpus lacustris* along the elevation gradient in the Marsh Ecology Research Program cells during the pre-disturbance (baseline) and post-disturbance (fifth year of reflooding) coenoclines.

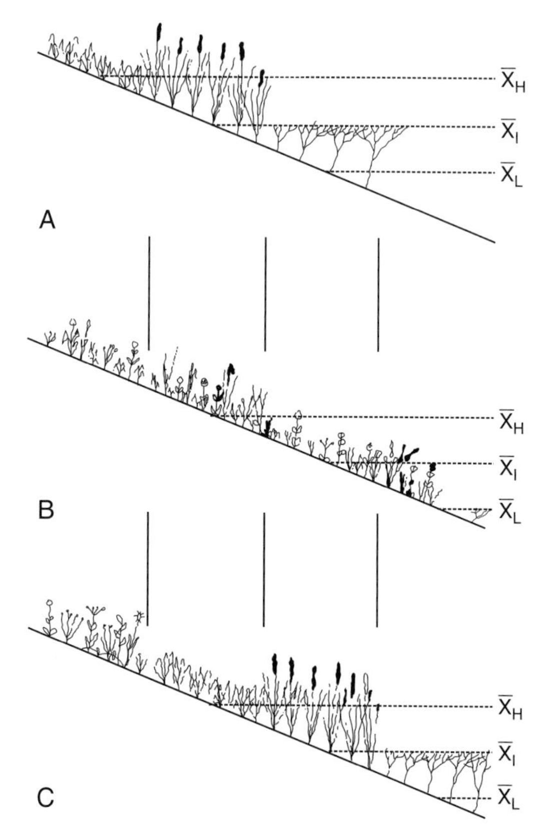

FIG. 7 Effect of permanently lowering mean interannual (x_i) water level on a wetland coenocline: (**A**) the original coenocline, (**B**) the first year or two after water levels are lowered, (**C**) the coenocline under the new water-level regime. (From van der Valk [1991].)

optimal water depths as is generally predicted to occur. There were two major reasons for their inability to adjust their distributions. One, the seeds of these four emergent species do not normally germinate under water (van der Valk, 1981), and two, there was no indication that the seeds of any emergent germinated in the field during the reflooding years in the MERP cells. Thus, these species could not become established at the higher or lower elevations by recruitment from the seed bank. Initially, the clonal spread of these species at elevations at which they became established was restricted to areas that were free of other emergent species. It was only in the last few years of the reflooding period, when species growing at water levels that they could not tolerate began to die, that any significant expansion of emergent species upslope

(*Phragmites*, Table 2) or downslope (*Typha*, Table 3) occurred. In summary, recruitment patterns of emergent species along newly forming coenoclines had lasting effects, and the distribution of species along coenoclines was not simply a function of their water depth tolerances.

COENOCLINE DEVELOPMENT: DIFFERENT PRE- AND POST-DISTURBANCE WATER LEVELS

In prairie wetlands, post-disturbance water levels may not necessarily return to pre-disturbance water levels. Water levels in wetlands are also sometimes permanently altered to improve the wetland as waterfowl or fish habitat, to reduce flooding in adjacent areas, to reduce the area of unwanted wetland species, and for many other reasons. It is often assumed that raising or lowering water levels will cause the various vegetation zones (i.e., the entire coenocline) in a wetland to shift up- or downslope (van der Valk, 1994). However, there is evidence that wetland coenoclines may not respond in the same way to raising water levels as to lowering them (Bukata *et al.*, 1988). In other words, wetland coenoclines can shift downslope when water levels are permanently lowered (Fig. 7), but may not be able to shift upslope when water levels are permanently raised (Fig. 8). When water levels are lowered, coenoclines can shift downslope largely because the emergent species can move downslope either clonally or by becoming established from seed at lower elevations that are not flooded. As seen in the MERP cells in which water levels were returned to pre-disturbance levels, emergent species cannot become established from seed at flooded higher elevations and they seem to have only a limited ability to shift their position through clonal growth.

When water levels are permanently raised, coenoclines have been predicted to respond in two ways (van der Valk, 1994): (1) the entire coenocline shifts upslope (the migration model) and (2) the coenocline is significantly altered in composition (the extirpation model). The latter model implies little or no shift in the location of species along the elevation gradient except their extirpation at elevations at which they cannot tolerate the water depth. Specifically, the extirpation model predicts that some species may be eliminated or significantly reduced in abundance and that the post-disturbance coenocline will be compressed or shortened when compared to the pre-disturbance coenocline. There were no barriers within or at the upper ends of the cells that would prevent emergent species from moving into more optimal water depths upslope. In other words, there were no dikes at the upper ends of these cells.

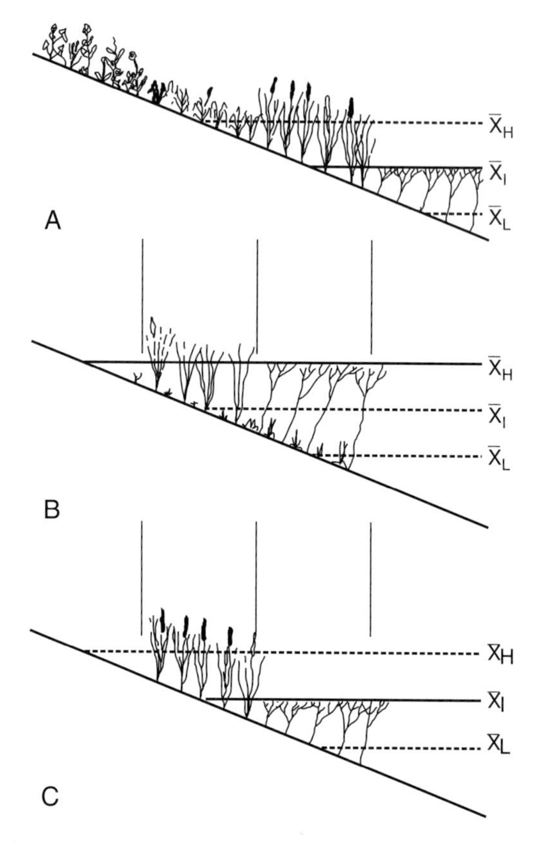

FIG. 8 Effect of permanently raising mean interannual (x_i) water level on a wetland coenocline: (**A**) the original coenocline, (**B**) the first year or two after water levels are raised, (**C**) the coenocline under the new water-level regime. (From van der Valk [1991].)

In the medium and high treatments of MERP, post-disturbance water levels were raised permanently by 30 and 60 cm, respectively, above the pre-disturbance water level of 247.5 m. Because the cells in the medium and high treatment experienced the same deep flooding period (disturbance) that eliminated emergents and the same drawdown period, the initial post-disturbance coenoclines in cells in these treatments were identical to those previously described in the normal treatment.

An experimental study of the water depth tolerances of the dominant emergent species in the Delta Marsh (Squires and van der Valk, 1992) indicated that these emergents fell into two fairly distinct groups. Some species (*Scirpus lacustris* ssp. *glaucus* and *Typha glauca*) grew best in deep water (ca. +45 cm), while others (*Scolochloa festucacea, Phragmites australis*, and *Scirpus lacustris* ssp. *validus (Vahl) T. Koyama*) grew best in shallow water (+20 cm or less). This pattern is consistent with their distribution in the MERP cells during the baseline period (Tables 1–4) when only *Scirpus lacustris* ssp. *glaucus* was present in the cells. To illustrate the response of shallow and deep water species to the medium and high treatments, Tables 6 and 7 give the mean water depth and frequency of occurrence of *Scolochloa*, a shallow water species, and *Typha*, a deep water species, in all three treatments during the reflooding period (1985–1989).

TABLE 6 Mean Water Depth and Frequency of *Scolochloa festucacea* in the Normal, Medium, and High Water-Level Treatments from 1985 to 1989*

	Year					
Treatment	**1985**	**1986**	**1987**	**1988**	**1989**	**Mean**
Mean water depth (m)						
Normal	−0.02	0.01	−0.05	−0.08	−0.06	−0.04
Medium	0.15	0.23	0.08	0.04	0.06	0.11
High	0.49	0.45	0.20†	0.09†	0.19†	0.33
Frequency‡						
Normal	149	246	212	169	151	185
Medium	186	240	77	73	62	128
High	115	155	48†	30†	20†	74

*A negative number indicates that the species was growing at an elevation above 247.5 m.
†Species absent from one or more cells in 1987, 1988, and 1989.
‡These are absolute frequencies (i.e., the number of times a species was recorded on a grid sampling of vegetation maps of the relevant cells in a given year. See de Swart *et al.* (1994) for details.

TABLE 7 Mean Water Depth and Frequency of *Typha glauca* in the Normal, Medium, and High Water-Level Treatments from 1985 to 1989

	Year					
Treatment	**1985**	**1986**	**1987**	**1988**	**1989**	**Mean**
Mean water depth (m)						
Normal	0.13	0.10	0.07	0.07	0.10	0.09
Medium	0.25	0.34	0.27	0.25	0.25	0.27
High	0.44	0.57	0.52	0.51	0.42	0.49
Frequency*						
Normal	23	137	146	186	194	137
Medium	66	152	119	123	117	115
High	104	191	129	130	181	147

*These are absolute frequencies (i.e., the number of times a species was recorded on a grid sampling of vegetation maps of the relevant cells in a given year. See de Swart *et al.* (1994) for details.

In the medium and high treatments, as in the normal treatment, *Scolochloa* was more abundant in 1986 than 1985. In 1987, however, *Scolochloa* frequency declined in all three treatments; this decline continued through 1989. In the medium and high treatments, *Scolochloa* frequency in 1987 was nearly 70% lower than in 1986, while in the normal treatment it declined only about 15% (Table 6). During 1985 and 1986, this species was able to survive in areas that were too deep for it to tolerate permanently by mobilizing stored reserves of carbohydrates from rhizomes. These reserves were exhausted by 1987. The decline of *Scolochloa festucacea*, *Scirpus lacustris* ssp. *validus,* and other shallow water species after several years of high water has also been recorded in other studies (Millar, 1973; Neckles, 1984) and is consistent with experimental studies of water-depth tolerances of these emergents (Neil, 1990; 1993; Waters and Shay, 1990; Squires and van der Valk, 1992). Because of the disappearance of *Scolochloa* from deeper parts of the cells, the mean water depth at which it was found declined after 1986 (Table 6). Its mean water depth in the medium treatment was 0.15 m in 1985 and 0.06 m in 1989 while in the high treatment it went from 0.49 m to 0.19 m.

Typha glauca (Table 7) also increased in frequency from 1985 to 1986. It was eliminated from some areas with very deep water in 1987, but only in the medium and high treatments did its frequency decline by about 20% and 30%, respectively, in 1987. Its frequency increased again in 1988 and 1989 because of its clonal spread, mostly into areas where shallow water species had been eliminated. The mean water depth at which *Typha* was found in any treatment, however, changed little from 1985 to 1989. In the medium treatment, it was 0.25 m in 1985 and 1989; in the high treatment, it was 0.44 m in 1985 and 0.42 m in 1989 (Table 7).

By 1989, the coenoclines in the medium and high treatments were significantly different from those in the normal treatment (Fig. 9). *Scirpus lacustris* was undetectable in the medium and high treatments in 1989. The other three emergents were growing in deeper water and were less frequent than in the normal treatment. By 1989, between 35% and 45% of the emergent vegetation in the medium and high treatment cells had been eliminated (van der Valk *et al.*, 1994). Although some emergent losses (ca. 5%) were caused by herbivory by muskrats (Clark and Kroeker, 1993; Clark, 2000), most were caused by the loss of two species (*Scolochloa festucacea* [Table 6] and *Scirpus lacustris* ssp *validus* and *glaucus*). The emergent sections of the coenocline in the normal treatment in 1989 covered an elevation range of 70 cm while those in the medium and high treatments covered a range of 60 cm and 40 cm, respectively (Fig. 9). During the baseline period, it covered a maximum range of 90 cm in the MERP cells. Assuming an equal slope, the coenoclines in the medium and

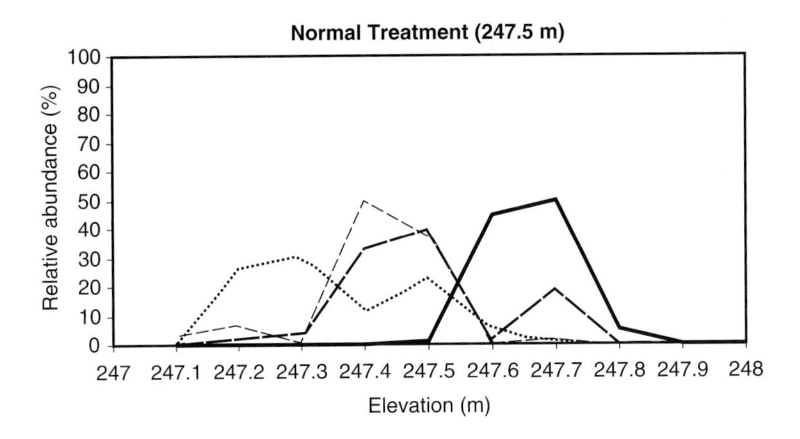

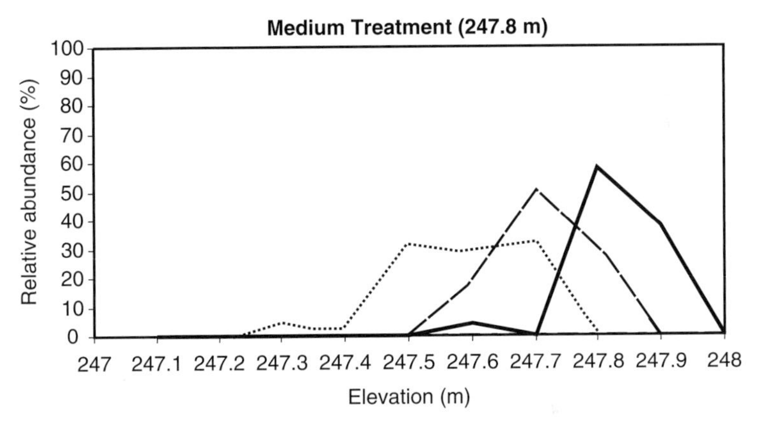

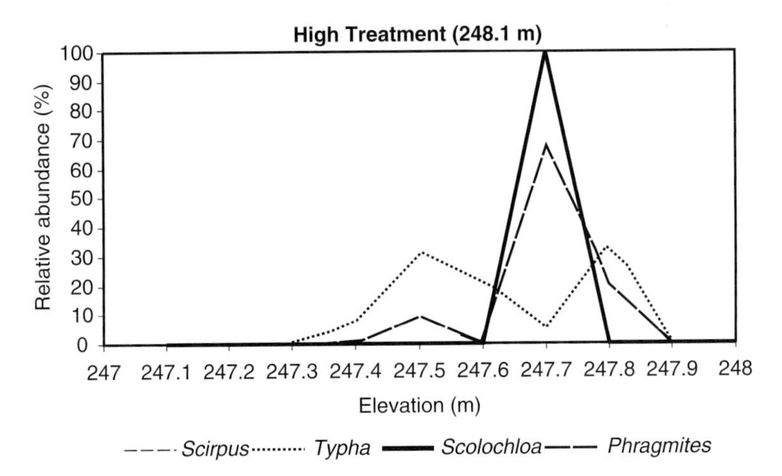

FIG. 9 Mean relative abundance of emergent species along post-disturbance coenoclines in the Marsh Ecology Research Program cells in the normal, medium, and high water treatments during the fifth year of the reflooding period (1989).

high treatments were shorter than those in the normal treatment and during the baseline period. In summary, the medium and high treatment coenoclines had fewer species, lower species frequencies, and smaller elevation ranges than the coenoclines in the normal treatment and during the baseline period.

Species growing along the new coenoclines in the medium and high treatment were unable to shift their positions along the water depth gradient. In areas too deep to support their growth, emergent species were extirpated. Emergents did spread by clonal growth into areas with water depths that they could tolerate, especially after emergents previously growing in these areas had disappeared. All the data from the MERP study support the extirpation model (Fig. 8).

MODELS OF COENOCLINE DEVELOPMENT

The first conceptual model of wet-dry cycles was published by van der Valk (1981). This simple model, based primarily on plant life history characteristics (seed dispersal syndromes, seed germination requirements, and life expectancy), could only predict changes in the overall composition of a wetland's flora during various stages of wet-dry cycles (Fig. 1). For emergent species, as noted, both seed germination and life expectancy are highly dependent on water depth, with the absence of standing water, a drawdown, being a particularly crucial time for the recruitment of emergent species. Because it was not a spatially explicit model, it could not predict the distribution of species along coenoclines.

During MERP, two models of coenocline development were constructed and tested. The simpler of these models is the niche model of de Swart *et al.* (1994). The niche model assumes that water depth alone is sufficient to predict the distribution of emergent species along coenoclines in the MERP cells. The niche model uses logistic regression equations developed from data on the distribution of emergent species along elevation gradients during the pre-disturbance (baseline) years to predict the probability of finding a given emergent species at a given location in the MERP cells in the post-disturbance years. The second model is the spatially explicit model of Seabloom *et al.* (2001). In this cellular automaton model, the cells of the MERP complex are divided into 3 m by 3 m units. In each unit, emergent species can exist in four stages: dormant seeds, seedlings, first-year adults, and mature adults with clonal growth. Five sets of rules determine the probability that a given species would go from one stage to another in a unit (Fig. 10). Mature adults of only a single species can persist in a unit. A species can become dominant in a unit because it

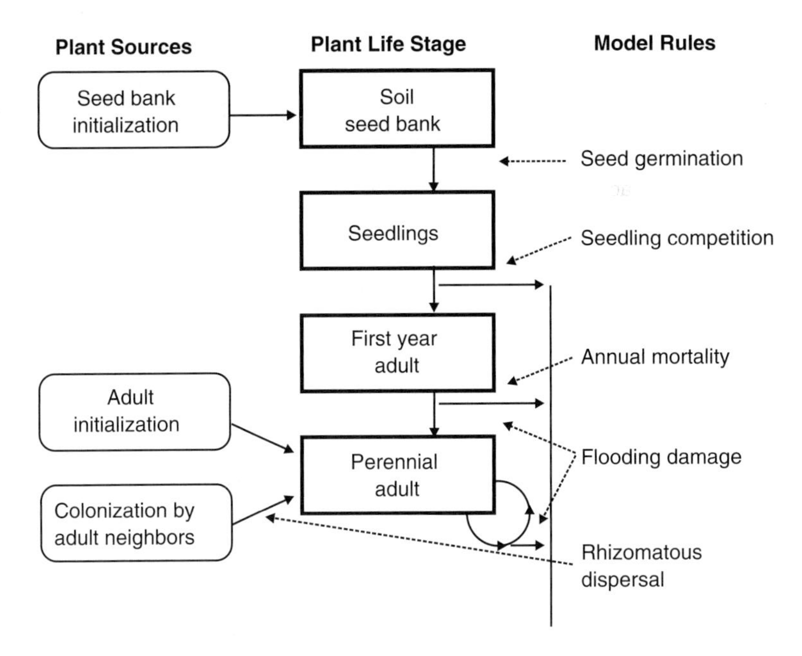

FIG. 10 The structure of the spatially explicit model of coenocline development. From Seabloom *et al.* (2001).

was the dominant species at the start of a simulation run, it colonized an empty adjacent unit, or it germinated from the seed bank and survived as a seedling and first-year adult in the unit (Fig. 10). These and other rules incorporated into the spatially explicit models were derived from a variety of studies of vegetation dynamics in prairie wetlands, not from the MERP studies.

Both the niche and spatially explicit models were used to generate vegetation maps of the MERP cells by predicting the dominant emergent species in each 3 m by 3 m unit in each cell. The elevation of each unit was derived from a detailed topographic map of each MERP cell. This enabled the water depth of every unit in each cell in each year to be estimated. For more model details and procedures used to test these models, see de Swart *et al.* (1994) and Seabloom *et al.* (2001). The simulated vegetation maps generated by each model were compared to actual vegetation maps derived from low-level aerial photographs. Both the simulated maps and actual vegetation maps had the same resolution (3 m by 3 m).

Fig. 11 shows the mean proportion of the MERP cells predicted by both models to be covered with each emergent species and the actual

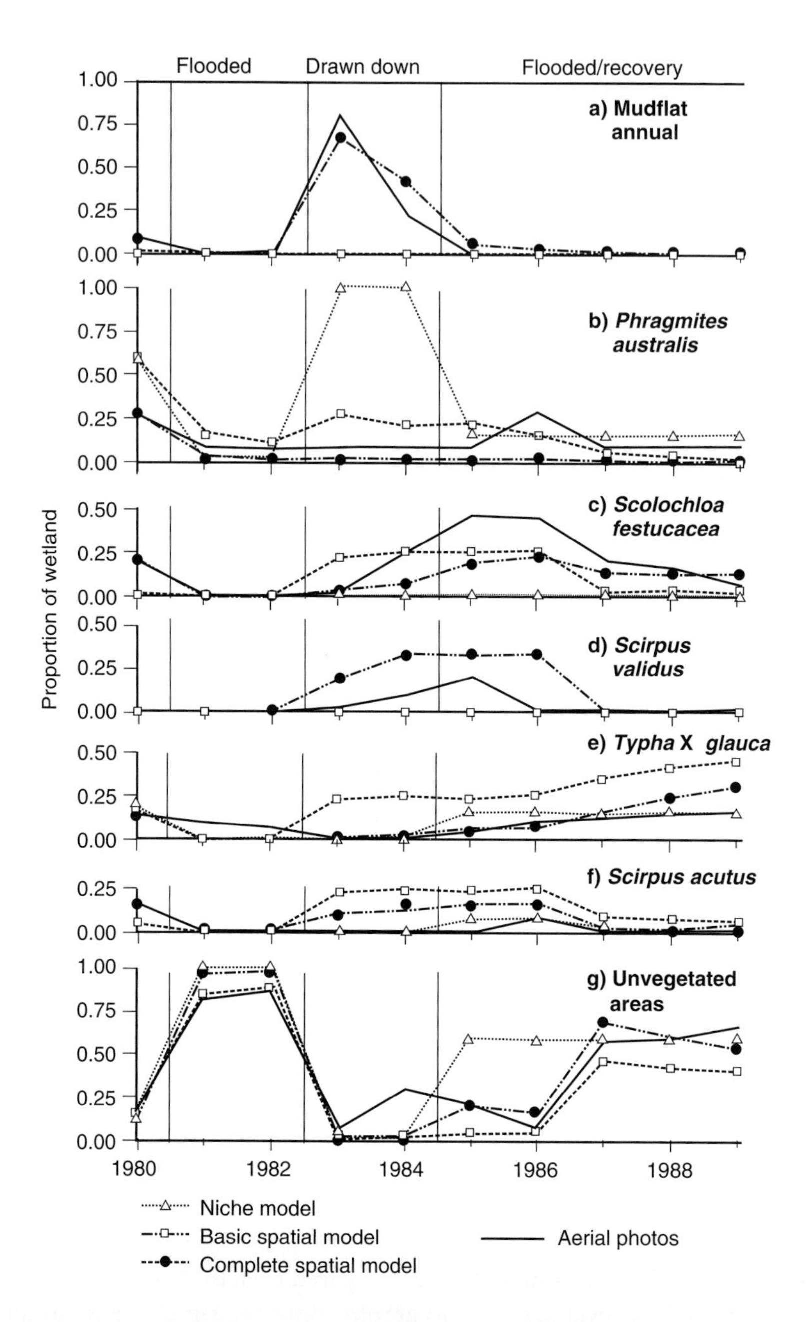

FIG. 11 Proportion of the Marsh Ecology Research Program cells covered by dominant emergent species and mudflat annuals predicted by the niche and spatially explicit models and the actual proportions estimated from aerial photographs. *Scirpus acutus* = *Scirpus lacustris* ssp. *glaucus*. (From Seabloom *et al.* [2001].)

proportions of each species estimated from aerial photographs. The accuracy of both models was high during periods when the cells were flooded, but low during the drawdown period (Fig. 11). During the reflooding period (1985 through 1989), the accuracy of both models increased with time. The niche model was the least accurate during the drawdown period because it greatly overestimated the amount of *Phragmites australis*. It was also unable to predict the presence of species that had not been part of the pre-disturbance coenocline. Because it does not incorporate time lags, the niche model also did more poorly predicting the abundance of emergent species during the first couple of years of the reflooding period (Fig. 11).

An examination of the model results confirms that constraints on species' ability to colonize areas creates both short- and long-term deviations from predicted distribution patterns based solely on water depth. Species were constrained because of seedling recruitment patterns, because they could not move upslope in response to an increase in water level, and because of landscape geometry in some cases. In prairie wetlands with their wet-dry cycles, the composition of coenoclines cannot necessarily be accurately predicted solely from local water depths at any given time. In the MERP cells, recruitment constraints were minimal because all the emergent species were present in the seed bank. This eliminated dispersal problems that could affect the initial composition of the coenocline during a drawdown, and thus establishment effects caused by preemption (i.e., the order in which species arrived at a site). Consequently, dispersal and establishment constraints will be much more important in the development of most other coenoclines. For example, in recently restored prairie potholes in which there is no seed bank of any consequence (Wienhold and van der Valk, 1989), the development of coenoclines is highly dependent on seed dispersal from other sites. Consequently, coenoclines vary widely among restored wetlands (Galatowitsch and van der Valk, 1996; Seabloom and van der Valk, 2003).

Neither the niche nor the spatially explicit model incorporated herbivory, plant pathogens, exploitative competition, or many other factors known to influence zonation patterns in wetlands (Spence, 1982). These did not have a detectable role in the development of the new coenoclines in the MERP cells. For example, muskrats, whose feeding and lodge building activities can have a significant impact on the abundance and distribution of emergents in some prairie wetlands (Weller and Spatcher, 1965), were never common in the MERP cells (Clark, 2000). In wetlands subject to less frequent and extreme water level fluctuations, including other kinds of prairie wetlands (Euliss *et al.*, 2004), these other factors may play a role in coenocline development.

CONCLUSIONS

The almost complete destruction of coenoclines in some prairie wetlands occurs periodically because of several years of above-normal water levels. Post-disturbance coenoclines, however, are not necessarily identical to pre-disturbance coenoclines, even when water levels under which the post-dispersal coenoclines develop are identical to those during the pre-disturbance period. Although pre- and post-disturbance coenoclines had the same species composition, the elevation ranges over which emergent species were found along these coenoclines were not similar. Differences in initial patterns of seed germination and seedling survival that resulted in the development of the pre-disturbance and post-disturbance coenocline seem to be responsible for these distribution differences. Thus, differences in soil moisture and temperature during the early stages of coenocline development can have lasting effects. When post-disturbance water levels are higher than pre-disturbance water levels, post-disturbance coenoclines are very different from pre-disturbance coenoclines. Emergent species were not able to shift their location upslope. Instead, they died out in areas too deep for them. This effectively compresses the coenoclines by eliminating emergents at their deep ends and even resulted in the extirpation of species along some post-disturbance coenoclines. Spatially explicit models of coenocline development confirmed the lasting impacts of constraints on post-disturbance coenocline development.

Disturbances over time result in changes in the plant communities found along environmental gradients. These changes can be very minor when pre- and post-disturbance environmental conditions are similar and much greater when pre- and post-disturbance environmental conditions are dissimilar. In the Delta Marsh, field observations and other data (e.g., aerial photographs) document vegetation changes over three wet-dry cycles. The post-disturbance vegetation that developed during the drawdown and regenerating stages of these cycles was in all three cases different from the pre-disturbance vegetation. Recruitment of species along newly developing coenoclines in the Delta Marsh is from the seedbank. Because dispersal uncertainties were not a factor, the results of the MERP study should not be extrapolated to situations where post-disturbance dispersal of species is required for coenocline development. In such cases (e.g., recently restored prairie wetlands), coenocline composition and structure are highly variable from wetland to wetland.

REFERENCES

Batt, B. D. J. (2000). The Delta Marsh. In: *Prairie Wetland Ecology*. (Murkin., H. R., van der Valk, A. G., and Clark, W. R., Eds.), pp. 17–33. The Contribution of the Marsh Ecology Research Program, Iowa State University Press, Ames, IA.

Bukata, R. P., Bruton, J. E., Jerome, J. J., and Harris, W. S. (1988). An evaluation of the impact of persistent water level changes on the areal extent of Georgian Bay/North Channel marshland. *Environ Manage* 12, 359–368.

Cittadino, E. (1990). *Nature as the Laboratory: Darwinian Plant Ecology in the German Empire, 1880–1900*. Cambridge University Press, Cambridge, UK.

Clark, W. R. (2000). Ecology of muskrats in prairie wetlands. In: *Prairie Wetland Ecology*. (Murkin, H. R., van der Valk, A. G., and Clark, W. R., Eds.), pp. 287–313. The Contribution of the Marsh Ecology Research Program, Iowa State University Press, Ames, IA.

Clark, W. R., and Kroeker, D. W. (1993). Population dynamics of muskrats in experimental marshes at Delta, Manitoba. *Can J Zool* 71, 1620–1628.

de Swart, E. O. A. M., van der Valk, A. G., Koehler, K. J., and Barendregt, A. (1994). Experimental evaluation of realized niche models for predicting responses of plant species to a change in environmental conditions. *J Veg Sci* 5, 541–552.

Euliss, Jr., N. H., and Mushet, D. M. (1996). Water-level fluctuation in wetlands as a function of landscape condition in the prairie pothole region. *Wetlands* 16, 587–593.

Euliss, Jr., N. H., Labaugh, J. W., Fredrickson, L. H., Mushet, D. M., Laubhan, M. K., Swanson, G. A., Winter, T. C., Rosenberry, D. O., and Nelson, R. D. (2004). The wetland continuum: a conceptual framework for interpreting biological studies. *Wetlands* 24, 448–458.

Galatowitsch, S. M., and van der Valk, A. G. (1996). Characteristics of recently restored wetlands in the prairie pothole region. *Wetlands* 16, 75–83.

Galinato, M. I., and van der Valk, A. G. (1986). Seed germination traits of annuals and emergents recruited during drawdowns in the Delta Marsh, Manitoba, Canada. *Aquat Bot* 26, 89–102.

Grime, J. P. (1979). *Plant Strategies and Vegetation Processes*. Wiley, Chichester, UK.

Johnson, W. C., Boettcher, S. E., Poiani, K. A., and Guntenspergen, G. (2004). Influence of weather extremes on the water levels of glaciated prairie wetlands. *Wetlands* 24, 385–398.

Kantrud, H. A., Krapu, G. L., and Swanson, G. A. (1989a). *Prairie Basin Wetlands of the Dakotas: A Community Profile*. Biological Report 85, U.S. Fish and Wildlife Service, Washington, DC.

Kantrud, H. A., Millar, J. B., and van der Valk, A. G. (1989b). Vegetation of wetlands of the prairie pothole region. In: *Northern Prairie Wetlands*. (van der Valk, A. G., Ed.), pp. 132–158. Iowa State University Press, Ames, IA.

LaBaugh, J. W., Winter, T. C., Adomaites, V. A., and Swanson, G. A. (1987). *Hydrology and Chemistry of Selected Prairie Wetlands in the Cottonwood Lake Area, Stutsman County, North Dakota, 1979–1982*. U.S. Geological Survey Professional Paper 1431, Denver, CO.

LaBaugh, J. W., Winter, T. C., and Rosenberry, D. O. (1998). Hydrologic functions of prairie wetlands. *Great Plains Res* 8, 17–37.

McIntosh, R. P. (1967). The continuum concept of vegetation. *Bot Rev* 33, 130–187.

Meredino, M. T., and Smith, L. M. (1991). Influence of drawdown date and reflood depth on wetland vegetation establishment. *Wildlife Soc B* 19, 143–150.

Meredino, M. T., Smith, L. M., Murkin, H. R., and Pederson, R. L. (1990). The response of prairie wetland vegetation to seasonality of drawdown. *Wildlife Soc B* 18, 245–251.

Millar, J. B. (1973). Vegetation change in shallow marsh wetlands under improving moisture regime. *Can J Bot* 51, 1443–1457.

Murkin, H. R., van der Valk, A. G., and Clark, W. R., (Eds.) (2000). *Prairie Wetland Ecology: The Contribution of the Marsh Ecology Research Program.* Iowa State University Press, Ames, IA.

Neckles, H. A. (1984). *Plant and Macroinvertebrate Response to Water Regime in a Whitetop Marsh.* M.S. thesis, University of Minnesota, St. Paul, MN.

Neil, C. (1990). Effects of nutrients and water levels on species composition in prairie whitetop (*Scolochloa festucacea*) marshes. *Can J Bot* 68, 1015–1020.

Neil, C. (1993). Growth and resource allocation of whitetop (*Scolochloa festucacea*) along a water depth gradient. *Aquat Bot* 46, 235–246.

Pederson, R. L. (1981). Seed bank characteristics of the Delta Marsh, Manitoba: applications for wetland management. In: *Selected Proceedings of the Midwest Conference on Wetland Values and Management.* (Richardson, B., Ed.), pp. 61–69. Freshwater Society, Navarre, MN.

Pederson, R. L., and van der Valk, A. G. (1985). Vegetation change and seed banks in marshes: ecological and management implications. *T N Am Wildl Nat Res* 49, 271–280.

Scoggan, H. J. (1978–1979). *The Flora of Canada. Parts 1-4.* National Museum of Natural Science, Ottawa, Canada.

Seabloom, E. W., and van der Valk, A. G. (2003). The development of vegetative zonation patterns in restored prairie pothole wetlands. *J Appl Ecol* 40, 92–100.

Seabloom, E. W., van der Valk, A. G., and Moloney, K. A. (1998). The role of water depth and soil temperature in determining initial composition of prairie coenoclines. *Plant Ecol* 138, 203–216.

Seabloom, E. W., Moloney, K. A., and van der Valk, A. G. (2001). Constraints on the establishment of plants along a fluctuating water-depth gradient. *Ecology* 82, 2216–2232.

Shay, J. M., and Shay, C. T. (1986). Prairie marshes in western Canada, with specific reference to the ecology of five emergent macrophytes. *Can J Bot* 64, 443–454.

Shay, J. M., de Geus, P. M., and Kapinga, M. R. M. (1999). Changes in shoreline vegetation over a 50-year period in the Delta Marsh, Manitoba in response to water levels. *Wetlands* 19, 413–425.

Spence, D. H. N. (1982). The zonation of plants in freshwater lakes. *Adv Ecol Res* 12, 37–125.

Squires, L., and van der Valk, A. G. (1992). Water-depth tolerances of the dominant emergent macrophytes of the Delta Marsh, Manitoba. *Can J Bot* 70, 1860–1867.

Stewart, R. E., and Kantrud, H. A. (1971). *Classification of Natural Ponds in the Glaciated Prairie Region.* U.S. Fish and Wildlife Service Resource Publication No. 92, Washington, DC.

Stewart, R. E., and Kantrud, H. A. (1972). *Vegetation of Prairie Potholes, North Dakota in Relation to Quality of Water and Other Environmental Factors.* U. S. Geological Survey Professional Paper 585-D, Washington, DC.

Swanson, G. A., Euliss, Jr., N. H., Hanson, B. A., and Mushet, D. M. (2003). Dynamics of a prairie pothole wetland complex: implications for wetland management. In: *Hydrological, Chemical, and Biological Characteristics of a Prairie Pothole Wetland Complex under Highly Variable Climate Conditions – The Cottonwood Lake Area, East-central North Dakota.* (Winter, T. C., Ed.), pp. 55–94. U.S. Geological Survey Professional Paper 1675, Denver, CO.

van der Kamp, G., and Hayashi, M. (1998). The groundwater recharge functions of prairie wetlands. *Great Plains Res* 8, 39–56.

van der Kamp, G., Hayashi, M., and Gallen, D. (2003). Comparing the hydrology of a grassed and cultivated catchment in the semi-arid Canadian prairies. *Hydrol Process* 17, 559–575.

van der Kamp, G., Stolte, W. J., and Clark, R. G. (1999). Drying out of small prairie wetlands after conversion of their catchments from cultivation to permanent brome grass. *J Hydrol Sci* 44, 387–397.

van der Valk, A. G. (1981). Succession in wetlands: a Gleasonian approach. *Ecology* 62, 688–696.

van der Valk, A. G. (1986). The impact of litter and annual plants on recruitment from the seed bank of a lacustrine wetland. *Aquat Bot* 24, 13–26.

van der Valk, A. G. (Ed.) (1989). *Northern Prairie Wetlands.* Iowa State University Press, Ames, IA.

van der Valk, A. G. (1992). Establishment, colonization and persistence. In: *Plant Succession: Theory and Prediction.* (Glenn-Lewin, D. C., Peet, R. K., and Veblen, T. T., Eds.), pp. 60–102. Chapman and Hall, London, UK.

van der Valk, A. G. (1994). Effects of prolonged flooding on the distribution and biomass of emergent species along a freshwater wetland coenocline. *Vegetatio* 110, 185–196.

van der Valk, A. G. (2000). Vegetation dynamics and models. In: *Prairie Wetland Ecology: The Contribution of the Marsh Ecology Research Program.* (Murkin, H. R., van der Valk, A. G., and Clark, W. R., Eds.), pp. 125–161. Iowa State University Press, Ames, IA.

van der Valk, A. G. (2005a). The prairie pothole region of North America. In: *The World's Largest Wetlands: Ecology and Conservation.* (Fraser, L. H., and Keddy, P. A., Eds.), pp. 393–423. Cambridge University Press, Cambridge, UK.

van der Valk, A. G. (2005b). Water-level fluctuations in North American prairie wetlands. *Hydrobiologia* 539, 171–188.

van der Valk, A. G., and Davis, C. B. (1976). Seed banks of prairie glacial marshes. *Can J Bot* 54, 1832–1838.

van der Valk, A. G., and Davis, C. B. (1978). The role of the seed bank in the vegetation dynamics of prairie glacial marshes. *Ecology* 59, 322–335.

van der Valk, A. G., and Pederson, R. L. (1989). Seed banks and the management and restoration of natural vegetation. In: *Ecology of Soil Seed Banks.* (Leck, M. A., Parker, V. T., and Simpson, R. L., Eds.), pp. 329–346. Academic Press, New York.

van der Valk, A. G., and Welling, C. H. (1988). The development of zonation in freshwater wetlands: an experimental approach. In: *Diversity and Pattern in Plant Communities.* (During, H. J., Werger, M. J. A., and Willems, J. H., Eds.) pp. 145–158. SPB Academic Publishing, The Hague, the Netherlands.

van der Valk, A. G., Squires, L., and Welling, C. H. (1994). An evaluation of three approaches for assessing the impacts of an increase in water level on wetland vegetation. *Ecol Appl* 4, 525–534.

Walker, J. M. (1959). *Vegetation Studies on the Delta Marsh, Manitoba.* M.Sc. Thesis, University of Manitoba, Winnipeg, MB, Canada.

Walker, J. M. (1965). *Vegetation Changes with Falling Water in the Delta Marsh, Manitoba.* Ph.D. Dissertation, University of Manitoba, Winnipeg, MB, Canada.

Waters, I., and Shay, J. M. (1990). A field study of the morphometric response of *Typha glauca* shoots to a water depth gradient. *Can J Bot* 68, 2339–2343.

Weller, M. W., and Fredrickson, L. H. (1974). Avian ecology of a managed glacial marsh. *Living Bird* 12, 269–291.

Weller, M. W., and Spatcher, C. S. (1965). *Role of Habitat in the Distribution and Abundance of Marsh Birds.* Special Report 43, Iowa Agriculture and Home Economics Experiment Station, Ames, IA.

Welling, C. H., Pederson, R. L., and van der Valk, A. G. (1988a). Recruitment from the seed bank and the development of zonation of emergent vegetation during a drawdown in a prairie wetland. *J Ecol* 76, 483–496.

Welling, C. H., Pederson, R. L., and van der Valk, A. G. (1988b). Temporal patterns in recruitment from the seed bank during drawdowns in a prairie wetland. *J Appl Ecol* 25, 999–1007.

Whittaker, R. H. (1967). Gradient analysis of vegetation. *Biol Rev* 42, 197–264.

Wienhold, C. E., and van der Valk, A. G. (1989). The impact of duration of drainage on the seed banks of northern prairie wetlands. *Can J Bot* 67, 1878–1884.

Winter, T. C. (1989). Hydrological studies of wetlands in the northern prairie. In: *Northern Prairie Wetlands.* (van der Valk, A. G., Ed.), pp. 16–54. Iowa State University Press, Ames, IA.

Winter, T. C. (2003). Geohydrological setting of the Cottonwood Lake Area. In: *Hydrological, Chemical, and Biological Characteristics of a Prairie Pothole Wetland Complex under Highly Variable Climate Conditions—The Cottonwood Lake Area, East-central North Dakota.* (Winter, T. C., Ed.), pp. 1–24. U.S. Geological Survey Professional Paper 1675, Denver, CO.

Winter, T. C., and Rosenberry, D. O. (1995). The interaction of ground water with prairie pothole wetlands in the Cottonwood Lake area, east-central North Dakota, 1979–1990. *Wetlands* 15, 193–211.

Winter, T. C., and Rosenberry, D. O. (1998). Hydrology of prairie pothole wetlands during drought and deluge: a 17-year study of the Cottonwood Lake wetland complex in North Dakota in the perspective of longer term measured and proxy hydrological records. *Climatic Change* 40, 189–209.

Woo, M.-K., and Rowsell, R. D. (1993). Hydrology of a prairie slough. *J Hydrol* 146, 175–207.

12

Modeling Heating Effects

Geoffry N. Mercer and Rodney O. Weber

University of New South Wales at the Australian Defence Force Academy

INTRODUCTION

The main physical processes involved in wildland fires are heat and mass transfer and chemical and physical change. The consequent ecological effects upon individual plants can be characterized according to the tissue response of live vegetation. This chapter considers models that enable us to relate these physical processes to the ecological response of vegetation in as fundamental a way as currently possible. In particular, we discuss heat transfer because this is the main source of damage to vegetation that is not impinged upon directly by flames. To do this, we begin with the fundamental equations for energy and mass conservation and show how these can be applied to selected idealized situations, such as spherical and cylindrical vegetation samples. We then discuss the application of this type of modeling to more realistic scenarios and provide a detailed case study of the survival of seeds in a woody fruit subjected to an experimental heating regime.

Many plants rely on various survival mechanisms for their post-fire regeneration. Mathematical models of some of these can be constructed with partial differential equations (PDEs). For example, seed survival in woody fruits can be modeled by using the heat conduction equation in a spherical coordinate system, subject to time-dependent heating on the boundary. Impact upon stem and thick branches can be modeled by using cylindrical coordinates or one-dimensional slab geometry for large-diameter trees. These models can then be used to try to predict plant survival after exposure to either laboratory heating or field fires. Temperature time traces within the vegetation can be determined for a given temperature time exposure at the surface of the vegetation.

The aim here is not to cover all specific fire or vegetation possibilities but rather to show the possibilities of this type of modeling. Specific cases need to consider the predicted (or measured) temperature time exposure at various heights, which is very fire dependent, and the types and sizes of vegetation being studied. A case study of the survival of seeds in woody fruits is used to show the application of the models.

CONSERVATION LAWS

The chemical reactions and physical processes that occur in combustion and fire situations can liberate stored chemical energy and allow for changes in physical form, such as solid matter to gaseous matter, but they do not cause any net changes to the total energy and mass in the complete system.

Our basis for creating a mathematical description of the physical and chemical process are the conservation laws for energy and mass. These conservation laws ensure that the total energy and mass in the system are kept constant. The most compact and convenient way to express these laws is by mathematical equations that include all the possibilities for the transfer of energy and the change of forms of energy and mass. These have been written down previously by many authors (e.g., Williams, 1985; several of the chapters in Johnson and Miyanishi, 2001).

For the energy, the fundamental law can be expressed as a word equation:

$$\begin{Bmatrix}\text{rate of accumulation}\\ \text{of heat energy}\end{Bmatrix} = \begin{Bmatrix}\text{rate at which}\\ \text{energy flows in}\end{Bmatrix} - \begin{Bmatrix}\text{rate at which}\\ \text{energy flows out}\end{Bmatrix} + \begin{Bmatrix}\text{rate at which}\\ \text{energy is released}\end{Bmatrix} \quad (1)$$

This can then be written as a differential equation, provided that we assume that heat is conducted in a homogeneous, isotropic solid (Ozisik, 1980):

$$\gamma \rho C_p \frac{\partial T}{\partial t} = \nabla \cdot (k \nabla T(r,t)), \quad (2)$$

where ρ is the density of the solid, C_p is the specific heat, k is the thermal conductivity, and $T(r,t)$ is the temperature at position r and time t.

The symbol ∇ represents the first-order spatial derivative operator in whatever coordinate system is appropriate, and ∇^2, the second-order one. In writing down this equation in this form, several assumptions have been made. First, it assumed that the only form of heat transfer mechanism is heat conduction, so effects such as sap movement transferring heat are explicitly not modeled. Second, it is assumed that there is no mass transfer such as occurs with pyrolysis or dessication of the material. To model effects such as these, additional differential equations representing the woody mass and the water content, respectively, are needed. A study of this is beyond the scope of this chapter. If the thermal conductivity is assumed constant (independent of position and temperature), then Equation (2) reduces to

$$\frac{\partial T}{\partial t} = \alpha \nabla^2 T, \tag{3}$$

where

$$\alpha = \frac{k}{\rho C_p} \tag{4}$$

is the thermal diffusivity which has units of $m^2\ s^{-1}$. In reality, α is not constant throughout any living material, with varying properties in different layers and different types of vegetation. Sometimes these variations are important, but often they can be neglected because the variation is minimal; for example, the variation is minimal among wood, live bark, and dead bark (Dickinson and Johnson, 2001).

SIMPLE EXAMPLES

In this section we develop the general problem of heating an arbitrarily shaped vegetation sample and then apply it to three simple examples that can be solved analytically. These examples are the slab approximation (for studies of the exposure of leaves and large diameter tree trunks, as in Dickinson and Johnson, 2004), the external heating of an idealized "cylindrical tree" (for studies of tree stems and thick branches, as in Potter and Andresen, 2002), and the heating of a sphere (for studies of spherical fruits, as in Mercer *et al.*, 1994).

General Statement of the Problem

We are considering a general shape, Ω, with boundary denoted by $\partial\Omega$.

Inside the shape, if the thermal diffusivity (α) is taken to be a constant throughout, then the temperature is governed by the partial differential equation (e.g., Carslaw and Jaeger, 1959; Ozisik, 1980)

$$\frac{\partial T}{\partial t} = \alpha \nabla^2 T, \tag{5}$$

an appropriate value for α when considering woody fruits or bark is $\alpha \approx 10^{-7}$ m^2 s^{-1} (Spalt and Reifsnyder, 1962).

Let the initial temperature within the shape be a uniform temperature of T_0 so that we have

$$T = T_0 \quad \text{in} \quad \Omega \quad \text{for} \quad t = 0. \tag{6}$$

At time $t = 0$ the boundary temperature is raised to T_B (e.g., when a fire is present near the sample) so that we have

$$T = T_B \quad \text{on} \quad \partial\Omega \quad \text{for} \quad t > 0. \tag{7}$$

For simplicity here, we will not let this temperature T_B be time-dependent. Time-dependent external temperatures are considered later in this chapter.

To simplify the analysis, introduce a new temperature variable $u = T_B - T$. Then, in terms of this new variable, the problem can be stated as

$$\frac{\partial u}{\partial t} = \alpha \nabla^2 u \quad \text{in} \quad \Omega, \tag{8}$$

$$u = 0 \quad \text{on} \quad \partial\Omega, \tag{9}$$

$$u = T_B - T_0 \quad \text{for} \quad t = 0. \tag{10}$$

This is now a familiar, homogeneous boundary condition, PDE problem. This can be solved by using the well-known technique of separation of variables.

Let u be separable such that $u = F(r)G(t)$ and substitute into the PDE to obtain

$$\frac{1}{\alpha}\frac{G'}{G} = \frac{\nabla^2 F}{F}.$$

The left side is a function of t only, and the right side is a function of x only; hence, they both must be equal to a constant for this to be possible. Introduce a constant, $-\lambda^2$ and obtain

$$\frac{1}{\alpha}\frac{G'}{G} = -\lambda^2 \quad \text{and} \quad \frac{\nabla^2 F}{F} = -\lambda^2$$

These rearrange to the ordinary differential equations

$$G' + \alpha\lambda^2 G = 0 \quad \text{and} \quad \nabla^2 F + \lambda^2 F = 0 \tag{11}$$

The first of these has solution given by

$$G(t) = e^{-\alpha\lambda^2 t} \tag{12}$$

The problem is then to solve

$$\nabla^2 F + \lambda^2 F = 0 \quad \text{in} \quad \Omega, \tag{13}$$

subject to $F = 0$ on $\delta\Omega$. Now we will consider the three cases of slab, cylinder, and spherical geometries. In these three idealized geometries the equation can be written

$$F'' + \frac{n-1}{r} F' + \lambda^2 F = 0 \quad \text{in} \quad \Omega, \tag{14}$$

where $n = 1$ for slab, $n = 2$ for cylinder, and $n = 3$ for sphere geometries.

One-Dimensional Cases

One-Dimensional Cartesian Case

Consider a cartesian slab geometry of width $2R$. The boundary condition ($u = 0$) is then expressed as

$$u = 0 \quad \text{at} \quad r = \pm R \tag{15}$$

Hence, in the separated variables we now need to solve

$$F'' + \lambda^2 F = 0, \quad F(\pm R) = 0 \tag{16}$$

Solving this gives

$$F(r) = \cos(\lambda_n r)$$

with $\lambda_n = (n - \frac{1}{2})\frac{\pi}{R}$. Summing over all the possible solutions (because the problem is linear) and including the t variable part of the solution give the general solution as

$$u(r,t) = \sum_{n=1}^{\infty} C_n \cos(\lambda_n r) e^{-\alpha\lambda_n^2 t}, \tag{17}$$

where the unknown constants C_n have to be determined from the initial condition. The initial condition is $u(r,0) = T_B - T_0$, which means the C_n must satisfy

$$T_B - T_0 = \sum_{n=1}^{\infty} C_n \cos(\lambda_n r) \tag{18}$$

Using a standard Fourier series analysis [or by multiplying the above equation by $\cos(\lambda_m r)$ integrating over the range and using the orthogonality of $\cos(\lambda_m r)$] gives the constants as

$$C_n = \frac{4(-1)^{n-1}(T_B - T_0)}{(2n-1)\pi}, \tag{19}$$

and our final solution is

$$u(r,t) = (T_B - T_0)\sum_{n=1}^{\infty}\frac{4(-1)^{n-1}}{(2n-1)\pi}\cos(\lambda_n r)e^{-\alpha\lambda_n^2 t}, \tag{20}$$

with $\lambda_n = \left(n - \frac{1}{2}\right)\frac{\pi}{R}$.

One-Dimensional Cylindrical Case

Consider an infinite cylinder of radius R (e.g., a simple model of a branch or stem). The boundary condition is that $u = 0$ at $r = R$. In cylindrical coordinates we therefore need to solve

$$F'' + \frac{1}{r}F' + \lambda^2 F = 0, \quad F(R) = 0. \tag{21}$$

Solving this gives

$$F(r) = J_0(\lambda_n r) \tag{22}$$

where $J_0(r)$ is the Bessel function of the first kind of order zero and λ_n = (zeros of J_0)/R. The first three values are given by $\lambda_1 = 2.405/R$, $\lambda_2 = 5.520/R$, $\lambda_3 = 8.653/R$. Summing over all the possible solutions (because the problem is linear) and including the t variable part of the solution gives the general solution as

$$u(r,t) = \sum_{n=1}^{\infty} C_n J_0(\lambda_n r)e^{-\alpha\lambda_n^2 t}, \tag{23}$$

where the unknown constants C_n have to be determined from the initial condition. The initial condition is $u(r, 0) = T_B - T_0$, which means the C_n must satisfy

$$T_B - T_0 = \sum_{n=1}^{\infty} C_n J_0(\lambda_n r). \tag{24}$$

Using a standard Fourier-Bessel series analysis gives the constants as

$$C_n = \frac{2(T_B - T_0)}{R^2 J_1^2(\lambda_n R)}\int_0^R rJ_0(\lambda_n r)\,dr \tag{25}$$

which, on using the identity

$$\int_0^R rJ_0(\lambda_n r)\,dr = \frac{R}{\lambda_n}J_1(\lambda_n R), \tag{26}$$

can be rewritten as

$$C_n = \frac{2(T_B - T_0)}{\lambda_n R J_1(\lambda_n R)} \tag{27}$$

So our final solution is

$$u(r,t) = (T_B - T_0)\sum_{n=1}^{\infty}\frac{2}{\lambda_n R J_1(\lambda_n R)}J_0(\lambda_n r)e^{-\alpha\lambda_n^2 t} \tag{28}$$

with λ_n = (zeros of J_0)/R.

One-Dimensional Spherical Case

Consider a sphere of radius R (e.g., a simple model of a seed). The boundary condition is that $u = 0$ at $r = R$. In spherical coordinates we therefore need to solve

$$F'' + \frac{2}{r}F' + \lambda^2 F = 0, \quad F(R) = 0 \tag{29}$$

Solving this gives

$$F(r) = \frac{\sin(\lambda_n r)}{r}$$

with $\lambda_n = n\pi/R$. Summing over all the possible solutions (since the problem is linear) and including the t variable part of the solution give the general solution as

$$u(r,t) = \sum_{n=1}^{\infty} C_n \frac{\sin(\lambda_n r)}{r} e^{-\alpha\lambda_n^2 t} \tag{30}$$

where the unknown constants C_n have to be determined from the initial condition. The initial condition is $u(r,0) = T_B - T_0$, which means the C_n must satisfy

$$T_B - T_0 = \sum_{n=1}^{\infty} C_n \frac{\sin(\lambda_n r)}{r}. \tag{31}$$

Using a standard Fourier analysis (by multiplying the above equation by $r\sin(\lambda_m r)$ integrating over the range and using the orthogonality of $\sin(\lambda_m r)$ and $\sin(\lambda_n r)$) gives the constants as

$$C_n = \frac{(T_B - T_0)2(-1)^{n+1}}{\lambda_n} \tag{32}$$

So our final solution is

$$u(r,t) = (T_B - T_0)\sum_{n=1}^{\infty} \frac{2(-1)^{n+1}}{\lambda_n} \frac{\sin(\lambda_n r)}{r} e^{-\alpha\lambda_n^2 t} \tag{33}$$

with $\lambda_n = n\pi/R$.

Example for a Finite One-Dimensional Case

In each case we note that we can recover the physical temperature from $T = T_B - u$ as

$$T = T_B - \sum_{n=1}^{\infty} C_n h(\lambda_n r) e^{-\alpha\lambda_n^2 t}, \tag{34}$$

where $h(\lambda_n r)$ are the appropriate basis functions.

We now show how these formulae can be applied to determine the temperature inside idealized vegetation samples. Consider the example of a sample initially at the ambient temperature of 20°C, which is then subjected to a fire that results in an exterior temperature of 500°C. Let $\alpha = 1.35 \times 10^{-7}$ m^2 s^{-1} (Spalt and Reifsnyder, 1962), which is a value appropriate for wood and bark. For a slab of half width $R = 0.1$ m, the above formula can be applied to obtain Fig. 1, which shows the temperature in the slab as time increases. Each line is at a 300-second interval in time and the direction of increasing time is shown by the arrow. Clearly, as time increases, the interior of the slab increases in temperature as the heat diffuses inwards. Similar calculations can be done for the case of a cylinder of radius $R = 0.1$ m and a sphere of radius $R = 0.1$ m using the preceding expressions. In these scenarios, the result is similar although the interior of the sample heats up quicker because the ratio of surface area to volume is larger.

Fig. 2 shows the temperature histories 0.02 m (2 cm) below the surface for the slab, cylinder, and sphere geometries of size 0.1 m when subjected to a temperature of 500°C at the surface. The spherical sample heats up the quickest, followed by the cylindrical one, and finally by the slab. This is expected because of the decreasing ratio of surface area to volume in progressing from spherical to cylindrical to slab geometries. Also plotted, as the dashed line, is the mortality temperature, taken here to be 70°C (discussed later in this chapter). From this, it is possible to read off the graph the time to death at this depth within the sample. For example, at 0.02 m into the sample, the times to reach 70°C are approximately 480, 520, and 560 seconds for the sphere, cylinder, and slab, respectively.

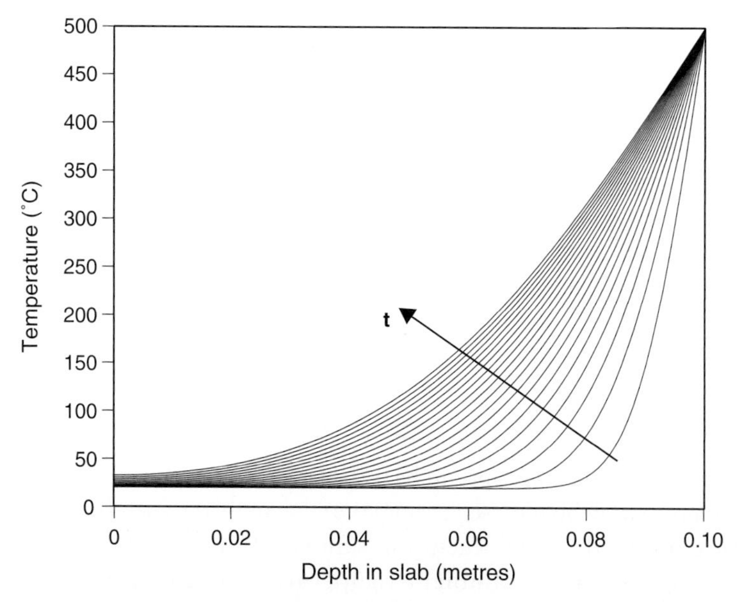

FIG. 1 Temperature throughout the slab at 300-second time intervals with $T_b = 500$°C and $T_0 = 20$°C. The arrow shows the direction of increasing time.

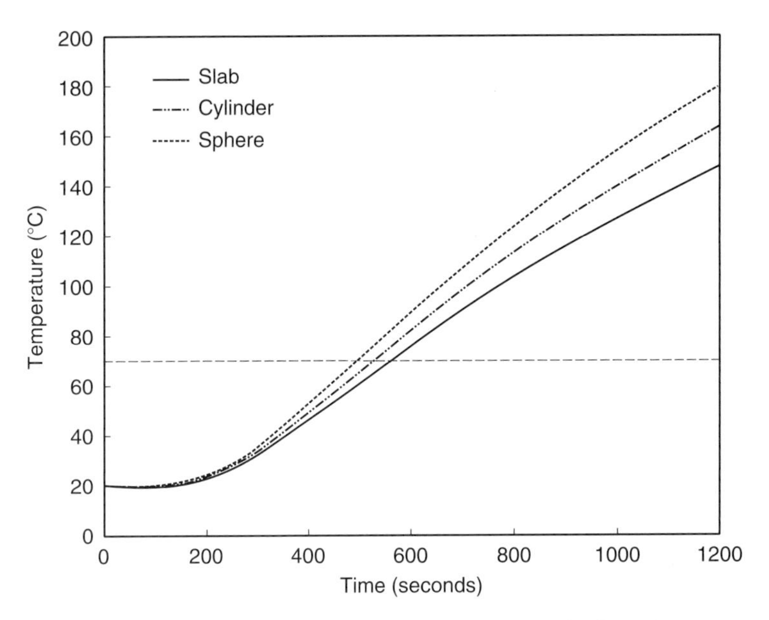

FIG. 2 Comparison of temperature history in slab, cylinder, and spherical geometries at a depth of 0.02 m below the surface when subjected to a burning temperature of 500°C at the surface. Also shown as the dotted line is the mortality temperature (70°C) from which it is possible to read off time to death of the vegetation at that depth.

By using a more realistic temperature time history at the surface, such as those obtained by modeling real fires (Mercer and Weber, 2001), it is possible to use this method to determine the temperature time history within vegetation that is modeled by slab, cylinder, or sphere.

Dimensional Analysis and Similarity Solution

We can also apply the principle of dimensional analysis to this problem in any regular geometry. Begin from the equation

$$\frac{\partial T}{\partial t} = \alpha \nabla^2 T \quad \text{in} \quad \Omega \tag{35}$$

There is a characteristic scale for length, R, and for time, t, and there are temperatures T_0, T_B, and T_i (temperature of interest, such as necrosis temperature) as well as the thermal diffusivity α (units of $m^2\ s^{-1}$).

Hence, the dimensionless parameters in the problem can be expressed as

$$\frac{T_i - T_B}{T_0 - T_B} \quad \text{and} \quad \frac{\alpha t}{R^2} \tag{36}$$

Assuming we have values for T_0 (20°C), T_i (70°C), and T_B (500°C) in mind (as per the previous example), we can then infer that the characteristic time is given by

$$t = \kappa R^2 / \alpha, \tag{37}$$

where κ is a dimensionless constant that depends only upon our choices for the temperature and the specific details of the geometry.

We conclude that the time for the temperature to rise by a given amount should be proportional to the square of the distance into the sample from the exterior edge. The constant of proportionality depends upon the geometry of the sample.

The preceding dimensional analysis suggests that we can define a similarity variable, x, according to

$$x = \frac{r}{\sqrt{\alpha t}}, \tag{38}$$

and seek solutions of the heat equation of the form

$$u(r, t) = f(x) \tag{39}$$

This entails finding the derivatives

$$\frac{\partial u}{\partial t} = -\frac{r}{2t\sqrt{\alpha t}} f'(x) \tag{40}$$

$$\frac{\partial u}{\partial r} = \frac{1}{\sqrt{\alpha t}} f'(x) \tag{41}$$

$$\frac{\partial^2 u}{\partial r^2} = \frac{1}{\alpha t} f''(x) \tag{42}$$

and substituting them into the heat equation which can be written generally as

$$\frac{\partial^2 u}{\partial r^2} + \frac{(n-1)}{r}\frac{\partial u}{\partial r} = \frac{1}{\alpha}\frac{\partial u}{\partial t}, \tag{43}$$

where n depends on the geometry under consideration. The result is an ordinary differential equation for $f(x)$

$$f'' + (n-1)\frac{\sqrt{\alpha t}}{r} f' = -\frac{1}{2}\cdot\frac{r}{\sqrt{\alpha t}} f', \tag{44}$$

which can be rearranged to give

$$f'' + \left(\frac{(n-1)}{x} + \frac{x}{2}\right) f' = 0 \tag{45}$$

Here the prime (′) denotes ordinary differentiation with respect to the variable x, and we have used the definition of x to show how the result is indeed an ordinary differential equation for $f(x)$, and how the similarity variable ensures that r and t occur only in their combined form.

This equation can be solved for the three main geometries of interest:

slab, $n = 1$: $f(x) = c_1 + c_2 \operatorname{erf}\left(\frac{x}{2}\right)$

cylinder, $n = 2$: $f(x) = c_1 + c_2 E_i\left(1, \frac{x^2}{4}\right)$

sphere, $n = 3$: $f(x) = c_1 + \frac{c_2}{x}\left(2\exp\left(\frac{-x^2}{4}\right) + x\sqrt{\pi}\operatorname{erf}\left(\frac{x}{2}\right)\right)$

where erf(x) is the Gauss error function and $E_i(x)$ is the exponential integral function. Also note that the natural domain for each of these solutions is $0 < x < \infty$, which can be referred back to the original space and time variables to give a domain of $0 < r < \infty$ and $t > 0$. This is characteristic of similarity solutions, as they are usually applicable only over semi-infinite (or infinite) domains; this restricts their use in applications since the samples are not semi-infinite. However, for slabs, cylinders, and spheres that are rather large (or, equivalently, when the time frame of interest is relatively short), these similarity solutions can provide an excellent approximation near the physical boundary.

Semi-infinite Slab

A related method of solution in the slab geometry, equivalent to the method of similarity solutions just discussed, has been described by Dickinson and Johnson (2001). It uses the method of Laplace transforms to determine that

$$\frac{T - T_B}{T_0 - T_B} = \operatorname{erf}\left(\frac{r}{2\sqrt{\alpha t}}\right) \tag{46}$$

where erf(x) is the Gauss error function as tabled in mathematical handbooks, an extract of which is given in Table 1. To illustrate the application of this solution, consider the now familiar case T = 70°C, T_0 = 20°C, T_B = 500°C. Then the scaled temperature is

$$\frac{T - T_B}{T_0 - T_B} = \frac{70 - 500}{20 - 500} = 0.8958 \tag{47}$$

and so, using Equation (46) and the tables, we can determine that

$$\frac{r}{2\sqrt{\alpha t}} \approx 1.15 \tag{48}$$

TABLE 1 Some Values of the Error Function erf(x)

x	erf(x)
0.95	0.8209
1.00	0.8427
1.05	0.8624
1.10	0.8802
1.15	0.8961
1.20	0.9103
1.25	0.9229

Rearranging and choosing a representative value for the thermal diffusivity, such as $\alpha = 1.35 \times 10^{-7}$ m^2 s^{-1} (Spalt and Reifsnyder, 1962), we obtain a critical time of

$$t_{cr} \approx 1.40 \times 10^6 \, r^2. \tag{49}$$

If we adopt the criterion that mortality occurs when the inside of the bark reaches a temperature of 70°C and assume a bark thickness of 0.02 m (2 cm), then the critical time for exposure of the outside of the tree to a temperature of 500°C is 560 seconds.

This value is in excellent agreement with the values found for the same scenario using the slab solution earlier. This excellent agreement is caused by the fact that the finite nature of the slab has had little impact at this temperature, depth, and time interval. If a deeper level or a higher temperature is used, then the finite slab results and the semi-infinite slab solution used here will differ more substantially.

APPLICATION TO MORE REALISTIC SCENARIOS

Although the models presented in the preceding section give us a good insight into the modeling of fire effects on vegetation in general, it is not as straightforward for real fire scenarios and different vegetation types. Many compounding factors need to be investigated for particular applications. Effects such as varying physical properties with temperature, moisture content, and vegetation structure; time and directional varying temperature exposures; and cell mortality from different heating regimes can all be modeled to some degree. Generally, when these types of factors are included, analytic solutions to the models are no longer possible and one must resort to numerical solutions of the governing equations.

Numerical Solution Techniques

A complete discussion of the numerical techniques used to solve these types of problems is beyond the scope of this chapter, but a brief outline follows. Numerical solutions to these types of problems can generally be classified into two categories: finite difference methods and finite element methods. Finite difference methods are best suited to simple geometries, such as slabs (Rego and Rigolot, 1989; Dickinson and Johnson, 2004) and circles (Potter and Andresen, 2002). Finite element methods are typically used for more complicated geometries or for multiple layers or regions of differing thermal properties and complicated boundary conditions.

When finite difference methods are used, the region of interest is gridded with usually a uniform grid and the derivatives are replaced by finite differences between neighboring points. So, for example, if i refers to stepping in the spatial direction and j the time direction, then the simplest approximation to the standard constant diffusivity heat equation

$$\frac{\partial T}{\partial t} = \alpha \nabla^2 T \tag{50}$$

is

$$\frac{T(i,j+1) - T(i,j)}{\Delta t} = \alpha \left(\frac{T(i-1,j) - 2T(i,j) + T(i+1,j)}{\Delta x^2} \right) \tag{51}$$

where Δt is the time step and Δx the spatial step. This can be rearranged to give the temperature at the next time level $(j + 1)$ in terms of values at the current time level (j) as

$$T(i,j+1) = T(i,j) + \frac{\alpha \Delta t}{\Delta x^2} \left(T(i-1,j) - 2T(i,j) + T(i+1,j) \right) \tag{52}$$

This is known as the forward time centered space (FTCS) method and is the simplest (and least accurate) of the finite difference methods. By marching across the x direction and forward in time, the temperature can be determined at all space and time points. This is called an explicit method because the new time can be written explicitly in terms of the current time. Explicit methods are plagued by stability problems where numerical errors can build up; they often require very small time steps to be stable and hence can be very inefficient. In other methods, known as implicit methods, values at the new time are given by a matrix equation as a function of the old time. These do not have stability problems but are more difficult to implement.

Finite element methods do not use a regular grid over the space-time domain. Instead, they use a triangular mesh in space, which means they can more easily incorporate complicated boundaries and layers of different physical parameters. The equation to be solved is expressed at each point of the mesh in terms of its neighbors. This is similar to the implicit finite difference methods because a matrix equation then needs to be solved to obtain the solution. Because of the complexity involved, most models solved by using finite element methods use commercially available software packages. FlexPDE (PDE Solutions, Inc., Sunol, Calif) is one such relatively simple-to-use product.

Variable Thermophysical Parameters

In general, parameters such as the thermal conductivity (k), specific heat (C_p), and density (ρ) vary with the temperature and moisture content of the material. Unfortunately, no data are available in the literature to be able to make accurate estimates for many of these dependencies. Jones *et al.* (2004) give a good summary of the available relationships, many of which are based on oven-dried wood. Typically, for the range of temperature of interest here, the variation is minimal and less than the uncertainty in the measured values, so the use of a constant value is reasonable (Jones, 2003). The type of material (e.g., wood, live bark, dead outer bark) also varies, but again the lack of available data means that a generic constant value is often used (Dickinson and Johnson, 2001).

Moisture content is important not only for its effect on the thermophysical parameters but also in terms of the latent heat needed to drive out and evaporate moisture. Thus, the process of desiccation can be an important effect in reducing the temperature rise in vegetation. Modeling this process correctly entails the addition of complicated mass conservation equations for the water content in both liquid and gaseous form and their transport through the material. Fortunately, simpler approximate methods act as good bounds on the process. The simplest is to restrict the maximum temperature possible to 100°C (see Rego and Rigolot, 1989). More complex methods involve tracking the energy input at each point and restricting the temperature to 100°C until the latent heat of vaporization has been met at that point for the moisture content there (Jones *et al.*, 2004). This method does not model the transport of the vapor, which may be a limiting process in the heating of the vegetation and hence may give a lower temperature than in reality.

Other factors, such as pyrolysis of the material, charring, and swelling, can also affect the thermophysical parameters (Jones *et al.*, 2004; Butler *et al.*, 2005). Unfortunately, these processes have received little attention in the literature and are difficult to model at present. Generally, they occur at temperatures greater than 180°C (Jones *et al.*, 2004) and so can often be overlooked if the main interest is at lower temperatures deeper into the vegetation, although their effect at the outer edges of the vegetation will somewhat affect the inner temperatures.

Spatially Dependent Heating

In general, the heating of a vegetation element is not uniform over its surface, and so one-dimensional models may not be adequate. For instance, the heating of a tree stem during a fire is not uniform around its circumference. Depending on the diameter of the tree, this may be an important feature. Typically with nonuniform surface heating, one must resort to numerical solutions. As an example, shown in Fig. 3 are temperature contours (in increments of 30°C starting from 30°C) of a circular tree trunk when exposed to an external temperature that decreases around the radius of the tree for a total of 1200 seconds. This is a simplified model of the temperature exposure of a tree during a fire that is known to be higher on the leeward side of the tree trunk. See Dickinson and Johnson (2001) for a more detailed description of this scenario. The parameter values are 500°C at the leeward edge of the tree, reducing to 20°C at the leading edge. These parameter values are for illustrative purposes only; more detailed fluid flow considerations are necessary to accurately predict the temperature exposure

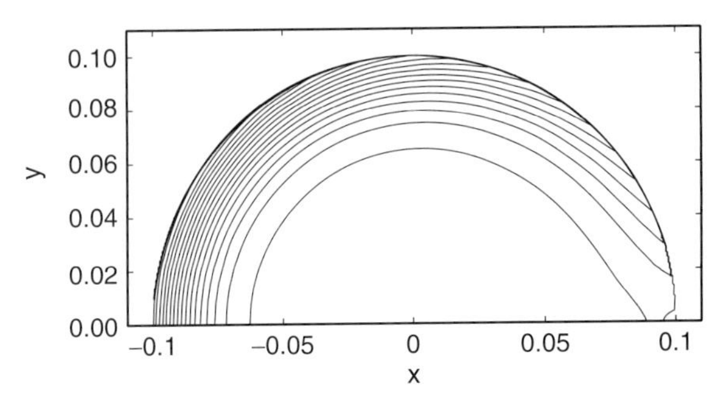

FIG. 3 Contours of temperature at 30° increments for a circular tree heated with an external temperature that varies with the radius of the tree at 1200 seconds of heating.

of a tree trunk during a fire in a wind. It can be seen from these calculations that the leeward side does indeed suffer a higher temperature within the trunk, leading to the well-known fire scars on the leeward sides of trees.

By using finite element methods and software such as FlexPDE, it is a relatively simple matter to solve for specific cases, such as the example described above. Other factors, such as layering within the vegetation sample (e.g., bark and wood with varying thermal diffusivity), are also easily included in this type of software.

Time-Dependent Heating

The boundary condition used previously in the simple one-dimensional examples was that there is a constant high temperature at the surface of the material. In reality, the heating exposure of vegetation during the passage of a fire is time dependent. As the fire approaches, the temperature rapidly increases and, after it has passed, there is a more gradual decrease. Models of temperature versus time above a fire, such as those presented in Mercer and Weber (2001), could be used as inputs to time-dependent temperature boundary conditions. Similarly, methods used by Dickinson and Johnson (2004), which assume a given flame residence time and then exponential cooling, are also valid.

A more realistic boundary condition on the surface of the vegetation is not that there is a given temperature there but rather that there is a given heat flux at the surface (Jones *et al.*, 2004). Mathematically, this is expressed as

$$\frac{\partial T}{\partial r} = H(T - T_e) \quad \text{at the surface} \tag{53}$$

where T_e is the external gas temperature and H is the heat transfer coefficient, which consists of free and forced components that depend on the wind at the surface (Potter and Andresen, 2002). While T_e is relatively easy to measure, determining H is more complicated and often only simple correlations are available. The often used approximation of a given temperature on the surface is just the limit of very large heat transfer coefficient. Although not ideal, this is a reasonable approximation for highly forced (high wind) scenarios, but is not appropriate in many other cases.

Mortality Models

One of the main reasons for modeling the heating of vegetation subject to fire is to determine the effect upon the vegetation. In particular, the depth of the mortality caused by the heating is often of interest. Cell mortality does not occur at one specific temperature but rather is a function of the temperature regime to which the plant has been exposed (Dickinson *et al.*, 2004) and depends on the type of vegetation involved (Dickinson and Johnson, 2004). A frequently used value for mortality is 60°C (Vines, 1968), but this was found to be too low for seeds (Bradstock *et al.*, 1994). Combining the modeling outlined here to determine temperature time exposures within vegetation with the recent advances in tissue response models such as in Dickinson *et al.* (2004) gives exciting new potential in determining lethal exposures from fires. This is an area of active current research.

CASE STUDY: A MODEL OF SEED SURVIVAL

Introduction

Here we consider the application of these types of models to a particular situation. Of interest is the temperature exposure of fruits above an idealized moving ground fire and the corresponding temperature exposure of the seeds within the fruit. We compare experimental observations and theoretical model calculations. This is an example of the use of the previously mentioned models in a real situation where we are trying to investigate some property (in this case seed survival).

Little is known of the fate of seeds in woody fruits subjected to fire (Lamont *et al.*, 1991), even though many plants rely on such seed for regeneration. Few studies have measured the temperature exposure of seeds within fruits or determined what constitutes a lethal exposure. Judd and Ashton (1991) subjected the mature woody fruits of three different Myrtaceae species to experimental heating at 250°C in a laboratory muffle furnace for various times up to 5 minutes and then determined the time of seed survival. Although this heating is not representative of the heating caused by fire, it does give an indication of seed survival. We compare their results to the model presented here. More realistic experiments on the heating of seeds in fruits were undertaken by Bradstock *et al.* (1994). They measured the internal and external temperatures of fruits of six serotinous *Hakea* species when exposed to laboratory heating in a muffle furnace and fire conditions in the field. In a similar manner, Judd (1993) subjected mature capsules of four small-fruited Myrtaceae species to various temperatures (200–750°C) in a muffle furnace for various lengths of time (15–300 seconds) and determined seed survival.

Mathematical Formulation

The fruit will be modeled as a sphere of radius b with a temperature that is dependent only upon the radial distance, r, and time. Defining the temperature above ambient (T_0) to be T', hence the governing differential equation is

$$\frac{\partial T'}{\partial t} = \alpha \nabla^2 T' \tag{54}$$

which in spherical coordinates can be rewritten as

$$\frac{\partial^2 (rT')}{\partial r^2} = \frac{1}{\alpha}\frac{\partial (rT')}{\partial t} \tag{55}$$

This is subject to the initial condition that the initial temperature within the shape be a uniform temperature of T_0 so that we have

$$T'(r, 0) = 0. \tag{56}$$

There is some prescribed time-dependent external temperature, $T_e(t)$, so that on the boundary of the fruit we have

$$T'(b,t) = T_e(t) - T_0 \tag{57}$$

Making the substitution

$$U(r,t) = rT'(r,t), \tag{58}$$

the statement of the problem reduces to

$$\frac{\partial^2 U}{\partial r^2} = \frac{1}{\alpha}\frac{\partial U}{\partial t} \tag{59}$$

$$U(r,0) = 0 \tag{60}$$

$$U(b,t) = b(T_e(t) - T_0) = g(t). \tag{61}$$

By the use of Green's functions or Duhamel's theorem (Ozisik, 1980), this system has a solution given by

$$U(r,t) = \frac{r}{b} g(t) + \frac{2}{b} \sum_{m=1}^{\infty} (-1)^m \frac{\sin(\lambda_m r)}{\lambda_m} \left[g(0) e^{-\alpha \lambda_m^2 t} + \int_0^t e^{-\alpha \lambda_m^2 (t-\tau)} dg(\tau) \right] \tag{62}$$

where

$$\lambda_m = \frac{\pi m}{b}, \tag{63}$$

the function $g(t)$ is defined in Equation (61), and τ is the time-like integration variable. Hence, on returning to the original temperature variable, the time-dependent temperature within the fruit can be determined as

$$T(r,t) = T_0 + \frac{1}{r} U(r,t). \tag{64}$$

So, for a given time-dependent external temperature, the temperature time dependence within a model fruit can be determined.

Comparisons are made between the present model and the experimental work of Judd and Ashton (1991). Unfortunately, as discussed earlier, the thermal diffusivity of the fruits of plants is not available in the literature. For many woody fruits, such as those of *Eucalyptus regnans*, an estimate can be made by using the thermal diffusivity of the bark of the plant. A commonly quoted value for the thermal diffusivity of bark is 1.3×10^{-7} $m^2 s^{-1}$ (Vines, 1968; Rego and Rigolot, 1989) although values as low as 0.65×10^{-7} $m^2 s^{-1}$ have been used (Costa *et al.*, 1991). Reifsnyder *et al.* (1967) determined the thermal diffusivity of bark of three species of pine for various densities and moisture contents and found "with the range of bark density and moisture content likely to be found in nature, there is a twofold difference between lowest and highest values of thermal diffusivity." Their values for the thermal diffusivity of the bark of pines were in the range 0.6 to 1.6×10^{-7} $m^2 s^{-1}$. Values of 0.65×10^{-7} $m^2 s^{-1}$ or 1.3×10^{-7} $m^2 s^{-1}$ will be used here; these are only estimates, however, and work is needed to determine this parameter for woody fruits. Three different radii of the fruits will be considered: $b = 3$ mm, $b = 5$ mm, and $b = 10$ mm, to demonstrate the effect the radius has upon the seed survival.

Here we take thermal death to occur when the inner 20% of the radius reaches a temperature of 70°C (see earlier discussion). That is, the time to thermal death of the seed is taken to be the time t when

$$T\,(0.2b,t) = 70°\text{C} \tag{65}$$

It is really the temperature time regime that is important to survival (Dickinson *et al.*, 2004), not the outright temperature, but unfortunately those types of models have not yet been developed sufficiently for use here. The experimental results of Judd (1993) indicate that the lethal temperature is indeed greater than 60°C and more likely to be in the range 70°C to 100°C, depending on the seed type, fruit type, and heating used. The chosen lethal threshold of 70°C is therefore at the lower end of this lethal threshold temperature range. Choosing other values for the lethal threshold temperature (e.g., 90°C) is simply a matter of reading off the appropriate lethal exposure times from the figures that are presented later.

Results and Discussion

Shown in Fig. 4(A–C) are the temperature versus time profiles for $r = 0.2b$ ($b = 3$, 5, and 10 mm, respectively) with $\alpha = 1.3 \times 10^{-7}$ $m^2 s^{-1}$ (the solid

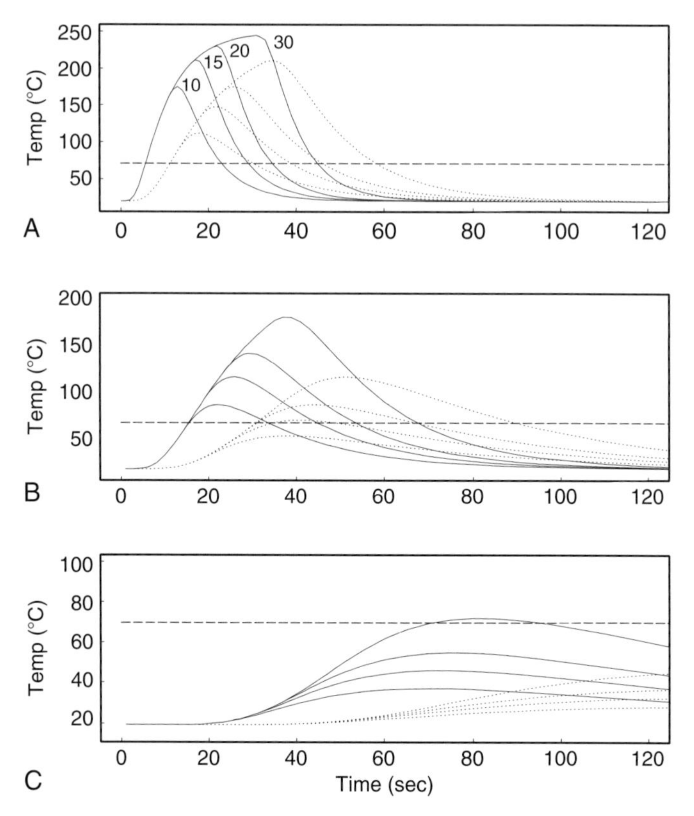

FIG. 4 Temperature versus time profiles at a depth of $r = 0.2b$, for radius of (**A**) $b = 3$, (**B**) $b = 5$, (**C**) $b = 10$ mm, with $\alpha = 1.3 \times 10^{-7}$ m^2 s^{-1} (solid lines) and $\alpha = 0.65 \times 10^{-7}$ m^2 s^{-1} (dotted lines). Fruits are exposed to an external temperature of 250°C for 10, 15, 20, and 30 s and T_a = 20°C. Dashed line is the lethal threshold of 70°C.

lines) and $\alpha = 0.65 \times 10^{-7}$ m^2 s^{-1} (the dotted lines). The ambient (room) temperature is taken to be 20°C (as in Judd and Ashton, 1991), and the fruits are exposed to an external temperature of 250°C for 10, 15, 20, and 30 seconds. Also shown as the dashed line is the lethal threshold of 70°C.

As expected, the fruit radius has a critical bearing on the internal temperature of the fruit. This is obvious when comparing the figures for b = 3 mm (Fig. 4A) or b = 5 mm (Fig. 4B) with a substantially larger fruit (b = 10 mm) (Fig. 4C). The larger fruit has a lower maximum temperature and a longer time lag before that maximum temperature is reached. With $\alpha = 1.3 \times 10^{-7}$ m^2 s^{-1} (the solid lines), the smaller fruits have easily reached the thermal death of the seed for even the shortest time exposure to the heating, whereas the largest fruit has only just reached thermal seed death for the longest time exposure considered. The change in the radius from b = 3 mm to b = 5 mm can significantly alter the maximum temperature reached within the fruit, from 147.5°C (b = 3 mm) to 72.1°C (b = 5 mm) for a 15-second exposure at a temperature of 250°C with $\alpha = 0.65 \times 10^{-7}$ $m^2\ s^{-1}$. This could have a significant effect upon seed survival.

The data provided in Judd and Ashton (1991) give a survival time for *Eucalyptus regnans* seeds of 15 to 30 seconds when exposed to an external temperature of 250°C. The present method gives a survival time of less than 10 seconds for both $\alpha = 1.3 \times 10^{-7}$ $m^2\ s^{-1}$ and $\alpha = 0.65 \times 10^{-7}$ $m^2\ s^{-1}$ for fruits approximately the size of E. *regnans* (3 mm in radius). Nevertheless, given the variability in many of the parameters (thermal diffusivity, moisture content, seed radius, thermal death temperature regime) and the relatively coarse time intervals available in the literature for experimental results, the present result is encouragingly close to those experimental results.

In formulating this model, many simplifying assumptions have been made and several physical features of the fruits have been overlooked. Woody fruits are usually not spherical in shape, although some are close to spherical. By using a few representative radii, ranges for seed survival can be determined; hence, the assumption of a spherical fruit is not necessarily too simplistic. Seeds are often not centrally positioned within the fruit, but this is not a critical factor in the seed survival if the seeds are located in the central core of the fruit. Fruits are not homogeneous, isotropic solids but are instead composed of many layers, presumably with differing thermal properties. This model is in many respects an averaging of these many layers, and better estimates of the average thermal properties of the fruits are clearly needed. Using the range of values of the thermal diffusivity for bark is only a first approximation to the appropriate range to be used. Comparing thermal diffusivities that differ by a factor of 2 (the solid lines versus the dotted lines in Fig. 4) demonstrates the effect the choice of the thermal diffusivity can have. The moisture contents of the fruit and seeds can be critical to seed survival for many reasons. The thermal diffusivity of fruits depends on their moisture content (Reifsnyder *et al.*, 1967), as does the lethal temperature exposure of the enclosed seeds (Judd, 1993). In general, the exposure of a fruit to heat in a fire is not uniform as supposed here. Clumping of fruits or shielding from stems is known to help protect seeds (Judd and Ashton, 1991); hence, this model is most relevant to single fruits that are not shielded by stems or other fruits.

The desiccation process mentioned earlier is evident in the experimental results of Judd (1993) (Figure 6 in particular), where there is an obvious plateau in the intracapsule temperature just below 100°C when the fruits are subjected to an external temperature of 250°C. As would be expected, this phenomenon is more evident in moister fruits. Here this could easily be modeled by truncating the external temperature to 100°C which would give an upper bound on the time to seed death. This assumes that thermal arrest is always present at the exterior edge of the fruit and thus the interior will only be exposed to a maximum temperature of 100°C. This upper bound will invariably be an overestimate of the actual time to death since it is clearly a larger reduction in the temperature exposure than would be expected physically.

The relatively simple model outlined here for the calculation of the internal temperatures of fruits provides a predictive tool for the determination of the survival of seeds within fruits subjected to fire conditions. Clearly more experimental work is necessary in determining many of the parameters involved before this model can be fully used. What it does provide is a link between experimental work on seed survival (a prescribed temperature for various times, or measured internal temperature during fire conditions) and seed survival from fire exposure in the field. Given sufficient information (from experimental work) on parameters such as thermal diffusivity and lethal seed temperatures to calibrate the model, it could then be applied to fire conditions. Models of temperature versus time above a fire such as those presented in Mercer and Weber (2001) could be used as inputs to the time-dependent temperature used on the exterior of the fruit in these models.

CONCLUSION

We have presented models for the heat transfer in three one-dimensional scenarios (slab, cylinder, and sphere) that have applications to different aspects of the heating of vegetation (leaf and large-diameter tree, stem, and seed, respectively). For a given temperature time input as obtained from fire models, it is possible to use these models to determine the temperature time profile within the vegetation sample and thereby infer the effect upon the vegetation. To use these models in a given situation requires knowledge of not only the temperature time history but also the type of vegetation involved (e.g., size, physical properties). A case study into the survival of different sized seeds within a woody fruit when subjected to a typical experimental heating regime shows the application of these models.

ACKNOWLEDGMENTS

The authors would like to thank A. M. Gill, E. A. Johnson, K. Miyanishi, and M. Dickinson for enlightening discussions and ongoing encouragement to continue with these modeling attempts.

APPENDIX NOTATION

b	radius, m
C_n	summation constants
C_p	specific heat at constant pressure, J kg^{-1} K^{-1}
E_i	exponential integral function
erf	error function
F	separation function
f	similarity function
G	separation function
g	boundary function
h	basis function
H	heat transfer coefficient, W m^{-2} K^{-1}
i	finite difference spatial index
J_0	Bessel function of the first kind of order zero
J_1	Bessel function of the first kind of order one
j	finite difference time index
k	thermal conductivity, W m^{-1} K^{-1}
n	coordinate index, dimensionless
R	spatial dimension, m
r	spatial coordinate, m
$\mathbf{r}$	spatial position vector
T	temperature, K
t	time, s
u	temperature, dimensionless
x	similarity variable

Greek symbols

α	thermal diffusivity, m^2 s^{-1}
Δt	finite difference time step
Δx	finite difference spatial step
κ	dimensionless constant
λ	separation constant

∇ spatial derivative operator
Ω arbitrary shape
$\partial\Omega$ boundary of Ω
π 3.1415926 . . .
ρ density, kg m^{-3}

Subscripts

B boundary property
cr critical property
e external property
0 initial or ambient property
i value of interest

REFERENCES

Bradstock, R. A., Gill, A. M., Hastings, S. M., and Moore, P. H. R. (1994). Survival of serotinous seedbanks during bushfires: comparative studies of *Hakea* species from south-eastern Australia. *Aust J Ecol* 19, 276–282.

Butler, B. W., Webb, B. W., Jimenez, D., Reardon, J. A., and Jones, J. L. (2005). Thermally induced bark swelling in four North American tree species. *Can J Forest Res* 35, 452–460.

Carslaw, H. S., and Jaeger, J. C. (1959). *Conduction of Heat in Solids*. Oxford University Press, London.

Costa, J. J., Oliveira, L. A., Viegas, D. X., and Neto, L. P. (1991). On the temperature distribution inside a tree under fire conditions. *Int J Wildland Fire* 1, 87–96.

Dickinson, M. B., and Johnson, E. A. (2001). Fire effects on trees. In: *Forest Fires: Behavior and Ecological Effects*. (Johnson, E. A., and Miyanishi, K., Eds.), pp. 477–525. Academic Press, San Diego, CA.

Dickinson, M. B., and Johnson, E. A. (2004). Temperature-dependent rate models of vascular cambium cell mortality. *Can J Forest Res* 34, 546–559.

Dickinson, M. B., Jolliff, J., and Bova, A. S. (2004). Vascular cambium necrosis in forest fires: using hyperbolic temperature regimes to estimate parameters of a tissue-response model. *Aust J Bot* 52, 757–763.

FlexPDE. www.pdesolutions.com.

Johnson, E. A., and Miyanishi, K. (2001). *Forest Fires: Behavior and Ecological Effects*. Academic Press, San Diego, CA.

Jones, J. L. (2003). *Development of an Advanced Stem Heating Model*. M.Sc. Thesis. Brigham Young University, Provo, UT.

Jones, J. L., Webb, B. W., Jimenez, D., Reardon, J., and Butler, B. (2004). Development of an advanced one-dimensional stem heating model for application in surface fires. *Can J Forest Res* 34, 20–30.

Judd, T. S. (1993). Seed survival in small myrtaceous capsules subjected to experimental heating. *Oecologia* 93, 576–581.

Judd, T. S., and Ashton, D. H. (1991). Fruit clustering in the Myrtaceae: seed survival in capsules subjected to experimental heating. *Aust J Bot* 39, 241–245.

Lamont, B. B., Le Maitre, D. C., Cowling, R. M., and Enright, N. J. (1991). Canopy seed storage in woody plants. *Bot Rev* 57, 277–317.

Mercer, G. N., Gill, A. M., and Weber, R. O. (1994). A time-dependent model of fire impact on seed survival in woody fruits. *Aust J Bot* 42, 71–81.

Mercer, G. N., and Weber, R. O. (2001). Fire plumes. In: *Forest Fires: Behavior and Ecological Effects* (Johnson, E. A., and Miyanishi, K., Eds.), pp. 225–255. Academic Press, San Diego, CA.

Ozisik, M. N. (1980). *Heat Conduction*. John Wiley and Sons, New York.

Potter, B. E., and Andresen, J. A. (2002). A finite-difference model of temperatures and heat flow within a tree stem. *Can J Forest Res* 32, 548–555.

Rego F., and Rigolot, E. (1989). Heat transfer through bark: a simple predictive model. In: *Proceedings of the 3rd International Symposium on Fire Ecology*. (Goldammer, J. G., and Jenkins, M. J., Eds.), pp. 157–161. S.P.B. Academic Publishing, The Hague, the Netherlands.

Reifsnyder, W. E., Herrington, L. P., and Spalt, K. W. (1967). *Thermophysical Properties of Bark of Shortleaf, Longleaf and Red Pine*. Bulletin No. 70, School of Forestry, Yale University, New Haven, CT.

Spalt, K. W., and Reifsnyder, W. E. (1962). *Bark Characteristics and Fire Resistance: A Literature Survey*. USDA Forest Service Occasional Paper, Southern Forest Experiment Station, New Orleans, LA.

Vines, R. G. (1968). Heat transfer through bark, and the resistance of trees to fire. *Aust J Bot* 16, 499–514.

Williams, F. A. (1985). *Combustion Theory*. 2nd edition. Benjamin Cummings, Menlo Park, CA.

13

Fire Effects on Grasslands

Paul H. Zedler

University of Wisconsin–Madison

INTRODUCTION

Grasses are one of the great successes of evolution. They occur in nearly every habitat from impoverished Antarctic wastes to the lush tropics and are dominant over vast stretches of landscape (Woodward *et al.*, 2004). The importance of grasslands in every sense—biological, economic, cultural, and aesthetic—has been repeatedly celebrated (e.g., Bryant, 1832).

The essential qualities of grasses and grasslands that have made them prominent in the biosphere and central to human history emerged as grasses adapted to tolerate drought and to exploit full-sun habitats. Grass progenitors were believed to have developed their basic features in wooded habitats, initially as subsidiary species, then perhaps as co-dominants in savanna-like situations (Linder and Rudall, 2005). But they then spread into drier and sunnier habitats where trees could not compete. In simplified terms, the outcome of this process has been that the uplands of the earth are divided between grasses, woody plants, and their mixtures—grasslands, forests, shrublands, savannas, and shrub steppes. Grasslands take control of landscapes when drought is annually recurrent and episodically severe but precipitation is usually sufficient to allow most of the growing space to be exploited in most years (Bredenkamp *et al.*, 2002; Woodward *et al.*, 2004). Where these conditions prevail, grasses are expert at seizing control of the –2 m to +2 m zone and holding it against the competition of other life forms.

Grasslands and grassy patches also occur in regions of higher and more predictable rainfall in situations that present difficulties for woody plants—steep topography, shallow or excessively drained soils subject to drought, or chemically deviant soils (Kruckeberg, 1984). Of special interest here are the cases where grasslands are created or maintained against competition by woody plants through disturbance, either massive, as in volcanic landscapes (Partomihardjo *et al.*, 1992), or recurrent, as with fires at short intervals (Curtis, 1959; Vogl, 1974). Another important class of grasslands, not considered in this chapter, includes the grass-dominated wetlands of both freshwater and saline environments (Archibold, 1995).

Situations involving fire disturbance are not rare (Bond *et al.*, 2005). As a consequence of their herbaceous growth form, their tendency to grow in dense aggregations, and the usual seasonal or episodic death and drying out of their stems and leaves, grasses are a near-perfect medium for fire propagation. The paleoecological and historical data show conclusively that grasslands can and did burn without the help of humans (Scott, 2000; Kershaw *et al.*, 2002) and that humans have only enhanced the capacity of fire to open up habitats. What is still debated, however, is whether the high frequency of fire was an accidental consequence of adaptation to other more stringent selective forces or if grasses specialized to promote and survive fire and used this scorched-earth strategy to wrest dominance from woody plants and less fire-tolerant herbaceous species.

Some researchers have emphasized a different strategic alliance. Continental grasslands generally have or have had in their recent past some complement of large grazing mammals. The long association of grazers and grasslands has prompted the hypothesis that grasses and their megaherbivore grazers are a highly co-evolved system (McNaughton, 1985). Flooding favors grasses in many situations, most notably in the seasonally flooded grasslands and savannas of the tropical regions (Bren, 1992; Piedade *et al.*, 1992) and freshwater and coastal marshes (Feldman and Lewis, 2005).

Chronic disturbances—those that may cause little immediate malaise or death—can nevertheless have long-term effects. Wind is a persistent force in most grasslands, propagating waves of bending stems but usually causing little breakage. Over time, however, wind by its erosive effects can modify or even destroy grasslands, as when dune movement is initiated (Forman *et al.*, 2005), or more locally, when fine soil materials are mobilized and differentially deposited beneath shrub canopies (Herbel *et al.*, 1972).

It is apparent that knowledge of disturbance is essential to understanding grasses and grassland ecosystems. All the most important management issues, whether motivated by purely economic considerations or by a desire to preserve natural systems for their own sake, require consideration of disturbance either as a means of effecting change or, when not subject to control, as a constraint on the attainment of optimal outcomes. A basic question in the practical management of grassland systems is, Should fire be allowed or imposed and, if so, with what frequency and intensity?

THE GRASS GROWTH FORM

The remarkable ability of grasses to thrive in so many ecological settings and their resilience to disturbance are largely attributable to their growth form, which is characterized by streamlined reduction and elegant simplicity. The tiller is the basic structural element. These originate from growing points (meristems) typically near, at, or below the surface of the soil. They expand upward through the production of repeating modules or "phytomers" (Clarke and Fisher, 1986; Briske, 1991) to form a jointed stem or "culm." The precise definition of phytomer varies. Following Clarke and Fisher (1986), the phytomer consists of a node, associated buds, an internode (essentially a stem), a sheath, and leaf blade (Fig. 1A). The sheath arises at the node and closely covers the internode. The leaf blade originates at the top of the sheath and generally diverges from the culm at a pronounced

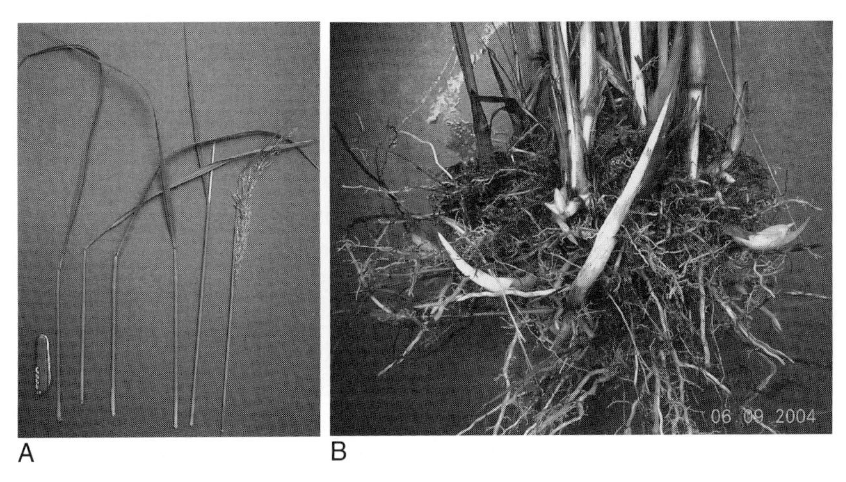

FIG. 1 (**A**) Flowering tiller of *Sorghastrum nutans* dissected into its component phytomers. The convention of considering the node to belong with the sheath above is followed. The lighter color at the base of the lower phytomers is caused by overlap by the sheath of the phytomer below. (**B**) Base of an *Andropogon gerardii* clump showing the typically dense tangle of roots, old tiller bases, and new tillers. Several actively growing rhizomes are clearly visible in this "sod" grass.

angle but generally less than 90°. Usually, the internode of one phytomer is covered by the uppermost part of the sheath of the phytomer below it. A bud is found at the base of the internode and is usually hidden by the sheath. These buds can become active and form new phytomers along the culm. This is a regular pattern in some grasses and primarily a response to damage to the distal phytomers in others. In an actively growing tiller before the formation of the inflorescence, the uppermost portion is an unexpanded leaf that is protected at its base by the sheaths of older phytomers. This vertically oriented "spear point" growth form is well suited for pushing upward through litter or the canopies of other plants (Campbell *et al.*, 1992).

The location of meristematic activity within the plant is key to the capacity of grasses to tolerate disturbance. The meristems that produce tillers are usually well protected by their location near or below the soil surface. As the tiller develops, meristematic activity comes to be restricted to intercalary meristems at the nodes and the base of the leaf blades and sheaths (Bell, 1992; Briske and Derner, 1998). This pattern restricts the energetically costly meristematic activity to points in the plant where it is needed to regenerate or reorient absorbing, photosynthetic, and reproductive elements. Despite this restriction, a typical grass nevertheless has considerable flexibility of growth response, allowing it to recover from a variety of disturbances and to adapt to changing physical and biological environments. Of particular importance, given the association of grasses with grazing faunas and fire, is that tissue lost from the distal portion of the leaf can be replaced by elongation of the leaf from the meristematic base. The same mechanism makes it possible to maintain dense artificial grasslands (lawns) by artificial grazing (mowing). The intercalary meristems at the nodes allow tillers to adjust the vertical angle of the stems by differential growth, and so to reestablish the vertical orientation after trampling or blowdown. If the defoliation is more severe, the buds located at the nodes can be activated so that secondary shoots can be formed from axillary buds. Under extreme grazing or after fires, a grass can mobilize new tillers from the buds at the base of the plant or located on underground rhizomes (Fig. 1B). Grasses also are capable of producing adventitious roots from nodes as needed to deal with flooding or burial.

The phytomer of a grass is built around the stems that are hollow (e.g., *Poa* spp.) or parenchyma-filled (e.g., *Andropogon* spp.) cylinders stiffened with structural tissues. The stem is further supported by the sheath, which closely surrounds it (Niklas, 1990). In nonflowering tillers, the internodes remain short and the sheaths may hide most or all of them (Fig. 1A). The sheath protects the developing phytomers while structural tissues are developing during the time they are susceptible to bending and wilting. The sheaths are photosynthetically active and may fix more carbon in the latter part of the growing season than the senescing leaves (Caldwell *et al.*, 1981). When a tiller produces a flowering stem, the internodes elongate greatly, presumably with the object of raising the flowers and seeds to facilitate pollination and seed dispersal. The appearance of the inflorescence terminates the growth of that tiller, but branches, flowering or not, may be produced from the axillary buds.

The narrow leaves of grasses are generally well supplied with supporting tissues in the form of fibrous and thick-walled cells (sclerenchyma). These features, along with a capacity to fold or roll in the lateral dimension, permit the leaf to endure periods of water stress without collapse (Abernethy *et al.*, 1998; Niklas, 1999). A notable feature of the grass leaf is the presence of siliceous deposits and silicified cells (phytoliths). Although silica deposits are present in many other plant families, they are characteristic of grasses (Bonnett, 1972). Phytoliths often have distinctive forms within taxonomic groups and because they persist in the soil they can be used to detect historical shifts in dominance from one grass tribe to another (Stromberg, 2004). The high silica content makes grass forage abrasive and, whether or not this evolved as a grazing deterrent, it is generally accepted that the evolution of high crown abrasion-resistant teeth (hypsodonty) characteristic of modern grazing animals was a response to tooth-wearing effects of a diet high in grass (Jernvall and Fortelius, 2002).

The entire grass structure is clearly designed to make efficient use of materials in the creation of a vertical structure that displays photosynthetic area and facilitates the production of flowers and fruit (Metcalfe, 1960). For example, detailed structural data of Anten and Hirose (1999) for a tall grass meadow showed that the grass, *Miscanthus sinesis* Anderss., was taller per unit stem biomass than any of the common forbs with which it occurs. A study of grasses and forbs in savannas in the central United States showed that grasses had superior leaf-level resource use efficiency compared with co-occurring forbs and woody plants (Tjoelker *et al.*, 2005).

Grasses show large variations in the individual structural elements (tillers, phytomers, flowers, and inflorescences) and in the way the tillers are aggregated as the genets expand from their origin as seedlings. The realized patterns of growth are continuously variable in several dimensions, but two extremes are commonly recognized—bunch forming (caespitose) and sod-forming (rhizomatous). In bunchgrasses, the tillers arise at steep angles from within the sheaths of previously existing tillers (intravaginally: Bell,

1992). As a result, genets form distinct clumps of ramets. As in many other clonal herbaceous plants, the vascular connections between successive generations are not permanent, even within a dense bunch. Typically, the parent ramet will die, severing any direct vascular connection between the offspring and other ramets within the clone (Briske and Derner, 1998). Thus, a bunchgrass will usually consist of a collection of autonomous "ramet hierarchies" (Briske and Derner, 1998). A common pattern among bunch-forming grasses is for the center of clumps to die out so that the genet appears as a well-defined ring (Samuel and Hart, 1995), though this phenomenon is also seen in a less dramatic form in rhizomatous grasses (Kobayashi and Yokoi, 2003). These rings in turn can break apart, giving rise to another level in the genet–ramet hierarchy (Liston *et al.*, 2003).

The caespitose habit is common and widely distributed, so it follows that there must be strong selective forces in many diverse settings acting to reduce spacing between ramets within the clone despite the overlap in resource acquisition that would result. Theories to explain the bunchgrass growth form have emphasized proximate mechanisms, with ultimate explanations tending towards concepts of adaptation to resource limitation, especially involving nitrogen (Tomlinson and O'Connor, 2004). It has been shown for some bunchgrass species that nutrient resources are concentrated beneath the plant (e.g., nitrogen and soil organic carbon; Derner and Briske, 2001), and therefore the caespitose form may be a means of sequestering resources in habitats in which these are in short supply. But disturbance factors may also have played a role because the accumulation of dead material in grasses deters grazers (Pfeiffer and Hartnett, 1995). This effect will be most pronounced in the bunch form and in unburned vegetation. Tillers that emerge separated from the main cluster may be more likely to be selected by grazers, especially where surrounding cover is sparse and tillers are very apparent.

In sod-forming grasses, new tillers emerge at some distance from the parent, the new tiller either pushing through the parental sheath and running a distance horizontally or arising from underground stems (rhizomes). Grasses that spread by rhizomes have the ability to adjust the depth at which buds are located, which might be partly determined by their need to survive fire or grazing. The resource-based explanation for sod-formers is that dispersing new tillers reduces resource competition among ramets while also allowing the genet to encounter and exploit resource-rich patches (Lovett Doust, 1981).

But "bunch" and "sod" are not discrete categories. Valverde and Pisanty (1999) studied *Schizachyrium scoparium* (Michx.) Nash var. *littoralis* on dunes. This strongly caespitose grass showed considerable plasticity by producing "daughter tussocks" at some distance from the parent tussock in stable but not in moving dunes. The perception of a particular species as "bunch" or "sod" is also partly dependent on the density and continuity of the cover of the grasslands in which they occur. *Andropogon gerardii* Vitman, the dominant species in the North American tallgrass prairie, is classified as a sod-forming grass in reference books, yet its ramets are generally highly aggregated. After fire removes litter and standing stems, this aggregation is very apparent yet, by the time of flowering, the grassland can have the appearance of a uniform sward.

The preceding description captures the major features of the grass species that are the dominants in the world's natural and anthropogenic grasslands. But included in the Poaceae are some species and groups that deviate markedly from this general pattern. The obvious examples are the woody bamboos, the largest of which can reach tree size and which, for the most part, are specialized for forest habitats in the tropics and subtropics (Renvoize and Clayton, 1992). *Poa flabellata* (Lam.) Hook. f. (or *Parodiochloa flabellata*[G]) of the South Atlantic Islands is one of a set of examples of grasses that functionally approximate shrubs. This giant tussock grass (up to 3.5 m) has a dense canopy of long-lived leaves topping a tall tussock base. Individuals are thought to live for hundreds of years (Woods, 1979 cited in Chapman, 1996). Some of the spinifex grasses of Australia have a strong tussock form with stiff leaves and persistent leaves with a shrub-like appearance. Chapman (1996) described *Triodia basedowii* E. Pritz. as resembling "a curled up porcupine with grey pointed leaves poking in all directions." The amphibious grasses of the Amazon basin represent another extreme, alternating between a terrestrial existence and a period as floating aquatics (Piedade *et al.*, 1992). A number of grasses are "resurrection plants" with leaves capable of desiccating to air dryness but resuming function on rewetting (Gaff, 1986).

REGENERATION FROM SEED

Grasses have a diverse range of dispersible propagules and seed and fruit types. The subject is too large to review here, but a few points deserve mention because they relate to the disturbance resilience of grasses. Most grass seeds are relatively small, as are the fruits (caryopses) that contain them, and with exceptions (e.g., *Guadua*: Londono and Peterson, 1991) are dry. The number of propagules produced per flowering stem is generally large. As one example, a single individual of the small annual grass

Agrostis avenacea J. F. Gmel. with 13 tillers was estimated to have produced more than 14,000 seeds (Zedler and Black, 2004).

Grass seeds and fruits generally lack the morphological and physiological traits that allow deep dormancy and long survival in the soil, though nondeep physiological dormancy is common (Baskin and Baskin, 1998). Species vary widely with respect to the transiency of seedbanks. Odgers (1999) followed the "germinable" seedbanks of 24 grasses found in frequently burned vegetation in Australia over five seasons. Nine of the 11 exotic species had persistent seedbanks (observed seed densities >0 for five samples), whereas only five of the 13 natives maintained seed densities above zero in all seasons. All seedbanks, including those that did not drop to zero, were still temporally variable, as would be expected if seeds were not capable of prolonged survival. This study did not isolate the effect of seed predators, usually ants, which is known to be significant in some instances (Schoning *et al.*, 2004). In general, grasses can have persistent seed banks but usually only because of continual input. The sheer number of seeds produced seems to be a primary mechanism to survive disturbance (e.g., Lonsdale *et al.*, 1998).

Most grasses lack obvious features to promote long distance dispersal other than small diaspore size. As grasslands are often windswept, "anemochory" might be expected, and it does exist in species such as the "feather grass" (*Stipa pennata* L.) of the Eurasian steppes and tumbleweed grasses such as "blown grass" (*Agrostis avenacea*) of Australia. Some diaspores are specialized for adherence to animals, such as the sand bur species (*Cenchrus* spp.) in which the inflorescence is modified to produce a spiny fruit. For most species, however, seed dispersal shadows are relatively small, suggesting that, in most situations, holding space locally has greater fitness benefits than seeking open sites at a distance.

A special case is that of grasses in which a hygroscopic appendage, usually an awn firmly attached to the caryopsis, functions to implant the seeds into the soil (Frost, 1984). Studies in Africa [*Hyparrhenia diplandra* (Hack.) Stapf: Garnier and Dajoz, 2001] and Australia (*Heteropogon, Hyperthalia, Tristachya,* and *Themeda* spp.: Tothill, 1969) have shown that planting can enhance survival through intense fires but possibly at the cost of reduced survival in years without fire (Garnier and Dajoz, 2001). Peart (1984) suggested a classification of grass diaspores as having "active awns" (as in *Heteropogon*), passive awns (ones that did not have strong hygroscopic movement, e.g., *Aristida* spp.), and those with no appendages (e.g., *Eragrostis* spp.). He observed that passive awns, though they did not drill the diaspores, did position them vertically and anchor them with antrorse hairs so that the developing root could penetrate the soil. But Peart (1984) found that unappendaged species had relatively greater establishment after a fire than appendaged species, a fact he attributed to the unappendaged diaspores being able to move through the litter to the soil. Seed burial by subterranean emplacement of flowers and fruits, common in some aquatic situations [e.g., *Lilaea scilbides* (Poiret) Hanman: Zedler, 1987], is also known in a terrestrial grass, *Amphicarpum purshii* Kunth of the northeastern United States. This robust annual produces both aerial and buried seeds, and Cheplick and Quinn (1982) have shown that seed burial is an effective means of surviving fire, though whether this was the primary selective force operating is open to question. There is also evidence that awns reduce the loss to seed harvesting ants both by burial and subsequent loss of the awn or because the awn makes seed handling difficult (Schoning *et al.*, 2004).

Some grass seeds survive quite high temperatures and have higher germination with heat and smoke. Gashaw and Michelsen (2002) tested the diaspores (palea, lemma, and glumes were not removed) of three grass species of Ethiopian savannas [*Eleusine coracana* (L.) Gaertn., *Hyparrehnia confinis* Anderss., and *Sorghum arundinaceum* (Desv.) Stapf]. Although all three were killed when heated for 1 minute at 150°C, all had good germination after being subjected to 1 minute at 120°C, and *S. arundinaceum* and *H. confinis* survived 5 minutes at 90°C. Clarke and French (2005) studied the germination of Australian grasses after heat and smoke treatments. Six of 22 species were stimulated by heat, two of them showing the best response to the highest temperature treatment of 1 minute at 120°C. Seven of 22 showed improved germination with application of "smoke water." A different set of seven species showed a smoke × temperature interaction. Clearly there is more to be learned about fire/germination interactions.

GRASSES AS FUEL, MULCH, AND FORAGE

Grasses as Fuel

Grasses, whatever the selective forces that shaped them, seem designed to burn. The fuel elements are finely divided and usually spaced far enough apart to allow good aeration but close enough to allow transfer of heat from one element to the next. The culm-leaf structure of a typical grass is also a kind of "kindling-log" system that favors both rapid spread and a high level of total heat release. Grass leaves have surface/volume ratios on the order of 190 to 380 cm^{-1}, whereas for stems the range is 45 to 75 cm^{-1} (Brown, 1970). Because of its high surface/volume ratio, a dead grass leaf can be thoroughly wetted by rain at night, yet dry to a point capable of propagating a fire with a few hours at moderate humidity levels when exposed to sun and wind (Cheney and Sullivan, 1997). The capacity to dry is important in maximizing the probability that ignition will correspond with conditions that allow the fire to spread. Before a fuel element can flame, the water it contains must first be heated and then evaporated so that the element temperature can be raised to 200°C, the temperature at which volatilization of the organic components begins (Nelson, 2001). The moisture level above which too much heat and too much time are required to allow propagation is commonly called the "moisture of extinction" and expressed as a percentage or fraction of oven dry weight. These values are determined empirically and vary with species and geographical region in the range of about 12% to 40%, though values as high as 110% are also reported (Biddulph and Kellman, 1998).

These fine and abundant fuel elements require very small inputs of heat energy to reach the point of ignition, allowing fire to propagate readily through the fuel bed. For example, 20 of 24 rates of spread recorded for a set of wildfires in southeastern Australia were over 100 $\mathrm{m\,min^{-1}}$ with the fastest rate of 383 $\mathrm{m\,min^{-1}}$ (Cheney *et al.*, 1998). Once the leaves are burning, the heat they generate can raise the stems (the culms with their enclosing sheaths) with their greater mass and density to the ignition point, increasing the fire intensity and prolonging the duration of high temperatures. Catchpole *et al.* (1993) have demonstrated this process in "mixed fuel" experimental simulations. The fine fuels (for their experiment, fine wood shavings—"excelsior") increased the rate of spread through the larger fuel elements (pine sticks), which, once ignited, remained burning after the fine fuels had been consumed. The positive effect of the fine fuels was greater for lower packing ratios (volume of fuel per unit volume of fuel bed) and higher wind speeds. Under some combinations of low wind speed and low stick packing, the fire would not propagate without the fine fuels being present. By analogy, the finer leaves of grasses perform the function of the experimental excelsior.

Burning leaves can propagate a fire ahead of the burning front. Leaves do not always burn from the tip to the base, but can ignite first at a lower point, for example, when the fire moves up the stem. As a result, an actively burning leaf can become detached from the stem. Since many of these leaves break free at some distance above the ground, they are readily lofted in the turbulence of the locally rising column of heated air and may be carried forward by the prevailing wind or by local "fire whirls"—vortices that form when a disruption of the rotation along a horizontal axis causes it to be stretched upward and intensified by the conservation of angular momentum (Jenkins *et al.*, 2001) or that form in local shear zones (Albini, 1993). These "floaters" are a problem in management burns. In an otherwise uneventful controlled grassland burn, such fire whirl–entrained burning grass leaves moved ahead of the fire and ignited a conifer crown on the opposite side of the narrow road that served as a control line (T. Anchor, personal communication).

Of key importance to grass as fuel is that the strengthening tissues that keep the stems upright and allow leaves to resist wilting and wind damage continue this support function after the death of the tissues. The dense packing of stems, especially in bunchgrasses, provides additional support. As a result, grasslands typically have high levels of standing dead material during the dormant seasons—over winter and into spring in cold climates and through drought periods in warmer regions. This drying-off of grass ("curing") is a fundamentally important transition for both fire ecology and the utilization of grass by grazing animals (Cheney and Sullivan, 1997). It is a gradual biophysical process in which the decline in moisture content follows a negative exponential trend that can be interrupted by dew formation and rainfall (Fig. 2).

The pattern of synchronized death and subsequent curing to produce a readily ignited fuel bed reaches an extreme in the puzzling case of the simultaneously flowering bamboos of tropical regions. Keeley and Bond (1999) postulated that this behavior may have evolved to promote widespreading and intense fires in an environment where such fires would otherwise be rare or less intense. Currently, humans are creating a similar situation by facilitating the expansion of the dreaded alang-alang grass [*Imperata cylindrical* (L.) Beauv.] into tropical forests in Indonesia (Riswan and Hartanti, 1995), New Guinea (Hartemink, 2001), and elsewhere.

Even the most persistent grass stems do eventually fall. By analogy with the process that afflicts the growers of cereal crops, this process may be called "fuel lodging." The rate at which standing dead biomass becomes litter through the fuel lodging process varies with the growth form and with the impact of disturbances such as windfall, rainfall, snowfall, and trampling by grazers. In dense swards the process can be slowed because early falling culms can be supported by other stems. In bunchgrasses, the clumped growth form provides support for dead stems, slowing or in extreme cases effectively preventing dead stems from falling away from the clump, as in Australian spinifex grasses (*Triodia*, *Symplectrodia*, and *Monodia*: Allan and Southgate, 2002). In such cases the stems disintegrate rather than fall.

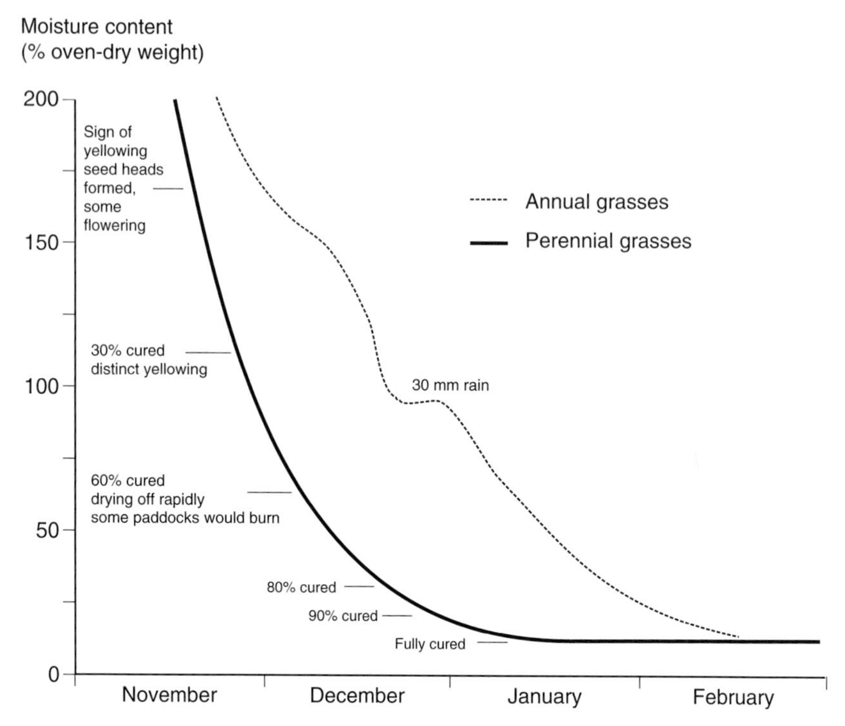

FIG. 2 The "curing curve" for annual and perennial grasslands in the Australian Capital Territory during the summer of 1964–1965. The annuals cure at a faster rate, a fact that can result in an early dormant season fire spreading though annuals while perennials still retain large amounts of green tissue. (From Luke and McArthur [1978], reproduced in Cheney and Sullivan, [1998]. Copyright Commonwealth of Australia, reproduced with permission.)

In grasslands with continuous cover, the fuel bed packs more densely as the fuel lodging process continues and, if unburned for a number of years, a dense litter layer forms on the soil surface that fluctuates seasonally but will eventually reach a quasi-steady state when each year's addition of litter is balanced by the loss of litter through decomposition. Thus, a typical fuel bed in a grassland unburned for a few years consists of some live tissue, mostly standing dead from the previous period of maximum growth, and a multiyear accumulation of litter, only a fraction of which may be dry enough or well enough aerated to burn in the event of fire. The proportion of the dead material that gets "hung up" or that remains standing is important because it will decompose more slowly than the material in contact with the soil and its biota. In pine-wiregrass (*Aristida stricta* Michx.) communities in the southeast United States, wiregrass that was placed on the soil surface had a decay constant (k) of –0.159 versus a value of –0.074 for material that was "elevated"—kept from contact with soil or compacted litter (Hendricks *et al.*, 2002).

Although one might expect rates of grass litter accumulation to vary substantially for different grasses and in different climatic regions, values reported in the literature seem to indicate that the times to reach quasi-equilibrium span a relatively small range. Daubenmire (1968) reported that about 3 to 5 years is required for litter biomass to accumulate to pre-fire levels, with no obvious correlation with climate or grassland type. In one detailed study of decomposition, the leaves of plains rough fescue [*Festuca hallii* (Vasey) Piper], a dominant species in the northern Great Plains, were shown to decompose rapidly compared to tree foliage (Trofymow *et al.*, 2002). Averaged across 18 sites, the fescue litter lost 98.2% of its mass in 6 years. Because this experiment placed the litter in direct contact with soil, it probably overestimates the actual rate of loss of litter in a natural prairie setting, where leaves could be above the soil for some period of time. Assuming exponential decomposition and constant annual productivity with a decay constant in the range of –1.5, the litter mass would be expected to equilibrate at roughly 5 years. Such calculations assume that production is constant whereas it often shows a trend after fire, either up or down. Pylypec and Romo (2003), working in a system in which fire tends to depress production, reported that litter accumulation leveled off at about 11 years. Aerts *et al.* (2003) reported that only 5% to 13% of grass litter remained after 3 months in a study in the Netherlands. Wright's (1974) results from *Hilaria* grasslands of Arizona showed that litter is about 90% rebuilt in 5 years. Kochy and Wilson (1997) reported 60% loss of mass over the summer for mixed litter from a *Stipa–Bouteloua–Koeleria* prairie in Saskatchewan.

Litter is important in grassland fires because it can be a significant part, even the greater part (e.g., Brye *et al.*, 2002), of the fuel load and because it can increase fuel homogeneity. Therefore, greater fire intensities and more uniform consumption patterns are expected in less frequently burned grasslands, an expectation confirmed in a study of grassland fires in Kansas and Florida (Gibson *et al.*, 1990). Lunt and Morgan (2002) also reported higher maximum fire temperatures in less frequently burned sites in Australia (Fig. 3).

The presence of litter can add to the complexity of outcomes because fuel moisture may differ between the litter and the standing dead biomass. If a burn occurs a few warm sunny days after a rain that has thoroughly wetted the litter and the soil, only the standing dead may burn, leaving the litter almost untouched. But if a fire were to occur in the same grassland

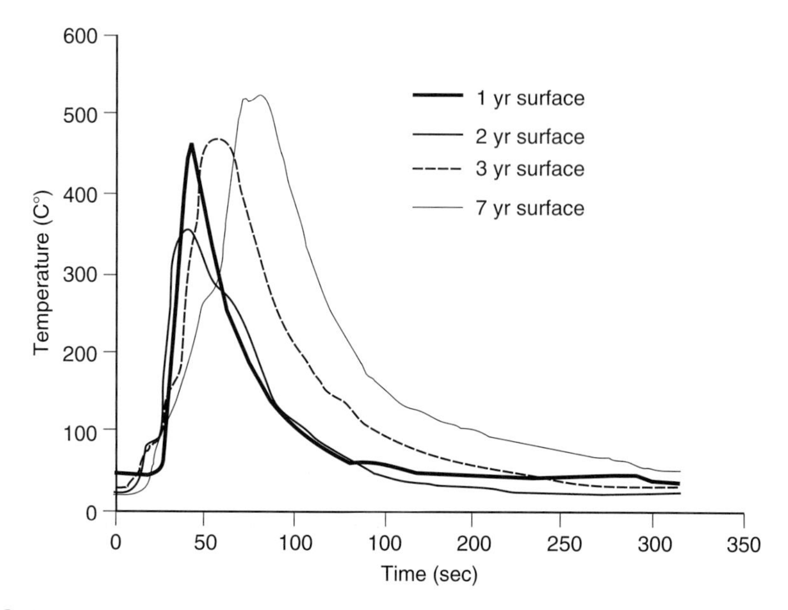

FIG. 3 Temperature curves recorded in *Themeda triandra*–dominated grasslands in eastern Australia. Fire intensities in these grasslands are moderate at 100 to 1200 kW m^{-1}. Note that peak temperature and the duration of temperatures in the lethal range for plant tissue (approximately >60°C) is greater where there has been a longer interval between fires. (From Morgan [1999], reproduced in Lunt and Morgan [2002]. Copyright Cambridge University Press, reproduced with permission.)

after a prolonged drought, most of the aboveground biomass would probably be consumed. In the former case, the unburned litter could be fuel in a subsequent fire, possibly as soon as in the next prolonged drought. In more arid grasslands there would probably rarely be large discrepancies between the fuel moisture levels of litter and standing dead. This underscores the importance of the antecedent and current climatic conditions in determining fire effects.

Effects of Grass Mulch and Standing Dead

The mulching effect of accumulated dead plant parts can be seen as an unavoidable consequence of the life form of grasses. [Mulch *sensu* Hedrick (1948): "the whole of the protective blanket after the forage has dried."] Structurally sturdy stems cannot easily be made to self-destruct to clear the way for the next generation of photosynthetic surfaces. The importance of this accumulation of dead material goes far beyond its role as fuel because of how it affects light, moisture, and nutrient relations. Facelli *et al.* (1988) found that litter accumulation exerts strong control on species composition of a grassland in Argentina. Similarly strong effects on local biodiversity have been reported from the Northern Hemisphere by Tilman (1993) and Foster and Gross (1998). In these cases, the strong effect of the litter seems largely attributable to shading, which makes it difficult for seedlings to establish and for smaller and low-growing plants to thrive. It also limits the amount of light reaching the newly emerging tillers of the dominant grasses. The negative effects can go beyond subsidiary species. Lunt and Morgan (2002) reported that *Themeda* grasslands in southeast Australia will decline at least in part because of self-shading from accumulated litter. In unburned productive tallgrass prairie, litter reduces the photon flux density at 20 cm to only 10% of full sunlight (Knapp *et al.*, 1998). It is estimated that litter blocks more than half of the photosynthetically active radiation in the first 30 days of growth, a reduction that cannot easily be made up in the remainder of the growing season (Knapp *et al.*, 1998). These strong litter effects are not universal. For example, Tunnell *et al.* (2004) reported no effect of litter in a disturbed Oklahoma grassland, and Hayes and Holl (2003) found no significant effects of litter on plant cover in a California coastal grassland. Lunt and Morgan (2002) noted that in the same region of Australia in which *Themeda* grasslands suffer negative effects of litter, *Poa*–dominated grasslands appear unaffected. Given that small mammals are very sensitive to cover and that small mammals can be a significant disturbance factor (e.g., Noy-Meir, 1988; Howe and Lane, 2004), it has also been suggested that litter accumulations could increase herbivory in grasslands generally and especially in the smaller gaps (Lunt and Morgan, 2002). Given this often marked effect of mulch, its removal by fire can be an important influence on ecosystem function, enhancing productivity by minimizing self-shading and opening the habitat for regeneration (Launchbaugh, 1973; Oesterheld *et al.*, 1999; Lunt and Morgan, 2002). Thus, it might be expected that a lack of fire could have negative consequences for local biodiversity by limiting opportunities for the establishment of smaller and shorter-lived species. Just such a pattern has been reported for prairie remnants by Leach and Givnish (1996). They attributed dramatic losses of legumes, small-seeded species, and species of low stature in small remnants to a lack of fire.

Grasses as Forage

Fire is not the only factor removing accumulated biomass. Grasslands present a vast and readily exploited resource for suitably adapted grazers. For mammals, the necessary adaptations were developed coincident with the expansion of grasslands in the Tertiary (Jacobs *et al.*, 1999). Of particular note is the emergence of mammals with high-crowned teeth (hypsodonts) able to deal with the abrasive silica-rich leaves of grasses during a period when browsing animals that utilized trees, shrubs, and broad-leaved herbs (mesodonts, brachydonts) were in gradual decline (Janis *et al.*, 2000).

There is no disagreement that large grazers have been a factor in grassland ecology since their origin. But the degree to which grasslands and large mammalian grazers constitute a tightly coupled mutualistic co-evolutionary system is a matter of debate. Grasses are also heavily utilized by small mammals (e.g., Bonham and Lerwick, 1976; Stromberg and Griffin, 1996), arthropods (Tscharntke and Greiler, 1995), and even mollusks (e.g., Hanley *et al.*, 1996; Silliman and Bertness, 2002).

The high silica content of grasses has been suggested as evidence of the evolutionary response of grasses to mammal grazing (McNaughton, 1985). But this idea has been challenged by the recent discovery of a diversity of grass phytoliths in Late Cretaceous dinosaur coprolites in India (Prasad *et al.*, 2005). This finding suggests that grasses were already substantially differentiated and in possession of abrasive phytoliths before the explosion of mammalian grazers in the Oligocene and Miocene epochs.

However, there is no doubt that the impact of natural grazers in natural grasslands can be substantial. McNaughton (1985) estimated that, in some areas of the East African Serengeti grasslands, between 15% and 94% of the annual aboveground net primary productivity was consumed by ungulate megaherbivores. Data from herbivorous small mammals (e.g., voles) suggest impacts nearly as great from these animals, with a 5% to 35% reduction in maximum standing crop in California grasslands (Batzli and Pitelka, 1970) and 40% to 50% higher standing crop in plots from which small mammals were excluded in Kenyan grasslands (Keesing, 2000). Grasses can absorb such high rates of herbivory (as well as regular mowing in suburban lawns) because of their ability to offset losses of leaves by the expansion from the intercalary meristem, sprouting of axillary buds, and the rapid recovery of photosynthetic area by the emergence of new tillers from the base. They can also exhibit "compensatory growth" that raises rates of carbon fixation of grazed plants above that of ungrazed individuals (McNaughton, 1979). But such subtle aspects of disturbance through faunal change are overwhelmed by the conspicuously disastrous consequences that follow from the overgrazing that has historically been imposed through human mismanagement of both wild (e.g., Birkett, 2002) and domesticated grazers (e.g., Sharp and Whittaker, 2003). Grazing by both native (Dublin, 1995) and domesticated (Savage and Swetnam, 1990) animals can alter fire regimes by reducing fuel amounts and fuel bed connectivity.

DROUGHT DISTURBANCE: A PRIMARY DRIVER

Because fire will not propagate in fully hydrated vegetation, fire and drought are highly and positively correlated over short periods of time. This makes it difficult to separate their independent contributions to grassland function. Further complicating the picture is that, over longer periods of time, protracted drought (i.e., lower than average rainfall for several or many years) can mean less, rather than more, fire (Brown *et al.*, 2005). This observation does not contradict the statement made above. Whether a grassland is in a drought cycle or a wet cycle, it will have to dry before it can burn. But when drought reduces the amount of fuel available to burn, fires may be less likely to start and may tend to be smaller when they occur because there is less fuel available to burn. Putting it simply, the data of Brown *et al.* (2005) suggest that grasslands have endured drought without much fire or have better growing conditions with frequent fire. This only underscores how the effect of these two factors are interrelated.

The value of aridity as a predictor of grass dominance is obvious from a consideration of the patterns at a global scale. Grasslands fade away on the dry end of the climatic gradient where surface moisture becomes highly unpredictable, so that the advantage shifts to species that can tap deeper and more dependable reserves of water or that are equally equipped to deal with temporally unpredictable surface moisture (Sala *et al.*, 1989). At the wet end of the aridity gradient grasses yield to forest and dense shrublands as precipitation increases and becomes more reliable. For example, study of the historical climate data for the triangular prairie peninsula that extends eastward through Illinois, Wisconsin, and into Indiana shows that drought as measured by the Palmer index is present in 10% to 15% of the years in the peninsula but only 5% to 10% of the years in the forested areas to the north and south (Changnon *et al.*, 2003). Disturbance and especially fire or other factors such as nutrient deficiency, shallow soils, or favorable radiation balance conditions permit grasslands to exist outside of the usual aridity envelope (Edwards and Armbruster, 1989). Likewise, a predominance of winter over summer precipitation favors woody plants as in the shrublands of the U.S. Great Basin, the Mediterranean-climate scrub regions, and South Africa (O'Connor and Roux, 1995).

The controlling importance of aridity has been amply demonstrated by many studies that have shown that average annual productivity of grasslands measured in various ways is highly correlated with annual precipitation (e.g., Walter, 1964; Rosenzweig, 1968; Lauenroth, 1979; Paruelo *et al.*, 1999; Huxman *et al.*, 2004). Given this, one would predict that interannual variation in rainfall, which generally increases with decreasing rainfall, would have a strong effect on productivity and all aspects of grassland function, a prediction confirmed by long-term studies in several regions of the world (Oesterheld *et al.*, 2001).

It is the modular growth pattern and annual renewal of aboveground elements of grasses that allow them to track variations in precipitation. They permit the individual grass to adjust its biomass and growth to the prevailing conditions by producing more or fewer units (tillers, phytomers, inflorescences). Bunch grasses, for example, will shrink in size and may fragment during drought years (Weaver, 1954). Annual grasses adjust to drought by scaling down with fewer tillers, reduced stature, and accelerated phenology. Although grasses are good at tracking variations in rainfall, some degree of lag is inevitable because of a loss of capacity that cannot instantly be compensated for (e.g., fewer buds, less carbohydrate reserve) or on the other end because of excess capacity (more tillers are launched than can be supported through the season). As a result, predictions of annual net primary productivity are often improved by including rainfall in prior years into the predictive equations (Wiegand *et al.*, 2004). The plasticity of grasses also explains the observation that the capacity of annual net primary productivity to track temporal variations in precipitation is best developed at intermediate levels of aridity—roughly 475 mm mean annual precipitation (Paruelo *et al.*, 1999). It is at this range on the aridity gradient that there is a high level of occupancy of the habitat but also unexploited space and other resources. The high occupancy means that the resources made accessible by removal of the moisture constraint can be exploited. With greater aridity, resources potentially released by higher than average precipitation cannot be fully exploited without a greater than annual lag. At the wetter end of the gradient, precipitation is great enough that there is very high occupancy and therefore other factors (light, nutrients) constraining response (Paruelo *et al.*, 1999).

Annual net primary productivity is fuel, and Gill *et al.* (2002) pointed out that therefore precipitation can affect fuel connectivity, which in arid systems (but not necessarily mesic systems) can be the difference between no fire or small fires and regional conflagrations if high connectivity resulting from a year of high net primary productivity is followed by drought and extreme weather.

DIRECT FIRE EFFECTS

As in any fire, the heat release from grassland fires is primarily a function of the amount of fuel, its moisture content, its structure, windspeed, and topography. Because the fuel elements are small, grass fires spread rapidly and are of short duration at any specific point. The main hazard in following behind a grass fire is not the residual flaming or the hot soil, but the smoke. Although data on grassland fires are limited, several generalities are well established. Because of the short duration and the fact that much of the fuel is held above the ground, temperatures peak rapidly and therefore soil heating into the range where biological damage is expected (usually considered to be >60°C) occurs only at the surface or a few centimeters into the soil. Roots and subterranean buds therefore have excellent protection against even the most intense fires. This would lead one to expect that grasses would simply move their growing points below the danger zone and so be immune to direct fire damage, in the manner of geophytes of fire-prone areas (Le Maitre and Midgley, 1992). But Tainton and Mentis (1984) point out that in South Africa it is the sod-forming grasses of rarely burned grasslands that have the deepest bud placement whereas the growing points of bunchgrasses of the frequently burned areas (e.g., *Themeda*, *Heteropogon*) are mostly not in the soil, though they are protected by surrounding live and dead tissues. This situation, which is not restricted to South Africa, seems evidence that bud mortality through fire has not been a strong selective factor or at least not strong enough to cause grasses in frequently burned areas to move their buds deeper into the soil.

Although the flaming mode of grass fires is of short duration, smoldering combustion also occurs and can have pronounced effects. It is expected in areas of high fuel density (high packing ratio), such as dense litter accumulations in long-unburned grasslands, especially when moist, and in the center of bunchgrasses. In bunchgrasses, the concentration of biomass presumably causes there to be more local heat release, perhaps in the flaming mode, but certainly with smoldering combustion that can allow heat to penetrate to the growing points. Wright (1971), using an experimental approach, showed that *Stipa comata* Trin. & Rupr., which retains more dead stems and leaves in its clumps than *Sitanion hystrix* (Nutt.) J. G. Smith, suffers greater damage from fire. He ascribed this to the longer duration burning in *Stipa* and the penetration of combustion deeper into the clump structure. This effect probably is part of the explanation for the more negative effect of fire on more arid grasslands as discussed below. The study by Bogen *et al.* (2002) of the tussock grass, *Festuca campestris* Rydb., showed that this grass would be highly susceptible to sustained burning within clumps since tiller mortality at 60°C is 50% at about 30 seconds and 100% above about 90 seconds. In another artificial experiment, Robberecht and Defossé (1995) looked at the fire response of two prairie species in the northwest United States, *Festuca idahoensis* Elmer and *Agropyron spicatum* (Pursh.) Scribn. & J. G. Sm. Although it is difficult to know the temperatures actually experienced by the growing points or whether the temperature-time curves would resemble those in natural fires, the experiments did confirm that both species were capable of mobilizing growing points to compensate for assumed death of buds and young tillers from the fire treatment. A similar study by Peláez *et al.* (1997) examined three species of tussock grasses native to a region of Argentina with mean annual precipitation of 344 mm. They reported significant bud mortality from their experimental heat treatments, which they attributed to the fact that the buds are located 1 to 2 cm above the soil surface. In both these studies the death of entire individuals was not significant, but this is not always the case. Management burning in California grasslands killed about 10% of the individuals of the bunchgrass *Nasella (Stipa) pulchra* (A. S. Hitchc.) Barkworth (Marty *et al.*, 2003). Busso *et al.* (1993) tested for fire effects on the bunchgrass *Stipa tenuis* Phil. by assaying for bud viability with tetrazolium. They found significant mortality but good recovery in terms of numbers of tillers produced by the end of the growing season. They attributed the mortality to the more damaging effect of fire in the dense clumps as described by Wright (1971). Tussock-forming *Calmagrostis* spp. in the high-elevation páramo vegetation of Ecuador was similarly sensitive to fire. The tussocks consist of 50% to 80% dead biomass, and this probably accounts for the significant bud mortality and the 3 to 5 years required to recover from a burn (Ramsay and Oxley, 1996).

The susceptibility of bunch grasses to direct fire mortality is perhaps most clearly shown in the spinifex grasslands of Australia. Burrows *et al.* (1991) noted that the spinifex grasses (genera *Pletrachne* and *Triodia* in Western Australia) have a fuel arrangement almost ideal for igniting and burning intensely because they tend to retain dead resinous leaves and stems. Therefore, when windspeeds are high enough to spread the flames from clump to clump (>12–17 kmh^{-1}), spinifex grasses can support fires of sometimes gargantuan proportions. For example, a single fire burned 36,000 km^2 in the Simpson Desert (Allan and Southgate, 2002). Given their propensity to burn, a clear set of fire-recovery traits might be expected but, in fact, the response of spinifex species to fire seems to defy any simple explanation. Rice and Westoby (1999) surveyed the fire response of 117 recently burned populations across Australia and found that in all of them the clumps were, as expected, intensely burned. Seventy-eight of the populations resprouted after fire. Resprouting was facilitated by the tendency for the grass clumps to accumulate wind-blown soil to form small mounds or columnar formations around their stems. The soil buildup offered protection to the plants, and regeneration was often seen arising below the surface of these soil accumulations. A surprising result, however, was that 39 of the populations were "fire killed" and were dependent on seeds for recovery. Seedling establishment after fire was observed in all burns, but seedling densities were generally higher in situations where the adult population had all been killed, as would be expected if, in these situations, the life history had adjusted towards the so-called "obligate seeder" mode. As noted above, this would seem a dangerous strategy for species that have seeds with limited longevity and that do not have the specialized structures to store seeds for release after fire (serotiny or bradyspory). An intriguing finding by Rice and Westoby (1999) is that the nonsprouting seeder response showed no clear taxonomic, geographic, edaphic, or climatic trends. Some species had local populations that were killed by fires and others that resprouted.

Annual grasses in fire-prone regions must survive fire as seeds and so there is the same question about fire resilience. The self-planting mechanisms discussed above (*Heteropogon*, *Amphicarpum*) are an effective way to survive fire, but most grasses, including a majority of those that thrive in high-fire situations, manage without highly specialized syndromes.

The robust *Sorghum* spp. of the Australian tropical savannas exemplify the resilience of annual grasses (Williams *et al.*, 1998). *Sorghum intrans* Benth. sustains seed mortality up to 38% from the frequent fires (Andrew and Mott, 1983 cited in Watkinson *et al.*, 1989) yet rebounds vigorously because of its large seed reserve and its ability to compensate for lower densities by higher seed production per plant. Dry season fires, therefore, do not normally threaten *Sorghum* populations, whereas early growing season fires after germination depletes the seed reserve can cause local extinction of populations (Watkinson *et al.*, 1989; Lonsdale *et al.*, 1998). *Andropogon brevifolius* Sw., an annual species of frequently burned Argentinean grasslands, apparently thrives with burning (Canales *et al.*, 1994). It is protected against fire-caused seedling loss by germinating only after the rains have begun. A matrix population model showed that the population in burned vegetation had strongly positive population growth ($\lambda = 2.8$) whereas unburned populations showed declining growth ($\lambda = 0.7$). Because the study was short term, we can't be certain that the population will go extinct without fire, but the expectation is that it will be significantly less abundant. Another example is the species of Flinders grass (*Iseilema* spp.) that occurs in the *Astrelba* grasslands of Queensland. Like the annual sorghums, these species will lose seeds in burns, especially where litter accumulations are high, but they respond positively to fire and will tiller more extensively than plants at similar densities in unburned situations (Scanlan and O'Rourke, 1982).

The propensity for annuals to carry and survive fire has also become a major element in the ecosystem function of arid and semi-arid areas in western North America because of the invasion of alien species. In northern California, burning for 3 successive years did not eliminate the exotic annuals, *Taeniatherum asperum* (Sim.) Nevski (colorfully called medusahead) and *Bromus tectorum* (less colorfully called cheat grass because of its brief window of palatability), where they co-occurred with native perennials (Young *et al.*, 1972). *B. tectorum* did decrease but *T. asperum* was little affected, apparently because its seeds were sufficiently protected merely by being on the soil surface. This same annual–grass fire syndrome has intruded into desert areas. In the Mojave Desert of western North America, exotic annual grasses have become a major component of the ephemeral flora in desert shrublands, threatening native biodiversity. Brooks (2002) performed experimental burns in a *Larrea-Ambrosia* desert site and found that the high annual plant biomass beneath *Larrea* canopies, which was mostly the introduced *Bromus madritensis* L., produced peak temperatures that killed most of the annual plant seeds on and below the surface. In contrast, fire in the interspaces burning in vegetation of lower biomass dominated by *Schismus* spp. had only moderate effects on the seed bank. Fire, in general, seemed to weaken the competitive edge of the introduced annual grasses and allow native annuals to recover, but the recovery was relative; fire did not eliminate the alien species that overall are as well or better adapted to fire as the native flora.

Managers in the western United States are frustrated by the difficulty of using fire to control alien annual grasses. Some of these grasses (e.g., *Bromus madritensis*) tend to retain their seeds into the dormant season and so, in theory, could be vulnerable to fire while the bulk of the seed production is still on the plants. But fires at such times often are not uniform or not uniformly intense enough to kill all of the seeds; therefore, the populations generally recover well from single fires. Vulnerability early in the growing season was discussed above for annual *Sorghum* species and could perhaps be exploited by carefully timed fire. But as in so many invasive control situations, the rate of reduction of species by such methods is slow relative to the rate at which they can expand when management is suspended.

Primacy of the Bud Bank

Although annual and short-lived grasses are important in many grasslands, long-lived perennials are usually the dominants. For these perennials, it is unusual for fire-caused mortality of individuals (genets or discrete clumps) to be more than moderate (but see discussion of spinifex above), and typically it is insignificant because of the resilience conferred by a large population of well-protected regenerative buds and their capacity to multiply stems from surviving vegetative parts. Because of the short duration of most grassland fires and the considerable insulating capacity of the soil, maximum temperatures in the soil decrease sharply and exponentially with depth (Valette *et al.*, 1994) so that lethal temperatures (approximately 60°C) are only reached in the upper few centimeters and in many grassland fires only in the upper few millimeters.

This resilience focuses attention on the "bud bank," the population of growing points from which the next generation of tillers will arise (Hartnett and Fay, 1998; Benson *et al.*, 2004). If the bud bank is large and well-distributed spatially, the tiller populations will be sustained. If the population of buds is shrinking and not well distributed, the grasses will yield dominance to other graminoids, forbs, or more usually, woody plants. For most dominant perennials, seed reproduction is a sideshow in ecological time. The opportunity to capture space by seedling establishment is usually limited, especially in dense, productive grasslands (Benson *et al.*, 2004).

The primacy of vegetative resilience is not a recent discovery, yet the difficulties of studying grasses in terms of tiller demography are such that relatively few studies have been done on this vital aspect of grassland ecology. Rather, studies of fire response have tended to look at aggregate measures such as cover and biomass. These studies have been extremely valuable, of course, in forming a picture of how fire and grasses interact. The main generality to emerge is that grasses in general tolerate fire extremely well and, more important, in many cases reach their maximum rates of net primary productivity and reproductive activity in the immediate post-fire years. This point was well established at the time of comprehensive reviews by Daubenmire (1968) and Vogl (1974). More recently, Oesterheld *et al.* (1999) have reviewed studies of the effect of fire on production. They concluded that the stimulatory effect of fire depends on the position of the grassland along the aridity gradient (Fig. 4). Highly productive

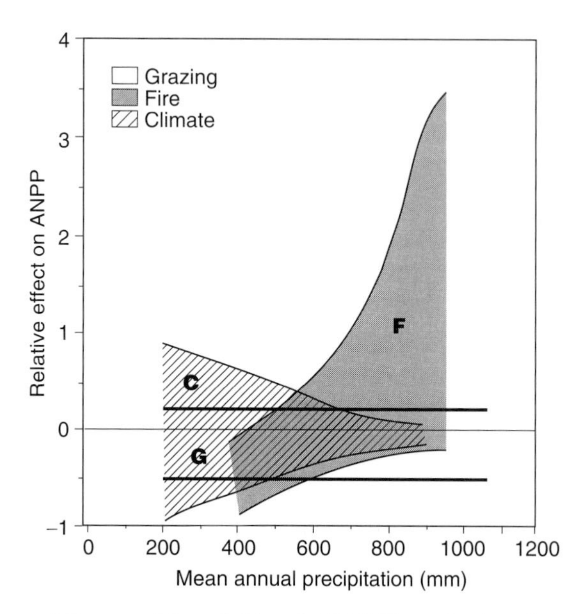

FIG. 4 Conceptual diagram of Oesterheld *et al.* (1999) illustrating how three major "disturbances"—fire, grazing, and climate (primarily drought)—affect the annual net primary productivity of grasslands across the mean annual precipitation gradient. Figure is based on a compilation of results from many studies. The response area of fire is indicated by a gray tint, of grazing by heavy lines, and that of climate is hatched. The graph illustrates the positive effect of fire on grasslands of high productivity and the often negative effect of fire in grasslands of drier climates. Note the rough correspondence with the zone in which fire effects become positive and the mean annual precipitation at which the limit on maximum woody cover is released from climatic control in Fig. 6. (From Oesterheld *et al.* [1999], copyright Elsevier Science Publishers, reproduced with permission.)

grasslands on the higher precipitation end of the rainfall gradient generally show moderate to strongly positive response to burning (Oesterheld *et al.*, 1999), whereas more arid grasslands, and bunchgrass grasslands specifically, show reduced productivity and standing crops in the first few years after fire (Tainton and Mentis, 1984; Busso *et al.*, 1993; Redmann *et al.*, 1993; Allan and Southgate, 2002; Pylypec and Romo, 2003).

The causes of stimulation of productivity are still debated. Tiller demography certainly is involved. (The contribution of nutrients is discussed below.) In woody plants, fire can release a large population of dormant buds, but accumulation of such "bud banks" between fires is not thought to be a factor in grasses. The results of Benson *et al.* (2004) seem to point in the opposite direction. They found that the bud banks were larger in annually burned tallgrass prairie than in otherwise comparable infrequently burned prairie (Fig. 5). But the bud population was not

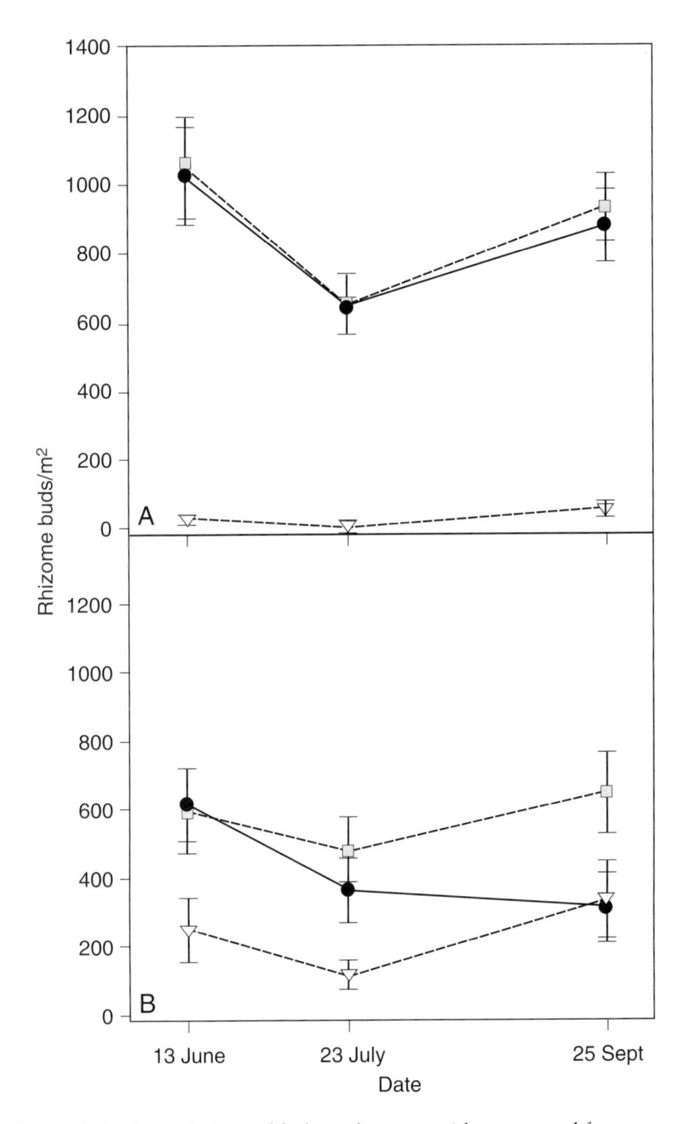

FIG. 5 Change in bud populations of forbs and grasses with season and for contrasting burn regimes in tallgrass prairie in Kansas. (**A**) Annually burned, (**B**) infrequently burned. Open squares = total bud density. Filled circles = grass bud density. Open triangles = forb bud density. The grass bud population is significantly higher in the annually burned prairie, while the reverse is true for the forb bud population. Both fluctuate seasonally. (From Benson *et al.* [2004], copyright Botanical Society of America, reproduced with permission.)

directly related to the ultimate stem population. The ratio of buds in annually burned to buds in infrequently burned prairie was about 2.5:1, whereas the stem density ratio was 1.3:1. The narrowing of the ratio was the result of an average stem production of 0.75 per bud in the annually burned versus 1.36 per bud in the infrequently burned area.

Silva *et al.* (1991) created a matrix population model of the perennial, *Andropogon semiberbis* (Nees) Kunth, in a frequently burned savanna in Venezuela. In results similar to the annual *Andropogon* studied by Canales *et al.* (1994), they found that burned populations in their stage-based model had a rapid rate of increase ($\lambda = 1.3$) whereas the unburned population was shrinking even faster ($\lambda = 0.3$). Elasticity analysis pointed to survival and growth of the smallest clumps as the important effect. They attributed the improved growing conditions to the removal of litter by fire.

It is tempting to make this into an adaptive story. A burned population with the resource pulse associated with burning (Blair, 1997) would benefit from extra investment in buds to ensure that the opportunity was fully exploited. According to Vesk and Westoby (2004), the cost of dormant buds is not high, and so investments could easily be returned in an environment favoring high production. But a grass clone in reduced resource circumstances would benefit by a more conservative strategy that begins the growth season with fewer buds and mobilizes additional stems as resources become available. Essentially, this is an argument for a plastic response designed to adjust to variable resources. This argument parallels that made earlier for the differing capacity of grasslands to track variation in precipitation.

At least for tallgrass prairie, the increase in the bud bank may be an important part of the positive response to fire. But a bud only represents the potential for growth, and therefore something more must be involved. Evidence seems to point to a fertilizing effect caused by the conversion of nitrogen, especially, from slowly available forms to readily available forms (nitrates, nitrites, ammonium) combined with an increase in light resulting from the removal of litter and standing dead material and increased early season soil temperatures (Blair, 1997) as the most critical abiotic factors. The nitrogen response occurs despite the well-documented fact that burning overall depletes the stock of nitrogen through volatilization (Blair *et al.*, 1998; Reich *et al.*, 2001).

The strongly positive response to fire seen in tallgrass prairie is not universal. As noted previously, burning depresses productivity in many grassland types. It is clear that in many cases this is caused in large part by the direct negative effect of fire on the regenerative capacity of the grasses (i.e., on the bud bank). Bunchgrasses (caespitose grasses) are especially susceptible to damage by fire.

Season of Burn

Most grasslands follow a strong seasonal cycle usually marked by a protracted dormant season. The probability that a fire will occur naturally or that a grassland will carry a fire if it is artificially ignited is dependent on the absolute and relative amount of dead biomass. Fires therefore are most likely to propagate during the dormant season or in the earliest and latest parts of the growing season. Fires that burn at different times during this seasonal cycle would not be expected to have uniform effects because both the physical factors (climate, fuel amount, and condition) and the physiological state of the plants will vary over time. Burns conducted when some of the species are dormant and others actively growing have the potential to disadvantage those caught in the more vulnerable growing condition. Also, the time at which litter and standing dead material are removed can determine the subsequent pattern of soil moisture and soil warming, especially in the early growth stages after the fire. But despite considerable interest in seasonal effects, we still lack a sophisticated understanding (Engle and Bidwell, 2001). The main problem is that "season" is not a quality of grasslands but only a rough predictor of the state of the vegetation and the physical condition at the time of the burn, neither of which is easily measured. The studies available suggest that most grasslands are resilient to variations in time of burn, though differences have been found. Benning and Bragg (1993), in an experiment that burned tallgrass prairie in Nebraska at 4-day intervals starting in early April, found that there was a strong stimulation of flowering of *Andropogon gerardii* in all burns compared to unburned grassland, but burns had about the same effect until the last two in mid-May, which produced a significantly stronger flowering response. Experiments in Wisconsin tallgrass prairie with summer burns showed large increases in subdominant grasses and forbs relative to controls, presumably because burning C_4 grasses during active growth reduced their competitive ability (Howe, 1995; Copeland *et al.*, 2002). In another tallgrass prairie experiment, Towne and Kemp (2003) burned in November, February, and April. The November and February burns produced similar results and slightly favored forbs relative to the dominant grasses. The response of two of the dominants differed; *A. gerardii* increased with all burns but *Sorghastrum nutans* (L.) Nash only with the spring burns. The effect of season on the annual *Sorghum* spp. of tropical Australia has been commented on previously.

Theory and data therefore both suggest that timing burns for particular stages in the growth cycle can alter grassland composition. But the great resilience of grasses ensures that these changes will mostly be subtle and often difficult to detect in short-term experiments. More work with better measurement of critical physical and biological features of the system is needed to clarify seasonal burn effects.

GRASSFIRE AND NUTRIENTS

One of the indirect effects of fire is to alter the abundance, form, and distribution of nutrients in the soil-vegetation system. Nutrients are lost in gaseous form (volatilization, gasification) or carried away in particulate form in smoke. Some are returned to the surface as ash (cations especially) or remain in place in unburned or partly burned elements of the pre-fire biomass. Although any element can be lost by one or more of these mechanisms, nitrogen is of particular interest because it is often the limiting nutrient in grasslands and because it volatilizes at about 200°C, whereas flame temperatures are in the range of 300 to 1400°C. Phosphorus, the other most likely limiting nutrient, volatilizes at about 500°C and thus is less susceptible to loss. Although the importance of the fire-related loss and change of state of nitrogen and phosphorus is widely appreciated, nutrient dynamics are difficult to study because of technical challenges in obtaining precision and accuracy (e.g., Brye *et al.*, 2003) and the inherent complexities of the fire-soil-plant-microbial system (Neary *et al.*, 1999). One of these is that nutrient loss is confounded with litter removal, which, as discussed previously, also can have major effects. Regional and global redistribution processes must also be considered (Burke *et al.*, 2002).

Some generalities emerge from the fog of complexity. First, fire results in significant local losses of nitrogen from the aboveground pool so that the nitrogen capital of any site is to some degree depleted by any fire. Although nitrogen released by a fire in one place may increase the nitrogen added by precipitation or dry fall in another, local balance sheets show a short-term net loss. The amount of loss is directly proportional to the amount of biomass consumed and, since grasslands typically lose 80% to 90% or more in a fire, the proportional loss of the aboveground pool is higher in grasslands than most other vegetation (Wan *et al.*, 2001). But most of the nitrogen in grasslands is in the soil, the aboveground losses are invariably small as a proportion of the total nitrogen pool (Neary *et al.*, 1999; Wan *et al.*, 2001). For example, in a semi-arid savanna, 93% of the nitrogen in the aboveground biomass was lost, but this was only 0.3% of the total nitrogen pool (Van de Vijver *et al.*, 1999). This means that a single fire is usually of little importance to long-term nutrient dynamics and it is only with very high fire frequencies that cumulative nitrogen loss can be large enough to impair function relative to less frequently burned sites. In tallgrass prairie, however, even annual burning did not seem to significantly reduce primary productivity and strongly favored grasses at the expense of other life forms (Collins and Steinauer, 1998).

In more productive grasslands, fire after a period without fire does have a strong positive effect on primary productivity (Old, 1969; Vogl, 1974; Oesterheld *et al.*, 1999). Such a productivity spike obviously means that the vegetation, and the grasses in particular, is in some way released from resource limitations. The question here is the degree to which this can be ascribed to fire-induced improvement in nutrient status. One possibility is that the fire converts nutrients, and especially nitrogen, in slowly available organic form to readily available mineral forms. But, as discussed, the maximum amounts that can be contributed from this source are relatively modest. Detailed study of tallgrass prairie shows that amounts of inorganic nitrogen in the soil are at least as great or greater in unburned vegetation and lower in annually burned prairie than in vegetation burned after a fire-free period (Blair, 1997). This suggests that the stimulatory effect of fire usually occurs despite a loss of nitrogen from the system and not because of it, and that it is probably due to a complex combination of factors, of which release from light limitations associated with litter buildup is a major element (Blair, 1997). Clearly, further study, including work across moisture and soil fertility gradients, is needed. But from a practical standpoint it seems that, at least in more productive grasslands, decisions to apply or withhold fire need not be constrained by concerns about the depletion of nutrient stocks. This is especially so in consideration of the huge increase in global nitrogen deposition from anthropogenic sources (Vitousek *et al.*, 1997). If there is to be a nutrient-based rationale for burning, it is likely to be aimed at reducing, not increasing, nitrogen pools.

GRASSES AND WOODY PLANTS

The capacity of woody plants for secondary growth allows them to accumulate capital that permits a fuller exploitation of the environment than is possible for herbaceous grasses. Where precipitation is higher and soil moisture generally favorable (wetlands excluded), woody plants shade out most grasses and, where soil water supply becomes too sparse and unpredictable, woody plants again assert control by their capacity to tap deeper reserves. But the two life forms broadly overlap. At a larger spatial scale, woody plants coexist with grasses in every biome and every vegetation type (Woodward *et al.*, 2004). Much of this coexistence is explained by local conditions that tip the balance, an example being the gallery forests that follow streams and rivers even in quite arid and frequently burned grasslands (e.g., Biddulph and Kellman, 1998).

Two situations, however, require some explanation beyond climate. The first is the existence of vast areas of savanna and shrub-steppe where the two life forms mingle in a fine-grained mosaic that can't be explained entirely by corresponding small-scale spatial variation in soil, topography, or local conditions affecting water balance. The existence of these mixtures has so far eluded simple explanation and in recent decades the debate about the "savanna problem" has generated a large literature and a wealth of hypotheses too large and too complex to review in detail here. The second is the readiness with which woody plants grow in some grassland areas when planted and the many places where woody plants have encroached onto formerly treeless grasslands (e.g., Grover and Musick, 1990; Titshall *et al.*, 2000; Silva *et al.*, 2001).

That fire can tip the balance toward grasses is beyond doubt and has been long known (Curtis, 1959; Daubenmire, 1968; Vogl, 1974). The basic mechanism is clear. Woody plants, unlike grasses, cannot avoid exposure of their buds and meristematic tissues to the direct effects of fires (Bond and van Wilgen, 1996), and the investment in secondary tissues means a longer period of recovery after damage. As discussed previously, although grasses often do suffer some direct harm from fire, in general they can recover full function, including reproduction, within one or a few years—well before most woody plants. Given this and the demonstrated capacity of grasses to persist once established in a wide range of settings, one might ask why grasses have not burned out all the forests of the world. A primary factor limiting the fire-driven advance of grasslands is the discontinuity in fuels and fuel moisture at the grassland–forest boundary. Although there is usually burnable biomass on the forest floor, there is a high probability that these fuels will be both moister and sparser than the fuels in adjacent grasslands (Biddulph and Kellman, 1998). Since the grassland is susceptible to burning earlier and for a longer period than the forest floor, fires will tend to stop at the grassland–forest edge. Humans can upset this natural quasi-equilibrium by their ability to ignite fires at times and places where the fuel discontinuity is less pronounced. Bond *et al.* (2005) pointed out another important factor: woody plants also have been shaped by fire and these fire-tolerant species have been able to extend woody plant dominance and presence into situations also favorable for grass. They cite the extensive radiation in fire-tolerant groups, of which the Australian eucalypts and rich woody flora of the Brazilian cerrado are two clear examples. Pines (*Pinus* spp.) are among those woody plants that have the capacity to coexist with and exploit the opportunities offered by fire (Keeley and Zedler, 1998). Perhaps the most dramatic examples are the "grass-stage" pines (e.g., *Pinus palustris* P. Mill., *P. michoacana* Mart.), elegantly designed to survive frequent, grass-driven understory fires. Even the depauperate flora of eastern North America has fire-tolerant specialists among the oaks, most notably *Quercus macrocarpa* (Abrams, 1992).

Bond *et al.* (2005) are among those who believe that fire is a critical factor in the woody plant–grass balance. Using climate–vegetation models, they predicted that if fire were (hypothetically) removed as a factor, the forested area of global vegetation (represented by grid cells classed as one or the other in their model) would increase from 27% to 56%. That is, they believe that natural grasslands dominate in many situations where a fire-free climate would favor forest or woodland. The long historical view encourages this kind of conclusion. The grass–fire nexus is widely believed to have fostered the expansion of grasslands beginning in the late Cretaceous epoch and through the Tertiary epoch (Bond *et al.*, 2005; Keeley and Rundel, 2005). Although, as Bond *et al.* argued, there was an evolutionary response by woody plants, the inherent limitations of the woody growth form could not be overcome entirely. The best-adapted woody plants have adjusted in the direction of grasses so as to be able to exploit post-fire conditions either by proliferating stems from dormant buds (such as grasses) or by forgoing vegetative regeneration in favor of specialization for seed survival, good establishment, and rapid early growth. But unless they become grass-like through an equal degree of reduction, woody plants carry the burden of their energetic commitment to oxidizable secondary tissues that are always at risk from fire. Proliferation of throwaway photosynthetically active units at ground level "cannot be beat" when fire or drought or both set strict limits to the capacity to support permanent leaf area and there is reasonably reliable moisture available in the surface soil for part of the year.

The Savanna Question

It has proven difficult to find a general model that explains the full range of savanna and shrub-steppe situations (Scholes and Archer, 1997). Some earlier views focused on what many might characterize as niche separation, with grasses being better able to exploit shallow moisture and woody plants having exclusive access to deep moisture (Walter, 1964). Because the resource is only partly shared (some water is assumed to move to depth beyond the reach of the grasses), a hypothetical equilibrium based only on this single resource is possible. In practice, this model leaves much unexplained. The other extreme is to explain savannas as inherently unstable mixtures that fluctuate but resist being pulled towards either state. Jeltsch *et al.* (2000) take the view that savannas are maintained by mechanisms that push back as the system approaches the boundaries. Thus, they see savannas as being repelled from "grass" or "tree" but free to drift around in an ample intermediate space. As examples of these buffering mechanisms they cite elephant browsing and fire pushing systems away from woodland and grazers and the dormancy and tree seedling banks pushing them away from grassland. Some of these mechanisms seem overparticular, and Jeltsch *et al.* acknowledge this by suggesting that the specific buffering mechanisms may differ among regions.

The search for workable generalizations requires large data sets. Sankaran *et al.* (2005) assembled data from hundreds of sites in Africa and were able to show a remarkably clear picture that provides a convincing case that mean annual precipitation (MAP) sets a limit to the woody component of mixed systems up to about 650 mm with a linear relation between the maximum woody cover and MAP (Fig. 6). Beyond 650 mm, the maximum values level off and there is no relation. Across the entire gradient studied (up to 1200 mm MAP), woody values range from zero to the maximum. The authors suggested that below 650 mm savannas are "climatically determined," whereas above this, reduced tree cover will occur only in "disturbance-driven savannas." A regression tree analysis of a subset of the data confirmed that MAP is the major explanatory variable, and that it is at higher MAP that fire return interval separates out more and less densely wooded savannas. Soils also enter below these factors, with greater woody plant cover on very sandy soils, as Walter's (1964) classic model predicts.

An interesting concordance, if only rough, may be found between the assertion that below 375 to 500 mm MAP grasslands have an immediate negative response to fire, and above this a generally positive response (Bennett *et al.*, 2003, citing Kucera, 1967 and Scanlan, 1980). If so, there is an approximate match with "fire-sensitive" grasslands dominating the part of climate space where trees are at a serious disadvantage, and "fire-responsive" grasslands in the part of climate space where they face the most serious competition from woody plants. In the fire-responsive part of the gradient, fire has the effect of stimulating the grasses while harming woody plants. At the risk of inappropriate anthropomorphization, it is as though where the climate protects against woody plant dominance, selection in grasses for fire survival and fire recovery is relaxed in favor of adaptation for resource scarcity.

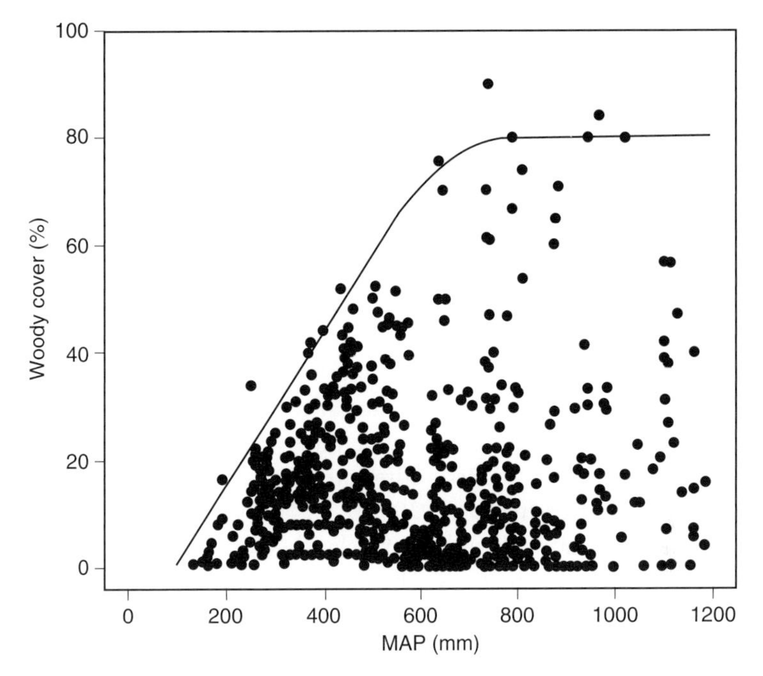

FIG. 6 Relation between woody cover from grassland/savanna sites across Africa with mean annual precipitation (Sankaran, 2005). Each point summarizes data from one site. Below about 650 mm mean annual precipitation the maximum percentage of woody cover has an upper level. Beyond this, woody cover varies over a wide range. Note, however, that across the entire range of mean annual precipitation many sites are more than 90% grass, indicating the action of factors other than climate. Fire is the likely cause for the abundance of grasses in many of these instances. (From Sankaran *et al* [2005], *Nature*, copyright 2005, reprinted by permission from Macmillan Publishers Ltd.)

Grasslands and Ecosystem Change

Another view of grassland stability and change is embodied in the concept of "state and transition" or "alternative stable state" models (Westoby *et al.*, 1989). Encouraged by mathematical models that suggest that complex systems can exhibit strongly nonlinear responses, threshold effects, jumps to new domains of attraction, and hysteresis in recovery, ecologists have contemplated the degree to which such change of state dynamics provide a more appropriate view of grassland and savanna ecosystems. This has become an active area for theorists, but here we consider only the role that fire might play in such complex dynamics. Fire needs to be considered because it can shift the relative biomass of species almost instantly, and because it can virtually eliminate vulnerable species. Yet, as we have seen, for the most part grasses are not these vulnerable species.

There have been dramatic changes of state in grasslands worldwide in the last few centuries. These have all been attributable to direct or indirect human action. In one well-documented case, the combination of grazing and the introduction of aggressive alien species has transformed landscapes in western North America (Mack, 1981; Mack and Thompson, 1982). In northern Australia an introduced perennial grass, *Andropogon gayanus* Kunth., produces fires that burn with eight times the intensity of fires in natural grasslands (Rossiter *et al.*, 2003). Such a large change in fire intensity cannot fail to modify nearly all aspects of ecosystem function and structure. These cases do not require complex alternative stable state explanations, beyond the rather obvious fact that the introduction of novel species preadapted to an ecosystem must always have the potential to modify that system. Invaded systems may definitely be in a new state, but there is fear that there is no realistic alternative attainable through existing management (e.g., Rossiter *et al.*, 2003).

The other dramatic change in grasslands, noted previously, is woody plant invasion and expansion. A change in fire regime is thought to have played at least a secondary role in most of these situations, but climatic variation and grazing and interactions among all three of these probably are involved. If those who believe in delicate balances between grazers and vegetation (e.g., Dublin, 1995) are correct, then the introduction of alien grazers would seem to be a sufficient explanation because of their multiple effects. They reduce grass vigor and therefore fire intensity and perhaps frequency, while also creating conditions favorable for the dispersal and growth of woody plants (Scholes and Archer, 1997). But removal of the grazers and reintroduction of fire often do little to alter the state of the vegetation, which is perhaps the best evidence for the alternative stable state theory. Definitive tests of the hypothesis are made difficult because when grasses contend with woody plants that are accommodated to (if not "adapted to") fire (such as *Prosopis* spp.), recovery may take a long time. Briske *et al.* (2003) suggested that the alternative stable state theory has applicability within a certain realm, but that it does not displace older ideas of directional change that can lead to recovery.

A FINAL CAUTION—GRASSES AND FIRES

The drama of a good grass fire encourages the belief that every feature of grasses and grasslands must have been selected to deal with this awesome force of nature. But careful study of the response of grasses to fires suggests a more nuanced view. Some grasses and grasslands seem to thrive on fire and to languish if left long unburned. In general, these are to be found on the wetter end of the grassland climatic gradient. Other grasses are resilient to fire, in the sense that they can suffer damage but readily recover from it. And burning will almost always result in an increase of grass relative to competing life forms almost everywhere, wet or dry, temperate or tropics. These facts should not be taken as endorsement of a maximally aggressive burning policy. First, we need also to consider overall biodiversity issues, not just grasses, and in many situations an inappropriate fire frequency or extreme fire intensity will reduce biodiversity or have negative effects on the abiotic environment. It also appears that many grasses and grasslands can do very well with few or no fires. The admonition of Keeley and Bond (2001) to consider fire in all its complexity and not as a simple qualitative yes–no factor is relevant here. The first branch in the management decision tree may be "fire–no fire" but, if we choose fire, the decision tree has many branches. Even without the fear of stumbling into an unexpected "alternative state," we have reason to believe that the choices matter. Although there is a wealth of information, it is difficult to extract guidance from it in specific situations. We have a great need for a more refined understanding of direct and indirect fire effects and how these interact with all the other influences that cause change in grassland ecosystems.

REFERENCES

Abernethy, G. A., Fountain, D. W., and McManus, M. T. (1998). Observations on the leaf anatomy of *Festuca novae-zelandiae* and biochemical responses to a water deficit. *N Z J Bot* 36, 113–123.

Abrams, M. (1992). Fire and the development of oak forests. *Bioscience* 42, 346–353.

Aerts, R., De Caluwe, H., and Beltman, B. (2003). Plant community mediated vs. nutritional controls on litter decomposition rates in grasslands. *Ecology* 84, 3198–3208.

Albini, F. A. (1993). Dynamics and modeling of vegetation fires: observations. In: *Fire in the Environment: The Ecological, Atmospheric, and Climatic Importance of Vegetation Fires.* (Crutzen, P. J., and Goldammer, J. G., Eds.), pp. 39–52. John Wiley and Sons, Chichester, UK.

Allan, G. E., and Southgate, R. I. (2002). Fire regimes in the spinifex landscapes of Australia. In: *Flammable Australia: The Fire Regimes and Biodiversity of a Continent.* (Bradstock, R. A., Williams, J. E., and Gill, M. A., Eds.), pp. 145–176. Cambridge University Press, Cambridge.

Anten, N. P. R., and Hirose, T. (1999). Interspecific differences in above-ground growth patterns result in spatial and temporal partitioning of light among species in a tall-grass meadow. *J Ecol* 87, 583–597.

Archibold, O. W. (1995). *Ecology of World Vegetation.* Chapman and Hall, New York.

Baskin, C. C., and Baskin, J. M. (1998). Ecology of seed dormancy and germination in grasses. In: *Population Biology of Grasses.* (Cheplick, G. P., Ed.), pp. 30–83. Cambridge University Press, Cambridge.

Batzli, G. O., and Pitelka, F. A. (1970). Influence of meadow mouse populations on California grassland. *Ecology* 51, 1027–1039.

Bell, A. D. (1992). *Plant Form: An Illustrated Guide to Flowering Plant Morphology.* Oxford University Press, Oxford.

Bennett, L. T., Judd, T. S., and Adams, M. A. (2003). Growth and nutrient content of perennial grasslands following burning in semi-arid, sub-tropical Australia. *Plant Ecol* 164, 185–199.

Benning, T. L., and Bragg, T. B. (1993). Response of big bluestem (*Andropogon-gerardii*-Vitman) to timing of spring burning. *Am Midl Nat* 130, 127–132.

Benson, E. J., Hartnett, D. C., and Mann, K. H. (2004). Belowground bud banks and meristem limitation in tallgrass prairie plant populations. *Am J Bot* 9, 416–421.

Biddulph, J., and Kellman, M. (1998). Fuels and fire at savanna gallery forest boundaries in southeastern Venezuela. *J Trop Ecol* 14, 445–461.

Birkett, A. (2002). The impact of giraffe, rhino and elephant on the habitat of a black rhino sanctuary in Kenya. *Afr J Ecol* 40, 276–282.

Blair, J. M. (1997). Fire, N availability, and plant response in grasslands: a test of the transient maxima hypothesis. *Ecology* 78, 2359–2368.

Blair, J. M., Seastedt, T. R., Rice, C. W., and Ramundo, R. A. (1998). Terrestrial nutrient cycling in tallgrass prairie. In: *Grassland Dynamics: Long-term Research in Tallgrass Prairie.* (Knapp, A., Briggs, J. M., Hartnett, D. C., and Collins, S. L., Eds.), pp. 222–243. Oxford University Press, New York.

Bogen, A. D., Bork, E. W., and Willms, W. D. (2002). Rough fescue (*Festuca campestris* Rydb.) response to heat injury. *Can J Plant Sci* 82, 721–729.

Bond, W. J., and van Wilgen, B. W. (1996). *Fire and Plants.* Chapman and Hall, London.

Bond, W. J., Woodward, F. I., and Midgley, G. F. (2005). The global distribution of ecosystems in a world without fire. *New Phytol* 165, 525–538.

Bonham, C. D., and Lerwick, A. (1976). Vegetation changes induced by prairie dogs on shortgrass range. *J Range Manage* 29, 221–225.

Bonnett, O. T. (1972). *Silicified Cells of Grasses: A Major Source of Plant Opal in Illinois.* Bulletin 742, Agricultural Experiment Station, University of Illinois at Urbana-Champaign, Urbana, IL.

Bredenkamp, G. J., Spada, F., and Kazmierczak, E. (2002). On the origin of northern and southern hemisphere grasslands. *Plant Ecol* 163, 209–229.

Bren, L. J. (1992). Tree invasion of an intermittent wetland in relation to changes in the flooding frequency of the River Murray, Australia. *Aust J Ecol* 17, 395–408.

Briske, D. D. (1991). Developmental morphology and physiology of grasses. In: *Grazing Management. An Ecological Perspective.* (Heitschmidt, R. K., and Stuth, J. W., Eds.), pp. 85–108. Timber Press, Portland, OR.

Briske, D. D., and Derner, J. D. (1998). Clonal biology of caespitose grasses. In: *Population Biology of Grasses.* (Cheplick, G. P., Ed.), pp. 106–135. Cambridge University Press, Cambridge.

Briske, D. D., Fuhlendorf, S. D., and Smeins, F. E. (2003). Vegetation dynamics on rangelands: a critique of the current paradigms. *J Appl Ecol* 40, 601–614.

Brooks, M. L. (2002). Peak fire temperatures and effects on annual plants in the Mojave Desert. *Ecol Appl* 12, 1088–1102.

Brown, J. K. (1970). Ratios of surface area to volume for common fine fuels. *Forest Sci* 16, 101–105.

Brown, K. J., Clark, J. S., Grimm, E. C., Donovan, J. J., Mueller, P. G., Hansen, B. C. S., and Stefanova, I. (2005). Fire cycles in North American interior grasslands and their relation to prairie drought. *Proc Natl Acad Sci U S A* 102, 8865–8870.

Bryant, W. C. (1832). *The Prairie: Poems by William Cullen Bryant.* E. Bliss, New York.

Brye, K. R., Norman, J. M., and Gower, S. T. (2002). Assessing the progress of a tallgrass prairie restoration in southern Wisconsin. *Am Midl Nat* 148, 218–235.

Brye, K. R., Norman, J. M., Gower, S. T., and Bundy, L. G. (2003). Methodological limitations and N-budget differences among a restored tallgrass prairie and maize agroecosystems. *Agr Ecosyst Environ* 97, 181–198.

Burke, I. C., Lauenroth, W. K., Cunfer, G., Barrett, J. E., Mosier, A., and Lowe, P. (2002). Nitrogen in the central grasslands region of the United States. *Bioscience* 52, 813–823.

Burrows, N., Ward, B., and Robinson, A. (1991). Fire behaviour in spinifex fuels on the Gibson Desert Nature Reserve, western Australia. *J Arid Env* 20, 189–204.

Busso, C. A., Boo, R. M., and Peláez, D. V. (1993). Fire effects on bud viability and growth of *Stipa tenuis* in semiarid Argentina. *Ann Bot* 71, 377–381.

Caldwell, M. M., Richards, J. H., Johnson, D. A., Nowak, R. S., and Dzurec, R. S. (1981). Coping with herbivory: photosynthetic capacity and resource allocation in two semiarid *Agropyron* bunchgrasses. *Oecologia* 50, 14–24.

Campbell, B. D., Grime, J. P., and Mackey, J. M. L. (1992). Shoot thrust and its role in plant competition. *J Ecol* 80, 633–641.

Canales, J., Trevisan, M. C., Silva, J. F., and Caswell, H. (1994). A demographic-study of an annual grass (*Andropogon-brevifolius* Schwarz) in burnt and unburnt savanna. *Acta Oecol* 15, 261–273.

Catchpole, E. A., Catchpole, W. R., and Rothermel, R. C. (1993). Fire behavior experiments in mixed fuel complexes. *Int J Wildland Fire* 3, 45–57.

Changnon, S. A., Kunkel, K. E., and Winstanley, D. (2003). Quantification of climate conditions important to the tall grass prairie. *T Illinois State Acad Sci* 96, 41–54.

Chapman, G. P. (1996). *The Biology of Grasses*. CAB International, Wallingford, UK.

Cheney, N. P., Gould, J. S., and Catchpole, W. R. (1998). Prediction of fire spread in grasslands. *Int J Wildland Fire* 8, 1–13.

Cheney, P., and Sullivan, A. (1997). *Grassfires: Fuel, Weather and Fire Behaviour*. CSIRO Publishing, Collingwood, Australia.

Cheplick, G. P., and Quinn, J. A. (1982). *Amphicarpum purshii* and the pessimistic strategy in amphicarpic annuals with subterranean fruit. *Oecologia* 52, 327–332.

Clarke, L. G., and Fisher, J. B. (1986). Morphology of grasses: shoots and roots. In: *Grass Systematics and Evolution*. (Soderstrom, T. R., Hilu, K. W., Campbell, C. S., and Barkworth, M. E., Eds), pp. 37–45. Smithsonian Institution Press, Washington, DC.

Clarke, S., and French, K. (2005). Germination response to heat and smoke of 22 Poaceae species from grassy woodlands. *Aust J Bot* 53, 445–454.

Collins, S. L., and Steinauer, E. M. (1998). Disturbance, diversity, and species interactions in tallgrass prairie. In: *Grassland Dynamics: Long-Term Research in Tallgrass Prairie*. (Knapp, A., Briggs, J. M., Hartnett, D. C., and Collins, S. L., Eds.), pp. 140–156. Oxford University Press, New York.

Copeland, T. E., Sluis, W., and Howe, H. F. (2002). Fire season and dominance in an Illinois tallgrass prairie restoration. *Restor Ecol* 10, 315–323.

Curtis, J. T. (1959). *The Vegetation of Wisconsin: An Ordination of Plant Communities*. University of Wisconsin Press, Madison, WI.

Daubenmire, R. F. (1968). The ecology of fire in grasslands. *Adv Ecol Res* 5, 209–266.

Derner, J. D., and Briske, D. D. (2001). Below-ground carbon and nitrogen accumulation in perennial grasses: a comparison of caespitose and rhizomatous growth forms. *Plant Soil* 237, 117–127.

Dublin, H. T. (1995). Vegetation dynamics in the Serengeti-Mara ecosystem: the role of elephants, fire, and other factors. In: *Serengeti II: Dynamics, Management, and Conservation of an Ecosystem*. (Sinclair, A. R. E., and Arcese, P., Eds.), pp. 71–90. University of Chicago Press, Chicago.

Edwards, M. E., and Armbruster, W. S. (1989). A tundra-steppe transition on Kathul Mountain, Alaska, USA. *Arctic Alpine Res* 21, 296–304.

Engle, D. M., and Bidwell, T. G. (2001). The response of central North American prairies to seasonal fire. *J Range Manage* 54, 2–10.

Facelli, J. M., Montero, C. M., and Leon, R. J. C. (1988). Effect of different disturbance regimen on seminatural grasslands from the subhumid pampa. *Flora* 180, 241–249.

Feldman, S. R., and Lewis, J. P. (2005). Effects of fire on the structure and diversity of a *Spartina argentinensis* tall grassland. *Appl Veg Sci* 8, 77–84.
Forman, S. L., Marín, L., Pierson, J., Gómez, J., Miller, G. H., and Webb, R. S. (2005). Aeolian sand depositional records from western Nebraska: landscape response to droughts in the past 1500 years. *Holocene* 15, 973–981.
Foster, B. L., and Gross, K. L. (1998). Species richness in a successional grassland: effects of nitrogen enrichment and plant litter. *Ecology* 79, 2593–2602.
Frost, P. G. H. (1984). The responses and survival of organisms in fire-prone environments. In: *Ecological Effects of Fire in South African Ecosystems* (Booysen, P. d. V. and Tainton, N. M., Eds.), pp. 274–309. Springer-Verlag, Berlin.
Gaff, D. F. (1986). Desiccation tolerant "resurrection" grasses from Kenya and West Africa. *Oecologia* 70, 118–120.
Garnier, L. K. M., and Dajoz, I. (2001). The influence of fire on the demography of a dominant grass species of West African savannas, *Hyparrhenia diplandra*. *J Ecol* 89, 200–208.
Gashaw, M., and Michelsen, A. (2002). Influence of heat shock on seed germination of plants from regularly burnt savanna woodlands and grasslands in Ethiopia. *Plant Ecol* 159, 83–93.
Gibson, D. J., Hartnett, D. C., and Merrill, G. L. S. (1990). Fire temperature heterogeneity in contrasting fire prone habitats: Kansas tallgrass prairie and Florida sandhill. *B Torrey Bot Club* 117, 349–356.
Gill, M. A., Bradstock, R. A., and Williams, J. E. (2002). Fire regimes and biodiversity: legacy and vision. In: *Flammable Australia: The Fire Regimes and Biodiversity of a Continent*. (Bradstock, R. A., Williams, J. E., and Gill, M. A., Eds), pp. 429–446. Cambridge University Press, Cambridge.
Grover, H. D., and Musick, H. B. (1990). Shrubland encroachment in southern New Mexico, U.S.A.: an analysis of desertification processes in the American southwest. *Climatic Change* 17, 303–330.
Hanley, M. E., Fenner, M., and Edwards, P. J. (1996). The effect of mollusc grazing on seedling recruitment in artificially created grassland gaps. *Oecologia* 106, 240–246.
Hartemink, A. E. (2001). Biomass and nutrient accumulation of *Piper aduncum* and *Imperata cylindrica* fallows in the humid lowlands of Papua New Guinea. *Forest Ecol Manag* 144, 19–32.
Hartnett, D. C., and Fay, P. A. (1998). Plant populations: patterns and processes. In: *Grassland Dynamics: Long-Term Research in Tallgrass Prairie*. (Knapp, A., Briggs, J. M., Hartnett, D. C., and Collins, S. L., Eds.), pp. 81–100. Oxford University Press, New York.
Hayes, G. F., and Holl, K. D. (2003). Site-specific responses of native and exotic species to disturbances in a mesic grassland community. *Appl Veg Sci* 6, 235–244.
Hedrick, D. W. (1948). The mulch layer of California annual grasslands. *J Range Manage* 1, 22–25.
Hendricks, J. J., Wilson, C. A., and Boring, L. A. (2002). Foliar litter position and decomposition in a fire-maintained longleaf pine-wiregrass ecosystem. *Can J Forest Res* 32, 928–941.
Herbel, C. H., Ares, F. N., and Wright. R. A. (1972). Drought effects on a semidesert grassland range. *Ecology* 53, 1084–1093.
Howe, H. F. (1995). Succession and fire season in experimental prairie plantings. *Ecology* 76, 1917–1925.
Howe, H. F., and Lane, D. (2004). Vole-driven succession in experimental wet-prairie restorations. *Ecol Appl* 14, 1295–1305.
Huxman, T. E., Smith, M. D., Fay, P. A., Knapp, A. K., Shaw, M. R., Loik, M. E., Smith, S. D., Tissue, D. T., Zak, J. C., Weltzin, J. F., Pockman, W. T., Sala, O. E., Haddad, B. M., Harte, J., Koch, G. W., Schwinning, S., Small, E. E., and Williams, D. G. (2004). Convergence across biomes to a common rain-use efficiency. *Nature* 429, 651–654.

Jacobs, B. F., Kingston, J. D., and Jacobs, L. L. (1999). The origin of grass-dominated ecosystems. *Ann Mo Bot Gard* 86, 590–643.

Janis, C. M., Damuth, J., and Theodor, J. M. (2000). Miocene ungulates and terrestrial primary productivity: where have all the browsers gone? *Proc Natl Acad Sci U S A* 97, 7899–7904.

Jeltsch, F., Weber, G. E., and Grimm, V. (2000). Ecological buffering mechanisms in savannas: a unifying theory of long-term tree-grass coexistence. *Plant Ecol* 150, 161–171.

Jenkins, M. A., Clark, T., and Coen, J. (2001). Coupling atmospheric and fire models. In: *Forest Fires: Behavior and Ecological Effects*. (Johnson, E. A., and Miyanishi, K., Eds.), pp. 257–302. Academic Press, San Diego, CA.

Jernvall, J., and Fortelius, M. (2002). Common mammals drive the evolutionary increase of hypsodonty in the Neogene. *Nature* 417, 538.

Keeley, J. E., and Bond, W. J. (1999). Mast flowering and semelparity in bamboos: the bamboo fire cycle hypothesis. *Am Nat* 154, 383–391.

Keeley, J. E., and Bond, W. J. (2001). On incorporating fire into our thinking about natural ecosystems: a response to Saha and Howe. *Am Nat* 158, 664–670.

Keeley, J. E., and Rundel. P. W. (2005). Fire and the Miocene expansion of C4 grasslands. *Ecol Lett* 8, 683–690.

Keeley, J. E., and Zedler, P. H. (1998). Evolution of life histories in pines. In: *Ecology and Biogeography of* Pinus. (Richardson, D. M., Ed.), pp. 219–250. Cambridge University Press, Cambridge.

Keesing, F. (2000). Cryptic consumers and the ecology of an African savanna. *Bioscience* 50, 205–215.

Kershaw, A. P., Clark, J. S., and Gill, M. A. (2002). A history of fire in Australia. In: *Flammable Australia: The Fire Regimes and Biodiversity of a Continent*. (Bradstock, R. A., Williams, J. E., and Gill, M. A., Eds.), pp. 3–25. Cambridge University Press, Cambridge.

Knapp, A. K., Briggs, J. M., Blair, J. M., and Turner. C. L. (1998). Patterns and controls of aboveground net primary production in tallgrass prairie. In: *Grassland Dynamics: Long-Term Research in Tallgrass Prairie*. (Knapp, A., Briggs, J. M., Hartnett, D. C., and Collins, S. L., Eds.), pp. 196–221. Oxford University Press, New York.

Kobayashi, K., and Yokoi, Y. (2003). Spatiotemporal patterns of shoots within an isolated *Miscanthus sinensis* patch in the warm-temperate region of Japan. *Ecol Res* 18, 41–51.

Kochy, M., and Wilson, S. D. (1997). Litter decomposition and nitrogen dynamics in aspen forest and mixed-grass prairie. *Ecology* 78, 732–739.

Kruckeberg, A. R. (1984). *California Serpentines: Flora, Vegetation, Geology, Soils, and Management Problems*. University of California Press, Berkeley, CA.

Lauenroth, W. K. (1979). Grassland primary production: North American grasslands in perspective. In: *Perspectives in Grassland Ecology: Results and Applications of the US/IBP Grassland Biome Study*. (French, N., Ed.), pp. 3–24. Springer-Verlag, New York.

Launchbaugh, J. L. (1973). Effect of fire on shortgrass and mixed prairie species. *P Tall Timbers Fire Ecol Conf* 12, 129–151.

Leach, M. K., and Givnish, T. J. (1996). Ecological determinants of species loss in remnant prairies. *Science* 273, 1555–1558.

Le Maitre, D. C., and Midgley, J. J. (1992). Plant reproductive ecology. In: *The Ecology of Fynbos: Nutrients, Fire, and Diversity*. (Cowling, R. M., Ed.), pp. 135–174. Oxford University Press, Cape Town, SA.

Linder, H. P., and Rudall, P. J. (2005). Evolutionary history of Poales. *Annu Rev Ecol Evol S* 36, 107–124.

Liston, A., Wilson, B. L., Robinson, W. A., Doescher, P. S., Harris, N. R., and Svejcar, T. (2003). The relative importance of sexual reproduction versus clonal spread in an aridland bunchgrass. *Oecologia* 137, 216–225.

Londono, X., and Peterson, P. M. (1991). *Guadua sarcocarpa* (Poaceae: Bambuseae), a new species of Amazonian bamboo with fleshy fruits. *Syst Bot* 16, 630–638.

Lonsdale, W. M., Braithwaite, R. W., Lane, A. M., and Farmer, J. (1998). Modelling the recovery of an annual savanna grass following a fire-induced crash. *Aust J Ecol* 23, 509–513.

Lovett Doust, L. (1981). Population dynamics and local specialization in a clonal perennial (*Ranunculus repens*): I. The dynamics of ramets in contrasting habitats. *J Ecol* 69, 743–755.

Lunt, I. D., and Morgan, J. W. (2002). The role of fire regimes in temperate lowland grasslands of southeastern Australia. In: *Flammable Australia: The Fire Regimes and Biodiversity of a Continent*. (Bradstock, R. A., Williams, J. E., and Gill, A. M., Eds.), pp. 177–196. Cambridge University Press, Cambridge.

Mack, R. N. (1981). Invasion of *Bromus tectorum* L. into western North America—an ecological chronicle. *Agro-Ecosystems* 7, 145–165.

Mack, R. N., and Thompson, J. N. (1982). Evolution in steppe with few large, hoofed mammals. *Am Nat* 119, 757–773.

Marty, J. T., Rice, K. J., and Collinge, S. K. (2003). The effects of burning, grazing and herbicide treatments on restored and remnant populations of *Nassella pulchra* at Beale Air Force Base, California. *Grasslands* 13, 1, 4–9.

McNaughton, S. J. (1979). Grazing as an optimization process: grass-ungulate relationships in the Serengeti. *Am Nat* 113, 691–703.

McNaughton, S. J. (1985). Ecology of a grazing ecosystem: the Serengeti. *Ecol Monogr* 55, 259–294.

Metcalfe, C. R. (1960). *Anatomy of the Monocotyledons*. Clarendon Press, Oxford.

Neary, D. G., Klopatek, C. C., Debano, L. F., and Ffolliot, P. F. (1999). Fire effects on belowground sustainability: a review and synthesis. *Forest Ecol Manag* 122, 51–71.

Nelson, R. M. J. (2001). Water relations of forest fuels. In: *Forest Fires: Behavior and Ecological Effects*. (Johnson, E. A., and Miyanishi, K., Eds.), pp. 79–149. Academic Press, San Diego, CA.

Niklas, K. J. (1990). The mechanical significance of clasping leaf sheathing in grasses: evidence from two cultivars of *Avea sativa*. *Ann Bot* 65, 505–512.

Niklas, K. J. (1999). A mechanical perspective on foliage leaf form and function. *New Phytol* 143, 19–31.

Noy-Meir, I. (1988). Dominant grasses replaced by ruderal forbs in a vole year in undergrazed Mediterranean grasslands in Israel. *J Biogeogr* 15, 579–587.

O'Connor, T. G., and Roux, P. W. (1995). Vegetation changes (1949-71) in a semi-arid, grassy dwarf shrubland in the Karoo, South Africa: influence of rainfall variability and grazing by sheep. *J Appl Ecol* 32, 612–626.

Odgers, B. M. (1999). Seasonal variation in buried germinable seed banks of grass species in an urban eucalypt forest reserve. *Aust J Bot* 47, 623–638.

Oesterheld, M., Loreti, J., Semmartin, M., and Paruelo, J. M. (1999). Grazing, fire, and climate effects on primary productivity of grasslands and savannas. In: *Ecosystems of Disturbed Ground*. (Walker, L. R., Ed.), pp. 287–306. Elsevier, Amsterdam.

Oesterheld, M., Loreti, J., Semmartin, M., and Sala, O. E. (2001). Inter-annual variation in primary production of a semi-arid grassland related to previous-year production. *J Veg Sci* 12, 137–142.

Old, S. M. (1969). Microclimate, fire, and plant production in an Illnois prairie. *Ecol Monogr* 39, 355–384.

Partomihardjo, T., Mirmanto, E., and Whittaker, R. J. (1992). Anak Krakatau's vegetation and flora circa 1991 with observations on a decade of development and change. *GeoJournal* 28, 233–248.

Paruelo, J. M., Lauenroth, W. K., Burke, I. C., and Sala, O. E. (1999). Grassland precipitation—use efficiency varies across a resource gradient. *Ecosystems* 2, 64–68.

Peart, M. H. (1984). The effects of morphology, orientation and position of grass diaspores on seedling survival. *J Ecol* 72, 437–453.

Peláez, D. V., Boo, R. M., Elia, O. R., and Mayor, M. D. (1997). Effect of fire intensity on bud viability of three grass species native to central semi-arid Argentina. *J Arid Environ* 37, 309–317.

Pfeiffer, K. E., and Hartnett, D. C. (1995). Bison selectivity and grazing response of little bluestem in tallgrass prairie. *J Range Manage* 48, 26–31.

Piedade, M. T. F., Junk, W. J., and De Mello, J. A. N. (1992). A floodplain grassland of the central Amazon. In: *Primary Productivity of Grass Ecosystems of the Tropics and Sub-Tropics*. (Long, S. P., Jones, M. B., and Roberts, M. J., Eds.), pp. 127–158. Chapman & Hall, London.

Prasad, V., Strömberg, C. A. E., Alimohammadian, H., and Sahni, A. (2005). Dinosaur coprolites and the early evolution of grasses and grazers. *Science* 310, 1177–1180.

Pylypec, B., and Romo, J. T. (2003). Long-term effects of burning *Festuca* and *Stipa-Agropyron* grasslands. *J Range Manage* 56, 640–645.

Ramsay, P. M., and Oxley, E. R. B. (1996). Fire temperatures and postfire plant community dynamics in Ecuadorian grass paramo. *Vegetatio* 124, 129–144.

Redmann, R. E., Romo, J. T., Pylypec, B., and Driver, E. A. (1993). Impacts of burning on primary productivity of *Festuca* and *Stipa-Agropyron* grasslands in central Saskatchewan. *Am Midl Nat* 130, 262–273.

Reich, P. B., Peterson, D. W., Wedin, D. A., and Wrage, K. (2001). Fire and vegetation effects on productivity and nitrogen cycling across a forest-grassland continuum. *Ecology* 82, 1703–1719.

Renvoize, S. A., and Clayton, W. D. (1992). Classification and evolution of the grasses. In: *Grass Evolution and Domestication*. (Chapman, G. P., Ed.), pp. 3–37. Cambridge University Press, Cambridge.

Rice, B., and Westoby, M. (1999). Regeneration after fire in *Triodia* R. Br. *Aust J Ecol* 24, 563–572.

Riswan, S., and Hartanti, L. (1995). Human impacts on tropical forest dynamics. *Plant Ecol* 121, 41–52.

Robberecht, R., and Defossé, G. E. (1995). The relative sensitivity of two bunchgrass species to fire. *Int J Wildland Fire* 5, 127–134.

Rosenzweig, M. L. (1968). Net primary productivity of terrestrial communities: prediction from climatological data. *Am Nat* 102, 67–74.

Rossiter, N. A., Setterfield, S. A., Douglas, M. M., and Hutley, L. B. (2003). Testing the grass-fire cycle: alien grass invasion in the tropical savannas of northern Australia. *Divers Distrib* 9, 169–176.

Sala, O. E., Golluscio, R. A., Lauenroth, W. K., and Soriano, A. A. (1989). Resource partitioning between shrubs and grasses in the Patagonian steppe. *Oecologia* 81, 501–505.

Samuel, M. J., and Hart, R. H. (1995). Observations on spread and fragmentation of blue grama clones in disturbed rangeland. *J Range Manage* 48, 508–510.

Sankaran, M., Hanan, N. P., Scholes, R. J., Ratnam, J., Augustine, D. J., Cade, B. S., Gignoux, J., Higgins, S. I., Le Roux, X., Ludwig, F., Ardo, J., Banyikwa, F., Bronn, A., Bucini, G., Caylor, K. K., Coughenour, M. B., Diouf, A., Ekaya, W., Feral, C. J., February, E. C., Frost, P. G. H., Hiernaux, P., Hrabar, H., Metzger, K. L., Prins, H. T. T., Ringrose, S., Sea, W., Tews, J., Worden, J., and Zambatis, N. (2005). Determinants of woody cover in African savannas. *Nature* 438, 846–849.

Savage, M., and Swetnam, T. W. (1990). Early 19th-Century fire decline following sheep pasturing in a Navajo ponderosa pine forest. *Ecology* 71, 2374–2378.

Scanlan, J. C., and O'Rourke, P. K. (1982). Effect of spring wildfires on *Iseilema* (Flinders grass) populations in the Mitchell grass region of northwestern Queensland Australia. *Aust J Bot* 30, 591–600.

Scholes, R. J., and Archer, S. R. (1997). Tree-grass interactions in savannas. *Ann Rev Ecol Syst* 28, 517–544.

Schoning, C., Espadaler, X., Hensen, I., and Roces, F. (2004). Seed predation of the tussock-grass *Stipa tenacissima* L. by ants (*Messor* spp.) in southeastern Spain: the adaptive value of trypanocarpy. *J Arid Environ* 56, 43–61.

Scott, A. C. (2000). The pre-Quaternary history of fire. *Palaeogeogr Palaeocl* 164, 281–329.

Sharp, B. R., and Whittaker. R. J. (2003). The irreversible cattle-driven transformation of a seasonally flooded Australian savanna. *J Biogeogr* 30, 783–802.

Silliman, B. R., and Bertness, M. D. (2002). A trophic cascade regulates salt marsh primary production. *Proc Natl Acad Sci U S A* 99, 10500–10505.

Silva, J. F., Raventos, J., Caswell, H., and Trevisan, M. C. (1991). Population responses to fire in a tropical savanna grass, *Andropogon-semiberbis*—a matrix model approach. *J Ecol* 79, 345–356.

Silva, J. F., Zambrano, A., and Farinas, M. R. (2001). Increase in the woody component of seasonal savannas under different fire regimes in Calabozo, Venezuela. *J Biogeogr* 28, 977–983.

Stromberg, C. A. E. (2004). Using phytolith assemblages to reconstruct the origin and spread of grass-dominated habitats in the great plains of North America during the late Eocene to early Miocene. *Palaeogeogr Palaeocl Sp Iss* 207, 239–275.

Stromberg, M. R., and Griffin, J. R. (1996). Long-term patterns in coastal California grasslands in relation to cultivation, gophers, and grazing. *Ecol Appl* 6, 1189–1211.

Tainton, N. M., and Mentis, M. T. (1984). Fire in grassland. In: *Ecological Effects of Fire in South African Ecosystems*. (Booysen, P. d. V., and Tainton, N. M., Eds.), pp. 115–147. Springer-Verlag, Berlin.

Tilman, D. (1993). Species richness of experimental productivity gradients: how important is colonization limitation? *Ecology* 74, 2179–2191.

Titshall, L. W., O'Connor, T. G., and Morris, C. D. (2000). Effect of long-term exclusion of fire and herbivory on the soils and vegetation of sour grassland. *Afr J Range Forage Sci* 17, 70–80.

Tjoelker, M. G., Craine, J. M., Wedin, D., Reich, P. B., and Tilman, D. (2005). Linking leaf and root trait syndromes among 39 grassland and savannah species. *New Phytol* 167, 493–508.

Tomlinson, K. W., and Connor, T. G. (2004). Control of tiller recruitment in bunchgrasses: uniting physiology and ecology. *Funct Ecol* 18, 489–496.

Tothill, J. C. (1969). Soil temperature and seed burial in relation to the performance of *Heteropogon contortus* and *Themeda australis* in burnt native woodland pastures in eastern Queensland. *Aust J Bot* 17, 269–275.

Towne, E. G., and Kemp, K. E. (2003). Vegetation dynamics from annually burning tallgrass prairie in different seasons. *J Range Manage* 56, 185–192.

Trofymow, J. A., Moore, T. R., Titus, B., Prescott, C., Morrison, I., Siltanen, M., Smith, S., Fyles, J., Wein, R., Camirt, C., Duschene, L., Kozak, L., Kranabetter, M., and Visser, S. (2002). Rates of litter decomposition over 6 years in Canadian forests: influence of litter quality and climate. *Can J Forest Res* 32, 789–804.

Tscharntke, T., and Greiler, H.-J. (1995). Insect communities, grasses, and grasslands. *Ann Rev Entomol* 40, 535–558.

Tunnell, S. J., Engle, D. M., and Jorgensen, E. E. (2004). Old-field grassland successional dynamics following cessation of chronic disturbance. *J Veg Sci* 15, 431–436.

Valette, J.-C., Gomendy, V., Maréchal, J., Houssard, C., and Gillon, D. (1994). Heat transfer in the soil during very low-intensity experimental fires: the role of duff and soil moisture content. *Int J Wildland Fire* 4, 225–237.

Valverde, T., and Pisanty, I. (1999). Growth and vegetative spread of *Schizachyrium scoparium* var. *littoralis* (Poaceae) in sand dune microhabitats along a successional gradient. *Can J Bot* 77, 219–229.

Van De Vijver, C., Poot, P., and Prins, H. H. T. (1999). Causes of increased nutrient concentrations in post-fire regrowth in an East African savanna. *Plant Soil* 214, 173–185.

Vesk, P. A., and Westoby, M. (2004). Funding the bud bank: a review of the costs of buds. *Oikos* 106, 200–208.

Vitousek, P. M., Aber, J. D., Howarth, R. W., Likens, G. E., Matson, P. A., Schindler, D. W., Schlesinger, W. H., and Tilman, D. G. (1997). Human alteration of the global nitrogen cycle: sources and consequences. *Ecol Appl* 7, 737–750.

Vogl, R. J. (1974). Effects of fire on grasslands. In: *Fire and Ecosystems*. (Kozlowski, T. T., and Ahlgren, C. E., Eds.), pp. 139–194. Academic Press, New York.

Walter, H. (1964). *Vegetation der Erde*. Band I. Gustav Fischer Verlag, Jena, Germany.

Wan, S., Hui, D., and Luo, Y. (2001). Fire effects on nitrogen pools and dynamics in terrestrial ecosystems: a meta-analysis. *Ecol Appl* 11, 1349–1365

Watkinson, A. R., Lonsdale, W. M., and Andrew, M. H. (1989). Modelling the population dynamics of an annual plant *Sorghum intrans* in the wet-dry tropics. *J Ecol* 77, 162–181.

Weaver, J. E. (1954). *North American Prairie*. Johnsen Publishing Co., Lincoln, NE.

Westoby, M., Walker, B., and Noy-Meir, I. (1989). Opportunistic management for rangelands not at equilibrium. *J Range Manage* 42, 266–274.

Wiegand, T., Snyman, H. A., Kellner, K., and Paruelo, J. M. (2004). Do grasslands have a memory: modeling phytomass production of a semiarid South African grassland. *Ecosystems* 7, 243–258.

Williams, R. J., Gill, A. M., and Moore, P. H. R. (1998). Seasonal changes in fire behaviour in a tropical savanna in northern Australia. *Int J Wildland Fire* 8, 227–239.

Woodward, F. I., Lomas, M. R., and Kelly, C. K. (2004). Global climate and the distribution of plant biomes. *Philos T Roy Soc B* 359, 1465–1476.

Wright, H. A. (1971). Why squirreltail is more tolerant to burning than needle-and-thread. *J Range Manage* 24, 277–284.

Wright, H. A. (1974). Effect of fire on southern mixed prairie grasses. *J Range Manage* 27, 417–419.

Young, J. A., Evans, R. A., and Robison, J. (1972). Influence of repeated annual burning on a medusahead community. *J Range Manage* 25, 372–375.

Zedler, P. H. (1987). *The Ecology of Southern California Vernal Pools: A Community Profile*. Fish and Wildlife Service, U.S. Department of the Interior, Washington, DC.

Zedler, P. H., and Black, C. (2004). Exotic plant invasions in an endemic-rich habitat: the spread of an introduced Australian grass, *Agrostis avenacea* J. F. Gmel., in California vernal pools. *Austral Ecol* 29, 537–546.

14

Wildfire and Tree Population Processes

Sheri L. Gutsell and Edward A. Johnson
University of Calgary

INTRODUCTION

Interest in the ecological effects of wildfires has a long history in ecology, going back to Clements' (1910) study of wildfire in lodgepole pine forests. Since that time, the main approach of fire ecological studies has been to describe statistically patterns of fire effects and correlate these to environmental variables. Unfortunately, this approach does not easily lead to an understanding of how fires cause their ecological effects. An example of the problem with this approach can be found in the use of the term "fire severity." Severity is often defined in terms of ecological effects (e.g., tree mortality) and is used to classify a fire (e.g., high-severity fire) (Pollet and Omi, 2002; Fulé *et al.*, 2003; Odion *et al.*, 2004). Unfortunately, when defined in this way, it is not clear what wildfire process caused the effect. In addition, because classifications are subjective and no units are associated with the term, making comparisons among studies is difficult.

Another approach to studying the ecological effects of wildfires has been the process–response model, in which the mechanism or action of a wildfire process (e.g., heat transfer from conduction, convection, and radiation) causing an ecological effect (e.g., tree mortality) is clearly identified and the causal connection between the wildfire process and ecological effect is clearly established (Johnson, 1985; Johnson and Miyanishi, 2001). In this chapter, we make a preliminary attempt at this approach by using local and regional tree population dynamics as the ecosystem processes of interest. We show how wildfire processes and characteristics affect individual trees and the recruitment and mortality of their local and regional populations.

Tree populations can be thought of as having a hierarchical structure. The smallest unit is the individual, a group of individuals make up a birth cohort, two or more birth cohorts make up a local population, and a collection of local populations make up a regional population. A birth cohort consists of individuals that were born (i.e., recruited) at roughly the same time and have similar mortality schedules. The recruits in a birth cohort may come from within the local population or from nearby local populations. In the regional population, the local populations can be thought of as "individuals" that recruit and die. When wildfire kills one or more of the local populations within a regional population, the mortality schedule of interest becomes the time-since-fire distribution, which is the survivorship curve (1–cumulative mortality) of the regional population. Recruitment into the exterminated local populations occurs via seeds that originate from aerial seed banks or sprouts that originate from belowground bud banks. Alternatively, recruitment into the exterminated local population may occur via seeds that are dispersed from nearby surviving local populations.

We begin our chapter by introducing wildfire processes and characteristics, discussing those shown to have specific ecological effects on individual trees and tree populations. Next, we define local populations and their cohorts and then describe what is known about tree population recruitment and mortality processes in a variety of ecosystems. We describe the causal connections between specific wildfire processes or characteristics and tree population recruitment and mortality processes where these connections are well understood. Finally, we discuss the regional processes of geomorphology, large-scale mortality by wildfire, and seed dispersal.

Our discussion is by no means complete, but is intended to show one possible approach to understanding how wildfires cause ecological effects. In addition, as our expertise is mainly in the boreal forest, readers may find that at times we are biased to that literature. We also know that a lot of work is going on as we write, so the reader is encouraged to consult the recent literature.

WILDFIRE PROCESSES AND CHARACTERISTICS

Wildfires consist of many processes and characteristics operating at a variety of temporal and spatial scales. In this section, we discuss those that are known to have specific effects on individual tree mortality and the tree population processes of recruitment and mortality.

Wildfire Processes

In this section, we discuss two important wildfire processes: combustion and heat transfer. These processes are discussed thoroughly in numerous textbooks (e.g., Drysdale, 1985; Liñán and Williams, 1993; Cox, 1995; Holman, 1997; Incropera and DeWitt, 2002). For a more specific discussion of heat transfer into plants, including the relevant heat transfer models, the reader is also referred to Chapter 12 by Mercer and Weber, and Dickinson and Johnson (2001).

Both combustion and heat transfer, which are related, directly affect the mortality of individual trees (or their components); however, they may do so in different ways. Combustion is a process that kills plants (e.g., seedlings) or their parts (e.g., leaves) by consuming them. In contrast, heat transfer is a process that may kill plants and their parts without consuming them; mortality may result if enough heat is transferred from the fire or its buoyant plume to their critical tissues (e.g., vascular cambium, apical meristem, and transport tissues).

Combustion

Fire is the common name we give to high-temperature gaseous combustion, which is the rapid transformation of the stored chemical energies in fuels into kinetic energies of heat and motion (Liñán and Williams, 1993; Saito, 2001). In a wildfire, combustion may be in one of two forms: flaming combustion or smoldering combustion. Each form of combustion is discussed in turn below.

Flaming Combustion Flaming combustion, which results in a visible flame, involves the conversion of solid fuels to gaseous form (Drysdale, 1985). It is the rapid oxidation of gaseous fuels (volatiles) released from solid fuels. In wildfires, surface fuels consist of the organic litter layer (i.e., recently fallen and partially decomposed tree leaves, fallen twigs, bark, and branches, dead herbaceous plants), as well as live grasses, forbs, and shrubs (Davis, 1959). Canopy fuels consist of the small-diameter branches, leaves, and buds within the canopies of most conifer and broadleaf deciduous trees and the chaparral and pocosin shrub types (Nelson, 2001). In flaming combustion, heat transfer from the flame to the solid fuels occurs by conduction, convection, and radiation. The heat transfer liberates volatiles that are mixed with oxygen from the air by laminar or turbulent processes; when they are in the right stoichiometry, combustion occurs (Liñán and Williams, 1993).

The physical properties of fuels and their arrangement influence flaming combustion by affecting the fuel's relative susceptibility to heating. Small-diameter fuels with high surface-to-volume ratios within loosely packed fuel arrays allow for more of the total fuel mass to be exposed to preheating, liberating extractives that will burn in the flaming zone. In contrast, larger fuel particles, such as fallen tree boles, with low surface-to-volume ratios, may absorb more heat than they liberate during the passage of a flame. This explains why small-diameter fuels are the main drivers of flaming combustion and larger fuels contribute relatively little to the process.

In flaming combustion, the ignition of volatiles from solid fuels often proceeds in a sequence, beginning with the fuel components that thermally decompose at relatively low temperatures and ending with those that require much higher temperatures to decompose (Simmons, 1995). Fuel components that ignite at the lowest temperatures and release high-energy gases are called extractives (terpenes, fats, oils, and waxes); these are low–molecular-weight compounds with high heats of combustion. The combustion of extractives provides the heat needed to drive off the moisture and ignite fuel components, such as cellulose, that require higher temperatures to decompose. Cellulose is a high–molecular-weight fuel component that produces large amounts of flammable gases. This sequence of increasingly higher temperatures means that more of the fuel load becomes available for flaming combustion. It also explains why living fuels, which contain high-energy extractives, can burn, despite their very high fuel moistures (80–200% of oven dry weight). The abundant high-energy gases released from living fuels provide a sufficiently large amount of heat to overcome the otherwise prohibitively large latent heat sink created by their high moisture content (Drysdale, 1985).

The flaming combustion of small-diameter surface fuels and, in some situations, crown fuels causes at least two important ecological effects: (1) the partial or complete mortality of the aboveground portion of grasses, forbs, shrubs, seedlings, and saplings; and (2) the partial or complete mortality of canopy trees (depending on fire intensity; see discussion later in this chapter). Both in turn affect the post-fire recruitment of herbaceous and woody plants. The removal of ground vegetation and tree canopies results in a period of reduced competition for soil resources (i.e., soil moisture and nutrients) and light, which can help facilitate the rapid post-fire recruitment of both sexually and asexually reproducing tree populations.

Smoldering Combustion Smoldering combustion is a self-sustaining, slowly propagating, low-temperature flameless combustion process in which solid fuel first undergoes thermal decomposition (pyrolysis), producing volatiles and carbonaceous char (Drysdale, 1985). Subsequently, the char oxidizes, providing the heat required for propagation of the smolder process. In a wildfire, smoldering occurs primarily after passage of the visible flame. Heat transfer in smoldering combustion is by conduction, convection, and radiation. See Miyanishi (2001) for an overview of the heat transfer models proposed for the propagation of smoldering combustion.

Smoldering combustion is an important process in only some ecosystems affected by wildfires. It occurs within the fermentation and humus (F and H) layers of the organic soil (i.e., duff), which lie between the litter layer and the A horizon of the mineral soil in some forests. Duff is present primarily within conifer-dominated forests with cool to cold climates, such as boreal and subalpine forests and Douglas fir forests of the Pacific Northwest (Miyanishi, 2001). Smoldering combustion may also occur in dead fallen tree boles and within thick layers of litter impregnated with roots, called root mats, in some humid tropical forests; see Miyanishi (2001) for a summary of this work. Smoldering is not an important process in most dry, warm savanna, woodland, and forest ecosystems (e.g., Mediterranean pine forests, ponderosa pine savannas). These ecosystems often have only a litter layer, which is consumed in flaming combustion.

Only porous materials that form a solid carbonaceous char when heated can undergo self-sustained smoldering combustion (Drysdale, 1985). Duff consists primarily of cellulose, hemicellulose, and lignin (Mason, 1976); of the three components, lignin is the most resistant to thermal decomposition and thus it is the main component in duff that produces char (Johnson, 1992; Miyanishi, 2001). The surface oxidation of char provides the heat necessary to cause further thermal degradation of adjacent virgin duff (Drysdale, 1985). The propagation of smoldering combustion requires that volatiles be progressively driven out ahead of the zone of active combustion (where char is oxidized) to expose fresh char that will then begin to oxidize. Heat losses from the reaction zone must not be too high because sufficient heat for the endothermic pyrolysis of the virgin duff must be supplied from this reaction zone (Drysdale, 1985).

Smoldering combustion is controlled by three properties of duff: bulk density, moisture content, and depth (Miyanishi, 2001). Bulk density ($kg\,m^{-3}$) affects the penetration of oxygen through duff, which affects the rate of oxidation of char (Palmer, 1957; Ohlemiller, 1985; 1990). The moisture content of duff affects how much heat is required to evaporate the water before it can start to raise the temperature of the virgin duff high enough to drive off volatiles and produce char (Palmer, 1957; Miyanishi, 2001). The depth of duff affects heat losses from the oxidation zone because the duff itself acts as insulation; the thinner the duff layer, the greater the proportional heat loss from the system (Palmer, 1957; Bakhman, 1993; Miyanishi, 2001). In addition, the effects of these properties interact; for example, the thinner the duff, the lower the moisture content threshold for the propagation of smoldering (Fig. 1) (Jones *et al.*, 1994; Miyanishi and Johnson, 2002). See Miyanishi and Johnson (2002) for a more detailed discussion of the significance of each of these three factors in smoldering combustion of duff.

Smoldering combustion of duff causes at least three important ecological effects: (1) mortality of seeds stored in the duff; (2) partial or complete mortality of the vegetative parts of plants (i.e., roots and rhizomes); and (3) exposure of mineral soil. All three in turn affect the post-fire recruitment of herbaceous and woody plants. Seeds stored in duff are typically

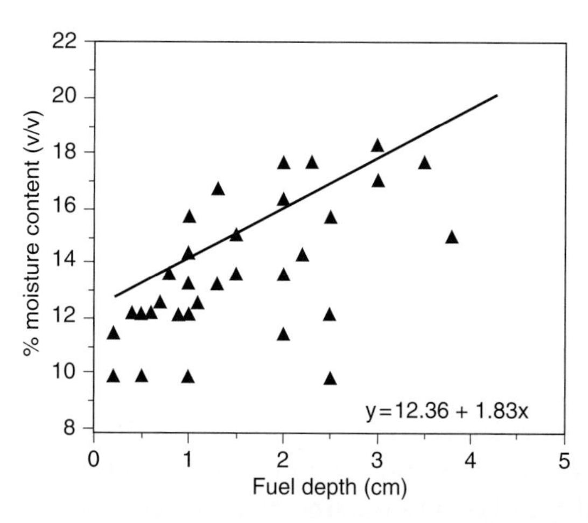

FIG. 1 The regression line determined from the fuel depth–moisture combinations that resulted in failure of smoldering propagation (above the line) and the data points for the depth–moisture combinations that resulted in successful smoldering propagation (below the line). (From Miyanishi and Johnson [2002], with permission from Academic Press.)

killed in the smoldering process; either the heat transferred through the duff is enough to kill the seeds or seeds are consumed in the process. See Chapter 12 for a detailed case study of the application of heat transfer models to seed survival within woody fruits. Smoldering combustion may also partially or completely kill the roots and rhizomes of trees, which can reduce or inhibit the post-fire recruitment of tree populations that sprout from basal buds. The partial or complete mortality of the tree roots of sexually reproducing tree populations may contribute to overall tree mortality. Finally, several studies have demonstrated the importance of exposed mineral soil to the post-fire recruitment of sexually reproducing tree populations. This is discussed in detail later in this chapter.

Heat Transfer

The transfer of heat from flaming and smoldering combustion and the plume above a flame to the components of a tree (roots, bole, canopy buds, leaves, and small branches) occurs by conduction, convection, and radiation. In what follows, we briefly review each mode of heat transfer and then describe each one's relevance to understanding and predicting the mortality of individual trees.

Conduction occurs when there is a temperature gradient in a stationary medium (solid, liquid, or gas) (Incropera and DeWitt, 2002). It is the transfer of energy from the more energetic (i.e., higher temperature) to the less energetic (i.e., lower temperature) molecules of a substance (e.g., tree bark) resulting from the collisions between adjacent molecules. This energy is related to the molecules' random translational motion and internal rotational and vibrational motion. In a solid, the transfer of energy also results from lattice waves and the translational motion of free electrons (Incropera and DeWitt, 2002). The further apart molecules are, the less likely they are to collide; therefore, of the three forms of matter, gases are the poorest conductors of heat. This explains why, for example, in the transfer of heat from the flames and buoyant plume of a surface fire to a tree canopy conduction is a relatively unimportant heat transfer process (compared to convection and radiation).

Convection is the transfer of heat within a fluid (e.g., air or water) at a nonuniform temperature by the bulk motion of the fluid molecules. Convection may be classified according to the nature of the fluid flow (Incropera and DeWitt, 2002). Free convection is induced by buoyancy forces, which arise from density differences caused by the temperature variations in the fluid. For example, on a windless, hot day in a desert, free convection occurs when the warm air within the sand's boundary layer rises because it is less dense than the cool air above it. Forced convection occurs when the flow of the fluid is caused by an external means, such as the wind or a fan. For example, on a hot day the wind will carry away the warm air that becomes trapped within the boundary layer of a tree cone and replace it with cooler ambient air.

Thermal radiation is the energy emitted by matter (solid, liquid, or gas) that is at any temperature above absolute zero (0 K) (Incropera and DeWitt, 2002). The emission may be attributed to changes in the electromagnetic configuration of the constituent atoms or molecules. The energy of the radiation field is transported by electromagnetic waves and does not require a material medium. The radiant energy flux (W m^{-2}) emitted by the surface of a radiating object is a function of emissivity (dimensionless), which is the radiative property of the object's surface, the Stefan–Boltzmann constant (5.669×10^{-8} W m^{-2} K^{-4}), and the fourth power of the absolute temperature (K) of the surface of the material. This is called the Stefan–Boltzmann law. Emissivity is the ratio of the emissive power of a body to that of an ideal emitter (blackbody) and is assumed to be a constant over all temperatures and wavelengths. Because the emitted radiant flux is directly proportional to temperature, an object that is heated above a fire reradiates a portion of the heat it gains from the flame or plume, causing a proportional reduction in its temperature (Dickinson and Johnson, 2001). In addition, the amount of radiant energy received by an object from a source (e.g., flame) is inversely proportional to the distance between them while the amount of radiant energy absorbed by an object depends on its albedo (reflectivity).

Heat transfer from a surface fire into the foliage, buds, cones, and small branches of a tree canopy involves primarily free convection (e.g., Johnson and Gutsell, 1993; Dickinson and Johnson, 2001; Michaletz, 2005). Heat transfer from a surface fire into a tree bole involves convective and radiative heat transfer between the flame and surface of the bark, and conduction through the bark and underlying wood (e.g., Fahnestock and Hare, 1964; Martin, 1963; Peterson and Ryan, 1986; Gutsell and Johnson, 1996; Dickinson and Johnson, 2001; Jones, *et al.*, 2004; Michaletz, 2005). Heat transfer into the forest floor organic material, mineral soil, and roots of a tree occurs primarily by conduction, although it may also occur by convection through porous organic material (e.g., DeVries, 1963; Oliveira *et al.*, 1997; Dickinson and Johnson, 2001; Miyanishi, 2001). See Chapter 12 (Mercer and Weber, 2007) and Dickinson and Johnson (2001) for specific examples of the modeling of heat transfer through various tree components, and Dickinson and Johnson (2001) for predictions of individual tree mortality based on these models. Later in this chapter, we provide an example of some recent work (Michaletz, 2005) that predicts tree mortality based on conduction and convection heat transfer theory and tree allometry models.

Wildfire Characteristics

In this section, we discuss two important wildfire characteristics: fire intensity and fire spatial patterns (i.e., area burned, fire shape, number and area of unburned residual islands). A fire's intensity and spatial patterns are examples of characteristics that have direct effects on the recruitment and mortality of tree populations. Fire intensity determines the extent of mortality of canopy trees within a burn, whereas fire spatial patterns determine whether some tree populations can recruit within a burn after fire. Specifically, the post-fire recruitment of tree populations whose exposed seeds are killed by fire (i.e., within open cones or catkins) require that seed be dispersed from live, reproductively mature trees around the edge of the burn or from unburned residual islands within the burn perimeter. Another important wildfire characteristic, fire frequency, is discussed later in this chapter.

Fire Intensity

Heat transfer from a flame to the solid fuels ahead of the flame is by conduction, convection, and radiation. In flaming combustion, the supply of volatiles from the surface of solid fuels is directly linked to the rate of heat transfer from the flame to the fuels ahead of the flame. The rate at which heat is released at the linear flaming front is called fire intensity (kW m^{-1}); it is the most important single factor that characterizes a fire's behavior (Drysdale, 1985). Intensity has a power law relationship to flame length (Byram, 1959; Thomas, 1971), which can be defined as the vertical distance from the flame base to the time-averaged yellow visible flame tip (Saito, 2001). One equation illustrating this relationship is

$$I = 259L^{2.2} \tag{1}$$

where I is intensity (kW m^{-1}) and L is length (m) (Byram, 1959). Therefore, as intensity increases, flame length increases faster.

Forest fires are generally characterized as one of two types based on their intensity: (1) surface fires, which are of relatively low intensity (e.g., 100 to 15 000 kW m^{-1}; Van Wagner, 1983) and burn strictly through surface fuels; and (2) crown fires (active or passive), which are of relatively high intensity (e.g., 8000–100,000 kW m^{-1}; Van Wagner, 1983) and burn through surface fuels and crown fuels. Depending on their intensity, surface fires kill the aboveground portion of herbaceous and woody plants and may partially kill canopy trees (see discussion later of the mechanisms of partial tree mortality). They may become passive crown fires, extending into single tree crowns and then dropping back to the surface fuels, resulting in the death of individual canopy trees (Van Wagner, 1977).

Active crown fires form a wall of flames extending from the surface fuels to above the tree canopies. They kill and consume the aboveground portion of understorey herbaceous and woody plants; they also kill canopy trees (at least their aboveground tissues) and consume their small-diameter canopy components. The occurrence of active crown fires is limited by three factors: (1) a critical surface fire intensity that ignites the crown fuels; (2) a critical horizontal heat flux that results primarily from the burning tree crowns, but may also come from the unburned buoyant gaseous fuels from the surface fire; and (3) a critical rate of spread below which the crown fire cannot be sustained (Van Wagner, 1977). The critical rate of spread ensures an adequate supply of new fuel to maintain the crown fire. It is dependent primarily on the bulk density of the crown space (kg m^{-3}) (Fig. 2). Closed forests (e.g., subalpine coniferous, boreal, and Mediterranean pine forests) have relatively high crown bulk densities and thus require a relatively lower rate of spread to sustain an active crown fire compared to open canopy ecosystems (i.e., savannas, woodlands). For example, in the boreal forest, canopy bulk densities generally are greater than 0.20 kg m^{-3} (Johnson, 1992), and thus a relatively low rate of spread (15 m min^{-1}) is required to maintain an active crown fire (Fig. 2). Rates of spread up to 60 m min^{-1} have been reported in many crown fire systems (McRae *et al.*, 1989). Given the importance of crown bulk density to rate of spread, and of rate of spread to the maintenance of active crown fires, it is not surprising that closed forests (particularly those that have trees with small leaves or needles and thus smaller branches, for example, conifer forests) are dominated by crown fires.

In contrast to closed forests, open canopy ecosystems, such as savannas and woodlands, typically have lower crown bulk densities, and thus the maintenance of active crown fires requires relatively higher rates of spread. High fire spread rates are possible in open ecosystems because there are often high winds as well as continuous surface fuels that dry rapidly because of the high winds and high solar radiation. However, the presence of trees in savannas and woodlands reduces the effective wind speed and its effect on fire spread rates (e.g., Fig. 3.2 in Catchpole, 2002). Spread rates of

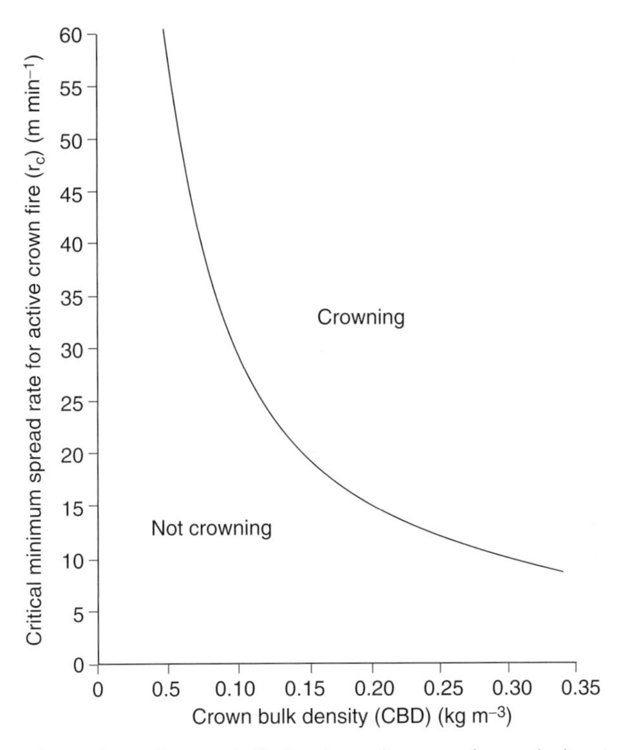

FIG. 2 The relationships of crown bulk density to fire rate of spread, showing the crown bulk density and fire rate of spread combinations required for active crown fires. (From Alexander [1988].)

approximately 6 m min^{-1} are typical of surface fires with steady state frontal fire spread in open ecosystems (McRae *et al.*, 1989). The connections between crown bulk density, fire rate of spread, and active crown fires in open ecosystems were demonstrated in a study by Alexander *et al.* (1991). They estimated the bulk density of canopy trees and conducted experimental burns in eight study plots in a black spruce [*Picea mariana* (Mill.) BSP]–lichen [*Stereocaulon paschale* (L.) Hoffm.] woodland in northern Canada. Active crown fires developed in two plots with crown bulk densities of 0.12 and 0.17 kg m^{-3} and fire rates of spread of 26.3 and 33.3 m min^{-1}. In all other plots where active crown fires did not develop, crown bulk density was higher (0.18–0.32 kg m^{-3}), but fire spread rates were much lower (1.2–6.1 m min^{-1}). It is important to note that the development of active crown fires in these plots was also facilitated by the foliage of black spruce trees that extends to the ground (i.e., provides ladder fuels that connect the surface and canopy fuels). In woodlands and savannas, where tree crowns are well above the ground (e.g., *Pinus ponderosa* Laws., *Pinus banksiana* Lamb.) and there are few or no ladder fuels, the development of an active crown fire is much more difficult (e.g., Quintilio *et al.*, 1977). Despite the potential for high wind speeds and fire spread rates in open ecosystems, the relatively lower crown bulk density of these ecosystems and lack of ladder fuels make it difficult to sustain an active crown fire. This probably contributes to the relatively high frequency of low intensity surface fires or passive crown fires in many open ecosystems, and low frequency (or lack) of high intensity active crown fires (e.g., Mast *et al.*, 1998; Veblen *et al.*, 2000; Ehle and Baker, 2003).

When forest fires are characterized by their intensity, the measure given generally describes the heat output where intensity is highest, at the advancing front of a fire (i.e., in the main direction of spread). However, the intensity at other points along the perimeter of a fire usually differs, with relatively lower intensities at the sides of the fire (i.e., flank fires) and even lower intensities at the back of the fire (i.e., back fire). Intensity variations are seen within all fires and may relate to spatial and temporal variations in wind speed, fuel moisture content, fuel load, and presence of barriers, such as rock outcrops (Catchpole *et al.*, 1982). Clearly, these varying intensities will have different effects on vegetation and therefore it must be clear where on the fire an ecological effect is being studied.

Fire Spatial Patterns

In this section we discuss the spatial patterns of fire, including the area burned, the shape of a burn, and the number and size of unburned residual islands within a burn. Each of these characteristics affects the post-fire recruitment of some tree populations whose seeds are unavailable from within a burn after fire.

Area Burned The total amount of area burned by a fire is determined primarily by the weather conditions before and during a fire, which influence fuel conditions, and by the fire's rate of spread. Fires that burn with low frontal rates of spread typically burn relatively small areas; the size of these fires is primarily controlled by the properties of the surface fuels through which they burn (e.g., Díaz-Delgado *et al.*, 2004) and the air movements within and a few meters above the surface fuels (Jenkins *et al.*, 2001). The spread of smaller fires occurs primarily through radiative and conductive heat transfer, with some additional heat transfer by convection at the fire front (Jenkins *et al.*, 2001).

Fires that burn at high frontal rates of spread usually burn large areas and occur when there has been a sustained period of dry, warm weather before the fire and dry, warm, or hot and windy weather during the fire (e.g., Quintilio *et al.*, 1977; Flannigan and Harrington, 1987; Hirsch and Flannigan, 1990). These weather conditions are associated with a critical synoptic weather pattern of upper level ridge build-up and breakdown (e.g., Nimchuk, 1983; Johnson and Wowchuk, 1993; Skinner *et al.*, 1999; Skinner *et al.*, 2002; Fauria and Johnson, 2006). The presence of an upper level ridge is associated with warm, dry conditions that dry out surface and canopy fuels; the breakdown of the ridge is often accompanied by increased lightning activity and strong and gusty surface winds [see Flannigan and Wotton (2001) for a more detailed discussion of the relationship of climate, weather, and area burned]. The frontal spread of these large, wind-driven fires is controlled by both radiative and forced convective heat transfer from the flames and their plumes to the unburned fuels ahead of the flames. The sides or flanks of these fires are partly driven by radiation from the smaller flames, made upright or curving inwards by the convection column (Catchpole, 2002). Thus, flank fire spread rates may increase with wind speed, but the wind has much less influence on flank fire spread compared to frontal fire spread. At the back of a fire the flames typically bend away from the unburned fuels, and the fire spreads slowly by radiation and conduction, at more or less the same spread rate as in zero wind speed (Catchpole, 2002).

Fire Shape The final shape of a burn may depend on many factors: fire intensity, wind speed, fuel types, fuel homogeneity and continuity, and topography (Catchpole, 2002). The shape of small fires appears to be controlled primarily by the fuels through which they burn (e.g., Díaz-Delgado *et al.*, 2004); small fires tend to form round shapes (Eberhart and Woodard, 1987). Larger fires, particularly those in homogeneous fuel, terrain, and environmental conditions, form roughly elliptical shapes because the frontal fire spread rate is much faster than the flank and back spread rates (Catchpole, 2002). Changes in wind direction and terrain over the duration of a fire cause many fires to have irregular shapes (Catchpole, 2002).

Unburned Residual Islands Many studies have reported the presence of unburned patches of trees (i.e., unburned residual islands) within the perimeter of burns. Attempts to explain their presence often are based on environmental variation, including spatial differences in edaphic water content and topography, changes in wind direction or speed, and the spatial location of humid, barren, or rocky areas, water bodies, and other landscape elements, resulting in the spatial discontinuity of fuels. Unfortunately, these explanations lack a process-based model. More recent work on coupled atmosphere-fire modeling (e.g., Clark *et al.*, 1996; Jenkins *et al.*, 2001; Jenkins, 2002; Coen, 2005) suggests that the presence of unburned residual islands in large fires might be better explained by the turbulent, coherent structure in a fire. Unfortunately, our understanding of this is only in the beginning stages (but readers should consult the recent literature).

A few studies have shown that the presence of residual islands varies with area burned (and thus fire behavior), with smaller area burned fires (e.g., 0.2–2 km^2) having few or no residuals (e.g., Eberhart and Woodard, 1987; Johnson *et al.*, 2003; Díaz-Delgado *et al.*, 2004). Unfortunately, there seems to be little agreement on the patterns of the number and area of residuals within larger burns. For example, Eberhart and Woodard (1987) examined the patterns of unburned residuals within 69 burns (0.2–200 km^2) in the boreal forest of Alberta. They found that the number of residuals per 1 km^2 of burn area was highest in the intermediate fire size classes (2.01–20 km^2) and lowest in both the smallest and largest size classes (0.41–0.2 km^2 and 20–200 km^2, respectively); the median area within residuals increased significantly as burn size increased. In another study of residuals within 445 burns (6.0–17.5 km^2) in *Pinus-Quercus*–dominated forests of northeastern Spain, Díaz-Delgado *et al.* (2004) found that both the number and area of unburned residuals increased significantly with burn size. Finally, Johnson *et al.* (2003) examined the unburned residuals in 27 burns (0.02–110 km^2) in the boreal forest of Saskatchewan and found that the number of unburned islands per unit area of burn decreased with burn area, whereas the amount of unburned forest per unit area of burn remained constant with increasing burn area. Some of the conflicting results of these studies can be explained by differences in the sizes of burns examined. The burn sizes examined by Díaz-Delgado *et al.* (2004) were within the intermediate range examined by the other two studies. Also, the different methods of these studies make comparisons difficult. Clearly, we need a better understanding of the processes that give rise to these patterns before we can draw meaningful conclusions about the number and sizes of residuals within burns.

As mentioned at the beginning of this section, area burned, fire shape, and the number and area of unburned residuals within burns have implications for some tree populations whose seeds are unavailable from within a burn after fire. The post-fire recruitment of these populations requires that seed be dispersed from live reproductive trees around the edge of the burn or from unburned residuals within the burn perimeter. By knowing the area burned, shape of a fire, and number and area of unburned residuals within a burn, one can determine the distribution of distances from the burn edge into the burn that seeds must travel in order to recruit back onto the burned substrate. This is discussed in more detail later in this chapter.

We've now discussed the principal wildfire processes and characteristics that affect individual trees and the tree population processes of recruitment and mortality. Our goal has been to explain some of the details that one should know when trying to understand how fires cause their ecological effects. Next, we discuss local population recruitment and mortality processes and their connections to specific wildfire processes and characteristics.

LOCAL POPULATIONS AND PROCESSES

In this section, we first define local populations and their birth cohorts, and then describe what is known about the recruitment and mortality of birth cohorts in a variety of ecosystems. We describe the causal connections between specific wildfire processes and characteristics and tree population processes where these connections are well understood. At the end of the section we discuss some recent work that gives the causal connections between heat transfer processes in low-intensity surface fires and the mortality of individual trees.

Local Populations and Their Cohorts

A local population can be defined as a group of individuals of the same species living adjacent to each other and with relatively homogeneous recruitment and mortality schedules (Ebert, 1999). If recruitment and mortality schedules are not homogeneous, then the local population is subdivided into birth cohorts. In tree populations initiated following fires, there are often two types of cohorts: the immediate post-fire cohort and the understorey cohort (Johnson and Fryer, 1989).

Local Population Recruitment

We define a recruit as an individual that germinates from an abscised seed or sprouts from surviving vegetative structures and survives through its first year. In this section, we discuss the recruitment of post-fire cohorts and understorey cohorts in turn.

Post-Fire Cohort Recruitment

A post-fire cohort consists of individuals that recruit within the first few years after a fire. An understanding of the early post-fire recruitment period of some tree populations is essential to understanding the future composition of many open (i.e., savannas and woodlands) and closed (i.e., forest) ecosystems because for many tree populations some of the highest mortality rates occur within the first year of seed germination or within the first few years of the onset of sprouting (e.g., Chrosciewicz, 1974; Thanos *et al.*, 1989; Thanos and Marcou, 1991; Greene and Johnson, 1999; Charron and Greene, 2002; Borchert *et al.*, 2003).

The post-fire recruitment period is relatively short partly because fires cause only short-term changes in environmental conditions that facilitate recruitment. Specifically, in both low- and high-intensity fires, flaming combustion kills and consumes the aboveground portion of live grasses, forbs, and shrubs, and consumes the organic litter layer. Heat transfer through flaming combustion in low-intensity fires may also partially or completely kill some canopy trees. High-intensity active crown fires will kill canopy trees and consume their small-diameter canopy components. The consumption of litter, ground vegetation, and tree canopy components results in a short period of reduced competition for soil resources (i.e., moisture, nutrients) and light, all of which facilitate the post-fire recruitment of both sexually and asexually reproducing tree populations. Other fire-caused, ecosystem-specific environmental changes that facilitate the early post-fire recruitment of sexually and asexually reproducing tree populations are explained below.

Recruitment of Sexually Reproducing Tree Populations

Conifer-dominated Forests in Cool to Cold Climates. In conifer-dominated forests with cool to cold climates (e.g., boreal and subalpine forests, Douglas fir forests of the Pacific Northwest), the consumption of duff by smoldering combustion also facilitates early post-fire cohort recruitment of sexually reproducing tree populations. In fact, several studies have found that the most favorable seedbeds for tree seed germination and germinant survival in these forests are thin humus or exposed mineral soil (e.g., St. Pierre *et al.*, 1992; Duchesne and Sirois, 1995; Charron and Greene, 2002). There are higher recruitment rates and lower mortality rates on these seedbeds, primarily because their higher bulk density means they are less subject to drying by wind and warm temperatures and therefore they retain moisture more effectively than the more porous litter and fermentation layers (Van Wagner, 1987). For seeds resting on a litter or fermentation layer, the distance to the humus or mineral soil below is lengthened, making it more difficult for a germinant's radicle to reach the moister humus or mineral soil (Williams *et al.*, 1990; Caccia and Ballaré, 1998; Charron and Greene, 2002). This is particularly true for the small-seeded tree populations [e.g., *Picea glauca* (Moench) Voss, *Pinus banksiana*] common in cool coniferous forests. The length of the hypocotyl is proportional to the mass of the seed, and thus germinant size is dependent on seed size for the first few weeks, until the deployed first true leaves replace the endosperm and cotyledons as the primary source of carbohydrates (Grime and Jeffrey, 1965; Charron, 1998). The germinants of seeds that fall through the litter layer also have low survival rates because the litter mechanically restricts the upward growth of the hypocotyl (Caccia and Ballaré, 1998; Bai *et al.*, 2000), especially when the litter consists of organic material with a large surface area, e.g., *Quercus prinus* L. leaves (Williams *et al.*, 1990) and *Populus tremuloides* Michx. leaves (e.g., Charron and Greene, 2002).

In conifer-dominated forests, the amount of duff consumed is one of the most important factors controlling early post-fire recruitment of sexually reproducing tree populations. This knowledge allows us to predict with reasonable certainty where on the ground within a burn we will find relatively high densities of recruits. Our reasonably good understanding of how smoldering combustion processes consume duff has allowed us to identify and determine the role of the variables (i.e., bulk density, moisture content, and depth) that control the amount of duff consumed. With a better understanding of how these variables change within and across local populations, we could make predictions about recruitment patterns at both the local population and landscape levels. This was demonstrated by Miyanishi and Johnson (2002) for *Pinus banksiana* and *Picea mariana* (Mill.) BSP–dominated stands in central Saskatchewan. They used an understanding of smoldering combustion processes to explain observed patterns of duff consumption within stands, among stands within fires, and between different fires. They found that the moisture content of duff and its interaction with duff depth were the two most important determinants of duff consumption patterns within stands. Within a stand, duff was consumed in patches, with relatively more duff consumed under trees alive during the fire where duff is relatively drier because of the interception of precipitation by tree crowns. Between stands, larger patches of consumed duff were found in *Picea*-dominated stands compared to *Pinus*-dominated stands, where duff was thinner. Across stands, there was a significantly greater area of duff removed by smoldering in a fire that occurred when the duff was drier compared to a fire that occurred under wetter duff conditions; such differences in duff moisture across stands were caused by differences in both pre-fire weather conditions and hill slope hydrology. The between- and across-stand patterns were partially explained by the positive relationship between duff depth and moisture content in the propagation of smoldering combustion (Fig. 1). The relationship indicates that, at low moisture contents, smoldering can be propagated in even very thin duff, whereas at higher moisture contents, a thicker duff depth is required to propagate smoldering. Or conversely, in thin duff a lower moisture content is required to propagate smoldering than in thicker duff. Miyanishi and Johnson (2002) also suggested patterns of duff consumption at a landscape level based on hill slope geomorphology (see later in this chapter for more detail on how geomorphic processes affect the distribution of tree populations). Briefly, in their study area, *Pinus banksiana*–dominated stands occupy the tops and middle of glaciofluvial hill slopes whereas *Picea mariana*–dominated stands occupy the bottoms of these hill slopes. This spatial distribution has been attributed to a moisture-nutrient gradient (Bridge and Johnson, 2000). Furthermore, the bottom-slope *Picea* stands have thicker duff than the top-slope *Pinus* stands. Thus, hill slope geomorphic processes determine the moisture-nutrient gradient which, in turn, affects the depth of duff within stands. Consequently, they conclude that the hill slope is the fundamental unit for understanding patterns of tree recruitment, not only because it creates the moisture-nutrient gradient, and hence the species composition gradient, but also because it determines the patterns of duff consumption which, in turn, determine tree seedling recruitment patterns. Therefore, to better predict patterns of duff consumption at the local population and landscape levels, we need better distributed hydrological models that predict patterns of duff moisture at varying scales.

The length of the post-fire recruitment period of some sexually reproducing tree populations in cool coniferous forests has been examined by using permanent plots. Studies from the boreal forest of Canada and conifer forests of Yellowstone National Park have found that the recruitment period of three conifer tree populations with serotinous cones, *Picea mariana*, *Pinus banksiana*, and *Pinus contorta* var. latifolia, occurs within 4 years after fire. Recruitment of *P. mariana* and *P. banksiana* peaks in either the first and second post-fire summers (Greene *et al.*, 2004) or the second and third post-fire summers (St. Pierre *et al.*, 1992; Charron and Greene, 2002). Recruitment of *P. contorta* in Yellowstone National Park also peaks in the first and second post-fire summers (Turner *et al.*, 1997).

In a longer-term (20–30 year) permanent plot study in the boreal forest of Yukon and Alaska, Johnstone *et al.* (2004) examined post-fire recruitment of *P. contorta*, *P. mariana*, and *Populus tremuloides*. They found that recruitment of all three species occurred within 10 years after fire, with most recruitment occurring 3 to 7 years after fire. These results are consistent with retrospective studies of the same species (e.g., Johnson *et al.*, 1994; Lavoie and Sirois, 1998; Gutsell and Johnson, 2002).

The length of the recruitment period for tree populations with serotinous cones appears to depend primarily on seed availability (Greene and Johnson, 1999; Charron and Greene, 2002; Greene *et al.*, 2004). The accrual of litter and aggradation of mosses may also play a role; however, we could find no studies that examined this directly after fire. Seed availability from serotinous cones involves both cone opening and seed abscission processes. In serotinous cones, there is resin between cone scales that holds the cones closed until they are heated, usually by fire. When the resin bonds are broken (i.e., the resin melts), the cone scales reflex, allowing seeds to be released. Thus, cone opening is a two-part process: (1) the resin bonds weaken and break, and (2) the cone scales reflex away from the cone axis. Johnson and Gutsell (1993) examined the heat transfer mechanisms that cause the resin bonds to weaken and break. They used a lumped capacitance model to show that the heat flux into a cone depends on a cone's thermal properties (e.g., convective surface area, volume, density, and specific heat capacity), the duration of heating, and the temperature in the flame or its plume to which the cones are exposed. From this work and previous studies (e.g., Beaufait, 1960), it is clear that the resin bonds are broken soon (seconds to minutes) after a fire passes, depending on the fire's intensity and rate of spread. The reflex mechanism of cone scales, which is affected primarily by cone moisture, appears to take relatively longer (hours to days), although we know of no studies that have documented this. Studies that have examined seed abscission schedules for some serotinous cone species indicate that seeds are released over a much longer period (days to a few years). For example, Greene and Johnson (1999) examined the seed abscission schedule for the serotinous cones of *P. mariana* and *P. banksiana*. They found that 35% of seeds remained in the cones of standing burnt *P. mariana* trees 2 years after fire and only 6% remained 4 years after fire. Only 3% and 0.8% of *P. banksiana* seeds remained in cones 2 and 4 years after fire, respectively. Similarly, Greene *et al.* (2004) observed that only approximately 2% to 5% of filled seeds of *P. mariana* and *P. banksiana* remained in cones 5 years after fire.

Seed availability has been shown to be a limiting factor in the post-fire recruitment of tree populations whose cones open upon maturation [e.g., *P. glauca*, *Pseudotsuga menziesii* (Mirbel) Franco; Taylor and Aarssen, 1989; Peters *et al.*, 2005]. For example, a study in the boreal forest of Alberta found significantly higher densities of *P. glauca* in stands initiated after fires that coincided with large seed crops (i.e., mast years) than after fires that coincided with years of low seed crops (Peters *et al.*, 2005).

Savannas, Woodlands, and Forests in Warm, Dry Climates. In drier and warmer savanna, woodland, and forest ecosystems (e.g., ponderosa pine savannas, Coulter pine forests of California, and Mediterranean pine forests), high recruitment of post-fire cohorts occurs rapidly after fire and is facilitated by the consumption of ground vegetation and the organic litter layer by flaming combustion. In many savanna ecosystems, the removal of the aboveground portion of ground vegetation, particularly grasses, is critical to the recruitment and survival of some tree populations (e.g., *P. ponderosa*). Studies of seedling-grass competition show that initially after fire *P. ponderosa* recruitment is high but mortality increases rapidly because grass roots quickly dominate the upper soil layers and compete with trees for moisture (Pearson, 1942; Kolb and Robberecht, 1996). However, the brief post-fire period of reduced competition with grasses gives seedlings enough of a competitive advantage to increase their survival.

Unlike in cooler conifer-dominated forests, the consumption of duff by smoldering combustion does not appear to be an important fire process limiting tree recruitment in warmer and drier savannas, woodlands, and forests. In fact, the mineral soil in these ecosystems is typically overlain by only a litter layer, which is consumed and converted to ash by flaming combustion (C. Thanos, personal communication). Several studies have shown that a litter layer of scorched needles that accumulates after a fire facilitates tree recruitment in some warm, dry ecosystems, such as *P. brutia* Ten. forests of Greece (Thanos *et al.*, 1989), *P. pungens* Lamb. growing on exposed ridge tops in the southern Appalachians (Williams *et al.*, 1990), and *P. ponderosa* Dougl. ex Laws. forests in South Dakota (Bonnet *et al.*, 2005), but see Grace and Platt (1995). The accumulation of scorched needles reduces the variation in subsurface soil temperature and increases soil moisture retention, both of which result in greater germinant survival rates. The post-fire availability of scorched needles clearly depends on the intensity of the fire. In high-intensity active crown fires, the small-diameter needles in the canopy are consumed in the flame, whereas in low-intensity surface fires, needles may only be scorched (killed) by heat transfer from the plume but not consumed.

In at least some of these warm, dry ecosystems, two additional factors appear to facilitate rapid post-fire recruitment: the availability of a large crop of seeds released after serotinous cones are opened by high-intensity fire and a post-fire wet season favorable for recruitment (e.g., Thanos *et al.*, 1989; Moravec, 1990; Thanos and Marcou, 1991; Thanos *et al.*, 1996; Herranz *et al.*, 1997; Spanos *et al.*, 2000; Borchert *et al.*, 2003). For example, recruitment of two Mediterranean pine species, *P. brutia* and *P. halepensis* Mill., and a coastal California pine species, *P. coulteri* (D. Don), appears to take place almost exclusively during the first post-fire wet season, after high-intensity fire opens their serotinous cones and strong winds blow their seeds onto the bare mineral soil (Thanos *et al.*, 1989, 1996; Thanos and Marcou, 1991; Borchert *et al.*, 2003).

Several studies in warm, dry ecosystems have also found that animal seed caching is an important factor controlling the post-fire recruitment of some tree populations (e.g., Acherar *et al.*, 1984; Castro *et al.*, 1999; Borchert *et al.*, 2003; Ordóñez and Retana, 2004). For example, in the dry *P. coulteri*–chaparral forests of coastal California, Borchert *et al.* (2003) found that seed caching by rodents and birds facilitated the rapid post-fire recruitment of *P. coulteri*. Both animals collect and cache all or nearly all available tree seeds in a short period after fire; seeds that are buried but not recovered then germinate during the first post-fire wet season. Seeds from caches are more likely to germinate than noncached seeds (e.g., Vander Wall, 1992).

In summary, warmer and drier savanna, woodland, and forest ecosystems appear to have many variables that control the post-fire recruitment of sexually reproducing tree populations: the consumption of ground vegetation and litter, the post-fire accumulation of scorched needle litter, the availability of a large crop of seeds from serotinous cones, a post-fire wet season, and animal seed caching. Knowledge of these variables allows us to predict with some certainty the specific situations that would result in high post-fire recruitment. Unfortunately, it is unclear from the studies described in this section whether these variables are ubiquitous across a burn or whether they vary within and across local populations. Furthermore, few of the studies make any causal connections between specific processes or characteristics and the observed recruitment patterns. This makes it difficult to generalize and make meaningful predictions about recruitment patterns within and across these local populations. For example, all studies stated that they examined recruitment after high-intensity crown fires. However, some of them noted the presence of scorched needle litter after fire, indicating that in some areas the fires must have been of low intensity. How much variation in fire intensity occurs within these ecosystems, and does this variation affect recruitment patterns? For tree populations whose recruitment is affected by animal seed caching, what effect do the population dynamics of each animal species have on post-fire recruitment?

Recruitment of Asexually Reproducing Tree Populations Many tree populations reproduce asexually after fire by sprouting from the basal buds (lignotubers, rhizomes, roots, or burls) of trees whose aboveground tissues are killed by fire. The mortality of aboveground tissues induces suckering by hormonal control (auxin production in the crown and cytokinin production in the roots). Strong apical dominance is exerted by a tree crown over the root system through the downward movement of auxins (Farmer, 1962; Steneker, 1974). Once the crown is killed, apical dominance is lost and sprouting is initiated, provided cytokinins have been produced in the roots (Schier, 1976).

In cool coniferous forests, the post-fire recruitment of tree populations that sprout from basal buds is facilitated by the consumption of litter, ground vegetation, and tree canopy components by flaming combustion, which increases light availability and soil temperatures (Maini and Horton, 1966; Steneker, 1974; Greene *et al.*, 2004). Sprouting is also initiated by the consumption of duff by smoldering combustion (e.g., Brown and Debyle, 1987; Schimmel and Granström, 1996) and depends on the characteristics of fire-affected trees. Relatively smaller and older trees appear to sprout less vigorously after fire (e.g., Sharma and Rikhari, 1997; Hoffman, 1998; Keeley *et al.*, 1998), apparently because they have a smaller bud bank and relatively less available carbohydrate reserves stored in their belowground tissues (e.g., Miyanishi and Kellman, 1986; Weber, 1990; Greene and Johnson, 1999).

We could find only two studies that used permanent plots to examine the length of the recruitment period of asexually reproducing tree populations. Brown and Debyle (1987) and Greene *et al.* (2004) examined post-fire recruitment of *Populus tremuloides* in Idaho and the boreal forest of Canada, respectively, and found that recruitment occurs primarily within 2 years after fire.

Most studies seem to agree that the key variables that control early post-fire recruitment of asexually reproducing tree populations are the consumption of duff by smoldering combustion and the amount of available carbohydrate reserves within fire-affected trees. Unfortunately, few studies have made causal connections between the fire processes and asexual recruitment processes, although there have been some attempts (e.g., Kaufmann, 1991; Schimmel and Granström, 1996; Sharma and

Rikhari, 1997; Broncana and Retana, 2004). For example, some studies have correlated fire severity to the number and size of sprouts (e.g., Kauffman, 1991; Sharma and Rikhari, 1997; Broncana and Retana, 2004). Unfortunately, because fire severity is defined in terms of an ecological effect (e.g., scorch height), the wildfire process causing the ecological effect (i.e., the number or size of sprouts that emerge) is not clearly identified. In addition, given that basal bud banks are located within the organic or mineral soil, an understanding of how the heat flux through smoldering combustion affects sprouting from basal buds is required. There have been a few studies of necrosis of roots and basal vascular cambium from soil heating and smoldering combustion (e.g., Swezy and Agee, 1991; Ryan and Frandsen, 1991). However, a heat flux model describing the mechanisms that either stimulate or inhibit sprouting from basal buds has yet to be developed.

As a final note on post-fire recruitment, we mentioned at the beginning of this section that an understanding of the early recruitment period of post-fire cohorts is essential to understanding the future composition of open and closed forested ecosystems because for many tree populations some of the highest mortality rates occur within the first year of seed germination or first few years of the onset of sprouting. Another reason is that individuals from post-fire cohorts are often the only ones to form the canopy. In some cooler coniferous forests, understorey cohorts have high mortality rates and slow height growth rates. The high mortality rates of understorey cohorts mean that few individuals live very long; their slow height growth rates mean that they have a low probability of making it into the canopy before the next high-intensity crown fire (e.g., Gutsell, 2001; Gutsell and Johnson, 2002). In drier and warmer savanna ecosystems, where frequent low-intensity fires occur between infrequent high-intensity fires, the individuals from cohorts that recruit after each low-intensity fire (for convenience we will refer to them as secondary post-fire cohorts) may reach the canopy of the post-fire cohorts that had initiated after high-intensity fire. Their elevation into the canopy can occur if their bark has increased in thickness and their buds have become elevated above the ground such that they avoid the most intense heat of the fire (see discussion earlier in chapter). Thus, the elevation of secondary post-fire cohorts into the canopy depends primarily on the length of time between low-intensity fires and the growth rates of individuals within secondary post-fire cohorts. Baker and Ehle (2001) suggested that *P. ponderosa* trees require a period of 50 years or more before they can survive low-intensity fire and make it into the canopy; Brown and Wu (2005) suggested they require a shorter period of 10 to 40 years.

Understorey Cohort Recruitment

The understorey cohort consists of individuals that recruit later than the post-fire cohort, when the post-fire environmental conditions change such that there is a significant change in cohort recruitment and mortality rates. Typically, conditions change to result in a decrease in the recruitment rate and/or increase in the mortality rate. However, conditions may improve for some species such that there is a significant increase in understorey cohort recruitment rates and decrease in their mortality rates. This seems to be the case for some plant populations (e.g., *Quercus agrifolia* Nee in oak savannas of California), particularly in warm, dry ecosystems, that require nurse plants before they can recruit (Callaway, 1995).

Tree populations in many conifer-dominated forests in cool climates appear to have understorey cohorts. They have relatively higher mortality rates than post-fire cohorts because of significant changes in environmental conditions. Specifically, there are rapid re-growth of ground vegetation (including post-fire cohort seedlings and saplings) and an accumulation of litter and duff on the forest floor, both of which can cause significant increases in germinant mortality rates (e.g., Williams *et al.*, 1990; Caccia and Ballaré, 1998; Charron and Greene, 2002). Retrospective studies in these forests that have carefully aged understorey tree stems (e.g., Johnson *et al.*, 1994; Gutsell and Johnson, 2002) have shown that the recruitment of some understorey cohorts is more or less continuous. However, few individuals within understorey cohorts live very long, indicating that these cohorts have high mortality rates. Static age distributions show a large number of young seedlings and saplings and much older canopy trees, with few or no trees of intermediate age (e.g., Johnson *et al.*, 1994; Gutsell and Johnson, 2002). Some tree populations in these forests do not have an understorey cohort, particularly those whose seeds are largely unavailable beyond the few years after fire (e.g., within serotinous cones) and/or are intolerant of shade (e.g., many *Pinus* populations).

In drier and warmer savannas, woodlands, and forests, it is unclear whether there are understorey cohorts. Several studies have shown that there is recruitment beyond the first few post-fire years (e.g., Keeley *et al.*, 1998; Ehle and Baker, 2003; Brown and Wu, 2005); however, it isn't clear whether individuals recruiting several years after fire have different recruitment or mortality rates compared to those that recruit in the first few post-fire years.

Sources of Seeds and Sprouts for Local Population Recruitment

After high-intensity crown fires, aerial seed banks (i.e., serotinous cones) of fire-killed trees within the burn provide a local source of seeds for post-fire cohorts. Local sources of sprouts would be the bud banks of trees whose aboveground tissues are killed by fire. Thus, the pre-fire community of local populations may more or less replace itself (e.g., Ne'eman, 1997; Greene and Johnson, 1999; Spanos *et al.*, 2000; Greene *et al.*, 2004).

Seeds will be deposited from aerial seed banks within a local population if trees in the fire-killed population have reached reproductive maturity and if the intensity and rate of spread of the fire are such that cones open but their seeds are not killed (Johnson and Gutsell, 1993). If seed is available from within the fire-killed local population, then these populations may not have a regional population structure (see discussion later in this chapter). However, if fire-killed trees are not mature, the intensity and rate of spread of the fire are such that the cones do not open (e.g., low fire intensity, high rate of spread), or seeds are killed by fire (e.g., high fire intensity, low rate of spread), then seed must come from reproductive trees with open cones and live seed from elsewhere in the burn. In these situations, tree populations with aerial seed banks would have a regional population structure.

Sprouts will emerge from the bud banks of asexually reproducing tree populations if the heat transfer from flaming combustion kills their aboveground tissues and consumes ground vegetation and litter, if their bud banks are not killed by heat transfer from smoldering combustion, and if fire-affected trees have adequate carbohydrate reserves in their belowground tissues. If sprouts emerge from within a fire-affected local population, then the population may not have a regional population structure. However, when asexual reproduction is not successful, recruitment of these populations depends on their ability to reproduce sexually. Many asexually reproducing tree populations are also capable of sexual reproduction (e.g., *Populus tremuloides*, *Betula papyrifera* Marsh.). However, the seeds of these populations typically are exposed (i.e., within cones that open at maturity or catkins) and are killed by fire. As with all tree populations whose seeds are exposed and killed by fire, post-fire recruitment depends on seeds being dispersed from reproductively mature live trees located around the burn edge. In these situations, tree populations would have a regional population structure.

Local sources of seeds for understorey cohorts will come from within a local population once post-fire cohorts have reached reproductive maturity. Until then, seeds must come from mature trees located outside of the local population. Local sources of sprouts for understorey cohorts may emerge from trees whose aboveground tissues have died (e.g., windthrow).

After low-intensity surface fires, local sources of seeds for secondary post-fire cohorts will come from canopy trees whose seeds survived fire. In these situations, tree populations may not have a regional population structure. However, if some canopy trees and their seeds or bud banks are killed by fire (e.g., passive crown fire), these populations may have a regional population structure. This will depend on the spatial extent of canopy tree mortality, which is a function of the intensity variations within a fire.

Local Population Mortality

Several studies have examined the mortality rates of post-fire cohorts, particularly within the first few years after high-intensity fire. Most have shown that the mortality is very high within the first year of germination and the first few years of the onset of sprouting (e.g., Chrosciewicz, 1974; Thanos *et al.*, 1989; Thanos and Marcou, 1991; Greene and Johnson, 1999; Charron and Greene, 2002; Borchert *et al.*, 2003). For example, Charron and Greene (2002) conducted sowing experiments for *P. banksiana*, *P. mariana*, and *P. glauca* seeds on different seedbeds in the boreal forest of Saskatchewan. They found that the period from the time of seed abscission through the first summer accounted for 63% to 100% of the mortality observed over a 3-year period. On mineral soil and humus seedbeds, roughly 50% to 75% of *P. banksiana*, *P. mariana*, and *P. glauca* seeds died in that period. On thick organic material, mortality was significantly higher (>90%). Similarly, in a warm, dry Mediterranean pine forest, Thanos *et al.* (1996) found that the highest period of mortality for *Pinus halepensis* was in the first post-fire summer, when nearly 35% of germinants died. From the end of the first summer to the end of the second post-fire year, 25% of the remaining seedlings died, with few dying thereafter.

We could find few studies that directly measured the mortality of post-fire cohorts beyond the first few years after fire. One study by Johnstone *et al.* (2004) examined the mortality of *P. contorta*, *P. mariana*, and *Populus tremuloides* in the boreal forest of Alaska and northeastern Yukon through repeated observations over a 20- to 40-year post-fire period. They found that *P. mariana* experienced very little mortality 10 to 30 years after fire, whereas *P. contorta* and *Populus tremuloides* experienced a significant increase in mortality 10 to 19 years after fire. The period of increased mortality for *P. contorta* and *Populus tremuloides* was attributed to intraspecific density-dependent thinning. The results for *P. contorta* are similar to that found by Johnson and Fryer (1989) for *P. contorta* in the Canadian Rockies. Using stand reconstructions, they found that the onset of thinning mortality begins in the second or third decade after fire. However, the results of Johnstone *et al.* (2004) for *Populus tremuloides* differed from those of two other studies. Both Zasada *et al.* (1992) and Greene and Johnson (1999) found that density-dependent thinning of *Populus tremuloides* begins almost immediately after the onset of recruitment. Johnstone *et al.* attributed these differences to the fact that the recruits in their study originated from seeds, whereas the recruits in Zasada *et al.* (1992) and Greene and Johnson (1999) were asexual in origin. In a study of a dry *P. ponderosa* savanna where frequent low-intensity surface fires occur between infrequent high-intensity fires, Ehle and Baker (2003) showed that most *P. ponderosa* mortality is associated with low-intensity fires.

In the next section, we discuss the mortality of individual trees. We discuss some recent work that shows how low-intensity fires actually kill individual trees, including the heat transfer processes into a tree bole and tree canopy.

Mortality of Individual Trees High-intensity active crown fires kill canopy trees, at least their aboveground tissues, and consume their small-diameter canopy components. In contrast, canopy trees often survive low-intensity fires but may be partially killed. Partial tree mortality may result if the heat from flaming combustion kills some of the bole or crown tissues and/or smoldering combustion kills some of the root tissues. Therefore, the amount of post-fire tree mortality from low-intensity surface fires depends on the characteristics of the fire (e.g., intensity, rate of spread, duff consumption) and of the trees (e.g., basal bark thickness, vertical bud distribution, foliage structure). The mortality of tree components depends on each component's physical dimensions and position relative to heat sources. Relative to other tree components, a tree's trunk (bole) is large in diameter and has a protective layer of bark; thus, it has considerable resistance to heating by a flame. The live tissues within a tree trunk, called vascular cambium, that are susceptible to necrosis from heating are located just beneath the bark. Crown components (branches and vegetative buds) are relatively small in diameter but are high above the ground, and thus are farther away from the hottest part of the flame and buoyant plume. However, their small diameter means that the heat transfer required for necrosis of live tissues is much lower than that required for bole necrosis. After flaming combustion consumes the organic litter layer, tree roots remain insulated by duff.

However, depending on its bulk density, moisture content, and depth, the duff may be subsequently consumed by smoldering combustion. Unfortunately, there have been few studies on necrosis of tree roots and basal vascular cambium from soil heating and smoldering combustion; therefore, we do not discuss them further here.

Tree mortality can be predicted by linking heat transfer processes to tree mortality mechanisms. Because fires and trees vary in their characteristics, tree mortality predictions have been based on a series of couplings between the fire and individual trees. Peterson and Ryan (1986) present a flow chart of some of the factors and couplings between fire behavior and their effects on tree mortality. Unfortunately, much of the work on modeling of tree mortality up to this point has used logistic regression to identify the significant independent variables that predict tree mortality (e.g., Beverly and Martell, 2003; Hély *et al.*, 2003; McHugh and Kolb, 2003). Generally, independent variables are characteristics or surrogates of fire behavior (e.g., fire intensity, duff consumption) and fire effects (e.g., crown scorch, bark char) (Dickinson and Johnson, 2001). However, the relationships among independent variables are chosen without considering the processes by which fire causes tissue necrosis or tissue necrosis causes tree mortality. By not including processes, most current models of tree mortality lack generality and cannot be applied beyond the specific conditions in which they were based. Given these problems, our understanding of how fires partially or completely kill a tree has been poor.

Dickinson and Johnson (2001) summarize some of the ways heat transfer between trees and low-intensity surface fires have been modeled. For example, some studies have considered the effects of fires on tree carbon budgets (e.g., Ryan *et al.*, 1988; Harrington, 1993; Glitzenstein, 1995). Trees die when they are unable to obtain enough resources to maintain their normal functions following injury or stress (Waring, 1987). More recently, Michaletz (2005) has derived a process model of tree mortality based on heat transfer theory and tree allometry models. He uses a buoyant line source plume model to drive a transient conduction model for vascular cambium necrosis and a lumped capacitance heat transfer model for vegetative bud necrosis (for a detailed discussion of lumped capacitance heat transfer models, see Chapter 12). In Michaletz's model, vascular cambium and vegetative bud necroses are linked to tree mortality by using a sapwood area budget derived from tree allometry models. The resulting model is general and predictive, and it clearly defines the causal mechanisms of tree mortality. Model results are illustrated for *Populus tremuloides*, *P. glauca*, and *P. contorta*, but the generality of the model means that it can be easily applied to other tree populations.

Michaletz's (2005) model predicts tree mortality resulting from either vascular cambium mortality around the entire circumference of the base of a tree bole or from necrosis of all the vegetative buds in the tree crown. Tree mortality is quantified by M, the proportion of necrotic vegetative buds in the crown. It is assumed that basal girdling or necrosis of all vegetative buds will result in tree mortality ($M = 1$); otherwise there is necrosis of some proportion of vegetative buds ($M \leq 1$). M is given by

$$M = (N_n/N_t)^{1-U(x_n/x_{ba})} \tag{2}$$

where N_n is the number of necrotic vegetative buds, N_t is the total number of vegetative buds, x_n is the maximum depth of necrosis in the bole, and x_{ba} is the basal bark thickness. $U(x_n/x_{ba})$ is called the Heaviside step function and is defined as

$$0 \text{ if } x_n/x_{ba} < 1$$

$$1 \text{ if } x_n/x_{ba} \leq 1$$

In other words, when the depth of necrosis at the base of a tree is greater than the basal bark thickness ($x_n/x_{ba} = 1$), the bole is girdled and tree mortality occurs ($M = 1$). When the necrosis depth is less than the basal bark thickness ($x_n/x_{ba} < 1$), the bole is not girdled. Tree mortality through vegetative bud necrosis will occur when $N_n = N_t$.

The pipe model assumption of a constant ratio of sapwood area to leaf area SA:LA (Shinozaki *et al.*, 1964) is used along with the assumption of a constant ratio of leaf area to number of buds, N_{bu} (Michaletz and Johnson, unpublished data) so that it is assumed that the ratio $SA{:}N_{bu}$ is constant. By substituting sapwood area, SA, as a surrogate for the number of buds, N, the tree mortality model can be rewritten as

$$M = (1 - SA_n/SA_{lcb})^{1-U(x_n/x_{ba})} \tag{3}$$

where SA_n is the bole sapwood area at the maximum height of bud necrosis (Michaletz and Johnson, 2006) and SA_{lcb} is the bole sapwood area at the height of the live crown base. Therefore, the above equation identifies basal bark thickness and sapwood area taper (i.e., vertical bud distribution) as variables linking meristem necroses to tree mortality.

Sapwood area is an appropriate framework for modeling tree form because resource distribution is considered a fundamental process driving allometry (e.g., Shinozaki *et al.*, 1964; West *et al.*, 1999; Dreyer and Puzio, 2001; Banavar *et al.*, 2002; Niklas and Spatz, 2004). Well-established allometric relationships describe the mutual scaling of basal bark thickness, x_n, and sapwood area taper, SA_n/SA_{lcb}. Basal bark thickness, x, scales linearly with basal bole diameter, D, as

$$x = aD + b \tag{4}$$

where the slopes a and intercepts b are species-specific constants. Bole sapwood area SA scales with bole diameter according to

$$SA = CD^m \tag{5}$$

where the normalization constants, C, and scaling exponent, m, are species-specific constants. Bole diameter at any given height is related to the diameter at breast height *(DBH)* by a bole taper model. Bole taper, d_z/D_i, is given by

$$d_z/\mathrm{D_i} = (Z-z/Z-1.3)^{1/(expA)} \tag{6}$$

where d_z is the diameter inside bark at height z, D_i is the diameter inside bark at 1.3 m, and Z is the bole height (Newnham, 1991). The species-specific scaling exponent, *1/(expA)*, varies continuously along Z to account for height-dependent changes in bole form (from neiloidal to paraboloidal to conical). Tree height scales with *DBH* according to

$$z = 1.3 + C(DBH^m) \tag{7}$$

where the normalization constant, C, and scaling exponent, m, are species-specific.

Model results relate the DBH, fire intensity, and the proportion of necrotic vegetative buds. Model simulations across a wide range of fire intensities indicate that tree mortality is dependent on the scaling relationships between basal bark thickness and the vertical bud distribution. Bole girdling is the primary mechanism of tree mortality for all tree populations; vegetative bud necrosis was a mortality mechanism in a limited number of cases. As fire intensity increases, the DBH of trees that can be girdled increases for all species.

We've discussed the local population processes of recruitment and mortality and the mechanisms of individual tree mortality. We have coupled local population processes to specific fire processes and characteristics, where these couplings are well understood. As readers will have noticed, many areas require more work before we will have a complete understanding of the couplings between fire and local population processes. In the next section, we discuss regional tree populations, including the regional processes of geomorphology, large-scale mortality by wildfire, and seed dispersal.

REGIONAL POPULATIONS AND PROCESSES

In ecosystems affected by wildfires, some tree populations have a regional population structure. A regional population is a group of local populations linked together by seed dispersal and in which some of the local populations suffer extinction by disturbance. In other words, the regional population has mortality processes (local extinction by disturbance) and recruitment processes (seed dispersal). The size of a regional population in fire-affected ecosystems will vary among tree populations, depending on the landscape distribution of local populations, the size and frequency of fires, and the seed dispersal curve of each population (e.g., Hanski and Gilpin, 1991; Eriksson, 1996). In this section, we first discuss the role of geomorphic processes in the landscape distribution of local populations. Then we discuss the frequency and size distribution of fires and the seed dispersal processes that link local populations into a regional population.

Geomorphology and the Landscape Distribution of Local Populations

The size of a regional population depends in part on the landscape distribution of local populations. As many studies have shown, plant populations are not randomly distributed over the landscape but respond to heterogeneity in environmental conditions and resources. This heterogeneity affects the distribution and abundance of plant populations because each population is able to recruit, grow, and reproduce only within a specific range of conditions and resources. Several studies have demonstrated that soil moisture and nutrients are the two most important resources explaining the distribution and abundance of plant populations (e.g., Chabot and Mooney, 1985; Barbour and Billings, 1988).

The distribution of soil moisture and nutrients has distinctive universal patterns that are governed by geomorphic processes. All landscapes consist of hill slopes, which are bounded by ridgelines and drainage channels, i.e., stream courses (Selby, 1982). Hill slopes form the fundamental upland hydrological gradient, and stream courses form the lowland hydrological gradient. Hill slope moisture is determined by regional precipitation patterns, contributing area, slope steepness, and soil transmissivity. Precipitation determines the amount of moisture added to soil. The contributing area of any given point on a hill slope is the area of excess saturation flow; it determines how much water will flow through a point (Selby, 1982). Slope steepness determines the moisture available at any point because it influences the rate at which water flows down the hill slope. Water flows more rapidly down steep slopes; thus, for any given slope position, steep slopes will have lower values of soil moisture than gentler slopes. Soil transmissivity is a product of the depth of the zone through which water flows and hydraulic conductivity (i.e., ease with which soil pores permit water movement), which depends on the size and configuration of soil pores. Soils with larger pores have higher transmissivity and thus relatively lower moistures than those with smaller pores.

The downward flow of water from ridgelines to stream courses also results in the downward flow of nutrients. Therefore, hill slopes create the gradients of both moisture and nutrients that explain the distribution of many, if not most, plant populations. Bridge and Johnson (2000) provide an example of two such hill slope gradients on glaciofluvial and glacial till substrates in the southern mixed-wood boreal forest of Saskatchewan (Fig. 3). Allometric equations describe each type of hill slope and the relationship between tree species position on the moisture and nutrient gradient and their distances down slope from the ridgeline.

Given an understanding of the general pattern of moisture and nutrients down hill slopes and the responses of tree populations to these gradients, it should be possible to predict a general landscape distribution of tree populations (e.g., Bridge and Johnson, 2000) and species richness (e.g., Chipman and Johnson, 2002). One way this has been accomplished is by calculating an index of hill slope moisture, called a wetness index, using a digital elevation model and surficial geology data in a hydrological model. The results of such studies have shown that most landscapes have a similar distribution of wetness values (e.g., Sivapalan *et al.*, 1987). The driest

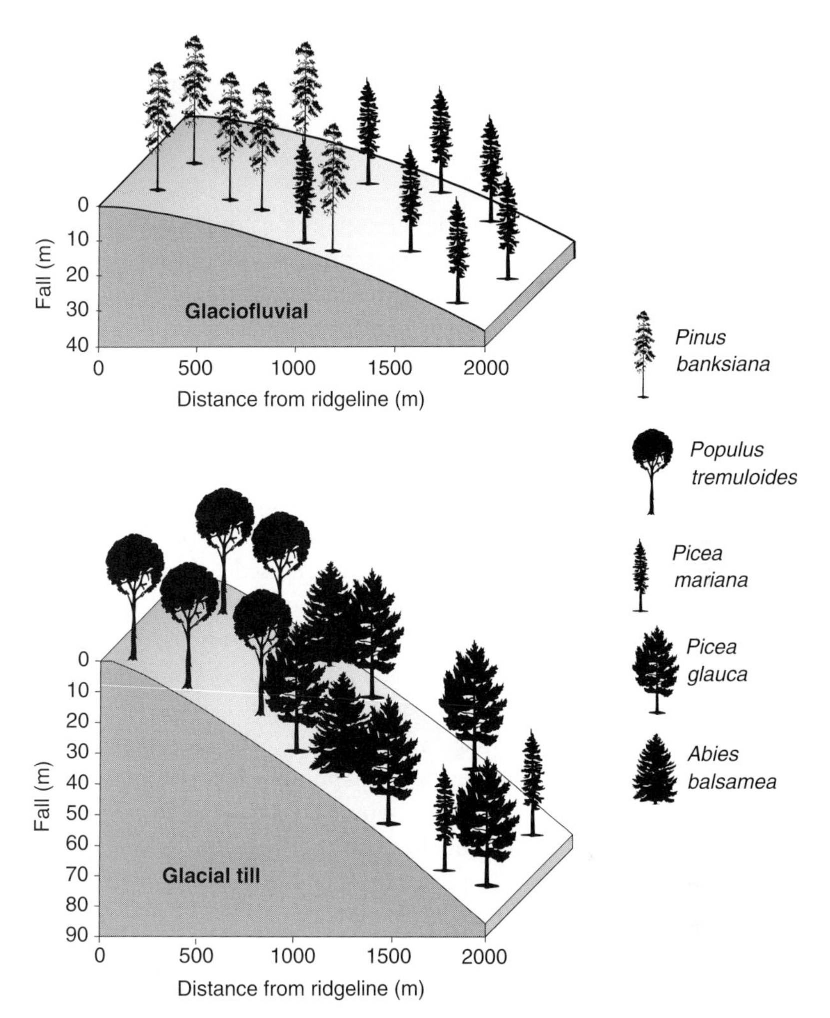

FIG. 3 Landscape patterns of canopy tree species composition in the mixedwood boreal forest in central Saskatchewan, based on the relationship between the stand position on the moisture and nutrient gradients and the stand distances from the ridgeline. (Adapted from Bridge and Johnson [2000], with permission from Opulus Press.)

(along ridgelines and hill tops) and wettest (along stream courses and bottom slopes) parts of the landscape are the least abundant on the landscape, whereas the areas just below the ridgelines are the most abundant on the landscape. Therefore, tree populations that tolerate and are competitive within the range of wetness values that are the most abundant on the landscape should be the most widespread in the area. Clearly, what is needed next is to apply this model in a variety of ecosystems.

Regional Population Mortality

An understanding of the size of regional populations requires that we know not only the spatial distribution of local populations but also the frequency with which fires occur and the sizes of fires that affect these populations. In this section, we discuss the frequency and size distribution of fires.

Fire Frequency Distribution

The mortality of local populations has been examined in studies of fire frequency (Johnson and Gutsell, 1994). Fire frequency is an estimate of the probability distribution of survival or mortality from fire in a regional population. Traditionally, in areas subject to high-intensity fires that kill all (or nearly all) canopy trees, time-since-fire data are used to construct the survivorship distribution, from which the fire frequency can be statistically estimated (Reed, 1998; 2000). In areas subject to low-intensity surface fires that kill only some canopy trees, fire interval data are used to estimate the fire frequency. However, the survivorship distribution must have other terms incorporated into it in order to account for the fact that only some parts (i.e., canopy trees) of the regional population are killed by fire. Unfortunately, this area of study remains largely unexplored, although some modest headway has been made recently (Reed and Johnson, 2004).

Many fire frequency studies have been conducted in closed canopied forests subject to crown fires (e.g., Masters, 1990; Bergeron and Archambault, 1993; Yarie, 1998; Weir *et al.*, 2000) and open ecosystems (i.e., savannas) subject to surface fires (e.g., Grissino-Mayer, 1999; Heyerdahl *et al.*, 2001). Most have found that the fire frequency has changed more than once over the past few hundred years and that these changes are a result of large-scale changes in climate. Between changes in frequency, the distributions also suggest that as forests get older there is no change in their probability of burning. Many fire frequency studies in closed forests have found that fires occur within the lifespan of post-fire cohorts (e.g., Masters, 1990; Bergeron and Archambault, 1993; Weir *et al.*, 2000), suggesting that understorey cohorts do not have time to replace post-fire cohorts in the canopy. Therefore, post-fire cohorts are the only ones to form the canopy and produce seeds for the next generation.

The fire frequency determines the age distribution of stands across the landscape, which determines whether post-fire cohorts have reached the age (or size) at which they reproduce. This clearly affects whether seeds are available to be dispersed after fire, and has important implications for the post-fire recruitment of tree populations. This is discussed in more detail later in this chapter.

Fire Size Distribution

The sizes of fires that occur in an area over time have been examined by several studies, with data presented in one of two ways: (1) as a probability density distribution (frequency histogram) of fire sizes or (2) as a proportion of total area burned by each fire size. Several probability density distributions have been constructed in different ecosystems, and most show that there are a large number of small fires and a decreasing number of fires in larger size classes (e.g., Johnson *et al.*, 1998; Malamud *et al.*, 1998; Reed and McKelvey, 2002; Díaz-Delgado *et al.*, 2004). Interestingly, despite the disproportionately large number of small fires, the proportion of total area burned by different fire sizes shows that it is the few large fires that account for most of the area burned (e.g., Strauss, 1989; Johnson *et al.*, 1998; Piñol *et al.*, 1998; Díaz-Delgado *et al.*, 2004) because it takes many small (e.g., 1 ha) fires to equal the area of a single large (e.g., 10,000 ha) fire. Therefore, these large fires have the greatest effect on local populations whose post-fire recruitment relies on dispersal of their seeds from live reproductive populations located near the burn edge. This is discussed in more detail below.

Seed Dispersal Processes

A dispersal curve summarizes the distribution of horizontal distances traveled by seeds (Levin *et al.*, 2003). Most empirical seed dispersal data fit a relatively simple, unimodal leptokurtic distribution characterized by a peak at or close to the source, followed by a rapid decline and a long, relatively "fat" tail (Kot *et al.*, 1996; Willson, 1993). Several mechanistic models have been used to describe empirical seed dispersal data. They have shown that the observed pattern of horizontal seed dispersal distances at a local scale is a function of the terminal velocity of seeds, the height of seed release, and the horizontal wind speed experienced by a seed during seed flight (e.g., Greene and Johnson, 1989; Okuba and Levin, 1989; Andersen, 1991; Greene and Johnson, 1996; Nathan *et al.*, 2001). Nathan *et al.* (2001) showed that of the three variables, the temporal variation in wind conditions has the greatest influence on horizontal dispersal distance. Other studies have examined patterns of long-distance seed dispersal (i.e., distances traveled by the 99th or 99.9th percentile of seeds). These studies have shown that seed uplifting and escape of seeds from the forest canopy, which are caused by the fine-scale turbulence structure within a forest, are the two most important variables explaining patterns of long-distance seed dispersal (Horn *et al.*, 2001; Nathan *et al.*, 2002; Katul, 2005).

An understanding of both the local and long-distance seed dispersal patterns is critical to an understanding of the post-fire recruitment of tree populations whose seeds or sprouts are unavailable from within a burned local population after fire. This includes populations with aerial seed banks whose cones do not open, whose cones open but seeds are killed, or that are killed by fire before they are reproductive. It also includes populations with exposed seeds (i.e., in open cones or catkins) whose seeds and bud banks are killed by fire.

The post-fire recruitment of tree populations with aerial seed banks whose seeds are unavailable from within a fire-killed local population depends on seed being dispersed from reproductive fire-killed populations located elsewhere in the burn. Thus, an understanding of the post-fire recruitment of these populations requires that we know their landscape spatial distribution within a burn as well as the reproductive status of each local population. The age of first reproduction of trees appears to depend on a minimum stem size (Sedgley and Griffin, 1989) and canopy position (i.e., the size of a tree's nearest neighbors). Thus, the capacity for sexual reproduction must be determined by the size of the carbohydrate pool available to sponsor the heavy investment in reproduction (Greene *et al.*, 1999). Greene *et al.* (1999) used the basal area of canopy stems as a surrogate predictor of this carbohydrate pool. If we could relate the basal area of post-fire cohorts to their age and we know the fire frequency, then we could determine the probability of fire occurring before post-fire cohorts reproduce (e.g., Gutsell, 2001; Johnson *et al.*, 2003). If, as suggested by fire frequency studies, most ecosystems have an average fire frequency close to the senescence age of canopy trees, then the post-fire recruitment of these populations is limited primarily by the timing of first reproduction. Clearly, the longer it takes canopy trees to reproduce, the more these populations are limited in terms of their ability to recruit after fire.

Seed dispersal processes are also critical for the post-fire recruitment of tree populations with exposed seeds in catkins or open cones whose seeds and/or bud banks are killed by fire. The post-fire recruitment of these populations depends on seed being dispersed from live reproductive populations located around the burn perimeter or within unburned residual islands. Therefore, for an understanding of the post-fire recruitment of these populations we need to know the spatial pattern of fires and the dispersal curve of the local population. The spatial pattern of a fire includes the area burned, the shape of the burn, and the number and size of unburned residual islands within the burn perimeter. Together these characteristics determine the distribution of distances that seeds must travel from the edge of live vegetation (i.e., the burn perimeter and unburned residual islands) into the burn. From this distribution and the seed dispersal curve of a local population, one can determine the likelihood that a particular seed dispersing tree population will be able to recruit into a burn, or parts of a burn (i.e., the percentage of the burn area with the potential for recruitment by that population).

Johnson *et al.* (2003) present data for the mean distance to nearest live edge as a function of nominal burn area for 27 fires in Saskatchewan. They found that despite differences in burn area, the mean distance to the nearest live edge did not exceed 150 m. This result is explained by the increasingly oblong shape of fires as they get bigger, the irregular convoluted shape of the burn edge, and the ubiquitous nature of residuals in large fires. Given that the median seed dispersal distance is 30 to 50 m for the two seed-dispersing tree populations in their study area [*P. glauca* and *Abies balsamea* (L.) Mill.], one can see that these populations are dispersal limited. In fact, these populations are more dispersal limited than their analysis suggests. As suggested by the landscape distribution of local populations, Johnson *et al.* make two assumptions that are unlikely to be met: the seed-dispersing tree populations are present at all burn edges and would be able to recruit anywhere in the burn. In addition, as with aerial seed bank populations, recruitment of these populations may be further limited by the fire frequency (which affects the landscape age distribution) and the timing of reproduction of post-fire cohorts. Clearly, with knowledge of the landscape distribution of local populations, the fire frequency, and the timing of reproduction of tree populations, we would have a better understanding of the real limitations these populations face in terms of their post-fire recruitment into burns.

CONCLUSIONS

The study of fire behavior has been largely developed by physical scientists, particularly engineers and physicists. Unfortunately, much of the literature on the physical aspects of fire behavior is underutilized by foresters and ecologists, who have seemed more interested in describing the ecological responses to fire rather than how specific fire processes cause ecological effects. This chapter, along with Chapter 12 by Mercer and Weber, has been an attempt to bring the two perspectives of physical scientists and natural scientists (foresters and ecologists) closer together by identifying the fire behavior processes and characteristics that have been shown to have specific ecological effects on individual trees, and their local and regional populations. As readers will have noticed, much more work is needed before we have a complete understanding of the causal mechanisms involved.

REFERENCES

Acherar, M., Lepart, J., and Debussche, M. (1984). La colonization des friches par le pin d'Alep (*Pinus halepensis* Mill.) en Languedoc méditerranéen. *Acta Ecol* 19, 179–189.

Alexander, M. E. (1988). Help with making crown fire hazard assessments. *Proceedings of the Symposium and Workshop on Protecting People and Homes from Wildfire in the Interior West*. pp. 147–56. U.S. Department of Agriculture Forest Service, General Technical Report INT-251. Intermountain Research Station, Ogden, UT.

Alexander, M. E., Stocks, B. J., and Lawson, B. D. (1991). Fire behavior in black spruce-lichen woodland: the Porter Lake project. Forestry Canada Information Report NOR-X-310. Northern Forestry Centre, Edmonton.

Andersen, M. (1991). Mechanistic models for the seed shadows of wind-dispersed plants. *Am Nat* 137, 476–497.

Bai, Y. G., Thompson, D., and Broersma, K. (2000). Early establishment of Douglas-fir and ponderosa pine in grassland seedbeds. *J Range Manage* 53, 511–517.

Baker, W. L., and Ehle, D. (2001). Uncertainty in surface-fire history: the case of ponderosa pine forests in the western United States. *Can J Forest Res* 31, 1205–1226.

Bakhman, N. N. (1993). Smoldering wave propagation mechanism. 1. Critical conditions. *Combust Explo Shock* 29, 14–17.

Banavar, J. R., Damuth, J., Maritan, A., and Rinaldo, A. (2002). Supply-demand balance and metabolic scaling. *Proc Natl Acad Sci U S A* 99, 10506–10509.

Barbour, M. G., and Billings, W. D. (Eds.). (1988). *North American Terrestrial Vegetation*. Cambridge University Press, Cambridge.

Beaufait, W. R. (1960). Some effects of high temperatures on the cones and seeds of jack pine. *Forest Sci* 6, 194–199.

Bergeron, Y., and Archambault, S. (1993). Decreasing fire frequency of forest fires in the southern boreal zone of Québec and its relationship to global warming since the end of the "Little Ice Age." *Holocene* 3, 255–259.

Beverly, J. L., and Martell, D. L. (2003). Modeling *Pinus strobus* mortality following prescribed fire in Quetico Provincial Park, northwestern Ontario. *Can J Forest Res* 33, 740–751.

Bonnet, V. H., Schoettle, A. W., and Sheppard, W. D. (2005). Postfire environmental conditions influence the spatial pattern of regeneration for *Pinus ponderosa*. *Can J Forest Res* 35, 37–47.

Borchert, M., Johnson, M., Schreiner, D. S., and Vander Wall, S. B. (2003). Early postfire seed dispersal, seedling establishment and seedling mortality of *Pinus coulteri* (D. Don) in central coastal California, USA. *Plant Ecol* 168, 207–220.

Bridge, S. R. J., and Johnson, E. A. (2000). Geomorphic principles of terrain organization and vegetation gradients. *J Veg Sci* 11, 57–70.

Broncana, M. J., and Retana, J. (2004). Topography and forest composition affecting the variability in fire severity and post-fire regeneration occurring after a large fire in the Mediterranean basin. *Int J Wildland Fire* 13, 209–216.

Brown, J. L., and DeByle, N. V. (1987). Fire damage, mortality, and suckering in aspen. *Can J Forest Res* 17, 1100–1109.

Brown, P. M., Kauffman, M. R., and Sheppard, W. D. (1999). Long-term, landscape patterns of past fire events in a montane ponderosa pine forest of central Colorado. *Landscape Ecol* 14, 513–532.

Brown, P. M., and Wu, R. (2005). Climate and disturbance forcing of episodic tree recruitment in a southwestern ponderosa pine landscape. *Ecology* 86, 3030–3038.

Byram, G. M. (1959). Combustion of forest fuels. In: *Forest Fire: Control and Use*. (Davis, K. P., Ed.), pp. 61–80. McGraw-Hill, New York.

Caccia, F. D., and Ballaré, C. L. (1998). Effects of tree cover, understorey vegetation, and litter on regeneration of Douglas-fir (*Pseudotsuga menziesii*) in southwestern Argentina. *Can J Forest Res* 28, 683–692.

Callaway, R. M. (1995). Positive interactions among plants. *Bot Rev* 61, 306–349.

Castro, J., Gómez, J. M., García, D., Zamora, R., and Hódar, J. A. (1999). Seed predation and dispersal in relict Scots pine forests in southern Spain. *Plant Ecol* 145, 115–123.

Catchpole, E. A., de Mestre, N. J., and Gill, A. M. (1982). Intensity of a fire at its perimeter. *Aust Forest Res* 12, 47–54.

Catchpole, W. (2002). Fire properties and burn patterns in heterogeneous landscapes. In: *Flammable Australia: The Fire Regimes and Biodiversity of a Continent*. (Bradstock, R. A., Williams, J. E., and Gill, A. M., Eds.), pp. 49–76. Cambridge University Press, Cambridge.

Chabot, B. F., and Mooney, H. A. (Eds.) (1985). *Physiological Ecology of North American Plant Communities*. Chapman and Hall, New York.

Charron, I. (1998). *Sexual Recruitment of Trees Following Fire in the Southern Mixedwood Boreal Forest of Canada*. M.Sc. thesis, Concordia University, Montreal.

Charron, I., and Greene, D. F. (2002). Post-wildfire seedbeds and tree establishment in the southern mixedwood boreal forest. *Can J Forest Res* 32, 1607–1615.

Chipman, S. J., and Johnson, E. A. (2002). Understory vascular plant species diversity in the mixedwood boreal forest of western Canada. *Ecol Appl* 12, 588–601.

Chrosciewicz, Z. (1974). Evaluation of fire-produced seedbeds for jack pine regeneration in central Ontario. *Can J Forest Res* 4, 455–457.

Clark, T. L., Jenkins, M. A., Coen, J., and Packham, D. (1996). A coupled atmosphere-fire model: convective feedback on fire-line dynamics. *J Appl Meteorol* 35, 875–901.

Clements, F. E. (1910). *The Life History of Lodgepole Pine Forests*. Bulletin 79, U.S. Department of Agriculture Forest Service, Washington, DC.

Coen, J. (2005). Simulation of the Big Elk Fire using coupled atmosphere-fire modeling. *Int J Wildland Fire* 14, 49–59.

Cox, G. (1995). *Combustion Fundamentals of Fire*. Academic Press, New York.

Davis, K. P. (1959). *Forest Fire Control and Use*. McGraw-Hill, New York.

DeVries, D. A. (1963). Thermal properties of soil. In: *Physics of Plant Environment* (Van Wijk, W. R., Ed.), pp. 210–235. North-Holland Publishing Company, Amsterdam.

Díaz-Delgado, R., Lloret, F., and Pons, X. (2004). Spatial patterns of fire occurrence in Catalonia, NE, Spain. *Landscape Ecol* 19, 731–745.

Dickinson, M. B., and Johnson, E. A. (2001). Fire effects on trees. In: *Forest Fires: Behavior and Ecological Effects* (Johnson, E. A., and Miyanishi, K., Eds.), pp.477–525. Academic Press, San Diego, CA.

Dreyer, O., and Puzio, R. (2001). Allometric scaling in animals and plants. *J Math Biol* 43, 144–156.

Drysdale, D. (1985). *An Introduction to Fire Dynamics*. John Wiley and Sons, Chichester, UK.
Duchesne, S., and Sirois, L. (1995). Phase initial de régénération après feu des populations conifériennes subarctiques. *Can J Forest Res* 25, 307–318.
Eberhart, K. E., and Woodard, P. M. (1987). Distribution of residual vegetation associated with large fires in Alberta. *Can J Forest Res* 17, 1207–1212.
Ebert, T. A. (1999). *Plant and Animal Populations: Methods in Demography*. Academic Press, New York.
Ehle, D. S., and Baker, W. L. (2003). Disturbance and stand dynamics in ponderosa pine forests in Rocky Mountain National Park, USA. *Ecol Monogr* 73, 543–566.
Eriksson, O. (1996). Regional dynamics of plants: a review of evidence for remnant, source-sink, and metapopulations. *Oikos* 77, 248–258.
Fahnestock, G. R., and Hare, R. C. (1964). Heating of tree trunks in surface fires. *J Forest* 62, 799–805.
Farmer, R. E. (1962). Aspen root sucker formation and apical dominance. *Forest Sci* 8, 403–410.
Fauria, M., and Johnson, E. A. (2006). Large-scale climatic patterns control large lightning fire occurrence in Canada and Alaska forest regions. *J Geophys Res* 111, G04008.
Flannigan, M. D., and Harrington, J. B. (1987). Synoptic conditions during the Porter Lake burning experiment. *Climatol Bull* 21, 19–40.
Flannigan, M. D., and Wotton, B. M. (2001). Climate, weather, and area burned. In: *Forest Fires: Behavior and Ecological Effects*. (Johnson, E. A., and Miyanishi, K., Eds.), pp. 351–373. Academic Press, San Diego, CA.
Fulé, P. Z., Crouse, J. E., Heinlein, T. A., Moore, M. M., Covington, W., and Verkamp, G. (2003). Mixed-severity fire regime in a high-elevation forest of Grand Canyon, Arizona, USA. *Landscape Ecol* 18, 465–486.
Glitzenstein, J. S., Platt, W. J., and Streng, D. R. (1995). Effects of fire regime and habitat on tree dynamics in north Florida longleaf pine savannas. *Ecol Monogr* 65, 441–476.
Grace, S. L., and Platt, W. J. (1995). Effects of adult tree density and fire on the demography of pregrass stage juvenile longleaf pine (*Pinus palustris* Mill.). *J Ecol* 83, 75–86.
Greene, D. F., and Johnson, E. A. (1989). A model of wind dispersal of winged or plumed seeds. *Ecology* 70, 339–347.
Greene, D. F., and Johnson, E. A. (1996). Wind dispersal of seeds from a forest into a clearing. *Ecology* 77, 595–609.
Greene, D. F., and Johnson, E. A. (1999). Modelling recruitment of *Populus tremuloides, Pinus banksiana,* and *Picea mariana* following fire in the mixedwood boreal forest. *Can J Forest Res* 29, 462–473.
Greene, D. F., Noel, J., Bergeron, Y., Rousseau, M., and Gauthier, S. (2004). Recruitment of *Picea mariana, Pinus banksiana,* and *Populus tremuloides* across a burn severity gradient following wildfire in the southern boreal forest of Québec. *Can J Forest Res* 34, 1845–1857.
Greene, D. F., Zasada, J. C., Sirois, L., Kneeshaw, D., Morin, H., Charron, I., and Simard, M.-J. (1999). A review of regeneration dynamics of North American boreal forest tree species. *Can J Forest Res* 29, 824–839.
Grime, J. P., and Jeffrey, D. W. (1965). Seedling establishment in vertical gradients of sunlight. *J Ecol* 53, 621–642.
Grissino-Mayer, H. D. (1999). Modeling fire interval data from the American Southwest with the Weibull distribution. *Int J Wildland Fire* 9, 37–50.
Gutsell, S. L. (2001). *Understanding Forest Dynamics Incorporating Both Local and Regional Ecological Processes*. Ph.D. dissertation, University of Calgary, Calgary, AB.
Gutsell, S. L., and Johnson, E. A. (1996). How fire scars are formed: coupling a disturbance process to its ecological effect. *Can J Forest Res* 26, 166–174.

Gutsell, S. L., and Johnson, E. A. (2002). Accurately ageing trees and examining their height growth rates: implications for interpreting forest dynamics. *J Ecol* 90, 153–166.

Haase, S. M. (1986). Effect of prescribed burning on soil moisture and germination of southwestern ponderosa pine seed on basaltic soils. U.S. Department of Agriculture Forest Service Research Note RM-462, Rocky Mountain Forest and Range Experiment Station, Fort Collins, CO.

Hanski, I., and Gilpin, M. (1991). Metapopulation dynamics: brief history and conceptual domain. *Biol J Linn Soc* 42, 3–16.

Harrington, M. G. (1993). Predicting *Pinus ponderosa* mortality from dormant season and growing season fire injury. *Int J Wildland Fire* 3, 65–72.

Heinselman, M. L. (1973). Fire in the virgin forests of the Boundary Waters Canoe Area. *Quaternary Res* 3, 329–382.

Hély, C., Flannigan, M., and Bergeron, Y. (2003). Modeling tree mortality following wildfire in the southeastern Canadian mixed-wood boreal forest. *Forest Sci* 49, 566–576.

Herranz, J. M., Martinez-Sanchez, J. J., Marin A., and Ferrandis, P. (1997). Postfire regeneration of *Pinus halepensis* Miller in a semi-arid area in Albacete province (southeastern Spain). *Ecoscience* 4, 86–90.

Heyerdahl, E. K., Brubaker, L. B., and Agee, J. K. (2001). Spatial controls of historical fire regimes: a multiscale example from the Interior West, U.S.A. *Ecology* 82, 660–678.

Hirsch, K. G., and Flannigan, M. D. (1990). Meteorological and fire behavior characteristics of the 1989 fire season in Manitoba, Canada. In: *International Conference on Forest Fire Research*, pp. B.06-1–B.06-16, Coimbra, Portugal.

Hoffman, W. A. (1998). Post-burn reproduction of woody plants in a neotropical savanna: the relative importance of sexual and vegetative reproduction. *J Appl Ecol* 35, 422–433.

Holman, J. P. (1997). *Heat Transfer*. McGraw-Hill, New York.

Horn, H. S., Nathan, R., and Kaplan, S. R. (2001). Long-distance dispersal of tree seeds by wind. *Ecol Res* 16, 877–885.

Incropera, F. P., and DeWitt, D. P. (2002). *Fundamentals of Heat and Mass Transfer*. John Wiley and Sons, New York.

Jenkins, M. A. (2002). An examination of the sensitivity of numerically simulated wildfires to low-level atmospheric stability and moisture, and the consequences for the Haines index. *Int J Wildland Fire* 11, 1–20.

Jenkins, M. A., Clark, T., and Coen, J. (2001). Coupling atmospheric and fire models. In: *Forest Fires: Behavior and Ecological Effects*. (Johnson, E. A., and Miyanishi, K., Eds.), pp.257–302. Academic Press, San Diego, CA.

Johnson, E. A. (1985). Disturbance: the process and the response. An epilogue. *Can J Forest Res* 15, 292–293.

Johnson, E. A., and Fryer, G. I. (1989). Population dynamics in lodgepole pine–Engelmann spruce forests. *Ecology* 70, 1335–1345.

Johnson, E. A., and Gutsell, S. L. (1993). Heat budget and fire behavior associated with the opening of serotinous cones in two *Pinus* species. *J Veg Sci* 4, 745–750.

Johnson, E. A., and Gutsell, S. L. (1994). Fire frequency models, methods, and interpretations. *Adv Ecol Res* 25, 239–287.

Johnson, E. A., Miyanishi, K., and Kleb, H. (1994). The hazards of interpretation of static age structures as shown by stand reconstructions in a *Pinus contorta—Picea Engelmannii* forest. *J Ecol* 82, 923–931.

Johnson, E. A., and Miyanishi, K. (2001). Strengthening fire ecology's roots. In: *Forest Fires: Behavior and Ecological Effects*. (Johnson, E. A., and Miyanishi, K., Eds.), pp. 1–11. Academic Press, San Diego, CA.

Johnson, E. A., Miyanishi, K., and Weir, J. M. H. (1998). Wildfires in the western Canadian boreal forest: landscape patterns and ecosystem management. *J Veg Sci* 9, 603–610.

Johnson, E. A., Morin, H., Miyanishi, K., Gagnon, R., and Greene, D. F. (2003). A process approach to understanding disturbance and forest dynamics for sustainable forestry. In: *Towards Sustainable Management of the Boreal Forest.* (Burton, P. J., Messier, C., Smith, D. W., and Adamowicz, W. L., Eds.), pp. 261–306. NRC Research Press, Ottawa.

Johnson, E. A., and Wowchuk, D. R. (1993). Wildfires in the southern Canadian Rocky Mountains and their relationship to mid-tropospheric anomalies. *Can J Forest Res* 23, 1213–1222.

Johnstone, J. F., Chapin III, F. S., Foote, J., Kemmett, S., Price, K., and Viereck, L. (2004). Decadel observations of tree regeneration following fire in boreal forests. *Can J Forest Res* 34, 267–273.

Jones, J. C., Goh, T. P. T., and Dijanosic, M. J. (1994). Smoldering and flaming combustion in packed beds of *Casuarina* needles. *J Fire Sci* 12, 442–451.

Jones, J. L., Webb, B. W., Jimenez, D., Reardon, J., and Butler, B. (2004). Development of an advanced one-dimensional stem heating model for application in surface fires. *Can J Forest Res* 34, 20–30.

Katul, G. G., Porporato, A., Nathan, R., Siqueira, M., Soons, M. B., Poggi, D., Horn, H. S., and Levin, S. A. (2005). Mechanistic analytical models for long-distance seed dispersal by wind. *Am Nat* 166, 368–381.

Kauffman, J. B. (1991). Survival by sprouting following fire in tropical forests of the eastern Amazon. *Biotropica* 23, 219–224.

Keeley, J. E., Keeley, M. B., and Bond, W. J. (1998). Stem demography and post-fire recruitment of a resprouting serotinous conifer. *J Veg Sci* 10, 69–76.

Kolb, P. F., and Robberecht, R. (1996). *Pinus ponderosa* seedling establishment and the influence of competition with the bunchgrass *Agropyron spicatum. Int J Plant Sci* 157, 509–515.

Kot, M., Lewis, M. A., and van den Driessche, P. (1996). Dispersal data and the spread of invading organisms. *Ecology* 77, 2027–2042.

Lavoie, L., and Sirois, L. (1998). Vegetation changes caused by recent fires in the northern boreal forest of eastern Canada. *J Veg Sci* 9, 483–492.

Levin, S. A., Muller-Landau, H. C., Nathan, R., and Chave, J. (2003). The ecology and evolution of seed dispersal: a theoretical perspective. *Annu Rev Ecol Syst* 34, 575–604.

Liñán, A., and Williams, F. A. (1993). *Fundamental Aspects of Combustion.* Oxford Engineering Science Series 34. Oxford University Press, Oxford.

Maini, J. S., and Horton., K. W. (1966). Vegetative propagation of *Populus* species. I. Influence of temperature and formation on initial growth of aspen suckers. *Can J Bot* 44, 1183–1189.

Malamud, B. D., Morein, G., and Turcotte, D. L. (1998). Forest fires: an example of self-organized critical behavior. *Science* 281, 1840–1841.

Martin, R. E. (1963). Thermal properties of bark. *Forest Prod J* 13, 419–426.

Mason, C. F. (1976). *Decomposition.* Edward Arnold (Publishers), London.

Mast, J. N., Veblen, T. T., and Linhart, Y. B. (1998). Disturbance and climatic influences on age-structure of ponderosa pine at the pine-grassland ecotone, Colorado Front Range. *J Biogeogr* 25, 743–755.

Masters, A. M. (1990). Change in forest fire frequency in Kootenay National Park, Canadian Rockies. *Can J Bot 68,* 1763–1767.

McHugh, C. W., and Kolb, T. E. (2003). Ponderosa pine mortality following fire in northern Arizona. *Int J Wildland Fire* 12, 7–22.

McRae, D. J., Stocks, B. J., and Ogilvie, C. J. (1989). Fire acceleration on large-scale convection burns. In: *Proceedings of the 10th Conference on Fire and Forest Meteorology.* (MacIver, D. C., Auld, H., and Whitewood, R., Eds.), pp. 101–107. Forestry Canada, Ottawa.

Mercer, G. N., and Weber, R. O. (2007). Modeling heating effects. In: *Plant Disturbance Ecology: The Process and the Response*. (Johnson, E. A., and Miyanishi, K., Eds.) pp. 371–395. Academic Press, San Diego, CA.

Michaletz, S. T. (2005). *Biophysical Processes of Tree Mortality in Surface Fires*. M.Sc. thesis, University of Calgary, Calgary.

Michaletz, S. T., and Johnson, E. A. (2006). A heat transfer model of crown scorch in forest fires. *Can J Forest Res* 36, 2839–2851.

Miyanishi, K. (2001). Duff consumption. In: *Forest Fires: Behavior and Ecological Effects*. (Johnson, E. A., and Miyanishi, K., Eds.), pp. 437–476. Academic Press, San Diego, CA.

Miyanishi, K., and Johnson, E. A. (2002). Process and pattern in duff consumption in the mixedwood boreal forest. *Can J Forest Res* 32, 1285–1295.

Miyanishi, K., and Kellman, M. (1986). The role of root nutrient reserves in regrowth of two savanna shrubs. *Can J Bot* 64, 1244–1248.

Moravec, J. (1990). Regeneration of N.W. African *Pinus halepensis* forests following fire. *Vegetatio* 87, 29–36.

Nathan, R., Katul, G. G., Horn, H. S., Thomas, S. M., Oren, R., Avissar, R., Pacala, S. W., and Levin, S. A. (2002). Mechanisms of long-distance dispersal of seeds by wind. *Nature* 418, 409–413.

Nathan, R., Safriel, U. N., and Noy-Meir, I. (2001). Field validation and sensitivity analysis of a mechanistic model for tree seed dispersal by wind. *Ecology* 82, 374–388.

Ne'eman, G. (1997). Regeneration of natural pine forest—review of the work done after the 1989 fire in Mount Carmel, Israel. *Int J Wildland Fire* 7, 295–306.

Nelson, R. M. Jr. (2001). Water relations in forest fuels. In: *Forest Fires: Behavior and Ecological Effects* (Johnson, E. A., and Miyanishi, K., Eds.), pp. 79–150. Academic Press, San Diego, CA.

Newnham, R. M. (1991). Variable-form taper functions for four Alberta tree species. *Can J Forest Res* 22, 210–233.

Niklas, K. J., and Spatz, H. C. (2004). Growth and hydraulic (not mechanical) constraints govern the scaling of tree height and mass. *Proc Natl Acad Sci U S A* 101, 15661–15663.

Nimchuk, N. (1983). *Wildfire Behaviour Associated with Upper Ridge Breakdown*. ENR Report No. T/50. Alberta Energy and Natural Resources, Forestry Service, Edmonton, AB.

Odion, D. C., Frost, E. J., Strittholt, J. R., Jiang, H., Dellasala, D. A., and Moritz, M. A. (2004). Patterns of fire severity and forest conditions in the western Klamath Mountains, California. *Conserv Biol* 18, 927–936.

Ohlemiller, T. J. (1985). Modeling of smoldering combustion propagation. *Prog Energ Combust* 11, 277–310.

Ohlemiller, T. J. (1990). Smoldering combustion propagation through a permeable horizontal fuel layer. *Combust Flame* 81, 341–353.

Okuba, A, and Levin, S. A. (1989). A theoretical framework for data analysis of wind dispersal of seeds and pollen. *Ecology* 70, 329–338.

Oliveira, L. A., Viegas, D. X., and Raimundo, A. M. (1997). Numerical predictions on the soil thermal effect under surface fire conditions. *Int J Wildland Fire* 7, 51–63.

Ordóñez, J. L., and Retana, J. (2004). Early reduction of post-fire recruitment of *Pinus nigra* by post-dispersal seed predation in different time-since-fire habitats. *Ecography* 27, 449–458.

Palmer, K. N. (1957). Smoldering combustion in dusts and fibrous materials. *Combust Flame* 1, 129–154.

Pearson, G. A. (1942). Herbaceous vegetation a factor in natural regeneration of ponderosa pine in the Southwest. *Ecol Monogr* 12, 315–338.

Peters, V. S., MacDonald, S. E., and Dale, M. R. T. (2005). The interaction between masting and fire is key to white spruce regeneration. *Ecology* 86, 1744–1750.

Peterson, D. L., and Ryan, K. C. (1986). Modeling postfire conifer mortality for long-range planning. *Environ Manage* 10, 797–808.

Piñol, J., Terradas, J., and Lloret, F. (1998). Climate warming, wildfire hazard, and wildfire occurrence in coastal eastern España. *Climatic Change* 38, 345–357.

Pollet, J., and Omi, P. N. (2002). Effect of thinning and prescribed burning on crown fire severity in ponderosa pine forests. *Int J Wildland Fire* 11, 1–10.

Quintilio, D., Fahnestock, G. R., and Dubé, D. E. (1977). Fire behavior in upland jack pine: the Darwin Lake project. Information Report NOR-X-174. Environment Canada, Canadian Forestry Service, Edmonton, AB.

Reed, W. J. (1998). Determining changes in historical forest fire frequency from a time-since-fire map. *J Agr Biol Envir St* 3, 430–450.

Reed, W. J. (2000). Reconstructing forest fire history—identifying hazard rate change points using the Bayes' information criterion. *Can J Stat* 28, 352–365.

Reed, W. J., and Johnson, E. A. (2004). Statistical methods for estimating historical fire frequency from multiple fire-scar data. *Can J Forest Res* 34, 2306–2313.

Reed, W. J., and McKelvey, K. S. (2002). Power-law behaviour and parametric models for the size–distribution of forest fires. *Ecol Model* 150, 239–254.

Ryan, K. C., and Frandsen, W. H. (1991). Basal injury from smoldering fires in mature *Pinus ponderosa* Laws. *Int J Wildland Fire* 1, 107–118.

Ryan, K. C., Peterson, D. L., and Reinhardt, E. D. (1988). Modeling long-term fire-caused mortality of Douglas-fir. *Forest Sci* 34, 190–199.

Saito, K. (2001). Flames. In: *Forest Fires: Behavior and Ecological Effects.* (Johnson, E. A., and Miyanishi, K., Eds.), pp. 12–54. Academic Press, San Diego, CA.

Schier, G. A. (1976). Physiological and environmental factors controlling vegetative regeneration of aspen. U.S. Department of Agriculture Forest Service General Technical Report RM-29, Rocky Mountain Forest and Range Experiment Station, Fort Collins, CO.

Schimmel, J., and Granström, A. (1996). Fire severity and vegetation response in the boreal Swedish forest. *Ecology* 77, 1436–1450.

Sedgley, M., and Griffin, A. R. (1989). *Sexual Reproduction of Tree Crops.* Academic Press, New York.

Selby, M. J. (1982). *Hillslope Materials and Processes.* Oxford University Press, Oxford.

Sharma, S., and Rikhari, H. C. (1997). Forest fire in the central Himalaya: climate and recovery of trees. *Int J Biometeorol* 40, 63–70.

Shinozaki, K., Yoda, K., Hozumi, K., and Kira, T. (1964). A quantitative analysis of plant form: pipe model theory. I. Basic analysis. *Jpn J Ecology* 14, 97–105.

Simmons, R. F. (1995). Fire chemistry. In: *Combustion Fundamentals of Fire.* (Cox, G., Ed.), pp. 408–473. Academic Press, London.

Sivapalan, M., Wood, E. F., and Beven, K. J. (1987). On hydrological similarity. 2. A scaled model of storm runoff production. *Water Resour Res* 23, 2266–2278.

Skinner, W. R., Flannigan, M. D., Stocks, B. J., Martell, D. L., Wotton, B. M., Todd, J. B., Mason, J. A., Logan, K. A., and Bosch, E. M. (2002). A 500 hPa synoptic wildland fire climatology for large Canadian forest fires, 1959–1996. *Theor Appl Climatol* 71, 157–169.

Skinner, W. R., Stocks, B. J., Martell, D. L., Bonsal, B., and Shabbar, A. (1999). The association between circulation anomalies in the mid-troposphere and area burned by wildland fire in Canada. *Theor Appl Climatol* 63, 89–105.

Spanos, I. A., Daskalakou, E. N., and Thanos, C. A. (2000). Post-fire, natural regeneration of *Pinus brutia* forests in Thasos Island, Greece. *Acta Œcol* 21, 13–20.

Steneker, G. A. (1974). Factors affecting the suckering of trembling aspen. *Forest Chron* 50, 32–34.

St. Pierre, H., Gagnon, R., and Bellefleur, P. (1992). Régénération après feu de l'épinette noire (*Picea mariana*) et du pin gris (*Pinus banksiana*) dans la forêt boréale, Québec. *Can J Forest Res* 22, 474–481.

Strauss, D., Bednar, L., and Mees, R. (1989). Do one percent of forest fires cause ninety-nine percent of the damage? *Forest Sci.* 35, 319–328.

Swezy, D. M., and Agee, J. K. (1991). Prescribed fire effects on fine-root and tree mortality in old-growth ponderosa pine. *Can J Forest Res* 21, 626–634.

Taylor, K., and Aarssen, L. (1989). Neighbor effects in mast year seedlings of *Acer saccharum*. *Am J Bot* 76, 546–554.

Thanos, C. A., Daskalakou, E. N., and Nikolaidou, S. (1996). Early post-fire regeneration of a *Pinus halepensis* forest on Mount Parnis, Greece. *J Veg Sci* 7, 273–280.

Thanos, C. A., Marcou, S., Christodoulakis, D., and Yannitsaros, A. (1989). Early post-fire regeneration in *Pinus brutia* forest ecosystems of Samos Island (Greece). *Acta Œcol* 10, 79–94.

Thanos, C. A., and Marcou, S. (1991). Post-fire regeneration in *Pinus brutia* forest ecosystems of Samos Island (Greece): 6 years after. *Acta Œcol* 12, 633–642.

Thomas, P. H. (1971). Rates of spread of some wind-driven fires. *Forestry* 44, 155–175.

Turner, M. G., Romme, H. W., Gardner, R. H., and Hargrove, W. W. (1997). Effects of fire size and pattern on early succession in Yellowstone National Park. *Ecol Monogr* 67, 411–433.

Vander Wall, S. B. (1992). The role of animals in dispersing a "wind-dispersed" pine. *Ecology* 73, 614–621.

Van Wagner, C. E. (1977). Conditions for the start and spread of a crown fire. *Can J Forest Res* 7, 23–34.

Van Wagner, C. E. (1983). Fire behavior in northern conifer forests and shrublands. In: *The Role of Fire in Northern Circumpolar Ecosystems*. (Wein, R. W., and MacLean, D. A., Eds.), pp. 65–80. John Wiley and Sons, New York.

Van Wagner, C. E. (1987). Development and structure of the Canadian forest fire weather index system. Forestry Technical Report No. 35. Canadian Forestry Service, Ottawa.

Veblen, T. T., Kitzberger, T., and Donnegan, J. (2000). Climatic and human influences on fire regimes in ponderosa pine forests in the Colorado Front Range. *Ecol Appl* 10, 1178–1195.

Waring, R. H. (1987). Characteristics of trees predisposed to die. *BioScience* 37, 569–574.

Weber, M. G. (1990). Response of immature aspen ecosystems to burning and cutting in relation to vernal leaf-flush. *Forest Ecol Manag* 31, 15–33.

Weir, J. M. H., Johnson, E. A., and Miyanishi, K. (2000). Fire frequency and the spatial age mosaic of the mixedwood boreal forest in western Canada. *Ecol Appl* 10, 1162–1177.

West, G. B., Brown, J. H., and Enquist, B. J. (1999). A general model for the structure and allometry of plant vascular systems. *Nature* 400, 664–667.

Williams, C. E., Lipscomb, M. V., Johnson, W. C., and Nilsen, E. T. (1990). Influence of leaf litter and soil moisture regime on early establishment of *Pinus pungens*. *Am Midl Nat* 124, 145–152.

Willson, M. R. (1993). Dispersal mode, seed shadows, and colonization patterns. *Vegetatio* 107/108, 261–280.

Yarie, J. (1998). Forest fire cycles and life tables: a case study from interior Alaska. *Can J Forest Res* 11, 554–562.

Zasada, J. C., Sharik, T. L., and Nygren, M. (1992). The reproductive process in boreal forest trees. In: *A Systems Analysis of the Global Boreal Forest*. (Shugart, H. H., Leemans, R., and Bonan, G., Eds.), pp. 85–125. Cambridge University Press, Cambridge.

15

Insect Defoliators as Periodic Disturbances in Northern Forest Ecosystems

Barry J. Cooke, Vincent G. Nealis, and Jacques Régnière
Natural Resources Canada, Canadian Forest Service

INTRODUCTION

Recurrent outbreaks of forest insect populations have been discussed extensively in the ecological literature (Myers, 1988; Berryman, 1996), but rarely from the perspective of disturbance ecology. The reason lies, in part, in the traditional focus of quantitative animal ecologists on the species of interest (e.g., Morris, 1959) contrasted with the traditional emphasis of plant ecologists on whole communities (e.g., Tilman, 1982). Further, historical research on forest insects has been motivated largely by the practical problem of insect pest management. While yielding a rich legacy of information on the biology and ecology of a few selected insects, the modeling literature that utilizes this information has generally tended to view disturbance as an economic disruption to be predicted and prevented (Berryman, 1991) rather than as an ecosystem process to be emulated. Emulation of natural disturbances, an approach to ecologically sustainable forest management (Bergeron and Harvey, 1997; Perera *et al.*, 2004), has rekindled interest in insect disturbance modeling. The challenge is how to bring the rich insect population dynamics literature to bear on the subject of disturbance in forest dynamics.

In this chapter, we argue that herbivorous insects constitute a class of forest disturbance that is distinct from fire, wind-throw, or flooding in that: (1) insect outbreaks tend to be spatially synchronized and temporally periodic (periods of high impact followed by periods of low impact, over extensive areas); and (2) insect outbreaks are predictably selective and therefore result in different legacies than do abiotic disturbances. As Holling (1992) suggested, insect disturbances, because they tend to be highly selective as a result of biological relationships between the insects and their hosts (Rausher, 1983), represent an agent of "creative destruction" in the boreal forest.

We suggest that recognition of the distinct nature of disturbances caused by insect outbreaks may shed light on the analysis of other disturbances that are, by nature, aperiodic and difficult to forecast (Pickett and White, 1985). In this chapter we review the dynamics of several outbreak systems. We argue that the population density of forest insects is determined by the ecological relationships among host plants, the herbivorous insects, and their natural enemies, and that these "tritrophic" interactions[1] are often lagged in space-time, and therefore tend to generate periodic, high-amplitude fluctuations, or "harmonic oscillations,"[2] in population densities that are synchronized over vast areas of forest. Forest disturbance in these systems happens when periodic population oscillations are amplified to the point that insect densities have a measurable impact on the structure and/or function of dominant elements of the forest ecosystem.

Where our argument differs from that typically stated in the literature is with regard to the degree of periodicity and synchrony of oscillations in

[1] Tritrophic interaction = set of all resource-consumer interactions spanning three trophic levels: plants, herbivores, and predators

[2] From physics, the term "harmonic oscillation" is used to distinguish it from the idea of a "relaxation oscillation," which was central to the phenomenological cusp-catastrophe models developed in the 1970s to describe the process of insect outbreak occurrence (e.g., Ludwig *et al.*, 1978; Rose and Harmsen, 1981; Hassell *et al.*, 1999).

defoliator populations. We argue that the complexity and nonlinear nature of tritrophic interactions result in oscillations that may be imperfectly periodic. Moreover, stochastic environmental variables may further influence these dynamics and contribute to spatiotemporal patterns of disturbance that are imperfectly synchronized. So, although we agree with the idea of a well-defined outbreak return interval, we think the variation about that mean is significant and is under the control of a number of biotic and environmental variables that are worth studying.

To study these relationships, we use a comparative, process-oriented approach. In this chapter we illustrate the approach by comparing the ecologies of four species of insect defoliators (Table 1). A major challenge in insect disturbance modeling is the fact that these systems tend to be modeled in detail as single-species systems. The comparative approach to insect disturbance ecology seeks to identify meaningful similarities among systems, which helps to reduce model complexity to a bare minimum. It focuses on critical, functional ecological relationships rather than species. It is thus ideally suited to disturbance modeling, where the goal is mimicking realistic patterns of natural disturbance caused by a variety of insect species.

The chapter is organized according to Fig. 1, which illustrates a process-oriented view of the way in which insect populations cause selective forest disturbance. At the bottom of Fig. 1, insect hazard modelers will recognize the familiar terms "susceptibility" and "vulnerability" as the component factors whose product is the risk or probability of "disturbance" (defined variously as growth loss, stem mortality, or stand volume loss). Susceptibility typically refers to the probability that herbivory reaches

TABLE 1 Foliage Grazers

Common Name	Latin Name	Acronym	Preferred Host	Secondary Hosts
Forest tent caterpillar	*Malacosoma disstria* [Hbn.] (Lasiocampidae)	FTC	Trembling aspen	Sugar maple, other *Populus* species
Spruce budworm	*Choristoneura fumiferana* (Clem.) (Tortricidae)	SBW	White spruce, balsam fir	Black spruce, red spruce
Jack pine budworm	*Choristoneura pinus pinus* Freeman (Tortricidae)	JPBW	Jack pine	None
Western spruce budworm	*Choristoneura occidentalis* Freeman (Tortricidae)	WSBW	Douglas-fir	White fir, spruce

a certain intensity over a certain time frame, while vulnerability refers to the probability of tree mortality (or growth loss) at a given level of herbivory. Whereas an empirical modeler might attempt to explain observed levels of disturbance (Fig. 1, *bottom*) as a direct function (Fig. 1, *dark-shaded area*) of static spatial features in the environment (Fig. 1, *top right*), a process modeler would seek to understand and quantify the most important ecosystem processes (Fig. 1, *middle*) that determine susceptibility and vulnerability. These ecosystems processes are ultimately governed by environmental and ecological factors (Fig. 1, *top*) that vary both temporally (Fig. 1, *top left*) and spatially (Fig. 1, *top right*) at a variety of scales.

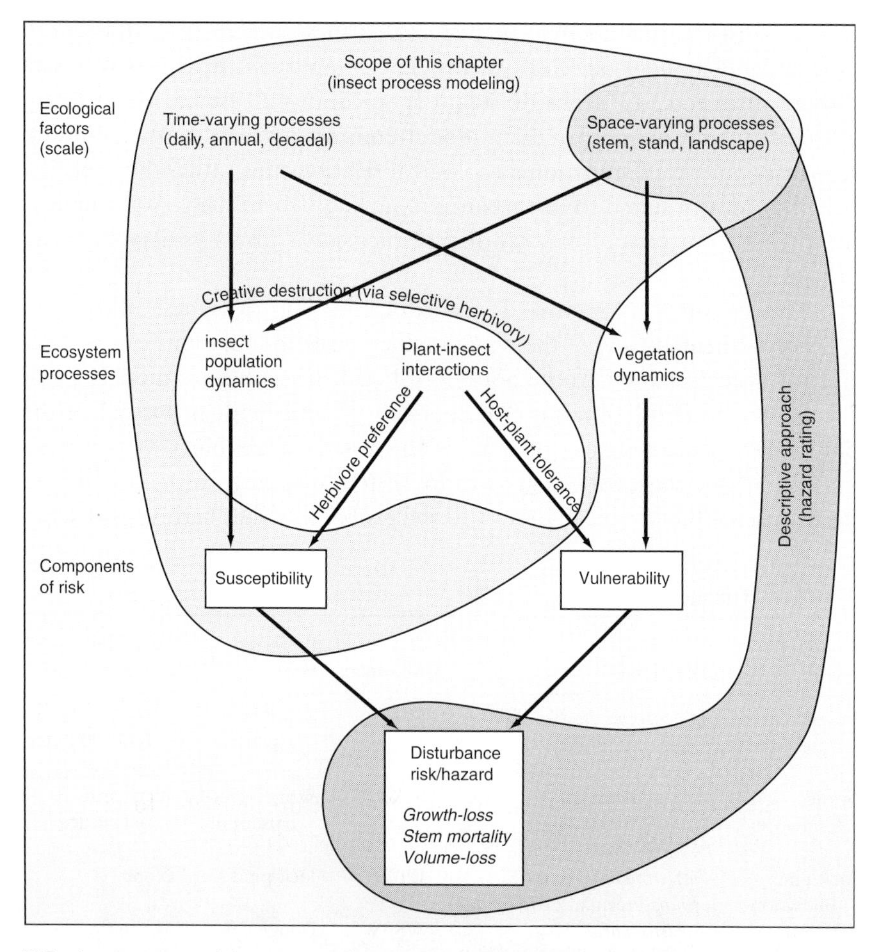

FIG. 1 Overview of the common framework underlying two approaches to insect disturbance ecology: the descriptive/phenomenological/correlative approach (i.e., hazard rating/risk analysis) and the process approach (i.e., process-oriented simulation modeling).

The scope of this chapter is limited to those elements relating most directly to insect population dynamics and plant–insect interactions (Fig. 1, *light-shaded area*). We focus in particular on the creative–destructive nature of disturbance caused by the highly selective process of insect herbivory (Fig. 1, *white area*). Although we are tackling a major piece of the insect disturbance ecology puzzle, we are relying on research in other domains to help bridge the gap between insect population dynamics and insect disturbance ecology. This chapter is thus designed to complement others in this volume, as well as existing reviews that examine insect population processes (Schowalter *et al.*, 1986) and forest insect disturbance (MacLean, 2004) from a more general taxonomic and ecological perspective. We do not go so far as to show quantitatively how different patterns of insect-caused disturbance can be fully explained by existing process-oriented models. Instead, our more modest objective is to convince the reader that the study of population processes is helping to solve many of the prevailing mysteries about insect-caused disturbance.

DEFOLIATING INSECTS AS A DISTINCT CLASS OF FOREST DISTURBANCE

Insects Compared to Fire

To understand the role of defoliating insects as a source of forest disturbance, it helps to compare the magnitude of insect disturbances to physical disturbances, such as fire. In Canadian forests, insects and fire are roughly equivalent in terms of the area disturbed. A regional scale analysis by Kurz and Apps (1999) for the period between 1920 and 1989 indicated that insects are the dominant disturbance in the East, while fire tends to dominate in the West. This distinction, however, is biased by the widespread outbreak of spruce budworm (*Choristoneura fumiferana* Clem.) in eastern Canada from 1970 to about 1990 (Hardy *et al.*, 1986) and the absence, in their analysis, of the more recent outbreaks of both spruce budworm and mountain pine beetle (*Dendroctonus ponderosae* Hopkins) in British Columbia, which engulfed more than 4 million ha in 2003 (British Columbia Ministry of Forests, 2004). Over the lifetime of a forest, both sources of disturbance are significant throughout Canada.

An important distinction between insects and fires, however, is the way the two disturbance regimes behave on the landscape over time. In western Canada, for example, insect-caused disturbance has been somewhat periodic, while fire occurrence is not (Fig. 2). Notably, there is no negative correlation between the fire disturbance and insect disturbance in western

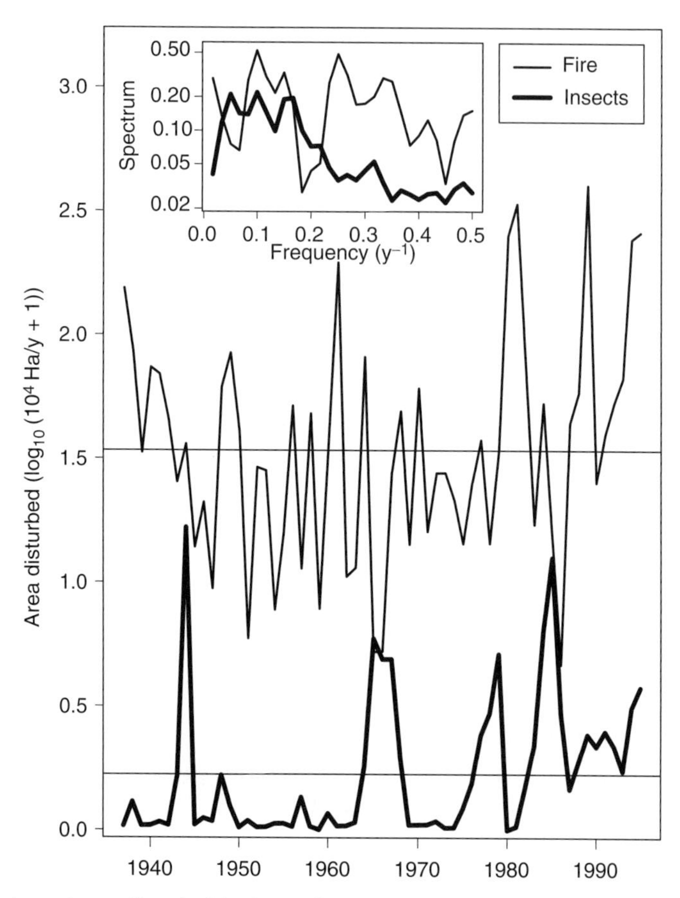

FIG. 2 Annual area disturbed by fire and insects in west central Canada (after Li *et al.*, 2003). Spectral analysis measuring periodicity is shown in inset.

Canada, despite the apparent correspondence between the bi-decadal insect outbreaks of the mid-1940s, 1960s, and 1980s and periods of reduced fire disturbance (Fig. 2). The data do reveal, however, a time-lagged, positive correlation between the fire and insect outbreak time-series ($r = 0.36$ for lag = 4 y; $p = .003$), indicating that high insect disturbance tends to precede high fire disturbance by about 4 years—a result similar to that reported in Ontario with regard to spruce budworm (Fleming *et al.*, 2002). Thus, fire and insect disturbance regimes are not entirely independent of each other. And yet, despite this interdependence, insect disturbance is distinguished by its periodicity and spatial synchrony.

Outbreaks as Regionally Synchronized Population Oscillations

The insects listed in Table 1 are among the most damaging insect species in northern forest ecosystems. All exhibit extensive, spatially synchronized oscillations of high amplitude. It is the extent of synchrony of these cycles, and not their local destructiveness, that creates the political will to intervene as populations go through epidemic phases. Fire, in contrast, is a relatively nonsynchronous disturbance, where the need to intervene stems not from its synchrony but from its enormous destructive capacity at the local scale, and its potentially rapid rate of spread. On the temporal scale, fire occurrence tends to be uncorrelated from one year to the next, but insect outbreaks tend to last for several years because of the strong relationship between insect densities in successive generations (Royama, 1992). Thus, fire ecologists do not use the term "cycle" in its trigonometric sense but instead use the notion of a fire-return interval that is a random variable whose mean might vary according to local circumstances (Van Wagner, 1978). In the entomological literature, the term "cycle" is used liberally but most often implies oscillations that are accompanied by a fair amount of noise. Insect ecologists may interpret a cycle as a "phase-forgetting quasi-cycle" (Nisbet and Gurney, 1982) because the phase, and to a lesser degree its periodicity, is influenced readily by stochastic inputs, such as weather and migration (Royama, 1992). This phase-forgetting aspect of insect population cycles is critical because this is what makes cycles vulnerable to large-scale synchronization via weak external forces, including spatially autocorrelated but temporally random perturbations, such as those caused by minor fluctuations in seasonal climate (Moran, 1953) and interpopulation dispersal of egg-bearing adults (Barbour, 1990).

Understanding the processes that promote synchronization of cycles is important because synchronization is the critical scaling mechanism that allows insect ecologists to be more categorical than fire ecologists in their estimates of disturbance return intervals. For example, in Ontario, forest tent caterpillar (*Malacosoma disstria* Hübner) populations cycle every 13.0 ± 0.95 SE years (Fleming *et al.*, 2000). If local population oscillations were not regionally synchronized, such precise generalizations would not be possible. Thus, a major topic of concern for forest entomologists is the origin of cycles and the mechanisms by which local-scale population oscillations are synchronized to form regional-scale outbreaks. For disturbance ecologists, the important point is that insect-caused disturbances are fundamentally different from physical disturbances, both in pattern and in genesis.

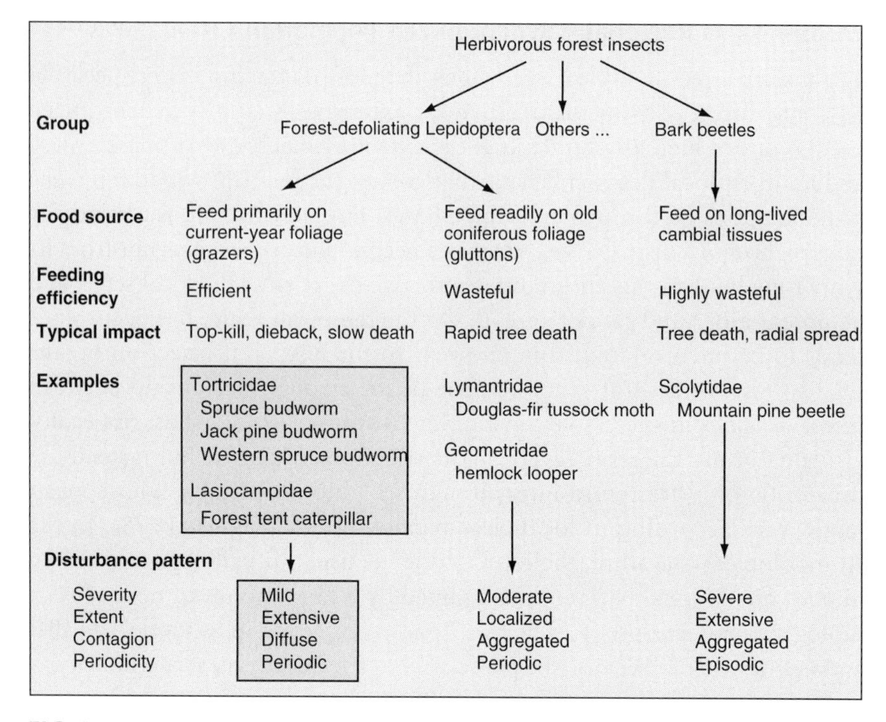

FIG. 3 The biology and dynamics of some foliage-grazing Lepidoptera (the focus of this chapter) contrasted with two other groups: the wasteful coniferophagous Lepidoptera and the bark beetles.

Foliage-Grazers versus Other Forest Insects

Most insects have little detectable impact on the forest; however, the destructive capacity of those in Table 1 is well documented. A general classification of herbivorous forest insects is illustrated in Fig. 3.

Many destructive insect species, such as bark beetles, feed on vital tissues such as the cambium and can directly kill their hosts. Some defoliators, such as the Douglas-fir tussock moth, *Orgyia pseudotsugata* McDunnough (Beckwith, 1978), and the hemlock loopers, *Lambdina* spp. (Rose and Lindquist, 1985), feed on all age classes of foliage (new and old) and so can completely defoliate their host tree in one or two seasons. With evergreen host species, complete defoliation usually results in death of the tree. The defoliating insect species considered in this chapter (Table 1), however, feed almost exclusively on current-year foliage of their host plant and rarely completely defoliate the tree. As such, they are more like grazers that exploit a renewable, and somewhat nonvital, host resource. The result is less immediate stress to their host. The single deciduous tree defoliator in Table 1, the forest tent caterpillar, may seem, at first, an exception to this generalization. However, its primary host, trembling aspen (*Populus tremuloides* Michx.), can produce a second flush of foliage, even after complete defoliation, either from reserves stored in roots and stems or from photosynthesis through the bark so that defoliated trees rarely remain unfoliated throughout the season.

THE PROCESS OF INSECT DISTURBANCE

Impact of Herbivory

During an outbreak of early-season defoliators, the tree's photosynthetic capacity is reduced. Repeated defoliation acts as a drain on the tree's nonstructural carbohydrates, leading to severe and sustained reductions in tree growth (Kulman, 1971). Repeated defoliation eventually leads to die-back of twigs, branches, roots, stems, and whole stands, as structural modules are sacrificed to reduce the demand for water and life-sustaining reserves. In the initial phase of an outbreak, however, the defoliators are actually less of a drain than one might expect because the younger, expanding tissues actually represent an early-season nutrient sink, and the photosynthetic efficiency of formerly shaded foliage of the lower crown increases following defoliation of the upper crown. Nevertheless, feeding by these insects eventually becomes less benign as the outbreak persists and the inability of the tree to renew photosynthetic material compromises vigor. For example, perennial defoliation reduces the ability of trembling aspen to produce a second flush of leaves as photosynthetic reserves (starches and storage proteins) are depleted. In the case of fir (*Abies*) and spruce (*Picea*), persistent budworm feeding leads to low recruitment of new foliage to the tree crown and increased retention of old needles (Baskerville and Kleinschmidt, 1981), which reduces the tree's overall photosynthetic capacity because old needles are less efficient at photosynthesis than are young needles (Clark, 1961).

Interaction with Other Agents

Insect-caused disturbances may manifest suddenly (e.g., over a few years) or gradually (e.g., over a decade or more). The ultimate disturbance may be attributed directly to the action of the defoliator outbreak, or it may be accelerated or modified by additional stressing agents. In trembling aspen and spruce and fir, tree health declines after persistent defoliation as trees become susceptible to other pathologies (Belyea, 1952; Churchill *et al.*, 1964). Other biotic factors, such as competing vegetation, pathogens, and bark beetles, can also be involved in decline. These agents typically act additively and in conjunction with abiotic forces, such as wind and drought. The total impact of all these disturbance agents is mediated by site conditions (e.g., soil depth, drainage, slope, aspect, and exposure).

The interactions among various disturbance agents adds complexity to spatial and temporal patterns of tree and stand death, such that the ultimate pattern of disturbance caused by an insect outbreak may bear little resemblance to the initial pattern of defoliation. Franklin *et al.* (1987) suggested that forest insects are part of a "Gordian knot of intertwined causality" where tree death is governed by a complex insect-pathogen-tree-site-climate interaction. Manion and Lachance (1992) suggested that large-scale forest decline is even more complex because of multiscale spatial variation in each of these agents. Such multicausal systems are notoriously difficult to analyze (Hilborn and Stearns, 1982) because causality is hard to ascertain, and quantifying the marginal impacts of individual agents is virtually impossible. For example, while there is little doubt that spruce budworm can kill trees directly (MacLean, 1984; Nealis and Régnière, 2004a), different hazard modeling studies conducted in different parts of eastern Canada have yielded differing results regarding risk factors influencing the severity of budworm damage (MacLean and Ostaff, 1989; Archambault *et al.*, 1990; Dupont *et al.*, 1991; Bergeron *et al.*, 1995; MacLean and MacKinnon, 1997; MacKinnon and MacLean, 2003). Similarly, in the case of the forest tent caterpillar on trembling aspen, although Witter *et al.* (1975) stated that "almost all the mortality [of trembling aspen during an outbreak in Minnesota] can be attributed to chronic, severe defoliation by the forest tent caterpillar," it is not entirely clear to what extent herbivory alone was to blame for recent large-scale aspen declines in Alberta (Hogg *et al.*, 2002) and Ontario (Candau *et al.*, 2002). Any defoliator can kill its host if the host is stressed and defoliation persists for long enough.

Defoliators Cause Selective Disturbance

A distinctive aspect of disturbances caused by insects is their highly selective nature. Selectivity arises from three sources: insect host plant preferences, insect population dynamics, and host tolerance to herbivory (Fig. 1).

All insects have, to a greater or lesser extent, host plant preferences. During an insect outbreak, nonhost plant species are often left undamaged.

With reduced interspecific competition, these undamaged nonhosts are left free to grow and to play an important role in stand-level succession and ecosystem-level maintenance of biodiversity. Even in so-called "stand-replacing" insect-caused disturbances, not all hosts are completely removed. Residual, surviving host trees and even dead trees and snags continue to have a significant impact on stand dynamics.

At the level of the individual insect, host plant preferences can manifest differently in the different life stages. Sometimes these stage-specific preferences reinforce one another, as when adults lay eggs in close proximity to the food source preferred by the larvae. But this is not always the case. The different life stages of spruce budworm, for example, show varying degrees of preference for black spruce (*Picea mariana* (P. Mill.) BSP). Spruce budworm moths find black spruce highly acceptable as a host on which to oviposit, and hatchlings find many suitable sites on the tree to use as overwintering sites. Young feeding larvae, however, will disperse actively in large numbers from black spruce in springtime as the late-flushing characteristic of black spruce renders the buds too difficult for early-season feeders, such as budworm, to establish feeding shelters. Once budburst has occurred, not only is black spruce an adequately nutritious host for spruce budworm larvae, but larvae actually perform better later in the season on black spruce than on either white spruce [*Picea glauca* (Moench) Voss] or balsam fir [*Abies balsamea* (L.) Mill.] (Nealis and Régnière, 2004a). In this case, preference is a function of relative phenology of host and insect, and this could vary across a landscape. Thus, generalizing host plant preferences can be misleading without specific knowledge of the mechanisms that govern this process.

Insects being relatively small organisms, the preferences discussed so far are manifest at very small spatial scales. However, the act of dispersing to fulfill a preference, when carried out by many individuals in a population, leads to enormous potential for richly textured patterns of herbivory and disturbance at larger scales. As fine-scale, subpopulation processes, insect aggregation, congregation, and dispersion are major sources of spatial variability in the selective process leading to disturbance. To the extent that preferred resources are clumped, insect aggregation can lead to consistently higher rates of immigration into some patches over others and thus link small- and large-scale patterns. Hardy *et al.* (1983) observed that spruce budworm "epicenters" tend to occur in old white spruce stands located in river valleys in the subboreal mixed-wood forest region of Quebec. Moth congregation and aggregation could play a critical role in the formation of these epicenters. What is interesting from a disturbance ecology perspective is that although outbreak epicenters may indicate a preferred habitat type for ovipositing moths, these are not the areas where outbreaks tend to last the longest or cause the most damage. The areas of greatest disturbance are those where outbreaks tend to linger the longest, and in spruce budworm these areas are located typically outside the epicenters (Gray *et al.*, 2000). A reasonable conjecture is that the points that are first to show signs of damage are those that are first to begin exporting moths in large numbers (Nealis and Régnière, 2004b). This would be consistent with key elements of the epicentric theory of spruce budworm outbreak spread (Greenbank *et al.*, 1980) while at the same time maintaining consistency with Royama's (1984) noneruptive theory of budworm outbreak cycles. In other words, if these so-called epicenters are also the initial points of outbreak collapse, then they may not function as mere sources of outbreaks. If they are also a source of predators, then they may be acting as relay points in what is a loosely coupled network of oscillating predator and prey populations (Régnière and Lysyk, 1995).

Overview of Insect Population Dynamics

When insect preferences for a particular species, size class, or condition of host plant are met, the result is increased rates of survival and reproduction on that host. Over time, this can lead to an increase in insect density and disturbance. For example, survival of jack pine budworm (*Choristoneura pinus pinus* Freeman) larvae is much higher on host trees that produce pollen cones than on those that do not (Nealis and Lomic, 1994). Consequently, disturbance tends to be more severe in stands with a high proportion of flowering trees (Nealis *et al.*, 2003).

While it is conceptually useful to distinguish between sub–population-level aggregative processes and population-level reproductive processes, that distinction is often blurred by the fact that preferred host plants tend to both boost population growth rates and foster aggregations, either through active recruitment or passive arrestment of individuals simultaneously. Three points arise. First, host plant preferences have a strong influence on the spatial pattern of disturbance because they are operating through two synergistic mechanisms, insect aggregation and population growth. The blurring that results from the strong interplay between individual-level host preferences, population dynamics, and host distributions is precisely the reason that the species-level insect–plant association is such a durable ecological concept. Second, different systems will exhibit different patterns of selective disturbance, depending on whether the aggregative or reproductive processes dominate. With bark beetles, where the aggregative process is key to overcoming strong plant defenses, disturbance tends to be locally patchy and aperiodic (see Chapter 16). With the defoliators discussed here, where the reproductive process often dominates over the aggregation processes, disturbance tends to be diffuse and periodic (see Chapter 17). Third, to the extent that preferred resources are clumped, so too will population growth be higher in some patches than others, and so therefore will be the pattern of disturbance.

Once insect populations have increased to a very dense state, the positive synergy between population growth and insect dispersal can become negative, as insects switch from aggregative to diffusive behavior, and as populations switch from a phase of unlimited exponential growth to a phase of resource-limited growth. The spillover effect that results from density-dependent dispersal often leads to seemingly paradoxical patterns of disturbance. For example, during a jack pine budworm outbreak, although stands with many pollen-producing trees are far more prone to recruitment of budworms, even the trees not producing pollen cones will be defoliated, as there is often spillover from the flowering trees. Similarly, spruce budworm oviposit in the upper canopy of large trees but spill over in large numbers onto smaller size classes of suppressed trees as they disperse there in springtime upon emerging, or in midsummer in search of food. That subdominant and understory balsam fir tend to be killed first in a spruce budworm outbreak indicates the preference of moths to oviposit in tall trees, the preference of larvae to avoid overcrowding (or starvation), and the frequency with which understory fir are found occupying canopy gaps in mature coniferous stands. As a result, during outbreaks, spruce budworm can sometimes act in the manner of a silvicultural "thinning from below."

The same spillover phenomenon has been reported in western spruce budworm (*Choristoneura occidentalis* Freeman; Fellin, 1976) and Douglas-fir tussock moth (Harris *et al.*, 1985), where, in shelterwood and partial cuts of Douglas-fir [*Pseudotsuga menziesii* (Mirb.) Franco], the absence of an intercepting subcanopy led to the movement of young larvae in large numbers from canopy trees to regenerating trees, ultimately killing them. These are cases where modified silvicultural practices stimulated, rather than emulated, natural disturbance. They are noteworthy because many defoliating insects tend to perform better in thinned forests (Bauce, 1996; Dobesberger, 1998) and in sunny, exposed, edge-dominated habitats (Moore *et al.*, 1988; Roland and Kaupp, 1995; Fortin and Mauffette, 2001).

Host Tolerance

Some host plants are more tolerant of herbivory than others because of their ability to cope with reduced foliage or their ability to recover lost foliage (Haukioja and Koricheva, 2000). These responses vary as functions of numerous factors, such as tree architecture, site condition, competition, and health status or vigor (reviewed in Tiffin, 2000). A tree's internal programs for resource acquisition and allocation are also important because they determine the tree's ability to re-establish apical dominance, to alter production of short and long shoots or leaf-level photosynthetic activity, to use stored reserves, or to shed and re-grow fine roots. Differences in this capacity can be manifest at any level, from individual (e.g., large vs. small crown) to species (e.g., white spruce vs. fir), and can be associated with a range of compensatory growth responses, such as epicormic shoot production, rootlet regrowth or enhanced flowering and seed production in conifers, or suckering in trembling aspen. Variability in tolerance to herbivory introduces spatial variability and patchiness in a tree's vulnerability to a given level of herbivory. Even if insect densities were uniform in space (i.e., no host preferences, no dynamics), impact would vary locally.

Modeling Herbivore Impact: Empirical versus Process Approach

There are many examples of herbivore impact studies reaching different conclusions under ecological circumstances that vary only slightly. One case in point concerns the spruce budworm in eastern Canada, where two correlative impact studies in the province of Quebec (Dupont *et al.*, 1991; Bergeron *et al.*, 1995) reached different conclusions, despite a fairly similar range of ecological circumstances. Dupont *et al.* (1991) emphasized local factors acting within stands and concluded that site conditions (hydrology, geology, soil moisture, and texture) were critical factors influencing tree death during an outbreak. In contrast, Bergeron *et al.* (1995) concluded that forest conditions (stand age and species composition) were more important and that the spatial context of stand composition in the surrounding forest matrix was significant. A third study from New Brunswick suggests both studies may be partially correct, and that the differing results could be a product of differing sampling schemes and ecological contexts, possibly emphasizing different aspects of susceptibility and vulnerability (MacLean and MacKinnon, 1997). Specifically, they argue that because outbreak intensity varies among regions, the risk factors that emerge as significant in any spatial analysis also tend to vary.

Rather than dwell on the failures of the correlative approach to insect disturbance modeling, we wish to highlight its close relationship with process modeling, showing how the two approaches can be pursued synergistically within a common framework (Fig. 1). The correlative approach, or empirical hazard modeling, is conceptually simple and more direct than the process approach in that it seeks to explain spatial variation in disturbance risk as a function of spatial variation in forest landscape characteristics. The problem is that hazard models are static, descriptive models and therefore limited in their predictive capacity, especially under conditions of system change. Because they downplay the importance of the dynamic temporal processes that mediate insect population growth and forest structure, hazard models that work well during one outbreak will tend to perform less well in subsequent outbreaks, where forest conditions and population growth processes may differ.

Process studies can help resolve unexpected differences between impact studies, and they may suggest more appropriate hypotheses worth testing. For example, where Bergeron *et al.* (1995) showed that conifer stands are less vulnerable to budworm disturbance when embedded in mixed-species forests, Su *et al.* (1996) showed that hardwoods located within the focal stand produce the same effect, only locally. They hypothesized that this was a result of the effects of forest composition on the budworm natural enemy community—a result that is now supported by more recent process studies showing that the natural enemy complex indeed varies as a function of forest attributes both within the focal stand and in the surrounding forest matrix (Cappuccino *et al.*, 1998; Quayle *et al.*, 2003).

Patterns of Disturbance

Spatial Patterns

Although some defoliators tend to be more destructive than others, all species exhibit substantial spatial and temporal variability in the extent of damage they cause. The extent and degree of disturbance during a defoliator outbreak depend largely on the intensity and duration of defoliation over the landscape. When herbivory is intense and prolonged, stand death and stand-replacement can occur. What is more common, under regimes of more moderate or brief herbivory, are milder disturbances, such as the formation of small canopy gaps, patches of accelerated self-thinning among suppressed stems, or a brief increase in light, nutrient, and water availability, and loss of productivity.

For herbivore population fluctuations to reach outbreak status, it is typically necessary that the size, density, and quality of preferred hosts exceed some level. For an outbreak to be extensive, the preferred hosts need to be distributed extensively, although not necessarily contiguously (Roland, 1993). When and where preferred hosts are less available, the herbivore must deal with suboptimal conditions, and this can influence the dynamics of local populations. Thus, the interplay between population dynamics processes and host plant preferences can be a source of both spatial and temporal variability in the pattern of disturbance.

Temporal Patterns

Boreal forest ecosystems are often dominated by relatively few tree species and, because of their genesis from previous disturbances, tend to be correlated over large spatial scales. In such homogeneous forest types, even a relatively specialist feeder, such as the jack pine budworm, can find sufficiently extensive, favorable habitats to support an increase in its density, leading to a significant disturbance at the landscape scale. Because of the relative homogeneity of northern forests, insect-caused disturbances can affect multiple regions simultaneously.

One of the most obvious results of forestry of the past century has been conversion of the forest toward more homogeneous or bimodal age classes dominated by a few tree species. Furthermore, the tree species favored have been those which tend to be shade intolerant, fast-growing, ruderal species: balsam fir over white spruce, jack pine (*Pinus banksiana* Lamb.) over white pine (*Pinus strobus* L.), poplar (*Populus* spp.) over everything. One of the urgent objectives of disturbance ecology is to determine whether these anthropogenic patterns have resulted in new patterns of natural disturbances, and whether this threatens the stability and sustainability of forest ecosystems.

Herein lies a major limitation in the correlative, historical approach to risk analysis: disturbance patterns of the past will not be a strong indicator of the future if environmental conditions have changed substantially. In contrast, process-oriented studies, because they model mechanisms, have greater predictive value.

POPULATION DYNAMICS OF FOLIAGE-GRAZERS

We have argued that insect defoliators constitute a distinct class of forest disturbance because they are selective and because they tend to exhibit cyclic and synchronized outbreaks. Next we discuss the nature of the mechanisms that generate cycles and synchronize these cycles across the landscape. We propose a template of intermediate complexity that could be used to structure a process-oriented insect disturbance model. We develop this template by comparing and contrasting the dynamics of the forest tent caterpillar and three of the Canadian conifer-feeding budworms, all cyclic defoliators.

Case Studies

Forest Tent Caterpillar, Malacosoma disstria *(Hübner)*

Forest tent caterpillar outbreaks in Ontario, Canada, are remarkably cyclic, with a periodicity of 13.0 ± 0.95 years SE during the 20th century (Fleming *et al.*, 2000). Longer-term reconstructions by Baird (1917) for all of central Canada suggest that outbreaks in the 19th century followed much the same pattern. While Baird found no reason to suspect any regional differences in outbreak periodicity, Hildahl and Reeks (1960), studying outbreak patterns in western Canada from 1920 onwards, were at odds with Sippell (1962) in Ontario as to whether forest tent caterpillar outbreaks were cyclic. Thus, the degree of cyclicity or synchrony in forest tent caterpillar outbreaks may vary by region. Such regional variability was confirmed by Cooke and Lorenzetti (2006), who reported a 9-year outbreak cycle in the aspen-dominated Abitibi region of northwestern Quebec and a 14-year cycle in the maple-dominated Appalachian region of southern Quebec, during the interval 1938–2002. They also noted that outbreak cycles in Quebec covered, on average, only 37% ± 13% SE of the insects' potential outbreak range.

A similar but less variable result is revealed in data from Ontario (Fig. 4). Of the six forest tent caterpillar cycles occurring during the period 1929–2003 (Fig. 4, *top panel*), the most extensive outbreak (cycle II) spanned 65% of the insects' outbreak range; the least extensive (cycle III) spanned only 22% of that range (Fig. 4, *bottom panels*). On average, individual outbreak cycles span only 43% ± 7% SE of the insects' outbreak range. Thus, despite the regularity of the cycle, it appears that there are always significant areas where populations do not get high enough to cause detectable damage. In fact, there are only two very small areas (near Dryden, northwest region, and Sudbury, northeast region) where all six outbreak cycles have occurred.

The spatial variation in cycle synchrony is an empirical fact which requires explanation. Many environmental factors, both stochastic and deterministic, could modulate cycle amplitude and vary sufficiently in space and time to effectively reduce observed levels of cycle synchrony. Cooke and Roland (2003), for example, discussed how fluctuations in winter

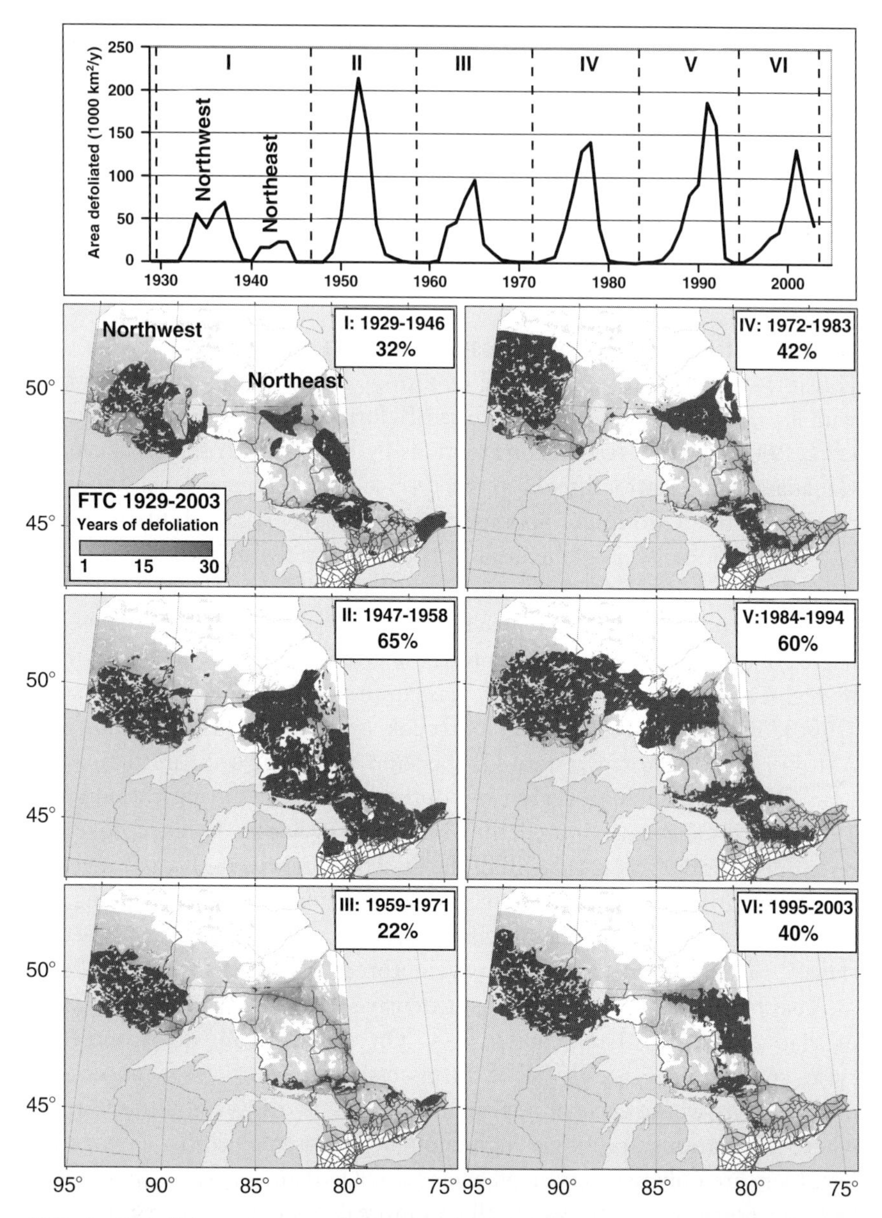

FIG. 4 Periodicity and extent of six cycles of forest tent caterpillar outbreaks in Ontario. The number of years of defoliation during the full period 1929–2003 (gradient) is overlain by the extent of defoliation during each cycle (*dark grey,* surface area measured as percentage of full extent).

temperature may play a complex role in both synchronizing and then desynchronizing forest tent caterpillar cycles in western Canada. We argue that process-oriented studies that explicitly address variable patterns of synchrony are needed to prevent uncritical acceptance or rejection of synchronization theory. Understanding how real-world population oscillations may be synchronized is key to understanding how large-scale insect disturbance occurs.

Jack Pine Budworm, Choristoneura pinus pinus *Freeman*

Analyses of jack pine budworm defoliation records in the Great Lakes region of North America indicate that outbreaks of this species recur at vaguely periodic intervals (Fleming *et al.*, 2000; McCullough, 2000). However, as with forest tent caterpillar, the degree of cyclicity and the estimates of periodicity vary among regions, such that it would be a drastic oversimplification to suggest that outbreaks are strictly cyclic. Although a 10-year cycle appears defensible in western Canada (Volney, 1988), outbreak periodicity in the pine barrens of Wisconsin varies geographically, with 10-year cycles in the northeast, 5- to 6-year cycles in the drier central region, and very weakly periodic 14-year cycles in the southwest (Volney and McCullough, 1994). The recent patterns in Ontario seem equally vague (Nealis *et al.*, 2003).

As with the forest tent caterpillar, there are always large subregions where expected cycles fail to materialize. Outbreaks in Ontario, for example, never expand to cover the entire range of the insect over the course of a decadal cycle (see Fig. 1 in McCullough, 2000) despite the presence of preferred host plants, favorable climate, and plenty of neighboring infestations that could act as sources of dispersing moths.

Spruce Budworm, Choristoneura fumiferana *(Clem.)*

Candau *et al.* (1998) found differences of 32, 45, and 36 years between the peaks of two successive outbreak cycles of spruce budworm in western, central, and eastern Ontario, respectively. With data from only two cycles, one cannot estimate interval means and variances. The estimate of low-frequency, approximately 30-year periodicity nevertheless agrees with similar short-term analyses of defoliation data from Quebec (Hardy *et al.*, 1986; Gray *et al.*, 2000) and of population data from New Brunswick (Royama, 1984)—all from the 20th century.

Longer-term tree ring studies of white spruce from Quebec, initiated by Blais (1983) and extended spatially by Jardon *et al.* (2003) and Boulanger and Arseneault (2004), confirm that spruce budworm outbreaks tend to occur every 32.0 ± 5.1 years SE although the six reconstructed cycles that are thought to have occurred since 1800 have varied substantially (see Chapter 17). A similar result has been reported from nearby New Brunswick (Royama, 1984).

Tree ring data from northern British Columbia indicate that spruce budworm outbreaks in the far west of its range are occurring at roughly 26-year intervals (Burleigh *et al.*, 2002). Thus, there is substantial long-term evidence for a low-frequency cycle in spruce budworm outbreaks over a wide geographic range of conditions. Spruce budworm cycles are not as regular as those of forest tent caterpillar, but neither is the forest tent caterpillar perfectly periodic.

It has been argued that the low-frequency outbreak cycle (approximately 30–40 years) of *C. fumiferana* in eastern North America is the result of the naturally short lifespan of the average balsam fir stand (approximately 60–80 years), which could be shortened to approximately 40 to 60 years in the presence of budworm (Ludwig *et al.*, 1978; Clark *et al.*, 1979; Hassell *et al.*, 1999). However, as discussed below, this argument clearly does not stand up for western spruce budworm (*C. occidentalis*). Although outbreaks of this species recur every 22 years, Douglas-fir forests are renewed on a much longer time frame (some individual trees may live up to 700 years). So, while *C. fumiferana* outbreaks in the east often result in significant forest disturbance, it does not follow that the low-frequency population cycle is dictated by the disturbance cycle. It may simply be that, for this particular system, the two processes are close enough to one another in frequency that they are often correlated without being causally linked.

Western Spruce Budworm, Choristoneura occidentalis *Freeman*

It is noteworthy that tree ring reconstructions of outbreaks of two other western North American budworms indicate highly periodic, low-frequency cycles. The 2-year cycle budworm (*C. biennis* Freeman), feeding on white spruce and subalpine fir (*Abies lasiocarpa* Nutt.) in northcentral British Columbia, appears to have an outbreak every 32 years, or 16 generations (Zhang and Alfaro, 2002; 2003). The western spruce budworm, *C. occidentalis*, on Douglas-fir in southern British Columbia appears to have an outbreak every 22 years (Alfaro *et al.*, 1982). In New Mexico, approximately 15 outbreak cycles of *C. occidentalis* have occurred in the past 400 years (Swetnam and Lynch, 1993). In Colorado, approximately 32 outbreak cycles have occurred in the past 700 years (Ryerson *et al.*, 2003). As a group it thus appears that the North American fir–feeding budworms do outbreak periodically, with a return interval that is much longer than a decade.

Many of the defoliated trees on which the Colorado C. *occidentalis* outbreak reconstructions are based exhibited top-kill, while relatively few were thought to have died as a result of defoliation (Ryerson *et al.*, 2003). These outbreaks are thus thought to be less disturbing to the forest than are outbreaks of the spruce budworm in the eastern spruce–fir forest.

This is strong evidence that not all defoliator population cycles translate into severe forest disturbances. Indeed, the cyclic western spruce budworm and the Douglas-fir forest may represent a resilient "equilibrium state"—even more resilient than the spruce budworm–fir forest ecosystem. This resilience implies that some process other than forest destruction must cause the primary western spruce budworm cycle. This may well be the case in other folivore-dominated forest systems, including spruce budworm (Royama, 1992).

Summary

Our goal was to show that much of the theoretical population dynamics literature on population cyclicity and spatial synchronization is highly relevant to disturbance modeling. At the same time, we also wanted to show that defoliator outbreaks are imperfectly cyclic and imperfectly synchronous.

If defoliator outbreaks were perfectly periodic and synchronized, and if the frequency of oscillation were fixed in time and in space, then, from a disturbance ecology perspective, there would be little need to model population processes; a simple mathematical description of the phase characteristics would be adequate. But as we have shown, there is systematic variability in systems behavior (e.g., trends in outbreak cycle amplitude, and areas where outbreak cycles do not emerge), suggesting that additional variables are at play. If these can be identified, then our ability to predict disturbance will increase. The key questions are the following:

1. How are cycles generated?
2. How are cycles synchronized?
3. What factors serve to modulate cycle amplitude and synchrony?

Volumes have been written on these complex questions, so a quick and lasting synthesis is simply not possible. But we believe it may be possible to make some gains in disturbance ecology by working with simpler, semi-mechanistic, data-rich models, as we outline next.

A Tritrophic Model of Cycle Induction

The specific reasons put forward for animal population cycles are many. Among forest insects, some of the most common biological mechanisms discussed are the following:

1. Predator–prey interactions (Nicholson and Bailey, 1935)
2. Host–disease interactions (Anderson and May, 1980)
3. Induction of host plant defenses or changes in plant nutritional quality (Denno and McClure, 1983; Haukioja, 1991)
4. Intrinsic changes in individual traits (Wellington, 1964; Chitty, 1967)
5. Changes in population allele frequencies (Campbell, 1966; Stenseth, 1981)
6. Induction of maternal effects (Ginzburg and Taneyhill, 1994)

The common consensus is that harmonic cycles in density of insect populations are the result of delayed, negative feedbacks between the insect and its environment (Berryman, 1978; 2002). Negative feedback results from the finite capacity of populations to grow in a limited environment. The delay is inherent in the relationship between the density of the population and the impact of that density on the environment when the negative effects of resource depletion or increased predation are experienced by future generations (i.e., the effect of present-day population densities is delayed until the future).

All the biological mechanisms listed above are, in theory, capable of generating cycle-inducing feedback. In practice, modern insect ecologists tend to focus on a subset of these: the so-called "top-down" effect of natural enemies (1 and 2) and the "bottom-up" effect of the host plants (3). The ensemble, referred to as a tritrophic interaction, is a marvelously complex ecological phenomenon, despite its conceptual simplicity.

The tritrophic model is a good conceptual aid but is too abstract to use for analytically understanding how disturbances arise in particular systems as a result of specific biological processes. A more concrete model, presented here in strictly graphical form, shows how our concept can be implemented and used to describe the dynamics of many defoliator species.

Model Formulation

Fig. 5 represents the cyclic behavior of a general tritrophic interaction as it might be expected to manifest in the systems considered here. The dark gray line represents herbivore population densities over a population cycle. The black line represents natural enemy populations. These natural enemies could be predators, parasites, or pathogens. Both curves are jagged,

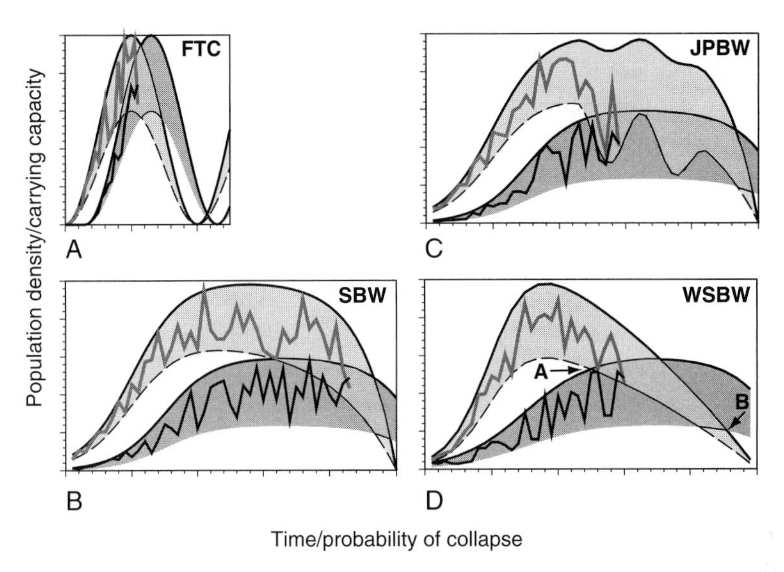

FIG. 5 The tritrophic herbivory model applied to four foliage-grazing Lepidoptera. (Symbols explained in text.)

rather than smooth, to represent the imposition of stochastic environmental influences (highly variable weather, generalist predators, and other density-independent effects). The shaded regions around these curves represent an envelope, or range, of possible population trajectories.[3] The top edge of the light gray region indicates the expected herbivore population trajectory in a favorable environment; the bottom edge of this region is the expected trajectory in an unfavorable environment. The dark gray region applies similarly to natural enemy populations.

When the trajectory of the natural enemy population (black line) intersects that of the herbivore population (dark gray line), a herbivore decline is inevitable. The decline results from the effects of both host plants and natural enemies, with natural enemies being the more proximal and fast-acting of the two agents. Note also how stochastic factors may help in bringing about decline, or in propping up a peak, while having no role in

[3] In effect, there are two ways to interpret these regions: as a probability field or as an environmental range. Under the first interpretation, the population cycle, as a stochastic process, may follow any path within an infinite ensemble of theoretical possibilities within the envelope. Under the second, cycle amplitude may be considered a variable parameter that changes (slowly in time, or predictably in space) according to varying ecological circumstances. The second interpretation is the one emphasized here.

causing the cycle—that is, the stochastic elements may affect the timing and amplitude, but not the nature, of the pattern.

When natural enemy population growth is favored over herbivore population growth (black trajectory near the top of dark gray region and the dark gray trajectory near the bottom of light gray region), a natural enemy–induced population collapse at point A (left portion of the medium gray zone of intersection) is likely. This would occur, for example, in forest types that are not rich in primary host plants but are rich in host plants for other species of herbivores on which these natural enemies can feed.

In contrast, where herbivore population growth is favored over natural enemy population growth (black line low, dark gray line high), herbivory lingers on long enough that there are some destruction of the host plant resource, some reduction in carrying capacity, and a decline in herbivore populations not attributable to natural enemies. In such a case, natural enemies will not appear to cause a population collapse until point B, at the right of the medium gray zone of intersection. Because this decline is retarded, the system suffers significant disturbance.

Model Behavior

As forest and natural enemy communities may vary, so may the shape of the regions and the behavior of the model. In this way the model can be made to fit a variety of systems. According to the scheme depicted in Fig. 5, the forest tent caterpillar, being more strongly limited by fast-acting natural enemies than are the budworms, exhibits a higher-frequency primary oscillation (Fig. 5A). The budworms, as a group, exhibit a lower-frequency primary oscillation because of the relative inefficiency of natural enemies and the relative importance of host plants.

The western spruce budworm, being the most oscillatory budworm system with the highest degree of synchrony[4] and the lowest amount of severe disturbance, exhibits the least distinct pattern of host plant responsiveness (Fig. 5D). It is like the forest tent caterpillar in that any host plant effects (i.e., ill effects of resource depletion) are manifest at the same spatial scale over which natural enemies operate (e.g., within stands); hence, a cycle that is regular in periodicity.

[4] Eastern spruce budworm outbreaks (in the 20th century) may have been more extensive than those of western spruce budworm (but not after adjustment for size of host range); however, tree ring studies indicate stronger periodicity and greater synchrony in western spruce budworm (especially when desynchronizing effects of terrain-related dispersal barriers and microclimatic variability in western North America are discounted).

In contrast, the spruce budworm (Fig. 5B) is a system where resource depletion has little impact early in the cycle (as larvae continually space themselves out and adjust their fecundity according to food availability) but a large impact later in the cycle (as damage accumulates, tree death becomes widespread, environmental carrying capacity drops, and moths emigrate en masse). At this late stage, a population collapse is inevitable and natural enemies will finally begin to take over.

The jack pine budworm is different still from the other two budworms in that larvae depend so heavily on pollen cones, which fluctuate periodically in availability. When pollen cone production is shut down by previous defoliation, local populations begin to decline as the impact of natural enemies is accentuated (Nealis, 1991). When flowering resumes, after defoliation has abated, local populations can rebound; however, the degree of rebound depends on natural enemies. The result is a scale-dependent multifrequential oscillation (wiggly region in Fig 5C). Jack pine budworm outbreaks are usually observed in areas where production of pollen cones is intense and natural enemies are lacking. Either of the two mechanisms is sufficient to bring about population decline, but reduction of pollen cone production, being a highly localized process, initiates the decline and natural enemies follow (Nealis *et al.*, 2003).

As in jack pine budworm, host plant–associated feedback in the spruce budworm system counteracts the cycle-inducing effect of negative feedback from natural enemies, with the result that oscillations are actually multifrequential rather than unifrequential. The difference is that the higher-frequency "sawtooth" oscillations are not as prominent a feature of spruce budworm outbreaks, and went largely unnoticed until a study by Royama (1984) was published.

Although all coniferous hosts probably reduce their rates of pollen cone production in response to budworm defoliation, it is in the jack pine system where this has a measurable impact on budworm spatial dynamics. Here, the reduction in pollen cone production leads to a spatially shifting mosaic of local-scale population eruptions that are nested within the landscape-scale outbreak cycle. For example, during the 1980s outbreaks of jack pine budworm in Ontario, population eruptions tended to last about 2 years at the stem level (approximately 10^0 m), about 4 years at the stand level (approximately 10^2 m), about 5 years at the regional level (approximately 10^5 m), and about 7 years at the inter-regional (provincial) level (approximately 10^6 m) (McCullough, 2000; Nealis *et al.*, 2003). In contrast, the high-frequency fluctuations in spruce budworm tend to be spatially correlated across whole regions (Royama, 1984). Indeed, this is the main reason that spruce budworm outbreak cycles are thought to be synchronized across whole regions (Royama *et al.*, 2005).

Model Summary

All four insect–forest systems can be interpreted within the context of the same tritrophic regulatory structure. What differs is the strength and nature of host plant feedback relative to the cycle-inducing effect of delayed feedback caused by natural enemies. Host plant feedback is relatively weak in the forest tent caterpillar system, hence the strongly periodic, natural enemy–related pattern of outbreaks. What differs among the individual budworm systems is the rate at which the cumulative effects of host damage are expressed and, therefore, the spatial scale over which damage and disturbance occur. The important point is that the tritrophic model accommodates both top-down (natural enemies) and bottom-up (host plant) forces explicitly; the functional relationships are the same but their speeds are different.

In all these systems, the amplitude of population oscillations is expected to vary randomly as a function of stochastic forces. Some cycles will be amplified to the level of evident disturbances; others will result in undetectable levels of defoliation, but cycles nonetheless. The nature of these stochastic forces will vary among cycles in time and space. Sometimes the modulating force will be more weather-related, other times more predation-related. Sometimes it will be a spell of consecutive years of favorable conditions; sometimes it will involve a single catastrophic event. A key challenge for insect ecologists is to identify in each system which stochastic elements are critical agents and how they influence other functional processes within the system.

Spatial Implementation

The collective dynamics of environments that are patchily structured in terms of forest and natural enemy communities can be represented by using a spatial mosaic of the various types. For example, we know that balsam fir is more like jack pine than white spruce in the low proportion of carbohydrate reserves held in storage. Fir declines more quickly in quality as a host (Nealis and Régnière, 2004a) and is more vulnerable to mortality (MacLean, 1980; Blais and Archambault, 1982). In the same way that jack pine is intolerant of and reproductively adapted to fire disturbance, balsam fir appears to be intolerant of and reproductively adapted to spruce budworm disturbance (Morin, 1994). Thus, we expect budworm disturbance dynamics to differ in spruce versus fir forests. This difference is not illustrated in Fig. 5 because in central Canada spruce and fir stands are often sufficiently intermingled that the distinctly different dynamic processes are not free to scale up from the stand level to the landscape level.

Fig. 5B can thus be considered as the average behavior of spruce budworm over a complex landscape in central Canada. In practice, the model would need to be parameterized for all the different stand types one encounters in the forest landscape, hence the need for spatially structured population models operating on realistic landscapes. For this purpose we expect that comparative studies in spruce-dominated western Canada versus fir-dominated eastern Canada to be particularly informative.

Spatial Synchronization

We have shown that periodic insect outbreaks are the result of local insect populations cycling in response to changes in both their resource and in the impacts of their natural enemies. A key question is how these local-scale oscillations are synchronized to form large-scale outbreaks. This is an area of intensive research, so a complete synthesis would be premature. However, a brief review is necessary to explain how such tiny organisms can inflict such widespread ecological disturbance. Three principal mechanisms could, in theory, account for cycle synchronization:

1. Spatially autocorrelated random perturbations, which act as a phase-locking mechanism (Moran, 1953)
2. Spatially autocorrelated nonrandom shocks, which act as a period-forcing mechanism (Berryman, 1981)
3. Interpopulation migration, which acts as both a frequency-homogenization and phase-locking mechanism (Barbour, 1990)

The literature has generally done a poor job discussing the various theories of cycle synchronization (but see Royama, 1997). Even now there is insufficient evidence with which to decide which of these mechanisms is most important. We cannot dismiss the possibility that all three processes are operating, perhaps at multiple spatial scales. The challenge for insect ecologists is to develop mature, inclusive hypotheses by looking at specific systems and using long-term data and formal analytical methods with well-designed, system-specific process models. This approach, for example, has led to a much clearer understanding of how spruce budworm population cycles are synchronized by multiple mechanisms (1 and 3) acting on a single life stage—the egg-carrying adult female moth (Royama *et al.*, 2005).

Synchronizing Mechanisms

Catastrophic shocks, periodic or otherwise, are "regional" in scale. Thus, only mechanisms 1 and 3 are capable of what is termed "global" synchronization: the ability to synchronize locally induced cycles across the entire range of a continentally distributed insect.

From a biological perspective, dispersal and spatially autocorrelated weather perturbations appear to be very different mechanisms. Mathematically, however, they operate in similar ways—with one resultant difference. The similarity is this: because the phase of a population oscillation is determined largely by the sum history of stochastic perturbations, any set of oscillators can be readily brought into phase with one another by ensuring they have a shared history of stochastic perturbation. As Bjornstad *et al.* (2002) put it: "cyclic systems yearn to align themselves." To understand the argument it is necessary to understand how delayed negative feedback is capable of generating oscillatory dynamics, and how stochastic perturbations imposed on this regulatory loop are incorporated into the system state, such that the system state at any point in time is largely determined by the sum history of perturbations. While the periodicity of oscillations is determined by the nature of the feedback loop, the phase of those oscillations is free to be determined by the accumulated history of perturbations, which eventually overwhelm any influence of initial conditions.

What Moran (1953) understood was that independently fluctuating populations at two locations will be quickly brought into a common phase when random perturbations, such as those caused by fluctuations in a climatic factor, are correlated between the two locations. The higher the correlation of these perturbations, the faster and more complete will be the synchronization of the phases of these populations.

What Barbour (1990) understood was that dispersal of moths could act not only as a source of correlated perturbation but also as a dynamic homogenization mechanism, whereby ecologically dissimilar populations regulated by different sets of factors could effectively be blended to form a single meta-population regulated by a common superset of factors. With sufficient dispersal rates between them, the dynamic similarity of populations is guaranteed, and any differences in periodicity or phasing will diminish to zero through time.

The main difference between Barbour's and Moran's proposed mechanisms of synchronization is that dispersal is a robust mechanism, whereas correlated perturbation is not. To be more precise, Moran's effect is constrained by the requirement that the oscillating populations be governed by a generating process that is homogeneous in its parameters (i.e., identical periodicity), whereas Barbour's is not. What this implies is that in real, heterogeneous landscapes, where predator and prey communities surely vary spatially in their composition, only the dispersal mechanism is robust enough to ensure landscape-wide spatially synchronized population oscillations.

Synchrony versus Synchronization

If the synchronizing forces of dispersal and correlated perturbation are strong and universal, then why is it rare to see systems that fluctuate in perfect synchrony? While dispersal and correlated perturbations may tend to pull systems toward a synchronized state, in real, natural systems, multiple uncorrelated perturbations are at play, so that precise and universal synchronization is by no means guaranteed. This would not be obvious from a reading of the ecological literature, which often refers to the pattern of population synchrony and rarely to the process of synchronization. The literature is so replete with examples of synchronized oscillatory systems that one might be tempted to conclude that all systems behaved this way. In fact, even the best-synchronized oscillatory systems, such as larch budmoth (*Zeiraphera diniana*) in Europe, do not exhibit synchrony over their full range (Bjornstad *et al.*, 2002). As we have shown with North American folivores, synchrony occasionally breaks down in time, in space, or under particular ecological circumstances.

The important message for disturbance ecologists is that synchronization is a critical scaling process that makes insect outbreaks very different from abiotic disturbance processes, both in pattern and in cause. Synchrony varies naturally, and can be made to vary artificially, by altering parameters such as insect dispersal rates, landscape connectivity, and homogeneity in regulatory processes. Once again, the insight is gained from the study of the processes that create the phenomenon, rather than analysis of the patterns the phenomenon creates.

Amplitude Modulation, Outbreak Trends, and Environmental Change

A mechanistic theory of insect-caused forest disturbance should be able to address the observation that outbreaks of several major periodic defoliators have increased in severity, extent, duration, or synchrony during the 20th century. Among these are spruce budworm (Blais, 1983), jack pine budworm (Volney, 1988), western spruce budworm (Swetnam and Lynch, 1993), and forest tent caterpillar (Roland, 1993).

The two theories most commonly invoked to explain the apparent dynamical changes in these insect outbreaks are:

1. Large-scale forest clearing, often with recolonization by pioneering tree species, is altering forest structure and species composition (Jackson *et al*., 2000; Carleton, 2000) in a way that may favor pioneering herbivores (Blais, 1983; Swetnam and Lynch, 1993) and disfavor their natural enemies (Roland, 1993).
2. A reduction in the frequency, duration, extent, and distribution of wildfires, caused by either climate change (Bergeron and Leduc, 1998; Carcaillet *et al*., 2001) or large-scale fire suppression (Swetnam and Lynch, 1993; Carleton, 2000), has created an older, more homogeneous forest, which may be more susceptible and more vulnerable to insect attack (Blais, 1983; Volney, 1988).

Of course, correlation does not imply causation, especially when it comes to shared trends in relatively short time-series. Thus, it is still a matter of some debate whether or not there is a causal link between environmental change and the increasing severity of outbreaks in any one of these systems.

That the evidence is equivocal is illustrated most clearly in the tree ring studies of western spruce budworm on Douglas-fir conducted by Swetnam and Lynch (1993) in New Mexico versus those of Ryerson et al. (2003) in Colorado. The New Mexico data prompted Swetnam and Lynch (1993) to speculate that the increased extent and severity of *C. occidentalis* outbreaks was a result of increased availability of susceptible hosts—echoing the argument made by Blais (1983) about increasing frequency and severity of *C. fumiferana* outbreaks in eastern Canada. However, the Colorado data of Ryerson *et al*. (2003) exhibited no such trend, despite a similar land use history. The parsimonious conclusion is that the western spruce budworm cycle is a highly stochastic process governed by random variables (such as weather and capricious, high-order interactions among members of the budworm community food web) that vary both spatially and temporally.

In summary, all four of these pests depend on disturbance-loving tree species (trembling aspen, balsam fir, jack pine, Douglas-fir) that are becoming more common across significant parts of the landscape. Yet there is little evidence of a consistent trend in population densities toward greater synchrony or cycle amplitude with increasing host plant abundance. Where such "trends" are reported, they often span as few as two to four population cycles, which is clearly insufficient for what we know, from process-oriented ecological studies, to be a highly stochastic process. This should not be construed as outright dismissal of the forest composition/land use hypothesis. Rather, what it indicates is healthy skepticism given limited data and limited opportunities to apply a meaningful test.

Proponents of this hypothesis often find it necessary to assume that the mechanism by which forest structure influences insect dynamics consists of a relaxation oscillation—that is, a periodically eruptive cusp-catastrophe event (Ludwig *et al.*, 1978; Hassell *et al.*, 1999). We wish to close by pointing out that this is not the case, that there are other more realistic mechanisms by which a trend toward more homogeneous host plant distributions could result in more extensive insect outbreaks. In the case of the spruce budworm/spruce-fir system, one scenario might operate as follows: 1) selective removal of white spruce from mixed coniferous stands leads to conversion toward fir-dominated stands; 2) reduced fire frequency increases the rate of retention of mature and "overmature" fir stands; 3) increased extent of a more vulnerable host leads to greater levels of defoliation and tree mortality earlier on in the budworm population cycle (Blais, 1958; MacLean, 1980); 4) increased levels of damage lead to higher rates of dispersal out of damaged stands (Royama, 1984); 5) decreased distances between acceptable host stands increase the survival rate, and realized fecundity, of dispersing moths; 6) increased rates of interpopulation migration lead to stronger synchronization of population oscillations (Barbour, 1990); 7) stronger density-dependence of dispersal leads to higher amplitude oscillation in population density (Royama, 1980); and 8) delayed density-dependent insect mortality factors act with increasing strength, further increasing the amplitude of harmonic oscillations (Royama, 1992). Over time, feedbacks between these various processes reinforce one another: increased cycle amplitude leads to increased damage, increased rates of dispersal, increased cycle synchrony, stronger collapse of budworm populations, stronger rebound of populations, and so on.

Although this scenario may seem complex, it is logical. And, as we have shown, several (though not all) of its presumptions and predictions are supported by data. To what extent the scenario may account for the apparent "increase in the frequency, extent and severity of [spruce budworm] outbreaks" reported by Blais (1983) we cannot say; however, it is consistent with his earlier (Blais, 1968) interpretation that the "extent of recent [spruce budworm] attacks was the result of a fortuitous synchronization of a number of independent outbreaks." What we can say for certain is that process-oriented theoretical and empirical studies of spruce budworm dynamics are necessary, and they are starting to reveal much about how disturbance arises and plays out across the landscape and through time.

As to the role of enviroclimatic change in modulating disturbance probabilities, time alone will not tell. In addition to large-scale, long-term monitoring, measurement and manipulative experiments are required to elucidate the independent roles of various environmental factors such as forest landscape structure and climate. These findings then need to be integrated in the form of process-based simulation models capable of predicting when and where disturbance is likely to occur. Once these models have been validated, we will be better able to emulate and manage for insect disturbances, which will bring us one step closer to the path of forest sustainability.

CONCLUSION

Defoliating insects constitute a special class of forest disturbance because the impact of outbreaks is selective, extensive, periodic, and spatially synchronized, albeit imperfectly. We have shown that in four systems, and maybe more, the ecological relationships between the defoliating insect, its resource, and its natural enemies are critical processes governing insect-caused disturbance. We know that the periodicity of outbreaks is caused by temporal lags in the interactions between herbivores, their host plants, and their natural enemies. But because these multiple processes act at multiple spatial scales, population fluctuations are loosely periodic rather than strictly cyclic. Stochastic influences such as weather further modify the detail of these patterns.

Population oscillations in different areas are synchronized through interpopulation migration and through the action of mild, spatially autocorrelated random perturbations, such as large-scale weather patterns. Departures from synchronous cyclicity are not well understood; however, we do know that this is an important determinant of the duration of outbreaks and of the amount of disturbance caused. Investigation of the causes of this variation requires focus on the common processes mentioned above through comparative analysis.

Understanding the causes of periodic fluctuations in insect populations helps in modeling insect-caused disturbance. But our imperfect knowledge of what modulates the amplitude of oscillations and the degree of synchrony identifies a gap that remains to be closed. Spatial analysis of outbreak patterns and impacts is revealing that the most severe budworm and forest tent caterpillar outbreaks tend to occur where local populations do not decline in synchrony with the collapse of surrounding populations or where the outbreak return interval is shorter than expected, and consequently trees do not have enough of a respite between individual outbreak cycles. We propose that this variability can best be understood by examining the dynamics of biotic interactions tempered by the influence of stochastic perturbations. Because the disturbance caused by insect outbreaks is itself a biotic interaction, casting insect population dynamics in terms of disturbance dynamics is a mutually beneficial context.

The other major requirement for closing the gap between insect population dynamics and insect-caused disturbance is a concerted effort to quantify the impacts of defoliators on tangible host plant parameters, such as foliage volumes, photosynthetic rates, carbohydrate levels, and growth rates. Because insects are highly selective, the variable pattern of tissue loss caused by insects leads to highly complex patterns of disturbance.

We suggest that the best way to simulate (and manage) insect-caused disturbance is through process-oriented modeling for at least four reasons:

1. The traditional, correlative approach to insect disturbance modeling (hazard modeling) is fast reaching its limit of effectiveness.
2. Major gains in predicting disturbance will come through a more careful consideration of the way local biotic processes are integrated across a variable forest mosaic to scale up to the landscape level.
3. Multivariate environmental change will continue to erode the capacity of the historical approach (using the past to forecast the future) to predict disturbance, and this at a time when the accuracy of inventory projection is becoming increasingly necessary to achieve sustainability.
4. Understanding what determines insect numbers will improve the usefulness of insect biodiversity studies that seek to explain spatial patterns of species abundance as a function of forest characteristics and land management practices.

We also suggest that the deepest insights in insect disturbance ecology are likely to come from model systems that have been studied for long periods of time, and where much of the basic biology has been worked out. The virtue of the detailed information gained to date is our current ability to design process-oriented research in a tightly structured, comparative framework, combining and interpreting results from a broad array of highly comparable process studies. This ensures a synthesis that applies to more than one forest–insect ecosystem. This is perhaps the most exciting aspect of process-based insect disturbance ecology: the possibility of having spatially explicit, multi-system models that can be used to examine broad questions about ecosystem resilience under multivariate enviroclimatic regime change.

ACKNOWLEDGMENTS

Thanks to Sylvie Gauthier for comments on an earlier draft.

REFERENCES

Alfaro, R. I., Van Sickle, G. A., Thomson, A. J., and Wegwitz, E. (1982). Tree mortality and radial growth losses caused by the western spruce budworm in a Douglas-fir stand in British Columbia. *Can J Forest Res* 12, 780–787.

Anderson, R. M., and May, R. M. (1980). Infectious diseases and population cycles of forest insects. *Science* 210, 658–661.

Archambault, L., Gagnon, R. R., Pelletier, G., Chabot, M., and Bélanger, L. (1990). Influence du drainage et de la texture du dépôt sur la vulnérabilité du sapin baumier et de l'épinette blanche aux attaques de la tordeuse des bourgeons de l'épinette. *Can J Forest Res* 20, 750–756.

Baird, A. B. (1917). An historical account of the forest tent caterpillar and of the fall webworm in North America. In: *47th Annual Report of the Entomological Society of Ontario*, pp. 73–87. Ontario Department of Agriculture, Toronto, ON.

Barbour, D. A. (1990). Synchronous fluctuations in spatially separated populations of cyclic forest insects. In: *Population Dynamics of Forest Insects*. (Leather, S. L., Watt, A. D., Hunter, M. D., and Kidd, N. A. C., Eds.), pp. 339–346. Intercept Press, Andover, UK.

Baskerville, G., and Kleinschmidt, S. (1981). A dynamic model of growth in defoliated fir stands. *Can J Forest Res* 11, 206–214.

Bauce, E. (1996). One and two years impact of commercial thinning on spruce budworm feeding ecology and host tree foliage production and chemistry. *Forest Chron* 72, 393–398.

Beckwith, R. (1978). Biology of the insect, introduction. In: *The Douglas-fir Tussock Moth: A Synthesis*. (Brookes, M. H., Stark, R. W., and Campbell, R. W., Eds.), p. 32. U.S. Department of Agriculture Forest Service Technical Bulletin No. 1585, Washington, DC.

Belyea, R. M. (1952). Death and deterioration of balsam fir weakened by spruce budworm defoliation in Ontario. *J Forest* 50, 729–738.

Bergeron, Y., and Harvey, B. (1997). Basing silviculture on natural ecosystem dynamics: an approach applied to the southern boreal mixedwoods of Québec. *Forest Ecol Manag* 92, 235–242.

Bergeron, Y., and Leduc, A. (1998). Relationships between change in fire frequency and mortality due to spruce budworm outbreak in the southeastern Canadian boreal forest. *J Veg Sci* 9, 493–500.

Bergeron, Y., Leduc, A., Morin, H., and Joyal, C. (1995). Balsam fir mortality following the last spruce budworm outbreak in northwestern Quebec. *Can J Forest Res* 25, 1375–1384.

Berryman, A. A. (1978). Population cycles of the Douglas-fir tussock moth (Lepidoptera: Lymantriidae): the time–delay hypothesis. *Can Entomol* 110, 513–518.

Berryman, A. A. (1981). *Forest Insects: Principles and Practice of Population Management*. Plenum Press, New York.

Berryman, A. A. (1991). Population theory: an essential ingredient in pest prediction management and policy-making. *Am Entomol* 37, 138–142.

Berryman, A. A. (1996). What causes population cycles of forest Lepidoptera? *Trends Ecol Evol* 11, 28–32.

Berryman, A. A. (2002). *Population Cycles: The Case for Trophic Interactions*. Oxford University Press, New York.

Bjornstad, O. N., Peltonen, M., Liebhold, A. M., and Baltensweiler, W. (2002). Waves of larch budmoth outbreaks in the European Alps. *Science* 298, 1020–1023.

Blais, J. R. (1958). The vulnerability of balsam fir to spruce budworm attack in northwestern Ontario, with special reference to the physiological age of the tree. *Forest Chron* 34, 405–422.

Blais, J. R. (1968). Regional variation in susceptibility of eastern North American forests to budworm attack based on history of outbreaks. *Forest Chron* 44, 17–23.

Blais, J. R. (1983). Trends in the frequency, extent, and severity of spruce budworm outbreaks in eastern Canada. *Can J Forest Res* 13, 539–547.

Blais, J. R., and Archambault, L. (1982). Vulnérabilité du sapin baumier aux attaques de la tordeuse des bourgeons de l'épinette au Québec. Centre Recherche Foret Laurentides, Ste-Foy, Québec. Rapp. Inf. LAU–X-51F.

Boulanger, Y., and Arseneault, D. (2004). Spruce budworm outbreaks in eastern Quebec over the last 450 years. *Can J Forest Res* 34, 1035–1043.

British Columbia Ministry of Forests. (2004). *2003 Summary of Forest Health Conditions in British Columbia*. Ministry of Forests, Forest Practices Branch, Victoria, BC.

Burleigh, J. S., Alfaro, R. I., Borden, J. H., and Taylor, S. (2002). Historical and spatial characteristics of spruce budworm *Choristoneura fumiferana* (Clem.) (Lepidoptera: Tortricidae) outbreaks in northern British Columbia. *Forest Ecol Manag* 168, 301–309.

Campbell, I. M. (1966). Genetic variation related to survival in Lepidopteran species. In: *Breeding Pest Resistant Trees*. (Gerhold, H. D., Schreiner, E. J., McDermott, R. E., and Winieski, J. A., Eds.), pp. 129–135. Pergamon Press, Oxford.

Candau, J.-N., Abt, V., and Keatley, L. (2002). Bioclimatic analysis of declining aspen stands in northeastern Ontario. Ontario Forest Research Institute, Forestry Research Report 154, Sault Ste-Marie, ON.

Candau, J.-N., Fleming, R. A., and Hopkin, A. (1998). Spatiotemporal pattern of large-scale defoliation caused by the spruce budworm in Ontario since 1941. *Can J Forest Res* 28, 1733–1741.

Cappuccino, N., Lavertu, D., Bergeron, Y., and Régnière, J. (1998). Spruce budworm impact, abundance and parasitism rate in a patchy landscape. *Oecologia* 114, 236–242.

Carcaillet, C., Bergeron, Y., Richard, P. J. H., Fréchette, B., Gauthier, S., and Prairie, Y. T. (2001). Change of fire frequency in the eastern Canadian boreal forests during the Holocene: does vegetation composition or climate trigger the fire regime? *J Ecol* 89, 930–946.

Carleton, T. J. (2000). Vegetation responses to the managed forest landscape of central and northern Ontario. In: *Ecology of a Managed Terrestrial Landscape: Patterns and Processes of Forest Landscapes in Ontario*. (Perera, A. H., Euler, D. L., and Thompson, I. D., Eds.), pp. 179–197. UBC Press, Vancouver, BC.

Chitty, D. (1967). The natural selection of self-regulatory behaviour in animal populations. *Proc Ecol Soc Aust* 2, 51–78.

Churchill, G. B., John, H. H., Duncan, D. P., and Hodson, A. C. (1964). Long-term effects of defoliation of aspen by the forest tent caterpillar. *Ecology* 45, 630–633.

Clark, J. (1961). *Photosynthesis and Respiration in White Spruce and Balsam Fir*. State University of New York, College of Forestry, Syracuse, NY.

Clark, W. C., Jones, D. D., and Holling, C. S. (1979). Lessons for ecological policy design: a case study of ecosystem management. *Ecol Model* 7, 1–53.

Cooke, B. J., and Lorenzetti, F. (2006). The dynamics of forest tent caterpillar outbreaks in Quebec, Canada. *Forest Ecol Manag* 226, 110–121.

Cooke, B. J., and Roland, J. (2003). The effect of winter temperature on forest tent caterpillar egg survival and population dynamics in northern climates. *Environ Entomol* 32, 299–311.

Denno, R. F., and M. S. McClure. (1983). *Variable Plants and Herbivores in Natural and Managed Systems*. Academic Press, New York.

Dobesberger, E. J. (1998). Stochastic simulation of growth loss in thinned balsam fir stands defoliated by the spruce budworm in Newfoundland. *Can J Forest Res* 28, 703–710.

Dupont, A., Bélanger, L., and Bousquet, J. (1991). Relationships between fir vulnerability to spruce budworm and ecological site conditions of fir stands in central Québec. *Can J Forest Res* 21, 1752–1759.

Fellin, D. G. (1976). Forest practices, silvicultural prescriptions, and the western spruce budworm. In: *Proceedings of a Symposium on the Spruce Budworm; November 11–14, 1974.* pp. 117–121. U.S. Department of Agriculture Forest Service Misc. Publ. 1327. Washington, DC.

Fleming, R., Candau, J.-N., and McAlpine, R. S. (2002). Landscape-scale analysis of interactions between insect defoliation and forest fire in central Canada. *Climatic Change* 55, 251–272.

Fleming, R. A., Hopkin, A. A., and Candau, J.-N. (2000). Insect and disease disturbance regimes in Ontario's forests. In: *Ecology of a Managed Terrestrial Landscape: Patterns and Processes of Forest Landscapes in Ontario.* (Perera, A. H., Euler, D. L., and Thompson, I. D., Eds.), pp. 141–162. UBC Press, Vancouver, BC.

Fortin, M., and Mauffette, Y. (2001). Forest edge effects on the biological performance of the forest tent caterpillar (Lepidoptera: Tortricidae) in sugar maple stands. *Ecoscience* 8, 164–172.

Franklin, J. F., Shugart, H. H., and Harmon, M. E. (1987). Tree death as an ecological process: the causes, consequences, and variability of tree mortality. *Bioscience* 37, 550–556.

Ginzburg, L. R., and Taneyhill, D. E. (1994). Population cycles of forest Lepidoptera: a maternal effect hypothesis. *J Anim Ecol* 63, 79–92.

Gray, D. R., Régnière, J., and Boulet, B. (2000). Analysis and use of historical patterns of spruce budworm defoliation to forecast outbreak patterns in Quebec. *Forest Ecol Manag* 127, 217–231.

Greenbank, D. O., Schaefer, G. W., and Rainey, R. C. (1980). Spruce budworm (Lepidoptera: Tortricidae) moth flight and dispersal: new understanding from canopy observations radar and aircraft. *Mem Entomol Soc Canada* 110, 1–49.

Hardy, Y. J., Lafond, A., and Hamel, L. (1983). The epidemiology of the current spruce budworm outbreak in Quebec. *Forest Sci* 29, 715–725.

Hardy, Y., Mainville, M., and Schmitt, D. M. (1986). *An Atlas of Spruce Budworm Defoliation in Eastern North America, 1938-80.* U.S. Department of Agriculture Forest Service Miscellaneous Publication No. 1449, Washington, DC.

Harris, J. W. E., Alfaro, R. I., Dawson, A. F., and Brown, R. G. (1985). *The Western Spruce Budworm in British Columbia 1909-1983.* Canadian Forest Service, Pacific Forestry Centre, Information Report BC-X-257, Victoria, BC.

Hassell, D. C., Allwright, D. J., and Fowler, A. C. (1999). A mathematical analysis of Jones's site model for spruce budworm infestations. *J Math Biol* 38, 377–421.

Haukioja, E. (1991). Cyclic fluctuations in density—interactions between a defoliator and its host tree. *Acta Oecol* 12, 77–88.

Haukioja, E., and Koricheva, J. (2000). Tolerance to herbivory in woody vs. herbaceous plants. *Evol Ecol* 14, 551–562.

Hilborn, R., and Stearns, S. C. (1982). On inference in ecology and evolutionary biology: the problem of multiple causes. *Acta Biotheor* 31, 145–164.

Hildahl, V., and Reeks, W. A. (1960). Outbreaks of the forest tent caterpillar *Malacosoma disstria* Hbn. and their effects on stands of trembling aspen in Manitoba and Saskatchewan. *Can Entomol* 92, 199–209.

Hogg, E. H., Brandt, J. P., and Kochtubajda, B. (2002). Growth and dieback of aspen forests in northwestern Alberta, Canada, in relation to climate and insects. *Can J Forest Res* 32, 823–832.

Holling, C. S. (1992). Cross-scale morphology, geometry and dynamics of ecosystems. *Ecol Monogr* 62, 447–502.

Jackson, S. M., Pinto, F., Malcolm, J. R., and Wilson, E. R. (2000). A comparison of pre-European settlement (1857) and current (1981–1995) forest composition in central Ontario. *Can J Forest Res* 30, 605–612.

Jardon, Y., Morin, H., and Dutilleul, P. (2003). Periodicité et synchronisme des épidémies de la tordeuse des bourgeons de l'épinette au Québec. *Can J Forest Res* 33, 1947–1961.
Kulman, H. M. (1971). Effects of insect defoliation on growth and mortality of trees. *Annu Rev Entomol* 6, 289–324.
Kurz, W. A., and Apps, M. J. (1999). A 70-year retrospective analysis of carbon fluxes in the Canadian forest sector. *Ecol Appl* 9, 526–547.
Li, Z., Apps, M. J., Kurz, W. A., and Banfield, E. (2003). Temporal changes of forest net primary production and net ecosystem production in west central Canada associated with natural and anthropogenic disturbances. *Can J Forest Res* 33, 2340–2351.
Ludwig, D., Jones, D. D., and Holling, C. S. (1978). Qualitative analysis of insect outbreak systems: the spruce budworm and the forest. *J Anim Ecol* 47, 315–332.
MacKinnon, W. E., and MacLean, D. A. (2003). The influence of forest and stand conditions on spruce budworm defoliation in New Brunswick, Canada. *Forest Sci* 49, 657–667.
MacLean, D. A. (1980). Vulnerability of fir spruce stands during uncontrolled spruce budworm outbreaks: a review and discussion. *Forest Chron* 56, 213–221.
MacLean, D. A. (1984). Effects of spruce budworm outbreaks on the productivity and stability of balsam fir forests. *Forest Chron* 60, 273–279.
MacLean, D. A. (2004). Predicting forest insect disturbance regimes for use in emulating natural disturbance. In: *Emulating Natural Forest Landscape Disturbance*. (Perera, A. H., Buse, L. J., and Weber, M. G., Eds.), pp. 69–82. Columbia University Press, New York.
MacLean, D. A., and MacKinnon, W. E. (1997). Effects of stand and site characteristics on susceptibility and vulnerability of balsam fir and spruce to spruce budworm in New Brunswick. *Can J Forest Res* 27, 1859–1871.
MacLean, D. A., and Ostaff, D. P. (1989). Patterns of balsam fir mortality caused by an uncontrolled spruce budworm outbreak. *Can J Forest Res* 19, 1087–1095.
Manion, P. D., and Lachance, D. (1992). *Forest Decline Concepts*. APS Press, St. Paul, MN.
McCullough, D. G. (2000). A review of factors affecting the population dynamics of jack pine budworm (*Choristoneura p*In:*us p*In:*us* Freeman). *Popul Ecol* 42, 243–256.
Moore, L. V., Myers, J. H., and Eng, R. (1988). Western tent caterpillars prefer the sunny side of the tree, but why? *Oikos* 51, 321–326.
Moran, P. A. P. (1953). The statistical analysis of the Canadian lynx cycle II. Synchronization and meteorology. *Aust J Zool* 1, 291–298.
Morin, H. (1994). Dynamics of balsam fir forests in relation to spruce budworm outbreaks in the boreal zone of Québec. *Can J Forest Res* 24, 730–741.
Morris, R. F. (1959). Single-factor analysis in population dynamics. *Ecology* 40, 580–588.
Myers, J. H. (1988). Can a general hypothesis explain population cycles in forest Lepidoptera? *Adv Ecol Res* 18, 179–242.
Nealis, V. G. (1991). Parasitism in sustained and collapsing populations of the jack pine budworm, Choristoneura pinus pinus Free. (Lepidoptera: Tortricidae) in Ontario 1985–1987. *Can Entomol* 132, 1065–1075.
Nealis, V. G., and Lomic, P. V. (1994). Host-plant influence on the population ecology of the jack pine budworm, *Choristoneura pinus pinus* (Lepidoptera: Tortricidae). *Ecol Entomol* 19, 367–373.
Nealis, V. G., and Régnière, J. (2004a). Insect-host relationships influencing disturbances by the spruce budworm in a boreal mixedwood forest. *Can J Forest Res* 34, 1870–1882.
Nealis, V. G., and Régnière, J. (2004b). Fecundity and recruitment of eggs during outbreaks of the spruce budworm. *Can Entomol* 136, 591–604.
Nealis, V. G., Magnussen, S., and Hopkin, A. A. (2003). A lagged, density-dependent relationship between jack pine budworm *Choristoneura pinus pinus* and its host *Pinus banksiana*. *Ecol Entomol* 28, 183–192.

Nicholson, A. J., and Bailey, V. A. (1935). The balance of animal populations. Part I. *Proc Zool Soc London* 3, 551–598.
Nisbet, R. M., and Gurney, W. S. C. (1982). *Modelling Population Fluctuations*. John Wiley and Sons, New York.
Perera, A. H., Buse, L. J., and Weber, M. G. (2004). *Emulating Natural Forest Landscape Disturbance*. Columbia University Press, New York.
Pickett, S. T. A., and White, P. S. (1985). *The Ecology of Natural Disturbance and Patch Dynamics*. Academic Press, New York.
Quayle, D., Régnière, J., Cappuccino, N., and Dupont, A. (2003). Forest composition, host-population density, and parasitism of spruce budworm *Choristoneura fumiferana* eggs by *Trichogramma minutum*. *Entomol Exp Appl* 107, 215–227.
Rausher, M. D. (1983). The ecology of host selection behavior in phytophagous insects. In: *Variable Plants and Herbivores in Natural and Managed Systems*. (Denno, R. F., and McClure, M. S., Eds.), pp. 223–257. Academic Press, New York.
Régnière, J., and Lysyk, T. J. (1995). Population dynamics of the spruce budworm, *Choristoneura fumiferana*. In: *Forest Insect Pests in Canada*. (Armstrong, J. A., and Ives, W. G. H., Eds.), pp. 95–105. Natural Resources Canada, Canadian Forest Service, Ottawa.
Roland, J. (1993). Large-scale forest fragmentation increases the duration of tent caterpillar outbreak. *Oecologia* 93, 25–30.
Roland, J., and Kaupp, W. J. (1995). Reduced transmission of forest tent caterpillar (Lepidoptera: Lasiocampidae) nuclear polyhedrosis virus at the forest edge. *Environ Entomol* 24, 1175–1178.
Rose, A. H., and Lindquist, O. H. (1985). *Insects of Eastern Spruces, Fir and Hemlock*. Natural Resources Canada, Canadian Forest Service, Forestry Technical Report No. 23, Ottawa, ON.
Rose, M. R., and Harmsen, R. (1981). Ecological outbreak dynamics and the cusp catastrophe. *Acta Biotheor* 30, 229–253.
Royama, T. (1980). Effect of adult dispersal on the dynamics of local populations of an insect species: a theoretical investigation. In: *Dispersal of Forest Insects: Evaluation, Theory, and Management Implications*. (Berryman, A. A., and Safranyik, L., Eds.), pp. 79–93. Washington State University, Pullman, WA.
Royama, T. (1984). Population dynamics of the spruce budworm *Choristoneura fumiferana*. *Ecol Monogr* 54, 429–462.
Royama, T. (1992). *Analytical Population Dynamics*. Chapman and Hall, New York.
Royama, T. (1997). Population dynamics of forest insects: are they governed by single or multiple factors? In: *Forests and Insects*. (Watt, A. D., Stork, N. E., and Hunter, M. D., Eds.), pp. 37–48. Chapman and Hall, New York.
Royama, T., MacKinnon, W. E., Kettala, E. G., Carter, N. E., and Hartling, L. K. (2005). Analysis of spruce budworm outbreak cycles in New Brunswick, Canada, since 1952. *Ecology* 86, 1212–1224.
Ryerson, D. E., Swetnam, T. W., and Lynch, A. M. (2003). A tree-ring reconstruction of western spruce budworm outbreaks in the San Juan Mountains, Colorado, USA. *Can J Forest Res* 33, 1010–1028.
Schowalter, T. D., Hargrove, W. W., and Crossley, D. A. (1986). Herbivory in forested ecosystems. *Annu Rev Entomol* 31, 177–196.
Shugart, H. H. (2003). *A Theory of Forest Dynamics: The Ecological Implications of Forest Succession Models*. Blackburn Press, Caldwell, NJ.
Sippell, L. (1962). Outbreaks of the forest tent caterpillar, *Malacosoma disstria* Hbn., a periodic defoliator of broad-leaved trees in Ontario. *Can Entomol* 94, 408–416.
Stenseth, N. C. (1981). On Chitty's theory for fluctuating populations: the importance of genetic polymorphism in the generation of regular density cycles. *J Theor Biol* 90, 9–36.

Su, Q., MacLean, D. A., and Needham, T. D. (1996). The influence of hardwood content on balsam fir defoliation by spruce budworm. *Can J Forest Res* 26, 1620–1628.

Swetnam, T. W., and Lynch, A. M. (1993). Multicentury, regional-scale patterns of western spruce budworm outbreaks. *Ecol Monogr* 63, 399–424.

Tiffin, P. (2000). Mechanisms of tolerance to herbivore damage: what do we know? *Evol Ecol* 14, 523–536.

Tilman, D. (1982). *Resource Competition and Community Structure*. Princeton University Press, Princeton, NJ.

Van Wagner, C. E. (1978). Age-class distribution and the forest fire cycle. *Can J Forest Res* 8, 220–227.

Volney, W. J. A. (1988). Analysis of jack pine budworm outbreaks in the prairie provinces of Canada. *Can J Forest Res* 18, 1152–1158.

Volney, W. J. A., and McCullough, D. (1994). Jack pine budworm population behaviour in northwestern Wisconsin. *Can J Forest Res* 24, 502–510.

Wellington, W. G. (1964). Qualitative changes in populations in unstable environments. *Can Entomol* 96, 436–451.

Witter, J. A., Mattson, W. J., and Kulman, H. M. (1975). Numerical analysis of a forest tent caterpillar (Lepidoptera: Lasiocampidae) outbreak in northern Minnesota. *Can Entomol* 107, 837–854.

Zhang, Q.-B., and Alfaro, A. I. (2002). Periodicity of two-year cycle spruce budworm outbreaks in central British Columbia: a dendro-ecological analysis. *Forest Sci* 48, 722–731.

Zhang, Q.-B., and Alfaro, R. I. (2003). Spatial synchrony of the two-year cycle budworm outbreaks in central British Columbia, Canada. *Oikos* 102, 146–154.

16

Dynamics of Mountain Pine Beetle Outbreaks

Justin Heavilin and James Powell
Utah State University

Jesse A. Logan
USDA Forest Service

INTRODUCTION

Native forest insects are the greatest forces of change in forest ecosystems of North America. In aggregate, insect disturbances affect an area that is almost 45 times as great as that affected by fire, resulting in an economic impact nearly five times as great (Dale *et al.*, 2001). Of these natural agents of ecosystem disturbance and change, the bark beetles are the most obvious in their impact, and of these, the mountain pine beetle (*Dendroctonus ponderosae* Hopkins) has the greatest economic importance in the forests of western North America (Samman and Logan, 2000). The primary reason for this impact is that the mountain pine beetle is one of a handful of bark beetles that are true predators in that they must kill their host to successfully reproduce, and they often do so in truly spectacular numbers.

Although the mountain pine beetle is an aggressive tree killer, it is a native component of natural ecosystems; in this sense, the forests of the

American West have co-evolved (or at least co-adapted) in ways that incorporate mountain pine beetle disturbance in the natural cycle of forest growth and regeneration. Such a relationship in which insect disturbance is "part and parcel of the normal plant biology" has been termed a normative outbreak by Mattson (1996). This normative relationship between native bark beetles and their host forests is undergoing an apparent shift, exemplified by an unusual sequence of outbreak events. Massive outbreaks of spruce beetle have recently occurred in western North America ranging from Alaska to southern Utah (Ross *et al.*, 2001; Munson *et al.*, 2004). A complex of bark beetles are killing ponderosa pine in the southwestern United States at levels not previously experienced during the period of European settlement. Pinyon pines are being killed across the entire range of the pinyon/juniper ecotype, effectively removing a keystone species in many locations. Mountain pine beetle outbreaks are occurring at greater intensity, and in locations where they have not previously occurred (British Columbia Ministry of Forests, 2003). Any one of these events is interesting; that they are occurring almost simultaneously is nothing short of remarkable. In many of these instances the outbreaks are anything but normative; they are occurring in novel habitats with potentially devastating ecological consequences (Logan and Powell, 2001; Logan *et al.*, 2003).

What is going on here? The root of these unprecedented outbreaks appears to be directly related to unusual weather patterns. Although drought, particularly in the Southwest, is playing an important role in some of these outbreaks, the dominant and ubiquitous factor at the continental scale is the sequence of abnormally warm years that began somewhere in the mid-1980s (Berg, 2003; Logan and Powell, 2004). Regardless of the underlying causes, the impact of warming temperatures on bark beetle outbreaks has resulted in a renewed research interest focused on understanding and responding to the economic and ecological threat of native insects functioning as exotic pests. Because of its ecological importance and economic impact, the mountain pine beetle is receiving much of this interest. Development of predictive models is an important component of this research effort. This chapter develops a minimally complex model that, on a landscape scale, describes the spatial and temporal interaction between the mountain pine beetle and the lodgepole pine forest.

We first briefly review the biology of the mountain pine beetle and one of its primary hosts, lodgepole pine (*Pinus contorta* Douglas). Lodgepole pine is a shade-intolerant species that opportunistically colonizes areas following large-scale disturbances (Schmidt and Alexander, 1985). It is therefore an early successional species that typically initiates a sequence of events that result in subsequent replacement by more shade-tolerant species. Over much of its distribution range, lodgepole pine would be replaced by spruce/fir forests without the intervening action of a major disturbance event. This disturbance is typically a stand-replacing crown-fire. Lodgepole pine reproduction is keyed to fire disturbance by producing a proportion of cones that release seeds only in the presence of high-heat (serotinous cones). The protected seeds inside the tightly closed cones remain viable for a prolonged time until the intense heat of a fire triggers their release. Seed establishment is also tied to conditions (exposed, mineral soil) created by stand-replacing fires (Muir and Lotan, 1985). Tree mortality caused by mountain pine beetles hypothetically plays a critical timing role in this reproductive cycle by creating the fuel conditions that predispose a stand to fire (Peterman, 1978).

The mountain pine beetle spends most of its lifecycle feeding in the protected environment of a host tree's phloem tissue (the nutrient rich inner bark). Adults emerge sometime in the summer (typically late July or early August) to attack new trees. If they are successful in overcoming the substantial host tree defensive chemistry and kill the host, the various life history events are subsequently carried out, resulting in continuation of the species' lifecycle. There are many subtle, and some not so subtle, nuances in the interplay between predator insect and prey tree. However, for a general synoptic model, the complex ecology of this beetle can be represented as a predator functional response curve. Simply stated, if enough beetles simultaneously attack a tree, tree defenses are overcome and the tree is killed; if not, the attack is unsuccessful and the tree lives. Beetle recruitment, in turn, is keyed to the number of trees that are killed during the previous attack cycle. Because beetle recruitment is keyed to the surrogate measure of trees killed (rather than actual beetle reproductive biology), this approach has been termed the "red-top" model because trees that are successfully attacked and killed begin to fade in color the summer following the attack, subsequently turning a characteristic bright orange or red.

Tree responses to beetle attack involve both constitutive and induced resin flow. When beetles attack a tree, the tree excretes pitch through the hole the beetle has chewed in the bark, creating "pitch tubes" (Amman and Cole, 1983). Resin flow contains toxic, defensive chemicals, and the resin flow induced by attack physically expels the attacking adult beetles. Not all trees have the same capacity to produce pitch for this defense because of varying size and fitness. Thus, not all trees are equally susceptible to attack. The lodgepole pine forest can be viewed as having three classes of trees from the perspective of susceptibility to beetle attack. Because the beetle feeds on the phloem of the tree, attacking a tree of adequate size to create egg galleries and sufficient nutrient supply is important. The first class of trees are juveniles, which have a diameter breast height (DBH) of less than 20 cm. Although juvenile trees provide enough nutrients, they generally do not provide enough clearance in the phloem for larvae to develop. Larger trees, with a DBH between 20 and 38 cm, constitute the second class of vigorous trees. Vigorous trees are large enough to house egg galleries and offer a suitable nutrient source. However, they have the strongest defenses against beetle attack. As an adult tree ages, the crown remains relatively constant while the diameter of the tree trunk increases. The defenses employed against attack are then spread over a greater surface area, reducing their effectiveness (Amman and Cole, 1983). This suggests a third class of susceptible trees, older with weaker defenses than the vigorous trees. This class also accounts for trees suffering from drought, crowding, and other stresses. The defenses of this third class of trees are more easily overcome by the mountain pine beetle and still offer sufficient nutrients for beetle development, although, generally, brood production will be less because phloem is spread across a larger surface.

The objective of this model is to describe the evolution of spatial patterns of beetle attacks in both endemic and epidemic states, as well as to predict the spread of the beetle population. The analysis will analytically demonstrate the potential for this model to emulate observed patterns of forest disturbance. The final model incorporates the basic three-tier demography of susceptible pine populations. The juvenile cohort increases by contributions from the two mature classes of trees via propagation and decreases due to maturation into the vigorous cohort. The vigorous cohort, in turn, increases by contributions from the juvenile cohort and decreases through maturation into the adult cohort. This model assumes that there is no death rate in juvenile and vigorous cohorts, independent of the mountain pine beetle. Finally, the adult cohort increases through contributions by the vigorous cohort and decreases through death. This is the simple demography of a healthy lodgepole pine forest. Timber inventory data provide values for the birth, maturation, and death parameters for the healthy forest dynamics. Beetle-caused mortality is represented by the removal of vigorous and adult trees from the forest, making a direct contribution to an additional cohort of infected lodgepole pine trees, named infectives or red tops.

By using data collected from aerial surveys and satellite imagery, the method of estimating functions can return parameter values that fit the model parameters governing attack dynamics to the data. Then the predictive capacity of the model can be tested against a decade of data collected in the Sawtooth National Recreation Area of central Idaho.

DERIVATION OF THE RED TOP MODEL

A disturbance model for lodgepole pine begins by constructing an age-structured model framework for the uninfected lodgepole pine forest. The age classes include a seed base (S_0), seedlings (S_1), and juvenile (J), vigorous (V), and adult (A) trees. Each of these classes subsists on contributions by their subordinate class, and contributes to the successive class in the hierarchy; the exception is the seed base, toward which both classes of reproducing trees (vigorous and adult) contribute. The following equations provide a starting point for this simplified model.

$$S_{0(t+1)} = (1 - s_0)\, S_{0t} + b_V V_t + b_A A_t \quad (1)$$

$$S_{1(t+1)} = s_0 S_{0t} + (1 - s_1)\, S_{1t} \quad (2)$$

$$J_{(t+1)} = s_1 S_{1t} + (1 - s_J)\, J_t \quad (3)$$

$$V_{(t+1)} = s_J J_t + (1 - s_V)\, V_t \quad (4)$$

$$A_{(t+1)} = s_V V_t + (1 - d)\, A_t \quad (5)$$

The discrete equations listed in Equations (1) to (5) describe the density of each class in the following time step, based on the current densities as well as fecundity, maturation, and death rates specific to each class. Equation (3), for example, represents the density of juvenile trees in the next time step based on the proportion of seedlings maturing to juveniles (s_1) and the proportion of juveniles that remain juveniles into the next time step ($1 - s_J$). Similarly, s_V is the proportion of vigorous trees that mature into adult trees. The contributions to the seed base are made by vigorous and adult trees, represented by b_V and b_A, respectively. Finally, the adult class is the only class that experiences mortality from aging. Because there is no age class to which the last class can transition, d represents the natural mortality in the adult class.

Availability of direct sunlight is important in the establishment of lodgepole pine stands. Shade intolerance makes it difficult for successive generations to mature beneath the canopy of adult lodgepole pines. In the absence of disturbance, the lodgepole pines are eventually replaced by more shade-tolerant species of conifers. The pressure caused by shading from larger trees can be modeled by a response function that retards the growth of one class of tree under the combined pressure of larger trees. Each understory class of lodgepole pine tree experiences shading from larger trees; the smaller the tree, the greater the shading. To address shading in the model, we look for response functions that are functions of the densities of all larger trees. In the case of the seedling class, the function $g_S(J_t, V_t, A_t) = \gamma_S\left(\frac{(J_t + V_t + A_t)}{J_t + V_t + A_t + \beta_S}\right)$ has these properties, where β_S is a parameter that relates the percent of canopy closure in terms of tree density, and γ_S is tuned to the sensitivity of the seedling class to the effects of shading. Similarly, $g_J(V_t, A_t) = \gamma_J\left(\frac{(V_t + A_t)}{(V_t + A_t) + \beta_J}\right)$ and $g_V(A_t) = \gamma_V\left(\frac{(A_t)}{(A_t) + \beta_V}\right)$ are response functions modeling the shading experienced by the juvenile and vigorous classes, respectively.

Including these response functions yields the following system of equations:

$$S_{0(t+1)} = (1 - s_0)\, S_{0t} + b_V V_t + b_A A_t \tag{6}$$

$$S_{1(t+1)} = s_0 S_{0t} + (1 - s_1) S_{1t} - \gamma_S\left(\frac{J_t + V_t + A_t}{(J_t + V_t + A_t) + \beta_S}\right) S_{1t} \tag{7}$$

$$J_{(t+1)} = s_1 S_{1t} + (1 - s_J) J_t - \gamma_J\left(\frac{V_t + A_t}{(V_t + A_t) + \beta_J}\right) J_t \tag{8}$$

$$V_{(t+1)} = s_J J_t + (1 - s_V) V_{1t} - \gamma_V\left(\frac{A_t}{A_t + \beta_V}\right) V_t \tag{9}$$

$$A_{(t+1)} = s_V V_t + (1 - d)\, A_t. \tag{10}$$

In the western North American lodgepole pine forests, the seedling class experiences shading by virtually any density of larger trees, which in the model implies that β_S is small. The remaining classes of larger trees are hardly affected by shading once they have passed the seedling class, $\gamma_V = \gamma_A \approx 0$. This allows us to approximate the effect of shading on the juvenile class by γ_S. Mountain pine beetle disturbance will not remove the juvenile class. Assuming β_S is small (i.e., seedlings are easily shaded out), the shading further is approximately a constant, γ_S. If we combine the three nonsusceptible classes, adding Equations (6) to (8) to yield a composite nonsusceptible class, $\hat{J}_t = S_{0t} + S_{1t} + J_t$, then we have a simplified equation,

$$\hat{J}_{(t+1)} = (1 - s_J)\, \hat{J}_t + b_V V_t + b_A A_t - \gamma_S S_{1t} + s_J (S_{0t} + S_{1t}). \tag{11}$$

In the absence of fire, the density of the juvenile class is significant in shading the seedling class. To further simplify, we will assume that the survivorship from the seedling class to the juvenile class, $s_J(S_{0t} + S_{1t})$, is roughly equal to the mortality caused by shading, $\gamma_S S_{1t}$, and therefore the last two terms in Equation (11) may be negated, given:

$$\hat{J}_{(t+1)} = (1 - s_J)\,\hat{J}_t + b_V\,V_t + b_A A_t, \tag{12}$$

$$V_{(t+1)} = s_J \hat{J}_t + (1 - s_V)\,V_t, \tag{13}$$

$$A_{(t+1)} = s_V\,V_t + (1 - d)\,A_t. \tag{14}$$

The simplified demographics may be written as the following Leslie matrix equation:

$$\begin{pmatrix} \hat{J}_{t+1} \\ V_{t+1} \\ A_{t+1} \end{pmatrix} = \begin{pmatrix} 1 - s_V & b_V & b_A \\ s_V & 1 - s_A & 0 \\ 0 & s_A & 1 - d \end{pmatrix} \begin{pmatrix} \hat{J}_t \\ V_t \\ A_t \end{pmatrix}.$$

Next we include the beetle's influence on the forest. Berryman *et al.* (1985) observed that the probability of a lodgepole pine tree being killed as a function of beetle attacks per square meter of tree surface fit a sigmoidally shaped curve such as that which results from the Holling III response function, $P(B) = \frac{B^2}{B^2 + a^2}$. This relationship can be read as the probability, P, as a function of the beetle density, B, with a parameter representing the effectiveness of the beetles, a, in attacking lodgepole pine trees. The units of a are beetles per hectare (the more effective the beetles are, the smaller the value of a). From this clue we can arrive at a model that describes the probability of successful attack by the mountain pine beetle as a function of red top density.

We let I_{t+1} represent a new class called infectives. These are the next year's density (in trees per hectare) of infected trees, resulting from beetle attacks, and let S_t be the density of susceptible trees in the current year, t; then

$$I_{t+1} = P(B) \cdot S_t = \frac{B^2}{B^2 + a^2}\,S_t. \tag{15}$$

Considering that the beetles attack a tree for the purpose of building egg galleries and thus rearing young, each infected tree can be considered as having a beetle fecundity, f, in units of beetles per tree. Density of beetles is then related to density of infected trees, $B_t = fI_t$.

Substituting this relationship into the response function in Equation (15) gives:

$$I_{t+1} = P(B = f_t I_t) \cdot S_t = \frac{(f_t I_t)^2}{(f_t I_t)^2 + a^2}\,S_t. \tag{16}$$

This results in an equation relating one year's density of infected trees in terms of the prior year's density of infectives and the density of susceptible trees, effectively removing the density of beetles from the equation.

Then, factoring the fecundity from the response function, we introduce a new parameter, $\alpha = \frac{a}{f}$, with units of trees per hectare,

$$\frac{I_t^2}{I_t^2 + \left(\frac{a_t}{f_t}\right)^2} S_t = \frac{I^2}{I^2 + \alpha_t^2} S_t \tag{17}$$

This new parameter, α, can be interpreted as the beetle fecundity per tree that will result in a 50–50 chance of susceptible trees in an area becoming infected, as illustrated in Fig. 1. This parameter can also be viewed from the perspective of the beetle as a level of the beetles' effectiveness in their effort to attack healthy trees and reproduce.

Because the susceptibility and potential beetle fecundity differ between vigorous and adult lodgepole pine trees, it is useful to consider the source of infected trees coming from two cohorts, $S_t = V_t + A_t$. This necessitates the assignment of distinct α^2 parameters to each susceptible class. The contribution to the infectives class in the following year is the sum of the trees killed by the mountain pine beetle from the two susceptible classes.

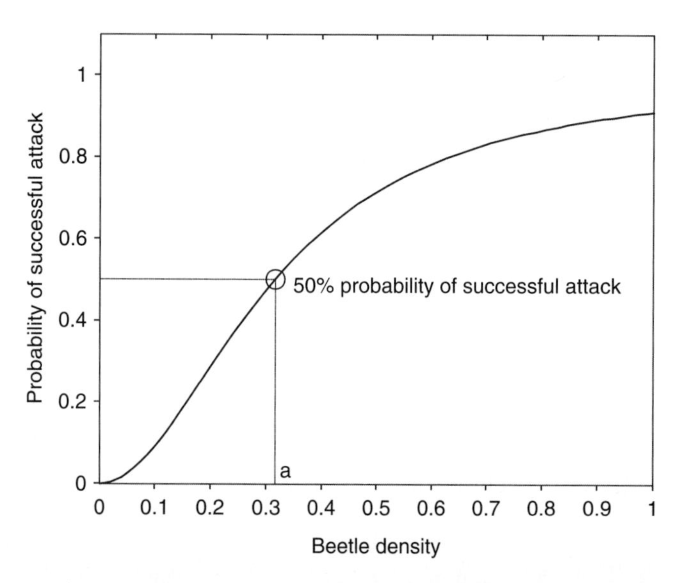

FIG. 1 Low densities of beetles are not able to overcome the defenses of the tree. Once enough beetles have arrived at the tree, the beetle in turn releases a nonaggregating pheromone, which repels beetles from the tree and forces them to search out another victim. The *a* value equates to the point at which 50% of the susceptible trees can be successfully attacked.

Using this relationship between the population of infected trees and the next year's population of newly emergent beetles, substituting these values into the probability function for a successful attack as a function of beetle density, and multiplying by the population of the two cohorts that are susceptible to beetle attack (V_t and A_t), we arrive at an equation for the next year's infectives in terms of the current year's density of infectives and susceptible trees

$$I_{t+1} = \frac{I_t^2}{I_t^2 + \alpha_{V,t}^2} V_t + \frac{I_t^2}{I_t^2 + \alpha_{A,t}^2} A_t. \tag{18}$$

The simplified demographics may be written as the following Leslie matrix equation (Equation 19). We can now assemble a complete model for mountain pine beetle disturbance in lodgepole pine forest. The final step is including the mortality exerted on the forest by the beetles. In our model, this is accomplished by subtracting a portion of the trees from the overall forest demographic model and adding them to the infectives cohort.

$$\begin{pmatrix} \hat{J}_{t+1} \\ V_{t+1} \\ A_{t+1} \end{pmatrix} = \begin{pmatrix} 1 - s_V & b_V & b_A \\ s_V & 1 - s_A & 0 \\ 0 & s_A & 1 - d \end{pmatrix} \begin{pmatrix} \hat{J}_t \\ V_t \\ A_t \end{pmatrix} - \begin{pmatrix} 0 \\ \dfrac{I_t^2}{I_t^2 + \alpha_{V,t}^2} V_t \\ \dfrac{I_t^2}{I_t^2 + \alpha_{A,t}^2} A_t \end{pmatrix}, \tag{19}$$

$$I_{t+1} = \frac{I_t^2}{I_t^2 + \alpha_{V,t}^2} V_t + \frac{I_t^2}{I_t^2 + \alpha_{A,t}^2} A_t. \tag{20}$$

The model presented in Equations (19) and (20) describes the growth and mortality of the lodgepole pine forest under pressure from a local beetle population, neglecting the effects of dispersal to and from adjacent patches of forest, and assuming that shading mortality on seedlings roughly balances their rate of survivorship.

Dynamic Analysis Without Dispersal

If we consider the behavior of this model locally, say in a small stand of trees within the forest, and make some simplifying assumptions, we recognize compelling behavior even at this point in the model's development. We will show that the model without dispersal exhibits a lack of an endemic state of infected trees but, instead, describes an increase in susceptibility from encroaching infected trees that leads to periodic waves of beetle outbreak.

This suggests that there is no endemic state per se but that outbreaks elsewhere in the forest serve as a population reservoir.

For simplicity and to clarify model behavior, we will assume that the adult and vigorous classes may be combined into one susceptible class (i.e., $V_t + A_t = S_t$). This is often the case in western North America, where fire management has compromised the age structure of forests, leaving a homogeneous stand of lodgepole pine and completely removing one class of trees from the equation. With this in mind, the system of Equations (19) and (20) can be written as

$$\begin{pmatrix} \hat{J}_{t+1} \\ S_{t+1} \end{pmatrix} = \begin{pmatrix} 1-s & b \\ s & 1-d \end{pmatrix} \begin{pmatrix} \hat{J}_t \\ S_t \end{pmatrix} - \begin{pmatrix} 0 \\ \dfrac{I_t^2 S_t}{I_t^2 + \alpha_t^2} \end{pmatrix}, \tag{21}$$

$$I_{t+1} = \frac{I_t^2}{I_t^2 + \alpha_t^2} S_t. \tag{22}$$

Let us examine how the model behaves locally in a relatively healthy stand of trees. Considering the density of infectives to be very low in the area under consideration, $I_t \ll 1$, implies $\frac{I_t^2}{I_t^2 + \alpha_t^2} \ll I_t$. This means that, at low den-sities of infectives, the loss of susceptible trees to infectives is negligible and the model with three classes, Equations (21) and (22), decouples into a matrix model of healthy forest and an equation modeling the change in density of the infectives.

The eigenvalues of the Leslie matrix $\begin{pmatrix} 1-s & b \\ s & 1-d \end{pmatrix}$ are $\lambda_{1,2} = 1 - \frac{1}{2}\left(s + d \pm (s-d)\sqrt{1 + \frac{4sb}{s-d^2}} \right)$.

The long-term behavior of the healthy forest is governed by the dominant eigenvalue, $\lambda_{\max}$, which can be interpreted as the intrinsic growth rate of the population. At least one eigenvalue is greater than one for $b > d$, and we observe that the determinant of the matrix in Equation (21) is $\lambda_1 \lambda_2 = 1 - d - s + sd - bs > 0$ for sufficiently small s and d, implying that both λ_1 and λ_2 must be positive. We can therefore conclude $\lambda_{\max} > 1$, which means the densities of both forest classes are increasing. If we consider S_0 to be the density of susceptible trees at time zero, then $S_t \approx \lambda_{\max}^t S_0$ describes the increase in density of the susceptible trees over time. For small numbers of I_t, the decoupled infective class is described by

$$I_{t+1} = \frac{I_t^2}{I_t^2 + \alpha^2} \lambda_{\max}^t S_0. \tag{23}$$

We find a fixed point analysis of Equation (23) very helpful in explaining the local dynamics of forest disturbance and recovery. The fixed points for low densities in the decoupled equation are found by setting $I_{t+1} = I_t$ and the associated stability of these particular points indicates trends for densities of infectives (e.g., increasing or decreasing). Fig. 2 is a bifurcation plot of the fixed points of Equation (23) as the density of susceptible trees, $S_t = \lambda^{t}_{\max} S_0$, increases. We see that for very low densities of S_t, there is only one fixed point. This fixed point is at $I_t = 0$ and is termed the "trivial fixed point," because an absence of beetles in the current time step implies that no beetles will be in the next time step. The trivial fixed point is attracting and is illustrated in Fig. 2 by the horizontal line of fixed points existing for infectives density of zero, $I_t = 0$. The local density of infectives appears to break away from the trivial fixed point through contributions from neighboring regions, but these small perturbations are quickly forced back to the extinction point.

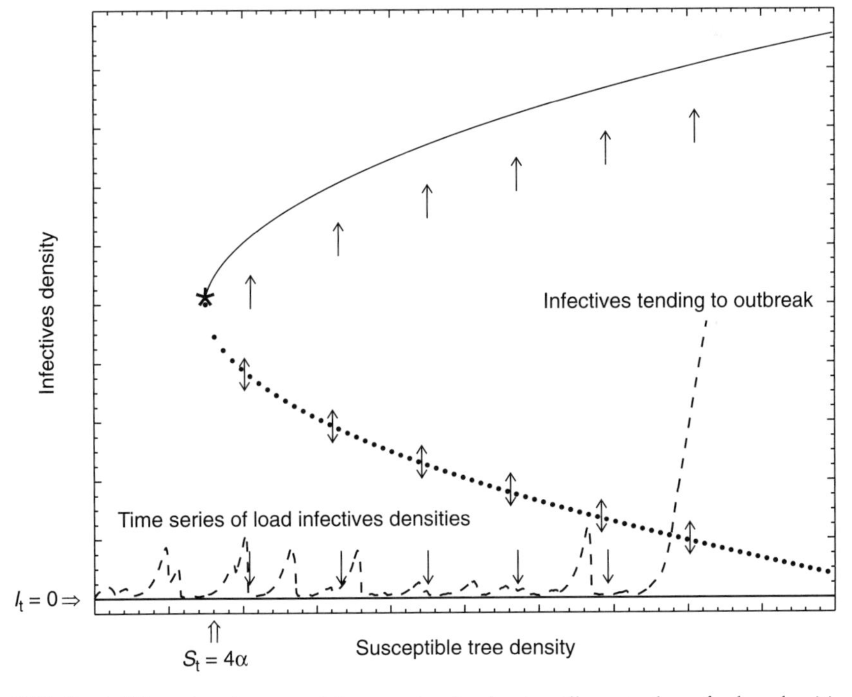

FIG. 2 A bifurcation diagram of the emerging fixed points illustrates how, for low densities of infected trees, the trivial fixed point (at $I_t = 0$) pulls the densities of these infected trees down to extinction. But as the susceptible tree density increases, the repelling fixed point moves closer to the trivial attracting fixed point. When by chance the density of infected trees shoots above this lowered threshold (through contributions by neighboring areas), the density of infected trees is thrust away from the low-density levels, permitting an outbreak to occur.

While the density of infectives is at this benign level, the healthy forest grows in accordance with the decoupled Leslie matrix model. This unbridled growth of the susceptible class eventually reaches a density level where a nontrivial fixed point emerges. This nontrivial fixed point first comes into existence through a saddle-node bifurcation when $\lambda^t_{max} S_0 = 4\alpha$. In Fig. 2, this appears at the point marked with an asterisk. The fixed point is found by setting $I_{t+1} = I_t$ and then through simple algebraic manipulation and use of the quadratic formula. When $S_0 > \frac{4\alpha}{\lambda^t_{max}}$, this nontrivial fixed point at $I_t = \frac{\lambda^t_{max} S_t}{2}$ bifurcates, or "splits" into two distinct fixed points, one attracting and the other repelling, represented by the solid and dotted curves, respectively. We ignore the nontrivial attracting fixed point because it violates the earlier assumption that I_t is small. It is the fixed point at the lower density that is of interest to us (the repelling point closer to the trivial fixed point). This fixed point near the extinction level of infectives is given by $I_t = \frac{\lambda^t_{max} S_t - \sqrt{(\lambda^t_{max} S_0)^2 - 4\alpha^2}}{2}$. This repelling fixed point pushes the density of infectives toward the extinction level. The model, therefore, predicts extinction for the infectives at this low level but, remembering that we are examining the behavior in a small stand of the forest, there are in fact contributions of infectives from the surrounding forest that continuously jump start the population. As time passes and the forest grows, the repelling fixed point moves closer and closer to the trivial fixed point. Eventually, the repelling fixed point moves so close to the trivial fixed point that even a small contribution to the infectives class from neighboring areas of the forest elevates the density of infectives above this repelling fixed point, at which time an epidemic occurs as the same repelling fixed point forces densities of infectives to increase. This leads to positive exponential growth and, consequently, the onset of an outbreak (Berryman *et al.*, 1984).

The picture that emerges is that of statistically periodic outbreaks. Starting with no red top trees, the forest grows increasingly vulnerable to ever-smaller invasions of mountain pine beetles. Eventually, some small external input of mountain pine beetles pushes the dynamics over the threshold and into an outbreak, which removes most of the susceptible class. A period of time must pass as juvenile trees grow to become susceptibles and eventually the cycle repeats. This explanation of the dynamics on a local scale depends on re-invasion from external sources of mountain pine beetles, as opposed to the existence of a stable, endemic population (which the model suggests does not exist). To complete this picture of periodic outbreak, we must explicitly include the effects of dispersal in space.

Including Dispersal in the Red Top Model

Thus far we have not addressed the spatial component of the beetle–forest interaction. But before we begin, let us consider two sensible options for investigating the spatial behavior of the beetle epidemic: one and two dimensions. Although it is obvious that the real world impact of the beetle outbreak can be protracted from a two-dimensional perspective, there are clear advantages to considering a lower dimensional interpretation of the phenomena (noting that a third dimension, height, is inconsequential compared to the landscape scale of the model).

When the region infected is large, as in the case of the Sawtooth National Recreation Area (SNRA), the spread of infected trees is essentially a one-dimensional event. Furthermore, in the particular case of the SNRA, the valley is relatively long and narrow, and patches of infected trees often span a significant breadth of the forest, leaving more or less only one dimension for an advancing infection. Protracting the forest in one dimension also allows for more simplified mathematical analysis, which can be directly applied to the higher order modeling. Stepping up from one dimension to two, we take with us the insights from the one-dimensional model and incorporate them into the two-dimensional model, thus allowing us to better interpret the results.

To this end, we now consider the beetles emerging from a host tree. To model the probability of a beetle at a point source (i.e., an infected tree) dispersing to a surrounding tree, we introduce a dispersal kernel. Neubert *et al.* (1995) propose a number of dispersal kernels that can describe dispersal behavior in one dimension. These kernels can be derived a priori from differential equations. If we assume that in a given year there is a density of flying beetles that are seeking a host tree and that these flying beetles find a host tree at some constant rate, we can model this by a system of equations that represent the density of dispersing beetles, $u\ (x, t)$, and the density of beetles that have settled on a tree, $v\ (x, t)$:

$$\frac{\partial u}{\partial t} = D\frac{\partial^2 u}{\partial x^2} - \mu u. \tag{24}$$

$$\frac{\partial v}{\partial t} = \mu u. \tag{25}$$

where D is the diffusion parameter governing the rate at which beetles search out a host tree and μ is the constant rate at which beetles settle on a tree. If we assume that the dispersing beetles originate at a single point source (i.e., an infected tree) and consider a sufficiently long period over which all the dispersing beetles can find host trees, we can define the dispersal kernel to be the final distribution of settled beetles, that is to say, the distribution of settled beetles as time goes to infinity (i.e., $\lim_{t \to \infty} v(x,t) = k(x)$).

Solving for the kernel under these assumptions results in the Laplacian distribution (or double exponential):

$$k(x) = \frac{1}{\sqrt{4\delta}} \exp^{-\sqrt{x^2/\delta}}. \qquad (26)$$

Here we have made the substitution $\delta = D/\mu$, which has the virtue of representing the mean dispersal distance as the single parameter, δ.

This kernel has a "tent" shape, with a maximum at the origin, sloping down exponentially as the distance from the origin increases. The solution to the two-dimensional diffusion equation with a constant failure rate takes the form of ratios of modified Bessel functions, which have singularities at the origin. For the purposes of simplicity in our two-dimensional model, we employ a function that is analogous to the one-dimensional kernel:

$$k(\bar{x}) = \frac{1}{2\pi\delta^2} \exp\left(\frac{-\sqrt{(x_1^2 + x_2^2)}}{\delta}\right) \qquad (27)$$

where a point in space is represented by the ordered pair $\bar{x} = (x_1, x_2)$. The behavior of this dispersal model is very similar to the correct, more complicated solution with Bessel functions. However, the advantage is that the simplified dispersal model is much easier to parameterize and understand in a biological context, not to mention removing the troublesome singularity. In the two-dimensional dispersal model, the mean dispersal distance is 2δ, and the units of the dispersal kernel are inverse hectares, $\frac{1}{ha}$.

Neubert *et al.* (1995) go on to describe how the convolution of the dispersal kernel with the initial population density function results in the population of organisms dispersed over space. One can think of the number of individuals located in a small interval dispersing according to the distribution kernel; thus, the probability that the individuals will be at $\bar{y} = (y_1, y_2)$ in the next time step, given that they were originally at $\bar{x} = (x_1, x_2)$ is $I(\bar{x}, 0)\, K(\bar{x} - \bar{y})\, d\bar{y}$, where the dispersal kernel is shifted so that it is centered at the original locus of individuals, $\bar{x}$. The total population after a time step would then be the sum over all such infinitesimal intervals containing populations. This leads us to the convolution, I^*, of two functions I and K as defined by

$$I^* = (I * K)(\bar{x}) = \int_{-\infty}^{\infty} \int_{-\infty}^{\infty} I(\bar{y})\, K(\bar{x} - \bar{y})\, d\bar{y}. \qquad (28)$$

This is the concept behind the use of convolutions of dispersal kernels with spatial population density functions. In short, I^* can be interpreted as the population density of infected trees after dispersing in accordance with the probability density function K. So the final model is

$$\begin{pmatrix} \hat{J}_{t+1} \\ V_{t+1} \\ A_{t+1} \end{pmatrix} = \begin{pmatrix} 1-s_V & b_V & b_A \\ s_V & 1-s_A & 0 \\ 0 & s_A & 1-d \end{pmatrix} \begin{pmatrix} \hat{J}_t \\ V_t \\ A_t \end{pmatrix} - \begin{pmatrix} 0 \\ \dfrac{I^{*2} V_t}{I^{*2} + \alpha_V^2} \\ \dfrac{I^{*2} A_t}{I^{*2} + \alpha_A^2} \end{pmatrix} \qquad (29)$$

$$I_{t+1} = \frac{I^{*2}}{I^{*2} + \alpha_V^2} V_t + \frac{I^{*2}}{I^{*2} + \alpha_A^2} A_t, \qquad (30)$$

where I^* is given by the convolution in Equation (28). Equations (29) and (30) form a stage-structured model of integrodifference equations, including a minimally complex description of forest recruitment, aging, and growth, with a realistic model for mountain pine beetle attack and dispersal on a stand scale (i.e., a scale larger than individual trees which allows for units of space to be comprised of similar vegetation). However, to compare the model to the real-world phenomenon of mountain pine beetle outbreaks, we need to find suitable values for the parameters in the model as they pertain to the epidemic in the SNRA. Only then can we begin to simulate and interpret the results of the model as compared to observations.

Parameter Estimation Based on Aerial Damage Survey Data

Nonlinear parameter estimation is recognized to be at best a challenging aspect of modeling real world phenomena. In the case of the Red Top Model, we are faced with the problem of fitting parameters to the nonlinear response functions used to model the growth of infected trees in the forest. The task is difficult because the response variable is defined on a stand scale, as opposed to individuals. This also involves the response of tree-stands to populations of dispersing mountain pine beetles [through I^* in Equation (28)], and these populations are impossible to measure directly either under the bark or in flight.

The data used to estimate parameters for the model are from the Aerial Damage Survey (ADS), collected from flights over the SNRA providing 30 m by 30 m resolution. Numbers of infected trees are detailed on a map and then the data are converted to densities on a GIS cover map in ARCVIEW. Fig. 3 is a map of the SNRA generated from ADS data obtained in 1991.

The accuracy of this data collection is inherently limited. It is very difficult to pick out individual infected trees from a dense stand of forest while flying in an airplane. Similarly, clumps of infectives might not be properly articulated. The cover map, which describes the distribution of trees, both healthy and infectives, is mostly homogeneous, also a shortcoming. There are patches of sagebrush, grass, and nonhost conifers throughout the SNRA, interspersed with the lodgepole pine forest.

Because there are many years of spatial data for the spread of infected trees around the forest, one possible approach would be to find a least-squares or maximum-likelihood solution to the problem. But aside from the computational intensity of a multidimensional parameter search, there is also the risk of arriving at a suboptimal solution. Alternatively, we can employ the method of estimating functions. The method used here is an

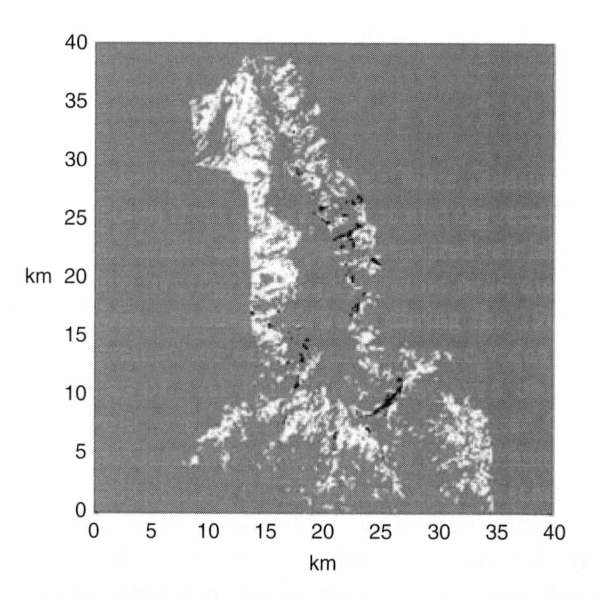

FIG. 3 A map of the Sawtooth National Recreation Area generated from aerial damage survey data taken in 1991. Each cell in the map represents a 30 m by 30 m square of the SNRA. The cell is classified as lodgepole pine or nonlodgepole pine forest. Over this forest cover map is superimposed a map of infectives, where each pixel contains the density of red top trees. In this map, the infectives were scaled to simply reflect presence or absence in a cell. The white represents the location of susceptible lodgepole pine trees and the black is the presence of infectives.

adaptation of the method described by Lele *et al.* (1998), who also used estimating functions to arrive at dispersal parameter values based on gypsy moth trap data.

The following is a brief overview of the method of estimating functions used to approximate parameter values for spatially related data. We begin by assuming that the data are in a spatial array where each element contains a density of individuals in that cell, i.e., $I_{i,j,t}$, where i and j are spatial indices and t is the time index. We construct the estimating function from the response function in Equation (30) containing the parameters to fit, δ and α^2. We assume that there is one cohort contributing to the infectives class and algebraically manipulate the equation

$$I_{i,j,t+1} = \frac{I^{*2}_{i,j,t}}{I^{*2}_{i,j,t} + \alpha^2} S_{i,j,t} \tag{31}$$

to read

$$g\,(I_{i,j,t}, S_{i,j,t}, \alpha^2, \delta) = I_{i,j,t+1}\,(I^{*2}_{i,j,t} + \alpha^2) - S_t I^{*2}_{i,j,t} = 0. \tag{32}$$

The result in Equation (32) is an estimating function that relates the parameters to the data, with expected value zero. Since we have two parameters to estimate, we need two equations. We want to combine them in a useful way that allows us to solve for parameters α^2 and δ. According to Lele *et al.* (1998), a near optimal combination of these functions can be found by introducing weighting functions that minimize the sensitivity of Equation (32) to the data. These are formed by taking the derivative of the estimating functions with respect to the parameter of interest (i.e., $w_{\alpha^2} = \frac{dg}{d\alpha^2}$ and $w_\delta = \frac{dg}{d\delta}$). The resulting system of equations is

$$\sum_t^{T-1} \sum_{i,j}^{n,m} W_{\alpha^2}[g\,(I_{i,j,t}, S_{i,j,t}, \alpha^2, \delta)] = H(\delta, \alpha^2) = 0, \tag{33}$$

and

$$\sum_t^{T-1} \sum_{i,j}^{n,m} W_{\delta}[g\,(I_{i,j,t}, S_{i,j,t}, \alpha^2, \delta)] = G(\delta, \alpha^2) = 0. \tag{34}$$

Now by stepping through an interval that is assumed to contain the best parameter value for δ and choosing candidates, $\hat{\delta}$, we can solve Equations (33) and (34) for the respective α^2 in terms of $\hat{\delta}$. This results in candidates for a solution of the form $(\hat{\alpha}^2, \hat{\delta})$. The ordered pairs represent possible parameter solutions constituting a one-dimensional curve. Since the candidates in turn need to be the zeros of the estimating function, $\sum g(I_t, S_t, \alpha^2, \delta) = 0$, we simply search for the root of this curve to arrive at the estimate for the parameters. Computationally, this one-dimensional search for optima is far less intensive than a search through two-dimensional parameter space, and all possible optima can be found and evaluated. Moveover, a one-dimensional root solving operation can be made arbitrarily accurate.

Because of heterogeneity in the SNRA in the form of water bodies, agricultural land, and a variety of vegetation classes, there are areas with zero host densities. To accommodate this patchiness, searches were carried out on a variety of subregions of varying sizes within the SNRA. There are also demographic differences between these subregions, such as stand age and density. The variability in demographics and composition results in differing parameter values throughout the SNRA. The parameter values listed in Table 1 are the averages of results of the estimating function procedure found on subregions of the forest using the two-dimensional Laplace kernel given in Equation (27).

Analysis with Dispersal in the One-Dimension Case

The spatial aspect of the model prompts the need to consider scale in the model. To begin with, let us consider a one-dimensional representation of the forest. It seems reasonable to assume that, for a sufficiently small forest, the spatial dynamics would not be evident, since it would emulate the non-spatial model. Consequently, we might think that there is a threshold area above which allows additional dynamics that result from dispersal. We therefore can postulate that although beetles cannot persist locally after killing all of the susceptible trees in a small area, a large enough forest might allow for beetles to find host trees in regenerated regions of the forest. In this manner, the beetles could persist in the forest despite decimating

TABLE 1 Average Values Resulting from the Estimating Function Procedure Applied to Patches of the Sawtooth National Recreation Area*

Year	δ_{ave}	${\alpha_{ave}}^2$
1990	5.9786	0.00008101
1991	4.6347	0.00581680
1993	4.6169	0.00581729
1995	4.8911	0.00537288
1997	4.9463	0.00537279
1998	5.2784	0.00469327
2000	5.2650	0.00469324
2001	5.3138	0.00452175

*Each year yields several solutions for δ and α^2 depending on the patch of forest over which the estimating function procedure is applied. For the years 1992, 1994, 1996, and 1999, the estimating function procedure did not converge to a set of solutions.

local populations. In light of this hypothesis, we consider the model's behavior in terms of the persistence of infectives throughout varying spatial scales. We run a sequence of 10 simulations at each forest size ranging in length from 1 km to 100 km at intervals of 5 km, fixing forest demographic parameters ($b = s = 0.06$) but selecting α^2 according to a uniform distribution from 0.004 to 0.005 for each year. A plot of the average number of years that the infectives class persists at each simulated forest size is shown in Fig. 4.

We see that for simulations with a forest length beyond this threshold near 40 km, there is qualitatively different behavior because of the spatial component of the model. For sufficiently large forests, we see the formation of waves of infectives that sweep back and forth across the simulation space as the forest regenerates. Fig. 5 illustrates a series of images taken at 10-year intervals as waves of infected trees swept across a simulation space of 50 km. Reading the frames from left to right and top to bottom, we can

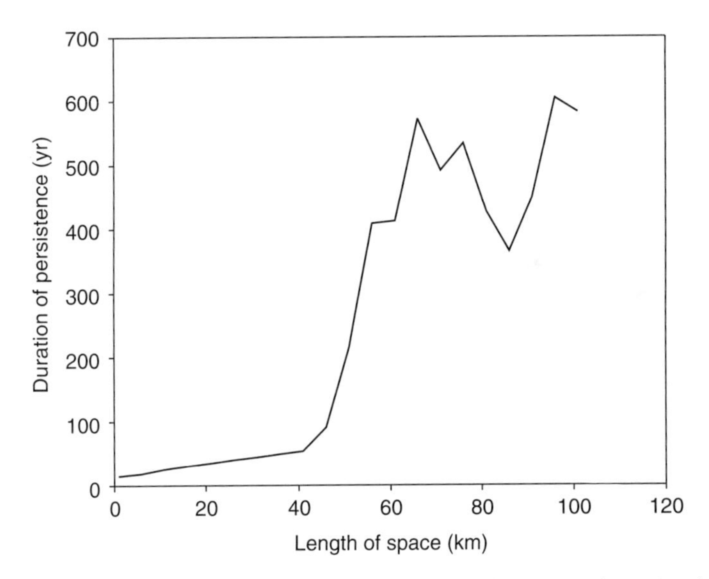

FIG. 4 Graph of the average persistence of the infectives class in one-dimensional simulations, using the one-dimensional simulations, using the one-dimensional Laplace kernel from Equation (26). Parameter values are α^2 varying uniformly between 0.004 and 0.005, the mean dispersal distance $\delta = 200$ meters, and fixed forest demographic parameters, $b = s = 0.06$, and $d = 0$. Observe that for simulated forest lengths greater than 40 km the persistence of infectives increases markedly. It is at this point that the spatial aspect of the model allows for wave formation that does not dissipate, increasing the duration of persistence. We observe fluctuations in the duration of persistence for larger forest sizes. This is caused by constructive resonance-like behavior resulting from combinations of particular forest lengths and wave speeds.

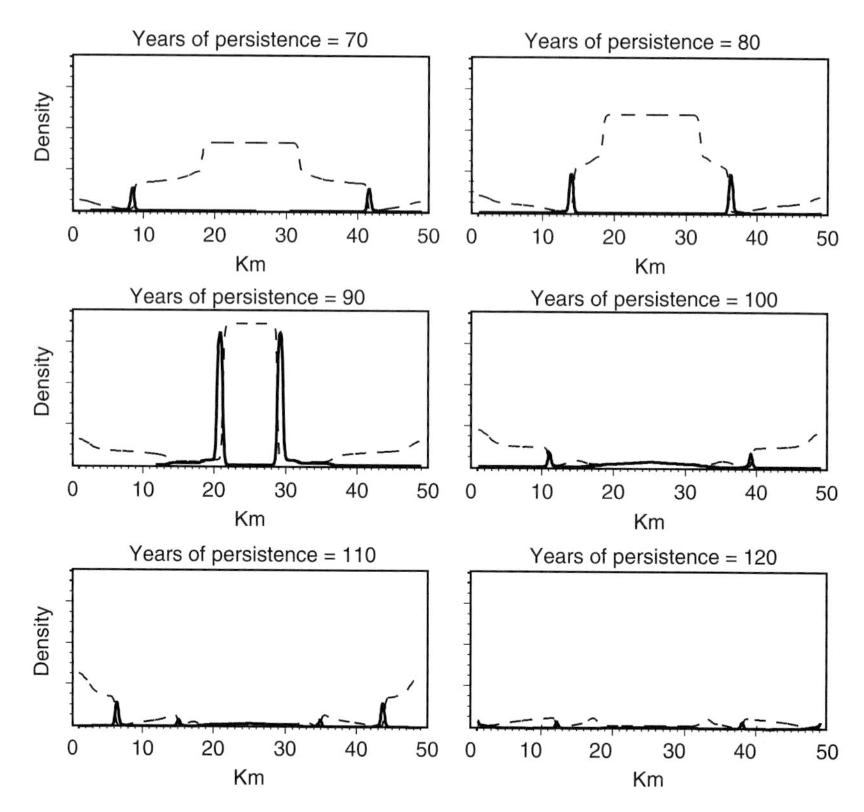

FIG. 5 This sequence of frames illustrates the advance of infestation through a forest of healthy trees. Reading left to right and top to bottom are six frames taken at 10-year intervals. This one-dimensional simulation depicts the waves of infected trees (*solid lines*) moving into regenerated stands of vigorous trees (*dashed lines*). As the waves collide they annihilate each other, having exhausted the available supply of susceptible trees. The simulation space is 50 km long and the parameter values are $b = s = 0.06$, $d = 0$, $\delta = 200$ meters, α^2 varies uniformly between 0.004 and 0.005.

see colliding waves of infectives that annihilate one another after exhausting the supply of hosts in the area. Successive waves follow as the forest regenerates, in turn exhausting the population of healthy susceptibles. This observation is highly compatible with the conjecture made during the analysis of the model in the absence of dispersal, where locally the forest regenerates, becoming increasingly susceptible to invasion until an external perturbation catalyzes an outbreak.

Fig. 6 is a time series of the results illustrating the percentage of the forest that is occupied by infective trees compared to the percentage of the forest occupied by susceptible trees. The pulses of infectives follow pulses of susceptible trees, as predicted earlier. We also see that the model simulated long-term persistence of the infectives within the forest and, over the course of time, the density of infectives varies greatly, demonstrating both endemic and epidemic population levels.

RESULTS OF THE FULLY DEVELOPED MODEL

Using the ADS data that provided parameter estimates, we apply the model to each year of data to predict the subsequent year's distribution of infectives. The coarse nature of the ADS data precludes low densities of infectives from being recognized. To address this, we set a lower threshold to ignore densities of trees that result from the model's smooth dispersal mechanism (the convolution) that are too low to be observed aerially. In Fig. 7 the threshold was set to 10 trees per hectare, because ADS surveyors are expected to note patches at greater than 10 trees per hectare. Using the α_{ave}

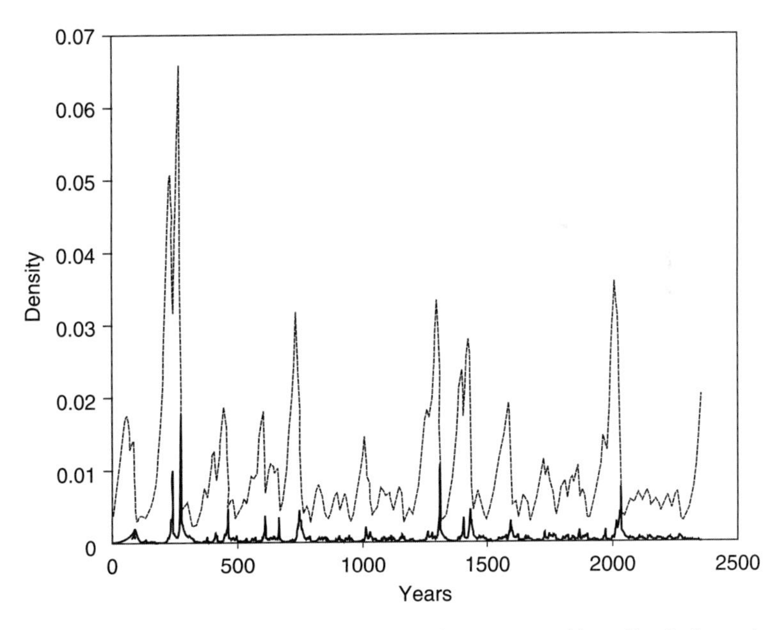

FIG. 6 This graph illustrates the cycles of outbreaks, represented by spikes in forest density (*dashed line*) followed by a spike in the infected tree density (*solid line*). This time series reveals the waves of infected trees following regenerated forest densities. In this 50-km-long one-dimensional simulation space, the infectives class persisted over 2000 years before going extinct. The parameter values in the simulation are $b = s = 0.06$, $d = 0$, and $\delta = 200$ meters, and α^2 varies uniformly between 0.004 and 0.005.

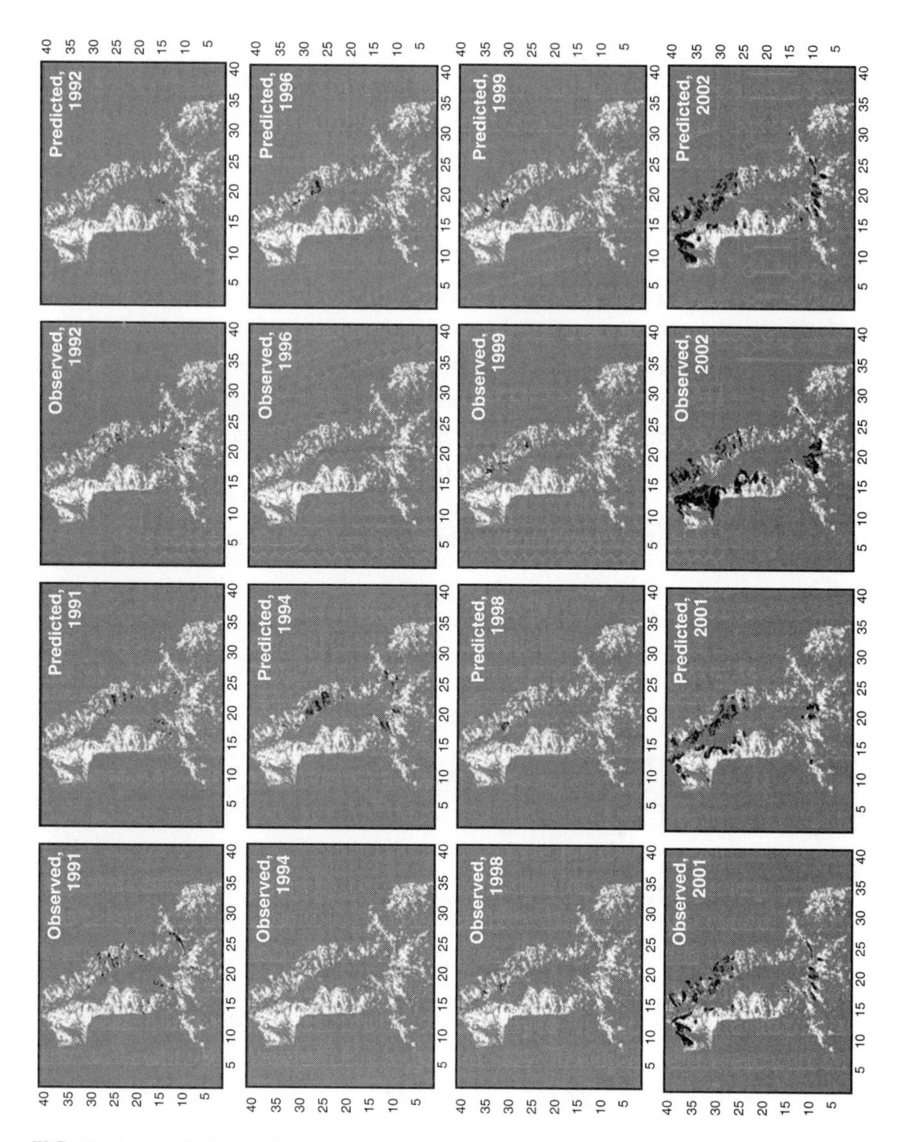

FIG. 7 Maps of observed and predicted infected trees in the Sawtooth National Recreation Area. The white regions represent the location of healthy lodgepole pine trees. The black regions represent the presence of infected lodgepole pines at densities above 10 trees per hectare. The infected trees' distribution is predicted by the two-dimensional model based on the previous year's observed distribution. Each prediction employs the respective parameter values for α^2 and δ listed in Table 1.

and δ_{ave} parameter values resulting from the estimating function procedure and applying the model to the entire SNRA forest yields predictions for infectives.

To gain another perspective on the predictions of the model, we compare year-to-year predictions of the proportion of the SNRA infected with the aerial damage survey data, as illustrated in Fig. 3. The ratio of cells that contain infectives over the total number of cells in the cover map that contain susceptible trees gives the percentage of the forest area that is infected with red tops. Fig. 8 is a graph of observed and predicted percentage of forest infected. Years of data that did not yield parameter values for α^2 and δ are omitted. Although this graph does not allow for examining the spatial distribution of the epidemic, it does provide a way of assessing the severity of the outbreak, as well as illustrating how well the model follows the outbreak's history.

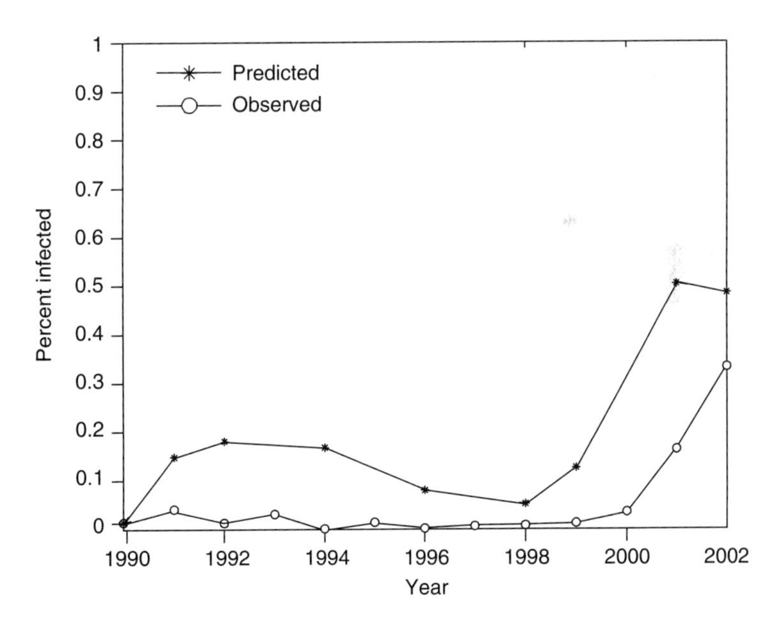

FIG. 8 Graph of the fraction of the total forested area of the Sawtooth National Recreation Area observed to contain densities of infectives (represented by o) and similarly the fraction of the forest predicted by the two-dimensional model using parameter values listed in Table 1 (represented by *). Predicted densities below 10 trees per hectare are ignored because it is believed that similar densities would not have been recorded by the aerial damage survey crew.

DISCUSSION AND CONCLUSION

In the preceding sections we used a variety of techniques and concepts to arrive at a model that, while relatively simple, still encapsulates the critical spatial and temporal mechanisms linking the mountain pine beetle distribution and forest recovery in space and time. Beginning with an understanding of bark beetle phenology and the attack dynamics observed in the forest, we proposed a relationship between the number of beetles and the number of infected trees in the forest. We also consider the probability of successful attack based on the density of infected trees in an area and use this to justify a Holling III response function that exhibited the desired Allee effect. Incorporating the notion of a dispersal kernel and a convolution to represent the spatial impact of current red tops on next year's fresh attacks, we arrived at a plausible heuristic for the spread of the mountain pine beetle through the lodgepole pine forest. These notions of density dependent attack dynamics and beetle dispersal in conjunction with a Leslie matrix describing the changing demographics of the forest form the integrodifference equations for the red top model.

Once we derived the red top model, we applied the theory of estimating functions to aerial damage survey data to find parameter values for beetle effectiveness and mean dispersal range. A range of parameter values resulted from this method, which is consistent with the understanding that beetle effectiveness is somehow temperature dependent, and in turn suggests that on a year-to-year basis the effectiveness of beetle attack on the lodgepole pine forest would fluctuate. There are clearly shortcomings of the data used to estimate parameter values. We were not able to ascertain densities of susceptible trees or infected trees and, as a consequence, we assumed a homogeneous standing timber and a homogeneous ratio of infected to susceptible trees in areas that contained infected trees. This is clearly a gross assumption and one that might be remedied by more accurate data. There are also years for which the estimating function procedure did not produce estimates for the parameters under investigation. This may be a consequence of too little beetle activity during those years or simply that the pattern of behavior did not lend itself to a solution.

Once we discovered what seemed to be reasonable parameter values to use in the model, we could investigate the behavior of the model and compare it to observed phenomena in the forest. We find that for small forest regions the model does not predict a nonzero stable equilibrium for the red top population, meaning that locally the forest cannot sustain an endemic population of mountain pine beetles. However, for a sufficiently large forested area, the reinvasion of beetles locally is facilitated by their presence in high densities elsewhere in the forest as waves of red tops move through susceptible stands of forest. We also observe that, just as in the real world situation, the size of the simulated forest plays a critical role in the persistence of the mountain pine beetle and forests that are too small to allow for regeneration after beetle attacks cannot sustain an endemic beetle population. Multiple simulations investigating persistence of the beetle population in a one-dimension forest found that, for an adequately sized simulation forest, the population of infected trees could persist for thousands of years. The effect of mountain pine beetle disturbance is to periodically reinfest forests with too many susceptible mature trees, removing this class and moving on. From the standpoint of our model, it seems quite reasonable that mountain pine beetle and lodgepole pine have co-adapted to maintain a dynamic self-regulation on large enough landscape scales.

In the case of the pine forests of western North America, in particular the SNRA, the mountain pine beetle has been presented with an undisturbed, contiguous, and mature forest structure that the model demonstrates is ripe for infestation. The advance of beetle attack is not mitigated by patches of previously disturbed forest regenerating from fire. Instead, crowding of susceptible trees has decreased the trees' defensive mechanisms and a sequence of unusually warm years has bolstered the beetle population above the unstable threshold described by the model. This eruptive outbreak observed in the SNRA is observed in the two-dimensional red top model.

Insect disturbance, as our model suggests, is important in maintaining a diverse age structure for lodgepole pine. Left to its own devices, lodgepole would develop into crowded and unhealthy forests of over-mature trees. With disturbances such as mountain pine beetles a certain homeostasis can be maintained, at least on sufficiently large spatial scales. As our model illustrates, insect disturbances can move at a self-limiting pace, balancing the rate of forest regeneration. Like fire (with which mountain pine beetle reforestation is associated), mountain pine beetle disturbance must be viewed as a normal and healthy part of ecosystem function on a sufficiently large scale. Our work helps establish on what scales, both in time and space, an insect disturbance such as that caused by mountain pine beetles can be expected to serve as a useful and normative disturbance.

ACKNOWLEDGMENTS

We are thankful for the significant contribution made by Leslie Brown for handling the imagery data used in this research. Additionally, portions of this research were funded by the U.S. National Science Foundation (grant DMS-0077663).

REFERENCES

Amman, G. D., and Cole W. E., (1983). *Mountain Pine Beetle Dynamics in Lodgepole Pine Forests: Part II Population Dynamics*. U.S. Department of Agriculture, General Technical Report INT-145, Intermountain Forest and Range Experiment Station, Ogden, UT.

Berg, E. E. (2003). Fire and spruce bark beetle disturbance regimes on the Kenai Peninsula, Alaska. In: *Second International Wildland Fire Ecology and Fire Management Congress*, Orlando, FL. pp. 1–4. American Meteorological Society, Boston, MA.

Berryman, A. A., Stenseth, N. C., and Wollkind, D. J. (1984). Metastability of forest ecosystems infested by bark beetles. *Res Pop Ecol* 26, 13–29.

Berryman, A. A., Dennis, B., Raffa, K. F., and Stenseth, N. C. (1985). Evolution of optimal group attack, with particular reference to bark beetles (Coleoptera: Scolyidae). *Ecology* 66, 989–903.

British Columbia Ministry of Forests. (2003). *Timber Supply and the Mountain Pine Beetle Infestation in British Columbia*. British Columbia Minister of Forests, Forest Analysis Branch, Victoria, BC.

Dale, V. H., Joyce, L. A., and McNulty, S. (2001). Climate change and forest disturbance. *BioScience* 51, 723–734.

Lele, S., Taper, M., and Gage, S. (1998). Statistical analysis of population dynamics in space and time using estimating functions. *Ecology* 79, 1489–1502.

Logan, J. A., and Powell, J. A. (2001). Ghost forests, global warming and the mountain pine beetle. *Am Entomol* 47, 160–173.

Logan, J. A., Régnière, J., and Powell, J. A. (2003). Assessing the impacts of global climate change on forest pest dynamics. *Front Ecol Environ* 1, 130–137.

Logan, J. A., and Powell J. A. (2004). Modelling mountain pine beetle phonological response to temperature. In: *Proceedings of a Mountain Pine Beetle Symposium: Challenges and Solutions*. (Brooks, J. E., and Shore, T. L., Eds.), pp 210–222. Pacific Forestry Centre, Victoria, BC.

Mattson, W. J. (1996). Escalating anthropogenic stresses on forest ecosystems: forcing benign plant-insect interactions into new interaction trajectories. In: *Caring for the Forest: Research in a Changing World* (Korpilahti E., Mikkelä, H., and Salonen, T., Eds.), pp. 338–342. IUFRO Secretariat, Vienna.

Munson, S. A., McMillin, J., Cain, R., and Allen, K. (2004). Spruce beetle in the Rockies—current status and management practices. In: *Effects of Spruce Beetle Outbreaks and Associated Management Practices on Forest Ecosystems in South-central Alaska*. (Burnside, R., Ed.), pp. 133–152. Kenai Peninsula Borough and the Interagency Forest Ecology Study Team.

Muir, P. S., and Lotan, J. E. (1985). Disturbance history and serotiny of *Pinus contorta* in western Montana. *Ecology* 66, 1658–1668.

Neubert, M. G., Kot, M., and Lewis, M. A. (1995). Dispersal and pattern formation in a discrete-time predator-prey model. *Theoret Pop Biol* 48, 7–43.

Peterman, R. M. (1978). *The Ecological Role of Mountain Pine Beetle in Lodgepole Pine Forests*. University of Idaho, Moscow, ID.

Ross, D. W., Daterman, G. E., Boughton, J. L., and Quigley, T. M. (2001). *Forest Health Restoration in South-Central Alaska: A Problem Analysis*. USDA Forest Service Gen. Tech. Rep. PNW-GTR-523, Pacific Northwest Research Station, Portland, OR.

Samman, S., and Logan, J. A. (2000). *Assessment and Response to Bark Beetle Outbreaks in the Rocky Mountain Area: A Report to Congress from Forest Health Protection*. USDA Forest Service, RMRS-GTR-62, Rocky Mountain Research Station, Fort Collins, CO.

Schmidt, W. C., and Alexander, R. R. (1985). Strategies for managing lodgepole pine. In: *Lodgepole Pine—The Species and Its Management*. (Baumgartner, D. M., Krebill, R. G., Arnott, J. T., and Weetman, G. F., Eds.), pp. 202–210. Office of Conferences and Institutes, Cooperative Extension, Washington State University, Pullman, WA.

17

Relationship Between Spruce Budworm Outbreaks and Forest Dynamics in Eastern North America

Hubert Morin, Yves Jardon, and Réjean Gagnon
Université du Québec à Chicoutimi

INTRODUCTION

In Chapter 15, Cooke *et al.* explain the periodic disturbances by native forest–defoliating insects by examining the population dynamics of the insects. They argue that the population density of forest insects is determined by the ecological relationships among host plants, the herbivorous insects, and their natural enemies in a so-called tritrophic interaction. The objective of this chapter is to discuss the effects of these disturbances by native forest–defoliators on the forest dynamics. However, we cannot understand the effects of forest-defoliating insects without understanding the reciprocal effect of forest dynamics on the population dynamics of the insects. Because both are so closely related and difficult to separate, this chapter focuses on the *relationship* between disturbances caused by native forest–defoliating insects and forest dynamics, which is recognized as an important part of the tritrophic interactions described in Chapter 15, especially for the spruce budworm system.

A good model to consider in this regard is the relationship between outbreaks of spruce budworm (*Choristoneura fumiferana* Clem.) and the dynamics of the boreal forest. This system is one of the best studied because the budworm is the insect that causes the most damage of any insect in eastern North America (MacLean, 1984; Morin, 1994). To give an idea of its importance, it is worth noting that the last outbreak, one of three major outbreaks that struck the North American boreal forest in the 20th century, destroyed 139 to 238 million m^3 of softwood on public land in Québec (Boulet *et al.*, 1996). This disturbance is more important than fire in many parts of eastern North America in terms of impact on the productivity of forests. Spruce budworm outbreaks have shown periodicity at the supra-regional scale, at least during the 20th century, occurring at an interval of approximately 33 years in eastern Canada (Candau *et al.*, 1998; Royama, 1984; Jardon *et al.*, 2003; Royama *et al.*, 2005). The impacts of the last four outbreaks are well registered in the boreal forest, and almost every natural stand of balsam fir [*Abies balsamea* (L.) Mill.] dates back to one or other of these outbreaks, following a cyclic mechanism of regeneration to which balsam fir is well adapted (MacLean, 1984; 1988; Morin, 1994). The balsam fir overstory is killed very rapidly by the insect, while some seedlings in the seedling bank survive and show a growth release to fill the gap created by the mortality of the trees (Morin, 1994; Morin and Laprise, 1997; Parent *et al.*, 2003). The seedling bank fills up again, and the process is ready to respond to a future outbreak (Johnson *et al.*, 2003). This is a simplification of a more complex process but, essentially, it represents what happens when a severe outbreak hits a mature balsam fir stand in eastern Canada (Morin, 1994; Jardon and Doyon, 2003; Johnson *et al.*, 2003).

There was a shift at the end of the 19th century in eastern Canada between two types of outbreaks affecting the landscape. What we actually know about outbreaks that occurred during the 19th century is that they were not synchronized at a supra-regional scale and that they took a long time to cover large areas in the landscape, affecting localized populations. In contrast, the 20th century outbreaks were better synchronized, with a great impact over very large areas (Blais, 1983; Jardon, 2002). This shift has been observed in other systems, such as the western spruce budworm (*Choristoneura occidentalis* Freeman) (Swetnam and Lynch, 1993). It has been linked to a shift in the forest mosaic, sometimes attributed to human impact on the forest (Blais, 1983; Swetnam and Lynch, 1993; Williams and Liebhold, 2000) or to natural phenomena, such as a change in fire frequency and impact (Jardon and Morin, submitted). The importance of the forest mosaic in explaining in part the oscillations of the insect populations in the system is not a new idea (Blais, 1954). For spruce budworm, it is now suggested as one of the major factors controlling the insect populations and the impact of budworm herbivory (see Chapter 15). However, the causes of the changes in the forest mosaic are still the subject of debate (Blais, 1983; Jardon and Morin, submitted).

Although this system has been well studied, it is not yet well understood. The factors that contribute to the periodicity, amplitude, and synchronicity of the outbreaks are not clearly identified, and the mechanisms are not clearly defined. Periodicity could be related to a predator–prey relationship (Royama, 1984; Royama *et al.*, 2005) and amplitude to food availability (Blais, 1983). These tritrophic interactions among host plants, insects, and natural enemies are well explained in Chapter 15. Synchronicity could be explained by migration and the Moran effect (Williams and Liebhold, 2000; Royama *et al.*, 2005; see also Chapter 15). The relationships between insect outbreaks and forest dynamics are just beginning to be understood for a limited number of outbreaks that occurred mainly in the 20th century, and these relationships are absolutely unknown for past outbreaks during the Holocene. We lack sufficient information on outbreaks that occurred before the 20th century to understand the changes that occurred at the end of the 19th century. In this chapter, we present some new results about the relationships between outbreaks of the last two centuries and boreal forest dynamics, and give an insight into what may have been the relationship during the Holocene, back to some 7000 years BP. Our hypothesis is that there is a close relationship between the structure of mature boreal forest stands and their occurrence within the forest and the periodicity, synchronicity, and amplitude of outbreaks. Structure here refers to species composition and distribution in the overstory of mature stands.

It is recognized that mature balsam fir stands are most susceptible to spruce budworm; they sustain the highest amount of defoliation during an outbreak (*sensu* MacLean and MacKinnon, 1997). Mortality is consistently higher in fir than in spruce (*Picea* spp.) stands and higher in mature than in immature stands (MacLean, 1980; MacLean and Ostaff, 1989). Balsam fir is the preferred host of the budworm, and it is also the most vulnerable species with the highest probability of mortality for a given level of budworm attack (Mott, 1963). The proportion of balsam fir in a stand remains the most consistent significant variable to explain the amount of defoliation (Bergeron *et al.*, 1995; MacLean and MacKinnon, 1997). White spruce [*Picea glauca* (Moench) Voss] would also be an important component in increasing the susceptibility of stands. In general, mature balsam fir stands in eastern Canada contain dispersed large white spruces. The stand structure that would be most susceptible would be one with a canopy of mature balsam firs and some dispersed white spruces protruding above the canopy. It has also been shown that the proportion of hardwood content has an impact on the level of defoliation by the budworm (Bergeron *et al.*, 1995; Su *et al.*, 1996; Cappuccino *et al.*, 1998; MacKinnon and MacLean, 2003; 2004). The higher the proportion of hardwood, the lower the level of defoliation of balsam fir. Red spruce (*Picea rubens* Sarg.) and black spruce [*P. mariana* (Mill.) B.S.P.] stands are also affected but to a lesser extent. We suggest that a forest mosaic that is mainly composed of balsam fir in the canopy is most likely to show periodic, synchronous, and severe outbreaks that have an impact over large areas. As stated in Chapter 15 by Cooke *et al.*, this would be a situation where herbivore population growth is favored over natural enemy population growth and where the decline of the herbivore population would be retarded, causing significant disturbance to the system. Conversely, a forest mosaic that is composed of alternate host and non–host tree species in the overstory is more likely to produce asynchronous outbreaks with less impact on restricted areas. The cyclicity, the pattern, or what we may call the entomological cycle caused by the budworm–natural enemy complex would remain the same but would be more difficult to identify in the landscape because of a moderate impact. This would be a situation where natural enemy population growth is favored over herbivore population growth. This hypothesis also implies that if there have been long periods in the past in which factors such as the climate or fire cycles did not allow balsam fir to occupy the canopy, there may have been long periods without any evident major impact of outbreaks, such as the ones we can identify in the growth rings of trees. This situation is completely different from the one we know something about, that is, the regular recurrence during the 20th century of major outbreaks that have a cyclic impact on the forest (Jardon *et al.*, 2003). This would be a new way to interpret the relationships between the forest and spruce budworm outbreaks that is in accordance with the tritrophic interactions described in Chapter 15 by Cooke *et al.* for the spruce budworm.

HISTORY OF SPRUCE BUDWORM OUTBREAKS OVER THE PAST 8600 YEARS

Dendrochronology Data

Since the pioneering work of Blais (1954), dendrochronology has been used extensively to show evidence of past outbreaks of numerous insects. Western and eastern spruce budworm (Morin and Laprise, 1990; Swetnam and Lynch, 1993; Krause, 1997), forest tent caterpillar (*Malacosoma disstria* Hbn.: Cooke and Roland, 2000), larch sawfly (*Pristiphora erichsonii* (Htg.): Jardon *et al.*, 1994), larch budmoth (*Zeiraphera diniana* Gn.: Weber, 1997), 2-year-cycle budworm (*Choristoneura biennis* Freeman: Zhang and Alfaro, 2002; 2003), and jack pine budworm (*Choristoneura pinus pinus* Free.: Volney, 1988) have been investigated, among others. Techniques for analyzing the data have been improved (Swetnam *et al.*, 1985), but the material used is still the same: the host trees. Thus, the maximum age that host trees reach still limits the analysis. This is particularly true for eastern spruce budworm, which regularly kills its host. Cross-dating has been helpful in using dead trees either in the field or in old buildings to extend the chronologies (Krause, 1997; Boulanger and Arseneault, 2004), but we are still limited to going back to the 17th century, and data available for extensive areas do not go back further than the 19th century (Fig. 1). Subfossil trees recovered from peatlands are now being used to go further back in time. Simard (2003) has studied a small peat bog surrounded by host trees of spruce budworm. Dendroecological analysis of subfossil spruce trees that covers a time period of about 4800 years BP has shown that the three well-known outbreaks of the 20th century are easily distinguishable in the growth rings but that previous outbreaks are difficult to identify. Characteristic growth reductions attributed to the budworm helped to cross-date the samples and to produce two floating chronologies between 4170 and 4740 years BP (Simard, 2003). Unfortunately, in many cases, the younger samples situated between 20 and 60 cm depth presented a complacent growth without significant reductions that would permit cross-dating. This result suggests that for that location, the impact of budworm on spruce growth before the 20th century was in general not as great as during the 20th century. But we are still lacking local and regional old

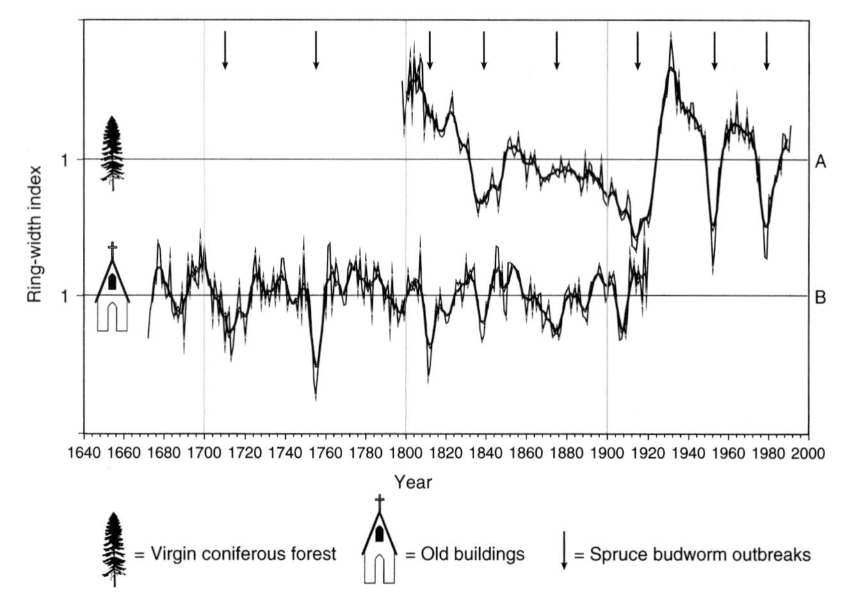

FIG. 1 Chronology of spruce budworm outbreaks in the boreal forest of Québec derived from dendroecological analysis from (**A**) Morin and Laprise (1990) and (**B**) Krause (1997).

chronologies over large areas. The most extensive dendroecological data set thus far for the eastern spruce budworm comes from the work of Yves Jardon (Jardon, 2002; Jardon *et al.*, 2003; Jardon and Morin, submitted). Jardon (2002) added 32 local chronologies to the existing chronologies in the meridional part of Québec and was able to study the periodicity, synchronism, and impact of the outbreaks that occurred during the last two centuries. This work has shown that budworm outbreaks occurred periodically at the landscape level every 25 to 38 years during at least the 20th century (Jardon *et al.*, 2003). It has also shown that the periodicity can change locally, perhaps because of migration or a complex local dynamic of the insect population, and that some sites escaped some infestations. Double waves of infestation were also present; they were similar and generally in phase, suggesting the influence of a common regional effect known as the Moran effect (Peltonen *et al.*, 2002).

Macrofossil Analysis

To identify outbreaks further back in time, we have developed a new method of analysis using macrofossils of the insect. Head capsules, pupae, and other insect remains have been used before for spruce budworm and other insects (Davis *et al.*, 1980; Bhiry and Filion, 1996). However, parts of the caterpillar or butterfly are fragile and possibly very rapidly recycled in the humus (Potelle, 1995). The first part of the work was to identify a macrofossil that would be produced in great quantity and that would preserve well in order to be retrieved easily in the sediments. Spruce budworm feces (frass pellets) were found to have all of these characteristics. During heavy infestations, they are so numerous that they fall to the ground like rain. Also, they have been shown to be identifiable to the species, parts of balsam fir leaves still being identifiable in the fecal matrix (Potelle, 1995). The method was first tested in fresh boreal humus under the canopy of balsam fir trees to ensure that the feces were from the budworm and to see whether some known outbreak episodes could be identified in the humus. Humus from balsam fir forests is not very deep (10–20 cm) and the turnover is rapid, but we were able to identify the three outbreaks of the 20th century in some stands. The outbreaks were represented by clear, easily distinguishable layers of feces (Potelle, 1995). The method was also tested successfully in deep humus (up to 1 m) in a humid maritime climate. In this environment, we were able to identify the 20th-century outbreaks and periods of budworm activity back to 1520 years BP (Simard *et al.*, 2002). An important reduction in the number of feces was observed below the upper 20 cm that corresponded approximately to the 20th century. This reduction was attributed to lower levels of budworm populations before the 20th century or to the intensification of decomposition with increasing humus depth. This study also showed the limits of the method. The resolution is not as high as dendrochronology, which is precise to the year. However, it is comparable to the resolution of many paleoecological methods. The macrofossil profile indicates periods of high budworm populations that may encompass several outbreaks.

The best environment for using this method is small peat bogs situated close to some balsam fir stands that were affected during known outbreaks. Unfortunately, this situation is not common, because balsam fir does not grow well in such an environment. The advantage of peat bogs is that, in such anaerobic and acid conditions, the feces, like other macrofossils, preserve well for a long time. Several such bogs were identified and sampled in the Saguenay area where some of the longest dendrochronologies were known (Isabelle Simard, Ph.D. thesis in prep.). We now have the longest budworm macrofossil profile available, covering more than 8000 calibrated (cal.) years BP (Simard *et al.*, 2005; Fig. 2). This profile from the Lac Désilets black spruce forested bog is 2.6 m deep and shows important periods of high insect populations. Spruce budworm feces started accumulating in the

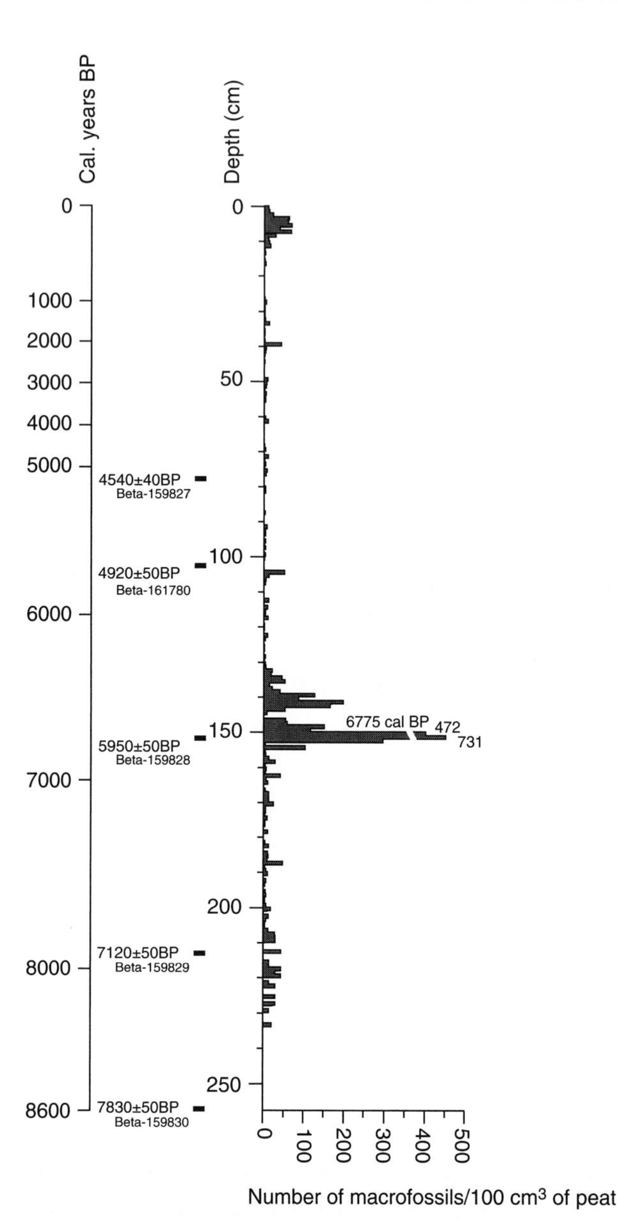

FIG. 2 Macrofossil profile showing the number of spruce budworm feces/100 cm^3 of peat in a small black spruce forested bog in the Saguenay area. (From Simard *et al.* [2005].)

study site around 8240 cal. years BP. They were observed throughout the profile, reaching important amounts at several depths, more particularly around 150 cm (6775 cal. years BP), at 140 cm (6550 cal. years BP), and in the last 10 cm (during the last century). We know that the 20th century saw high levels of spruce budworm populations in this area. Three well-known major outbreaks occurred during the 20th century (Blais, 1983; Morin and Laprise, 1990). If the number of feces is effectively a good indicator of the level of budworm populations, as we might expect, the profile suggests that there were only rare episodes during the Holocene when the populations were as high as or higher than in the 20th century. It also indicates that even if the environment permits better conservation of the macrofossils than in the deep humus, there is an important reduction in the number of feces below 20 cm that corresponds approximately to the 20th to 19th century limit, as shown in the deep humus of the Mingan archipelago (Simard *et al.*, 2002). Periods of low population levels seem to be the norm throughout the profile, at least at this local level. This pattern was also found in the three other sampled bogs where high numbers of feces were found for only two or three episodes during the Holocene (I. Simard, Ph.D. in prep). These periods do not necessarily correspond to the same dates as in the Lac Desilets bog, indicating that not every episode with high population levels is registered in one location as we might expect. Nevertheless, periods of high population levels were rare events during the Holocene. If we take into account that budworm populations have shown a cyclic pattern at the landscape level, these results suggest that episodes of high amplitude of the entomological cycle, such as occurred in the 20th century, were rare during the Holocene.

The paleophytogeographic context in which spruce budworm populations have evolved during the Holocene is of prime importance for interpreting these new results. An absence of budworm feces may have been caused by a momentary absence of coniferous host species in the landscape, especially balsam fir. However, studies of postglacial paleophytogeography of Québec and of the origin and dynamics of the mixed forest in this area show that balsam fir appeared very rapidly after the ice retreat which occurred between 11,000 and 8000 years BP in the regions considered here. Immediately after the ice retreat or after between ca. 1000 and 2000 years, depending on the region, balsam fir is present in the pollen and macrofossil diagrams and it maintains its presence throughout the diagrams (Richard, 1993; 1994). Of course, the representation of balsam fir did fluctuate in relation to other species. These fluctuations were mainly related to climate and fire occurrence, as indicated by the fluctuation of fire-adapted species, such as jack pine, and the analysis of charcoal. But host trees of the budworm, and especially balsam fir, never disappeared from the landscape situated south of the 52nd parallel in Québec since ca. 9000 to 8000 years BP (Delcourt and Delcourt, 1987).

VARIATION IN TEMPORAL AND SPATIAL DYNAMICS OF OUTBREAKS: REFLECTION OF CHANGES IN FOREST STRUCTURE

Outbreaks of the 20th Century

What was different during the Holocene to produce such markedly different amplitudes of the budworm population cycle? We use detailed dendroecological analysis to suggest some hypotheses. In a spatiotemporal study of spruce budworm outbreaks in southern Québec, Jardon (2002) has shown markedly different spatial outbreak patterns between the outbreak that occurred at the beginning of the 20th century (outbreak 3) and the one that occurred in the middle of the 20th century (outbreak 2). This difference was registered in the stands as well as in the entire study area. Outbreak 3 was very severe, and it appeared explosively across the study area without any clear spatial structure, whereas outbreak 2 was less severe and presented a spatial structure suggestive of a diffusion process. We are not suggesting here that outbreaks start from epicenters to spread across an area (Hardy *et al.*, 1983). Rather, in some situations, insect populations would increase more or less simultaneously so that their impact would rapidly be visible almost everywhere in the defoliation surveys as well as in the growth rings (outbreak 3 is a good example). In other situations, insect populations would increase more or less simultaneously so that the populations happened to reach a certain infestation level in some places slightly ahead of other areas (Royama 1984; Royama *et al.*, 2005). Their impact would then be detected in the defoliation surveys and in the growth rings at some places before others, giving a spatial pattern at a given level of defoliation or growth ring reduction (outbreak 2 is a good example). Seventy-three white spruce chronologies from northwestern and central Québec were added to the work of Jardon (2002) to cover an area extending north of the 50th parallel (Fig. 3). White spruce, when present, is generally chosen for such long-term studies because it is affected by the budworm, being one of the preferred hosts, and it can survive outbreaks so that we can find older trees than with balsam fir.

We show here the percentage of trees that were affected by the budworm during their life across the studied landscape (Fig. 4). Standard den-

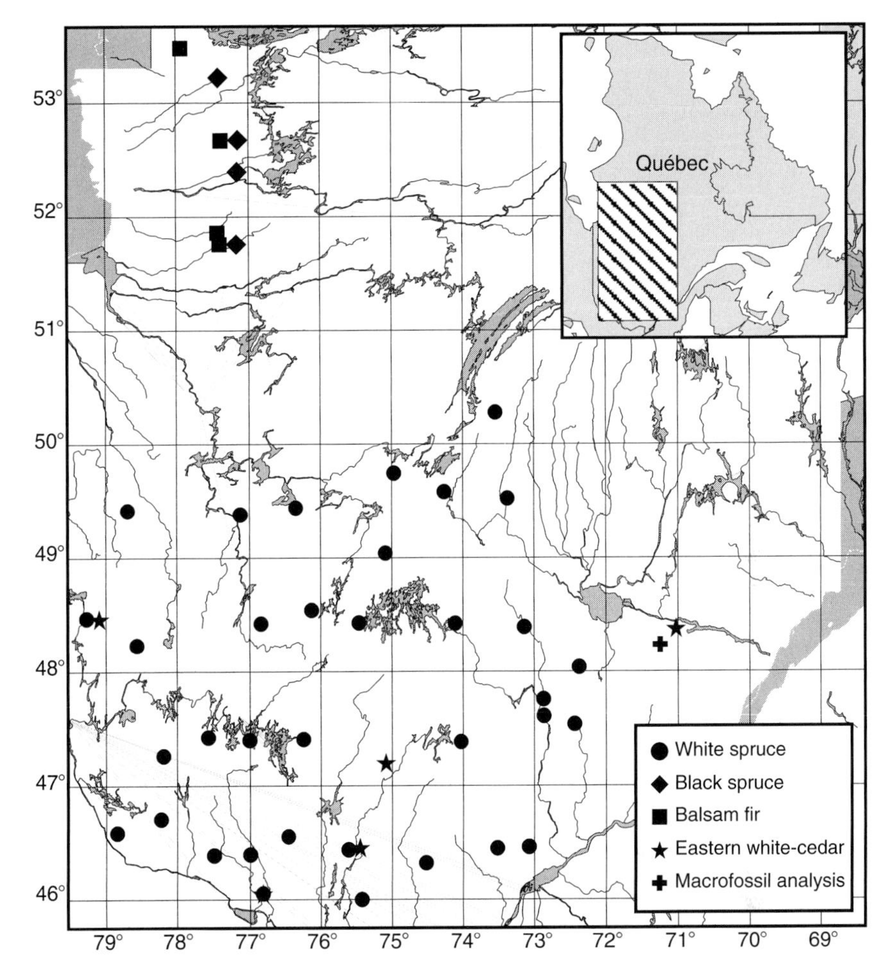

FIG. 3 Localization of host and nonhost chronologies for the Québec southern and boreal forests. (From Jardon [2002]; Morin and Laprise [1990]; Morin *et al.* [1993]; and Levasseur [2000].)

drochronological procedures (ARSTAN, ITRDB Lib V2.1; Holmes, 1999) were used to standardize the host and nonhost measured series using a cubic spline and a response frequency of 50% at 60 years. The OUTBREAK program (Holmes and Swetnam, 1996) was used to produce a corrected chronology by subtracting nonhost chronologies from host chronologies after an adjustment of the mean and standard deviations of the series to extract the regional climatic effect from the chronologies. To be considered as an outbreak period, a growth reduction must last at least 5 years and the

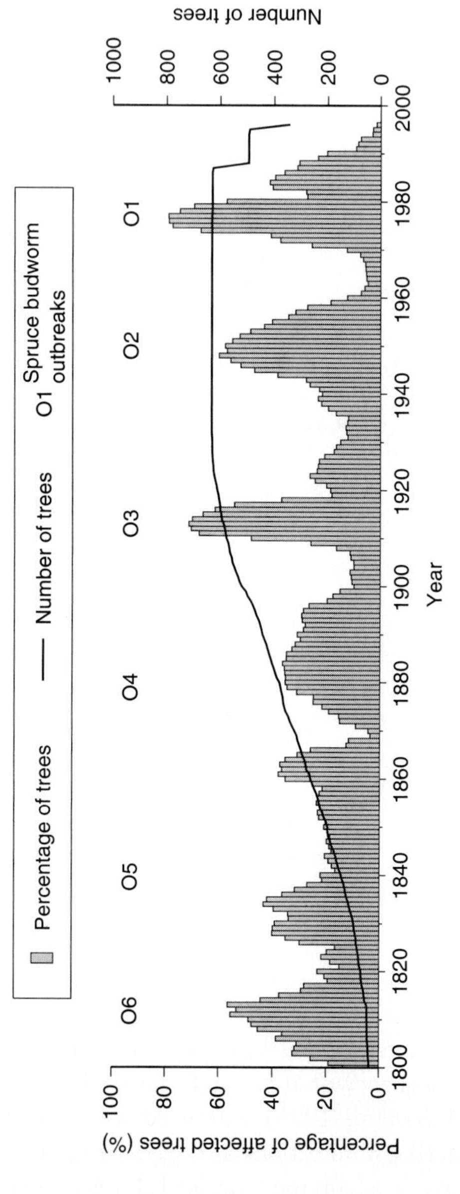

FIG. 4 Percentage of trees affected by spruce budworm defoliation in the southern part of Québec (south of the 50th parallel). Outbreak episodes are indicated above the histogram.

maximum growth reduction must be lower than one standard deviation of the corrected series (Swetnam *et al.*, 1985). The number of outbreak periods was compiled annually and presented as the percentage of the total number of trees present in the chronologies for a given year [see Jardon (2002) for details]. Because of the absence of nonhost series, black spruce and balsam fir series situated north of the 51st parallel were not corrected and the same criteria were used to identify outbreaks on the standardized, noncorrected series (Fig. 5). Outbreak O3 is known as a very severe outbreak that appeared very rapidly in each stand as well as in the entire province (Morin and Laprise, 1990; Jardon, 2002). As shown by white spruce trees, the number of affected trees rose very rapidly to attain a maximum close to 80% in 4 years. This outbreak killed more trees more rapidly than any other known outbreak in the province (Boulet *et al.*, 1996).

In contrast to this very explosive pattern is the pattern shown by outbreak O2. It took 12 years for the insect to defoliate the same area as outbreak O3, and only 60% of the trees were affected at the same time at the maximum of the infestation. What could have caused this difference? At the landscape level, the abundance of mature balsam fir stands was suggested as most important. Outbreak O3 was so intense that it killed almost every mature balsam fir stand (those that are the most susceptible). When outbreak O2 occurred following the very regular entomological cycle (Jardon *et al.*, 2003), mature balsam fir stands were rare. Some survivors were present, but the majority of the stands were young (less than 30 years old) and thus less susceptible, and were composed of released advanced growth following the opening of the canopy of killed stands (Blais, 1983; Morin, 1994). In contrast, outbreak O1, which came some 30 years later, affected more mature stands and showed a spatial pattern similar to outbreak O3, affecting 80% of the stands in 6 years. We can hypothesize that outbreak O3 occurred in a landscape that had mainly mature balsam fir in the canopy. The severity of this particular outbreak would have led to this pattern during the 20th century in a landscape dominated by balsam fir stands, where a severe outbreak is followed by a less severe one, which is followed by a severe one, and so on. But why didn't we find this pattern for the 19th century?

Comparing 20th-Century to 19th-Century Outbreaks

Outbreaks from the 19th century had a spatial pattern in each stand, as well as in the entire studied area, that was completely different from those of the 20th century. They all show a gradual pattern of occurrence, taking

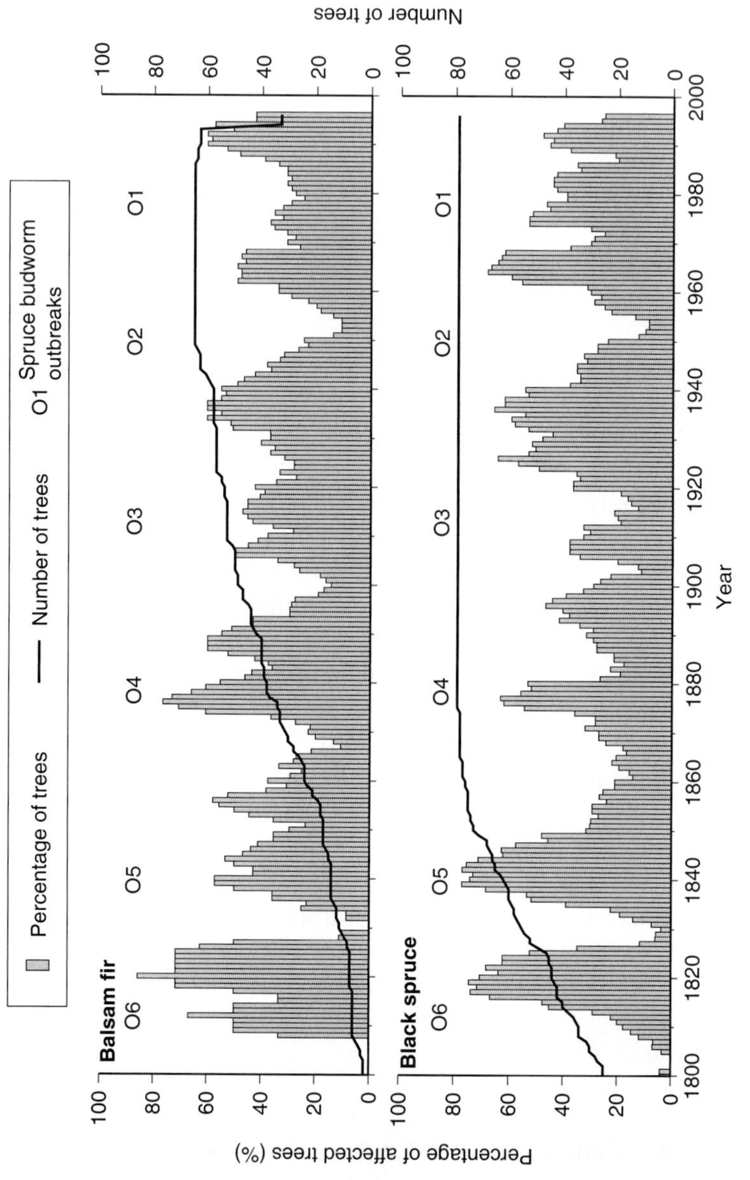

FIG. 5 Percentage of trees affected by spruce budworm defoliation in the northern part of Québec (between the 50th and the 53rd parallels). Balsam fir and black spruce were used because white spruce was absent. Outbreak episodes recognized for the southern part of Québec (cf. Fig. 4) are indicated for comparison.

from 10 to more than 20 years to attain the maximum number of affected trees. Furthermore, this maximum rarely exceeded 40% of affected trees at the same time. Thus, the impact on the forest was very gradual and defoliation was never present in the entire province at the same time. One can argue that as we go back in time, the sampled trees, which are survivors of past outbreaks, were younger and less susceptible. This is true and is a major limitation of the dendroecological techniques. But other results suggest that this limitation is not the only explanation for this particular pattern and that it is not merely an artifact. First, the outbreaks of the 19th century, and particularly outbreak O4, were always difficult to identify with dendroecological techniques (Blais, 1965; Morin and Laprise, 1990; Krause, 1997). In the oldest chronology made with living trees, Blais (1965) interpreted the growth reduction observed during the end of the 19th-century period (outbreak O4) as a drought and not as an outbreak because it coincided with a decrease in growth of pine trees, indicating a dryer period. Morin and Laprise (1990), Krause (1997), and Jardon *et al.* (2003) considered that there was an outbreak at that time but that it had a more local impact and never affected large areas at the same time. Working with trees from buildings that were older when this outbreak occurred, Krause (1997) and Boulanger and Arseneault (2004) were able to confirm this hypothesis. Second, these dendroecological data match markedly well the data from macrofossil analyses, which are not affected by this problem of tree age. Indeed, macrofossil analyses were always able to show the importance of the 20th-century outbreaks. They also showed that before the 20th century, there were no feces peaks that would indicate high population levels back to 6000 years BP (Simard *et al.*, 2005). These results suggest a major change in the amplitude of the fluctuations of budworm populations at the beginning of the 20th century.

The Forest Structure

The dendroecological and macrofossil results suggest that, most of the time, insect populations would show cyclic fluctuations that do not reach outbreak levels appearing in a stand or over large areas explosively. Local outbreaks would occur here and there and synchronicity would not be evident over large areas. As an example of this pattern, outbreak O2 suggests that the forest structure that would permit such outbreaks would present fewer host trees in the canopy of mature stands—and especially fewer balsam fir, the preferred host of the budworm. This interpretation is in accordance with the tritrophic interactions model of insect population dynamics presented by Cooke *et al.* in Chapter 15. In fact, studies comparing forests composed of less vulnerable species, such as white spruce–dominated western Canada or black spruce–dominated eastern Canada, against the more vulnerable fir-dominated eastern Canada forests are particularly informative in helping elucidate the mechanisms involved in the insect population dynamics. In eastern Canada, black spruce–dominated forest is the type of forest that we actually find in the middle of the boreal forest region, where black spruce is the dominant tree in the canopy throughout the landscape and balsam fir generally occurs in the understory and in the canopy of some rare isolated stands. This type of forest is maintained by a fire disturbance regime because black spruce, with its semi-serotinous cones, is better adapted for post-fire regeneration than balsam fir.

In a study of the importance of budworm outbreaks along a latitudinal gradient up to the 53rd parallel, the northern limit of balsam fir, Levasseur (2000) found that while budworm defoliation was present in such a boreal environment, the impact of the defoliation was more and more localized at both the stand and landscape scales the further north one went. Within a stand, not every tree was affected by every outbreak, and at the landscape level, not every stand was affected by every outbreak (Fig. 5). In this area, only balsam fir and black spruce were present in sufficient quantity to be sampled. The peaks in number of trees affected are not only generally asynchronous with those for the southern part of Québec but also more diffuse; the frequency of affected trees slowly rises to a maximum that rarely reaches 65% in more than 10 years. The spatial outbreak patterns and the impact of the 20th-century outbreaks in these stands and in the landscape in a black spruce–dominated environment are more similar to those depicted for the 19th century in the southern part of the province. At these latitudes, however, the climate may have influenced the regeneration capacity of the insect, and thus the forest structure may not be the only cause of outbreak patterns.

Insect populations would have reached outbreak level synchronized over large areas only in exceptional periods during the Holocene. The forest structure associated with this pattern would show balsam fir as a major component of the landscape forest structure. It has been proposed by Blais (1983) that the frequency, extent, and severity of outbreaks increased during the 20th century. He attributed this phenomenon to human influence, such as cutting practices, fire control, and spruce budworm spraying programs. These would have favored balsam fir because (1) it regenerates well after cutting, even in black spruce stands when it is present; (2) it does not regenerate very well after fire, so fire suppression would favor it in the long run; and (3) extensive spraying programs, especially during outbreak O1, would have permitted the conservation of mature balsam fir forests. This hypothesis was supported by very good surveys for the last outbreak of the 20th century (outbreak O1 in Fig. 4), partial to good surveys for the one that occurred in the middle of the century (outbreak O2), and very scattered and local surveys for the one that occurred at the beginning of the century (outbreak O3). Contrary to what was believed, it is now estimated that outbreak O2 was less severe than outbreak O3 and that outbreak O1 was less severe than outbreak O3. In fact, outbreak O3 would have been the most severe outbreak in terms of mortality of trees. It also covered the entire province the most rapidly. It is difficult to determine with precision the area covered by past outbreaks and the volume of wood that was killed, especially for outbreak O3, when observations were not as accurate. However, it is estimated that outbreaks O3, O2, and O1 would have covered an area of 300,000 km^2, 260,000 km^2, and 350,000 km^2, respectively, and that they would have caused the death of 360 to 540 million m^3, 66 to 180 million m^3, and 139 to 238 million m^3 of softwood, respectively (Boulet *et al.*, 1996). This interpretation is supported by dendroecological data on age structure of the boreal forest (Morin, 1994) and by many chronologies (Morin and Laprise, 1990; Krause, 1997; Jardon *et al.*, 2003).

Human impact is certainly involved in changes of the forest structure but probably mostly for outbreak O1, and perhaps O2, when logging operations were active throughout the province, including the boreal forest. At the beginning of the century, when outbreak O3 occurred, logging operations (using hand saws and horses) had very localized impact in the southern parts of Québec, mainly in the deciduous forest and the southern boreal forest within the balsam fir–white birch zone. It is unlikely that such practices would have influenced the entire area so drastically. Concerning the protection of the forest against spruce budworm, spraying with DDT was done in localized areas beginning in the 1950s. Spruce budworm spraying programs started effectively at the beginning of the 1970s in Québec. They were also generally localized and effective mostly during outbreak O1, when biological insecticide (B.t.) was used. The largest sprayed area in the province was 3,969,000 ha in 1973 (Armstrong and Cook, 1993; SOPFIM [Société de protection des forêts contre les insectes et les maladies], personal communication). Fire control started at the end of the 19th and beginning of the 20th century in Québec. At the time, it consisted mainly of a prevention campaign. The first fire observation tower dates back to 1910. Towers were gradually replaced by survey airplanes until the 1970s. Even if the fight against fire started around 1920, it was very localized and centered on human-caused fires. Organized firefighting started around the 1950s with the creation in 1947 of the first forest protection school and the gradual introduction of the airplane for material and fighter transport and aerial spraying. Water bombers were introduced at the beginning of the 1960s (Blanchet, 2003). At present, the evidence that firefighting has had any influence on the fire frequency, fire size distribution, or the forest structure at the landscape level has not been convincing (Johnson *et al.*, 2001; Miyanishi and Johnson, 2001; Ward *et al.*, 2001; Bridge *et al.*, 2005).

There must have been other factors acting at a very large scale on the forest that would have permitted the occurrence of such an explosive outbreak at the beginning of the 20th century. If we accept the hypothesis that a higher proportion of mature firs in the canopy could be the triggering factor, it would suggest that the event that could lead to this increase is a lower fire frequency. Thus, we propose that a change in the fire frequency would have permitted the buildup of such forests in the eastern boreal during the 19th century and the occurrence of major outbreak episodes during the 20th century. Indeed, a change in fire frequency at the end of the Little Ice Age (circa 1850) has been proposed for eastern Canada (Bergeron and Archambault, 1993, Bergeron *et al.*, 2001). The Little Ice Age climate in the southern boreal forest would have been under the influence of an atmospheric circulation bringing a greater frequency of cold and dry polar air masses. With the end of the Little Ice Age and the migration of the polar front towards higher latitudes, the southern fringe of the boreal forest could have seen a greater penetration of warm and humid air masses, a situation conducive to a reduced frequency of drought periods and fire. Initial fire ignitions and the total area burned have decreased, but it is mainly the absence of large fires that characterizes the recent fire regime in the area studied by Bergeron and Archambault (1993). The fact that important fire years were encountered in many parts of the southern boreal forest, and that the growth pattern of white cedar, indicative of drought periods in the area studied by Bergeron and Archambault (1993), fits the growth pattern of white cedar from the Saguenay area (Morin, unpublished data), suggests a general phenomenon for all the Québec southern boreal forest fringe. The relationship between this decrease in fire frequency and a possible change in the proportion of host trees in the canopy of the boreal forest remains speculative. However, it has been shown that this decrease in fire frequency is responsible for an increase in balsam fir and white cedar abundance during the 20th century because the proportion of balsam fir increases as the time since last fire increases in the southern boreal forest (Bergeron, 1998). Mortality due to spruce budworm outbreaks also increases with time since last fire (Bergeron and Leduc, 1998). Thus, considering the dynamics of balsam fir forests and the susceptibility of mature balsam fir forests to spruce budworm, it is proposed that this event is possibly responsible, at least in part, for the differences in forest structure and outbreak patterns observed between the 19th and the 20th centuries.

Predicting the Next Outbreak

If our hypothesis is confirmed, the upcoming outbreak should be less severe and show a spatial pattern much like outbreak O2 (i.e., gradual and less synchronized over the landscape) because mature balsam fir stands were severely damaged by the very severe O1 outbreak. Several models were developed to predict the severity and spatial pattern of the coming outbreak. Gray *et al.* (2000) have developed a model using defoliation data from outbreak O1 to predict where and when the coming outbreak would show different patterns and lengths of defoliation period and to propose some general spatial patterns. The patterns show the lag period of one region versus another once the expected outbreak begins. It is possible that the relative difference of patterns proposed might still apply in the next outbreak, but the model has the limitation that it is based on one outbreak situation occurring in a given landscape. There is actually no sign of significant budworm population increase in the boreal forest. Only some localized areas of severe defoliation are seen in southwestern Québec and in some areas in the east and center of the province (Gouvernement du Québec, 2003; 2004). We think that outbreak O2 would be a good example of what might happen during the next outbreak: a less severe outbreak that succeeds to a severe one and shows a very gradual spatial pattern across the landscape. In fact, this pattern does appear to be the one that is occurring, because defoliation is known until 1989 (Boulet *et al.*, 1996) in the southwest part of the province and other localized impacts have gradually appeared with year to year population fluctuations.

It should be noted that predictions are based on our comprehension of past events under the assumption that known variables such as climate will vary within some recognized range of fluctuations. In the context of recent climate changes that may result in unprecedented changes in temperature and aridity, things could be completely different (Fleming and Volney, 1995; Fleming and Candau, 1998; Volney and Fleming, 2000). However, our growing knowledge and understanding of the population dynamics of the insects (Chapters 15 and 16) and of the relationships between outbreaks and forest dynamics should lead to better predictions of what may occur under changing environmental conditions.

REFERENCES

Armstrong, J. A., and Cook, C. A. (1993). *Traitements Aériens des Forêts Canadiennes de 1945 à 1990*. Forêt Canada, Rapport d'information ST-X-2F, Ottawa.

Bergeron, Y. (1998). Les conséquences des changements climatiques sur la fréquence des feux et la composition forestière au sud-ouest de la forêt boréale québécoise. *Géogr Phys Quatern* 52, 167–173.

Bergeron, Y., and Archambault, S. (1993). Decreasing frequency of forest fires in the southern boreal zone of Québec and its relation to global warming since the end of the "Little Ice Age." *Holocene* 3, 255–259.

Bergeron, Y., and Leduc, A. (1998). Relationships between change in fire frequency and mortality due to spruce budworm outbreak in the southeastern Canadian boreal forest. *J Veg Sci* 9, 493–500.

Bergeron, Y., Leduc, A., Morin, H., and Joyal, C. (1995). Balsam fir mortality following the last spruce budworm outbreak in northwestern Quebec. *Can J Forest Res* 25, 1375–1384.

Bergeron, Y., Gauthier, S., Kafka, V., Lefort, P., and Lesieur, D. (2001). Natural fire frequency for the Canadian boreal forest: consequences for sustainable forestry. *Can J Foest Res* 31, 384–391.

Bhiry, N., and Filion, L. (1996). Mid-Holocene hemlock decline in eastern North America linked with phytophagous insect activity. *Quaternary Res* 45, 312–320.

Blais, J. R. (1954). The recurrence of spruce budworm infestations in the past century in the Lac Seul area of northwestern Ontario. *Ecology* 35, 62–71.

Blais, J. R. (1965). Spruce budworm outbreaks in the past three centuries in the Laurentide Park, Québec. *Forest Sci* 11, 130–138.

Blais, J. R. (1983). Trends in the frequency, extent, and severity of spruce budworm outbreaks in eastern Canada. *Can J Forest Res* 13, 539–547.

Blanchet, P. (2003). *Feux de Forêt. L'histoire d'une Guerre*. Éditions Trait d'union, Montréal.

Boulanger, Y., and Arseneault, D. (2004). Spruce budworm outbreaks in eastern Quebec over the last 450 years. *Can J Forest Res* 34, 1035–1043

Boulet, B., Chabot, M., Dorais, L., Dupont, A., and Gagnon, R. (1996). Entomologie forestière. In: *Manuel de Foresterie*. (Ordre des Ingénieurs Forestiers, Ed.), pp. 1008–1043. Les Presses de l'Université Laval, Québec.

Bridge, S. R. J., Miyanishi, K., and Johnson, E. A. (2005). A critical evaluation of fire suppression effects in the boreal forest of Ontario. *Forest Sci* 51, 41–50.

Candau, J.-N., Fleming, R. A., and Hopkin, A. (1998). Spatiotemporal patterns of large-scale defoliation caused by the spruce budworm in Ontario since 1941. *Can J Forest Res* 28, 1733–1741.

Cappuccino, N., Lavertu, D., Bergeron, Y., and Régnière, J. (1998). Spruce budworm impact, abundance and parasitism rate in a patchy landscape. *Oecologia* 114, 236–242.

Cooke, B. J., and Roland, J. (2000). Spatial analysis of large-scale patterns of forest tent caterpillar outbreaks. *Ecoscience* 7, 410–422.

Davis, R. B., Anderson, R. S., and Hoskins, B. R. (1980). A new parameter for paleoecological reconstruction: head capsules of forest-tree defoliator Microlepidopterans in lake sediment. In: *Abstracts and Program of the 6th Biennial Meeting of the American Quaternary Association, 18-20 August 1980, Institute of Quaternary Studies, University of Maine*, p. 62. Orono, Maine.

Delcourt, P. A., and Delcourt, H. R. (1987). *Long-term Forest Dynamics of the Temperate Zone. A Case Study of Late-Quaternary Forests in Eastern North America*. Ecological studies # 63, Springer-Verlag, New York.

Fleming, R. A., and Candau, J. N. (1998). Influences of climatic change on some ecological processes of an insect outbreak system in Canada's boreal forests and the implications for biodiversity. *Environ Monit Assess* 49, 235–249.

Fleming, R. A., and Volney, W. J. A. (1995). Effects of climate change on insect defoliator population processes in Canada's boreal forest: some plausible scenarios. *Water Air Soil Poll* 82, 445–454.

Gouvernement du Québec. (2003). *Insectes, Maladies et Feux dans les Forêts Québécoises, en 2002.* Direction de la Conservation, MRNFP, Québec.

Gouvernement du Québec. (2004). *Défoliation de la Pousse Annuelle en 2004, Tordeuse des Bourgeons de l'Épinette.* Ministère des Ressources Naturelles de la Faune et des Parcs, Direction de la Conservation des Forêts, Gouvernement du Québec, Québec.

Gray, D. R., Régnière, J., and Boulet, B. (2000). Analysis and use of historical patterns of spruce budworm defoliation to forecast outbreak patterns in Quebec. *Forest Ecol Manag* 127, 217–231.

Hardy, Y., Lafond, J. A., and Hamel, L. (1983). The epidemiology of the current spruce budworm outbreak in Quebec. *Forest Sci* 29, 715–725.

Holmes, R. L. (1999). *ITRDB, Dendrochonology Program Library Users Manual.* Laboratory of Tree-Ring Research, University of Arizona, Tucson.

Holmes, R. L., and Swetnam, T. W. (1996). *Dendroecology Program Library, Program OUTBREAK Users Manual.* Laboratory of Tree-Ring Research, University of Arizona, Tucson.

Jardon, Y. (2002). *Analyses Temporelles et Spatiales des Épidémies de la Tordeuse des Bourgeons de l'Épinette au Québec.* Ph.D. thesis, Université du Québec à Chicoutimi, Chicoutimi, PQ.

Jardon, Y., and Doyon, F. (2003). *Balsam Fir Stand Dynamics after Insect Outbreak Disturbances in Western Newfoundland Ecoregion (Corner Brook Subregion).* Report for the Model Forest Network, Corner Brook, NF.

Jardon, Y., Filion, L., and Cloutier, C. (1994). Tree-ring evidence for endemicity of the larch sawfly in North America. *Can J Forest Res* 24, 742–747.

Jardon, Y., Morin, H., and Dutilleul, P. (2003). Périodicité des épidémies de la tordeuse des bourgeons de l'épinette au cours des deux derniers siècle. *Can J Forest Res* 33, 1947–1961.

Johnson, E. A., Miyanishi, K., and Bridge, S. R. J. (2001). Wildfire regime in the boreal forest and the idea of suppression and fuel buildup. *Conserv Biol* 15, 1554–1557.

Johnson, E. A., Morin, H., Miyanishi, K., Gagnon, R., and Greene, D. F. (2003). A process approach to understanding disturbance and forest dynamics for sustainable forestry. In: *Towards Sustainable Management of the Boreal Forest.* (Burton, P. J., Messier, C., Smith, D. W., and Adamowicz, W. L., Eds.), pp. 261–306. NRC Research Press, Ottawa.

Krause, C. (1997). The use of dendrochronological material from buildings to get information about past spruce budworm outbreaks. *Can J Forest Res* 27, 69–75.

Levasseur, V. (2000). *Analyse Dendroécologique de l'Impact de la Tordeuse des Bourgeons de l'Épinette* (Choristoneura fumiferana) *Suivant un Gradient Latitudinal en Zone Boréale au Québec.* M.Sc. thesis, Université du Québec à Chicoutimi, Chicoutimi, PQ.

MacKinnon, W. E., and MacLean, D. A. (2003). The influence of forest and stand conditions on spruce budworm defoliation in New Brunswick, Canada. *Forest Sci* 49, 657–667.

MacKinnon, W. E., and MacLean, D. A. (2004). Effects of surrounding forest and site conditions on growth reduction of balsam fir and spruce caused by spruce budworm defoliation. *Can J Forest Res* 34, 2351–2362.

MacLean, D. A. (1980). Vulnerability of fir-spruce stands during uncontrolled spruce budworm outbreaks: a review and discussion. *Forest Chron* 56, 213–221.

MacLean, D. A. (1984). Effects of spruce budworm outbreaks on the productivity and stability of balsam fir forests. *Forest Chron* 60, 273–299.

MacLean, D. A. (1988). Effects of spruce budworm outbreaks on vegetation, structure, and succession of balsam fir forest on Cape Breton Island, Canada. In: *Plant Form and Vegetation Structure: Adaptation, Plasticity and Relation to Herbivory.* (Werger, M. J. A., van der Aart, P. J. M., During, H. J., and Verhoeven, J. T. A., Eds.), pp. 253–261. SPB Academic Publishing bv, The Hague, the Netherlands.

MacLean, D. A., and MacKinnon, W. E. (1997). Effects of stand and site characteristics on susceptibility and vulnerability of balsam fir and spruce budworm in New Brunswick. *Can J Forest Res* 27, 1859–1871.

MacLean, D. A., and Ostaff, D. P. (1989). Pattern of balsam fir mortality caused by an uncontrolled budworm outbreak. *Can J Forest Res* 19, 1087–1095.

Miyanishi, K., and Johnson, E. A. (2001). Comment—a reexamination of the effects of fire suppression in the boreal forest. *Can J Forest Res* 31, 1462–1466.

Morin, H. (1994). Dynamics of balsam fir forests in relation to spruce budworm outbreaks in the boreal zone of Quebec. *Can J Forest Res* 24, 730–741.

Morin, H., and Laprise, D. (1990). Histoire récente des épidémies de la tordeuse des bourgeons de l'épinette au nord du lac St-Jean (Québec): une analyse dendrochronologique. *Can J Forest Res* 20, 1–8.

Morin, H., and Laprise, D. (1997). Seedling bank dynamics in boreal balsam fir forests. *Can J Forest Res* 27, 1442–1451.

Morin, H., Laprise, D., and Bergeron, Y. (1993). Chronology of spruce budworm outbreaks near Lake Duparquet, Abitibi region, Quebec. *Can J Forest Res* 23, 1497–1506.

Mott, D. G. (1963). The analysis of the survival of small larvae in the unsprayed area. *Mem Entomol Soc Can* 31, 42–52.

Parent, S., Simard, M. J., Morin, H., and Messier, C. (2003). Establishment and dynamics of the balsam fir seedling bank in old forests of northeastern Quebec. *Can J Forest Res* 33, 597–603.

Peltonen, M., Liebhold, A. M., Bjornstad, O. N., and Williams, D. W. (2002). Spatial synchrony in forest insect outbreaks: roles of regional stochasticity and dispersal. *Ecology* 83, 3120–3129.

Potelle, B. (1995). *Potentiel de l'Analyse des Macrorestes pour Détecter les Épidémies de la Tordeuse des Bourgeons de l'Épinette dans des Sols de Sapinières Boréales*. M.Sc. thesis, Université du Québec à Chicoutimi, Chicoutimi, PQ.

Richard, P. J. H. (1993). Origine et dynamique postglaciaire de la forêt mixte au Québec. *Rev Palaeobot Palyno* 79, 31–68.

Richard, P. J. H. (1994). Postglacial paleophytogeography of the eastern St. Lawrence River watershed and the climatic signal of the pollen record. *Palaeogeogr Palaeocl* 109, 137–161.

Royama, T. (1984). Population dynamics of the spruce budworm *Choristoneura fumiferana. Ecol Monogr* 54, 429–462.

Royama, T., MacKinnon, W. E., Kettela, E. G., Carter, N. E., and Harting, L. (2005). Analysis of spruce budworm outbreak cycles in New Brunswick, Canada, since 1952. *Ecology* 86, 1212–1224.

Simard, S. (2003). *La Tordeuse des Bourgeons de l'Épinette à Travers les Arbres Subfossiles*. M.Sc. thesis, Université du Québec à Chicoutimi, Chicoutimi, PQ.

Simard, I., Morin, H., and Lavoie, C. (2006). A millenial-scale reconstruction of spruce budworm abundance in Saguenay, Québec, Canada. *The Holocene* 16, 31–37.

Simard, I., Morin, H., and Potelle, B. (2002). A new paleoecological approach to reconstruct long-term history of spruce budworm outbreaks. *Can J Forest Res* 32, 428–438.

Su, Q., MacLean, D. A., and Needham, T. D. (1996). The influence of hardwood content on balsam fir defoliation by spruce budworm. *Can J Forest Res* 26, 1620–1628.

Swetnam, T. W., and Lynch, A. M. (1993). Multicentury, regional-scale patterns of western spruce budworm outbreaks. *Ecol Monogr* 63, 399–424.

Swetnam, T. W., Thompson, M. A., and Sutherland, E. K. (1985). *Using Dendrochronology to Measure Radial Growth of Defoliated Trees*. U.S. Department of Agriculture Forest Service Agriculture Handbook, Washington, DC.

Volney, W. J. A. (1988). Analysis of historic jack pine budworm outbreaks in the Prairie provinces of Canada. *Can J Forest Res* 18, 1152–1158.

Volney, W. J. A., and Fleming, R. A. (2000). Climate change and impacts of boreal forest insects. *Agr Ecosyst Environ* 82, 283–294.

Ward, P. C., Tithecott A. G., and Wotton B. M. (2001). Reply—a re-examination of the effects of fire suppression in the boreal forest. *Can J Forest Res* 31, 1467–1480.

Weber, U. M. (1997). Dendroecological reconstruction and interpretation of larch budmoth (*Zeiraphera diniana*) outbreaks in two central Alpine valleys of Switzerland from 1470-1990. *Trees Struct Funct* 11, 277–290.

Williams, D. W., and Liebhold, A. M. (2000). Spatial scale and the detection of density dependence in spruce budworm outbreaks in eastern North America. *Oecologia* 124, 544–552.

Zhang, Q. B., and Alfaro, R. I. (2002). Periodicity of two-year cycle spruce budworm outbreaks in central British Columbia: a dendro-ecological analysis. *Forest Sci* 48, 722–731.

Zhang, Q. B., and Alfaro, R. I. (2003). Spatial synchrony of the two-year cycle budworm outbreaks in central British Columbia, Canada. *Oikos* 102, 146–154.

18

Impact of Beaver (*Castor canadensis* Kuhl) Foraging on Species Composition of Boreal Forests

Noble T. Donkor

Canadian University College

INTRODUCTION

Herbivores can greatly influence plant community composition through selective foraging (Naiman *et al.*, 1988; Pastor *et al.*, 1988; Donkor and Fryxell, 1999). Local vegetation diversity is produced and maintained not only by what and how much herbivores eat but also by what they do not eat. For example, browsed and unbrowsed species differentially alter nutrient flows through soils, and hence the ultimate response of the ecosystem may be related to changes in the nutrient cycles imposed by herbivory (Bryant and Chapin, 1986). The effects of herbivores on community composition are particularly acute in boreal forests, where limited light, temperature, and nutrient resources restrict the capacity of woody plants to replace tissues eaten by browsing animals (Bryant and Chapin, 1986). It is well established that herbivory by large mammals can significantly influence boreal forest ecosystem structure and dynamics (Beals *et al.*, 1960; Snyder and Janke, 1976; Keith *et al.*, 1984; Baker, 2003; Baker *et al.*, 2004). However, of all the large herbivores, the beaver (*Castor canadensis* Kuhl) has a much greater potential to alter these ecosystems through herbivory because most of the biomass they harvest is composed of mature trees.

Although numerous studies have shown that beavers prefer to cut and eat the leaves and bark of deciduous trees and shrubs while avoiding conifers (e.g., Brenner, 1962; Northcott, 1971; Gill, 1972; Jenkins, 1980; Belovsky, 1984; McGinley and Whitham, 1985; Fryxell and Doucet, 1991), these studies have largely ignored how herbivory by beavers alters forest composition and structure. Such studies on the impact of beavers are limited (e.g., Johnston and Naiman, 1990; Donkor and Fryxell, 1999); moreover, most of their conclusions are based on the Clementsian notion of succession as a deterministic developmental sequence of plant species replacements following disturbance (Clements, 1916). As argued in Chapter 1, the empirical evidence (e.g., Chapin *et al.*, 1994; Fastie, 1995; Gutsell and Johnson, 2002) does not support this idea of boreal forest succession. Particularly lacking in the previous studies of beaver impacts is the use of an integrative approach to explore the role of beaver herbivory and causative edaphic factors in shaping the composition and structure of the plant community.

Biological systems are structured by the individual and combined effects of multiple biotic and abiotic ecological factors that can result in nonlinear dynamics and multiple community domains of attraction. In this chapter, I review herbivory in boreal forests, discuss temporal dynamics of beaver populations in these forests, summarize the current traditional understanding of beaver foraging impacts in boreal forests, and offer an alternative approach to gain understanding of how beaver herbivory causes the ecological effects seen in the composition and structure of boreal forests. I discuss both direct and indirect vegetation responses to beaver activity in boreal forests. Direct effects of beavers on vegetation would involve their actual browsing and cutting of preferred plant species, both mature trees and saplings, that would then affect regeneration of nonpreferred species. Indirect effects would involve beaver impacts on hydrology, thus influencing moisture–nutrient gradients along which plant species are distributed, depending on their tolerances and requirements.

HERBIVORY IN BOREAL FORESTS

There are numerous examples of the importance of herbivory in the boreal region by rodents, ungulates, insects, and birds. Some major mammal herbivores, such as snowshoe hare (*Lepus americanus* Erxleben) (Keith *et al.*, 1984), moose (*Alces alces* L.) (Snyder and Janke, 1976), and elk (*Cervus elaphus* L. *canadensis* Erxleben) (Baker *et al.*, 2004), have been shown to influence plant community composition through selective foraging over several growing seasons. In at least some areas of the boreal region, snowshoe hares drive many ecological changes; if they were eliminated, the boreal vertebrate community would be greatly affected. When moose populations are high, they can be important agents of ecosystem disturbance, not only through their selective feeding habits but also through trampling, defecation, and urination. Where elk compete with moose for beaked willow (*Salix bebbiana* Sarg.), their combined herbivory can cause long-term alteration of stand composition (Pastor *et al.*, 1988), although it has no immediate effect on the forest overstory.

Moose and beaver are two important and interacting herbivores of the North American boreal forest. Although they browse on the same plant species, their foraging strategies and consequently their effect on ecosystems differ substantially (Pastor and Naiman, 1992). Moose browse young *Populus tremuloides* Michx. (trembling aspen) and other hardwoods and avoid *Picea* (spruce) (Peek *et al.*, 1976; Belovsky, 1981). This selective foraging seems to be related to the high nitrogen and low lignin content of *P. tremuloides* leaves and twigs (Bryant and Kuropat, 1980). Moose browse *P. tremuloides* year-round, stripping leaves in summer and eating current twigs in winter (Peterson, 1955). Consequently, their continuous heavy browsing often kills the young trees (Krefting, 1974). In contrast, the ability of beavers to harvest mature trees can affect the overstory (Johnston and Naiman, 1990). Stumps left from beaver harvesting are easily discernible and distinguished from that of other mammals by their conical shape, height above the ground, and size and arrangement of tooth marks (Johnston and Naiman, 1990).

There are several reasons why beavers have a much greater potential than other large animals to alter ecosystems through herbivory (Johnston and Naiman, 1990). First, as the only animals in North America other than humans that can fell mature trees, their ability to decrease forest biomass is much greater than that of other herbivores. For example, in a study in Canada, Donkor and Fryxell (1999) reported that the diameter of large trees felled by beavers, mostly *P. tremuloides* and *Betula papyrifera* Marsh. (paper birch), averaged 15.1 cm and the maximum stem diameter cut was 45.3 cm. In Minnesota, Johnston and Naiman (1990) found that the maximum diameter of *P. tremuloides* trees cut by beavers was 43.5 cm and the average stem diameters cut at two beaver ponds were 13.9 and 10.2 cm. Also, beavers removed more than 40% of the stem density, basal area, and aboveground biomass. Second, beaver foraging is restricted to a relatively narrow band of forest surrounding their ponds (up to 60 m from the edge of ponds), concentrating the impact of their foraging in a small area (Johnston and Naiman, 1987; Donkor and Fryxell, 1999; 2000). Anderson (1978) pointed out that central place foragers are expected to concentrate feeding activity close to their lodge or denning sites because both time and energy needed to provide the central place with food items increase with distance. Field studies of beavers (e.g., Hall, 1960; Jenkins, 1980; McGinley and Whitham, 1985; Basey *et al.*, 1988; Donkor and Fryxell, 1999) have shown that the spatial pattern of foraging by beavers usually exhibits a sharp decline over 60 m from the edge of their ponds. Third, beavers remove a much higher proportion of biomass within their foraging range than any other herbivore. Each individual of one colony of six beavers felled an average of about 1.3 $Mg\,y^{-1}$ of wood biomass within the 0.96-ha browse zone (Johnston and Naiman, 1990). Howard (1982) reported a similar value of 1 $Mg\,y^{-1}$ for beavers in Massachusetts. In comparison, individual moose browse only 0.0003 to 0.091 $Mg\,ha^{-1}\,y^{-1}$ of woody biomass within their foraging area (McInnes, 1989), and a whole herd of Serengeti ungulates consumes 1 $Mg\,ha^{-1}y^{-1}$ (McNaughton, 1985). Of course, the amount of wood harvested by beavers far exceeds the amount actually ingested; beavers consume approximately one third of the edible biomass of *P. tremuloides* that they fell (Aldous, 1938). These factors confirm the impact of beavers on ecosystem structure and composition far beyond their immediate requirements for food and space (Naiman and Melillo, 1984).

TEMPORAL CHANGES IN BEAVER POPULATIONS

The beaver was effectively extirpated from most regions of the North American continent because of intense hunting and trapping pressure by the mid- to late 1800s (Jenkins and Busher, 1979; Larson and Gunson, 1983). During the 1900s, management programs designed to protect the beaver from trapping and hunting, coupled with a decline in the market for their fur, allowed beavers to recolonize many parts of their former range. Active reintroduction programs were also instituted in many regions (Busher and Lyons, 1999). With a decline in trapping and hunting intensity, and low to nonexistent predation pressure, beaver populations in North America have rebounded from widespread extirpation to high abundance (Schulte and Muller-Schwarze, 1999).

A typical example of beaver population dynamics is that of the unexploited population on Prescott Peninsula, Quabbin Reservation, in western Massachusetts, over a 44-year period (1952–1996). Trapping and hunting are prohibited in the area. Beavers on the Prescott Peninsula are found on small streams, larger streams, and ponds and along the shore of a reservoir. The history of this population can be described by five phases (Fig. 1): the initial phase (1952–1968) was a time of very slow population growth; the second phase (1968–1975) was one of rapid increase; the third phase (1975–1983) was a time of fluctuating population levels at a high density; the fourth phase (1983–1988) represented decline in local colonies, possibly because of a decline in local food sources; and the fifth phase (1988–1996) indicated a period of relative stability at low population levels.

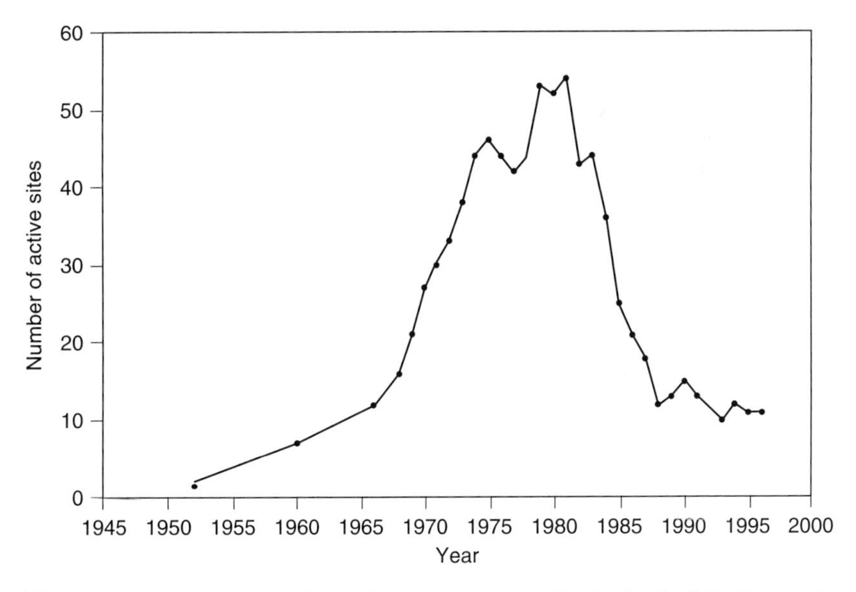

FIG. 1 *Castor canadensis* population dynamics on Prescott Peninsula, Quabbin Reservation, western Massachusetts, 1952–1996. (From Busher and Lyons [1999], redrawn with kind permission of Springer Science and Business Media.)

The basic life history characteristics of beavers have been summarized in detail elsewhere (e.g., Bradt, 1938; Jenkins and Busher, 1979; Allred, 1986; Hilfiker, 1990). Because of their variability in reproductive output, beavers have the potential to rapidly take advantage of high-quality habitats, increasing their population numbers in a relatively short time (Payne, 1984; Smith and Petersen, 1988). Although general principles of furbearer population management, including those for beavers, are well described (e.g., Novak, 1987), few attempts have been made to model beaver population dynamics. Yearling beavers are often abundant in a population and can significantly contribute to the overall reproductive output of a population in a given year (Payne, 1984; Smith and Petersen, 1988). In heavily trapped populations, beaver colonies are composed of more single individuals and pairs, while in untrapped populations there are more typically family units (an adult pair and progeny from two successive years, averaging three to eight per site). As the population density increases, the composition of beaver colonies is altered. Two-year-olds often delay dispersal, resulting in more than two adults per colony. In addition, yearlings may disperse and reproduce (Jenkins and Busher, 1979). Some beavers may live for a time as transients or floaters (Allred, 1981), and this number is likely to increase with population size and with a concomitant decline in local food resources. Fryxell (2001) reported that the mean colony size and probability of recurrence from year to year were associated with local food availability (Fig. 2). Fryxell estimated that the abundance of beavers at the study sites (previously exploited) in Algonquin Provincial Park, Ontario,

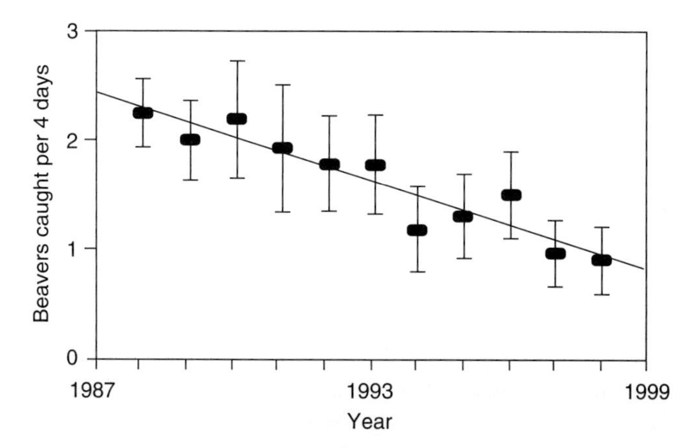

FIG. 2 Annual changes in *Castor canadensis* density averaged over 14 sites in Algonquin Provincial Park, Ontario, during 1988–1998. (From Fryxell [2001] 7, redrawn with permission from Blacewell publishing.)

during 1988–1998 dropped over 50%, ranging from a maximum of 2.3 to a minimum of 0.9 animals captured per trapping season.

TRADITIONAL UNDERSTANDING OF BEAVER FORAGING IMPACT ON PLANT COMMUNITY STRUCTURE

Beavers are the classical example of ecosystem engineers whose disturbance checks and reverses plant compositional changes and determines vegetational development (Vogl, 1981; Pollock *et al.*, 1995). Beavers are generalist herbivores that feed on a variety of woody and herbaceous vegetation, including grasses, forbs, ferns, shrubs, and trees (Northcott, 1971; Jenkins, 1975; Svendsen, 1980; Novak, 1987). However, they are highly selective in their choice of woody plants (Aldous, 1938; Hall, 1960; Brenner, 1962; Jenkins, 1975; Fryxell and Doucet, 1991). They harvest woody plant species in a manner consistent with an optimal "central place" foraging strategy (Orians and Pearson, 1979; Schoener, 1979; Jenkins, 1980). With this strategy, beavers concentrate more of their foraging effort and harvest on a wider range of plant sizes near the edge of the pond (up to 25 m from the water). They generally feed preferentially on a small number of deciduous species. The foraging activity by beavers in the lowland boreal forests at Algonquin Provincial Park in Ontario, Canada, was concentrated on six species, although more than 20 species of trees and shrubs were cut at least once (Donkor and Fryxell, 1999; 2000). Seventy-eight percent of the 1841 stems cut by beavers around 15 beaver ponds consisted of *Alnus rugosa* (L). Moench, *Corylus cornuta* Marsh., *Acer rubrum* L., *Populus tremuloides, Betula papyrifera*, and *Salix bebbiana* (Table 1). The spatial pattern of foraging by beavers at Algonquin Park exhibited a sharp decline over 60 m, similar to patterns recorded in field studies elsewhere (e.g., McGinley and Whitham, 1985; Basey *et al.*, 1988). The effect of beavers on forest structure is possibly greatest in boreal forests, where most of the biomass cut by beavers is mature trees (Johnston and Naiman, 1990). However, when their chief food supply of *P. tremuloides* has been depleted, the beavers switch over to shrubs such as *S. bebbiana* (Smith, 1997).

Traditionally, the effect of beaver foraging on the structure and composition of boreal forests has been based on Clements' (1916) theory of plant succession, which involved a series of dominant plant species taking over from one another. For forest development following disturbance, this would mean that different plant species recruited and became dominant in the canopy at different times after stand initiation (Gutsell and Johnson, 2002). On the basis of this notion, ecologists working in many different

TABLE 1 Species Composition of Standing and Browsed Woody Plants More than 1 cm in Diameter Recorded in Transects at 15 Sites in Algonquin Provincial Park, Ontario

	Standing Vegetation*		Cut Vegetation†	
Species	% stems	% area	% stems	% area
Alnus rugosa	25.6	5.0	49.2	36.6
Abies balsamea	22.3	10.1	0.5	1.9
Picea glauca	11.0	15.2	1.5	1.9
Picea mariana	9.6	18.1	1.7	0.5
Acer rubrum	5.3	2.1	8.0	5.6
Corylus cornuta	4.8	0.1	13.8	1.7
Betula papyrifera	3.1	12.4	2.4	27.5
Pinus resinosa	2.2	12.2	0.2	0.2
Salix bebbiana	1.9	0.2	2.4	0.4
Populus tremuloides	1.6	7.1	2.4	10.7
Acer saccharum	1.5	1.1	2.2	1.1
Pinus strobus	1.5	11.1	0.0	0.0
Acer spicatum	1.4	0.2	1.4	1.1
Amelanchier arborea	0.7	0.1	0.5	0.2
Populus balsamifera	0.4	0.1	0.1	<0.1
Tsuga canadensis	0.3	0.1	0.1	<0.1
Prunus pensylvanica	0.3	0.1	0.0	0.0
Others	6.6	4.7	13.7	10.4

*Total standing stems = 7438, total basal area of standing vegetation = 48.9311 m^2. Each transect measured 2 m × 60 m, and there were 130 transects in all. Hundred percent mensuration was done on each transect.

†% stems = percentage of stems cut out of the 1841 total number of stems cut by *Castor canadensis* at the 15 sites; % basal area = percentage of basal area cut out of the total basal area harvested by *C. canadensis* at the 15 sites (3.1974 m^2).

types of forests have classified species of the older (usually inferred from larger size) trees in a stand as early-successional and species of the younger (i.e., smaller) trees as mid- or late-successional (i.e., such classification was based on the inferred timing of recruitment or dominance in the canopy) (e.g., Kneeshaw and Burton, 1997; Linder *et al.*, 1997; Frelich and Reich, 1999).

In this regard, it has been traditionally thought that beavers cut large-diameter trees of early successional species, particularly *Populus* and *Salix*, near their ponds. The canopy openings that such cutting created would allow sufficient light penetration to permit regeneration of shade-intolerant species (Pastor and Naiman, 1992). It was thus suggested that beavers could convert mid-successional stands to early successional stands (Aldous, 1938; Gill, 1972), although it has also been argued that their cutting could release shade-tolerant understory *Abies* and *Pinus* from competition and thus hasten succession (Naiman *et al.*, 1988; Johnston and Naiman, 1990).

Another characteristic that was used to distinguish successional status of plants, and hence their response to herbivory, is nutrient cycling. Boreal forests are composed of a few tree species that differ in nutrient cycling properties and landscape distribution (Bridge and Johnson, 2000). Recent theories of nutrient cycling in boreal forests emphasize the different nutrient requirements and decay rates of litter from various tree species (Van Cleve and Viereck, 1981; Flanagan and Van Cleve, 1983; Gordon, 1983; Van Cleve *et al.*, 1983; Chapin *et al.*, 1986; Pastor *et al.*, 1987). The nitrogen cycle has been emphasized in these ecosystems because it is considered the most limiting nutrient for tree growth (Van Cleve and Zasada, 1976; Van Cleve and Oliver, 1982).

On the basis of the Clementsian view of succession, it has been believed that early-successional deciduous species, such as *Populus tremuloides*, *P. balsamifera* L. (balsam poplar), and *Betula papyrifera*, have easily decomposable, nitrogen-rich litter that enhances soil nitrogen availability (Flanagan and Van Cleve, 1983), but they also require large amounts of nitrogen for production because they grow rapidly and are deciduous (Van Cleve and Oliver, 1982; Pastor and Bockheim, 1984). These species are then believed to be succeeded by late-successional conifer species, such as *Picea glauca* (Moench) Voss (white spruce), *P. mariana* (Mill.) B.S.P. (black spruce), and *Abies balsamea* (L.) Mill. (balsam fir), whose slowly decomposing litters depress soil nitrogen availability (Flanagan and Van Cleve, 1983; Pastor *et al.*, 1987). These conifers are able to survive the depression of soil nitrogen availability because of low nitrogen requirements, low inherent growth rates, and retention of nitrogen for several years in evergreen photosynthetically active needles (Chapin *et al.*, 1986), although eventually severe nitrogen stress and even dieback may occur (Pastor *et al.*, 1987).

To varying degrees, most boreal species produce carbon-based compounds that have multiple consequences, one of which may be deterrence of herbivores (Bryant and Kuropat, 1980; Palo *et al.*, 1985; Tahvanainen *et al.*, 1985; Basey *et al.*, 1988). Fast-growing species seem to produce some of these compounds, such as phenolics, only during juvenility; once beyond the reach of most herbivores, these plants decrease their production of secondary compounds (Bryant *et al.*, 1983; Chapin *et al.*, 1985; Basey *et al.*, 1988). In contrast, slower-growing species, such as *Picea glauca* and *Picea mariana*, maintain production of phenolic polymers that form lignin and other secondary compounds throughout their lives (Bryant *et al.*, 1983).

These secondary compounds are carbon-based rather than nitrogen-based because nitrogen is a scarce resource for boreal plants (Corey *et al.*, 1985). Low nitrogen and high lignin contents both reduce digestibility (Corey *et al.*, 1985) and make litter difficult to decompose (Flanagan and Van Cleve, 1983) because the processes of both ruminant digestion and decomposition are microbially mediated. Therefore, the carbon and nitrogen cycles of boreal forests are tightly linked by positive and negative feedback loops between decomposers, plants, and herbivores: nitrogen availability controls net carbon fixation, but the types of carbon compounds produced control nitrogen availability and browsing intensity (Pastor and Naiman, 1992).

Despite this traditional understanding given in the limited studies of beaver impacts on boreal forests, there is now a sound body of evidence in the refereed literature that the Clementsian deterministic sequential replacement of dominants and communities does not take place in the boreal forest and certainly that "late successional" species do not require facilitation through habitat modification in order to establish (e.g., Chapin *et al.*, 1994; Fastie, 1995). These studies showed that when seeds and seedlings of "late successional" species are planted into the initial bare substrate ("pioneer" sites), they are as capable of germinating and establishing as "early successional" species or even have similar seedling mortality rates. It is obvious, therefore, that constraints on natural establishment have more to do with distance to seed source (seed dispersal) or disturbances (providing suitable seedbed and light conditions) than with any habitat alteration or facilitation by "early successional" species. Furthermore, Gutsell and Johnson (2002) conclude that there is no valid justification for the categorization of boreal forest species as early or late successional. They argue that the interpretation of sequential species invasion following disturbance (e.g., fire) is largely based on inaccurate aging of trees. Therefore, in the rest of this chapter, I will provide community ecologists with an alternative approach to understanding beaver foraging impacts on vegetation dynamics without invoking the traditional Clementsian notion of a developmental sequence of plant communities.

UNDERSTANDING BEAVER FORAGING IMPACTS ON COMPOSITION AND DYNAMICS OF THE BOREAL FOREST

The foraging strategies by other mammals and hence their effects on the ecosystem differ from that of beavers. Studies of the ways these mammals, including beavers, have an effect on forest structure and composition often ignore the potential effects of gaps created by environmental factors such as wind, lightning, diseases, and insects. It is important to know whether the process of beaver foraging alone is responsible for the vegetation composition observed or whether other biotic or abiotic variables are operating in conjunction with herbivory. Recently, many authors have emphasized the need for an integrative approach to exploring the role of both biotic and abiotic factors in shaping plant community composition (Chapin *et al.*, 1987; Dunson and Travis, 1991). Because of a paucity of information, I relied on literature on the effect of beaver herbivory on community composition of boreal forests (e.g., Johnston and Naiman, 1990; Donkor and Fryxell, 1999; Wright *et al.*, 2002) and integrative effects of beaver herbivory and edaphic conditions on forest composition (e.g., Donkor and Fryxell, 2000) to offer an alternative explanation as to how herbivory by beavers causes the ecological effects seen in the composition and dynamics of boreal forests.

Direct Vegetation Responses to Beaver Activity

Donkor and Fryxell (1999) reported that the community composition of woody plants varied in relation to a pronounced beaver foraging gradient. Species richness and diversity increased from the pond margin to peak about 25 m from ponds and declined to an intermediate condition at greater distances. If reduced species richness and diversity near ponds are caused by selective removal of plants by beavers, then abundance of preferred species farther from ponds should be increased, with no such pattern for nonpreferred species. Johnston and Naiman (1990) had reported that selected foraging by beavers decreased the relative importance (as measured by basal area) of preferred species (i.e., *Populus tremuloides*) and increased the importance of avoided species (i.e., *Alnus rugosa*, *Picea glauca*). Donkor and Fryxell (1999) found that total stem abundance and basal area of four of the seven preferred species (*Populus tremuloides*, *Acer rubrum*, *Acer saccharum*, and *Corylus cornuta*) increased significantly with distance from ponds as expected (Table 2). However, *Alnus rugosa* and *Salix bebbiana*, although also preferred by beavers, declined significantly with distance from ponds while *Acer spicatum* showed no significant trend. With respect to nonpreferred conifers, four of the five species (*Pinus strobus* L., *Pinus resinosa* Ait, *Picea glauca*, and *Picea mariana*) showed no significant pattern of total stem abundance and basal area cover with distance from ponds as expected while one species (*Abies balsamea*) exhibited a significant decline in stem density and a highly significant increase in basal area with distance from ponds (Table 3). Thus, it appears that in the case of

TABLE 2 Distance-Dependent Stem Abundance and Basal Area ($m^2 \times 10^{-4}$) (in Parentheses) of Seven Species of Woody Plants Preferred by Beavers (*Castor canadensis*) at the 15 Sites in Algonquin Provincial Park, Ontario*

	Distance from Pond (m)						
Species	5	15	25	35	45	55	Pa†
Alnus rugosa	682 (5814)	491 (6914)	347 (8834)	132 (1179)	152 (1108)	98 (413)	0.009 (0.097)
Populus tremuloides	4 (609)	7 (274)	32 (4475)	20 (8174)	25 (7936)	29 (13131)	0.012 (0.038)
Salix bebbiana	35 (240)	45 (272)	30 (235)	24 (93)	18 (79)	9 (45)	0.012 (0.010)
Acer rubrum	25 (82)	61 (479)	66 (2314)	65 (3404)	99 (1110)	79 (6325)	0.006 (0.009)
Acer saccharum	1 (9)	8 (17)	12 (371)	20 (661)	33 (1818)	31 (2673)	0.001 (0.004)
Corylus cornuta	3 (5)	74 (140)	51 (96)	62 (119)	80 (151)	84 (189)	0.022 (0.019)
Acer spicatum	5 (20)	18 (62)	23 (118)	27 (558)	30 (160)	12 (40)	0.402 (0.674)

*Totals were summed over all sites.

† Probability that the observed regression slope could have differed from zero by chance.

TABLE 3 Stem Density and Basal Area ($m^2 \times 10^{-4}$) (in Parentheses) of Five Coniferous Species versus Distance at 15 Sites in Algonquin Provincial Park, Ontario*

	Distance from Pond (m)						
Species	5	15	25	35	45	55	P[a†]
Pinus strobus	14 (5826)	16 (256)	22 (2591)	18 (4062)	21 (9286)	18 (32,494)	0.088 (0.412)
Pinus resinosa	7 (1566)	23 (5787)	33 (5083)	30 (34,912)	40 (7254)	33 (5028)	0.006 (0.184)
Abies balsamea	183 (10202)	277 (9275)	316 (7988)	311 (7449)	277 (7648)	294 (6625)	0.038 (0.003)
Picea glauca	64 (9430)	126 (11,155)	199 (22,404)	136 (9206)	169 (7494)	124 (14,515)	0.087 (0.834)
Picea mariana	152 (6982)	156 (10,057)	144 (52,840)	96 (6805)	93 (6375)	70 (5694)	0.049 (0.892)

*[*] *Castor canadensis* select against conifers. Distance-dependent total basal area for conifers averaged between 65% and 75% of total basal area for all species.

[†] Probability that the observed regression slope could have differed from zero by chance.

some species, factors other than beaver herbivory are determining spatial abundance patterns.

In addition, if plant community structure were determined by selective removal of plants by beavers, then sapling recruitment of preferred species should be negatively correlated with foraging intensity whereas sapling recruitment of nonpreferred species should be positively correlated with foraging intensity, because these species should have increased opportunity for establishment. Donkor and Fryxell (1999) reported that sapling recruitment of four preferred species (*Populus tremuloides, Acer rubrum, Acer saccharum, and Corylus cornuta*) did increase farther from ponds and that sapling recruitment of nonpreferred species was greater near ponds, as expected.

The traditional approach to explaining vegetation dynamics presumes that a more even mix of individuals of different plant species would replace plants cut by beavers (Novak, 1987). The opening of the canopy caused by the removal of large trees has many effects on forest composition and structure. Johnston and Naiman (1990) suggested that increasing light penetration and decreasing competition for soil moisture and nutrients could increase net primary productivity of existing subcanopy woody species. This was supported by their finding that the loss of *Populus tremuloides* resulted in small increases in the importance of *Alnus* spp., *Betula* spp., *Picea glauca,* and *Populus balsamifera.*

Recall from Table 2 that *Alnus rugosa* and *Salix bebbiana*, while preferred by beavers, increased in abundance close to the ponds. The most likely explanation for this pattern is that beaver cutting creates light gaps and frees up soil nutrients, thereby producing favorable conditions for regeneration of *A. rugosa* and *S. bebbiana*. This enhanced regeneration may more than compensate for any mortality caused by herbivory. Furthermore, because of their shrubby growth habit, these two species are more likely to respond to cutting by sprouting. Thus, beaver herbivory could stimulate regeneration and increase clone size (and hence basal area) in these plants. Finally, the mud piles that beavers create along pond banks are conducive to germination of *S. bebbiana* seedlings. In summary, beaver foraging would directly affect forest structure and composition from their actual browsing and cutting of preferred plant species, which would then affect regeneration of both preferred and nonpreferred species.

Indirect Vegetation Responses to Beaver Activity

In the study by Donkor and Fryxell (2000), there was a pronounced gradient in soil moisture as a function of distance from ponds: from 0 to 10 m distance from water, soil samples contained more than 25% moisture, while beyond 40 m soil samples had less than 10% moisture. Donkor and Fryxell also observed that richness of mesic species decreased with increasing distance from ponds, while richness of xeric species increased with distance from ponds. Thus, their observation that overall plant species richness was highest at intermediate levels of soil moisture found at intermediate distance from ponds is not surprising because this would be the region where substantial numbers of both more mesic (e.g., *A. rugosa*) and more xeric (e.g., *C. cornuta*) species would tend to occur. These observations are also consistent with Terborgh's (1973) assertion that the general pattern in temperate North American forests is highest species richness at intermediate moisture levels. The importance of both moisture and nutrient gradients in explaining species distribution (Whittaker, 1956) was shown by Bridge and Johnson (2000) for the boreal forest of central Saskatchewan; they found that the patterns of dominance by *Pinus banksiana* Lamb, *Picea mariana*, *Populus tremuloides*, *Picea glauca,* and *Abies balsamea* were explained by these gradients along hill slopes of different substrates (glaciofluvial and glacial till), rather than by stand age (again indicating little support for the traditional successional explanation for these species distributions).

If variation in species richness is caused primarily by variation in an abiotic factor (soil moisture), then sapling recruitment of mesic species should decline with distance from ponds, whereas recruitment of xeric species should increase with distance from ponds. This was supported by Donkor and Fryxell's (2000) study. Hence, both species richness and recruitment data were consistent with the edaphic factor influence.

Soil characteristics may also be important to plant distributional patterns. Studies on vegetation along environmental gradients have shown that factors related to temperature, moisture, available nutrients, and mechanical stress often correlate with observed vegetation distribution patterns (e.g., Terborgh, 1971; Whittaker and Niering, 1975; Keddy, 1985; Wilson and Keddy, 1985). Differences in sapling recruitment of mesic versus xeric species along the gradient suggest that moisture influences community organization through contrasting effects on different members of the community. Most environmental factors that affect organisms do so along a gradient. In the study by Donkor and Fryxell (2000), the other abiotic variables (phosphorus, magnesium, potassium, pH, and organic matter) did not vary consistently with distance from ponds, indicating that they are not primary factors determining the spatial variation in woody plant community structure, irrespective of their importance to plant growth. These results are consistent with those of Wilson and Keddy (1985), who found that the distributions of plant species were influenced by a gradient of substratum organic content, among other abiotic variables measured (nitrogen, phosphorus, and calcium). Whittaker and Niering (1965) also reported that in some systems, some soil differences are important only at a secondary level in accounting for the vegetation pattern.

Johnston and Naiman (1990) observed that long-term foraging by beavers significantly altered tree density and basal area; where beavers had been foraging uninterrupted for 6 years, the density and basal area of stems ≥ 5 cm in diameter were reduced by 43%. Donkor and Fryxell (1999; 2000) conducted vegetation sampling on one pond abandoned by beavers for 3 years (the only one of this status in the area). Though fresh stems cut by beavers were rare on this pond, stumps from previous cutting were still evident. Three years after the beavers left, patterns of species richness and diversity had decreased significantly. However, sapling recruitment of preferred and nonpreferred species on the abandoned pond was similar to that of the occupied ponds. To this end, indirect effects of beaver foraging on forest composition would involve their impacts on hydrology and thus on creating moisture-nutrient gradients along which plant species are distributed, depending on their tolerances and requirements.

Combined Direct and Indirect Effects of Beaver Herbivory on Forest Composition

Both biotic and abiotic factors seem to be co-influencing the woody plant community composition in boreal forests. In Donkor and Fryxell's (2000) study, *A. rugosa* and *S. bebbiana*, though preferred by beavers, declined in abundance with distance from ponds although their recruitment increased towards ponds. This pattern of abundance and sapling recruitment of these two species could be explained by the influence of selective foraging and the covarying moisture gradient that tends to enhance sprouting of these two woody species close to water (Donkor, 1993). Both *A. rugosa* and *S. bebbiana* are mesic species showing more tolerance to high moisture conditions. Although the study of the impacts of beaver abandonment involved only one pond (in comparison to the 15 beaver-occupied ponds studied), it does demonstrate that removal of beavers may reverse some of the effects of beaver foraging and edaphic factors on species richness and diversity. Remillard *et al.* (1987) found that vegetation within areas where beavers feed only changed with fluctuating water levels associated with cyclic abandonment and reoccupation of beaver sites.

Flooding of terrestrial areas is another big impact on plant community structure and composition. Flooding impacts result from a combination of direct and indirect effects of beavers. Dam-building beavers are clear examples of ecosystem engineers that are abundant throughout the boreal regions of North America. The ponds they create by damming streams increase the soil moisture, which influences community structure and ecosystem functioning (Naiman *et al.*, 1988). When ponds are abandoned and the associated dams are breached, extensive meadows form that can persist for many years (Ives, 1942). Beavers set the stage for the creation of meadows by building dams that trap nutrient-rich sediment and by both directly and indirectly killing woody vegetation in the riparian zone via herbivory and flooding (Wright *et al.*, 2002). In contrast to forested riparian zones, beaver meadows have high light penetration and elevated soil moisture and nutrient levels (Naiman *et al.*, 1994).

Long-term Effects of Beaver Herbivory on the Landscape

Because beavers concentrate feeding activity around ponds, the effect of herbivory is large at the scale of the beaver pond, and may be insignificant at the landscape level (Johnston and Naiman, 1987; 1990). However, there are many reasons why beaver foraging should also concern boreal forest ecologists and managers at the landscape scale. Beaver harvests in Ontario declined from 1919–1920 to the mid-1930s. Beavers became rare in the mid-1930s, and trapping seasons were closed. Through management, populations began to recover, and numbers peaked around 1970 (Novak, 1987; Ingle-Sidorowicz, 1982). Against the background that the North American beavers have rebounded from widespread extirpation to high abundance, with their swift increases in number and range expansion, I hypothesize that beaver browsing will have long-term impacts on the forest composition and structure for several reasons. First, most biomass harvested by beavers in these ecosystems is mature woody plants; therefore, where beaver dams and beaver densities are very high, their impacts (e.g., on carbon and nitrogen nutrient cycles) could be substantial. Second, when beavers deplete their food resource close to their ponds, they are likely to build more dams and extend the water perimeter to get closer to food sources and also to expand their refuge from predators. It cannot be simply said that their influence is concentrated within a narrow band of their feeding range. Large tracts of land are inundated each year by damming activities of beavers. Third, long-term beaver herbivory will cause substantial changes in forest composition because of their preferential selection of some browse species over others. Long-term browsing could shift the composition of some boreal forest stands from palatable deciduous tree species to unpalatable shrub and coniferous species (Pastor *et al.*, 1988). Thus, preferential central place foraging by beavers scattered throughout the landscape could establish a patchwork of shrubs and conifers. Given the increasing numbers of beavers in North America over the past 50 years (Ingle-Sidorowicz, 1982) and the increase in availability of prime beaver habitat as human logged areas regenerate to *Populus tremuloides* and other preferred foods, future beaver influence on boreal forests may be significant.

Surprisingly, little has been published on the long-term effects of beavers on forest composition and structure. In a more recent study, Wright *et al.* (2002) compared beaver-modified riparian sites and riparian sites with no history of beaver occupation in the Adirondack region of New York to determine the degree of similarity in species composition. Visual analysis of historical aerial photographs indicated that beaver activity is the only large-scale form of disturbance in that region. Beaver-modified sites were selected only if historical photographs, taken at 10-year intervals between 1942–2002, demonstrated that they had been forested at one point. Selected undisturbed riparian sites showed no evidence of beaver modification at any point during the 60-year period. Although beavers had no predictable effect on species richness at the patch scale, by increasing habitat heterogeneity, they increased the number of species of herbaceous plants by over 33% at the landscape scale. Also, ordination of plots indicated a striking separation in species composition between plots in beaver-engineered and nonengineered sites with very low levels of species overlap between the two habitat types (Fig. 3). The large increase in species richness caused by beavers appears to be caused by the creation of novel habitat types in the riparian zone and the presence of a large number of species capable of exploiting the resources provided in these engineered patches. Previous research on ecosystem engineers in a wide range of natural ecosystems has indicated that engineered patches can have both higher (Martinsen *et al.*, 1990) and lower (Bratton, 1975) species richness than nonengineered patches. An increase in richness within engineered patches is thought to occur when disturbance increases resource availability by eliminating competitively dominant species or ameliorating stressful conditions. Alternatively, a decrease in species richness could occur if the conditions created by the engineer either facilitate the growth of a competitive dominant or are so harsh as to eliminate most species (Wright *et al.*, 2002). The effect of the engineer on richness at the landscape scale will be

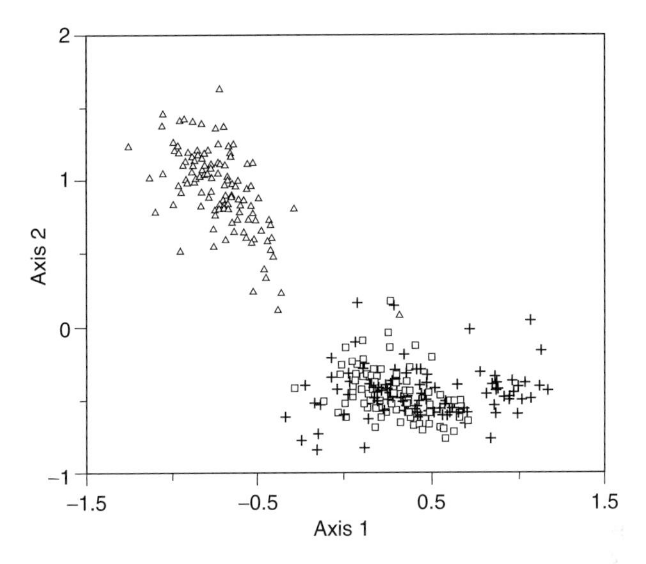

FIG. 3 Ordination of plant community composition between plots in beaver-modified and riparian forested sites in central Adirondack region of New York. Ordination of plots based on presence of species using nonmetric multidimensional scaling. Δ Forested riparian zone habitat, □ alder habitat, + meadow habitat. (From Wright *et al.* [2002], redrawn with kind permission of Springer Science and Business Media.)

negligible unless there are species found in engineered patches that are not found elsewhere in the landscape. Thus, even in systems where engineered patches have lower species richness than nonengineered patches, the existence of species uniquely present in the engineered patches will result in ecosystem engineering having a positive effect on richness at the larger scale.

CONCLUSION

Selective foraging by beavers influences the richness and diversity of plant communities surrounding beaver ponds in boreal forests by increasing the importance of nonpreferred plant species (conifers) and facilitating the regeneration of stems of both preferred and nonpreferred species. Additionally, variation in one abiotic factor, soil moisture, influences the woody plant community organization of these boreal forests through variable responses of different species to hydrid conditions. Therefore, the ecological effect seen in the structure and composition of boreal forests can be attributed to the process and response effects of both beaver foraging and edaphic conditions, although beavers encourage the abiotic effect.

ACKNOWLEDGMENTS

I thank T. K. Nunifu and D. Nanang and two other anonymous reviewers for insightful discussions and helpful comments on earlier drafts of the manuscript.

REFERENCES

Aldous, S. A. (1938). Beaver food utilization studies. *J Wildife Manage* 2, 215–222.

Allred, M. (1981). The potential use of beaver population behavior in beaver resources management. *J Idaho Acad Sci* 17, 14–24.

Allred, M. (1986). *Beaver Behavior: Architect of Fame and Bane*. Nature Graph, Happy Camp, CA.

Anderson, M. (1978). Optimal foraging: size and allocation of search effort. *Theor Popul Biol* 13, 397–409.

Baker, B. W. (2003). Beaver (*Castor canadensis*) in heavily browsed environments. *Lutra* 46, 173–181.

Baker, W. W., Mitchell, D. C. S., Ducharme, H. C., Stanley, T. R., and Peinetti, H. R. (2004). Why aren't there more beaver in Rocky Mountain National Park? In: *Wildlife and Riparian Areas*. (Colorado Riparian Association, Eds.), pp. 85–90. Colorado Riparian Association, Boulder, CO.

Basey, J. M., Jenkins, S. H., and Busher, P. E. (1988). Optimal central-place foraging by beavers: tree-size selection in relation to defensive chemicals of quaking aspen. *Oecologia* 76, 278–282.

Beals, E. W., Cottam, G., and Vogl, R. J. (1960). Influence of deer on vegetation of the Apostle Islands, Wisconsin. *J Wildlife Manage* 24, 68–80.

Belovsky, G. E. (1981). Food selection by a generalist herbivore: the moose. *Ecology* 62, 1020–1030.

Belovsky, G. E. (1984). Summer diet optimization by beaver. *Am Midl Nat* 111, 209–222.

Bradt, G. W. (1938). A study of beaver colonies in Michigan. *J Mammal* 19, 139–162.

Bratton, S. P. (1975). The effect of the European wild boar, *Sus scrofa*, on gray beech forest in the Great Smoky Mountains. *Ecology* 56, 1356–1366.

Brenner, F. J. (1962). Food consumed by beavers in Crawford County, Pennsylvania. *J Wildlife Manage* 26, 104–107.

Bridge, S. R. J., and Johnson, E. A. (2000). Geomorphic principles of terrain organization and vegetation gradients. *J Veg Sci* 11, 57–70.

Bryant, J. P., and Chapin, F. S. (1986). Browsing-woody plant interactions during boreal forest plant succession. In: *Forest Ecosystems in the Alaska Taiga: A Synthesis of Structure and Function*. (Van Cleve, K., Chapin, F. S., Flanagan, P. W., Viereck, L. A., and Dyrness, C. T., Eds.), pp. 213–225. Springer-Verlag, New York.

Bryant, J. P., and Kuropat, P. J. (1980). Selection of winter forage by subartic browsing vertebrates: the role of plant chemistry. *Annu Rev Ecol Syst* 11, 261–285.

Bryant, J. P., Chapin, F. S., and Klein, D. R. (1983). Carbon/nutrient balance of boreal plants in relation to vertebrate herbivory. *Oikos* 40, 357–368.

Busher, P. E., and Lyons, P. J. (1999). Long-term population dynamics of the North American beaver, *Castor canadensis*, on Quabbin Reservation, Massachusetts, and Sagehen Creek, California. In: *Beaver Protection, Management, and Utilization in Europe and North America*. (Busher, P. E., and Dzieciolowski, R. M., Eds.), pp. 147–160. Kluwer Academic/Plenum Publishers, New York.

Chapin, F. S., Bloom, A. J., Field, C. B., and Waring, R. H. (1987). Plant responses to multiple environmental factors. *BioScience* 37, 49–67.

Chapin, F. S., Bryant, J. P., and Fox, J. F. (1985). Lack of induced chemical defense in juvenile Alaskan woody plants in response to simulated browsing. *Oecologia* 67, 457–459.

Chapin, F. S., Vitousek, P. M., and Van Cleve, K. (1986). The nature of nutrient limitation in plant communities. *Am Nat* 127, 48–58.

Chapin, F. S., Walker, L. R., Fastie, C. L., and Sharman, L. C. (1994). Mechanisms of primary succession following deglaciation at Glacier Bay, Alaska. *Ecol Monogr* 64, 149–175.

Clements, F. E. (1916). *Plant Succession*. Publication 242. Carnegie Institute, Washington, DC.

Corey, P. D., Bryant, J. P., and Chapin, F. S. (1985). Resource availability and plant antiherbivore defense. *Science* 230, 895–899.

Donkor, N. T. (1993). *Influence of Beaver Foraging and Edaphic Factors on the Woody Plant Community Structure in Lowland Boreal Forest*. M.Sc. Thesis, Department of Zoology, University of Guelph, Guelph, ON.

Donkor, N. T., and Fryxell, J. M. (1999). Impact of beaver foraging on structure of lowland boreal forests of Algonquin Provincial Park, Ontario. *Forest Ecol Manag* 118, 83–92.

Donkor, N. T., and Fryxell, J. M. (2000). Lowland boreal forests characterization in Algonquin Provincial Park relative to beaver (*Castor canadensis*) foraging and edaphic factors. *Plant Ecol* 148, 1–12.

Dunson, W. A., and Travis, J. (1991). The role of abiotic factors in community organization. *Am Nat* 138, 1067–1091.

Fastie, C. L. (1995). Causes and ecosystem consequences of multiple pathways of primary succession at Glacier Bay, Alaska. *Ecology* 76, 1899–1916.

Flanagan, P. W., and Van Cleve, K. (1983). Nutrient cycling in relation to decomposition and organic matter quality in taiga ecosystems. *Can J Forest Res* 13, 795–817.

Frelich, L. E., and Reich, P. B. (1999). Neighborhood effects, disturbance severity, and community stability in forests. *Ecosystems* 2, 151–166.

Fryxell, J. M. (2001). Habitat suitability and source-sink dynamics of beavers. *J Anim Ecol* 70, 310–316.

Fryxell, J. M,, and Doucet, C. M. (1991). Foraging time and central-place foraging in beavers. *Can J Zool* 69, 1308–1313.

Gill, D. (1972). The evolution of a discrete beaver habitat in the Mackenzie River delta, Northwest Territories. *Can Field Nat* 86, 223–239.

Gordon, A. G. (1983). Nutrient cycling dynamics in differing spruce and mixedwood ecosystems in Ontario and the effects of nutrient removals through harvesting. In: *Resources and Dynamics of the Boreal Zone*. (Wein, R. W., Riewe, R. R., and Methven, I. R., Eds.), pp 97–118. Association of Canadian Universities for Northern Studies, Ottawa, ON.

Gutsell, S. L., and Johnson, E. A. (2002). Accurately ageing trees and examining their height-growth rates: implications for interpreting forest dynamics. *J Ecol* 90, 153–166.

Hall, J. G. (1960). Willow and aspen in the ecology of beaver on Sagehen Creek, California. *Ecology* 41, 484–494.

Hilfiker, E. L. (1990). *Beavers: Water, Wildlife and History*. Windswept Press, Interlaken, NY.

Howard, R. J. (1982). *Beaver Habitat Classification in Massachusetts*. M.S. thesis, University of Massachusetts, Amherst, MA.

Ingle-Sidorowicz, H. M. (1982). Beaver increase in Ontario: result of changing environment. *Mammalia* 46, 167–175.

Ives, R. L. (1942). The beaver-meadow complex. *J Geomorph* 5, 191–203.

Jenkins, S. H. (1975). Food selection by beavers: a multidimensional contingency table analysis. *Oecologia* 21, 157–173.

Jenkins, S. H. (1980). Seasonal and year-to-year differences in food selection by beavers. *Oecologia* 44, 112–116.

Jenkins, S. H., and Busher, P. E. (1979). *Castor canadensis*. *Mamm Species* 120, 1–8.

Johnston, C. A., and Naiman, R. J. (1987). Boundary dynamics at the aquatic-terrestrial interface: the influence of beaver and geomorphology. *Landscape Ecol* 1, 47–57.
Johnston, C.A. and Naiman, R.J. (1990). Browse selection by beavers: effects on riparian forest composition. *Can J Forest Res* 20, 1036–1043.
Keddy, P. A. (1985). Lakeshores in the Tusket River Valley, Nova Scotia: distribution and status of some rare species including *Coreopsis rosea* Nutt. and *Sabatia kennedyana* Fem. *Rhodora* 87, 309–320.
Keith, L. B., Cary, J. R., Rongstad, O. J., and Brittingham, M. C. (1984). Demography and ecology of a snowshoe hare population decline. *Wildlife Monogr* 90, 1–43.
Kneeshaw, D. D., and Burton, P. J. (1997). Canopy and age structures of some old sub-boreal *Picea* stands in British Columbia. *J Veg Sci* 8, 615–626.
Krefting, L. W. (1974). *The Ecology of the Isle Royale Moose with Special Reference to the Habitat*. Forestry Series 15, Technical Bulletin 297. University of Minnesota Agricultural Experiment Station, St. Anthony Park, MN.
Larsen, J. A. (1980). *The Boreal Ecosystem*. Academic Press, New York.
Larson, J. S., and Gunson, J. R. (1983). Status of the beaver in North America. *Acta Zool Fenn* 174, 91–93.
Linder, P., Elfving, B., and Zackrisson, O. (1997). Stand structure and successional trends in virgin boreal forest reserves in Sweden. *Forest Ecol Manag* 98, 17–33.
Martinsen, G. D., Cushman, J. H., and Whitman, T. G. (1990). Impact of pocket gopher disturbance on plant species diversity in a shortgrass prairie community. *Oecologia* 83, 132–138.
McGinley, M. A., and Whitham, T. G. (1985). Central Place foraging by beavers (*Castor canadensis*): a test of foraging predictions and the impact of selective feeding on the growth of cottonwoods (*Populus fremontii*). *Oecologia* 66, 558–562.
Mclnnes, P. F. (1989). *Moose Browsing and Forest Vegetation Dynamics*. M.S. thesis, University of Minnesota, St. Paul, MN.
McNaughton, S. J. (1985). Ecology of a grazing ecosystem: the Serengeti. *Ecol Monogr* 55, 259–294.
Naiman, R. J., and Melillo, J. M. (1984). Nitrogen budget of a subartic stream altered by beaver (*Castor canadensis*). *Oecologia* 62, 150–155.
Naiman, R. J., Johnston, C. A., and Kelley, J. C. (1988). Alteration of North American streams by beaver. *BioScience* 38, 753–762.
Northcott, T. H. (1971). Feeding habits of beaver in Newfoundland. *Oikos* 22, 407–410.
Novak, M. (1987). *Wild Furbearer Management and Conservation in North America*. Ontario Trappers Association, Toronto, ON.
Orians, G. H., and Pearson, N. E. (1979). On the theory of central place foraging. In: *Analysis of Ecological Systems*. (Horn, D. J., Stairs, G. R., and Mitchell, R. D., Eds.), pp. 155–177. Ohio State University Press, Columbus, OH.
Palo, R. T., Sunnerheim, K., and Theander, O. (1985). Seasonal variation of phenols, crude protein, and cell wall content of birch (*Betula pendula* Roth.) in relation to ruminant in vitro digestibility. *Oecologia* 65, 314–318.
Pastor, J. P., and Bockheim, J. G. (1984). Distribution and cycling of nutrients in an aspen-mixed hardwood-spodosol ecosystem in northern Wisconsin. *Ecology* 65, 339–353.
Pastor, J., and Naiman, R. J. (1992). Selective foraging and ecosystem processes in boreal forests. *Am Nat* 139, 690–705.
Pastor, J., Naiman, R. J., and Dewey, B. (1987). A hypothesis of the effects of moose and beaver foraging on soil carbon and nutrient cycles, Isle Royale. *Alces* 23, 107–124.
Pastor, J., Naiman, R. J., Dewey, B., and Mclnnes, P. (1988). Moose, microbes, and the boreal forest. *BioScience* 38, 770–777.
Payne, N. F. (1982). Colony size, age, and sex structure of Newfoundland beaver. *J Wildlife Manage* 26, 272–278.

Payne, N. F. (1984). Reproductive rates of beaver in Newfoundland. *J Wildlife Manage* 48, 912–917.
Peek, J. M., Ulrich, D. L., and Mackie, R. J. (1976). Moose habitat selection and relationships to forest management in northeastern Minnesota. *Wildlife Monogr* 48, 1–65.
Peterson, R. L. (1955). *North American Moose*. University of Toronto Press, Toronto, ON.
Pollock, M. M., Naiman, R. J., Erickson, H. E., Johnston, C. A., Pastor, J., and Pinay, G. (1995). Beaver as engineers: influence on biotic and abiotic characteristics of drainage basins. In: *Linking Species and Ecosystems*. (Jones, C. G., and Lawton, J. H., Eds.), pp. 117–126. Chapman and Hall, New York.
Remillard, M. M., Guendling, G. K., and Bogucki, D. J. (1987). Disturbance by beaver (*Castor canadensis*) and increased landscape heterogeneity. *Ecol Stud Anal Synthesis* 64, 103–122.
Schoener, T. W. (1979). Generality of the size-distance relation in models of optimal foraging. *Am Nat* 114, 902–914.
Schulte, B. A., and Muller-Schwarze, D. (1999). Understanding North American beaver behavior as an aid to management. In: *Beaver Protection, Management, and Utilization in Europe and North America*. (Busher, P. E., and Dzieciolowski, R. M., Eds.), pp. 109–128. Kluwer Academic/Plenum Publishers, New York.
Smith, D. W. (1997). *Dispersal Strategies and Cooperative Breeding in Beavers*. Ph.D. Dissertation, University of Nevada, Reno, NV.
Smith, D. W., and Peterson, R. O. (1988). *The Effects of Regulated Lake Levels on Beaver at Voyageurs National Park, Minnesota*. Research/Resources Management Report MWR-11. U.S. Department of the Interior, National Park Service, Midwest Regional Office, Omaha, NE.
Snyder, J. D., and Janke, R. A. (1976). Impact of moose browsing on boreal forests of the Isle Royal National Park. *Am Midl Nat* 95, 79–92.
Svendsen, G. E. (1980). Seasonal change in feeding patterns of beaver in southern Ohio. *J Wildlife Manage* 44, 285–290.
Tahvanainen, J. E., Hell, E., Julkunen-Titto, R., and Lavola, A. (1985). Phenolic compounds of willow bark as deterrents against feeding by mountain hare. *Oecologia* 65, 319–323.
Terborgh, J. (1971). Distribution on environmental gradients: theory and a preliminary interpretation of distributional patterns in the avifauna of the Cordillera Vilcabamba, Peru. *Ecology* 52, 23–40.
Terborgh, J. (1973). On the notion of favourableness in plant ecology. *Am Nat* 107, 481–501.
Van Cleve, K., and Oliver, L. K. (1982). Growth response of postfire quaking aspen (*Populus tremuloides* Michx.) to N, P, and K fertilization. *Can J Forest Res* 12, 160–165.
Van Cleve, K., and Viereck, L. A. (1981). Forest succession in relation to nutrient cycling in the boreal forest of Alaska. In: *Forest Succession: Concepts and Application*. (West, D. C., Shugart, H. H., and Botkin, D. B., Eds.), pp 185–211. Springer-Verlag, New York.
Van Cleve, K., and Zasada, J. (1976). Response of 70-year-old white spruce to thinning and fertilization in interior Alaska. *Can J Forest Res* 6, 145–152.
Van Cleve, K., Oliver, L. K., Schlentner, R., Viereck, L. A., and Dyrness, C. T. (1983). Productivity and nutrient cycling in taiga forest ecosystems. *Can J Forest Res* 13, 746–766.
Vogl, R. J. (1981). The ecological factors that produce perturbation-dependent ecosystems. In: *Forest Succession: Concepts and Application*. (West, D. C., Shugart, H. H., and Botkin, D. B., Eds.), pp. 63–94. Springer-Verlag, New York.
Whittaker, R. H. (1956). Vegetation of the Great Smoky Mountains. *Ecol Monogr* 26, 1–80.
Whittaker, R. H., and Niering, W. A. (1965). Vegetation of the Santa Catalina Mountains, Arizona II. A gradient analysis of the south slope. *Ecology* 46, 429–452.
Whittaker, R. H., and Niering, W. A. (1975). Vegetation of the Santa Catalina Mountains, Arizona V. Biomass, production, and diversity along the elevation gradient. *Ecology* 56, 771–790.

Wilson, S. D., and Keddy, P. A. (1985). Plant zonation on a shoreline gradient: physiological response curves of component species. *J Ecol* 73, 851–860.
Wright, J. P., Jones, C. G., and Alexander S. F. (2002). An ecosystem engineer, the beaver, increase species richness at the landscape scale. *Oecologia* 132, 96–101.

19

Beaver, Willow Shrubs, and Floods

J. Dungan Smith
U.S. Geological Survey

INTRODUCTION

In the semi-arid parts of western North America, the headwater tributaries of many rivers were populated by beaver-maintained willow carrs before 1800. Despite heavy trapping of beaver between 1810 and 1840 and their near extirpation from these areas, the well-established willow carrs typically remained intact for another century. Before and during this period, the carrs protected stream systems that had evolved their morphologies under a hydrological regime in which the beaver-maintained wetlands mitigated (1) the effects of heavy rainfall in the headwater drainage basins and (2) the consequences of upstream and hillslope erosion. Once the beaver had been removed and the bottomlands had been cleared of shrubs, these former beaver meadows often made excellent ranchland. As a result, in the late 1800s and early 1900s there was substantial encroachment of ranchlands onto former beaver-occupied bottomlands. Regulation of livestock grazing in the 1930s by the Taylor Act decreased grazing in riparian areas, and, since the middle of the past century, beaver have reoccupied many headwater streams in the western United States, rejuvenating the willow carrs. Beginning in the 1990s, however, human populations have been rapidly expanding into these headwater areas. As a result, the anthropogenic impacts on many western streams are again increasing.

Like human communities, beaver communities exert an important influence on the landscape (e.g., see Marston, 1994; Gurnell, 1998; Hillman, 1998; Butler and Malanson, 2005). Many other organisms depend on their ponds; consequently, their fate indirectly controls the fates of many wetland, pond, and stream inhabitants, and, to a lesser degree, their activities also affect the upland ecosystem (Wright *et al.*, 2002; Muller-Schwarze and Sun, 2003). Beaver pond complexes maintain an elevated water table over an expanded area. This substantially increases ground water recharge to the aquifers fed from these upland areas and changes the nature of the floodplain vegetation. Moreover, in many stream networks the wetlands upstream have a large impact on the geomorphology, hydrology, and water quality of the rivers farther downstream. Extensive beaver-maintained shrub carrs on low-order tributaries are particularly important with respect to flood mitigation. Locally, willow shrub ecology, beaver activity, stream geomorphology, and hydrology are tightly connected. As a result, some of the anthropogenic impacts on beaver populations are indirect, complex, and unexpected, and their ecological and geomorphic consequences can be extensive and profound.

Many low-order western streams have a morphology that developed over a long period of time when their meander belts and floodplains were densely populated with willow and beaver. When impacted by heavy rainfall, healthy beaver-maintained willow carrs provide good protection from floods, both at the site of the carr and further downstream. This flood protection is accomplished by (1) distributing the flood water across shrub-protected floodplains and (2) similarly retaining sediment and debris that could fill or destabilize the stream channels and rivers further downstream. By drastically reducing and variably delaying peak flows from tributaries, the beaver pond complexes and their associated shrub carrs broaden and greatly decrease the flood peaks further downstream in the river networks. In addition, the shrub-protected ponds provide refuge for many aquatic organisms during floods and, as long as the beaver also are able to remain in their ponds during the high flow events, any damage to their dams is quickly repaired as the stage drops. The latter is made possible because in well-developed shrub carrs most of the damage to beaver dams during a flood occurs as narrow breaches in areas that originally were channels, as is shown subsequently in this chapter. Only when a flood moves through an area that has a marginal supply of trees and shrubs are the dams likely to be washed out and the beaver forced to leave for lack of food and building materials. Examples of these situations are provided subsequently in this chapter. Before human interference with streams and stream networks, overall damage from large floods was generally limited by the dense carrs of shrubs that lined streams from their headwaters on downstream to the point where the streams became too wide and too deep for beaver to maintain permanent dam complexes across the floodplain (for example, see Smith, 2004.).

Even on rivers too wide for permanent beaver-dam complexes, temporary beaver dams on individual threads and backwater channels help maintain an elevated water table in the meander belt or beneath the floodplain and promote the growth of cottonwood forests with dense understories of shade-tolerant riparian species. Typically, large cottonwood trees are spaced too far apart to be effective at flood mitigation, but dense understory shrubs provide the same protection of the river and floodplain as do the willow carrs further upstream. Moreover, the beaver pond–controlled hydrographs from the tributaries help reduce the erosive fluid forces in these downstream reaches of the river and on the downstream floodplains. In this manner, the geomorphic integrity of western stream and river systems used to be well protected against floods for long distances downstream. In contrast, when the willow carrs and understory shrubs are removed along streams and rivers that evolved over many millennia in the presence of these riparian features, the streams and rivers are often too steep to maintain their geomorphic integrity during large floods and become vulnerable to floodplain erosion. Geomorphic alteration from narrow, single-threaded systems to wide straight systems (i.e., floodplain unraveling) even becomes possible.

To understand the interplay of floods, shrubs, and beaver, it is necessary first to understand and quantify the relation between shrubs and overbank flows. For flow through dense shrub carrs, the simple hydraulic ideas and formulae used for water conveyance canals and for in-channel flow in rivers are not appropriate. In open-channel flow formulations, water depth is the key hydraulic variable, whereas in wetlands full of stiff reeds and shrubs, on shrub-covered floodplains, and in cottonwood forests with dense shrub understories, stem density partially to completely replaces flow depth as the controlling hydraulic variable (Smith, 2001). For example, in shrub carrs with 100% canopy coverage in which the water depth is less than the shrub height, water velocity no longer increases with flow depth; rather, it becomes constant with respect to this variable and depends solely on stem density (Smith, 2004). Consequently, as overbank discharge increases, flow depth rises linearly with it. Boundary shear stress similarly becomes approximately constant with respect to flow depth and discharge, as shown later in this chapter, typically preventing floodplain and wetland erosion by even the largest floods. These results have profound hydraulic implications for floodplain response to large overbank flows and, specifically, they explain why beaver-maintained willow carrs are so effective in mitigating large floods and in preventing geomorphic unraveling of stream systems. Conversely, these results suggest that removal of beaver-maintained willow carrs might, in some cases, lead to catastrophic changes in stream morphology during large floods. It is important to evaluate under what conditions removal of willow carrs poses a threat to the integrity of a stream system because the consequences of unraveling are not just local and can be very severe physically, chemically, and biologically.

If a narrow meandering stream with a broad floodplain, which was once protected by shrub carrs along its banks, were to unravel into a wide braided stream, there inevitably would be major ecological changes. A narrow, deep, sinuous stream with a complex, well-armored, stable bed and with trees and shrubs nearby along the banks maintains a cool, protected, biologically productive environment for benthic organisms and fish. In particular, this type of stream provides an environment that can support a large number of fish per unit length of channel. In contrast, once such a stream has unraveled and become a broad, shallow, straight-flow conveyance with a simple, nonarmored, frequently mobile bed and remote banks, the streambed becomes hostile to many of the original benthic organisms and the water temperature is likely to rise to a point where the stream becomes unsuitable for many types of fish. Moreover, the shallow flow, without complexity and cover, makes the few fish that can inhabit the new environment vulnerable to birds of prey. From a fisheries perspective, a single flood in a geomorphically vulnerable stream system can change the in-stream environment from a morphologically stable, biologically productive one to a morphologically unstable, species-impoverished one. Eventually the unraveled stream may return to its original sinuous form, but this recovery is likely to take on the order of a century (Friedman *et al.*, 1996; Smith *et al.*, 1998).

In the past, beavers, not humans, have maintained a riparian environment conducive to willow growth and flood mitigation in the headwater tributaries of streams in the semi-arid North American West. One goal of this chapter is to demonstrate the geomorphic importance of beaver-maintained willow carrs on headwater tributaries of western rivers. Another is to point out the vulnerability of these willow carrs to the encroachment of managed herds of grazing ungulates and other anthropogenic activities. This task will be accomplished in a five-step manner. Field data procured by the author from three separate areas will be used. These areas are (1) the Clark Fork of the Columbia River in the Deer Lodge Valley, Montana, (2) an unnamed headwater tributary of East Plum Creek (south of Denver, Colorado), and (3) Rocky Mountain National Park, Colorado.

The first step is to present the salient hydrological, geomorphic, and ecological aspects of the history of the Clark Fork in the Deer Lodge Valley during the past two centuries and to use this information to show how dense shrubs can protect floodplains from eroding during very large floods. The second step is to discuss how a stream channel can unravel by using the geomorphic response of East Plum Creek to a major flood in 1965. The third step is to derive a simple but appropriate and accurate mathematical model for the interaction of overbank flow and woody vegetation on a floodplain. Boundary shear stresses from this model can be used to determine whether sediment will be deposited on a floodplain, such as occurred along the Clark Fork in the Deer Lodge Valley during its flood of record in 1908, or whether the overbank flow will cause the floodplain to unravel, as occurred along East Plum Creek downstream of a shrub-carr–lined unnamed tributary during its flood of record in 1965. The fourth step is to apply the overbank-flow model to the unnamed tributary of East Plum Creek with the aim of using it to investigate why the unnamed tributary did not unravel except just upstream of its mouth. An important part of this step is to examine the role of the beaver dams along the unnamed tributary and their response to the 1965 flood. Finally, in the fifth step, information on Carpenter Creek (a headwater tributary of East Plum Creek) and data procured by the author on streams that were recently abandoned by beaver in the eastern part of Rocky Mountain National Park will be used to demonstrate the detrimental impact that cattle and elk can have on beaver and willow populations. The nonlinear mechanism through which the beaver-maintained wetlands are converted to uplands when invaded by large herds of ungulates is delineated. The discussions in this chapter rely, in part, on previously reported, comprehensive investigations and reconstructions of the floods of record on the two above-mentioned rivers, namely the 1908 flood on the Clark Fork in the Deer Lodge Valley (Smith *et al.*, 1998; Smith and Griffin, 2002; Smith, 2004) and the 1965 flood on Plum Creek (Osterkamp and Costa, 1987; Griffin and Smith, 2001; Smith, 2001; Griffin and Smith, 2004).

BACKGROUND

Clark Fork of the Columbia River

The Clark Fork of the Columbia River begins with the confluence of Silver Bow and Warm Springs Creeks in the upper Deer Lodge Valley of Montana

(Fig. 1). It first flows approximately north through the Deer Lodge Valley to the town of Garrison, where it turns west-northwest and flows past Missoula, Montana. Silver Bow Creek begins in Butte, and Warm Springs Creek begins west of Anaconda. Owing to very heavy mining impacts along Silver Bow Creek and a major flood in 1908, which eroded tailings from the floodplain of this river and then deposited several decimeters of metals-contaminated tailings on the floodplain of the Clark Fork upstream of Garrison (Smith *et al.*, 1998), the meander belt of the Clark Fork in Deer Lodge Valley has been part of a large Environmental Protection Agency

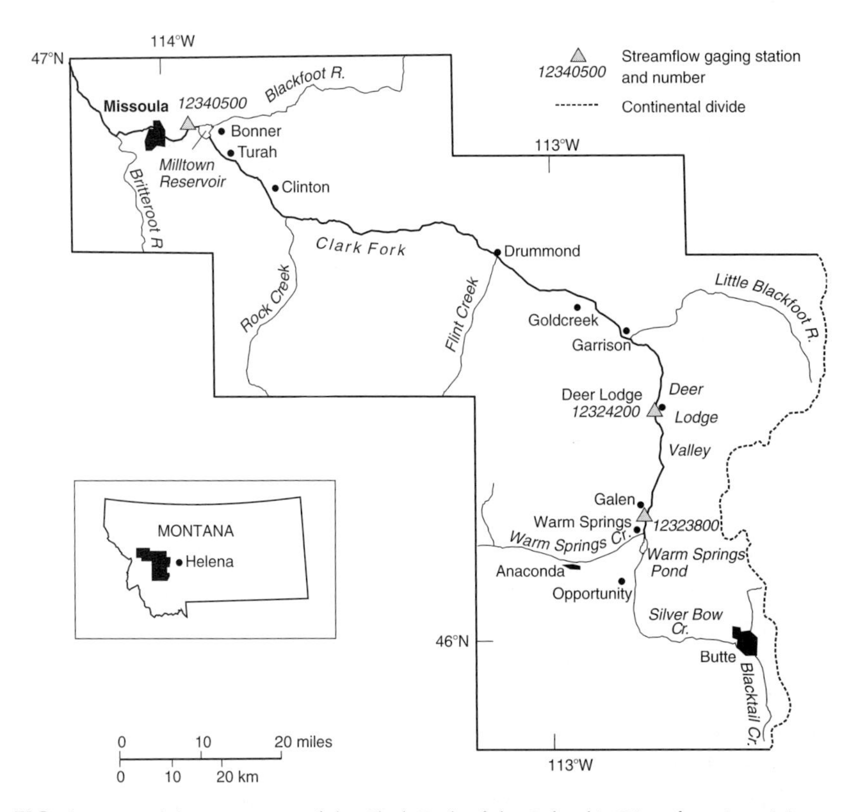

FIG. 1 Map of the upper part of the Clark Fork of the Columbia River from its origin near Warm Springs, Montana, to its confluence with the Bitterroot River just west of Missoula, Montana (modified from Smith *et al.*, 1998). Also shown are its major tributaries and the USGS stream gauge at Galen. The Little Blackfoot River enters the Clark Fork at Garrison and nearly doubles its discharge. Downstream of this junction, the Clark Fork is too large to have been permanently dammed by beaver. The Deer Lodge Valley extends from Warm Springs to Garrison, and historical evidence indicates that this reach of the Clark Fork was heavily impacted by beaver activity before the early 1800s.

(EPA) Superfund clean-up effort. Because of the loss of stream-bank vegetation after the 1908 flood and more recent ranchland encroachment into the meander belt, bank erosion is common and suspended-sediment fluxes in the river during spring flows are high. Commensurate with this bank erosion are high fluxes of metals, particularly copper, into the river, and associated with the high suspended-sediment transport rates are high suspended-sediment–load-controlled downstream fluxes of sorbed metals. This situation is in stark contrast to that in the early 19th century.

Before 1820 or so, the floodplain of the Clark Fork upstream of its confluence with the Little Blackfoot River near Garrison was occupied by a beaver-maintained shrub carr (Smith *et al.*, 1998; Smith and Griffin, 2002). A trapper named Warren Ferris visited the Clark Fork in the Deer Lodge Valley in 1831 and described the river as "clear, deep, rapid, and not fordable at high water" (Horstman, 1984). Ten years later, the wandering Jesuit priest, Father Pierre Jean DeSmet, also observed the Clark Fork and its tributaries in the Deer Lodge Valley (presumably Silver Bow Creek, Warm Springs Creek, and the Little Blackfoot River) and wrote "In no part of the world is the water more limpid or pure, for whatever may be the depth of the rivers, the bottom is seen as if there were nothing to intercept the view" (Horstman, 1984). The conclusion that can be gleaned from these statements is that the rivers were exceptionally narrow and deep and that they were devoid of even small amounts of fine sediment, which would have gone into suspension during the observed high flows and obstructed the view of the stream bed. There could have been no measurable amount of fine sediment or organic matter being transported down the river and, thus, no erosion of the riverbanks during the high flows that these travelers encountered. A geomorphic interpretation of either observation by itself requires that the rivers be lined with a dense shrub flora and that there be many beaver ponds upstream of the observation sites. This is particularly true of the Clark Fork, where beaver ponds were essential to filter out all of the fine sediment that must have been entering this river from the ephemeral streams and highly erodible hillslopes on the east side of the Deer Lodge Valley.

After the beaver were trapped out of the Deer Lodge Valley, the beaver pond–supported water table apparently dropped slowly enough for the roots of the willow shrubs to follow it downward, because the shrub flora in the meander belt of the Clark Fork remained healthy until the 1908 flood. Moreover, remnants of this dense shrub carr remain even now. Only after the willows were buried by several decimeters of tailings in 1908 did they become dormant and eventually begin to die. Even after they died, however, the dead stems remained as effective protection against floods until the dead material was removed in recent decades.

Beaver prefer aspen to cottonwood and cottonwood to willow for both building materials and food (personal observations; Muller-Schwarze and Sun, 2003); therefore, in areas occupied by beaver ponds for an extended period of time, most, if not all, of the large (seed-producing) aspen and cottonwood trees have been removed, leaving only willows. In addition, willows can resprout more readily and are more tolerant of extended inundation than cottonwoods (J. M. Friedman, personal communication). This was the situation in the upper Deer Lodge Valley before the 1908 flood, and it is generally the situation there at present. The cottonwood trees that now are growing in the Deer Lodge Valley are mostly located along post-1908 irrigation ditches rather than being along the river or on its floodplain (Griffin and Smith, 2002). Consequently, the 1908 flood can be presumed to have flowed straight down a meander belt that was densely covered for several tens of miles with a relatively uniform stand of large willow shrubs (Griffin and Smith, 2002). Whether or not there were occasional cottonwood trees in the meander belt of the Clark Fork in Deer Lodge Valley before the 1908 flood, there is no indication that they were numerous enough to crush or plough up the willows, produce debris dams, or have any other deleterious effect during the 1908 flood.

Overbank flow depths adjacent to the river in 1908, determined from the elevations of the edges of the tailings deposits, were from 1.0 to 1.5 m, but the willow shrubs were then and are now many times that in height (Smith, 2004). Consequently, the resistance to the overbank flow can be calculated from a model that treats the stems as random fields of vertically oriented circular cylinders. The many-day near-peak discharge during the 5-day 1908 flood was determined locally from that measured at Missoula and partitioned by the drainage area. Having flow depth, reconstructed cross-sectional geometry, and discharge made possible an inverse calculation to determine stem density. In this manner, Smith (2004) calculated shrub density in the meander belt during the 1908 flood by using a unique flow model that contained no empirically adjusted coefficients, and he found the canopy coverage to be approximately 100%. At the stem densities associated with 100% canopy coverage, the overbank flow is controlled entirely by the average stem diameter and the number of stems per unit area, as will be shown later in this chapter. In this "dense shrub" situation, depth has no effect on flow velocity. Assuming that the stem density in 1908 was uniform across the meander belt, therefore, gives a uniform velocity across the floodplain of about $0.22\ \mathrm{m\,s^{-1}}$ for the upper Deer Lodge Valley.

Using a depositional model involving downstream advection and vertical settling, Smith (2004) calculated a down-valley deposit-thickness profile, averaged over the varying width of the meander belt. In the upper Deer Lodge Valley, this was then compared to the data of Nimick (1990). Further downstream, the model result was compared to the less dense data obtained by Schafer and Associates (1997). When the field data were filtered to smooth out considerable spatial variability, the results of this simple predictive model were surprisingly good (Fig. 2). This depositional model is relatively sensitive to the velocity averaged across the wetted floodplain, and the close agreement between predicted and measured floodplain deposition, using a velocity of 0.20 m s^{-1} for the upper Deer Lodge Valley, confirmed that the flow velocity calculated from the model described in the previous paragraph and the commensurate stem density had to be fairly accurate. Form drag on the stems, therefore, reduced the flow velocity by an order of magnitude to approximately 0.20 m s^{-1} in the upper Deer Lodge Valley. It also reduced the boundary shear stress exerted by the overbank flow on the floodplain surface (τ_η) to less than the critical shear stresses for (1) resuspension of the fine sediment being transported by the flow (τ_{css}); (2) the material making up the floodplain surface ($\tau_{c\eta}$);

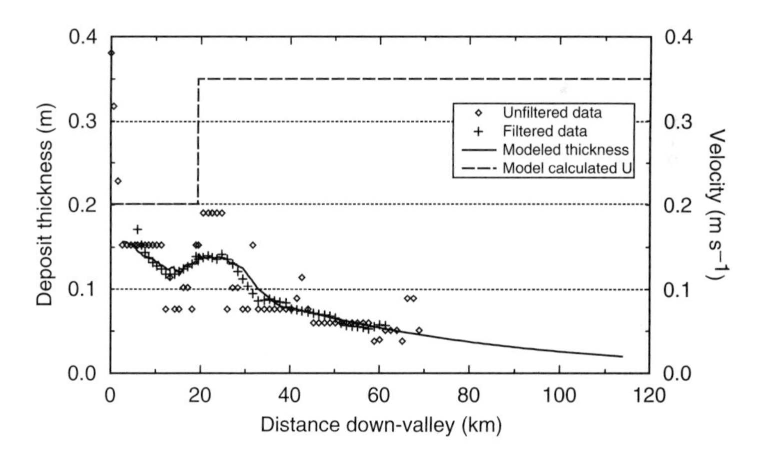

FIG. 2 Comparison of calculated and measured tailings thicknesses as a function of downvalley distance from Warm Springs to Missoula (from Smith [2004]). The diamonds represent cross-valley averaged tailings thicknesses approximately every mile. The plus signs represent a downvalley profile that has been filtered using a 13-point running mean to reduce the considerable scatter in the measurements resulting from sampling errors. The solid line represents model calculations of the deposit thickness by using the model calculated velocity profile shown by the dashed line.

and (3) bare, noncohesive soil (Wiberg and Smith, 1987). Further downstream in the region where the river was lined with cottonwood trees and the understory shrubs (primarily red osier dogwood) were thinner, the flow velocity through the shrubs increased to approximately 0.35 m s^{-1}, but the boundary shear stress still remained below the critical value for resuspension of the sediment in transit and for erosion of the floodplain soil (Smith, 2004). As a result of the low advective velocities in both of these areas, the suspended tailings settled rapidly on the floodplain in the meander belt; consequently, because of the low boundary shear stress on the floodplain surface caused by the shrubs, the material that was deposited could not be resuspended.

After the 5-day 1908 flood, the several-decimeter–thick tailings deposit on the floodplain of the Clark Fork in the Deer Lodge Valley capped the organic matter on the previous floodplain surface. The decay of this organic matter in the absence of oxygen caused the underlying soil to become anoxic. As a result, many of the willows upstream of Garrison became dormant and eventually died. These fields of dormant and then dead stems mostly remained in place, protecting each other from minor floods until recent decades. Unfortunately, many of the dead willow stems have now been removed for various anthropogenic reasons. As a consequence, the floodplain of the Clark Fork in the Deer Lodge Valley has become vulnerable over the past several decades to rapid surface erosion and eventual unraveling during a multidecadal recurrence-interval flood. It is worth noting here that there have not been any significant overbank flows on the Clark Fork in the Deer Lodge Valley since most of the dead stems in the upper reaches were removed. Thus, experimental confirmation of the Smith (2004) prediction that the system will unravel is still awaiting the next multidecade recurrence-interval flood.

The slope of the meander belt of the Clark Fork through the Deer Lodge Valley is 0.0025. Calculations showing how the average shear stress on the irregular, shrub-studded floodplain (τ_η) varies with stem density and flood recurrence interval for this slope and the cross-sectional geometry at the Galen gauge (located in Fig. 1) are presented in Fig. 3. The critical shear stress values for noncohesive soil (0.3 N m^{-2}) and bare tailings (1.0 N m^{-2}) along the Clark Fork are shown along with the boundary shear stresses on the actual soil surface in Fig. 3. This figure clearly indicates the transition from depth-independent flow (on the left side of the figure) to stem density–independent flow (on the right side of the figure). For reference, the nondimensional shrub spacing during the 1908 flood was estimated by Smith (2004) to be about 2.0, which is equivalent to a mean stem spacing

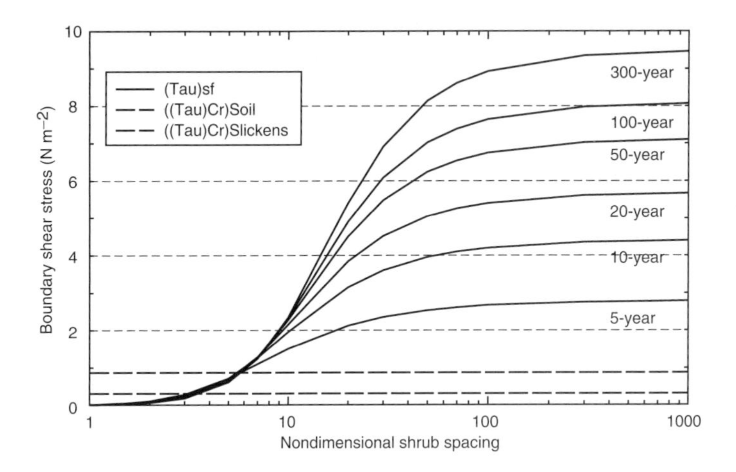

FIG. 3 Average shear stress on the actual boundary as a function of nondimensional shrub spacing for floods with recurrence intervals ranging from 5 to 300 years (from Smith, 2004). Shrub spacing is nondimensionalized by the mean shrub diameter in the meander belt. For nondimensional shrub spacings of less than 8 and flood recurrence intervals of more than 10 years, the average shear stress on the actual flow boundary (as opposed to that on the average flow boundary, which includes the form drag on the boundary irregularities) is essentially independent of discharge and flow depth. The horizontal lines on the lower part of the diagram represent the critical shear stresses for floodplain soil (lower) and tailings (upper). The former is exceeded once the nondimensional shrub spacing becomes larger than 3.5, and the latter is exceeded once this parameter becomes greater than 5.3.

of 0.17 m. Note that for nondimensional shrub spacings (average shrub spacing divided by average single shrub stem group diameter minus 1) of less than about 8 and flood-frequency recurrence intervals longer than 10 years, the boundary shear stresses become nearly independent of recurrence interval (discharge and flow depth) and almost completely dependent on stem density. In the case of overbank velocity, the dependence on stage and discharge is completely replaced by a dependence on stem density in the high-stem-density and high-flow asymptotes. Stated in a different way, for a given stem density, the vertically averaged overbank flow velocity becomes constant and flow depth increases directly with unit discharge. The free surface elevations over the channel and over the floodplain must be nearly the same; therefore, this effect concentrates more of the overall discharge in the channels and increases the boundary shear stress on the channel bottom. An analogous but lesser concentration of flow in the channel occurs whenever a floodplain is shallower for the same roughness or rougher for the same depth than the channel.

Using input parameters of 1.35 m for the overbank flow depth next to the river at the current location of the Galen gauge and 130 m^3 s^{-1} for the discharge at this location, Smith (2004) found the calculated nondimensional shrub spacing to be somewhat less than he had sampled beneath the tailings at several sites. Moreover, Smith's local measurements of pre-flood shrub spacings were found to be similar to the present shrub spacings in areas of dense old shrubs with interacting canopies. Consequently, he allowed for possible errors in the estimated input parameters and concluded that the overbank flow depth next to the river probably was only 1.20 m and that the flood discharge during the multiday peak flow was more likely 147 $m^3 s^{-1}$. For this new case, the nondimensional shrub spacing of 2.0 is in agreement with the field measurements. As mentioned previously, the average stem spacing for this shrub spacing is 0.17 m. The vertically averaged velocity is 0.22 $m s^{-1}$, and the average shear stress on the irregular boundary (skin friction) is 0.034 $N m^{-2}$. The increase in discharge from 130 $m^3 s^{-1}$ to 147 $m^3 s^{-1}$ increases the recurrence interval for the 1908 flood at the Galen gauge site from 270 years to 490 years.

In summary, our investigation of the Clark Fork through the Deer Lodge Valley shows that several major changes have occurred during the past 200 years. From a geomorphic point of view, the dominant changes have been the loss of the beaver in the early 1800s, the mortality of the shrubs in the early 1900s, and the more recent loss of the dead stems. From an environmental point of view, the loss of the shrubs and dead stems has transformed a robust fluvial geomorphic system into one that is vulnerable to unraveling during a major flood. The dense shrub carr promoted deposition of a thick layer of metals-contaminated mine tailings during the 1908 flood. In contrast, that same flood now would unravel the entire meander belt owing to the current sparse riparian shrub flora in it. Once unraveled, the only way to transform the fluvial system back to a single threaded meandering geometry would be to reestablish a dense willow carr on the braid plain, a process that unaided would take many decades (Smith, personal observations on Plum Creek; Friedman *et al.*, 1996). Once willows started to encroach onto the braid plain, beaver inevitably would, if uninhibited by human interference, inhabit the new fluvial system, build dam complexes, and accelerate the recovery process by distributing water and willow shoots across the braid plain during the low flow times of year. The specific question with regard to recovery of the Clark Fork in the Deer Lodge Valley, however, is whether willow growth would be drastically inhibited by the metals-contaminated sediment that would have been brought into the braided river channel as the floodplain unraveled and that would thereafter reside in high concentrations in the bed of the river. If willow growth were inhibited by the high concentrations of metals in the bed of the river, then it could take a very long time indeed for the system to recover.

Good agreement between measured and modeled tailings-deposit thicknesses and reasonable predicted shrub densities during the 1908 flood demonstrate that the floodplain modeling procedure using measured or estimated stem densities is likely to be quite accurate. Consequently, the predicted unraveling of the Clark Fork in the Deer Lodge Valley appears to be very likely in a period that is shorter than the period that will be required for recovery of the fluvial system under the best of conditions. This history of the Clark Fork in the Deer Lodge Valley shows that a fluvial system, which once was so robust that it caused several decimeters of fine sediment to be deposited during a single 5-day-long, multicentury recurrence–interval flood can be made vulnerable to unraveling by the removal of its riparian shrub flora. The next issues to address are first whether unraveling has been observed in a fluvial system, and second, if so, whether a model of the type used to predict unraveling of the Clark Fork in the Deer Lodge Valley can predict the location (shrub density) at which this unraveling began. In the early 19th century, the Clark Fork of the Columbia River in the Deer Lodge Valley was bordered by a beaver-maintained willow carr, and its ability to withstand the 1908 flood was a direct result of this history. A third issue, therefore, is the role that beaver-maintained willow carrs play in preventing this type of geomorphic unraveling. A fourth issue is what happens to healthy beaver pond complexes during large floods, and a fifth issue is what role grazing ungulates play in promoting the initiation of unraveling.

Plum Creek

Plum Creek is a small stream that flows approximately northward into the South Platte River south of Denver, Colorado (Fig. 4). It has two major branches, called East and West Plum Creek. On June 16, 1965, a very heavy, several-hour long rainfall event in the Plum Creek drainage basin caused a major flood on this stream. The flood affected both branches, but it was more severe on East Plum Creek. The Front Range of the Rocky Mountains in the vicinity of Plum Creek is aligned south-southeast to north-northwest, but there is an east-west trending ridge protruding from the Front Range between Denver and Colorado Springs called the Palmer Divide (Fig. 4, *bottom right*). This small topographical feature has a strong impact on the local meteorology, and particularly on precipitation. During the June 16 flood, the fluvial geomorphology of both branches of Plum

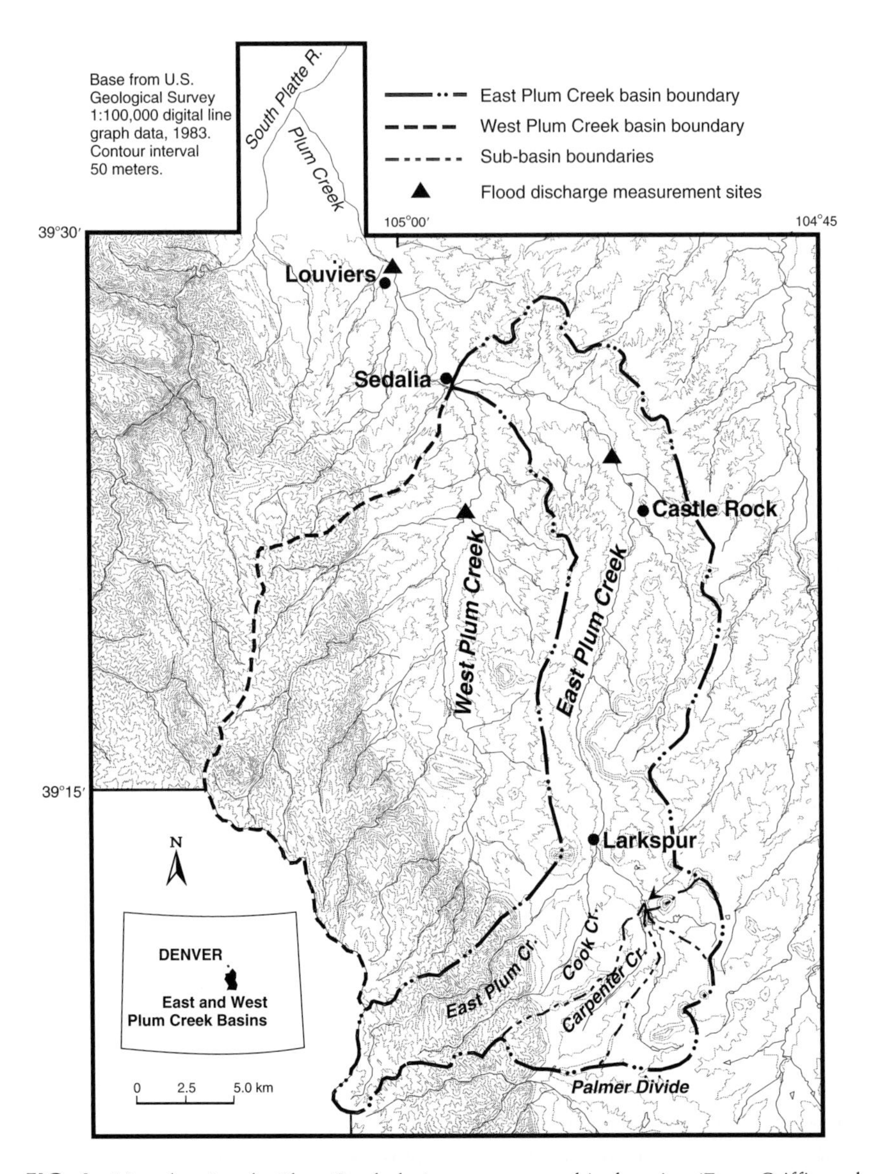

FIG. 4 Map showing the Plum Creek drainage system and its location (From Griffin and Smith [2004]). Of greatest interest in this chapter are East Plum Creek upstream (south) of Castle Rock and the headwater tributaries of East Plum Creek south of Larkspur. These headwater tributaries flow northward from high on the Palmer Divide. The Rocky Mountain Front is shown by a change in the closeness of the contours.

Creek was altered from that of a narrow, sinuous stream with a broad floodplain to a broad straight stream bounded by old terraces (Osterkamp and Costa, 1987; Griffin and Smith, 2004). The change of a stream or river from one geomorphic classification to another over the course of a single flow event is called unraveling (Smith, 2001). During this short but extreme unraveling event, the width of East Plum Creek increased by a factor of about 10. This increase in width was caused by erosion of the floodplain surface and associated filling of the original channel with sand from the floodplain, not by bank erosion.

The headwater tributaries of East Plum Creek begin on the Palmer Divide and flow northward toward Denver. One of these tributaries of East Plum Creek, Carpenter Creek, flows for a short distance next to Interstate Highway 25 (I-25). Moreover, Carpenter Creek has an unnamed headwater tributary that also flows along I-25. Two days after the 1965 flood on Plum Creek, a contractor for the Colorado Department of Transportation obtained a set of aerial photographs to facilitate assessment of the considerable flood damage to the highways and bridges. This set of aerial photographs shows that Carpenter Creek and this unnamed tributary to the east (Fig. 4) had both eroded their floodplains upstream of the confluence but that the floodplain of the unnamed tributary remained intact only a short distance farther upstream. A second unnamed tributary, with a floodplain that remained intact, is also shown in Fig. 4 coming into the first one just upstream of where the first one began to unravel. Consequently, the aerial photographs showing both the intact and unraveled segments of the unnamed tributaries provided an ideal opportunity for Smith (2001) and Griffin and Smith (2004) to test a process-based, predictive model of floodplain stability.

A meandering stream system unravels when the boundary shear stress on the floodplain, caused by the overbank flow, exceeds the erosion threshold of the floodplain ($\tau_\eta > \tau_{c\eta}$). When this occurs, down-valley oriented channels form on the tabs of floodplain that are inside the curving river channel. As the floodplain channels form and deepen, the flow depth (h) over them increases; therefore, the shear stresses on the floodplain-channel bottoms increase. (In the more or less straight, down-valley-oriented floodplain channels there are no shrubs, so $\tau_\eta \approx \rho g h S$, where ρ is the fluid density, g is the acceleration caused by gravity, h is the flow depth, and S is the down-valley slope of the floodplain tab.) The increase in shear stress on the channel bottom increases the rates of erosion and the sediment export ($Q_s \sim [\tau_\eta - \tau_{c\eta}]^\mu$, where $\mu \geq 3/2$), further widening and deepening the floodplain channels. In addition, the sediment transported out of these floodplain channels is added to the main river channel at a rate that exceeds the capacity of the river to transport it downstream, causing most of it to be deposited on the riverbed. As the river channel fills with sediment, more of the floodwater goes overbank, further increasing the depth of the overbank flow and, hence, the boundary shear stress on the floodplain and in the evolving floodplain channels. On true floodplains, which are composed of fine sediment deposited from weak overbank flows, this process goes on until the new, shallower river channel is the width of the meander belt. Once this new, geomorphically stable condition is reached, the river widens only slowly through erosion of its new banks, which are composed of old channel and floodplain deposits or of old terrace deposits. Bank erosion is a much slower process than the catastrophic floodplain erosion that leads to unraveling, particularly when the banks are against terraces and large amounts of material have to be removed from the toes of the banks by the flow for each increment of increase in width.

The effects of floodplain unraveling during a large flood can be seen clearly in Fig. 5. This figure shows two aerial photographs of the same section of East Plum Creek, located between 5.7 and 7.3 km south of Castle Rock (Fig. 4). The photograph on the left was taken in 1956, whereas the one on the right was taken on June 18, 1965, two days after the flood. The solid lines on the post-flood aerial photograph indicate the locations of the banks of the pre-flood channel. On the left photograph, this same channel is seen as a whitish sinuous line surrounded by a darkened tree- and shrub-covered meander belt. The trees, mostly medium-sized cottonwoods, were spaced too far apart, and the understory shrubs were too small and too far apart to protect the meander belt from unraveling during the flood peak. As the sandy floodplain eroded, most of these trees and shrubs were washed downstream. Outside the meander belt are terraces that are mostly covered with grass. Note that the width of the channel increased by an order of magnitude (an average for the reach of 7.3–96 m) and that the sinuosity decreased substantially relative to that of the pre-flood channel (1.25–1.05) during this several-hour long high-flow event. At one site near the bottom of the aerial photograph and at another near its top, the post-flood channel has even cut across large grassy tabs of floodplain and terrace, leaving the sediment-filled pre-flood channel as part of the new floodplain. This can be seen more clearly in Fig. 6, which magnifies the lower part of Fig. 5. The sandy channel bed at these two locations, as well as in most of the channel, has become flat because of the action of antidunes, the surface expressions of which are

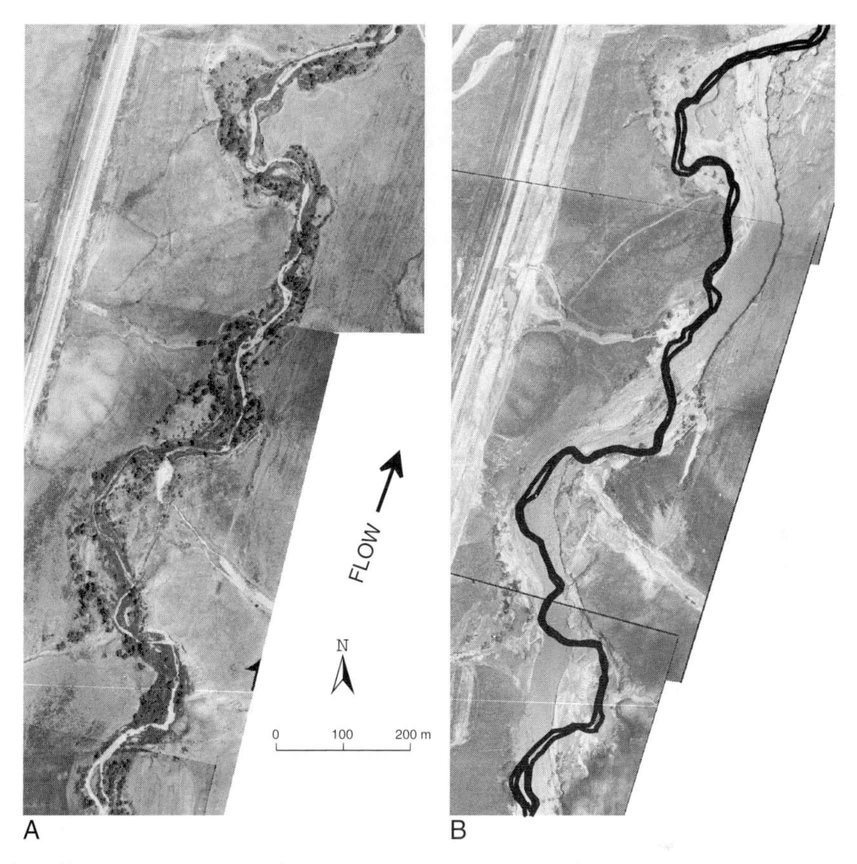

FIG. 5 Aerial photographs showing the geomorphic effects of the 1965 flood on East Plum Creek 7.3 to 5.7 km south of Castle Rock. The aerial photograph on the left was taken in 1956, nearly a decade before the flood, whereas the one on the right was taken 2 days after the flood. This figure demonstrates the concept of floodplain unraveling. The original channel is shown as a sinuous white line on the left panel meandering through a tree- and shrub-studded floodplain. On the right panel, two approximately parallel solid black lines show this channel. The actual channel after the flood fills the entire floodplain and has eroded into the terraces. Note that the original channel has been completely filled with sediment and, in two places, is not even near the post-flood channel, which is less sinuous and an order of magnitude wider. Figs. 6 and 7 show the southern part of this reach in greater detail. (From Griffin and Smith [2004].)

still clearly visible in the new channel on the left side of Fig. 6. Fig. 7 shows the same area as Fig. 6 before the flood. The dashed lines in Fig. 7 show the area affected by the flood. Note that there were over a dozen moderate to large cottonwood trees on the floodplain tab now cut by the channel, most of which were removed from the floodplain and transported out of the area by the flood.

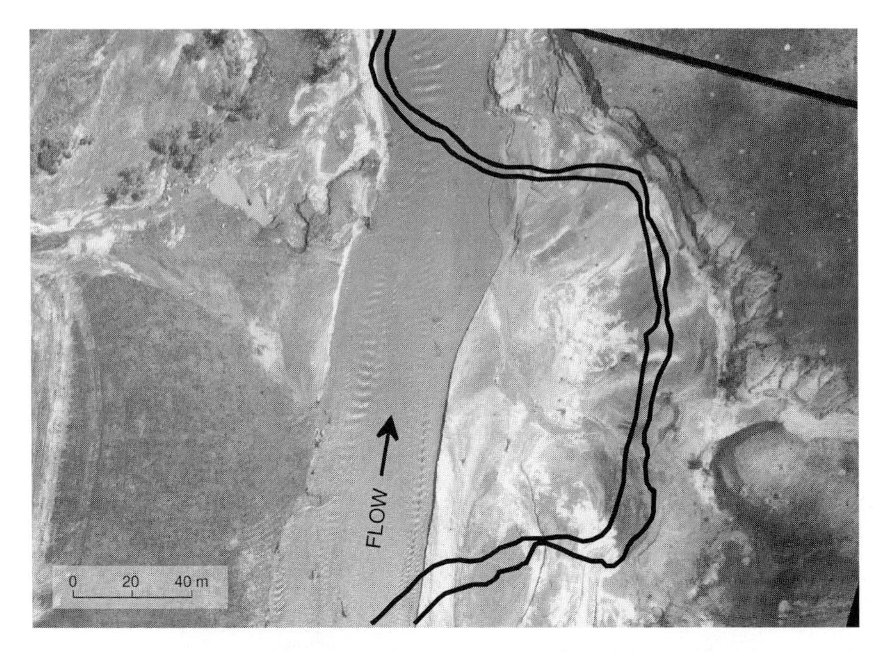

FIG. 6 Section of the southern part of Fig. 5 showing where the new channel cut straight through a low terrace during the June 16, 1965, flood. Note the surface expressions of antidunes in the new channel, despite the waning flow. The old channel has been completely filled with sand, and large flow-sculpted sand deposits have obliterated its trace, while the new channel has evolved a new hydraulic geometry in equilibrium with the high flow. Also note that all the trees and shrubs are gone from the old floodplain and have been removed from the area.

The 1908 flood on the Clark Fork of the Columbia River caused massive deposition of fine sediment through the Deer Lodge Valley, whereas the 1965 flood on Plum Creek caused almost complete erosion of its floodplain. The difference between the two cases lies in the nature and state of the vegetation on the two different floodplain surfaces. In the Clark Fork case, a dense shrub carr protected the floodplain in the Deer Lodge Valley from erosion, causing the shear stress on the floodplain surface to be much less than the critical shear stress for erosion of that surface ($\tau_\eta < \tau_{c\eta}$). In contrast, along Plum Creek, the trees alone were not dense enough to protect the floodplain surface from erosion during the 1965 flood, and the understory shrubs were overtopped by several meters of water, producing boundary shear stresses that were so high that the added flow resistance of the shrubs was unable to reduce these shear stresses below the critical value for erosion of the floodplain surface. That is, along Plum Creek, the boundary shear stresses resulting from the overbank flow exceeded the critical shear stress

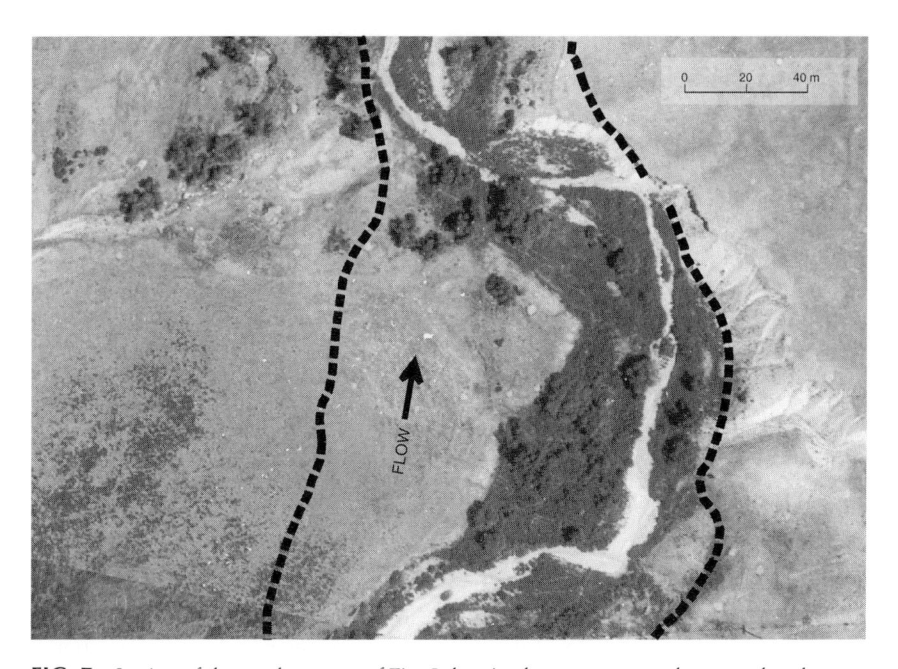

FIG. 7 Section of the southern part of Fig. 5 showing large cottonwood trees and understory shrubs on the pre-1965 floodplain. Owing to the sparseness of the cottonwood trees and the small stature of the understory shrubs, the woody vegetation on this part of the floodplain was completely removed by the 1965 flood. This figure also shows the low grass-covered terrace that was eroded by the flood and that can be observed in the previous figure to be occupied by the post-flood channel.

for erosion of the floodplain surface ($\tau_\eta > \tau_{c\eta}$) in spite of the vegetation. Fig. 7 clearly shows that there were trees and understory shrubs present in the meander belt, whereas Fig. 6 shows that after the flood they were gone. Once uprooted, these trees became tools for crushing and ripping up shrubs on the floodplain further downstream and for producing debris dams.

To quantify the differences between these two situations, an accurate mathematical model for overbank flow through shrub carrs that have been overtopped by the flow is necessary. Such a model was developed by Smith (2001). In order to test this model, sites both upstream and downstream of the location of the initiation of unraveling had to be examined and modeled by using pre-flood shrub characteristics. This was made possible by the unique set of aerial photographs of an unnamed headwater tributary of East Plum Creek that, for the most part, did not unravel. The unnamed tributary that did not unravel is bordered entirely by a sandbar-willow carr, which was maintained by beaver. Comparisons of these aerial photographs with field data collected in 2000 confirmed that the shrub densities and geomorphic situations were essentially the same now as they were before the flood (Griffin and Smith, 2004). The original analyses (Smith, 2001; Griffin and Smith, 2004) treated the reaches with beaver ponds as ones with reduced average shrub densities and approximated the bed roughness caused by the beaver dams using large dune-shaped obstructions on the surface of the floodplain. These papers did not examine the interplay of the flow, the beaver ponds, and the beaver dams in detail.

THEORY FOR INTERACTION OF FLOW AND SHRUBS

Rigid Stems that Protrude Through the Surface of the Water

Many of the species of willow shrubs that are found adjacent to beaver ponds and on beaver dams have a clump of stems emanating from a single root crown. The Bebb, Geyer, and Yellow willows that populate the floodplain of the Clark Fork in the Deer Lodge Valley are of this type. Typically these types of willows occupy areas where the overbank flow spreads laterally across a broad floodplain or over a beaver pond complex; therefore, they are not likely to be overtopped by even very large floods. Consequently, these "clumped willows" have evolved relatively rigid stems and are protected from floods by growing in large patches and distributing the force on the stems caused by high flows more or less uniformly over the whole patch. When the flow depth is less than one third or so of the height of these shrubs, the drag on their stems can be well approximated by that on a rigid field of circular cylinders with an appropriate spacing and average diameter. Smith (2004) presented a comprehensive model for this case, and the equations for it are a special case of those presented in the next section. Calculations for the hydraulic conditions of the unnamed tributary of Carpenter Creek using both a one-layer model and a three-layer model are compared later in this chapter, before I apply the three-layer model to the 1965 Plum Creek flood.

Flexible Stems That Do Not Protrude Through the Surface of the Water

Formulation of the Problem of Flow through a Field of Bent-over Shrubs

To determine the characteristics of deep overbank flows through dense carrs dominated by sandbar willows (*Salix exigua* Nutt.), such as was the situation in the headwater tributaries of East Plum Creek in 1965, a more general flow model is required than was used for the Clark Fork in the Deer Lodge Valley. In contrast to the clumped willows that are described in the previous section, willow species that are frequently and substantially overtopped by floods produce sprouts from long rhizomes and often produce flexible stems that bend over when subjected to high flows. In this manner, both the cross-sectional area exposed to the flow and the reference velocity for the drag force are decreased, substantially reducing the drag on each clone. Sandbar willows are of this type. They can be spaced relatively far apart, or they can grow in close proximity to each other. When in a patch, these willows bend over as a group. This produces a low-velocity zone near the bed and low boundary shear stresses on the bed, thereby protecting the sediment bed from erosion and the willows and understory plants from being uprooted. To model the flow through a patch of sandbar willows requires three primary layers: (1) a bottom layer in which the flow velocity is substantially reduced by the drag on all of the quasi-vertically oriented stems, (2) a middle layer containing the bent-over stems in which there is a very high drag force and a resulting large change in velocity with distance from the stream bed, and (3) a high-velocity outer flow over the tops of the bent-over plants.

The broad theoretical approach used here to develop a mathematical model for overbank flow through sandbar willow carrs is to (1) derive equations for each of the three vertical layers in the flow and couple them appropriately; (2) integrate the coupled set of equations to get the unit discharge ($Q = \langle u \rangle h$, where $\langle u \rangle$ is the vertically averaged flow velocity and h is the local flow depth); (3) calculate a slowly varying, spatially dependent friction coefficient ($\beta_f = \langle u \rangle / u_*$, where $u_* = (\tau_b/\rho)^{1/2}$ is the local shear velocity, τ_b is the boundary shear stress on the average floodplain surface, as calculated from streamwise-averaged equations of motion, and ρ is the fluid density); and (4) substitute $\beta_f u_*$ for $\langle u \rangle$ in a quasi-steady, hydrostatic, cross-sectionally integrated, one-dimensional numerical model or in a quasi-steady, hydrostatic, vertically integrated two-dimensional numerical model. The coupled equations also can be used in a reach-averaged manner, either with or without spatial accelerations. This set of procedures produces a closed set of equations that provides estimates of all flow variables at every point in the moving fluid, as is often required for sediment transport and ecological calculations. Smith and McLean (1984) used an analogous approach, with β_f based on quasi-logarithmic velocity profiles, to close their model for flow in meandering rivers.

The most important topographical features on the floodplain of the unnamed tributary are beaver dams; they are too steep to be treated explicitly with a hydrostatic flow model. Instead, either a fully three-dimensional, nonhydrostatic model has to be used at great computational expense, or the beaver dams have to be parameterized as roughness elements. The latter approach is pursued here. The effects of the shrubs on the friction coefficient β_f are very large in the overbank flow of concern in this chapter; therefore, friction dominates the force balance. Once a decision to treat the beaver dams as roughness elements, rather than as discrete topographical features in a three-dimensional model, has been made, local accelerations become minor components of the pertinent fluid mechanics and the flow can be approximated at most locations as quasi-uniform. This is consistent with the original formulation of the problem in which the vertically averaged flow velocity is related to the local boundary shear stress. By streamwise averaging of the momentum equations, the water surface and the floodplain become parallel, both being tilted in the down valley direction through a small angle, α. This means that form drag on all the irregularities of the floodplain surface are included in the boundary shear stress (τ_b). Dividing the net form drag on the floodplain surface by the area of that surface produces a shear stress–like variable that we might call "the component of shear stress on the boundary caused by form drag," or more simply the "drag stress." The latter situation arises even if the averaging process is considered a filter and weak accelerations are retained in the analysis. It should be emphasized here, however, that the present approach makes strong use of the slowly varying nature of the hydrostatic component of natural overbank flows, and it is not appropriate for flows that vary rapidly in the streamwise direction.

The three distinct layers of the model extend from the floodplain to a level denoted s_1, from s_1 to s_2, and from s_2 to a free surface at s_3. The variables in each layer are denoted by the subscript appropriate for the top of the layer, for example, $<u_1>$, $<u_2>$, and $<u_3>$ are the velocities averaged over the thicknesses of the first ($h_1 = s_1 - s_0$), second ($h_2 = s_2 - s_1$), and third ($h_3 = s_3 - s_2$) layers, respectively. The lowest layer has in it the measured number of stems per unit area in each size class of the measured stem–diameter probability distribution. We measured these diameters from 0.10 to 0.20 m above the mean floodplain surface. These stems are bent in circular arcs, with the tops of the willows being pulled directly downstream from the center of the middle layer. The circular arc, R_k, made by a bent live stem in the Plum Creek sandbar willow carr, was measured to be $\gamma_{RS}D_k$, where D_k is the diameter of a stem in a particular size class denoted by the subscript k and $\gamma_{RS} \cong 20$. The coefficient, γ_{RS}, in this expression depends, to some degree, on the force exerted on the top of the plant and the maturity of the plant. In a flow, if the larger stems bent only at $R_k = 20D_k$, they would stick further up in the fluid and have a higher drag on them. This would cause them to be bent further down until submerged in the main group. Similarly, the smaller diameter stems in that flow would reside lower in the fluid and have less drag on them. Therefore, they would bend over only enough to be submerged in the main group. As a consequence, the center of the middle layer was modeled as $20D_s$, where D_s is the mean diameter of the stems in the size distribution.

The drag on each of these stems depends on its projected area perpendicular to the flow; therefore, the stems can be modeled as circular cylinders extending vertically from the floodplain to s_1. The Reynolds numbers for the sandbar willow stems and branches subjected to the 1965 flood are such that they fall in the Reynolds similarity range, so the drag coefficient for them is independent of Reynolds number and has been set at the value for a circular cylinder (1.2). For smaller stems and branches in other situations an iterative correction can be made, starting with $C_D = 1.2$, to include any Reynolds number dependence of the drag coefficient. Owing to the small scale of the wake turbulence in the bottom layer, the eddy viscosity in the first layer is still that given by the "law of the wall." This could be matched at $z_m = 0.200$ h to a constant eddy viscosity–derived velocity profile for the outer part of the first layer, but the simplest and most convenient "shrub drag model" is to employ the law of the wall throughout the lowest layer. The hybrid approach used by Smith (2001) also might be employed, but it is inconsistent with the way the problem is posed in this paper, and it is no better than using the law of the wall throughout the lower layer. Smith (2001) used the square of the matched velocity in the bottom layer ($\langle u_1 \rangle^2 = ((u_*)_1/k)^2((\ln(h_1/z_0)) - 0.743)^2)$ instead of $\langle (u_1)^2 \rangle = ((u_*)_1/k)^2(\ln^2(h_1/z_0) - 2\ln(h_1/z_0) + 2)$ in the drag equation as employed in this chapter. The eddy viscosity for the second layer is essentially that for the interior region of a steady horizontally uniform channel flow. Note that the processes in the middle layer effectively cap the bottom layer at s_1 and require a surface stress to be imposed on the middle layer from the top layer at s_2.

As the stem spacing gets large, the velocity profile from this three-layer model will not become the same as that for an undisturbed flow because the layering and matching in the three-layer situation remain the same for all shrub spacings and that layering and matching are different from the layering and matching in the proper undisturbed flow velocity profile. This problem must be corrected by forcing the velocities from the constant eddy viscosity parts of the three-layer velocity field to match the velocities from the more appropriate one-layer model at each of the layer boundaries. Furthermore, the multiplicative corrections for the constant eddy viscosity parts of the velocity field generated in this manner must be applied to calculations for all stem spacings. In essence, this correction reduces the β's in the constant eddy viscosities to values appropriate for the more accurate velocity profile.

The middle layer contains an assumed uniformly distributed group of streamline-parallel stems from which protrude an assumed uniformly distributed group of branches. This layer also contains leaves that are assumed to be parallel to the stream surfaces and flowers (catkins). Owing to their prescribed orientation, there is only skin friction on most of the lengths of the stems [the height of an average sandbar willow ($100D_s$) minus its radius of curvature ($20D_s$)] and on the leaves, but there is substantial form drag on the numerous branches and on the catkins. There also is form drag on the tops of the circular arcs made by the bent stems. Like the stems in the lowest layer, the branches are projected onto a plane perpendicular to the flow and a drag coefficient of 1.2 is assigned to them. The thickness of the middle layer depends on the distance to which the branches of the sandbar willows extend above and below the stems when in the flow. This varies to some degree with flow velocity, but depends mostly on the diameters of the primary branches. For the unnamed tributary of Carpenter Creek, the diameters of the primary and secondary branches of the sandbar willows were measured 0.05 m above their bases and the mean primary branch diameter was found to be $D_{B1} = \gamma_{RB1} D_S$, where $\gamma_{RB1} = 0.21$. For the secondary branches measured in the same way, $D_{B2} = \gamma_{RB2} D_S$, where $\gamma_{RB2} = 0.04$. These values give $R_B = 5Ds$ for half the thickness of the second layer; therefore, $s_1 = 15\ D_s$ and $s_2 = 25\ D_s$. The mass of branches in this layer acts as a grid, producing substantial small-scale turbulence; however, much of this small-scale turbulence is damped by the leaves. Consequently, the dominant mixing still results from eddies the scale of the layer thickness. Owing to the large surface stress exerted on the top of the middle layer (s_2) by the flow in the top layer and the low velocities and shear stresses in the bottom layer, the stress divergence across the middle layer dominates over the gravitational driving force per unit volume in that layer, but the drag force in the layer caused by the stems, branches, catkins, and leaves produces what acts like a negative driving force.

The top layer contains no drag in its interior. The eddy viscosity for this layer is that of the outer part of a channel flow of depth $h_3 = s_3 - s_2$, with a shear velocity $(u_*)_3$ calculated from the shear stress at the base of the layer. Velocities and shear stresses have to be matched at s_1 and s_2. There is no velocity at s_0, but the shear stress (i.e., the boundary shear stress on the average floodplain surface for the spatially averaged flow) must be calculated at this level. In contrast, there is no stress at s_3, but the velocity (i.e., the surface velocity) may need to be calculated at this level.

In a spatially averaged (or spatially filtered) flow the stress divergence balances the downstream component of the fluid weight per unit volume. In the present case, this force per unit volume balance needs to be modified in the bottom and middle layers by subtracting the spatially varying drag forces per unit volume on the pertinent anatomical features of the trees and shrubs, as described above. In addition, form drag on the topographical features of the bed, including the beaver dams, needs to be added to the model. For the topographical features on the bed that are associated with the shrubs, this is done by using a long-spacing modification of the method of Kean and Smith (2005), which is an extension of the method of Smith and McLean (1977) developed for ripples and dunes. The method of Kean and Smith (2005) employs Gaussian shapes rather than dune shapes so that simple geometric features of the bed can be modeled using three geometric parameters (widths and heights of the features and the spacings between features) rather than two (height and wavelength) as needed for a ripple or dune field. Soil, organic debris, and leaf litter build up under the groups of sandbar willow stems and are scoured out from between these stem clumps; therefore, the spacing of the topographical elements caused by the stem clumps and, thus, comprising the bed roughness for this model is taken as the distance between clump centers. It is assumed that the leaf litter washes off the tops of the more solid piles of organic debris and soil during overbank flows, and, therefore, it is not part of the high-flow topographical elements. The geometry of the solid debris piles under the stem clumps at Plum Creek was such that the widths of the piles (equivalent to two standard deviations on a normal probability distribution and denoted 2σ) averaged about one half of the average stem clump diameter (that is, $2\sigma_{TE} = D_G/2$). The heights (H_{TE}) of the topographical elements averaged about two tenths of the stem-clump diameter ($H_{TE} = 0.2D_G = 0.8\sigma_{TE}$). Therefore, assuming a Gaussian shape, $\sigma_{TE}/H_{TE} = 1.2$ and, according to Kean and Smith (2005), the undisturbed-flow drag coefficient for these topographical elements is 0.8. An undisturbed-flow drag coefficient can be used properly in such a calculation only when the correct reference velocity is used in the drag equation. This correct reference velocity is the velocity that would be present at the site of the bump if it were removed. In general, such a calculation requires inclusion of the wake and the two relevant internal boundary layers from the upstream topographical element. The small vertical extent of the innermost internal boundary layer permits it to be neglected, and the velocity profile in the wake is approximated in this chapter as is done by Smith and McLean (1977) in their dune-form-drag model.

The geometry of a sequence of beaver dams makes use of a model for flow over a series of asymmetric saw-tooth–shaped structures spaced like a train of fluvial dunes, a suitable approximation for their effects on the flow. Consequently, the discussion in this paragraph follows that of Smith and McLean (1977). A typical beaver dam has an apron that protects its upstream face from high flows and keeps the lower part of the dam from being eroded by tunneling. The apron is located between the dam and an upstream pool from which the mud was excavated, producing a ramp-like approach to the dam. The downstream face of a well-made beaver dam is very steep. The shear stress (τ_T) on the bottom of a flow with beaver dams of height H removed, but with the force on these dams included in the roughness parameter (z_{0T}), is

$$\tau_T = \tau_{AS} + \tau_D = \tau_{AS}(1 + \sigma_{SM}). \quad (1a)$$

Therefore,

$$\tau_{AS} = \frac{\tau_T}{(1+\sigma_{SM})}. \quad (1b)$$

In Equation (1), τ_{AS} is the streamwise average shear stress on the actual bottom of the flow including the beaver dams, τ_D is the drag stress (drag force on a beaver dam, F_{DD}, divided by the spacing, λ_D, of the dams), and σ_{SM} is the "drag function" for the beaver dams. Writing the drag function in terms of a drag coefficient for an asymmetrical saw-tooth ($C_{DD} = 0.4$) and the square of the velocity that would be present at the location of the beaver dam, were the dam removed, divided τ_{AS}, gives

$$\sigma_{SM} = \frac{C_{DD}}{2k^2}\left[\ln^2\left(\frac{H}{z_0}\right) - 2\ln\left(\frac{H}{z_0}\right) + 2\right]\left(\frac{H}{\lambda_D}\right). \quad (1c)$$

Using this notation the drag force on the dam per unit width of the flow is

$$F_{DD} = \tau_{AS}\sigma_{SM}\lambda_D = \tau_T\frac{\sigma_{SM}}{(1+\sigma_{SM})}\lambda_D \quad (2)$$

and can be calculated once τ_T has been calculated.

The downstream beaver dam is in an internal boundary layer resulting from flow separation at the crest of the upstream dam and flow reattachment somewhat downstream of the base of the upstream dam. According to Elliot (1958), the thickness of the internal boundary layer (δ_I) grows as $z_{0AS}((x - x_0)/z_{0AS})^{4/5}$ where $(x - x_0)$ is the distance downstream of the point of reattachment and z_{0AS} is the roughness parameter for the actual surface excluding the drag on the beaver dams. The thickness of the internal boundary layer averaged over the spacing between the dams is

$$\delta_{av} \cong c_0 \left[(z_0)_{AS} \left(\frac{\lambda_D}{(z_0)_{AS}} \right)^{4/5} \right], \tag{3}$$

where $c_0 = 0.1$ is the coefficient that Smith and McLean (1977) found for dunes. Matching logarithmic velocity profiles for the internal and outer boundary layers at δ_{av}, gives $(z_0)_T$ in terms of $\gamma_{SM} = (1+\sigma_{SM})^{-1/2}$ and $(z_0)_{AS}$ as

$$(z_0)_T = (z_0)_{AS} \left(\frac{\delta_{av}}{(z_0)_{AS}} \right)^{\gamma SM} . \tag{4}$$

The spatially averaged velocity in the internal boundary layer ($z < \delta_I$) caused by flow separation at the crest of each beaver dam is

$$u_I = \frac{(u_*)_{AS}}{k} \ln \left(\frac{z}{(z_0)_{AS}} \right). \tag{5}$$

And the spatially averaged flow above the internal boundary layer ($z > \delta_I$) is

$$u_o = \frac{(u_*)_T}{k} \ln \left(\frac{z}{(z_0)_T} \right) \tag{6}$$

General Equations for a Multilayer Flow Problem

The equations used here are derived from a simpler but more consistent turbulence model than those employed by Smith (2001). Also, the analysis in this chapter is more complete and the equations are written in a simpler, more concise notation. However, the results from this new set of equations are not substantially different from those given by Smith (2001) and Griffin and Smith (2004).

In this section, ρ is the density of the fluid, g is the acceleration caused by gravity, θ is the upslope angle of the average bottom of the flow relative to the horizontal (so that the downhill component of the gravitational force per unit volume is $-\rho g(\sin\theta)$), z is the distance above the average bottom of the flow, z_i is the distance above the bottom of layer i, h_i is the thickness of layer i, $h = h_1 + h_2 + h_3$ is the flow depth, $(\tau_{zx})_i$ is the shear stress in layer i on planes parallel to the average bottom of the flow, $(\tau_{ib}) \equiv (\tau_{zx})_{ib} \equiv ((\tau_{zx})_i)_b$ is the shear stress on a plane at the base of layer i parallel to the average bottom of the flow, $(\tau_{ib}) \equiv (\tau_{zx})_{ib} \equiv ((\tau_{zx})_i)_b$ is the shear stress on the average bottom of the flow, $(\tau_{is}) \equiv (\tau_{zx})_{is} \equiv ((\tau_{zx})_i)_s$ is the shear stress on a plane at the top of layer i parallel to the average bottom of the flow, $k = 0.408$ is von Karman's constant, z_0 is the roughness parameter for the mean bed, $(\sigma_{Dj})_i$ is a drag function for anatomical component j of the woody vegetation in layer i, $(F_{Dj})_i$ is the drag force on anatomical component j of the woody vegetation in layer i, and $(\lambda_j)_i^2$ is the mean area affected by an anatomical component j of the woody vegetation in layer i. For example, $(\sigma_{DS})_1$ is the drag function for stems in layer 1, $(F_{DS})_1$ is the drag force on a stem in layer 1, and $(\lambda_S)_1^2$ is the mean area affected by a stem in layer 1. Similarly, $(\sigma_{DB})_2$ is the drag function for branches in layer 2 and $(F_{DB})_2$ is the drag force on a branch in layer 2. The drag coefficient for an anatomical component j of the woody vegetation is $(C_D)_j$.

The mean diameter of anatomical element j in layer i is denoted D_{ji}. The key geometric parameter associated with the shrubs in the drag function for layer i is D_{ji}/λ_{ji}^2. Even after the stems have been bent over by the flow, all the anatomical elements associated with one stem can be assigned to the layer in which they end up, and this can be done for each layer above the λs^2 for the bottom layer. Therefore, $D_{ji}/\lambda_{ji}^2 = N_{ji}D_{ji}/\lambda_s^2$, where N_{ji} is the number of anatomical elements j in layer i. For example, in layer 2 the "shrub geometry factor" for branches becomes $N_{B2}(D_{B2})/(\lambda_S)^2$, where N_{B2} is the number of branches in the second layer from a single stem. This also can be subdivided so that it represents the primary branches from a single stem in the second layer, in which case there have to be similar additional terms representing the secondary, tertiary, and further branches. The sum of these ratios then becomes the "shrub geometry factor" in $(\sigma_{DB})_2$.

Using the notation and concepts introduced in the previous two paragraphs, the drag function for anatomical element j in layer i is

$$(\sigma_{Dj})_i \equiv \frac{N_{ji}(F_{Dj})_i}{\lambda_s^2 h_i\left(\frac{(\tau_b)}{h_i}\right)}. \tag{7a}$$

Also, it is useful to write an expression for the sum of all anatomical elements in layer i. It is

$$(\sigma_T)_i \equiv \sum_{j=1}^{M} \frac{N_{ji}(F_{Dj})_i}{\lambda_s^2 h_i\left(\frac{(\tau_b)_i}{h_i}\right)} = \sum_{j=1}^{M}\left(\frac{C_{Dj}}{2k^2}\right)\left(\frac{k^2 < u_i^2 >}{(u_*)_i^2}\right)\left(\frac{N_{ji}D_{ij}}{\lambda_s^2}\right)(h_i). \tag{7b}$$

In Equation (7b), the square of the velocity averaged over the layer is denoted $<u_i^2>$.The second part of Equation (7b) contains four factors. The first is a product of coefficients. The second is the nondimensional average of the square of the velocity and can be denoted the "velocity factor." The third is the "shrub geometry factor" discussed in the previous paragraph. The fourth factor is the thickness of the layer.

When the flow is steady and horizontally uniform in layer i, the force balance per unit volume reduces to

$$\frac{\partial (\tau_{zx})_i}{\partial z} = \rho g (\sin\theta) + (\sigma_T)_i \frac{\rho (u_*)_i^2}{h_i}. \tag{8}$$

Recalling from a previous paragraph that the value of $(\tau_{zx})_i$ at the bottom surface of layer i is denoted by $(\tau_{zx})_{ib}$, the shear velocity $(u_*)_i$ for layer i can be defined as $((\tau_{zx})_{ib}/\rho)^{1/2}$. This shear velocity for layer i is the velocity scale for the turbulence in layer i. Integrating Equation (8) from z_i to s_i and noting that the constant of integration is $(\tau_{zx})_{is} - \rho g h_i(\sin\theta) - \rho(u_*)_i^2(\sigma_T)_i = (\tau_{zx})_{ib}$, gives

$$(\tau_{zx})_i = (\tau_{ib} - \tau_{is})\left(1 - \frac{z_i}{h_i}\right) + \tau_{is}, \tag{9a}$$

and

$$\tau_{ib} - \tau_{is} = -\rho g h_i (\sin\theta) - \rho(u_*)_i^2 (\sigma_T)_i. \tag{9b}$$

Equating τ_{ib} to $\rho(u_*)_i^2$ and rearranging Equation (9b) gives

$$\tau_{ib} = \frac{-\rho g h_i (\sin\theta) + \tau_{is}}{1 + (\sigma_T)_i}, \tag{10a}$$

and

$$(u_*)_i = \left(\frac{-g h_i (\sin\theta) + (\tau_{is}/\rho)}{1 + (\sigma_T)_{ib}}\right)^{1/2}. \tag{10b}$$

In Equation (9a) we see that the shear stress field is composed of two parts. The first term represents the gravitationally driven part of the stress field and is scaled by the stress difference across the layer. Owing to the decreasing thickness of the layer of fluid between z_i and s_i, as z_i increases, the shear stress resulting from this component of the flow decreases linearly to zero at s_i. The second term represents the top surface shear stress–driven component of the stress field. In contrast to the gravitationally driven component, the top surface shear stress–driven component remains the same throughout the layer. In a viscous flow with $(\sigma_T)_i = 0$, the first term would lead to a plane Poiseuille flow and the second would lead to a plane Couette flow (Kundu, 1990). In the turbulent flow of interest in this chapter it is also useful to consider briefly the resulting flow being composed of a gravitationally driven part and a top surface shear stress–driven part. Note here that the Couette-flow–producing term is independent of z_i; consequently, it gives a linear velocity profile for a constant eddy viscosity.

When $(\sigma_T)_i \neq$ zero, however, the term containing $(\sigma_T)_i$ produces what acts like a negative driving force per unit volume that resists the top surface shear stress–driven component of the flow. This term is represented by $\tau_{is}(1 - z_i/h_i)$ in Equation (9a). As a result, when $(\sigma_T)_i$ is nonzero, there can be no component of the shear stress field that is independent of z_i and no analogy to the plane Couette flow. That this is the case can be seen by rearranging Equation (3a) to separate the terms containing τ_{is} and τ_{ib} to get Equation (11) and then noting the linear z dependence of the τ_{is}–containing term.

$$(\tau_{zx})_i = \tau_{ib}\left(\left(1-\frac{z_i}{h_i}\right)+\left(\frac{\tau_{is}}{\tau_{ib}}\right)\left(\frac{z_i}{h_i}\right)\right) = \rho K_i \frac{\partial u_i}{\partial z_i}. \quad (11)$$

Stated in a different way, when $(\sigma_T)_i$ is zero $\tau_{ib} = \tau_{is}$ in the Couette flow analogy, but when $(\sigma_T)_i$ is nonzero $\tau_{is} > \tau_{ib}$ in the purely surface shear stress driven flow.

The shear stress $(\tau_{zx})_i$ in Equation (11) also has been equated to the rate of shear $(\partial u_i/\partial z_i)$ using an eddy viscosity (K_i) appropriate for the layer. For a steady, horizontally uniform flow in an infinitely wide channel the eddy viscosity can be expressed in two parts as follows:

$$K = ku_* z\left(\frac{\tau_{zx}}{\tau_b}\right), \quad \text{when} \quad z \leq z_m = 0.200h, \quad (12a)$$

and

$$K = \frac{ku_* h}{\beta}\left(\frac{\tau_b + \tau_s}{\tau_b}\right), \quad \text{where} \quad \beta = 6.24, \quad \text{when} \quad z \geq z_m = 0.200h. \quad (12b)$$

That the eddy viscosity can be best represented by a two-part profile of this form can be derived from a carefully crafted dimensional analysis and an assumption that the unknown functions of the pertinent shear stress ratios are linear. In both of these expressions, the velocity scale for the turbulent diffusion of momentum is represented by u_* times the appropriate shear stress ratio. The length scale for the turbulent diffusion of momentum differs between Equation (12a) and Equation (12b). In the near wall region represented by Equation (12a), it varies with distance from the solid boundary as kz, as in the law of the wall, whereas, further from the solid boundary, it must scale with h rather than z and is represented here as kh/β. Extracting h from the coefficient to get k/β is done for convenience. It is this difference in parameterization of the turbulent length scale that has required separation of the bottom layer of the flow into two parts when $\sigma_T = 0$. Rattray and Mitsuda (1974), using experimental data and a dimensional analysis that did not include the shear stress ratios, found β/k to be 15.6 using ($k = 0.40$, not 0.408) from which they calculated a matching level for the two eddy viscosity profiles of $z_m = 0.200h$. To get this value they examined a steady, horizontally uniform, gravitationally driven channel flow and matched the shear in the two resulting velocity profiles. The stress ratio must be included in Equation (12a) when considering density-stratified flows, and it had to be introduced into Equation (12b) in order to investigate the sandbar willow issue in this chapter owing to the top-surface shear stresses on the middle layers. Note that it disappears from (12b) when there is no top-surface stress. The value of von Karman's constant (k) was first determined experimentally by Nikuradse (1933) and later found by Long *et al.* (1993) to be 0.408 for hydraulically transitional and hydraulically rough flows.

Rattray and Mitsuda (1974) found that the logarithmic and parabolic velocity profiles resulting from the two-part eddy viscosity, discussed in the previous paragraph, should be matched at 0.200 times the flow depth in an unaccelerated flow in order for the shear rates to be continuous. This matching is not appropriate for the three-layer model discussed here because the layers and their flow properties are dictated by the characteristics of the shrubs. Consequently, in this chapter the shear rates will be allowed to be discontinuous at the layer boundaries in the three-layer model but, as mentioned in the previous section, a constraint on the velocity at each matching level and at the levels where the velocity is equivalent to the vertically averaged velocity for that layer ($0.42h_i$) must be applied to the three-layer model to replace this condition on the shear rates and to prevent the three-layer velocity field from going to the wrong asymptote as the shrub density goes to zero. The constraint is generated by forcing the velocities at the appropriate levels of the three-layer model in the no-shrub asymptote to agree with those calculated by using the 0.200h-matched velocity profile. This is accomplished after the layers of the three-layer model are established by letting the shrub density go to zero, then comparing the velocities at the appropriate levels in that particular uncorrected three-layer model to those calculated using the 0.200h matched-profile model. At each of these layers, a nearly depth-independent correction factor, which adjusts the values of velocity in the three-layer model to those given by the matched-profile model in this asymptote, can be generated for the constant eddy viscosity part of the velocity profile. These multiplicative corrections are specific to the layering in the three-layer model but apply to all shrub densities in that model. Typically the correction factor at the top of the first layer is zero or very small because most of the first layer has a logarithmic velocity structure in both profiles. However, the correction factor for the top of the second layer is very important, particularly as it affects the velocity in the interior of the second layer and the vertically averaged velocity in the third layer. The correction factor for the top of the third layer affects only the surface velocity.

For all but the bottom layer we can substitute Equation (12b), with the appropriate subscripts, into Equation (11) to get

$$(\tau_{zx})_i = \tau_{ib}\left(\left(1-\frac{z_i}{h_i}\right)+\left(\frac{\tau_{is}}{\tau_{ib}}\right)\left(\frac{z_i}{h_i}\right)\right) = \rho\frac{k}{\beta}(u_*)_i h_i \frac{(\tau_{ib}+\tau_{is})}{\tau_{ib}}\frac{\partial u_i}{\partial z_i}. \quad (13a)$$

Integrating Equation (13a) gives the velocity field for the layer.

$$u_i = \frac{(u_*)_i}{k}\left(\beta\left(\left(\frac{\tau_{ib}}{\tau_{ib}+\tau_{is}}\right)\left(\xi_i - \frac{\xi_i^2}{2}\right)+\left(\frac{\tau_{is}}{\tau_{ib}+\tau_{is}}\right)\left(\frac{\xi_i^2}{2}\right)\right)+\frac{ku_{ib}}{(u_*)_i}\right) \quad (13b)$$

In Equation (13b), $\xi_i \equiv z_i/h_i$ and u_{ib} is the velocity at the bottom of the layer, which must be equated to the velocity at the top of the layer below. The velocity at the top of layer i is

$$u_{is} = \frac{(u_*)_i}{k}\left(\frac{\beta}{2}\left(\left(\frac{\tau_{ib}}{\tau_{ib}+\tau_{is}}\right)+\left(\frac{\tau_{is}}{\tau_{ib}+\tau_{is}}\right)\right)+\frac{ku_{ib}}{(u_*)_i}\right). \quad (14a)$$

The vertically averaged velocity for layer i´th is

$$<u_i> = \frac{(u_*)_i}{k}\left(\frac{\beta}{3}\left(\left(\frac{\tau_{ib}}{\tau_{ib}+\tau_{is}}\right)+\frac{1}{2}\left(\frac{\tau_{is}}{\tau_{ib}+\tau_{is}}\right)\right)+\frac{ku_{ib}}{(u_*)_i}\right). \quad (14b)$$

The vertical average of the square of the velocity in layer i divided by the square of the shear velocity for that layer is

$$\frac{<u_i^2>}{(u_*)_i^2} = \left(\frac{1}{k}\right)^2\left(\beta^2\left(\frac{2}{15}\left(\frac{\tau_{ib}}{\tau_{ib}+\tau_{is}}\right)^2+\frac{3}{20}\left(\frac{\tau_{is}}{\tau_{ib}+\tau_{is}}\right)\left(\frac{\tau_{ib}}{\tau_{ib}+\tau_{is}}\right)+\frac{1}{20}\left(\frac{\tau_{is}}{\tau_{ib}+\tau_{is}}\right)^2\right)+\frac{2\beta}{3}\left(\left(\frac{\tau_{ib}}{\tau_{ib}+\tau_{is}}\right)+\frac{1}{2}\left(\frac{\tau_{is}}{\tau_{ib}+\tau_{is}}\right)\right)\left(\frac{ku_{ib}}{(u_*)_i}\right)+\left(\frac{ku_{ib}}{(u_*)_i}\right)^2\right). \quad (14c)$$

Equation (14a) is required to match the velocity at the base of the next layer upward; Equation (14b) times the thickness of the layer (h_i) gives the discharge per unit width (Q_i) for layer i, and Equation (14c) is needed to calculate the drag functions for the layer. The values of τ_{ib}, which are required to match the shear stresses at the tops of the next layers down, and $(u_*)_i$ are found from Equations (10a) and (10b).

The first layer is the only one with a solid boundary beneath it; therefore, it is the only one in which the turbulence structure is given by Equation (11). Substituting Equation (11), with the appropriate subscripts for its application to the first layer, into Equation (10) and then integrating the resulting expression from $z_1 = z_0$ to $z_1 \leq s_1$, gives

$$u_1 = \frac{(u_*)_1}{k}\left(\ln \frac{z_1}{z_o}\right). \tag{15}$$

The velocity at the top of the bottom layer is

$$u_1 = \frac{(u_*)_1}{k}\left(\ln \frac{h_1}{z_o}\right). \tag{16a}$$

The vertically averaged velocity for the first layer is

$$<u_1> = \frac{(u_*)_1}{k}\left[\left(\ln \frac{h_1}{z_o}\right) - 1\right]. \tag{16b}$$

Assuming $z_0/h_1 << 1$, the average of the square of the velocity for this lowest layer of thickness h_1 is

$$<u_1^2> = \frac{(u_*)_1^2}{k^2}\left[\left(\ln \frac{h_1}{z_o}\right)^2 - 2\left(\ln \frac{h_1}{z_o}\right) + 2\right] = \left(\frac{(u_*)_1}{k}\left[\left(\ln \frac{h_1}{z_o}\right) - 1\right]\right)^2 + 1 = <u_1>^2 + 1 \tag{16c}$$

Therefore, for the first layer, where there are only stems and $(\sigma_T)_1 = (\sigma_{DS})_1$,

$$(\sigma_{DS})_1 = \frac{C_{DS}}{2k^2}\left[\left(\ln \frac{h_1}{z_o}\right)^2 - 2\left(\ln \frac{h_1}{z_o}\right) + 2\right]\frac{h_1 D_S}{\lambda_S^2}. \tag{17}$$

Note from Equation (16c) that, when $\ln(h_1/z_0) >>1$, $<u_1>^2$ is a good approximation for $<u_1^2>$.

From Equation (10a) we get

$$\tau_b = \tau_{1b} = \frac{-\rho g h_1(\sin\theta) + \tau_{1s}}{1 + (\sigma_{DS})_1}, \tag{18a}$$

and from Equation(10b) we get

$$(u_*)_1 = \left(\frac{-g h_1(\sin\theta) + (\tau_{1s}/\rho)}{1 + (\sigma_{DS})_1}\right)^{1/2}. \tag{18b}$$

These two equations, however, are incomplete until $\tau_{1s} = \tau_{2b}$ is substituted into them. To find τ_{2b} we must carefully examine the mechanics of the second layer.

Application of the General Equations Derived above to a Three-Layer Flow Problem

In this model for flow in the unnamed tributary of Carpenter Creek during the 1965 Plum Creek flood, there are no shrubs in the top layer so σ_{T3} is zero. In addition, the flow is very deep, so the top layer is thick compared to the other two ($h_3 >> h_2, h_1$). This combination of conditions results in a high velocity and a large shear stress at the top of a relatively thin second layer. This middle layer contains the tops of the bent-over shrubs, which are composed of stems, branches, catkins, and leaves. The high velocities above this layer and the large number of anatomical elements in it produce a large drag stress in it. Consequently, the shear stress at its top is much greater than the shear stress at its bottom. The velocity similarly decreases substantially across the middle layer. The bottom layer contains only stems; because of the large shear across the middle layer, the bottom layer has a relatively low velocity and a lower drag stress in it.

In the second layer, where $(\sigma_T)_2$ is dominantly $(\sigma_{DB})_2$ and the length scale for the turbulent diffusion of momentum is kh/β, the shear stress at the bottom of the layer is, from Equation (10a),

$$\tau_{2b} = \frac{-\rho g h_2 (\sin\theta) + \tau_{2s}}{1 + (\sigma_T)_2}. \tag{19}$$

To specify τ_{2s} fully the shear stress at the bottom of the third layer must be used. Fortunately, the third layer contains no drag-producing elements (so $(\sigma_T)_3 = 0$) and has no shear stress on its top (so $\tau_{3s} = 0$); therefore,

$$\tau_{2s} = \tau_{3b} = -\rho g h_3 \, (\sin\theta). \tag{20}$$

Substituting Equation (20) into Equation (19) gives

$$\tau_{2b} = \frac{-\rho g (\sin\theta)(h_2 + h_3)}{1 + (\sigma_T)_2}. \tag{21}$$

The form of τ_{2b} given by Equation (21) is not conducive to the required iterative solution for the three layer problem and, thus, Equation (21) needs to be rearranged. This is done in Equation (22) by multiplying the left side of Equation (21) by the denominator of the right side, factoring τ_{1b} out of the second term, redefining $(\tau_{2b}/\tau_{1b})(\sigma_T)_2$ as $(\sigma'_T)_2$, and then substituting for τ_{1b} using Equation (18a). The advantage of the rearranged form will become obvious later in this paragraph.

$$\begin{aligned}
\tau_{2b}(1 + (\sigma_T)_2) &= \tau_{2b} + \tau_{2b}(\sigma_T)_2 = \tau_{2b} + \tau_{1b}\frac{\tau_{2b}}{\tau_{1b}}(\sigma_T)_2 \\
&= \tau_{2b} - \rho g (\sin\theta) h_1 \frac{(\sigma'_T)_2}{(1+(\sigma_{DS})_1)} + \tau_{2b}\frac{(\sigma'_T)_2}{(1+(\sigma_{DS})_1)} \\
&= \tau_{2b}\left(1 + \frac{(\sigma'_T)_2}{(1+(\sigma_{DS})_1)}\right) - \rho g (\sin\theta) h_1 \frac{(\sigma'_T)_2}{(1+(\sigma_{DS})_1)} \\
&= -\rho g (\sin\theta)(h_2 + h_3)
\end{aligned} \tag{22}$$

From the rearranged form of τ_{2b} in Equation (22) we can now obtain an equation in which $(\sigma'_T)_2$, and the shear velocity ratios imbedded in it explicitly cancel as their values get large. This equation is

$$\tau_{2b} = \left(1 + \frac{(\sigma'_T)_2}{(1+(\sigma'_{DS})_1)}\right) = -\rho g (\sin\theta)\left[(h_2 + h_3) - h_1 \frac{(\sigma'_T)_2}{(1+(\sigma'_{DS})_1)}\right]. \tag{23a}$$

This equation for τ_{2b} now becomes the focal point of the solution. The shear velocity for the second layer is

$$(u_*)_2 = \left[-g(\sin\theta) \left[(h_2 + h_3) - h_1 \frac{(\sigma'_T)_2}{(1 + (\sigma_{DS})_1)} \right] \right]^{1/2} \left(1 + \frac{(\sigma'_T)_2}{(1 + (\sigma_{DS})_1)} \right)^{-1/2}, \tag{23b}$$

and the expression for τ_{2b} from Equation (23a) can be substituted into Equations (18a,b) to get τ_{1b} and $(u_*)_1$.

For the second layer, the shear stress can be equated to the eddy viscosity times the rate of shear and integrated from $z_2 = s_2$ to $z_2 \leq s_3$ to get

$$u_2 = \frac{(u_*)_2}{k} \left(\beta \left(\left(\frac{\tau_{2b}}{\tau_{2b} + \tau_{2s}} \right) \left(\xi_2 - \frac{\xi_2^2}{2} \right) + \left(\frac{\tau_{2s}}{\tau_{2b} + \tau_{2s}} \right) \left(\frac{\xi_2^2}{2} \right) \right) + \frac{(u_*)_1}{(u_*)_2} ln \left(\frac{h_1}{z_0} \right) \right) \tag{24}$$

See Equation (13b). In Equation (24), $\xi_2 \equiv z_2/h_2$. As shown by Equation (14a), the velocity at the surface of the second layer is

$$u_{2s} = \frac{(u_*)_2}{k} \left(\frac{\beta}{2} \left[\left(\frac{\tau_{2b}}{\tau_{2b} + \tau_{2s}} \right) + \left(\frac{\tau_{2s}}{\tau_{2b} + \tau_{2s}} \right) \right] + \frac{(u_*)_1}{(u_*)_2} ln \left(\frac{h_1}{z_0} \right) \right). \tag{25a}$$

The vertically averaged velocity for the second layer, from Equation (14b), is

$$<u_2> = \frac{(u_*)_2}{k} \left(\frac{\beta}{3} \left(\left(\frac{\tau_{2b}}{\tau_{2b} + \tau_{2s}} \right) + \frac{1}{2} \left(\frac{\tau_{2b}}{\tau_{2b} + \tau_{2s}} \right) \right) + \frac{(u_*)_1}{(u_*)_2} ln \left(\frac{h_1}{z_0} \right) \right) \tag{25b}$$

The vertical average of the square of the velocity divided by the square of the shear velocity for the second layer is given by Equation (14c) to be

$$\frac{<u_2^2>}{(u_*)_2^2} = \left(\frac{1}{k} \right)^2 \left(\beta^2 \left(\frac{2}{15} \left(\frac{\tau_{2s}}{\tau_{2b} + \tau_{2s}} \right)^2 + \frac{3}{20} \left(\frac{\tau_{2s}}{\tau_{2b} + \tau_{2s}} \right) \left(\frac{\tau_{2b}}{\tau_{2b} + \tau_{2s}} \right) + \right. \right.$$

$$\left. \frac{1}{20} \left(\frac{\tau_{2s}}{\tau_{2b} + \tau_s} \right)^2 \right) + 2\frac{\beta}{3} \left(\left(\frac{\tau_{2b}}{\tau_{2b} + \tau_{2s}} \right) + \frac{1}{2} \left(\frac{\tau_{2s}}{\tau_{2b} + \tau_{2s}} \right) \right) \left(\frac{(u_*)_1}{(u_*)_2} ln \left(\frac{h_1}{z_0} \right) \right) +$$

$$\left. \frac{(u_*)_1}{(u_*)_2} ln \left(\frac{h_1}{z_0} \right) \right)^2. \tag{25c}$$

The average square of the velocity divided by the square of the shear velocity, given by Equation (25c), now must be substituted into

$$(\sigma_{DB})_2 = \frac{C_{DB}}{2k^2} \left(\frac{k^2 < u_2^2 >}{(u_*)_2^2} \right) \frac{(N_B D_{DB} + D_{DS}/2)}{\lambda_s^2} h_2, \tag{26}$$

which is obtained from Equation (7b), to get the drag function for the second layer. Smith (2001) has shown that σ_{SFS}, σ_{SFB}, and σ_{SFL} are all small compared to σ_{DB}; therefore, these terms have been dropped in this chapter. Similarly, because they are short and found only on the smallest twigs, drag on the catkins is small relative to that on the branches. Note that the drag function $(\sigma_{DB})_2$ is contained in and contains $(u_*)_2$, so Equation (26) must be determined as part of an iterative solution. Once a solution has been obtained it can be checked by calculating $(\sigma_T)_2$ from $(\sigma_T)_2'$ and then using Equation (21) to check the value of τ_{2b}.

The lower boundary of layer 3 is not solid; rather it is free to adjust to pressure and stress fluctuations caused by the turbulence field. This freedom eliminates the solid surface–caused turbulence represented by Equation (12a) near the bottom boundary of layer 3 and requires Equation (12b) to be used throughout the entire layer. The resulting velocity field, obtained from Equation (13b) with $\tau_{3s} = 0$, is

$$u_3 = \frac{(u_*)_3}{k}\beta\left(\xi_3 - \frac{\xi_3^2}{2}\right) + (u_3)_b. \tag{27}$$

The vertically averaged velocity for the third layer is

$$\langle u_3 \rangle = \frac{\beta (u_*)_3}{3k} + (u_3)_b. \tag{28}$$

The shear velocity for the third layer is obtained from Equation (22). It is

$$(u_*)_3 = (-gh_3(\sin\theta))^{1/2}. \tag{29}$$

MODEL RESULTS

Flow width, flow depth, and valley slope were measured by Griffin (Griffin and Smith, 2004) from available maps and the rectified aerial photographs. In addition, Griffin used maps of measured rainfall amounts and estimated stream flow hydrographs to determine peak discharges in the links of the Plum Creek drainage network during the 1965 flood. This information was obtained for each of the three individual southeastern most drainage basins of East Plum Creek displayed on Fig. 4. These drainage basins, from west to east, are (1) Carpenter Creek, (2) the unnamed tributary of Carpenter Creek, and (3) the unnamed eastern tributary of the unnamed tributary of Carpenter Creek. Discharge data from this analysis together with canopy cover measurements are presented in Table 1 for: (1) the unnamed tributary of Carpenter Creek just downstream of its unnamed tributary, (2) the unnamed tributary of Carpenter Creek upstream of its unnamed tributary, and (3) Carpenter Creek a short distance upstream of the mouth of the unnamed tributary. Also included in the table are stem spacings calculated by using Equation 30, which relates the fraction of floodplain covered by shrub canopy (F_{cc}) to stem spacing (λ_s) for $F_{cc} < 0.8$.

TABLE 1 Model Input Parameters and Calculated Flow Velocities and Boundary Shear Stresses along Carpenter Creek and the First Unnamed Tributary

Variable	Unnamed Tributary Downstream from Confluence with 2nd Tributary	Unnamed Tributary Upstream from Confluence with 2nd Tributary	Carpenter Creek
Canopy coverage (%)	21	55	13
Average stem spacing (m)	0.48	0.30	0.60
Nondimensional shrub spacing (λ_G/D_G)	4.4	2.2	5.2
Discharge ($m^3\ s^{-1}$)	309	253	571
Flow depth (m)	2.5	2.8	3.8
Flow velocity ($m\ s^{-1}$)	3.3	2.7	4.8
Calculated boundary shear stress ($N\ m^{-2}$)	2.2	0.28	17.0

$$\lambda_s = \left(\frac{\pi/4}{N_s F_{cc}}\right)^{1/2}(D_c). \tag{30}$$

In Equation (30), N_s is the number of stems in a cluster and D_c is the diameter of a cluster. Measurements made on the floodplain of the unnamed tributary give $N_s = 36$ and $D_c = 1.5$ m. On the average, the shrubs begin to intermingle when $F_{cc} = 0.8$, which is approximately the area of an inscribed circle (projected canopy area) divided by the area of a square (area of floodplain per shrub). Shrub spacing (λ_G) is given by Equation (31).

$$\lambda_G = \left(\left(\frac{\pi/4}{F_{cc}}\right)^{1/2} + 1\right)(D_G). \tag{31}$$

Nondimensional shrub spacing is defined by Smith (2004) as λ_G / D_G.

Smith (2004) determined the diameter and number of stems in an average old willow in the Deer Lodge Valley and used average shrub spacing instead of average stem spacing as a measure of shrub density. Griffin and Smith (2004) similarly grouped the stems into clumps because when the stems are closely spaced, the canopies of individual sandbar willows intermingle and it is the canopy of a clump that can be resolved in an aerial photograph. In both of these cases, however, form drag on the stems was modeled as if the stems were dispersed randomly on the floodplain. Average stem spacing as opposed to average shrub spacing is used in this chapter section because it is the primary shrub-based parameter in the model owing to its dominant role in the shrub geometry factor. Its meaning is unambiguous physically, and its use makes the shrub geometry factor easier to modify for other types of shrubs when applying the results presented in this chapter to other cases. It also is more appropriate for sandbar willows when they are sparse. Equations (30) and (31) can be used to relate canopy cover, shrub spacing, and stem spacing. For example, at 55% canopy cover, the nondimensional shrub spacing is 2.2 and the stem spacing is 0.30. In addition, Smith (2004) and Griffin and Smith (2004) used boundary shear stress on the irregular floodplain surface rather than that on the average floodplain surface because they were interested in erosion of or deposition on the floodplain, not in comparing the fluid mechanical effects in one case to those in the other. To display the effects of the stems alone and to facilitate comparison between cases where the shrubs are different types and diameters, the boundary shear stresses used in this section are for the average floodplain surface and do not have the form drag of the irregularities of the actual floodplain surface extracted from them. This added drag would decrease the values of the boundary shear stresses in the shrub carrs of the unnamed tributary of Carpenter Creek by a factor of 0.10 at stem spacing of 0.25 m, by a factor of 0.15 at a stem spacing of 0.30 m, and by a factor of 0.21 at a stem spacing of 0.35 m. Boundary shear stresses on the average floodplain surface and vertically averaged velocities from the three-layer model for this location are shown in Figs. 8 and 9, respectively. Similar one-layer model calculations for this location are displayed in Figs. 10 and 11. The two flow parameters are analogously graphed as functions of depth and stem spacing.

Boundary shear stresses are plotted on a logarithmic scale in Figs. 8 and 10 so that the full ranges of values can be displayed on the same graphs. Were they plotted on a linear scale, the top traces would both be straight lines, representing the same depth-slope-product boundary shear stress. This asymptote is reached within a few percent in each case at a stem spacing of 10 m. The boundary shear stress from the bent-over shrub model (Fig. 8) increases with flow depth for all stem densities and depths. In contrast, in the dense stem situation when the plants penetrate the water surface, the boundary shear stress first decreases with increasing flow depth (from 3.0–1.3 N m^{-2} at $\lambda_S = 0.28$) and then becomes nearly constant (Fig. 10). Moreover, this pattern of variation with depth in the one-layer case persists until the stem spacing exceeds 1.6 m, that is, when there is only one 0.03-m-diameter stem in a square 1.6 m on a side. At this value of λ_S, σ_{DS} varies with depth from 4 to 12 in the one-layer case. At $\lambda_S = 1.6$ m in the three-layer case, $\sigma_{DS1} \cong 0.48$ and σ_{T2} varies from 1.3 to 1.4 as the depth increases from 1.6 m to 5.0 m. Of course, a stiff shrub 0.03 m in diameter and several meters tall standing vertically in the 1965 flow is impossible, but a stiff tree four times that diameter at stem spacing of 3.2 m is not.

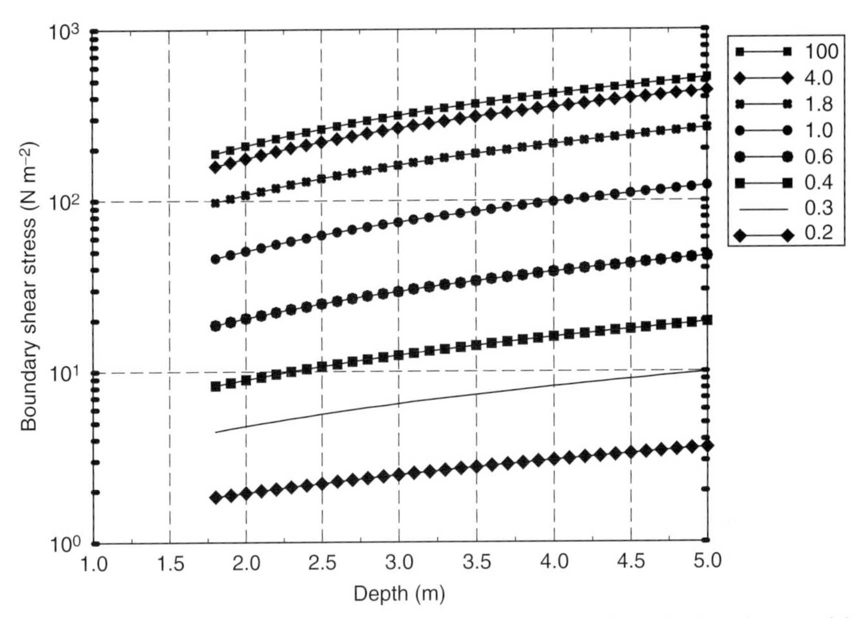

FIG. 8 Boundary shear stress on the average floodplain surface from the three-layer model as a function of flow depth and stem spacing. The boundary shear stress scale is graphed logarithmically because of the large range of values that are represented. At a stem spacing of 100 m, form drag on the shrubs is negligible, so the top curve represents the boundary shear stress from the depth-slope-product. This asymptote is reached within a few percent at a stem spacing of 10 m. The top line would be straight if the y-axis were arithmetic and not logarithmic.

Note that these examples both have the same "shrub geometric factor" (D_S/λ_S^2). Fig. 10 shows that the magnitude of the boundary shear stress at h = 3.0 m increases by nearly two orders of magnitude (0.8–40 N m^{-2}) as the stem spacing increases from 0.2 m to 1.6 m, even though it remains essentially independent of flow depth. This asymptote is reached because, when s_{DS} gets large compared to one, the depth in it cancels that in the depth-slope product in the numerator of the equation for boundary shear stress [Equation (18a) when $\tau_{is} = 0$]. In the three-layer case, the boundary shear stress also increases by nearly two orders of magnitude over this same range of stem spacing; however, at a stem spacing of 0.2 m its value is three times higher than it was for the one-layer case (2.4 N m^{-2} vs. 0.8 N m^{-2}). This example demonstrates that until a flow gets deep relative to the shrub height, the stiff shrubs are more effective at reducing the boundary shear stress than are flexible ones. At greater depths, however, the stiff shrubs are likely to break, particularly if there is debris in the flow.

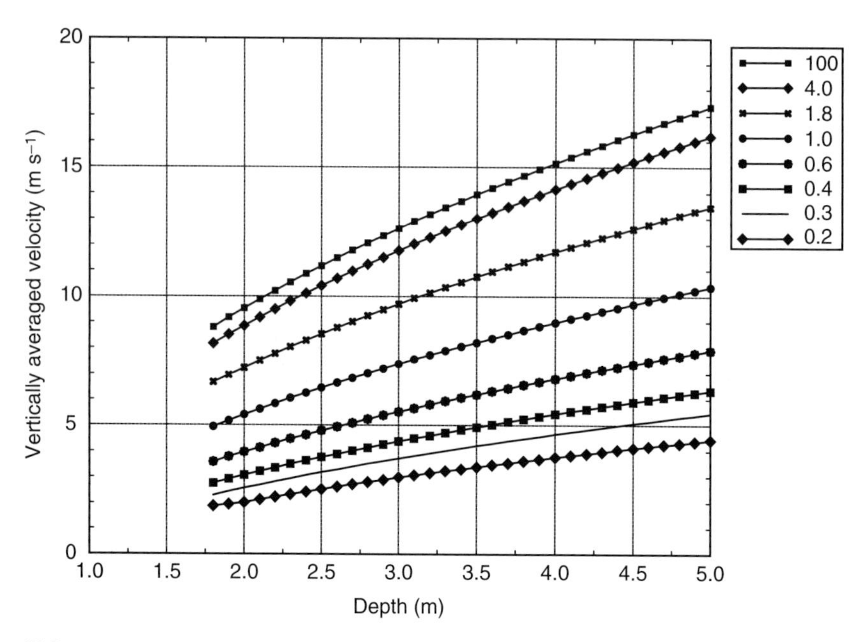

FIG. 9 Vertically averaged velocity from the three-layer model as a function of flow depth and stem spacing. The top curve represents the vertically averaged velocity that is derived from the depth-slope product shear stress and a fixed roughness parameter of 0.0123 m. Unlike the velocity in a shrub carr that is not overtopped, the vertically averaged velocity for a three-layer flow increases with flow depth for all stem spacings. This occurs because the top layer has no shrubs in it, so the discharge from this layer always increases with flow depth no matter how dense the stems are.

In the case of flexible shrubs, the vertically averaged velocity increases with increasing depth for all stem spacings (Fig. 9). In contrast, when the shrubs are stiff, moderately tall, and fairly dense, the vertically averaged velocity is essentially independent of flow depth and varies only with stem spacing (Fig. 11). This is because for large σ_{DS}, not only does the depth cancel out of both the numerator and the denominator of the equation for flow velocity but the velocity factor, which also contains the depth as h/z_0, nearly does as well. [Recall from Equation (16c) that $\langle u^2 \rangle = \langle u \rangle^2 + 1$.] For depths greater than 1.0 m, Fig. 11 shows that this asymptote is obtained for stem spacings less than 1.0 m. As mentioned in the discussion of tailings deposition on the Clark Fork floodplain, this asymptotic situation in a shrub carr causes the velocity to be altered only by stem density, making the flow velocity independent of both cross stream and streamwise-averaged downstream floodplain topography.

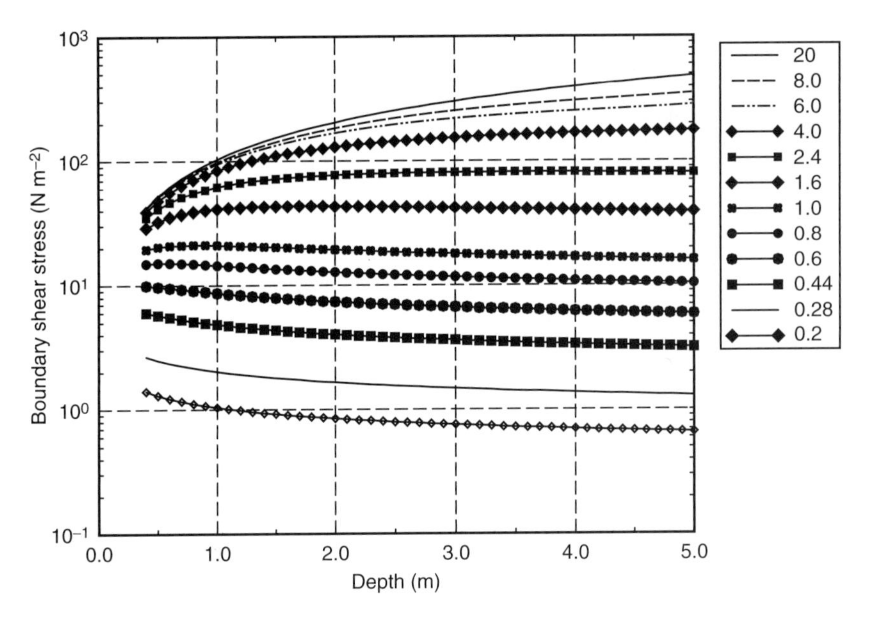

FIG. 10 Boundary shear stress on the average floodplain from a one-layer model as a function of flow depth and stem spacing. At a stem spacing of 20 m, the form drag on the shrubs is negligible, so the top curve represents the same boundary shear stress from the depth-slope product as did the top curve on Fig. 8. This line would be straight if the y-axis were arithmetic and not logarithmic. When the shrubs affect the flow all the way to the surface, the boundary shear stress decreases with flow depth at first, then becomes nearly independent of it. In fact, it does not increase with flow depth for flows deeper than 2 m until there is only one stem in a square 1.6 m on a side. At a stem spacing of 1.0 m the boundary shear stress does not increase with flow depth for depths greater than 1.0 m. The magnitude of the boundary shear stress in a carr of stiff shrubs, therefore, is controlled by stem spacing, not flow depth, as long as the depth is not less than the stem spacing.

When floodplains are covered with willow carrs, the stem spacing–caused reduction in boundary shear stress results in excellent protection of the floodplains from erosion, as long as the willows are not ripped up by debris. If the entire floodplain of the upper East Plum Creek drainage basin had been populated with a similarly dense sandbar willow carr (>55% canopy cover) in 1965, the shrubs would have provided sufficient protection to prevent large woody debris from entering the river and the floodplain from unraveling. Once erosion of the floodplain somewhere upstream along a river starts and large trees begin to be transported downstream by the flow, the floodplain-protecting shrubs can be ploughed up and ripped out by the roots of these floating trees and floodplain erosion then can occur along these paths of destruction. This is the main reason

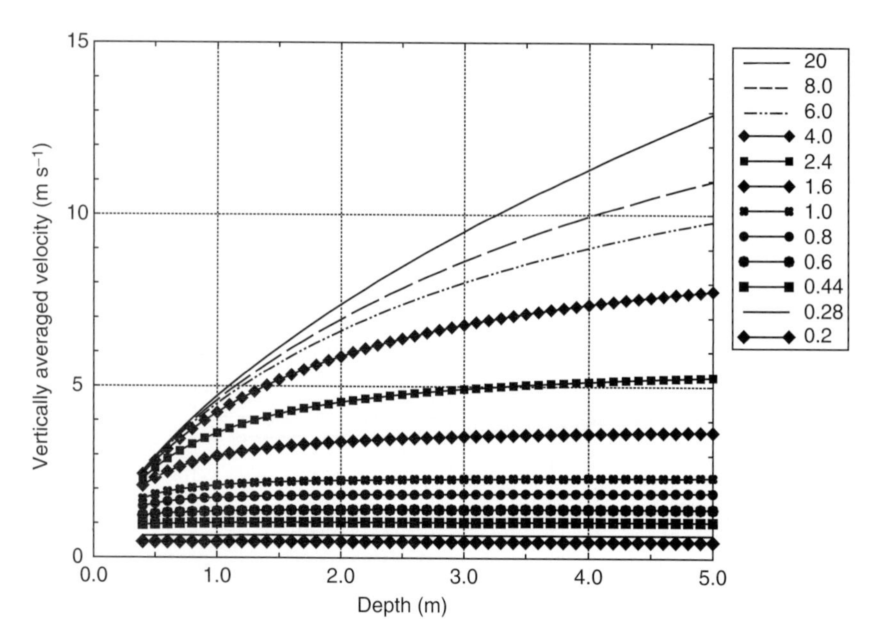

FIG. 11 Vertically averaged velocity from a one-layer model as a function of flow depth and stem spacing. For depths greater than the stem spacing and for stem spacings less than 1.6 m, the vertically averaged velocity is essentially independent of flow depth and bed roughness. This makes the vertically averaged velocity independent of floodplain topography and dependent primarily on stem spacing for deep overbank flows.

why well but intermittently protected floodplains can completely unravel. The calculations shown in Figs. 8 and 9 from the three-layer model demonstrate both why the sandbar willows would have protected the floodplain unless large woody debris was added to the flow further upstream and why the sparser, lower, and stiffer shrubs, displayed in Fig. 7 on the floodplain that was removed, were unable to protect that floodplain from erosion during the 1965 flood. The cottonwood trees at this East Plum Creek location provided less floodplain protection than would one 0.03-m diameter sandbar willow per square meter, despite their large diameters. In fact, the broad canopies of cottonwood trees result in stem spacings that usually are too large to make them effective at significantly reducing boundary shear stresses on floodplain surfaces. In a stand of cottonwood trees, it is the understory shrubs, which often are removed by grazing, that protect the floodplain from eroding during floods and that protect the cottonwood trees on the floodplain from being uprooted and carried downstream by the flood.

Along the unnamed tributary of Carpenter Creek (see upstream, that is, south, of the arrow near the bottom of Fig. 4) studied by Griffin and Smith (2004) and along its small tributary, beaver-maintained sandbar willow carrs with 55% or greater net canopy cover ($\lambda_S < 0.3$ and $D_S/\lambda_S^2 > 0.33$ m^{-1}) were sufficient to protect the general morphology of these two geomorphic systems, despite being subjected to overbank flows from 2 to 3 m deep (Fig. 12). The effects of the sandbar willow alone at greater than 55% canopy cover reduced the shear stresses on the average floodplain surface from more than 200 N m^{-2} to less than 6.0 N m^{-2} and the vertically averaged velocity from 9.5 m s^{-1} to 2.3 m s^{-1} (Figs. 12, then 8, and 9). These low boundary shear stresses and near bed velocities allow sediment and debris to accumulate on the floodplain among the stems of a cluster and the form drag

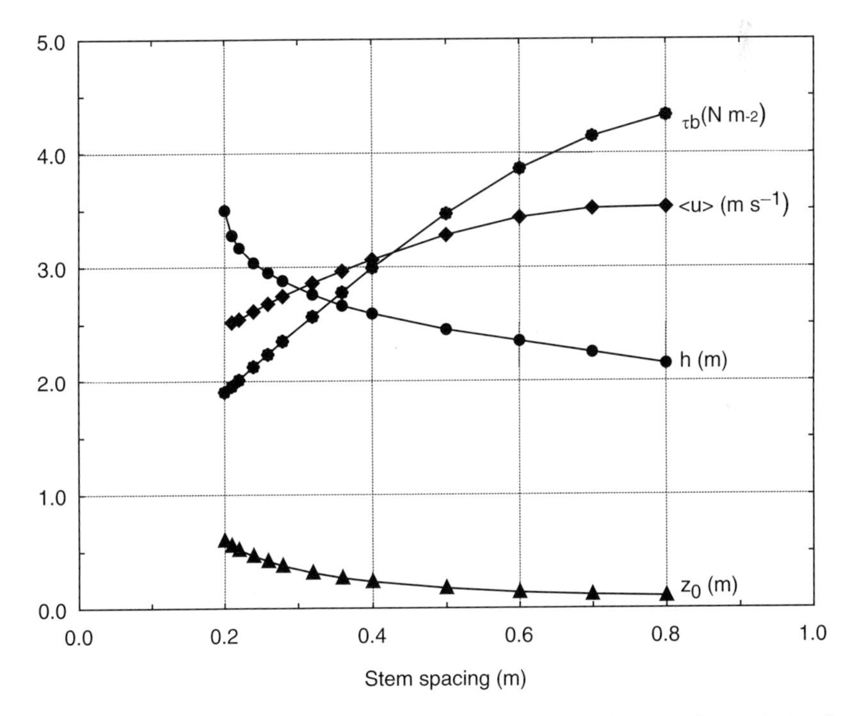

FIG. 12 Graphs of boundary shear stress on the average floodplain surface, velocity, flow depth, and bulk roughness parameter (which includes the effects of the shrubs). As functions of stem spacing for the larger unnamed tributary upstream of the smaller unnamed tributary. In this graph, all the flow parameters are for a fixed discharge per unit width of 7.9 m^2 s^{-1}. The canopy cover for this reach was 55%, giving a stem spacing of 0.30, as indicated in Table 1. Note that for a fixed discharge the flow depth and bulk roughness parameter (z_0) rise as the stem spacing decreases. In contrast, the boundary shear stress and velocity rise as the stem spacing increases.

on these topographic features further reduces the boundary shear stress on the actual floodplain surface.

Commensurate with the decreases in boundary shear stress and velocity with increasing stem density, shown on Fig. 12, the flow depth increases from 1.9 m to 2.8 m. Fig. 12 shows the important flow properties for the first unnamed tributary upstream of its unnamed tributary at a fixed discharge of 253 $m^3 s^{-1}$. Taking account of the form drag on the irregular floodplain surface, produced as described in the previous paragraph, the average shear stress on the actual boundary was reduced to 0.28 $N m^{-2}$ and, in addition, the elevated water table permitted growth of a dense herbaceous understory that further reduced the shear stresses on the mineral soil beneath. In contrast, below the second unnamed tributary the discharge on the first unnamed tributary was 309 $m^3 s^{-1}$, the canopy cover was 21%, and the stem spacing had increased from 0.30 m to 0.48 m (Table 1). At this location, the boundary shear stress on the average floodplain surface was 15 $N m^{-2}$ and that on the actual floodplain was 2.2 $N m^{-2}$. As shown at F in Fig. 13, the floodplain at this location had started to unravel. On the borders of the ranchland along Carpenter Creek, just upstream of the unnamed tributary, the sandbar willows were small and sparse (average stem diameter < 1.0 cm, canopy cover < 13%, and stem spacing > 0.60 m), and the understory vegetation was grass. At this location, the boundary shear stress on the average floodplain surface exceeded 70 $N m^{-2}$ and that on the actual floodplain surface exceeded 17 $N m^{-2}$, almost an order of magnitude above the critical shear stress for erosion that Smith (2004) found for floodplains covered with nonirrigated grass in the Deer Lodge Valley. The floodplain at this location on Carpenter Creek not only unraveled but the resulting channel appears to have achieved a new hydraulic geometry in equilibrium with the flood discharge (Griffin and Smith, 2004).

Although the discharge at the downstream location on the unnamed tributary had increased from 253 $m^3 s^{-1}$ to 309 $m^3 s^{-1}$ as a result of water entering from the second unnamed tributary, the valley had broadened and the flow depth had decreased slightly. Consequently, it had to be the

FIG. 13 Aerial photographs showing Carpenter Creek and the two unnamed tributaries that are discussed in the text (modified from Griffin and Smith, 2004). The confluence of the larger unnamed tributary and Carpenter Creek is seen just north of Reach F. The junction of the larger and smaller unnamed tributaries is about a third of the way from Reach E to Reach F. Upstream of this junction, both unnamed tributaries remained intact geomorphically, whereas downstream of the junction the larger unnamed tributary has begun to unravel, and it has completely unraveled by Reach F. The visible section of Carpenter Creek also has unraveled and has evolved a new equilibrium hydraulic geometry. The following figures focus on the smaller unnamed tributary and on the beaver ponds and dams in Reaches A through F.

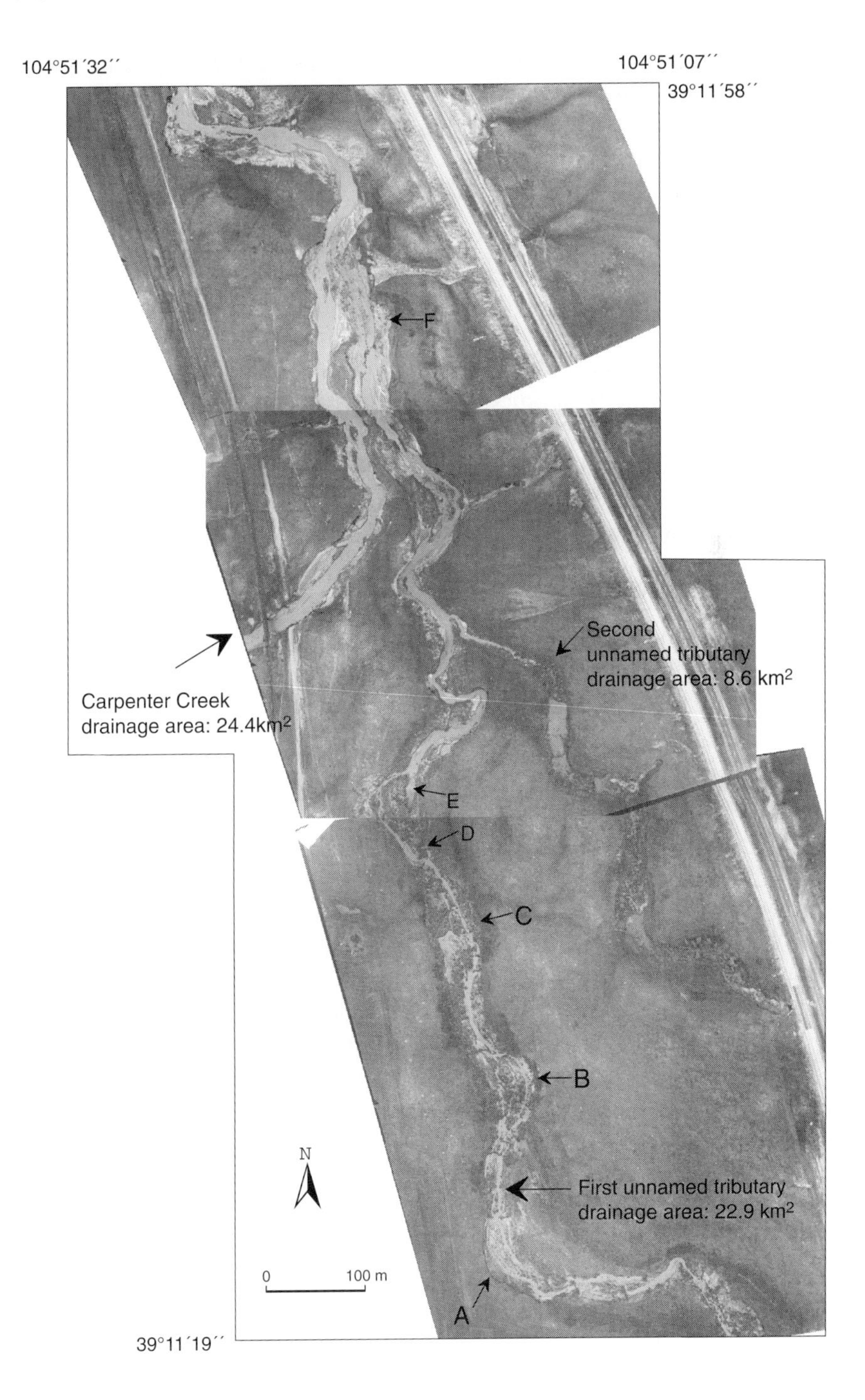
104°51′32′′
104°51′07′′
39°11′58′′
←F
Second
unnamed tributary
drainage area: 8.6 km2
Carpenter Creek
drainage area: 24.4km2
←E
←D
←C
←B
N
First unnamed tributary
drainage area: 22.9 km2
0
100 m
A
39°11′19′′

reduction in stem density rather than the increase in discharge, which acts by increasing the flow depth and, thus, the boundary shear stress that caused the unraveling to begin here. Owing to the reduction in stem diameters and the competition between the beaver and the cattle north of this site, it was about as far downstream as the beaver could maintain, under present conditions, a dense and healthy sandbar willow carr. It is likely, however, that before 1820 or so beaver and willows populated most of the drainage basin of upper East Plum Creek. The calculations and observations presented in this chapter suggest that the critical stem spacing for unraveling of the unnamed tributary during the 1965 flood was 0.48 m. Owing to submergence of the sandbar willows and, thus, a depth dependence in the boundary shear stress, this value of λ_S is applicable only for the 1965 flood at this site. Nevertheless, agreement between the calculated shear stress for this site and the critical shear stress for erosion of a floodplain covered by nonirrigated grass in a semi-arid climate indicates that the model presented in this chapter, combined with a critical shear stress for floodplain erosion of 2 $\mathrm{N\,m^{-2}}$ can be used with confidence to investigate floodplain stability in cases where the floodplain is protected by flexible shrubs that are overtopped by the flow.

DISCUSSION

Interplay of Floods and Beaver Ponds

In previous sections of this chapter, the importance of shrubs to stream stability has been investigated. A cursory examination of land-use patterns for upper East Plum Creek in available aerial photographs indicates that the unraveled areas were ranchlands covered with grass and small moderately to widely spaced shrubs. Further downstream these shrubs were under sparsely to moderately spaced cottonwood trees. In contrast, all the floodplain areas that did not unravel appeared to be covered by beaver-pond-supported willow carrs. This apparent pattern thus makes the origin and the past and present characteristics of the shrub carrs of fundamental importance in understanding the geomorphic stability of these and other headwater streams. The woody component of these shrub carrs was in 1965, and is at present, dominantly composed of sandbar willow and the aerial photographs of these willow carrs display dams and ponds indicating that they were all populated by beaver. In this section of the chapter the effects that floods have on beaver dams and ponds and the role that beaver play in maintaining willow carrs are investigated.

Beaver dams typically are not flimsy piles of sticks; rather, when the proper construction materials are present they are carefully made structures that are surprisingly resistant to floods. Many researchers have described beaver dams in the areas that they have worked, and there are clear variations from place to place, often related to depth of freezing of ponds in the winter time. Relevant recent investigations of dam construction, dam failure during floods, and their connection include the following: Zurowski (1992), Gurnell (1998), Hillman (1998), Meentemeyer and Butler (1999), and Butler and Malanson (2005). Nevertheless, a general description of the beaver dams in the Colorado Front Range and the features of these dams that are pertinent to their resistance to floods based on this author's observation is a necessary foundation for the rest of this chapter.

The largest available sticks are trimmed and laid parallel to the flow direction on the downstream side of a beaver dam. On the upstream side, smaller sticks with more branches are usually laid at a gentler angle and with a more random orientation. The upstream side of a beaver dam is well-packed with mud if it is available from the bottom of the pond. A well-made beaver dam is protected on the upstream side from the pressure of high flows by a gently upsloping apron and on the downstream side from undercutting by the plunging flow at its base by the mat of large, upstream-sloping, streamwise oriented logs. Near the threads of high flow, these aprons tend to contain both more sticks and more mud. On the unnamed tributaries, they also typically have blocks of sod with herbaceous plants growing in them at the top on the upstream side. Owing to the high proportions of mud in the apron and the use of willow shoots and mud to repair small breaches and to add to the apron, willows often begin to grow along the tops of the dams. Consequently, active but older beaver dams can be recognized from above as lines of willows. These willows, as they mature, aid in anchoring the dams and, thus, are particularly important components of long dams and of the dams across the main threads of a stream. When the dams are across rivers with gravel beds and sod is hard to get from the banks, the aprons are sometimes constructed of inter-bedded sticks and riverbed cobbles. Impressively, beaver appear to be able to carry medium sized cobbles for relatively long distances. The cobble and stick structure has the advantage of using local construction materials, which makes it possible to construct a longer apron. The cobble and stick structure is relatively strong, particularly when the apron is gentle. Under high flows, a cobble and stick dam with a long apron simulates a cobble dune, which is a stable geomorphic form on the bed of a river, and is more likely to grow than erode. Nevertheless, cobble and log dams are relatively porous and are vulnerable to tunneling, making mud and log dams more resistant to large floods. Along Carpenter Creek and its unnamed tributary, only in the vicinity of the confluence was there a sufficient source of cobbles for significant numbers to be incorporated into the beaver dams.

Field measurements made in both 2000 and 2005 along the first unnamed tributary of Carpenter Creek demonstrated that beaver dams well over 1 m in height and many tens of meters in extent were being constructed almost completely of these small-diameter willows. No cottonwood trees were seen anywhere near the study reaches in either 2000 or 2005, and only a few cottonwood stumps were observed at these times. The stumps were of two types. First, there were two very large and very old stumps in the bottom of a dry beaver pond. These probably had washed into the pond from upstream during the 1965 flood. Second, there were a few small cottonwood stumps, all well above the floodplain in elevation, located some distance from the beaver ponds. The latter were less than 80 mm in diameter. No cottonwood logs were found on the dam faces, and there was not a sufficient number of cottonwood stumps in the entire area to have made cottonwood logs a significant component of the dam cores. Similarly, the 1965 aerial photographs show no evidence of larger woody plants, and it is likely that beaver activity has prevented growth of significant-sized cottonwood trees along the unnamed tributary of Carpenter Creek for a long time by cutting them down before they grow to a significant size.

In 2005, the structures of the dams along the first unnamed tributary of Carpenter Creek were examined in detail. The beaver apparently had compensated for the paucity of large logs by using many somewhat entangled small sandbar willow stems, together with large amounts of mud in the dam cores, and by constructing a longer apron than the author has seen elsewhere. These stick and mud structures with long mud aprons were of a nature and shape that is likely to be very resistant to erosion, particularly from direct fluid-dynamical pressure on the dam face and by erosion of the tops of the structures. Once they are abandoned, beaver ponds often drain by flow tunneling under the dams and the superstructure collapsing into the tunnel, rather than by erosion from overtopping, but the robust apron of the dams on the unnamed tributary pushed the deep pool well upstream of the dam and appeared to provide good protection from this common mode of dam erosion. The author has observed that the beaver take particular care with the aprons of large dams on ponds with large lodges. It may just be that the beaver spend more time in these ponds and, hence, work on the dams more often or that the effort is intended primarily to deepen the pond beneath the winter ice, but the result is that these large dams are usually very resistant to high flows. In contrast, initial low dams often are not particularly carefully made: the stems and branches are not as well pruned or as consistently oriented, and the dams contain less mud and more herbaceous vegetation in them, blocking the leaks and evening the tops. These low dams often become the cores of higher ones and, if construction proceeds, the effort seems to become better organized and the amount of mud in them increases. As mentioned above, measurements made on aerial photographs of the diameters and heights of bent-over sandbar willow clumps indicated that they were about the same in 1965 as measured in 2000 and 2005; consequently, it is assumed here that the dams were similarly constructed on the unnamed tributaries before the 1965 flood.

Fig. 13 shows an aerial photograph of the entire area of interest. The flow is to the north (upward) toward East Plum Creek. The reaches displayed in greater detail in subsequent figures are labeled with capital letters, A through F, from upstream to downstream. Near the top of the figure, labeled F, is the confluence of the unnamed tributary of primary interest in this investigation (on the right) with Carpenter Creek (on the left). All of Carpenter Creek and the lower portion of the unnamed tributary have unraveled, whereas upstream of the junction of the first and second (on the right) unnamed tributaries, both of these two streams have remained geomorphologically intact despite the 2 m to 3 m of rapidly moving water that flowed down these two valleys. The unraveled area is shown by the broad swath of lighter color (suspended sediment transporting water and sand) covering most of the floodplain and an absence of darker colored shrubs. The willow carrs and the beaver ponds are clearly visible in the upstream areas along the unnamed tributaries. Between the first and second junctions, the floodplain of the larger unnamed tributary has been seriously eroded, suggesting that the critical shear stress for erosion has been exceeded here (although perhaps not by much). Currently there are active beaver ponds and shrub carrs all the way to and beyond F. On Carpenter Creek, both above and below its junction with the larger unnamed tributary, there were active beaver ponds in 2000 and 2005, but the sandbar willow in this area are sparse and small in diameter (<10 mm in 2000 and 2005). In addition, there were numerous ungulate clipped stem tips in May 2005, providing good evidence that cattle or deer were grazing the new shoots on Carpenter Creek in the vicinity of F, thereby preventing them from growing to a size useful to the beaver for construction of large dams and lodges. Where the fields come smoothly down to Carpenter Creek just north-northwest of F, small willows were absent in both 2000 and 2005, whereas in the stream valleys between low terraces south of F along both unnamed tributaries and upstream of F along Carpenter Creek, there were many small willows. Along Carpenter Creek, these smaller willows thinned noticeably toward the above mentioned area of easy access to the fields. Further downstream on Carpenter Creek, there were a few poor-quality low dams constructed of small sandbar willow stems (harvested from upstream along Carpenter Creek judging by the size) augmented by large amounts of mud and herbaceous vegetation and a few cobbles. Despite the mud and cobbles, it is doubtful that these dams would be able to survive even a moderate overbank flow. Fig. 13 shows that the floodplain along Carpenter Creek unraveled during the 1965 flood, that the stream filled the valley from terrace to terrace 2 days after the peak flow, and that the bed of the stream has been flattened by the activity of antidunes.

Fig. 14 provides an expanded view of the area from A to E on the larger unnamed tributary and of its unnamed tributary. From B through E on Fig. 14, the terrace is covered with grass and the stream flows through a carr of sandbar willow between the terraces. In this area, the stream is narrow compared to the floodplain. The valley of the first unnamed tributary through this reach is similar to that of the second unnamed tributary and can be compared to it quantitatively. Arrows on the smaller unnamed tributary denote beaver dams that survived the flood despite being overtopped by 1 m of water. In contrast to many of the beaver dams on the larger unnamed tributary, these dams have not been breached by the flood. At first it might be thought that the lower discharge on the second unnamed tributary and the associated lower flow depth and lower boundary shear stresses during the flood permitted what are probably similarly constructed

FIG. 14 Expanded view of the southern part of Fig. 13 showing the beaver dams and beaver ponds on the two unnamed tributaries. Arrows indicate the major dams on the smaller unnamed tributary. Note that none of the dams on the smaller tributary have been breached by the flood. In general, the beaver dams on both streams are aligned in a cross-valley direction and extend from terrace to terrace, thereby distributing the water across the entire floodplain and permitting dense sandbar willow carrs to form. On the smaller unnamed tributary, the willows form broad bands across the valley upstream of the dams, whereas on the larger unnamed tributary the channel provides a continuous line of high flow most of the way from A to E. This more or less continuous channel allowed the floodwaters to apply large forces on the dams where they crossed this channel, causing most of them to breach.

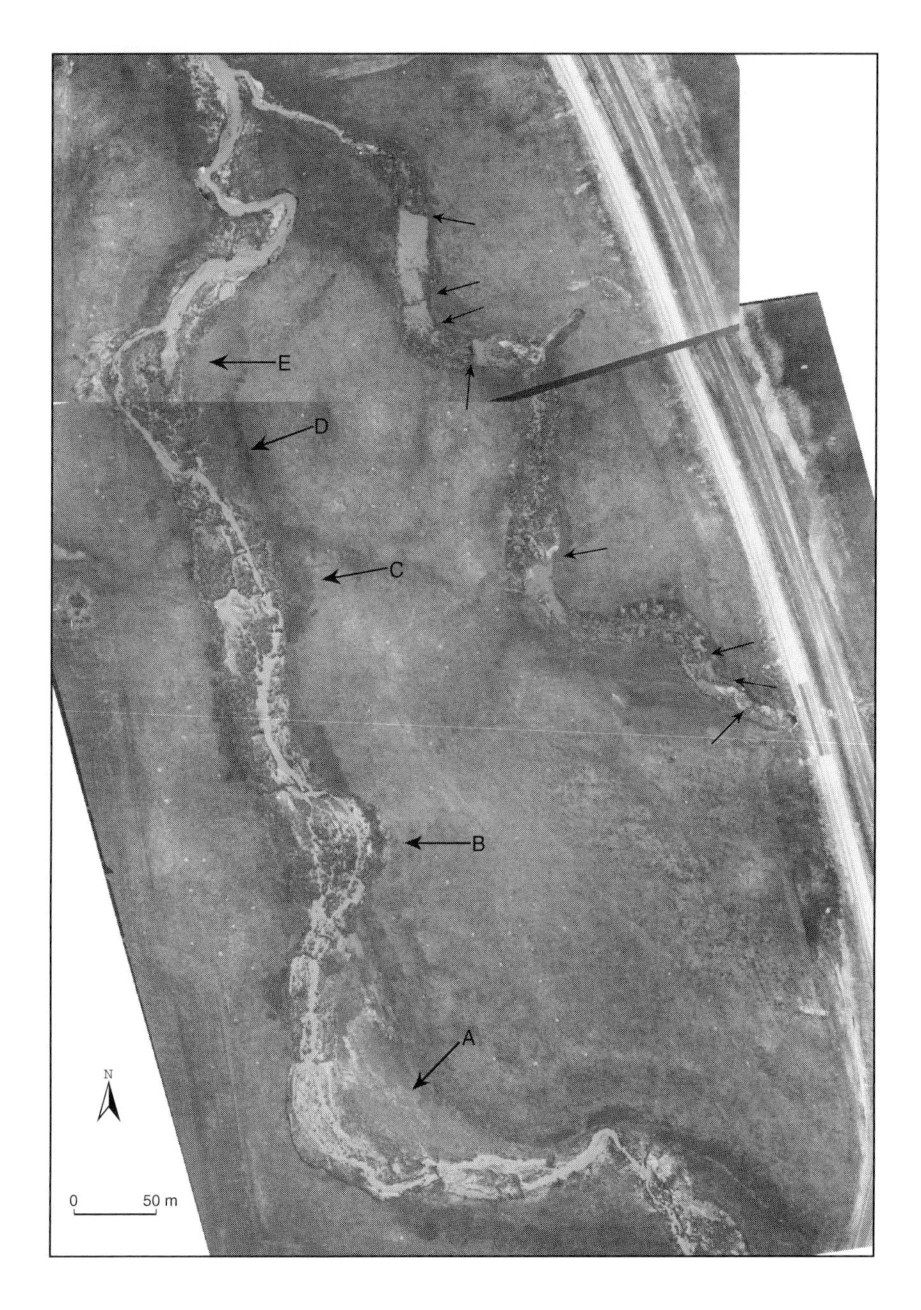
E
D
C
B
A
N
0 50 m

beaver dams to survive the flood without significant damage in one case and to be breached in the other. The difference in water depth probably played an important role, but this difference has occurred primarily because the high-flow channel is better defined in the first unnamed tributary than in the second (Fig. 14), making the shrubs less effective along that flow path.

Assuming the beaver dams to have sawtooth shapes and using the procedure described earlier in the chapter for calculating the drag on a regular sequence of them, their interaction with the flood can be determined for the various situations shown on the first unnamed tributary in Fig. 14. This also can be done in the presence of shrubs of varying densities. Referring to Fig. 8, we see that, were the shrubs removed, the flow depth for what otherwise was a 64% canopy cover in the larger unnamed tributary would be somewhat less than 2.0 m. For a 1.9-m-deep flow in a valley with a slope of 0.0107 and with no shrubs, the depth-slope product boundary shear stress (τ_{hS}) is 199 N m^{-2}. Estimating the beaver dams to be 0.8 m high and spaced 20 m apart gives a shear stress ratio ($R\tau = \tau_T / \tau_{AS}$) of 4.0, leading to an average shear stress on the actual floodplain (τ_{AS}) of 50 N m^{-2}. This same reduction in boundary shear stress would be produced by a stem spacing of 1.0 m. A τ_{AS} of 50 N m^{-2} would result in massive erosion of the floodplain and complete removal of the beaver dams. If the dams were infinitely strong and well anchored, however, the spatially averaged flow velocity would be 4.2 m s^{-1} and the average Froude number would be 0.98. Owing to this near critical Froude number there would be antidunes in the ponds. The velocity over the top of the dams would be 7.3 m s^{-1} and the Froude number there would be 2.3, resulting in shooting flow downstream of each dam and very large hydraulic jumps as the flow went back to tranquil before the next dam. The shear stress ratio is not particularly sensitive to the distance between dams, but it does have a measurable effect. The distance between dams taken in this example is on the short side, and a greater spacing would make the situation even worse. For example, at a 30-m spacing between dams the shear stress ratio is 3.0, giving τ_{AS} = 66 N m^{-2}, but even assuming the beaver dams are only 10 m apart and the stress ratio is 7.0, the average shear stress on the actual floodplain surface is 29 N m^{-2}. Moreover, in this scenario, the vertically averaged velocity would be 3.8 m s^{-1} and the Froude number would be 0.90. Therefore, the dams still would be eroded and there still would be antidunes on the bottoms of the ponds. In this latter case, the equivalent stem spacing is 0.8 m. By assuming that the shrubs upstream of the beaver ponds extend all the way across the channel, as shown in Fig. 14 for the second unnamed tributary, they can be included in the calculation using the three-layer model. Taking account of the open water, the stem spacing in this case is 0.38, which gives a flow depth of 2.7 m and a shear stress on the average floodplain surface of 2.9 Nm^{-2}. Including form drag on beaver dams spaced 10 m apart, τ_{AS} = 0.43 Nm^{-2}. Using a more realistic spacing of 30 m for the second unnamed tributary (Fig. 14), τ_{AS} = 0.97 Nm^{-2}. Given that there probably also are small topographical elements in the shrubs and on the bottom of the ponds, these values are likely to be too high. If the value of τ_{AS} = 0.97 Nm^{-2} were three times smaller, then it would give a shear stress on the actual bottom of the flow that is below the critical shear stress for the transport of the sand on the floodplain and on the bottoms of the beaver ponds.

For the smaller unnamed tributary, using the estimated discharge of 53 $m^3 s^{-1}$ and a valley width of 20 m gives a discharge per unit width of 2.7 $m^3 m^{-1} s^{-1}$. The valley slope is about 0.011 and the net canopy cover is approximately 68%. The calculated flow depth is 1.09 m and the boundary shear stress on the average surface is 3.1 N m^{-2}. Taking account of the drag stress resulting from the beaver dams, the shear stress on the floodplain surface is 0.44 N m^{-2}. This value is just above that for erosion of noncohesive sand on the floodplain or in the beaver ponds, and it is about half of that calculated in the previous paragraph for the larger unnamed tributary. Any significant irregularity in the floodplain surface or in the bottoms of the ponds would reduce the boundary shear stress to where sediment could not be transported by the flood on this second unnamed tributary. Although the boundary shear stresses on the actual floodplain surface were about twice this value for the larger unnamed tributary, this difference is too small to have been the cause for the beaver dams being breached on the larger tributary and not on the smaller one. Instead, the floodplain of the smaller tributary did not unravel because the shrubs extended across the entire valley upstream of each of the beaver dams; therefore, the floodwater had to go through a band of shrubs losing its momentum before it got to each dam.

In general, the larger unnamed tributary has also retained its pre-flood morphology; however, most of the beaver dams crossing the dominant channel have been breached and a few sites dominated by particularly large beaver ponds have started to unravel. Bend A provides a good example of this latter situation. This bend can be seen more clearly in Fig. 15. In this figure, five dams are numbered and the five associated ponds are labeled. There is a large grass-covered terrace to the right of the large pond, labeled P5. This terrace is too high at present for water to be diverted over it by the beaver, but the 1965 floodwaters clearly swept over it and deposited sand

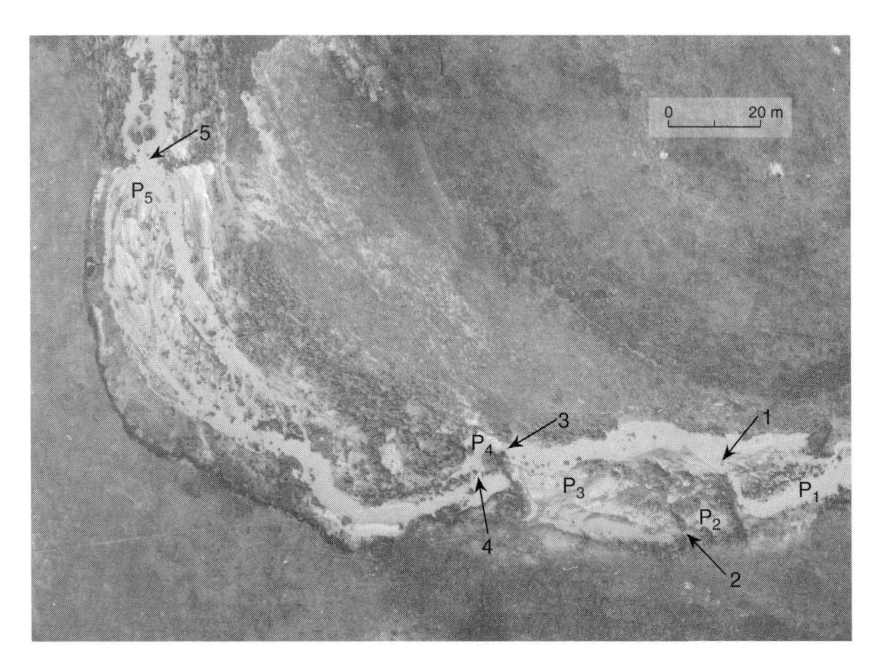

FIG. 15 Expanded view of Reach A. The major dams on this figure are labeled with numbers from 1 to 5, starting upstream. The pre-flood ponds are labeled in association with the dams as P1 through P5. This reach suffered from having been depleted of mature willows, and it would have unraveled during the flood were it not in a sharp bend. Note that the straight reach upstream of the bend did not fare too well. On the inside of the bend there is a low, grass-covered terrace, which was overtopped by the flood and which has sand deposits on it. Between the low terrace and the large pond labeled P5 there is a band of small shrubs that also were preserved as a result of their location on the inside of a bend. The bottoms of P3 and P5 show significant sand deposition to the left of the channel. This is a result of flow deceleration as it approached the dams at 3 and 5. All of the beaver dams in this reach have major breaches and all of the ponds have been drained.

on it to the right of #5. The group of small sandbar willows between the pond and the channel has remained in place despite the unimpeded flow on both sides. The bottom of the large pond (P5) has been scoured, and the dam (denoted 5) has been broadly breached, permitting the pond to drain. Sand has been deposited on the floodplain to the left of the channel between 3 and 5. Despite this activity, dams 3 and 4, upstream of the large pond, remained mostly intact, with the breach directing flow toward the band of small willows between the pond and the terrace and causing it to return to the channel through a small cut in dam-4. The bed of the pond (P3) upstream of dam-3 shows signs of dune formation on the south side (just below the label P3), providing a rough constraint on the boundary shear stress exerted by the flood on that surface. The furthest dam upstream (dam-1) has been seriously breached on the north side where the main channel was located before and during the flood. All in all, the flow was close to unraveling the floodplain in the vicinity of this bend, and it would have done so were it not for the resistance of the three main beaver dams upstream and the shrubs downstream of 5. Deceleration of the flow as it entered the shrubs downstream of 5 would have caused a decreasing boundary shear stress in the downstream direction and a decrease in the sediment transport rate. As a result, deposition of sediment had to occur upstream of 5. A slightly larger or longer flood probably would have started to scour around the shrubs and remove them, causing more serious channel filling and debris damage further downstream. Therefore, the boundary shear stress in this reach in 1965 was very close to the shear stress for floodplain unraveling.

This near catastrophe for the beaver in bend A was a direct result of the impoverished nature of the willow carrs next to ponds P1 and P2, and the large size of P5. Interestingly, downstream of P5 the willows on the floodplain were larger and more effective at overbank flow mitigation. At bend B the beaver dams across the main channel were protected by the larger and denser shrubs, and the dams were breached by cuts of small lateral extent (see locations 1, 2, 3, 4, and 5 on Fig. 16). Moreover, breaches in 3, 4, and 5 were along the main channel. At the upstream end of this reach the larger and denser shrubs forced the flow into numerous small channels through a dense willow carr rather than into a single large channel, as occurred at the upstream end of bend A. Most of the breaches in bend B were sufficiently small and the willows sufficiently large to permit quick repairs of the dams after the floodwaters had receded a bit more and the sediment transport rates had decreased. Bend A provides a good example of what can happen in a shrub impoverished area in a pond complex whereas bend B provides a good example of how shrub carrs can effectively protect the floodplain from erosion. Had there been large cottonwood trees in bend A and had they been eroded, the situation in bend B might have ended differently. The beavers' tendency to eliminate cottonwood trees in willow carrs, therefore, is an important accidental contribution to flood mitigation.

In the next reach downstream, shown in Fig. 17, the shrubs were dense and there similarly was no significant erosion. Nevertheless, there remained a well defined channel and the beaver dams across it were all breached (1a, 1b, 2, 4, and 5). The cuts were mostly small and could have been easily repaired once the flow depth and sediment transport decreased. The large

FIG. 16 Expanded view of Reach B. The major dams on this figure are labeled with numbers from 1 to 5, starting upstream. The pre-flood ponds are labeled in association with the dams as P1 through P5. Despite being just downstream of Reach A, the relatively dense woody vegetation associated with the numerous beaver ponds protected the floodplain. Although most of the beaver dams are breached, these breaches are small and would have been quickly repaired once the floodwaters receded a little further. Note that the flow resistance was so high in this bend from the presence of the mature willows that none of the geomorphic structures typical of bends are displayed at this location. The flow did not even erode the "cut" bank. There is some deposition of sand in the willow carr at the downstream end of the bend resulting from flow deceleration through the shrubs.

FIG. 17 Expanded view of Reach C. The major dams on this figure are labeled with numbers from 1 to 6, starting upstream. The pre-flood ponds are labeled in association with the dams as P1 through P6. Despite the large beaver pond (P3) near the middle of this reach and the relatively straight path that the stream channel takes through it, the dense willow carr in this area has fully protected the floodplain. Most of the breaches in the dams along the channel in this area are small and easily repaired. To the left of the ponds labeled P6a and P6b is an abandoned beaver pond with shrubs growing on its bottom. Note the sand deposition in the old channel leading into the large pond to the left. This deposition is a result of flow deceleration by the beaver dam and the dense willow carr beyond it.

pond (P3) at C was very effectively protected on both the upstream and downstream sides by sandbar willows. Particularly effective was the dense willow carr on the downstream side of the long dam. This long dam was breached with a very minor cut at 3b, but most of the water would have drained out of the pond through 3a. The dams along the main channel, even with the cuts in them, would have provided considerable resistance to the flow, causing the water to move through and over the areas of dense woody vegetation adjacent to the channel. Areas of sand deposition are shown in the channel leading to the pond (P3) on its west side, presumably a consequence of deceleration of the flow moving toward the dam and the shrub carr just downstream of the dam (thereby causing a decrease in the boundary shear stress). Despite the open pond, the net canopy cover at this location exceeded 60%, the average spacing between stems was less than 0.30 m, and τ_{AS} averaged about 0.30 Nm^{-2}. The vertically averaged flow velocity for the 3.1 m deep flow here was 2.6 m s^{-1}. The long east-west–oriented dam at 6 is for a large but previously abandoned pond with new willows growing on its bed. The breach in this dam is the width of the channel, and it probably was there long before the 1965 flood.

Downstream of reach C, the channel flows along the right terrace and then along the left terrace, as shown in Fig. 18. It is bordered by a dense sandbar willow carr. After dam-5 in reach C, the main channel flows for some distance before encountering another dam. As a result, dam-1 on Fig. 18 has been broadly breached and the deep accelerating flow in the main channel has washed out dam-3. Downstream of dam-3, the converging boundaries of the channel have forced the flow and suspended sediment back onto the floodplain. This has caused considerable deposition on the floodplain. Convergence of the channel also has caused the dam at the downstream end of P3 to breach broadly.

Examination of the sequence of aerial photographs shown in Figs. 15 through 18 indicates that a dominant channel formed in the first unnamed tributary. This did not happen in the second unnamed tributary (Fig. 14). The calculations for the two unnamed tributaries that included shrubs, which were carried out previously in this section, assumed that the willow carrs extended across the entire valley. This was the situation on the second unnamed tributary but not on the first. As a result, the calculations in that paragraph are appropriate for the former situation but not for the latter. Those calculations suggest that the beaver dams were not breached as long as τ_{AS} was less than the critical shear stress for erosion of the floodplain. In the case of the first unnamed tributary, flow resistance caused by the shrubs had a much smaller effect on the channel, and the original calculation using τ_{hS} probably provides a better estimate of the flow situation in that channel than does the one that includes shrub form drag. One result of the calculation using τ_{hS} is that in the channel the crests of the beaver dams in the larger tributary become hydraulic controls; therefore, in 1965 at these locations there was shooting flow on their downstream sides until the dams were breached. These high boundary shear stresses also could have caused

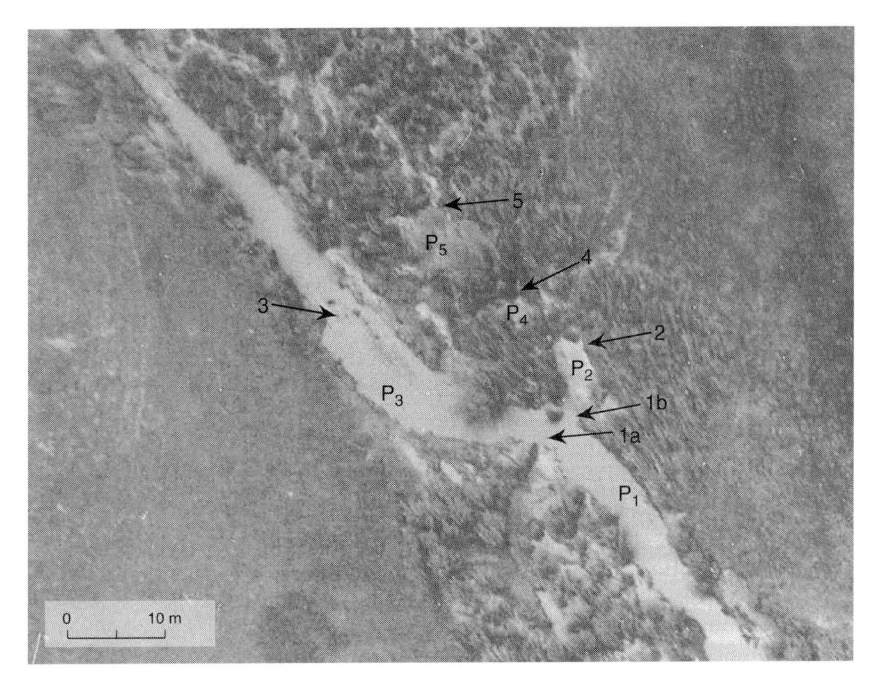

FIG. 18 Expanded view of Reach D. The major dams on this figure are labeled with numbers from 1 to 5, starting upstream. The pre-flood ponds are labeled in association with the dams as P1 through P5. The floodplain in this reach is covered with dense shrubs and is well protected. The channel suffers from being straight and from the absence of frequent, well-maintained beaver dams as in the previous reach. The dams at 1 and 3 appear to have been old and to not have been maintained. Note that the convergent channel downstream of 3 has forced flow and sediment laterally onto the floodplain where the decelerating flow caused the sand to be deposited among the shrubs.

scour of the floodplain near the ends of some of less well-protected dams. See, for example, breaches 1 and 3 in Fig. 15.

The drag force per unit width on a beaver dam is given by Equation (2). Applying this equation to the smaller unnamed tributary gives a drag force of 27 N per meter of cross flow width on a typical dam. For the larger tributary, F_{DD} is about the same when shrubs of equivalent density are present, but when they are absent, such as when the dams are across a channel, this value increases by an order of magnitude to 248 N m^{-1}. Owing to the high velocity in the channel of the larger tributary, the Froude number attained its critical value at the crest of the beaver dams, and owing to the shallow depth it did the same at the crests of the dams on the smaller unnamed tributary. Beaver dams are not well founded, and the shooting flow probably caused scour just downstream of the dams, making them vulnerable to the high pressures produced by the form drag on them. Apparently even under this condition, 27 N m^{-1} was not a sufficient force to breach the dams but 248 N m^{-1} was.

The effects of unraveling are clearly seen at F in Fig. 19. The bed of Carpenter Creek (on the left) at this location is sandy whereas the bed of the unnamed tributary has cobbles and boulders in it from cutting into a gravel terrace. Upstream of this cut bank, the bed of the unnamed tributary is sandy, but downstream of the confluence with Carpenter Creek the bed of Carpenter Creek also is covered with cobbles and boulders. Where the bed is sandy, antidunes have made it flat, so the water surface in these locations is smooth and of uniform reflectivity. Although the water in these aerial photographs is relatively shallow, the sediment in suspension is sufficient to obscure any bed forms. In contrast, if bar forms were present they would be visible in the aerial photographs. Bar forms are not seen; consequently, one can conclude that the sediment is too fine to move as bedload even 2 days after the flood and, thus, to produce bars.

In summary, flow in the two unnamed tributaries was slowed and deepened by beaver-maintained sandbar willow carrs. The shear stresses on the floodplain were reduced by several orders of magnitude where the shrubs were dense. In the smaller unnamed tributary, the discharge was much less than in the larger one and the flow depth was much lower. Were it not for the bands of willows upstream of the beaver dams on this tributary, the dams would have been washed out but, owing to the shrubs, none were even breached. The same would have been the case for the larger unnamed tributary had there not been a well-defined channel devoid of shrubs in the streamwise direction there. The channel allowed high boundary shear stresses to develop and caused most of the beaver dams across it to be badly breached. Also, dams in areas with sparse shrubs were breached (for example, in bend A). Even the breached beaver dams, however, retarded the flow to some extent and forced water out of the channels, onto the floodplain, and through the shrubs. In densely wooded areas, even large ponds did not lead to more than local damage to the beaver dams (for example, bend C) whereas, in sparsely wooded areas (for example, bend A), the dams were severely breached and the ponds drained. In major beaver pond complexes, there are many sites like bend A mixed with sites like bend B. The net result is some local erosion of the more sparsely populated parts of the floodplain and some sediment deposition in the ponds or on the more densely vegetated parts of the floodplain, but maintenance of the integrity of the overall system. Of considerable importance

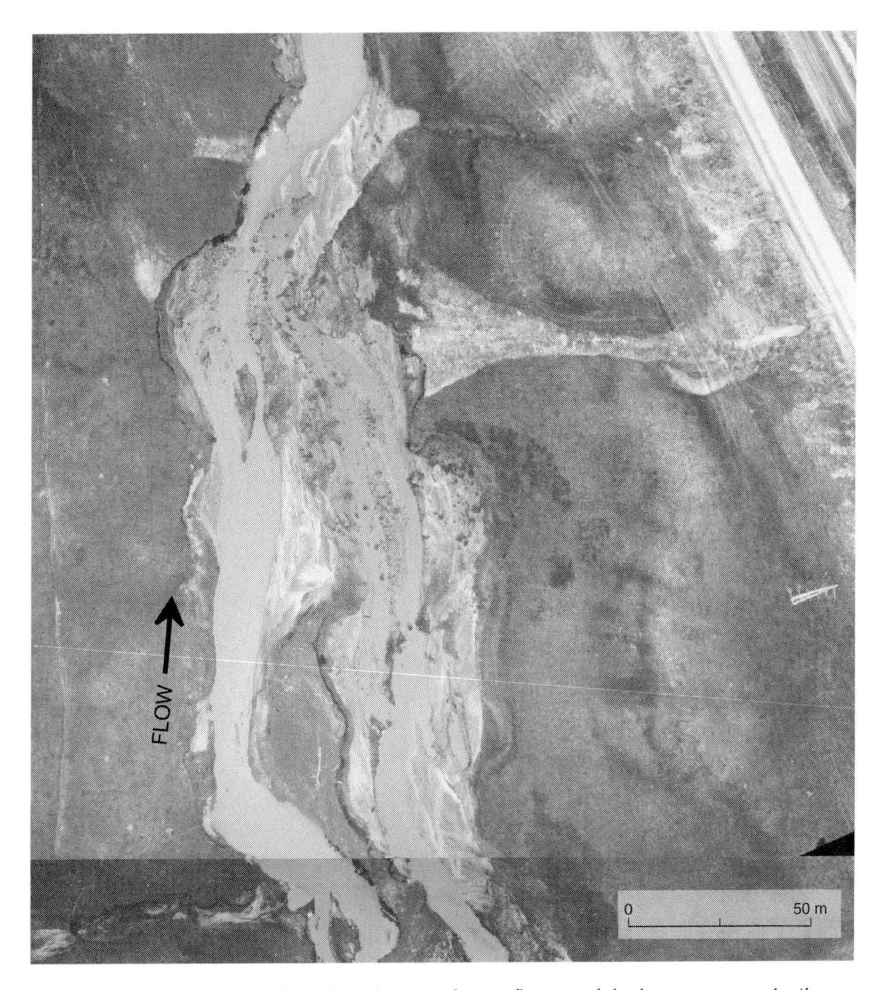

FIG. 19 Expanded view of Reach F showing the confluence of the larger unnamed tributary and Carpenter Creek. Gravel can be seen emanating from an eddy-caused cut in the bank at the center of the figure. Upstream of this location there is evidence of only sand on the bottom of the channel. Similarly, there is only sand in the channel of Carpenter Creek upstream of the confluence. This reach of the unnamed tributary has fully unraveled, but the gravel has prevented it from evolving a flood-related hydraulic geometry. Also, the gravel has disrupted the flood-related hydraulic geometry of Carpenter Creek for a short distance downstream of the confluence. The notches in the banks of both creeks are a result of recirculation zones and that the fluid mechanics of the confluence was complex.

to the protection of the floodplains along these unnamed tributaries was the absence of cottonwood trees that could have been eroded in the areas that were sparsely populated with willows. These large trees could have become ploughs capable of ripping up the otherwise shrub-protected floodplain and removing sequences of beaver from the channels.

The primary role that beaver dams play is to maintain ponds and, thus, to maintain a high water table. This, in turn, causes growth of wetland vegetation and shrub carrs. Although the beaver dams do provide physical roughness, it has been shown in this section that in the shrub carrs the dams promote, the shrubs are much more effective than the beaver dams at causing flow resistance. When the dams across the main thread of flow are close together relative to their height, they become effective roughness elements and slow the flow in the channel, forcing more water onto the floodplain than would have been the case in their absence. In this mode, they also produce a significant drag stress and, hence, reduce the shear stress on the actual boundary. This significantly reduces the sediment transport, no matter what is going on in the flow above the bed. It also reduces the capacity of the flow to erode any one of structures in a sequence of dams. The effectiveness of the beaver dams as roughness elements increases as the relative distance between them decreases. At relative spacings greater than 40, the dams act essentially independently and their primary effect on the flow is to constrict the cross-section locally and force more water onto the floodplain just upstream of the site of the dam. This has the opposite effect on the sediment transport from that described above because the shear stress associated with the locally accelerating flow dominates over the drag stress and the sediment transport rates increase locally. Similarly, the capacity of the flow to erode further the breached structure increases. The difference is a consequence of whether the dam acts as a single barrier to the flow or whether the dams act on the flow as a coherent series of roughness elements. Small breaches in the dams have a negligible hydrodynamic effect; large breaches allow a disproportionately greater amount of water to pass through the cross-section. As a rough estimate for a deep overbank flow, the effectiveness of a solitary dam depends on the percentage constriction of the channel cross–section, and the loss of effectiveness of a dam because of having been breached depends on the percentage increase in cross-sectional area relative to the cross-sectional area of the same structure when not breached. Only when the breach becomes large is there a significant added loss in effectiveness from acceleration of the flow through the cut.

Connection between Beaver and Large Ungulates

In previous sections of this chapter, the interplay between the beaver ponds, shrubs, and large ungulates has been mentioned briefly. Observations of branch tips by the author suggest that cattle and elk typically do not enter ponds and wetlands to graze the willow shoots and young branches as do moose; rather, they focus on eating the willow shoots and young branches that are in dry areas. In western streams, beaver usually do not maintain specific dams and ponds indefinitely. Rather, they move their dams around to maximize water access to the best food supplies. If the beaver have severely pruned the willow shrubs in one area, the family occupying that pond or pond complex is likely to construct one or more new dams in an area that provides better water access to a carr with more mature stems. The area being abandoned temporarily is unattractive to other beaver, so it is relatively safe for the departing family to leave it unused. Without maintenance, the ponds drain, usually by flow tunneling under or through the abandoned dam. This pond moving has a solid evolutionary foundation because it lets new willow shoots in the worked-out area mature over several years into a stem field that eventually is suitable for using. In this manner, a long-term winter food supply and a good supply of construction materials are maintained for a limited population of beaver. In general, well-fed beaver maintain healthy, more active ecosystems. Pond moving also helps maintain a regionally elevated water table and distributes the ground water beneath the carr relatively efficiently. By periodically shifting the locations of the ponds, the locations of the dominant threads of flow and their associated threads of gravel also are periodically changed. This has the ecologically desirable consequence of scouring the fine material from channels through old beaver ponds and maintaining an effective hydrologic connection between the pond complexes and the aquifer beneath. The opposite is true for recharge from beaver ponds that start as small lakes. In this situation, fine sediment and organic debris close the connection at the pond bottom between the pond and the subsurface aquifer. In addition, pond shifting permits recruitment of new shrubs along the edges and in the interior of the abandoned ponds (recall the new shrub growth in the abandoned pond at the northern end of Reach C) and, in the long term, provides an efficient means of expanding the shrub carr. Maintaining a larger carr results in better protection of the entire pond complex from large floods.

Unfortunately, when large grazing ungulates become overcrowded, they move into the abandoned pond areas and eat both the new plants and the young shoots off old root crowns. This also happens to the willow root crowns on the floodplain. As long as the grazing pressure is maintained, the willows cannot produce a significant number of mature shoots. The result is loss of beaver habitat because beaver can no longer return to the abandoned sites. As beaver habitat is lost, the water table inevitably drops under the willow carr. If the marginally functional roots of the overpruned root crowns cannot follow the dropping water table downward fast enough, the root crowns lose their water source, the carr constricts, and the upland area expands at its expense. Willows can remain dormant for many years when stressed by flooding. In contrast, they die fairly quickly when the water table drops faster than their roots can follow it downward. The ultimate result of this grazing pressure is local elimination of the beaver, the habitat that beaver ponds provided, and the ecological system that was based on the beaver ponds. The location of the initiation of unraveling on the unnamed tributary of Carpenter Creek is very near the boundary between unfettered willow growth along the unnamed tributaries and use of the riparian zone by ungulates for grazing along Carpenter Creek. It is possible that deer also are grazing the willow shoots in this area in late winter, but they are not numerous enough to limit the extent of the willow carr. It is more probable that the boundary beyond which significant willow growth occurs is associated with cattle ranching. Where the willow carrs are dense along the unnamed tributaries, access for cattle to the creek from the grassy terraces is more difficult than near the junction of the unnamed tributary with Carpenter Creek and further downstream. In addition, the grass near the Creek is more abundant and lusher than that on the high terraces farther upstream.

Ungulate-caused loss of beaver habitat is not restricted solely to ranching. It can also occur in national parks when deer and elk populations are allowed to increase without significant predation. In the Kawuneeche Valley of Rocky Mountain National Park and elsewhere in the Park, an excessively large population of elk currently is putting significant pressure on beaver populations. The reduction in area of the willow carrs, in turn, is limiting the moose habitat and may be causing a growing moose population to disperse across the Continental Divide. Moose are browsing ungulates with long legs and, when possible, tend to feed high up on the willow shrubs rather than clipping off new shoots near the ground, as the grazing ungulates prefer to do. Moose also are less social and more aggressive than cattle and elk. They do not tolerate an abundance of neighbors, so their population densities tend to be lower. Although sometimes blamed for willow shrub eradication, moose are more likely to coexist with the beaver and willow ecosystems than to destroy them. When this is not the case, it is a sign that the ecosystem already is badly out of balance. Interestingly, this complex ecological scenario in these national parks ultimately is the result of eliminating both wolves and hunting from the ecosystems, and in Yellowstone National Park reintroduction of wolves has pushed the elk out of the bottomlands. This has allowed extensive renewed growth of willows from old root crowns in many riparian zones (Smith and Ferguson, 2005).

One important geomorphic consequence of this ungulate incursion is an increased vulnerability to unraveling of the affected headwater streams. Another perhaps more important consequence is the potential to cause geomorphic alteration further downstream during large floods. Rivers form tree-like networks. If only a few headwater tributaries are affected, then the downstream impact of heavy rainfall and local unraveling may be minor, but when many of the headwater tributaries all are adversely affected, as is becoming all too frequent, then heavy rainfall on these vulnerable headwater drainage basins can result in catastrophic floods further downstream. For example, both the East and West Forks of Plum Creek unraveled during the 1965 flood, and this caused serious damage along the South Platte River all the way through Denver. Ultimately, the Plum Creek flood caused the construction of a major flood-protection project (Chatfield Reservoir) on the South Platte River.

SUMMARY AND CONCLUSIONS

This chapter has focused on (1) the role that shrub carrs play in protecting upland stream systems from damage caused by floods, (2) the role that beaver play in maintaining willow carrs, and (3) the interplay of beaver ponds and beaver dams with floods. These issues, however, cannot be addressed effectively without considering their ecological settings and ramifications and, particularly, the effects that humans and overpopulations of ungulates have in disrupting these ecological settings. From a human point of view, the issue may be phrased as an inquiry into whether we are making streams more vulnerable to catastrophic flooding by continually encroaching onto the floodplains of headwater tributaries.

To examine the first issue quantitatively, a mathematical model for flow through shrub carrs was essential. Moreover, for this model to be able to deal with extreme events, it had to be devoid of empirically adjusted parameters such as Manning's coefficient. Clearly, calibrated models cannot be used reliably for either extreme events or events that have occurred in the past. Rather, a model that takes explicit account of the salient fluid-mechanical characteristics of the flows of interest had to be constructed.

These flows of interest were floods through shrub carrs of varying densities, and the important issue was the interaction of the flows and the plants. Shrub density clearly had to be an important and continuously adjustable variable. Two types of models were required because two different types of shrub carrs were to be examined. The first type of model permitted the stems to be rigid and tall relative to the flow depth. The second type of model allowed the shrubs to be overtopped by the flow. This caused the shrubs to bend over. The author had previously constructed and used both types of models, but an improved version of the second type was developed for this chapter. The biological input to both models is the anatomical structure of the various shrubs of interest as it pertains to the salient characteristics of form drag on these plants.

The two types of models were previously applied to two separate field situations, and these applications were discussed in detail in this chapter. In particular, they were carefully compared and contrasted and general conclusions were extracted from these discussions. The background for each application is important and was provided in sufficient detail to support the conclusions. It is the general conclusions from these two separate studies, however, that are of concern in this chapter. The first application was to the flood of record on the Clark Fork of the Columbia River in the Deer Lodge Valley of Montana. The second application was to the flood of record on Plum Creek, located south of Denver, Colorado. In both cases, sufficient field data to permit quantitative tests of the modeling procedure had been or was obtained. In both cases, the predictions of the model were confirmed.

It was argued, based on several lines of evidence, that the Clark Fork in the Deer Lodge Valley originally flowed through a complex of beaver ponds and was lined by a dense willow carr. Calculations pertaining to the flood of record in 1908 on this river confirmed that the meander belt of the river had a willow flora that approached 100% canopy cover. Moreover, the overbank velocities produced by flow through a willow carr of this density agreed with a model for the net deposition of tailings in the meander belt of the river. Consequently, it was concluded that the flow model was accurate and could be used to examine the present state of the floodplain of the Clark Fork in the Deer Lodge Valley. Owing to the present impoverished state of the riparian vegetation, it was concluded that the floodplain would unravel during the first multidecadal recurrence interval flood. In addition, the model was used to show that for sparse woody vegetation, the average shear stress on the actual boundary depended on flow depth, but for dense woody vegetation, this boundary shear stress decreased somewhat as flow depth increased. The model also showed that the magnitude of the shear stress depended primarily on stem spacing. The vertically averaged velocity was shown to be nearly independent of flow depth and primarily dependent on stem spacing in the case of dense shrubs.

The present Clark Fork situation was compared to the 1965 Plum Creek situation. In the former case, the woody vegetation on the floodplain has been impoverished by tailings deposition during the 1908 flood and other human activities, and the floodplain has been predicted to unravel, although it has not yet done so. In the Plum Creek case, the floodplain had unraveled and the location of the place where unraveling began was both predicted and observed. Good agreement was again obtained between prediction and observation. The Plum Creek flood analysis, however, went further. The unraveled reaches were poorly protected by woody vegetation, whereas the areas that remained intact geomorphically were well protected by beaver-maintained sandbar willow carrs. Unlike the situation in the Deer Lodge Valley, where the willows were so tall and the floodplain so wide that overtopping the willow carr was nearly impossible, in the Plum Creek case the sandbar willows were overtopped. This means that in the Plum Creek case, the boundary shear stresses and vertically averaged velocities always maintained a dependence on flow depth and always increased with that variable. Unlike the situation for the floodplain of the Clark Fork, these variables never reached an asymptotic state where they depended only on stem spacing. Consequently, the deeper the flow became, the higher the vertically averaged velocities and boundary shear stresses.

Finally, the interplay of the flood and the beaver dams and ponds was examined carefully and quantitatively. It was shown that well-maintained, frequently occurring beaver dams provided good protection to the stream system in two ways. First, they maintained an elevated water table that made possible dense willow carrs. Second, when close together, they produced sufficient roughness to help protect the channels between the ponds from major bed and bank erosion. In essence, the beaver dams across the main channels complemented the woody floodplain vegetation by protecting the primary flow paths from serious scour. However, it was found that old beaver dams were substantially less effective in physically protecting the streams and that beaver dams across well-defined channels were vulnerable to being breached. As a result of this work, means for quantifying both the overbank and in-channel components of flow in beaver pond complexes and willow carrs have been developed, tested, and presented in a manner that permits similar calculations to be made easily in other situations.

REFERENCES

Butler, D. R., and Malanson, G. P. (2005). The geomorphic influences of beaver dams and failures of beaver dams. *Geomorphology* 71, 48–60.

Elliot, W. P. (1958). The growth of the atmospheric internal boundary layer. *EOS T Am Geophys Union* 38, 1048.

Friedman, J. M., Osterkamp, W. R., and Lewis, Jr., W. M., (1996). Channel narrowing and vegetation development following a Great Plains flood. *Ecology* 77, 2167–2181.

Griffin, E. R., and Smith, J. D. (2001). Computation of bankfull and flood-generated hydraulic geometries in East Plum Creek, Colorado. In: *Proceedings of the Seventh Federal Interagency Sedimentation Conference, March 25-29, 2001, Reno, Nevada, USA*. Vol. 1, Section II, pp 50–56.

Griffin, E. R., and Smith, J. D. (2002). *State of Flood Plain Vegetation within the Meander Belt of the Clark Fork of the Columbia River, Deer Lodge Valley, Montana*. U.S. Geological Survey Water-Resources Investigations Report 02-4109, U.S. Department of the Interior, U.S. Geological Survey, Boulder, CO.

Griffin, E. R., and Smith, J. D. (2004). Floodplain stabilization by woody riparian vegetation during an extreme flood. In: *Riparian Vegetation and Fluvial Geomorphology*. (Bennett, S. J., and Simon, A., Eds.), pp. 221–236. Water Science and Application 8, American Geophysical Union, Washington, DC.

Gurnell, A. M. (1998). The hydrogeomorphological effects of beaver-dam building activity. *Prog Phys Geog* 22, 167–189.

Hillman, G. R. (1998). Flood wave attenuation by a wetland following a beaver-dam failure on a second order boreal stream. *Wetlands* 18, 21–24.

Horstman, M. C. (1984). *Historical Events Associated with the Upper Clark Fork Drainage*. Montana Department of Fish, Wildlife, and Parks, Missoula, MT.

Kean, J. W., and Smith, J. D. (2005). Generation and verification of theoretical rating curves in the Whitewater River Basin, Kansas. *J Geophys Res* 110 (F4), Art. No. F04012.

Kundu, P. K. (1990). *Fluid Mechanics*. Academic Press, San Diego.

Long, C. E., Wiberg, P. L., and Nowell, A. R. M. (1993). Evaluation of von Karman's constant from integral flow parameters. *J Hydrol Eng* 119, 1182–1190

Marston, R. A. (1994). River entrenchment in small mountain valleys of the western USA: influence of beaver grazing and clearcut logging. *Rev Geogr Lyon* 69, 11–15.

Meentemeyer, R. K., and Butler, D. R. (1999) Hydrogeomorphic effects of beaver dams in Glacier National Park, Montana. *Phys Geogr* 20, 436–446.

Muller-Schwarze, D., and Sun, L. (2003). *The Beaver: Natural History of a Wetlands Engineer*. Comstock Pub. Associates, Ithaca, NY.

Nikuradse, J. (1933). *Laws of Flow in Rough Pipes*. (translated from German in 1950) National Advisory Commission, Aeronautical Technical Memo 1292, 1–62.

Nimick, D. A. (1990). *Stratigraphy and Chemistry of Metal-contaminated Flood Plain Sediments, Upper Clark Fork River, Montana*. M.Sc. thesis, University of Montana, Missoula, MT.

Osterkamp, W. R., and Costa, J. E. (1987). Changes accompanying an extraordinary flood on a sand-bed stream. In: *Catastrophic Flooding*. (Mayer, L., and Nash, D., Eds.), pp. 201–223. Allen and Unwin, Boston, MA.

Rattray, Jr., M., and Mitsuda, E. (1974). Theoretical analysis of conditions in a salt wedge. *Estuar Coast Mar Sci* 2, 373–394.

Schafer and Associates. (1997). *Soil and Tailings Map of a Portion of the Clark Fork River Floodplain*. Clark Fork Fluvial Geomorphology Committee, Environmental Protection Agency, Bozeman, MT.

Smith, D. W., and Ferguson, G. (2005). *Decade of the Wolf: Returning the Wild to Yellowstone*. Lyons Press, Guilford, CT.

Smith, J. D. (2001). On quantifying the effects of riparian vegetation in stabilizing single threaded streams. In: *Proceedings of the Seventh Federal Interagency Sedimentation Conference, March 25–29, 2001, Reno, Nevada, USA*. Vol. 1, Section IV, pp 22–29.

Smith, J. D. (2004). The role of riparian shrubs in preventing floodplain unraveling along the Clark Fork of the Columbia River in the Deer Lodge Valley, Montana. In: *Riparian Vegetation and Fluvial Geomorphology* (S. J. Bennett and A. Simon, Eds.), pp. 71–85. Water Science and Application 8, American Geophysical Union, Washington, DC.

Smith, J. D., and Griffin, E. R. (2002). *Relation between Geomorphic Stability and the Density of Large Shrubs on the Floodplain of the Clark Fork of the Columbia River in the Deer Lodge Valley, Montana*. U.S. Geological Survey Water-Resources Investigations Report 02-4070, Department of the Interior, U.S. Geological Survey, Helena, MT.

Smith, J. D., and McLean, S. R. (1977). Spatially averaged flow over a wavy surface. *J Geophys Res* 82, 1735–1746.

Smith, J. D., and McLean, S. R. (1984). A model for flow in meandering streams. *Water Resources Res* 20, 1301–1315.

Smith, J. D., Lambing, J. H., Nimick, D. A., Parrett, C., Ramey, M., and Schafer, W. (1998). *Geomorphology, Flood-plain Tailings, and Metal Transport in the Upper Clark Fork Valley, Montana*. U.S. Geological Survey Water-Resources Investigations Report 98-4170, U.S. Department of the Interior, U.S. Geological Survey, Helena, MT.

Wiberg, P. L., and Smith, J. D. (1987). Calculations of the critical shear stress for motion of uniform and heterogeneous sediments. *Water Resources Res* 23, 1471–1480.

Wright, J. P., Jones, G. C., and Flecker, A. S. (2002). An ecosystem engineer, the beaver, increases species richness at the landscape scale. *Oecologia* 132, 96–101.

Zurowski, W. (1992). The building activity of beavers. *Acta Theriologica* 37, 403–411.

INDEX

G

H

J

K

L

N

T

Z